Geometry from Dynamics, Classical and Quantum

José F. Cariñena · Alberto Ibort
Giuseppe Marmo · Giuseppe Morandi

Geometry from Dynamics, Classical and Quantum

Springer

José F. Cariñena
Departamento de Física Teórica
Universidad de Zaragoza
Zaragoza
Spain

Alberto Ibort
Departamento de Matemáticas
Universidad Carlos III de Madrid
Madrid
Spain

Giuseppe Marmo
Dipartimento di Fisiche
Universita di Napoli "Federico II"
Napoli
Italy

Giuseppe Morandi
INFN Sezione di Bologna
Universitá di Bologna
Bologna
Italy

ISBN 978-94-017-9219-6 ISBN 978-94-017-9220-2 (eBook)
DOI 10.1007/978-94-017-9220-2

Library of Congress Control Number: 2014948056

Springer Dordrecht Heidelberg New York London

Printed on acid-free paper

Springer is part of Springer Science+Business Media (www.springer.com)

Foreword

The Birth and the Long Gestation of a Project

Starting a book is always a difficult task. Starting a book with the characteristics of this one is, as we hope will become clear at the end of this introduction, even harder. It is difficult because the project underlying this book began almost 20 years ago and, necessarily, during such a long period of time, has experienced ups and downs, turning points where the project changed dramatically and moments where the success of the endeavor seemed dubious.

However the authors are all very grateful that things have turned out as they did. The road followed during the elaboration of this book, the innumerable discussions and arguments we had during preparation of the different sections, the puzzling uncertainties we suffered when facing some of the questions raised by the problems treated, has been a major part of our own scientific evolution and have made concrete contributions toward the shaping of our own thinking on the role of geometry in the description of dynamics. In this sense we may say with the poet:

Caminante, son tus huellas[1]
el camino y nada más;
Caminante, no hay camino,
se hace camino al andar.
Al andar se hace el camino,
y al volver la vista atras
se ve la senda que nunca
se ha de volver a pisar.
Caminante no hay camino
sino estelas en la mar.

Antonio Machado, *Proverbios y Cantares.*

[1] Wanderer, your footsteps are// the road, and no more;// wanderer, there is no road,// the road is made when we walk.// By walking the path is done,// and upon glancing back// one sees the path// that never will be trod again.// Wanderer, there is no road// only foam upon the sea.

Thus, contrary to what happens with other projects that represent the culmination of previous work, in this case the road that we have traveled was not there before this enterprise was started. We can see from where we are now that this work has to be pursued further to try to uncover the unknowns surrounding some of the beautiful ideas that we have tried to put together. Thus the purpose of this book is to share with the reader some of the ideas that have emerged during the process of reflection on the geometrical foundations of mechanics that we have come up with during the preparation of the book itself. In this sense it would be convenient to explain to the reader some of the major conceptual problems that were seeding the milestones marking the evolution of this intellectual adventure.

The original idea of this book, back in the early 1990s, was to offer in an accessible way to young Ph.D. students some completely worked significant examples of physical systems where geometrical and topological ideas play a fundamental role. The consolidation of geometrical and topological ideas and techniques in Yang-Mills theories and other branches of Physics, not only theoretical, such as in Condensed Matter Physics with the emergence of new collective phenomena or the fractional quantum Hall effect or High T_c superconductivity, were making it important to have a rapid but well-founded access to geometry and topology at a graduate level; this was rather difficult for the young student or the researcher needing a fast briefing on the subject. The timeliness of this idea was confirmed by the fact that a number of books describing the basics of geometry and topology delved into the modern theories of fields and other physical models that had appeared during these years. Attractive as this idea was, it was immediately clear to us that offering a comprehensive approach to the question of why some geometrical structures played such an important role in describing a variety of significant physical examples such as the electron-monopole system, relativistic spinning particles, or particles moving in a non-abelian Yang-Mills field, required us to present a set of common guiding principles and not just an enumeration of results, no matter how fashionable they were.

Besides, the reader must be warned that because of the particular idiosyncrasies of the authors, we were prone to take such a road. So we joyously jumped into the oceanic deepness of the foundations of the Science of Mechanics, trying to discuss the role that geometry plays in it, probably believing that the work that we had already done on the foundations of Lagrangian and Hamiltonian mechanics qualified us to offer our own presentation of the subject. Most probably it is unnecessary to recall here that, in the more than 20 years that had passed since publication of the books on the mathematical foundations of mechanics by V.I. Arnold [Ar76], R. Abraham and J. Marsden [Ab78] and J.M. Souriau [So70], the use of geometry, or better, the geometrical approach to Mechanics, had gained a widespread acceptation among many practitioners and the time was ripe for a second wave of the literature on the subject. Again our attitude was timely, as a number of books deepening and exploring complementary avenues in the realm of mechanics had started to appear.

When trying to put together a good recollection of the ideas embracing Geometry and Mechanics, including our own contributions to the subject, a feeling of uneasiness started to come over us as we realized that we were not completely satisfied with the various ways that geometrical structures were currently introduced into the description of a given dynamical system. They run from the "axiomatic" way as in Abraham and Marsden's book *Foundations of Mechanics* to the "constructive" way as in Souriau's book *Structure des Systèmes Dynamiques* where a geometrical structure, the Lagrange form, was introduced in the space of "movements" of the system, passing through the "indirect" justification by means of Hamilton's principle, leading to a Lagrangian description in Arnold's *Méthodes mathématiques de la mécanique classique*. All these approaches to the geometry of Mechanics were solidly built upon ideas deeply rooted in the previous work of Lagrange, Hamilton, Jacobi, etc. and the geometric structures that were brought to the front row on them had been laboriously uncovered by some of the most brilliant thinkers of all times. Thus, in this sense, there was very little to object in the various presentations of the subject commented above. However, it was also beginning to be clear at that time, that some of the geometrical structures that played such a prominent role in the description of the dynamical behavior of a physical system were not univocally determined. For instance there are many alternative Lagrangian descriptions for such a simple and fundamental system as the harmonic oscillator. Thus, which one is the preferred one, if there is one, and why? Moreover, the current quantum descriptions of many physical systems are based on either a Lagrangian or Hamiltonian description of a certain classical one. Thus, if the Lagrangian and/or the Hamiltonian description of a given classical system is not unique, which quantum description prevails? Even such a fundamental notion as linearity was compromised at this level of analysis as it is easy to show the existence of nonequivalent linear structures compatible with a given "linear" dynamics, for instance that of the harmonic oscillator again.

It took some time, but soon it became obvious that from an operational point of view, the geometrical structures introduced to describe a given dynamics were not a priori entities, but they accompanied the given dynamics in a natural way. Thus, starting from raw observational data, a physical system will provide us with a family of trajectories on some "configuration space" Q, like the trajectories photographed on a fog chamber displayed below (see Fig. 1) or the motion of celestial bodies during a given interval of time. From these data we would like to build a differential equation whose solution will include the family of the observed trajectories. However we must point out here that a differential equation is not, in general, univocally determined by experimental data. The ingenuity of the theoretician regarding experimental data will provide a handful of choices to start building up the theory. At this point we stand with A. Einstein's famous quote:

> *Physical concepts are free creations of the human mind, and are not, however it may seem, uniquely determined by the external world. In our endeavor to understand reality we are somewhat like a man trying to understand the mechanism of a closed watch. He sees the face and the moving hands, even hears its ticking, but he has no way of opening the case. If he is ingenious he may form some picture of a mechanism which could be responsible for all the things he observes, but he may never be quite sure his picture is the only one which could explain his observations. He will never be able to compare his picture with the real mechanism and he cannot even imagine the possibility or the meaning of such a comparison. But*

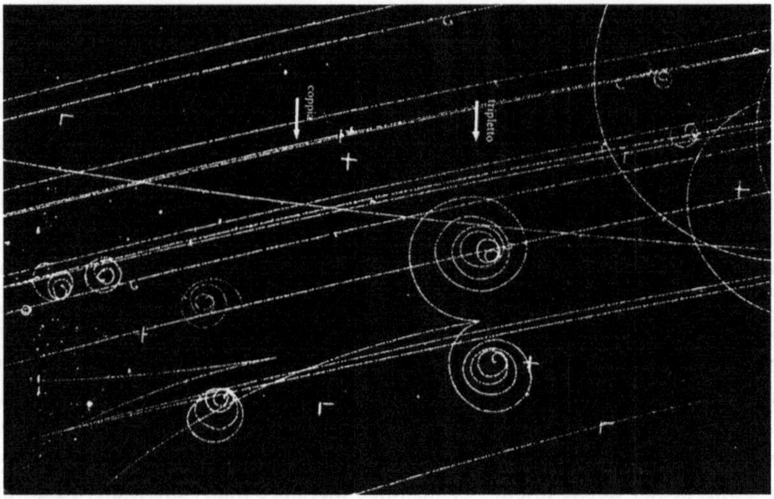

Fig. 1 Trajectories of particles on a fog chamber

> *he certainly believes that, as his knowledge increases, his picture of reality will become simpler and simpler and will explain a wider and wider range of his sensuous impressions. He may also believe in the existence of the ideal limit of knowledge and that it is approached by the human mind. He may call this ideal limit the objective truth.*
>
> A. Einstein, *The Evolution of Physics* (1938) *(co-written with Leopold Infeld).*

For instance, the order of the differential equation will be postulated following an educated guess of the theoretician. Very often from differential equations we prefer to go to vector fields on some (possibly) larger carrier space, so that evolution is described in terms of one parameter groups (or semigroups). Thus a first initial geometrization of the theory is performed.

At this point we decided to stop assuming additional structures for a given description of the dynamics and, again, following Einstein, we assumed that all geometrical structures should be considered equally placed with respect to the problem of describing the given physical system, provided that they were compatible with the given dynamics, *id est*[2] with the data gathered from it. Thus this notion of operational compatibility became the Occam's razor in our analysis of dynamical evolution, as geometrical structures should not be postulated but accepted only on the basis of their consistency with the observed data. The way to translate such criteria into mathematical conditions will be discussed at length throughout the text; however, we should stress here that such emphasis on the subsidiary character of geometrical structures with respect to a given set of data is already present, albeit in a different form, in Einstein's General Relativity, where the geometry of space–time is dynamically determined by the distribution of mass and energy in the universe. All solutions of Einstein's equations for a given energy–

[2] i.e., 'which is to say' or 'in other words'.

Fig. 2 The picture shows the movements of several planets over the course of several years. The motion of the planets relative to the stars (represented as unmoving points) produces continuous streaks on the sky. (Courtesy of the Museum of Science, Boston)

momentum tensor are acceptable geometrical descriptions of the universe. Only if there exists a Cauchy surface (i.e., only if we are considering a globally hyperbolic space–time) we may, after fixing some initial data, determine (locally) the particular solution of equations compatible with a given energy–momentum tensor Fig. 2.

From this point on, we embarked on the systematic investigation of geometrical structures compatible with a given dynamical system. We have found that such a task has provided in return a novel view on some of the most conspicuous geometrical structures already filling the closet of mathematical tools used in the theory of mechanical and also dynamical systems in general, such as linear structures, symmetries, Poisson and symplectic structures, Lagrangian structures, etc. It is apparent that looking for structures compatible with a given dynamical system constitutes an "Inverse Problem" a description in terms of some additional structures. The inverse problem of the calculus of variations is a paradigmatic example of this. The book that we present to your attention offers at the same time a reflection on the geometrical structures that could be naturally attached to a given dynamical system and the variety of them that could exist, creating in this way a hierarchy on the family of physical systems according with their degree of compatibility with natural geometrical structures, a system being more and more "geometrizable" as more structures are compatible with it. Integrable systems have played a key role in the development of Mechanics as they have constituted the main building blocks for the theory, both because of their simple appearance, centrality in the development of the theories, and their ubiquity in the description of the physical world. The avenue we follow here leads to such a class of systems in a natural way as the epitome of extremely geometrizable systems in the previous sense.

We may conclude this exposition of motives by saying that if any work has a motto, probably the one encapsulating the spirit of this book could be:

> *All geometrical structures used in the description of the dynamics of a given physical system should be dynamically determined.*

What you will Find and What you will not in This Book

This is a book that pursues an analysis of the geometrical structures compatible with a given dynamical system, thus you will not find in it a discussion on such crucial issues such as determination of the physical magnitudes relevant for description of mechanical systems, be they classical or quantum, or an interpretation of the experiments performed to gain information on it, that is on any theoretical description of the measurement process. Neither will we extend our enquiries to the domain of Field Theory (Fig. 3) (even though we included in the preparation of this project such key points but we had to discard them to keep the present volume at a reasonable size) where new structures with respect to the ones described here are involved. It is a work that focuses on a mathematical understanding of some fundamental issues in the Theory of Dynamics, thus in this sense both the style and the scope will be heavily determined by these facts.

Chapter 1 of the book will be devoted to a discussion of some elementary examples in finite and infinite dimensions where some of the standard ideas in dealing with mechanical systems like constants of motion, symmetries, Lagrangian, and Hamiltonian formalisms, etc., are recalled. In this way, we pretend to help the reader to have a strong foothold on what is probably known to him/her with respect to the language and notions that are going to be developed in the main part of the text. The examples chosen are standard: The harmonic oscillator, an electron moving on a constant magnetic field, the free particle on the finite-dimensional side, and the Klein–Gordon equation, Maxwell equations, and the Schrödinger equation as prototypes of systems in infinite dimensions. We have said that field theory will not be addressed in this work, that is actually so because the examples in infinite dimensions are treated as evolution systems, i.e., time is a privileged variable and

Fig. 3 Counter rotating vortex generated at the tip of a wing. (American Physical Society's 2009 Gallery of Fluid Motion)

no covariant treatment of them are pursued. Dealing with infinite-dimensional systems, already at the level of basic examples, shows that many of the geometrical ideas that are going to appear are not restricted by the number of degrees of freedom. Even though a rigorous mathematical treatment of them in the case of infinite dimensions will be out of the scope of this book, the geometrical arguments apply perfectly well to them as we will try to show throughout the book.

Another interesting characteristic of the examples chosen in the first part of Chap. 1 is that they are all linear systems. Linear systems are going to play an instrumental role in the development of our discourse because they provide a particularly nice bridge between elementary algebraic ideas and geometrical thinking. Thus we will show how a great deal of differential geometry can be constructed from linear systems. Finally, the third and last part of the first chapter will be devoted to a discussion of a number of nonlinear systems that have managed to gain their own relevant place in the gallery of dynamics, like the Calogero-Moser system, and that all share the common feature of being obtained from simpler affine systems. The general method of obtaining these systems out of simpler ones is called "reduction" and we will offer to the reader an account of such procedures by example working out explicitly a number of interesting ones. These systems will provide also a source of interesting situations where the geometrical analysis is paramount because their configuration/phase spaces fail to be open domains on an Euclidean space. The general theory of reduction together with the problem of integrability will be discussed again at the end of the book in Chap. 7.

Geometry plays a fundamental role in this book. Geometry is so pervasive that it tends very quickly to occupy a central role in any theory where geometrical arguments become relevant. Geometrical thinking is synthetic so it is natural to attach to it an a priori or relatively higher position among the ideas used to construct any theory. This attitude spreads in many occasions to include also geometrical structures relevant for analysis of a given problem. We have deliberately subverted this approach here considering geometrical structures as subsidiaries to the given dynamics; however, geometrical thinking will be used always as a guide, almost as a metalanguage, in analyses of the problems. In Chap. 2 we will present the basic geometrical ideas needed to continue the discussion started here. It would be almost impossible to present all details of the foundations of geometry, in particular differential geometry, which would be necessary to make the book self-consistent. This would make the book hard to use. However, we are well aware that many students who could be interested in the contents of this book do not possess the necessary geometrical background to read it without introducing (with some care) some of the fundamental geometrical notions that are necessarily used in any discussion where differential geometrical ideas become relevant; just to name a few: manifolds, bundles, vector fields, Lie groups, etc. We have decided to take a pragmatic approach and try to offer a personal view of some of these fundamental notions in parallel with the development of the main stream of the book. However, we will refer to standard textbooks for more detailed descriptions of some of the ideas sketched here.

Linearity plays a fundamental role in the presentation of the ideas of this book. Because of that some care is devoted to the description of linearity from a geometrical perspective. Some of the discourse in Chap. 3 is oriented toward this goal and a detailed description of the geometrical description of linear structures by means of Euler or dilation vector fields is presented. We will show how a small generalization of this presentation leads naturally to the description of vector bundles and to their characterization too. Some care is also devoted to describe the fundamental concepts in a dual way, i.e., from the set-theoretical point of view and from the point of view of the algebras of functions on the corresponding carrier spaces. The second approach is instrumental in any physical conceptualization of the mathematical structures appearing throughout the book; they are not usually treated from this point of view in standard textbooks.

After the preparation offered by the first two chapters we are ready to start exploring geometrical structures compatible with a given dynamics. Chapter 4 will be devoted to it. Again we will use as paradigmatic dynamics the linear ones and we will start by exploring systematically all geometrical structures compatible with them: zero order, i.e., constants of motion, first order, that is symmetries, and immediately after, second-order invariant structures. The analysis of constants of motion and infinitesimal symmetries will lead us immediately to pose questions related with the "integrability" of our dynamics, questions that will be answered partially there and that will be recast in full in Chap. 8. The most significant contribution of Chap. 4 consists in showing how, just studying the compatibility condition for geometric structures of order two in the case of linear dynamics, we arrive immediately to the notion of Jacobi, Poisson, and Hamiltonian dynamics. Thus, in this sense, standard geometrical descriptions of classical mechanical systems are determined from given dynamics and are obtained by solving the corresponding inverse problems. All of them are analyzed with care, putting special emphasis on Poisson dynamics as it embraces both the deep geometrical structures coming from group theory and the fundamental notions of Hamiltonian dynamics. The elementary theory of Poisson manifolds is reviewed from this perspective and the emerging structure of symplectic manifolds is discussed. A number of examples derived from group theory and harmonic analysis are discussed as well as applications to some interesting physical systems like massless relativistic systems.

The Lagrangian description of dynamical systems arises as a further step in the process of requiring additional properties to the system. In this sense, the last section of Chap. 5 can be considered as an extended exposition of the classical Feynman's problem together with the inverse problem of the calculus of variations for second-order differential equations. The geometry of tangent bundles, which is reviewed with care, shows its usefulness as it allows us to greatly simplify exposition of the main results: necessary and sufficient conditions will be given for the existence of a Lagrangian function that will describe a given dynamics and the possible forms that such a Lagrangian function can take under simple physical assumptions (Fig. 4).

Once the classical geometrical pictures of dynamical systems have been obtained as compatibility conditions for $(2, 0)$ and $(0, 2)$ tensors on the corresponding carrier

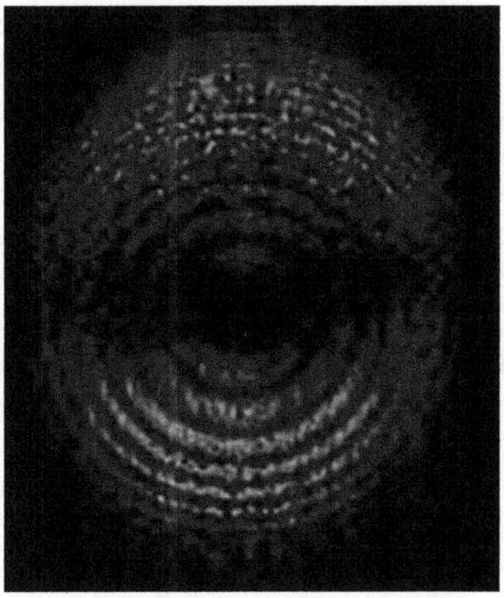

Fig. 4 Quantum stroboscope based on a sequence of identical attosecond pulses that are used to release electrons into a strong infrared (IR) laser field exactly once per laser cycle

space, it remains to explore a natural situation where there is also a complex structure compatible with the given dynamics. The fundamental instance of this situation happens when there is an Hermitean structure admissible for our dynamics. Apart from the inherent interest of such a question, we should stress that this is exactly the situation for the dynamical evolution of quantum systems. Let us point out that the approach developed here does not preclude their being an a priori given Hermitean structure. But under what conditions there will exist an Hermitean structure compatible with the observed dynamics. Chapter 6 will be devoted to solving such a problem and connecting it with various fundamental ideas in Quantum Mechanics. We must emphasize here that we do not pretend to offer a self-contained presentation of Quantum Mechanics but rather insist that evolution of quantum systems can be dealt within the same geometrical spirit as other dynamics, albeit the geometrical structures that emerge from such activity are of diverse nature. Therefore no attempt has been made to provide an analysis of the various geometrical ideas that are described in this chapter regarding the physics of quantum systems, even though a number of remarks and observations pertinent to that are made and the interested reader will be referred to the appropriate literature.

At this point we consider that our exploration of geometrical structures obtained from dynamics has exhausted the most notorious ones. However, not all geometrical structures that have been relevant in the discussion of dynamical systems are covered here. Notice that we have not analyzed, for instance, contact structures that play an important role in treatment of the Hamilton–Jacobi theory or Jacobi structures. Neither have we considered relevant geometrical structures arising in field theories or the theory of integrable systems (or hierarchies to be precise) like Yang–Baxter equations, Hopf algebras, Chern-Simons structures, Frobenius manifolds, etc. There is a double reason for that. On one side it will take us far beyond the purpose of this book and, more important, some of these structures are characteristic of a very restricted, although extremely significant, class of dynamics.

However we have decided not to finish this book without entering, once we are in possession of a rich baggage of ideas, some domains in the vast land of the study of dynamics, where geometrical structures have had a significant role. In particular we have chosen the analysis of symmetries by means of the so-called reduction theory and the problem of the integrability of a given system. These issues will be covered in Chap. 7 were the reduction theory of systems will be analyzed for the main geometrical structures described before.

Once one of the authors was asked by E. Witten, "how does it come that some systems are integrable and others not?" The question was rather puzzling taking into account the large amount of literature devoted to the subject of integrability and the attitude shared by most people that integrability is a "non-generic" property, thus only possessed by a few systems. However, without trying to interpret Witten, it is clear that the emergence of systems in many different contexts (by that time Witten had realized the appearance of Ramanujan's τ-function in quantum 2D gravity) was giving him a certain uneasiness on the true nature of "integrability" as a supposedly well-established notion. Without oscillating too much toward V. Arnold's answer to a similar question raised by one of the authors: "An integrable system is a system that can be integrated", we may try to analyze the problem of the integrability of systems following the spirit of these notes: given a dynamics, what are the fundamental structures determined by the structural characteristics of the flow that are instrumental in the "integrability" problem?

Chapter 8 will be devoted to a general perspective regarding the problem of integrability of dynamical systems. Again we do not pretend to offer an inclusive approach to this problem, i.e., we are not trying to describe and much less to unify, the many theories and results on integrability that are available in the literature. That would be an ill-posed problem. However, we will try to exhibit from an elementary analysis some properties shared by an important family of systems lying within the class of integrable systems and that can be analyzed easily with the notions developed previously in this book. We will close our excursion on the geometries determined by dynamics by considering in detail a special class of them that exhibit many of the properties described before, the so-called Lie–Scheffers systems which provide an excellent laboratory to pursue the search on this field.

Finally, we have to point out that the book is hardly uniform both in style and content. There are wide differences among its different parts. As we have tried to explain before a substantial part of it is in a form designed to make it accessible to a large audience, hence it can be read by assuming only a basic knowledge of linear algebra and calculus. However there are sections that try to bring the understanding of the subject further and introduce more advanced material. These sections are marked with an asterisk and their style is less self-contained. We have collected in the form of appendices some background mathematical material that could be helpful for the reader.

References

Abraham, R, Marsden, J.E.: Foundations of mechanics, (2nd ed.). Benjamin, Massachussetts (1978)

Arnol'd, V.I.: Méthodes mathématiques de la mécanique classique (Edition Mir), 1976. Mathematical Methods of Classical Mechanics. Springer, New York (1989)

Souriau, J.-M.: Structure des systemes dynamiques. Dunod, Paris (1970)

Acknowledgments

As mentioned in the introduction, we have been working on this project for over 20 years. First we would like to thank our families for their infinite patience and support. Thanks Gloria, Conchi, Patrizia and Maria Rosa.

During this long period we discussed various aspects of the book with a lot of people in different contexts and situations. We should mention some particular ones who have been regular through the years.

All of us have been participating regularly in the "International Workshop on Differential Geometric Methods in Theoretical Mechanics"; other regular participants with whom we have interacted the most have been Frans Cantrjin, Mike Crampin, Janusz Grabowski, Franco Magri, Eduardo Martinez, Enrico Pagani, Willy Sarlet and Pawel Urbanski.

A long association with the Erwin Schrödinger Institute has seen many of us meeting there on several occasions and we have benefited greatly from the collaboration with Peter Michor and other regular visitors.

In Naples we held our group seminar each Tuesday and there we presented many of the topics that are included in the book. Senior participants of this seminar were Paolo Aniello, Giuseppe Bimonte, Giampiero Esposito, Fedele Lizzi and Patrizia Vitale and of course, for even longer time, Alberto Simoni, Wlodedk Tulczyjew, Franco Ventriglia, Gaetano Vilasi and Franco Zaccaria.

Our long association with A.P. Balachandran, N. Mukunda and G. Sudarshan has influenced many of us and contributed to most of our thoughts.

In the last part of this long term project we were given the opportunity to meet in Madrid and Zaragoza quite often, in particular in Madrid, under the auspices of a "Banco de Santander/UCIIIM Excellence Chair", so that during the last 2 years most of us have been able to visit there for an extended period.

We have also had the befit of ongoing discussions with Manolo Asorey, Elisa Ercolessi, Paolo Facchi, Volodya Man'ko and Saverio Pascazio of particular issues connected with quantum theory.

During the fall workshop on Geometry and Physics, another activity that has been holding us together for all these years, we have benefited from discussions with Manuel de León, Miguel Muñoz-Lecanda, Narciso Román-Roy and Xavier Gracia.

Contents

Chapter 1
Some Examples of Linear and Nonlinear Physical Systems and Their Dynamical Equations

An instinctive, irreflective knowledge of the processes of nature will doubtless always precede the scientific, conscious apprehension, or investigation, of phenomena. The former is the outcome of the relation in which the processes of nature stand to the satisfaction of our wants.

Ernst Mach, *The Science of Mechanics* (1883).

1.1 Introduction

This chapter is devoted to the discussion of a few simple examples of dynamics by using elementary means. The purpose of that is twofold, on one side after the discussion of these examples we will have a catalogue of systems to test the ideas we would be introducing later on; on the other hand this collection of simple systems will help to illustrate how geometrical ideas actually are born from dynamics.

The chosen examples are at the same time simple, however they are ubiquitous in many branches of Physics, not just theoretical, and they constitute part of a physicist's wardrobe. Most of them are linear systems, even though we will show how to construct non-trivial nonlinear systems out of them, and they are both finite and infinite-dimensional.

We have chosen to present this collection of examples by using just elementary notions from calculus and the elementary theory of differential equations. More advanced notions will arise throughout that will be given a preliminary treatment; however proper references to the place in the book where the appropriate discussion is presented will be given.

Throughout the book we will refer back to these examples, even though new ones will be introduced. We will leave most of the more advanced discussions on their structure for later chapters, thus we must consider this presentation as a warmup and also as an opportunity to think back on basic ideas.

J.F. Cariñena et al., *Geometry from Dynamics, Classical and Quantum*,
DOI 10.1007/978-94-017-9220-2_1

1.2 Equations of Motion for Evolution Systems

1.2.1 Histories, Evolution and Differential Equations

A physical system is primarily characterized by histories, histories told by observers: trajectories in a bubble chamber of an elementary particle, trajectories in the sky for celestial bodies, or changes in the polarization of light. The events composing these histories must be localized in some carrier space, for instance the events composing the trajectories in a bubble chamber can be localized in space and time as well as the motion of celestial bodies, but the histories of massless particles can be localized only in momentum space.

In the Newtonian approach to the time evolution of a classical physical system, a configuration space Q is associated with the system, that at this moment will be assumed to be identified with a subset of $\mathbb{R}^N$, and space-time is replaced by $Q \times \mathbb{R}$ that will be the carrier space where trajectories can be localized. Usually, temporal evolution is determined by solving a system of ordinary differential equations on $Q \times \mathbb{R}$ which because of experimental reasons combined with the theoretician ingenuity freedom, are chosen to be a system of second-order differential equations:

$$\frac{d^2q^i}{dt^2} = F^i\left(q^1, \ldots, q^N, \frac{dq^1}{dt}, \ldots, \frac{dq^N}{dt}\right), \qquad i = 1, \ldots, N\,. \tag{1.1}$$

How differential equations are arrived at from families of 'experimental trajectories' is discussed in [Mm85]. Assuming evolution is described by a second-order differential equation was the point of view adopted by Joseph-Louis Lagrange and it led him to find for the first time a symplectic structure on the space of motions [La16].

The evolution of the system will be described by solving the system of Eq. (1.1) for each one of the initial data posed by the experimentalist, i.e., at a given time t_0, both the positions and velocities q_0 and v_0 of the system must be determined. The solution $q(t)$, that will be assumed to exist, of the initial value problem posed by Eq. (1.1) and $q(t_0) = q_0$, $\dot{q}(t_0) = v_0$, will be the trajectory described by the system on the carrier space Q. The role of the theoretician should be quite clear now. We started from a necessarily finite number of 'known' trajectories and we have found a way to make previsions for an infinite number of them, for each initial condition.

If we are able to solve the evolution Eq. (1.1) for all possible initial data q_0, v_0, then we may alternatively think of the family of solutions obtained in this way as defining a transformation φ_t mapping each pair (q_0, v_0) to $(q(t), \dot{q}(t))$ (for each t such that the solution $q(t)$ exists). The one-parameter family of transformations φ_t will be called the flow of the system and knowing it we can determine the state of the system at each time t provided that we know the state of the system (described in this case by a pair of points (q, v)) at a time t_0.

To turn the description of evolution into a one-parameter family of transformations we prefer to work with an associated system of first-order differential equations. In this way there will be a one-to-one correspondence between solutions and initial data. A canonical way to do that is to replace our Eq. (1.1) by the system of $2N$ equations

$$\frac{dq^i}{dt} = v^i, \qquad \frac{dv^i}{dt} = F^i(q, v), \qquad i = 1, \ldots, N, \tag{1.2}$$

where additional variables v^i, the velocities of the system, have been introduced.

If we start with a system in the form given by Eq. (1.1) we can consider as equivalent any other description that gives us the possibility to recover the trajectories of our starting system. This extension however has some ambiguities. The one we are describing is a 'natural one'. However other possibilities exist as has been pointed out in [Mm85]. In particular we can think of, for instance, a coordinate transformation that would transform our starting system into a linear one where such a transformation would exist. We will consider this problem in depth in relation with the existence of normal forms for dynamical systems at various places throughout the text. A large family of examples fitting in this scheme is provided by the theory of completely integrable systems.

A completely integrable system is characterized by the existence of variables, called action-angle variables denoted by (J_i, ϕ^i), such that when written in this new set of variables our evolution Eq. (1.2) look like:

$$\frac{d\phi^i}{dt} = \nu^i(J), \quad \frac{dJ^i}{dt} = 0, \quad i = 1, ..., N. \tag{1.3}$$

The general solution of such a system is given by:

$$\phi^i = \phi^i_0 + \nu^i_0 t, \qquad J_i = J_i(t_0).$$

where $\nu^i_0 = \nu^i(J(t_0))$ and $J_i(t_0)$ is the value taken by each one of the variables J_i at a given initial time t_0.

If $\det(\partial\nu^i/\partial J_j) \neq 0$, this system can be given an equivalent description as follows:

$$\frac{d}{dt}\begin{pmatrix}\phi^i\\ \nu^i\end{pmatrix} = \left(\begin{array}{c|c}\mathbf{0} & \mathbb{I}\\ \hline \mathbf{0} & \mathbf{0}\end{array}\right)\begin{pmatrix}\phi^i\\ \nu^i\end{pmatrix}, \tag{1.4}$$

and the $2N \times 2N$ matrix in Eq. (1.4) is nilpotent of index 2. We will discuss the classical theory of completely integrable systems and we will offer a general view on integrability in Chap. 8.

The family of completely integrable systems is the first of a large class of systems that can be written in the form of a homogeneous first-order linear differential equations:

$$\frac{dx}{dt} = A \cdot x, \tag{1.5}$$

where A is an $n \times n$ real matrix and $x \in \mathbb{R}^n$. Here and hereafter use is made of Einstein's convention of summation over repeated indices. The Eq. (1.5) is then the same as:

$$\frac{dx^i}{dt} = A^i{}_j x^j . \tag{1.6}$$

Then the solution of Eq. (1.5) for a given Cauchy datum $x(0) = x_0$ is given by:

$$x(t) = \exp(tA)\, x_0 , \tag{1.7}$$

where the exponential function is defined as the power series:

$$\exp tA = \sum_{k=0}^{\infty} \frac{t^k}{k!} A^k,$$

(see Sect. 2.2.4 for a detailed discussion on the definition and properties of $\exp A$).

1.2.2 The Isotropic Harmonic Oscillator

As a first example let us consider an m-dimensional isotropic harmonic oscillator of unit mass and proper frequency ω. Harmonic oscillators are ubiquitous in the description of physical systems. For instance the electromagnetic field in a cavity may be described by an infinite number of oscillators. An LC oscillator circuit consisting of an inductor (L) and a capacitor (C) connected together is described by the harmonic oscillator equation. In classical mechanics, any system described by kinetic energy plus potential energy, say $V(q)$, assumed to have a minimum at point q_0, may be approximated by an oscillator in the following manner. We Taylor expand $V(q)$ around the equilibrium point q_0, provided that V is regular enough, and on taking only the first two non-vanishing terms in the expansion we have: $V(q) \approx V(q_0) + \frac{1}{2}V''(q_0)(q - q_0)^2$. By using

$$\omega = \sqrt{\frac{V''(q_0)}{m}},$$

m being the mass, and removing the constant value $V(x_0)$ one finds and approximating potential $U(q) = \frac{1}{2}m\omega^2 q^2$ for our system. It is this circumstance that places the harmonic oscillator in a pole position in the description of physical systems.

Then, extending the previous considerations to systems with m degrees of freedom, we could write the equations of motion for an m-dimensional isotropic harmonic oscillator as the system of second-order differential equations on $\mathbb{R}^m$:

$$\frac{d^2 q^i}{dt^2} = -\omega^2 q^i, \qquad i = 1, \ldots, m.$$

We may write this system as a first-order linear system:

$$dx/dt = A \cdot x,$$

by introducing the vectors $q, v \in \mathbb{R}^m$, $x = (q, v)^T \in \mathbb{R}^{2m}$ and the $2m \times 2m$ matrix:

$$A = \left(\begin{array}{c|c} \mathbf{0} & \mathbb{I}_m \\ \hline -\omega^2 \mathbb{I}_m & \mathbf{0} \end{array}\right) . \tag{1.8}$$

Then the equations of motion read in the previous coordinates:

$$\frac{dq^i}{dt} = v^i , \qquad \frac{dv^i}{dt} = -\omega^2 q^i , \qquad i = 1, \ldots, m . \tag{1.9}$$

A direct computation shows:

$$A^{2j} = (-1)^j \omega^{2j} \mathbb{I}_N , \qquad A^{2j+1} = (-1)^j \omega^{2j} A , \qquad j \geq 0 , \tag{1.10}$$

and we find at once:

$$\exp(tA) = \cos(\omega t)\, \mathbb{I}_N + \frac{1}{\omega} \sin(\omega t)\, A , \tag{1.11}$$

as well as the standard solution for (1.9):

$$x(t) = e^{tA} x_0 .$$

given explicitly by:

$$q(t) = q_0 \cos(\omega t) + \frac{v_0}{\omega} \sin(\omega t), \quad v(t) = v_0 \cos(\omega t) - \omega q_0 \sin(\omega t) . \tag{1.12}$$

1.2.3 Inhomogeneous or Affine Equations

Because of external interactions very often systems exhibit inhomogeneous terms in the evolution equations. Let us show now how we can deal with them in the same setting as the homogeneous linear ones. Thus we will consider an inhomogeneous first-order differential equation:

$$\frac{dx}{dt} = A \cdot x + b . \tag{1.13}$$

First of all we can split off b in terms of its components in the range of A, and a supplementary space, i.e., we can write:

$$b = b_1 + b_2 , \tag{1.14}$$

where $b_2 = A \cdot c$ form some $c \in \mathbb{R}^n$. Then Eq. (1.13) becomes

$$\frac{d\widetilde{x}}{dt} = A \cdot \widetilde{x} + b_1 , \tag{1.15}$$

with $\tilde{x} = x + c$. Note that the splitting of b is not unique but depends on the choice of a supplementary space to the range of A. If $b_1 = 0$, we are back to the previous homogeneous case. If not, we can define a related dynamical system on $\mathbb{R}^{n+1}$ by setting:

$$\frac{d\xi}{dt} = \widetilde{A} \cdot \xi \tag{1.16}$$

with $\xi = (\widetilde{x}, a)^T$, $a \in \mathbb{R}$, and,

$$\tilde{A} = \left(\begin{array}{c|c} A & b_1 \\ \hline 0 & 0 \end{array}\right) . \tag{1.17}$$

Explicitly, this leads to the equations of motion

$$\frac{d\widetilde{x}}{dt} = A \cdot \widetilde{x} + b_1, \qquad \frac{da}{dt} = 0 , \tag{1.18}$$

and the solutions of Eq. (1.15) will correspond to those of Eq. (1.16) with $a(0) = 1$ (and vice versa). The latter will be obtained again by exponentiating $\tilde{A}$. Note that we can write

$$\tilde{A} = \tilde{A}_1 + \tilde{A}_2, \qquad \tilde{A}_1 = \left(\begin{array}{c|c} A & \mathbf{0} \\ \hline \mathbf{0} & \mathbf{0} \end{array}\right), \qquad \tilde{A}_2 = \left(\begin{array}{c|c} \mathbf{0} & b_1 \\ \hline \mathbf{0} & \mathbf{0} \end{array}\right) , \tag{1.19}$$

and that the commutator of $\tilde{A}_1$ and $\tilde{A}_2$ is given by:

$$[\tilde{A}_1, \tilde{A}_2] = \left(\begin{array}{c|c} \mathbf{0} & A \cdot b_1 \\ \hline \mathbf{0} & \mathbf{0} \end{array}\right) . \tag{1.20}$$

Hence an important case happens when the kernel of A is a supplementary space for the image of A, because had we chosen such space as the supplementary space giving raise to Eq. (1.14), in such case $A \cdot b_1 = 0$, and then,

$$\exp(t\tilde{A}) = \exp(t\tilde{A}_1) \exp(t\tilde{A}_2) \tag{1.21}$$

where, explicitly

$$\exp(t\tilde{A}_1) = \exp t \left(\begin{array}{c|c} A & \mathbf{0} \\ \hline \mathbf{0} & \mathbf{0} \end{array}\right) = \left(\begin{array}{c|c} e^{tA} & \mathbf{0} \\ \hline \mathbf{0} & \mathbf{I_n} \end{array}\right) ;$$

and

$$\exp(t\tilde{A}_2) = \exp t \left(\begin{array}{c|c} \mathbf{0} & b_1 \\ \hline \mathbf{0} & \mathbf{0} \end{array}\right) = \left(\begin{array}{c|c} \mathbf{I_n} & tb_1 \\ \hline \mathbf{0} & \mathbf{I_n} \end{array}\right).$$

Hence the solution of Eq. (1.15) would be written as:

$$\widetilde{x}(t) = e^{tA}\left(\widetilde{x}_0 + tb_1\right), \tag{1.22}$$

or,

$$x(t) = e^{tA}\left(x_0 + c + tb_1\right) - c, \tag{1.23}$$

with initial value $x(0) = x_0$, and only the explicit exponentiation of A is required. Very often this particular situation is referred to as the 'composition of independent motions'. Note that the fact that b can be decomposed into a part that is in the range and a part that is in the kernel of A is only guaranteed when $\ker A \oplus \operatorname{ran} A = \mathbb{R}^n$. Let us consider now two examples that illustrate this situation.

1.2.4 A Free Falling Body in a Constant Force Field

We start by considering the case of the motion of a particle in constant force field, a simple example being free fall in a constant gravitational field. As for any choice of the initial conditions the motion takes place in a plane, we can consider directly $Q = \mathbb{R}^2$ with the acceleration g pointing along the negative y-axis direction.

Then, again we take $x = (q, v)^T \in \mathbb{R}^4$, $q, v \in \mathbb{R}^2$ and Newton's equations of motion are:

$$\frac{dq^1}{dt} = v^1, \quad \frac{dq^2}{dt} = v^2, \quad \frac{dv^1}{dt} = 0, \quad \frac{dv^2}{dt} = -g. \tag{1.24}$$

If the initial velocity is not parallel to g, i.e., $v_0^1 \neq 0$, then the solutions of Eq. (1.24) will be the family of parabolas:

$$q^1(t) = q_0^1 + v_0^1 t, \qquad q^2 = q_0^2 + \left(\frac{v_0^2}{v_0^1}\right)(q^1 - q_0^1) - \left(\frac{g}{2(v_0^1)^2}\right)(q^1 - q_0^1)^2,$$

where $q_0 = (q_0^1, q_0^2)$ and $v_0 = (v_0^1, v_0^2)$ are the initial data. Equation (1.24) can be recasted in the form of Eq. (1.13) by setting, in terms of 2×2 blocks

$$A = \begin{pmatrix} \mathbf{0} & \mathbb{I}_2 \\ \mathbf{0} & \mathbf{0} \end{pmatrix}, \qquad b = -g \begin{pmatrix} 0 \\ 0 \\ 0 \\ 1 \end{pmatrix}. \tag{1.25}$$

A free particle would have been described by the previous equation with $g = 0$. In such a case the solutions would have been a family of straight lines.

Now, $\ker A = \operatorname{ran} A$ and it consists of vectors of the form $(x, \mathbf{0})^T$. Hence b is neither in $\ker A$ nor in $\operatorname{ran} A$, and in order to obtain the solution the simple method discussed in the previous section cannot be used. We would use the general procedure outlined above and would be forced to exponentiate a 5×5 matrix. In this specific case, using the decomposition Eq. (1.14), one might observe that, due to the fact that A is nilpotent of order 2 ($A^2 = 0$), both $\tilde{A}_1$ and $\tilde{A}_2$ commute with $[\tilde{A}_1, \tilde{A}_2]$, and this simplifies greatly the procedure of exponentiating $\tilde{A}$. However, that is specific to this case, that can be solved, as we did, by direct and elementary means, so we will not insist on this point.

1.2.5 Charged Particles in Uniform and Stationary Electric and Magnetic Fields

Let us consider now the motion of a charged particle in an electromagnetic field in $\mathbb{R}^3$. Denoting the electric and magnetic fields respectively by $\mathbf{E}$ and $\mathbf{B}$, and by $\mathbf{q}$, and $\mathbf{v}$ the position and velocity of the particle (all vectors in $\mathbb{R}^3$), we have that the equations of motion of the particle are given by Lorentz equations of motion:

$$\frac{d\mathbf{q}}{dt} = \mathbf{v}, \quad \frac{d\mathbf{v}}{dt} = \frac{e}{m}\mathbf{E} + \frac{e}{m}\mathbf{v} \times \mathbf{B}, \tag{1.26}$$

where m and e are the mass and charge of the particle respectively, and we have chosen physical units such that $c = 1$, with c the speed of light. Of course, we can consider the second equation in (1.26) as an autonomous inhomogeneous equation on $\mathbb{R}^3$. Let us work then on the latter.

We begin by defining a matrix $\mathcal{B}$ by setting $\mathcal{B} \cdot \mathbf{u} = \mathbf{u} \times \mathbf{B}$ for any $u \in \mathbb{R}^3$, i.e., $\mathcal{B}_{ij} = \epsilon_{ijk} B_k$. The matrix $\mathcal{B}$ is a 3×3 skew–symmetric matrix, hence degenerate. Its kernel, $\ker \mathcal{B}$, is spanned by $\mathbf{B}$ (we are assuming that $\mathbf{B}$ is not identically zero), and $\operatorname{ran} \mathcal{B}$ is spanned by the vectors that are orthogonal to $\mathbf{B}$. Hence, $\mathbb{R}^3 = \ker \mathcal{B} \oplus \operatorname{ran} \mathcal{B}$ and we are under the circumstances described after Eq. (1.20). We can decompose $\mathbf{E}$ along $\ker \mathcal{B}$ and $\operatorname{ran} \mathcal{B}$ as follows:

$$\mathbf{E} = \frac{1}{B^2}\left[(\mathbf{E} \cdot \mathbf{B})\,\mathbf{B} + \mathbf{B} \times (\mathbf{E} \times \mathbf{B})\right], \quad B^2 = \mathbf{B} \cdot \mathbf{B}. \tag{1.27}$$

In the notation of Eq. (1.13), we have:

$$x = \mathbf{v}, \quad A = \frac{e}{m}\mathcal{B}, \quad b_1 = \frac{e}{mB^2}\,(\mathbf{E} \cdot \mathbf{B})\,\mathbf{B},$$

and,

$$b_2 = A \cdot \mathbf{c} = \frac{e}{mB^2}\mathbf{B} \times (\mathbf{E} \times \mathbf{B}) \quad \text{with} \quad \mathbf{c} = \frac{1}{B^2}\mathbf{B} \times \mathbf{E}.$$

The equations of motion (1.15) become:

$$\frac{d}{dt}\left(\mathbf{v} - \frac{\mathbf{E}\times\mathbf{B}}{B^2}\right) = -\frac{e}{m}\mathbf{B}\times\left(\mathbf{v} - \frac{\mathbf{E}\times\mathbf{B}}{B^2}\right) + \frac{e}{mB^2}\left(\mathbf{E}\cdot\mathbf{B}\right)\mathbf{B},$$

and we find from Eq. (1.23)

$$\mathbf{v} - \frac{\mathbf{E}\times\mathbf{B}}{B^2} = e^{et\mathcal{B}/m}\left(\mathbf{v}_0 - \frac{\mathbf{E}\times\mathbf{B}}{B^2} + \frac{et}{mB^2}\left(\mathbf{E}\cdot\mathbf{B}\right)\mathbf{B}\right),$$

or, noticing that $\exp\left(et\mathcal{B}/m\right)\cdot\mathbf{B} = \mathbf{B}$,

$$\mathbf{v} - \frac{\mathbf{E}\times\mathbf{B}}{B^2} = e^{et\mathcal{B}/m}\left(\mathbf{v}_0 - \frac{\mathbf{E}\times\mathbf{B}}{B^2}\right) + \frac{et}{mB^2}\left(\mathbf{E}\cdot\mathbf{B}\right)\mathbf{B}.$$

If S is the matrix sending $\mathcal{B}$ into its normal form, i.e., S is the matrix defining a change of basis in which the new basis is made up by an orthonormal set with two orthogonal vectors to $\mathbf{B}$ such that $\mathbf{e}_1, \mathbf{e}_2, \mathbf{B}/\|\mathbf{B}\|$ is an oriented orthonormal set,

$$S\mathcal{B}S^{-1} = \begin{pmatrix} 0 & 1 & 0 \\ -1 & 0 & 0 \\ 0 & 0 & 0 \end{pmatrix},$$

then we have

$$S\left(\mathbf{v} - \frac{\mathbf{E}\times\mathbf{B}}{B^2}\right) = e^{\left(etS\mathcal{B}S^{-1}\right)}S\left(\mathbf{v}_0 - \frac{\mathbf{E}\times\mathbf{B}}{B^2} + \frac{et}{mB^2}\left(\mathbf{E}\cdot\mathbf{B}\right)\mathbf{B}\right),$$

and we find that $e^{et\mathcal{B}/m}$ is a rotation around the axis defined by $\mathbf{B}$ with angular velocity given by the cyclotron (or *Larmor*) frequency $\Omega = eB/m$ (recasting the light velocity c in its proper place we would find $\Omega = eB/mc$).

Proceeding further, the first of Eq. (1.26) can be rewritten as

$$\frac{d}{dt}\left(\mathbf{q} - \frac{\mathbf{E}\times\mathbf{B}}{B^2}t - \frac{et^2}{2mB^2}\left(\mathbf{E}\cdot\mathbf{B}\right)\mathbf{B}\right) = e^{et\mathcal{B}/m}\left(\mathbf{v}_0 - \frac{\mathbf{E}\times\mathbf{B}}{B^2}\right).$$

We can decompose $\mathbf{v}_0$ as well along $\ker\mathcal{B}$ and $\operatorname{ran}\mathcal{B}$, i.e.,

$$\mathbf{v}_0 = \frac{1}{B^2}\left[(\mathbf{v}_0\cdot\mathbf{B})\mathbf{B} + \mathbf{B}\times(\mathbf{v}_0\times\mathbf{B})\right]. \tag{1.28}$$

Hence,

$$\mathbf{v}_0 - \frac{\mathbf{E}\times\mathbf{B}}{B^2} = (\mathbf{v}_0\cdot\mathbf{B})\frac{\mathbf{B}}{B^2} + \left(\frac{\mathbf{B}\times\mathbf{v}_0}{B^2} - \frac{\mathbf{E}}{B^2}\right)\times\mathbf{B} = \frac{\mathbf{v}_0\cdot\mathbf{B}}{B^2}\mathbf{B} + \frac{1}{B^2}\mathcal{B}\left(\mathbf{B}\times\mathbf{v}_0 - \mathbf{E}\right).$$

By applying $e^{et\mathcal{B}/m}$ to this decomposition we get

$$e^{et\mathcal{B}/m}\left(\mathbf{v}_0-\frac{\mathbf{E}\times\mathbf{B}}{B^2}\right)=\frac{\mathbf{v}_0\cdot\mathbf{B}}{B^2}\mathbf{B}+\frac{1}{B^2}e^{et\mathcal{B}/m}\cdot\mathcal{B}\left(\mathbf{B}\times\mathbf{v}_0-\mathbf{E}\right) \tag{1.29}$$

which can be written as

$$e^{et\mathcal{B}/m}\left(\mathbf{v}_0-\frac{\mathbf{E}\times\mathbf{B}}{B^2}\right)=\frac{d}{dt}\left[t\frac{\mathbf{v}_0\cdot\mathbf{B}}{B^2}\mathbf{B}+\frac{m}{eB^2}e^{et\mathcal{B}/m}\cdot\mathcal{B}\left(\mathbf{B}\times\mathbf{v}_0-\mathbf{E}\right)\right], \tag{1.30}$$

and we find from Eq. (1.29) and (1.30) that the solution is:

$$\mathbf{q}(t)-\left[\frac{\mathbf{E}\times\mathbf{B}}{B^2}+\frac{\mathbf{v}_0\cdot\mathbf{B}}{B^2}\mathbf{B}\right]t-\frac{et^2}{2mB^2}\left(\mathbf{E}\cdot\mathbf{B}\right)\mathbf{B}-\frac{m}{eB^2}e^{et\mathcal{B}/m}\left(\mathbf{B}\times\mathbf{v}_0-\mathbf{E}\right)=\text{const.}$$

The constant can be determined in terms of the initial position q_0, and we find eventually for the general motion of a charged particle in some external electromagnetic fields $(\mathbf{E},\mathbf{B})$:

$$\begin{aligned}\mathbf{q}(t)=\mathbf{q}_0&+\left[\frac{\mathbf{E}\times\mathbf{B}}{B^2}+\frac{\mathbf{v}_0\cdot\mathbf{B}}{B^2}\mathbf{B}\right]t+\frac{et^2}{2mB^2}\left(\mathbf{E}\cdot\mathbf{B}\right)\mathbf{B}\\&+\frac{m}{eB^2}\left(\exp\left(et\mathcal{B}/m\right)-1\right)\left(\mathbf{B}\times\mathbf{v}_0-\mathbf{E}\right).\end{aligned} \tag{1.31}$$

We can examine now various limiting cases:

1. When $\mathbf{E}\times\mathbf{B}=0$, $\mathbf{E}$ is in $\ker\mathcal{B}$ and $(\exp(et\mathcal{B}/m)-1)\mathbf{E}=0$. So,

$$\mathbf{q}(t)=\mathbf{q}_0+\left[t\frac{\mathbf{v}_0\cdot\mathbf{B}}{B^2}+\frac{et^2}{2mB^2}\left(\mathbf{E}\cdot\mathbf{B}\right)\right]\mathbf{B}+\frac{m}{eB^2}(\exp\left(et\mathcal{B}/m\right)-1)\mathbf{B}\times\mathbf{v}_0,$$

   and the motion consists of a rotation around the B axis composed with a uniformly accelerated motion along the direction of $\mathbf{B}$ itself.
2. As a subcase, if $\mathbf{E}=0$, we have a rotation around $\mathbf{B}$ plus uniform motion along $\mathbf{B}$:

$$\mathbf{q}(t)=\mathbf{q}_0+t\frac{\mathbf{v}_0\cdot\mathbf{B}}{B^2}\mathbf{B}+\frac{m}{eB^2}(\exp\left(et\mathcal{B}/m\right)-1)\mathbf{B}\times\mathbf{v}_0.$$

3. $\mathbf{B}\approx 0$. In this case we can expand

$$\exp(et\mathcal{B}/m)-1\cong\frac{et\mathcal{B}}{m}+\frac{1}{2}\left(\frac{et\mathcal{B}}{m}\right)^2+O(B^3).$$

   Now,

$$\mathcal{B}(\mathbf{B}\times\mathbf{v}_0-\mathbf{E})=B^2\mathbf{v}_0-((\mathbf{v}_0\cdot\mathbf{B})\mathbf{B}+\mathbf{E}\times\mathbf{B})$$

while

$$\mathcal{B}^2(\mathbf{B} \times \mathbf{v}_0 - \mathbf{E}) = -(\mathbf{E} \times \mathbf{B}) \times \mathbf{B} + O(B^3) = B^2\mathbf{E} - (\mathbf{E} \cdot \mathbf{B})\mathbf{B} + O(B^3).$$

We see that there are quite a few cancellations and that we are left with

$$\mathbf{q}(t) = \mathbf{q}_0 + \mathbf{v}_0 t + e\,\mathbf{E}\,t^2/2m,$$

as it should be if we solve directly Eq. (1.26) with $\mathbf{B} = 0$.

4. $\mathbf{E} \cdot \mathbf{B} = 0$.

This is perhaps the most interesting case because this condition is Lorentz-invariant and this geometry of fields is precisely that giving raise to the Hall effect.

First of all, we notice that the motion along $\mathbf{B}$ will be a uniform motion. We will decouple it by assuming $\mathbf{v}_0 \cdot \mathbf{B} = 0$ (or by shifting to a reference frame that moves along $\mathbf{B}$ with uniform velocity $(\mathbf{v}_0 \cdot \mathbf{B})/B^2$). Then, we may write Eq. (1.31) as:

$$\mathbf{q}\,(t) = \mathbf{v}_D t + \mathbf{Q}_0 + \exp\left(et\mathcal{B}/m\right)\mathbf{R},$$

where

$$\mathbf{v}_D = \mathbf{E} \times \mathbf{B}/B^2, \qquad \mathbf{Q}_0 = \mathbf{q}_0 - \frac{m}{eB^2}(\mathbf{B} \times \mathbf{v}_0 - \mathbf{E}), \qquad \mathbf{R} = \frac{m}{eB^2}(\mathbf{B} \times \mathbf{v}_0 - \mathbf{E}).$$

The first term represents a uniform motion, at right angles with both $\mathbf{E}$ and $\mathbf{B}$, with the drift velocity $\mathbf{v}_D$. The second term represents a circular motion around $\mathbf{B}$ with center at Q_0, radius $R = \|\mathbf{R}\|$ and angular frequency Ω. Correspondingly, we have

$$\mathbf{v}\,(t) = \mathbf{v}_D + e^{et\mathcal{B}/m}\left(\mathbf{v}_0 - \mathbf{v}_D\right). \tag{1.32}$$

Remark 1.1 Here we can set $\mathbf{E} = 0$ too. Then $\mathbf{v}_D = 0$ and we recover case 2. The radius of the (circular) orbit is the Larmor radius

$$R_L = \|\mathbf{R}\| = v_0/\Omega\,. \tag{1.33}$$

More generally, the standard formulae for transformation of the fields under Lorentz boosts [Ja62] show that, if (in units $c = 1$): $\|\mathbf{v}_D\| = \|\mathbf{E}\|/\|\mathbf{B}\| < 1$, a Lorentz boost with velocity $\mathbf{v}_D$ leads to

$$\mathbf{E}' = 0, \quad \mathbf{B}' = \mathbf{B} + O\left((E/B)^2\right). \tag{1.34}$$

So, if $\mathbf{E} \cdot \mathbf{B} = 0$ and $\|\mathbf{E}\| < \|\mathbf{B}\|$ (actually under normal experimental conditions $\|\mathbf{E}\| \ll \|\mathbf{B}\|$) there is a frame in which the electric field can be boosted away.

1.2.5.1 Classical Hall Effect

If we have a sample with n charged particles per unit volume, the total electric current will be $\mathbf{j} = ne\mathbf{v}(t)$, with $\mathbf{v}(t)$ given by Eq. (1.32). Averaging over times of order Ω^{-1}, the second term in Eq. (1.32) will average to zero, and the average current $\mathbf{J} = \overline{\mathbf{j}(t)}$ will be given by

$$\mathbf{J} = ne\mathbf{v}_D. \tag{1.35}$$

Defining a conductivity tensor σ_{ij} as

$$J_i = \sigma_{ij} E_j, \quad i, j = 1, 2, \tag{1.36}$$

and taking the magnetic field in the z–direction, we find the (classical) Hall conductivity,

$$\sigma_{ij} = \frac{ne}{B}\epsilon_{ij}, \quad \epsilon_{12} = -\epsilon_{21} = 1, \quad \epsilon_{ii} = 0 \tag{1.37}$$

or

$$\sigma = \sigma_H \begin{pmatrix} 0 & 1 \\ -1 & 0 \end{pmatrix}, \qquad \sigma_H = \frac{ne}{B}. \tag{1.38}$$

1.2.6 Symmetries and Constants of Motion

A symmetry of the inhomogeneous equation (1.13) will be, for the moment, any smooth and invertible transformation: $x \mapsto x' = F(x)$ that takes solutions into solutions. Limiting ourselves to affine transformations, it can be easily shown that the transformation $x' = M \cdot x + d$, with M and d constant, will satisfy

$$\frac{dx'}{dt} = A \cdot x' + b$$

iff $[M, A] = 0$ and $A \cdot d = (M - I) \cdot b$.

Using Eq. (1.6) we can compute, for any smooth function f in $\mathbb{R}^n$ (the set of such functions will be denoted henceforth as $C^\infty(\mathbb{R}^n)$, $\mathcal{F}(\mathbb{R}^n)$, or simply as $\mathcal{F}$ if there is no risk of confusion):

$$\frac{df}{dt} = \frac{\partial f}{\partial x^j}\frac{dx^j}{dt} = \frac{\partial f}{\partial x^j} A^j{}_i x^i \,. \tag{1.39}$$

Then a constant of motion will be any (at least C^1) function $f(x)$ such that $df/dt = 0$. Limiting ourselves to functions that are at most quadratic in x, i.e., having the form:

$$f_N(x) = x^t N x + a^t x = N_{ij} x^i x^j + a_i x^i \tag{1.40}$$

where N is a constant symmetric matrix, we find that

$$\frac{df_N(x)}{dt} = x^t\left(A^t N + NA\right)x + \left(a^t A\right)x + a^t b + 2\left(b^t N\right)x.$$

Hence, f_N will be a constant of motion if and only if:

$$A^t N + NA = 0, \qquad a^t b = 0, \qquad a^t A + 2b^t N = 0.$$

If we want to apply these considerations to the motion of a charged particle for instance, we have to rewrite Eq. (1.26) in the compact form of Eq. (1.13). To this effect, let us introduce the six-dimensional vectors

$$x = \begin{pmatrix} \mathbf{q} \\ \mathbf{v} \end{pmatrix}, \quad b = \begin{pmatrix} \mathbf{0} \\ \mathbf{b} \end{pmatrix}, \quad \mathbf{b} = \frac{e\,\mathbf{E}}{m}, \tag{1.41}$$

and the 6×6 matrix (written in terms of 3×3 blocks)

$$A = \begin{pmatrix} \mathbf{0} & \mathbb{I} \\ \mathbf{0} & C \end{pmatrix}, \quad C \cdot \mathbf{u} = \frac{e}{m}\mathbf{u} \times \mathbf{B}\,. \tag{1.42}$$

Writing M and d in a similar form, i.e.,

$$M = \begin{pmatrix} \widehat{\alpha} & \widehat{\beta} \\ \widehat{\gamma} & \widehat{\delta} \end{pmatrix}, \quad d = \begin{pmatrix} \mathbf{d}_q \\ \mathbf{d}_v \end{pmatrix},$$

(with $\hat{\alpha}, \ldots, \hat{\delta}$ being 3×3 matrices), we find that the condition $[A, M] = 0$ becomes:

$$\hat{\gamma} = 0; \qquad \hat{\delta} = \hat{\alpha} + \hat{\beta}C, \qquad [\hat{\delta}, C] = 0,$$

while the condition $Ad = (M - 1)b$, yields:

$$\mathbf{d}_v = \hat{\beta}\mathbf{b}, \qquad C\mathbf{d}_v = (\hat{\delta} - I)\mathbf{b}.$$

There are no conditions on $\mathbf{d}_q$, and this corresponds to the fact that arbitrary translations of $\mathbf{q}$ alone are trivial symmetries of Eq. (1.26). Let us consider in particular the case in which M is a rotation matrix. Then the condition $M^T M = I$ leads to the additional constraints:

$$\widehat{\alpha}^T\widehat{\alpha} = \widehat{\delta}^T\widehat{\delta} + \widehat{\beta}^T\widehat{\beta} = I, \quad \widehat{\alpha}^T\widehat{\beta} = 0\,. \tag{1.43}$$

But then, as $\hat{\alpha}$ is an orthogonal matrix, $\hat{\beta} = 0$, and we are left with

$$\widehat{\gamma} = \widehat{\alpha}, \quad \widehat{\alpha}^t = \widehat{\alpha}^{-1}, \quad \mathbf{d}_v = 0, \quad \widehat{\alpha}^t\mathbf{b} = \mathbf{b}, \quad \left[\widehat{\delta}, C\right] = 0\,. \tag{1.44}$$

As C itself is proportional to the infinitesimal generator of rotations about the direction of $\mathbf{B}$, this implies that $\hat{\alpha}$ must represent a rotation about the same axis, and that $\mathbf{b}$

must be an eigenvector of $\hat{\alpha}$ with eigenvalue one. As $\mathbf{b}$ is parallel to $\mathbf{E}$, this implies, of course, that $\mathbf{E} \times \mathbf{B} = 0$. It is again pretty obvious that, if $\mathbf{E}$ and $\mathbf{B}$ are parallel, then rotations about their common direction are symmetries.

More generally, a simple counting shows that because of Eq. (1.43) the matrices $\hat{\alpha}$, $\hat{\beta}$, $\hat{\delta}$, generate in general a six–parameter family of symmetries. Special relationships between $\mathbf{E}$ and $\mathbf{B}$ (or vanishing of some of them) may enlarge the family.

The transformations determined by Eq. (1.44) (whether it is a symmetry or not) is an example of a point transformation, i.e., a transformation $\mathbf{q} \mapsto \mathbf{q}' = \mathbf{q}'(\mathbf{q})$, of the coordinates together with the transformation: $\mathbf{v} \mapsto \mathbf{v}' = d\mathbf{q}'/dt$, that preserve the relation between the position and the velocity. Such transformations are called also Newtonian.

For a given system of second-order differential equations one can permit also transformations (in particular, symmetries) that preserve the relationship between $\mathbf{q}$ and $\mathbf{v}$ without being point transformations (also-called 'Newtonoid' transformations).

Let us consider an example of the latter in the case of the motion of a charged particle. Consider, at the infinitesimal level, the transformation

$$\mathbf{q} \mapsto \mathbf{q}' = \mathbf{q} + \delta\mathbf{q}, \qquad \delta\mathbf{q} = \lambda \frac{e}{m} (\mathbf{v} \times \mathbf{B} + \mathbf{E}) = \lambda\, (C \cdot \mathbf{v} + \mathbf{b}) \;, \tag{1.45}$$

(with λ a small parameter) and the (infinitesimal) transformation:

$$\mathbf{v} \mapsto \mathbf{v}' = \mathbf{v} + \delta\mathbf{v}, \qquad \delta\mathbf{v} = \frac{d}{dt}\delta\mathbf{q}.$$

This transformation is clearly Newtonoid and is also a symmetry,

$$\frac{d}{dt}\mathbf{q}' = \mathbf{v}', \qquad \frac{d}{dt}\mathbf{v}' = C \cdot \mathbf{v}' + \mathbf{b}\,, \tag{1.46}$$

because

$$\delta\mathbf{v} = \lambda \frac{d}{dt} [C\mathbf{v} + \mathbf{b}] = \lambda C[C\mathbf{v} + \mathbf{b}]\,, \tag{1.47}$$

and then

$$\frac{d}{dt}\delta\mathbf{v} = \lambda C^2\, [C\mathbf{v} + \mathbf{b}] = C\delta\mathbf{v}\,. \tag{1.48}$$

1.2.6.1 Non-point Transformation Symmetries for Linear Systems

We describe now briefly a way of obtaining symmetries that are non-point transformations starting from a system of differential equations written in the form Eq. (1.5).

Let us first remark that if we start from a homogeneous linear system

$$\frac{d\xi}{dt} = A \cdot \xi \,, \tag{1.49}$$

where A is constant, we can consider for any natural number k, the infinitesimal transformation

$$\xi \mapsto \xi' = \xi + \delta\xi, \quad \delta\xi = \lambda A^k \xi \,, \tag{1.50}$$

then, it is easy to prove that

$$\frac{d\xi'}{dt} = A \cdot \xi' \,, \tag{1.51}$$

i.e., Eq. (1.50) is a symmetry. As long as taking powers generates independent matrices, this procedure will generate new symmetries at each step. Notice however that only a finite number of them, will be independent, because of the celebrated Hamilton–Cayley theorem [Ga59], according to which any matrix satisfies its characteristic equation.

In the particular case we are considering here,

$$A = \begin{bmatrix} 0 & \mathbb{I} & 0 \\ 0 & C & \mathbf{b} \\ 0 & 0 & 0 \end{bmatrix}, \tag{1.52}$$

and the infinitesimal symmetry Eq. (1.45) is precisely that generated by A^2, because

$$A^2 = \begin{bmatrix} 0 & C & \mathbf{b} \\ 0 & C^2 & C\mathbf{b} \\ 0 & 0 & 0 \end{bmatrix} \tag{1.53}$$

leads to $\delta\mathbf{q} = \lambda(C\mathbf{v} + \mathbf{b})$ and $\delta\mathbf{v} = \lambda C(C\mathbf{v} + \mathbf{b})$.

Let us discuss now briefly the constants of motion. A general symmetric matrix N can be written, in terms of 3×3 blocks, as

$$N = \begin{bmatrix} \widehat{\alpha} & \widehat{\beta} \\ \widehat{\beta}^t & \widehat{\gamma} \end{bmatrix}, \quad \widehat{\alpha}^t = \widehat{\alpha}, \quad \widehat{\gamma}^t = \widehat{\gamma} \,, \tag{1.54}$$

Writing the vector a of Eq. (1.40) as $a = \begin{bmatrix} \mathbf{a}_q \\ \mathbf{a}_v \end{bmatrix}$ and using Eqs. (1.41) and (1.42), we can spell out explicitly the condition for f_N to be a constant of motion as

1. $\hat{\alpha} + \hat{\beta} C = 0$.
2. $(\hat{\beta} + \hat{\gamma} C) + (\hat{\beta} + \hat{\gamma} C)^t = 0$.
3. $\hat{\beta} \cdot \mathbf{b} = 0$.
4. $\mathbf{a}_v \cdot \mathbf{b} = 0$.
5. $\mathbf{a}_q = C \cdot \mathbf{a}_v - 2\hat{\gamma} \cdot \mathbf{b}$.

Let us examine the case in which

$$\hat{\alpha} = \hat{\beta} = 0, \qquad \hat{\gamma} = \frac{1}{2} m I . \tag{1.55}$$

Then, as $C + C^t = 0$, Eq. (1.2.6) are automatically satisfied, and we are left with

$$\mathbf{a}_v \cdot \mathbf{b} = 0 \tag{1.56}$$

which implies $\mathbf{a}_v = \mathbf{E} \times \xi$ for an arbitrary vector ξ, and

$$\mathbf{a}_q = \frac{e}{m} \left[(\mathbf{E} \cdot \mathbf{B})\, \xi - (\xi \cdot \mathbf{B}) \mathbf{E} \right] - e\mathbf{E} . \tag{1.57}$$

Then,

1. $\xi = 0$ leads to

$$f_N = E_{\text{tot}} = \frac{1}{2} m \, [\mathbf{v}]^2 - e\mathbf{E} \cdot \mathbf{q} , \tag{1.58}$$

   where E_{tot} is the total energy.
2. For $\xi \neq 0$ we find then that

$$f_\xi = \frac{e}{m} \big[(\mathbf{E} \cdot \mathbf{B})\, (\xi \cdot \mathbf{q}) - (\xi \cdot \mathbf{B})\, (\mathbf{E} \cdot \mathbf{q}) \big] - (\mathbf{E} \times \mathbf{v}) \cdot \xi \tag{1.59}$$

   is another constant of motion.

We have discussed at length this example to show that the subsequent physical interpretation of the solutions of a given system contains much more than the general solution provided by Eq. (1.7), which from the mathematical point of view is rather exhaustive.

1.2.7 The Non-isotropic Harmonic Oscillator

Let us consider now in more detail the m-dimensional anysotropic harmonic oscillator If the oscillator is not isotropic, the system of equations (1.8) generalizes to:

$$A = \begin{pmatrix} \mathbf{0} & \mathbb{I}_m \\ -\Omega & \mathbf{0} \end{pmatrix}, \qquad \Omega = \begin{pmatrix} \omega_1^2 & & \\ & \ddots & \\ & & \omega_m^2 \end{pmatrix} \tag{1.60}$$

with different frequencies ω_k, and corresponds to the equations of motion:

$$\frac{dq^k}{dt} = v^k, \qquad \frac{dv^k}{dt} = -\omega_k^2 q^k, \qquad k = 1, \ldots, m , \tag{1.61}$$

whose general solution is the obvious generalization of Eq. (1.12).

It is more convenient to reorder the variables and denote by $\xi \in \mathbb{R}^{2m}$ the collective variable $\xi = (q_1, p_1, q_2, p_2, \ldots, q_m, p_m)$, where $p_i = v_i/\omega_i$, $i = 1, \ldots, m$. Then, the equations of motion still have the form (1.8) but now, in the new coordinates, the matrix A takes the form:

$$\tilde{A} = \begin{pmatrix} 0 & \omega_1 & 0 & 0 & . & . & . & 0 \\ -\omega_1 & 0 & 0 & 0 & . & . & . & 0 \\ 0 & 0 & 0 & \omega_2 & 0 & . & . & 0 \\ 0 & 0 & -\omega_2 & 0 & 0 & . & . & 0 \\ . & . & . & . & . & . & . & . \\ . & . & . & . & . & . & . & . \\ . & . & . & . & . & . & 0 & \omega_m \\ 0 & 0 & 0 & 0 & . & . & -\omega_m & 0 \end{pmatrix} \tag{1.62}$$

i.e., $\tilde{A}$ is block diagonal. The diagonal blocks are the 2×2 matrices

$$\begin{pmatrix} 0 & \omega_k \\ -\omega_k & 0 \end{pmatrix} = \omega_k J, \qquad J = \begin{pmatrix} 0 & 1 \\ -1 & 0 \end{pmatrix}.$$

Thus the matrix $\tilde{A}$ is a direct sum of 2×2 commuting matrices.

We can introduce complex variables

$$z_k = q_k + i p_k, \qquad k = 1, \ldots, m.$$

In terms of the latter the equations of motion (1.61) become simply:

$$\frac{dz_k}{dt} = -i\omega_k z_k\,, \tag{1.63}$$

or, in matrix form:

$$\frac{dz}{dt} = B \cdot z\,, \tag{1.64}$$

where $z = (z_1, \ldots, z_m)^T \in \mathbb{C}^n$, and B is a diagonal matrix

$$B = -i \begin{pmatrix} \omega_1 & 0 & \ldots & 0 \\ 0 & \omega_2 & \ldots & 0 \\ . & . & \ldots & . \\ . & . & \ldots & . \\ . & . & \ldots & . \\ 0 & 0 & \ldots & \omega_m \end{pmatrix}. \tag{1.65}$$

After an essentially trivial reshuffling of variables, it can be shown that the matrix $\tilde{A}$ is precisely the realified counterpart of the complex matrix B (see Sect. 2.2.3).

We can now adapt to the complex formalism the discussion of symmetries and constants of motion for the harmonic oscillator. A complex affine transformation,

$$z \mapsto z' = Mz + b\,, \tag{1.66}$$

will be a symmetry iff $[M, B] = 0$ and $B \cdot b = 0$, i.e., $b \in \ker B$. Of course, unless some of the frequencies vanish, $\ker B = 0$. Leaving aside this case, complex linear symmetries will be homogeneous transformations with

$$[M, B] = 0\,. \tag{1.67}$$

It is clear that symmetries will include those generated by all the powers of B, i.e., $M = B^k$ will generate a symmetry for all integer k's, and this will be true for any linear system. In the generic case (see below) the powers of B will exhaust all possibilities, i.e., any matrix M satisfying the condition (1.67) will be a linear combination thereof.

We can consider now two extreme cases, namely:

1. The generic case in which the frequencies are all pairwise different: $\omega_i \neq \omega_j$ for all $i \neq j$. Then M has to be diagonal, and the only symmetries are the dilations

$$z_k \mapsto \lambda_k z_k, \qquad \lambda_k \in \mathbb{C}^*,$$

which are obviously symmetries of the equation (1.63).
2. In the isotropic case, $\omega_1 = \cdots = \omega_m = \omega \neq 0$, instead, $B = -i\omega\,\mathbb{I}_{m\times m}$ and if we require the transformation to be invertible, then $M \in GL(m, \mathbb{C})$. The linear symmetry group of the m-dimensional isotropic oscillator is then the full general complex linear group $GL(m, \mathbb{C})$.

Intermediate cases can be worked out in a similar way.

Let us turn now to consider real quadratic constants of motion. In complex notation,

$$f_{a,N}(z) = z^\dagger N z + a^\dagger z + z^\dagger a, \qquad N^\dagger = N.$$

where $z^\dagger = (z_1^*, \ldots, z_m^*)$, and $N_{ij}^\dagger = N_{ji}^*$.

Again, $df_{a,N}/dt = 0$ leads to

$$B^\dagger N + NB = 0, \qquad B^\dagger a = 0.$$

But $B^\dagger = -B$ and, unless some of the frequencies vanish, this implies: $a = 0$. Quadratic constants of motion are therefore of the form

$$f_N(z) = z^\dagger N z, \qquad N^\dagger = N, \qquad [N, B] = 0\,. \tag{1.68}$$

We are here in a situation that is quite similar to the one we have just discussed in connection with symmetries, except that N is required now to be a Hermitean matrix. In the generic case, the requirement that N should commute with B forces N to be diagonal, and we have only the m independent constants of motion

$$E_i = |z_i|^2, \qquad i = 1, \ldots, m,$$

and any other quadratic constant of motion will be a linear combination with real coefficients of the functions E_i. The corresponding matrices N_i will of course commute with each other: $(N_i)_{jk} = \delta_{ij}\delta_{jk}$. Explicitly,

$$E_i = q_i^2 + v_i^2/\omega_i^2,$$

and will be proportional to the energy associated with the i-th coordinate of the oscillator, the proportionality constant being $m\omega_i^2/2$. In the isotropic case, any Hermitean matrix will generate a constant of the motion, and we will have a total of (as a basis of) m^2 independent constants of motion, namely,

$$R^{(ij)} = z_i z_j^* + z_j z_i^*, \tag{1.69}$$

$$I^{(ij)} = -i(z_i z_j^* - z_j z_i^*), \qquad \text{for } i \neq j \tag{1.70}$$

(notice that there are $m(m+1)/2$ $R^{(ij)}$'s and $m(m-1)/2$ $I^{(ij)}$'s), and again: $R^{(ii)} = 2E_i$, $I^{(ii)} = 0$. The constants of motion can be written as quadratic forms

$$R^{(ij)} = \sum_{k,l=1}^{m} z_k^* M_{kl}^{(ij)} z_l, \qquad I^{(ij)} = \sum_{k,l=1}^{m} z_k^* N_{kl}^{(ij)} z_l,$$

and the associated Hermitean matrices $M^{(ij)}$ and $N^{(ij)}$ will be given by:

$$M_{kl}^{(ij)} = \delta_k^i \delta_l^j + \delta_l^i \delta_k^j, \quad k \neq l, \qquad M_{kk}^{(ij)} = \delta_k^i \delta_k^i, \quad k = l, \qquad N_{kl}^{(ij)} = i(\delta_k^i \delta_l^j - \delta_l^i \delta_k^j,). \tag{1.71}$$

In particular the matrix associated with the total energy $E = z^\dagger z$ will be $M = \mathbb{I}$. The remaining constants of motion will be associated with traceless Hermitean matrices. A basis of $(m^2 - 1)$ such matrices will provide us with a basis in the Lie algebra of the special unitary group $SU(m)$, that has a natural action on $\mathbb{C}^m$. Recall that a matrix U is unitary if $U^\dagger U = I$ and the set of all unitary matrices is a subgroup of $GL(m, \mathbb{C})$ called the unitary group. The special unitary group $SU(m)$ is the subgroup of unitary matrices with determinant 1 (see Sect. 2.6.1 for a detailed discussion of these notions).

Hence, a basis of (quadratic) constants of motion will be provided by sesquilinear expressions of the form: $f_N = z^\dagger N z$, where N is either the identity or a generator of the Lie algebra of $SU(m)$. Let us remark that there is an obvious action of $U(m)$ on $\mathbb{C}^m$ by matrix multiplication. We recall that $U(m) \approx (U(1) \times SU(m))/\mathbb{Z}_m$, with $\mathbb{Z}_m$ being the cyclic group of order m, the correspondence given by $(e^{i\theta}, U) \mapsto e^{i\theta}U$.

Thus the $U(1)$ subgroup acts by multiplication through a phase, i.e.

$$U(1)\colon z \mapsto \exp(i\theta)z$$

and this action will be generated by $N = \mathbb{I}$. As the dynamics is described by

$$z(t) = \exp(-i\omega t)z_0\,, \tag{1.72}$$

it is clear that there is a one-to-one correspondence between constants of the motion and elements in a basis for the Lie algebra of $U(m)$, with the total energy being associated with the subgroup $U(1)$.

In view of this $U(1)$ action, it is clear that any function invariant under $U(1)$ (i.e., a constant of motion for $H = z^\dagger z$) must be at least a Hermitean quadratic invariant function. That is why there are no linear constants of motion.

We will discuss in Sect. 5.4.3 the topological features of the orbit space of a harmonic oscillator. It will define the Hopf fibration, that will be of great importance in discussing several examples throughout this book. Because of its interest, it deserves a special attention and careful analysis (see the cover picture where the trajectories of a two-dimensional non-isotropic harmonic oscillator with irrational ratio ω_1/ω_2, are depicted on a fixed energy surface).

There are other symmetries for our oscillator dynamics which are not linear. We first describe some of them and then make some general considerations. We have seen that the isotropic harmonic oscillator has constants of motion given by

$$f_N = N^{ij} z_i^* z_j\,.$$

Now we can perform any coordinate transformation of the kind

$$\xi_j = B_j{}^i z_i\,, \qquad \xi_j^* = B_j{}^i z_i^*\,,$$

with $B = (B_j{}^i)$ being now a matrix whose entries are real constants of the motion such that $\det B \neq 0$. The equations of motion in the new coordinates ξ_k are given by:

$$\dot{\xi}_k = -i\omega_k \xi_k; \qquad \dot{\xi}_k^* = i\omega_k \xi_k^*\,.$$

These equations are still linear but they are connected to the previous ones, Eq. (1.61), by a nonlinear coordinate transformation. It is clear however that symmetries that are now linear in the coordinates (ξ, ξ^*) will not come from linear symmetries in the coordinates (z, z^*). Thus, even though we are dealing with vector spaces, it is natural to allow also for nonlinear transformations. We have here an instance of a nonlinear transformation that connects to alternative linear descriptions of the same system.

If our interest is primarily on dynamical systems, we will be interested in all transformations that help visualizing the dynamics (think for instance of polar coordinates). As a matter of fact, a physical process will take place irrespective of which coordinate frame the observer may choose to observe it. For this reason it would be very useful to have an identification of our dynamical system with a mathematical entity that transforms naturally with respect to any transformation of coordinates. Of course these entities are known to be tensor fields on our space. We digress further in Chap. 2, Sect. 2.4.4. on how to associate matrices with tensor fields.

1.2.8 Lagrangian and Hamiltonian Descriptions of Evolution Equations

We will say that a given system whose evolution equations are of the form (1.1) or (1.2) admit of Lagrangian description if there exists a function $L = L(q, v)$ such that they may be written as

$$\frac{dq^i}{dt} = v^i, \quad \frac{d}{dt}\frac{\partial L}{\partial v^i} = \frac{\partial L}{\partial q^i}, \qquad i = 1, \ldots, n.$$

The function $L = L(q, v)$ will be called the Lagrangian of the system. If this happens, then the second equation in (1.2.8) implies that

$$\frac{\partial^2 L}{\partial v^i \partial v^j} F^j(q, v) = \frac{\partial L}{\partial q^i} - \frac{\partial^2 L}{\partial v^i \partial q^j} v^j, \qquad i = 1, \ldots, n, \tag{1.73}$$

We will discuss later on some of the foundational ideas behind the Lagrangian formalism and their geometrical form (see Chap. 5, Sect. 5.5) but now we will be only concentrated on discussing some consequences of the existence of such formalism for the systems presented before.

Whether or not Eq. (1.73) can be solved for L constitutes what is known as the Inverse Problem of Lagrangian Dynamics. It may admit no solutions at all or more than one solution. Notice that if L is a solution of (1.73), then λL is also a solution for any $\lambda \in \mathbb{R}$. Moreover, if L is a Lagrangian for (1.2), then $L' = L + \dot{q}^i \partial\alpha/\partial q^i$ with $\alpha(q)$ an arbitrary function, is also a solution of (1.73) in the sense that:

$$\frac{d}{dt}\frac{\partial \dot{\alpha}}{\partial v^i} = \frac{\partial \dot{\alpha}}{\partial q^i},$$

with $\dot{\alpha} = d\alpha/dt = v^i \partial\alpha/\partial q^i$ the total time derivative of the function α, is an identity. The latter case, when the solutions are not (trivially) proportional to each other or do not differ merely by a total time derivative, leads to the possibility of alternative Lagrangian descriptions for the same dynamical system. We will give examples of this later on.

We shall limit ourselves, although more general cases can be envisaged, to Lagrangians that are quadratic in the velocities. Also, we shall consider here only cases in which the Lagrangian is regular, i.e., such that

$$\det(W_{ij}) \neq 0, \qquad \text{with} \qquad W_{ij} = \frac{\partial^2 L}{\partial v^i \partial v^j} .$$

The matrix (W_{ij}) is called the Hessian matrix of the Lagrangian. If that is the case, and if $W^{-1} = (W^{ij})$ is the inverse of the Hessian matrix $W = (W_{ij})$, then Eq. (1.73) can be solved for the forces once the Lagrangian is given, i.e.,

$$F^i = W^{ij} \left(\frac{\partial L}{\partial q^j} - \frac{\partial^2 L}{\partial v^j \partial q^k} v^k \right), \qquad i = 1, \ldots, n.$$

Defining now the canonical momenta p_i as

$$p_i = \frac{\partial L}{\partial v^i}, \qquad i = 1, \ldots, n, \tag{1.74}$$

and using the implicit function theorem, one can invert this relation between the v_i's and the p_k's, $v^i = \phi^i(q, p)$, and pass over to the Hamiltonian description of the system (1.2), where the equations of motion are transformed into the well-known Hamilton equations:

$$\frac{dq^i}{dt} = \frac{\partial H}{\partial p_i}, \quad \frac{dp_i}{dt} = -\frac{\partial H}{\partial q^i}, \tag{1.75}$$

where

$$H = p_i \phi^i(q, p) - L(q, \phi(q, p)) = H(q, p)$$

is the Hamiltonian function of the system, and the velocities have been expressed as functions of the q^i's and p_k's by inverting Eq. (1.74). The space whose points are labelled by coordinates q^i's and p_i's is called the phase space of the dynamical system above and it will be denoted[1] as T^*Q if $Q \subset \mathbb{R}^n$ denotes the configuration space of the original system (1.1), hence $T^*Q \subset \mathbb{R}^{2n}$.

It is a well known fact, to be explained later on, that if Q is the configuration space of a mechanical system, there is a natural Poisson bracket structure on the phase space. Denoting collectively by ξ^i, $i = 1, \ldots, 2n$, the coordinates on T^*Q, ($\xi^i = q^i, \xi^{i+n} = p_i, i = 1, \ldots, n$), the fundamental Poisson brackets are defined as

$$\left\{\xi^i, \xi^j\right\} = J^{ij} \tag{1.76}$$

[1] For the time being that is just a convenient notation. The geometrical aspects of T^*Q and its geometrical definition will be considered later (see Sect. 5.3.2).

where $J = (J^{ij})$ is the $2n \times 2n$ symplectic matrix:

$$J = \left(J^{ij}\right) = \begin{pmatrix} \mathbf{0} & \mathbb{I} \\ -\mathbb{I} & \mathbf{0} \end{pmatrix}.$$

More explicitly,

$$\left\{q^i, q^j\right\} = \{p_i, p_j\} = 0, \quad \left\{q^i, p_j\right\} = \delta^i_j.$$

The Poisson bracket of any two functions f, g on T^*Q is defined as:

$$\{f, g\} = \sum_{ij} J^{ij} \frac{\partial f}{\partial \xi^i} \frac{\partial g}{\partial \xi^j}. \tag{1.77}$$

It is easy to show then that:

1. The Poisson bracket defined by Eq. (1.77) with J^{ij} given by (1.76), is an $\mathbb{R}$-bilinear skew-symmetric map $\{f, g\} = -\{g, f\}$, that satisfies the Jacobi identity,

$$\{\{f, g\}, h\} + \{\{h, f\}, g\} + \{\{g, h\}, f\} = 0,$$

for every triple of functions f, g, h, and that:
2. The Poisson bracket $\{f, gh\}$ can be evaluated using the Leibnitz rule, i.e.,

$$\{f, gh\} = \{f, g\}\, h + g\, \{f, h\}.$$

3. If $\{f, g\} = 0$ for any function g, then f is a constant function. We will say then that the Poisson bracket we have defined is nondegenerate.
4. In arbitrary coordinates, say $\{\eta^\alpha\}$, the Poisson bracket above can be written

$$\{f, g\} = \sum_{\alpha,\beta=1}^{2n} \frac{\partial f}{\partial \eta^\alpha} \left\{\eta^\alpha, \eta^\beta\right\} \frac{\partial g}{\partial \eta^\beta}$$

because

$$\{f, g\} = \sum_{\alpha,\beta,\gamma,\lambda=1}^{2n} \frac{\partial f}{\partial \eta^\gamma} \frac{\partial \eta^\gamma}{\partial \xi^\alpha} J^{\alpha\beta} \frac{\partial \eta^\lambda}{\partial \xi^\beta} \frac{\partial g}{\partial \eta^\lambda} = \sum_{\gamma,\lambda=1}^{2n} \frac{\partial f}{\partial \eta^\gamma} \left\{\eta^\gamma, \eta^\lambda\right\} \frac{\partial g}{\partial \eta^\lambda}.$$

Note that, in particular, under an arbitrary change of coordinates in Q: $q^i \mapsto \overline{q}^i = \overline{q}^i(q)$, the velocities will change as $v^k \mapsto \overline{v}^k = v^i \left(\partial \overline{q}^k / \partial q^i\right)$, and then, $\partial \overline{v}^k / \partial v^i = \partial \overline{q}^k / \partial q^i$. Consequently, the change in the coordinates p_i is given by

$$p_i = \frac{\partial L}{\partial v^i} = \frac{\partial v^k}{\partial v^i} \frac{\partial L}{\partial \overline{v}^k} = \frac{\partial \overline{q}^k}{\partial q^i} \overline{p}_k.$$

Then, it can be shown easily that, under such a change of coordinates, the fundamental Poisson brackets become

$$\left\{\overline{q}^i, \overline{q}^k\right\} = \left\{\overline{p}_i, \overline{p}_k\right\} = 0, \quad \left\{\overline{q}^i, \overline{p}_k\right\} = \delta^i_k, \tag{1.78}$$

and this proves that the Poisson bracket defined by Eq. (1.76) does not depend on the choice of coordinates on Q , i.e., it is an intrinsic object.

One of the first applications of the Lagrangian (and/or Hamiltonian) formalism is that they allow for procedures that associate constants of motion with infinitesimal symmetries, and *vice versa.* Let us start with the association of constants of motion with symmetries. We recall that the energy function

$$E_L = v^i \frac{\partial L}{\partial v^i} - L$$

associated with the Lagrangian L, is a constant of motion. Indeed

$$\frac{dE_L}{dt} = v^i \left(\frac{d}{dt}\frac{\partial L}{\partial v^i} - \frac{\partial L}{\partial q^i} \right) = 0.$$

Working at the infinitesimal level, let:

$$q^i \mapsto q^i + \delta q^i, \quad v^i \mapsto v^i + \delta v^i \tag{1.79}$$

be an infinitesimal symmetry. At this stage, we will not require the transformation (1.79) to be a point symmetry, i.e., one such that $\delta q^i = \delta q^i(q)$.

The condition for being a symmetry are

$$\frac{d}{dt}\delta q^i = \delta v^i = \delta \frac{dq^i}{dt}, \qquad \frac{d}{dt}\delta v^i = \delta F^i = \delta \frac{dv^i}{dt}$$

i.e., that the operations of taking variations and of taking time derivatives commute. This property being true for the variations of the fundamental functions q^i and v^i, it will be true for the variations of any function. If we consider then

$$\epsilon K = \delta E_L = \frac{\partial E_L}{\partial q^i}\delta q^i + \frac{\partial E_L}{\partial v^i}\delta v^i,$$

where ϵ is the infinitesimal parameter, it turns out that

$$\epsilon \frac{dK}{dt} = \frac{d}{dt}\delta E_L = \delta \frac{dE_L}{dt} = 0,$$

and hence K (if it is not trivially a constant) is a constant of motion.

As already said, it may happen that this procedure leads only to trivial results. Consider for example, the two-dimensional isotropic oscillator with the Lagrangian

$$L = \frac{1}{2}\mathbf{v}^2 - \frac{1}{2}\omega^2\mathbf{q}^2,$$

and then

$$E_L = \frac{1}{2}(\mathbf{v}^2 + \omega^2\mathbf{q}^2).$$

Let $\mathbf{q} = (q^1, q^2)$, $\mathbf{v} = (v^1, v^2)$, and denote by $\mathbf{k}$ a unit vector in the direction orthogonal to the plane, $\mathbf{k} = (0, 0, 1)$. Then,

$$\delta\mathbf{q} = \epsilon\,\mathbf{k} \times \mathbf{v}, \qquad \delta\mathbf{v} = \frac{d}{dt}\delta\mathbf{q} = -\epsilon\,\omega^2\mathbf{k} \times \mathbf{q},$$

is an infinitesimal symmetry, as indeed,

$$\frac{d}{dt}\delta\mathbf{v} = -\epsilon\omega^2\mathbf{k} \times \mathbf{v} = -\omega^2\delta\mathbf{q},$$

and we find

$$\epsilon\,K = \frac{1}{2}\delta(\mathbf{v}^2 + \omega^2\mathbf{q}^2) = 2\epsilon\,\omega^2(q^2v^1 - q^1v^2) = -2\epsilon\,\omega^2\mathbf{k}\cdot(\mathbf{q} \times \mathbf{v})$$

i.e., K is proportional to the angular momentum of the oscillator. If we take instead

$$\delta q^1 = \epsilon\,v^2, \qquad \delta q^2 = \epsilon\,v^1, \qquad \delta\mathbf{v} = \frac{d}{dt}\delta\mathbf{q},$$

$(\delta v^1 = -\epsilon\,\omega^2 q^2, \delta v^2 = -\epsilon\,\omega^2 q^1)$, we find $K = 0$, i.e., the above procedure turns out to be empty in this case.

The search for symmetries for systems admitting of a Lagrangian description is easier in the case of point symmetries. These symmetries correspond to symmetries for the Lagrangian. The standard way of associating constants of the motion with symmetries goes through Noether's theorem, that can be stated in a local language as follows.

Let Eq. (1.79) be a point symmetry, i.e., one such that: $\delta q^i = \delta q^i(q)$, for the system described by Eq. (1.2.8). If under Eq. (1.79),

$$L \mapsto L + \delta L, \qquad \delta L = \epsilon\,\frac{dg}{dt}, \qquad g = g(q),$$

then K given by

$$\epsilon\,K = \frac{\partial L}{\partial v_i}\delta q^i - \epsilon\,g$$

is a constant of motion.

As an example, let us consider again the two-dimensional isotropic oscillator. The equations of motion are clearly invariant under rotations in the plane, and therefore,

$$\delta \mathbf{q} = \epsilon \mathbf{k} \times \mathbf{q}, \quad \delta \mathbf{v} = \epsilon \mathbf{k} \times \mathbf{v}, \tag{1.80}$$

is a point symmetry. Under Eq. (1.80) $\delta L = 0$ and Noether's constant of motion is

$$\epsilon\, K = \frac{\partial L}{\partial \mathbf{v}} \cdot \delta \mathbf{q} = \epsilon\, \mathbf{v} \cdot \mathbf{k} \times \mathbf{q} = \epsilon\, (\mathbf{q} \times \mathbf{v}) \cdot \mathbf{k},$$

which gives us again the angular momentum. We see here that different procedures can associate the same constant of motion with different symmetries.

We can also try to work backwards, and see if and how one can associate symmetries with constants of motion. If the system happens to have a constant of motion K that is linear in the momenta, i.e., if

$$K = p_i \alpha^i - g, \qquad p_i = \frac{\partial L}{\partial v^i}, \qquad \alpha^i = \alpha^i(q), \qquad g = g(q),$$

is a constant of motion, i.e., using the equations of motion

$$0 = \frac{dK}{dt} = \frac{\partial L}{\partial q^i}\alpha^i + \frac{\partial L}{\partial v^i}\frac{d\alpha^i}{dt} - \frac{dg}{dt},$$

then, defining

$$\delta q^i = \epsilon \alpha^i, \quad \delta v^i = \frac{d}{dt}\delta q^i = \epsilon \frac{d\alpha^i}{dt} \tag{1.81}$$

it is easily seen that

$$\delta L = \epsilon\, \frac{dg}{dt}.$$

But, as the Lagrangian gets altered by a total time derivative, the equations of motion are unchanged and Eq. (1.81) is a (point and Noether) symmetry.

More generally, we can take advantage of the fact that (for regular Lagrangians) a Poisson bracket can be defined both at the Lagrangian and at the Hamiltonian level, and use it in the following manner. Let F be any function, and define the infinitesimal transformation

$$g \mapsto g + \delta g, \qquad \delta g = \epsilon\, \{g, F\},$$

(in particular, $g = q^i$ or p_i (on T^*Q) or $g = q^i$ or v^i). Considering now the time evolution we find, using the Jacobi identity,

$$\frac{d}{dt}\delta g = \epsilon\, \{\{g, F\}, H\} = \epsilon\, \{\{g, H\}, F\} + \epsilon\, \{g, \{F, H\}\}.$$

But when F is a constant of motion the last term vanishes, as being a constant of motion implies a vanishing Poisson bracket with the Hamiltonian, and we are left with

$$\frac{d}{dt}\delta g = \delta \frac{d}{dt} g,$$

which implies precisely that 'δ' commutes with the time evolution, i.e., that it is a symmetry.

1.2.9 The Lagrangian Descriptions of the Harmonic Oscillator

We have seen that different procedures for associating constants of motion with symmetries may lead to different kinds of associations. Even if we stick to point symmetries and to Noether's theorem, the association still depends in a crucial way on the Lagrangian, and it may be not be unique if the latter is not unique. To illustrate the ambiguities that may arise in such a case, let us consider again for simplicity the isotropic harmonic oscillator in m dimensions. Then, it is easy to prove that the equations of motion Eq. (1.61) (with $\omega_1 = \cdots = \omega_m = \omega$) can be derived from any Lagrangian of the form

$$L_B = \frac{1}{2} B_{ij}(v^i v^j - \omega^2 q^i q^j),$$

where (B_{ij}) is any constant nonsingular real symmetric matrix.

Let us consider, for example, the following two Lagrangians, both appropriate for the description of the two-dimensional isotropic harmonic oscillator, namely

$$L_1 = \frac{1}{2}[(v^1)^2 + (v^2)^2 - \omega^2((q^1)^2 + (q^2)^2)],$$

and

$$L_2 = v^1 v^2 - \omega^2 q^1 q^2 . \tag{1.82}$$

The Lagrangian function L_1 is invariant under rotations and Noether's theorem associates the angular momentum, $l = q^1 v^2 - q^2 v^1$, with this rotational symmetry. On the contrary, L_2 is not rotationally invariant. It is however invariant under the 'squeeze' transformations

$$(q^1, q^2) \mapsto (\lambda q^1, \lambda^{-1} q^2), \qquad (v^1, v^2) \mapsto (\lambda v^1, \lambda^{-1} v^2), \qquad \lambda \in \mathbb{R}^+.$$

Setting $\lambda = e^\tau$, we find, at the infinitesimal level:

$$\delta q^1 = \epsilon\, q^1, \qquad \delta q^2 = -\epsilon\, q^2 ,$$

and similarly for δv^1 and δv^2. The associated Noether constants of motion will be

$$\frac{\partial L}{\partial v^i}\delta q^i = (v^1\delta q^2 + v^2\delta q^1) = \epsilon\,(q^1v^2 - q^2v^1),$$

i.e., it is again the angular momentum. However, it is associated via the Lagrangian (1.82) with a symmetry that has nothing to do with rotations.

Similarly, for $m = 3$ we may consider the standard Lagrangian:

$$L_1 = \frac{1}{2}(\mathbf{v}^2 - \omega^2\mathbf{q}^2)$$

for the three-dimensional harmonic oscillator, together with, *exempli gratia*[2]

$$L_2 = \frac{1}{2}\left[\left(v^1\right)^2 + \left(v^2\right)^2 - \left(v^3\right)^2 - \omega^2\left(q^1\right)^2 - \omega^2\left(q^2\right)^2 + \omega^2\left(q^3\right)^2\right].$$

While L_1 is invariant under the orthogonal group $O(3)$, it is quite clear that L_2 will only be invariant under the pseudorthogonal group $O(2,1)$. Infinitesimal transformations of $O(2,1)$ corresponding to three generators of its Lie algebra are of the form:

$$\delta q^1 = \epsilon q^3, \quad \delta q^2 = 0, \quad \delta q^3 = \epsilon q^1$$

$$\delta q^1 = 0, \quad \delta q^2 = \epsilon q^3, \quad \delta q^3 = \epsilon q^2$$

$$\delta q^1 = \epsilon q^2, \quad \delta q^2 = -\epsilon q^1, \; \delta q^3 = 0$$

together with $\delta\mathbf{v} = d(\delta\mathbf{q})/dt$.

Notice that the two first infinitesimal symmetries correspond to 'boosts' in the q^1 and q^2 directions respectively, while the third one is a rotation in the (q^1, q^2) plane. It can be easily shown that the associated Noether constants are again the three components l_1, l_2 and l_3 of the angular momentum: $\mathbf{l} = \mathbf{q} \times \mathbf{v}$. While l_3 is still associated to a rotation, l_1 and l_2 are now associated with boosts.

1.2.10 Constructing Nonlinear Systems Out of Linear Ones

Up to now we have been discussing a variety of examples of linear systems. Our main motivation has been based, as in the rest of this book, on the idea that most, if not all, of the geometric structures associated with dynamical systems can be fully understood already at the (apparently) elementary level of linear systems.

However, a possible objection to this approach could be that there are quite a few dynamical systems (the Riccati equation, Calogero-Moser systems, Toda molecules,

[2] e.g.,'for example'.

to quote only a few of them) that are definitely nonlinear but are nonetheless (or perhaps just because of that) of a great interest.

All of these systems share with the linear ones the feature of being explicitly integrable. We will try to argue in this Section (conducted again on examples, and without pretension to full generality) that many nonlinear systems can be 'related' (in manners that will be specified immediately below) to linear ones, in such a way that (obvious) integrability of the latter will entail integrability of the former (which might look instead not so obvious). Many interesting geometrical structures can be unveiled in the study of these systems. In the spirit of this introductory chapter, we keep the exposition of the examples at the most possible elementary level, deferring a more detailed study of the same examples to the chapter on Integrability.

We will sketch here three different ways of constructing nonlinear systems which are associated with a linear one and that in addition are integrable in the same sense as the original dynamical system is, namely:

1. *Reparametrization.*
2. *Restriction to invariant surfaces.*
3. *Reduction and quotienting by equivalence relations.*

The first idea lies in the fact that changing the parametrization of a linear system sometimes preserves its explicit integrability. In this way the new system will not be linear anymore but still it will be integrable. The second idea consists simply in extracting a subsystem out of a linear system. Subsystems do not have to be linear, however they are obviously explicitly integrable if the total system is. Finally, the third procedure listed above relies in restricting the observable space of a system to a subset, to be more precise to a subalgebra, i.e., one associated with a quotient space, which need not to be linear.

A further way of constructing nonlinear systems out of linear ones by using nonlinear transformations will be discussed in Sect. 3.5.2, when discussing at length the possibility of constructing alternative (i.e. nonlinearly related) linear structures. This procedure will exploit the fact that a system can sometimes be brought to linear form using a nonlinear transformation, or in other words, that some systems are linear systems in disguise.

We illustrate next the use of the methods listed above with some elementary examples and we will leave a more detailed account of this subject to Chap. 7.

1.2.11 The Reparametrized Harmonic Oscillator

We will start by considering a very elementary example, namely the one-dimensional harmonic oscillator with unit mass $\ddot{q} = -\omega^2 q$. If u denotes the quotient $u = \dot{q}/\omega$ the equations of motion are:

$$\frac{dq}{dt} = \omega\, u, \qquad \frac{du}{dt} = -\omega\, q\,. \tag{1.83}$$

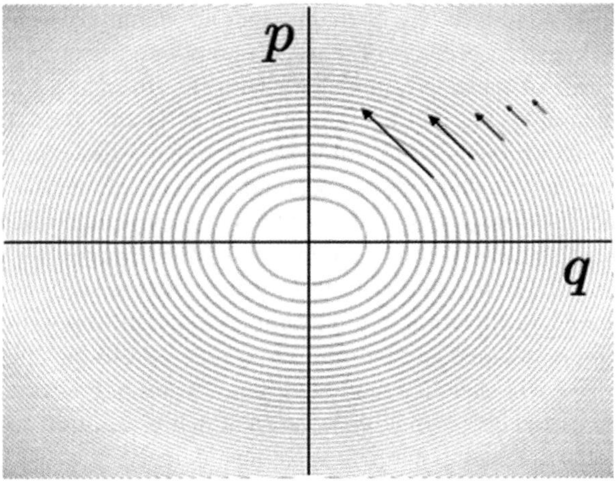

Fig. 1.1 Orbits of the 1D reparametrized harmonic oscillator. The angular velocity decays with the distance

In polar coordinates: $q = r\cos\theta$, $u = r\sin\theta$, we obtain

$$\frac{dr}{dt} = 0, \qquad \frac{d\theta}{dt} = -\omega \tag{1.84}$$

and the integral curves are obviously circles $r =$ constant, whose points are moving uniformly with angular velocity ω, i.e.,

$$\theta(t) = \theta_0 - \omega t.$$

Therefore the general solution of (1.83) is $(r_0\cos(\theta_0 - \omega t), r_0\sin(\theta_0 - \omega t))$.

If we look for instance to a spiral galaxy from a zenithal position, or to a viscous fluid rotating in a circular basin, the motion of its particles (or 'stars') is described to a first approximation by equations of motion similar to the previous ones, but the angular velocity of its integral curves is a function on the radius, where we assume the angular frequency ω to be a smooth function of r^2 at $r = 0$: $\omega = \omega(r^2)$, and thus the dynamics will be given by an equation like (Fig. 1.1):

$$\frac{d\theta}{d\tau} = -\omega\left(r^2\right). \tag{1.85}$$

The integral curves are again circles $r =$ constant, but now the particles on each circle 'move' with an angular velocity $\omega(r^2)$. Notice that r^2, and not r, is a smooth constant of motion, and any constant of the motion is a function of it. In the original coordinates (q, u) we obtain

$$\frac{dq}{d\tau} = \omega(q^2 + u^2)\,u, \qquad \frac{du}{d\tau} = -\omega(q^2 + u^2)\,q. \tag{1.86}$$

We say that the new dynamics is obtained from the previous one by time-reparametrization, and the 'time-parameter' t of the former is related to the 'time-parameter' τ of the latter by:

$$\frac{dt}{d\tau} = \frac{\omega(r^2)}{\omega},$$

or explicitly in integral form:

$$\tau = \int_0^t \frac{\omega}{\omega(r^2)} dt,$$

where the integral is computed along the integral curves of the initial dynamics.

The new dynamics is not linear anymore, even though we can integrate it explicitly. This phenomenon is general, as we will show in what follows. Before doing that, let us remark that the previous discussion can be carried out in the original linear coordinates and the ideas above do not depend on finding a 'clever' system of coordinates. Because $\omega(q^2 + u^2)$ is a constant of motion for the system , when we fix Cauchy data, say q_0, u_0, our equation can be evaluated on the curve

$$\Sigma = \left\{ \omega(q^2 + u^2) = \omega(q_0^2 + u_0^2) \right\} . \tag{1.87}$$

This restriction of a dynamics to a surface can be done every time the latter is an invariant surface under the associated flow. When restricted to the (one-dimensional) surface Σ, selected by the choice of the Cauchy data, the dynamics becomes

$$\frac{dx}{dt} = \omega(x_0^2 + y_0^2)\, y = \omega_0\, y\,, \qquad \frac{dy}{dt} = -\omega(x_0^2 + y_0^2)\, x = \omega_0\, x\,, \tag{1.88}$$

and we find now a linear system whose solution is

$$q(t) = a\cos(\omega_0 t + \alpha), \qquad u(t) = -a\sin(\omega_0 t + \alpha) \tag{1.89}$$

with $a^2 = q_0^2 + u_0^2$, $\tan\alpha = -u_0/q_0$.

1.2.11.1 The Free System in $\mathbb{R}^3$ and Invariant Surfaces

The second method consists in restricting a linear system to a nonlinear subspace. This procedure is conceptually very simple. We start with a linear system on a vector space E and restrict our Cauchy data to an invariant surface Σ which is not a linear subspace on E. Then it is obvious that if x_0 and x_1 are Cauchy data on Σ, it will not be true that $x_0 + x_1 \in \Sigma$. Therefore, the superposition rule does not hold and the restriction of Γ to Σ is not linear. In particular, if we adopt coordinates for Σ the differential equation in these coordinates will be a nonlinear differential equation.

A very simple but interesting example of this method is provided by the Newton equations of the free motion in $\mathbb{R}^3$ of a particle of unit mass:

$$\ddot{\mathbf{r}} = 0\,, \tag{1.90}$$

which can be rewritten in the form of a system

$$\begin{cases} \dot{\mathbf{r}} = \mathbf{v} \\ \dot{\mathbf{v}} = 0 \end{cases} \tag{1.91}$$

and is therefore associated to the second-order vector field Γ in $T\mathbb{R}^3$,

$$\Gamma = \mathbf{v}\frac{\partial}{\partial \mathbf{r}} \tag{1.92}$$

and has as constants of motion the velocity and the angular momentum:

$$\frac{d}{dt}\mathbf{v} = 0\,, \quad \frac{d}{dt}(\mathbf{r} \times \mathbf{v}) = 0.$$

Invariance under rotations suggests us introducing spherical polar coordinates in $\mathbb{R}^3$

$$\mathbf{r} = r\,\mathbf{n} \qquad \mathbf{n} \cdot \mathbf{n} = 1, \quad r > 0,$$

where $\mathbf{n} = \mathbf{r}/\|\mathbf{r}\| = \mathbf{r}/r$ is the unit vector in the direction of $\mathbf{r}$, and taking derivatives

$$\dot{\mathbf{r}} = \dot{r}\,\mathbf{n} + r\,\dot{\mathbf{n}}\,, \qquad \ddot{\mathbf{r}} = \ddot{r}\,\mathbf{n} + 2\,\dot{r}\,\dot{\mathbf{n}} + r\,\ddot{\mathbf{n}}\,.$$

Moreover, we have the identities

$$\mathbf{n} \cdot \mathbf{n} = 1\,, \qquad \mathbf{n} \cdot \dot{\mathbf{n}} = 0, \qquad \dot{\mathbf{n}}^2 = -\mathbf{n} \cdot \ddot{\mathbf{n}}\,,$$

and we see that $\dot{\mathbf{r}}{\cdot}\mathbf{n} = \dot{r}$, and $\mathbf{r} \cdot \dot{\mathbf{r}} = r\,\dot{r}$.

The equations $\ddot{\mathbf{r}} = 0$ are written as

$$\ddot{r} = -r\,\mathbf{n} \cdot \ddot{\mathbf{n}} = r\,\dot{\mathbf{n}}^2\,,$$

and we have for the angular momentum the expression

$$\mathbf{r} \times \mathbf{v} = r^2\mathbf{n} \times \dot{\mathbf{n}}.$$

Then $\ddot{r}$ cannot be expressed in terms of only the variable r and its derivative because of the term $\dot{\mathbf{n}}^2$.

However, by making use of constants of motion, we can choose a family of invariant surfaces Σ for Γ such that restricted to them we can associate with this equation an equation of motion involving only r, $\dot{r}$ and some "coupling constants" related to the values of the constants of motion determining the surface.

As a first instance we can consider the constant level sets of the energy function, which is a constant of motion. The leaves Σ_E are invariant under Γ and we take the restrictions of Γ on such subsets. Then

$$2E = \dot{\mathbf{r}} \cdot \dot{\mathbf{r}} = \dot{r}^2 + r^2 \dot{\mathbf{n}} \cdot \dot{\mathbf{n}} \Longrightarrow \dot{\mathbf{n}}^2 = \frac{1}{r^2}(2E - \dot{r}^2).$$

Therefore the equation of motion turns out to be

$$\ddot{r} = \frac{2E}{r} - \frac{\dot{r}^2}{r}.$$

We obtain in this way an equation of motion involving only r, $\dot{r}$ and some 'coupling constants' related to the values of the constants of motion.

We can proceed similarly with other constants of motion, for instance with the angular momentum, which is also rotationally invariant. We restrict ourselves to initial conditions with a fixed value of the angular momentum, and the corresponding surface Σ_ℓ say, for instance,

$$\ell^2 = r^4 \, \dot{\mathbf{n}}^2,$$

in order to get

$$\ddot{r} = \frac{\ell^2}{r^3}.$$

More generally, by selecting an invariant submanifold of $\mathbb{R}^3$ by means of a convex combination of energy and angular momentum, i.e. $\alpha(\mathbf{r} \times \dot{\mathbf{r}})^2 + (1-\alpha)\,\dot{\mathbf{r}}^2 = k$, with $0 \leq \alpha \leq 1$, we would find

$$\ddot{r} = \frac{\alpha\ell^2 + (1-\alpha)(2E - \dot{r}^2)r^2}{r^3} \tag{1.93}$$

which represents a completely integrable system interpolating between the two above mentioned completely integrable systems obtained for $\alpha = 0$ and $\alpha = 1$.

We can use other constants of motion to eliminate $\dot{\mathbf{n}}^2$. For instance we can use the invariant surfaces defined by

$$S = E\,\ell^2 = \frac{r^4}{2}\left(\dot{r}^2\dot{\mathbf{n}}^2 + r^2\dot{\mathbf{n}}^4\right), \tag{1.94}$$

and then we find

$$\ddot{r} = -\frac{\dot{r}^2}{2r} \pm \sqrt{r^2\,\dot{r}^2 + 8\,S}\,. \tag{1.95}$$

If we use a time-dependent surface Σ_t, or a time-dependent constant of the motion

$$r^2 + t^2\mathbf{v}^2 - 2\mathbf{r}\cdot\mathbf{v}\,t = k^2\,, \tag{1.96}$$

we find

$$\dot{\mathbf{n}}^2 = \frac{1}{r^2}[(k^2 + 2\mathbf{r}\cdot\mathbf{v}\,t - r^2)t^{-2} - \dot{r}^2]\,, \tag{1.97}$$

and replacing it in the equations of motion we find the time-dependent equation of motion,

$$\ddot{r} = \frac{k^2}{r}t^{-2} + 2\dot{r}t^{-1} - \frac{t^{-2}}{r} - \frac{\dot{r}^2}{r}\,. \tag{1.98}$$

Now it should be clear that to solve these equations it is enough to solve them on the velocity phase space $T\mathbb{R}^3$.

1.2.12 Reduction of Linear Systems

The previous discussions have shown that starting from linear systems we can obtain by elementary means nonlinear systems which are still explicitly integrable. In particular we have discussed reparametrized harmonic oscillator and generalized reparametrized systems. By restricting to surfaces we have found several nonlinear systems obtained from free motion in $\mathbb{R}^3$. In fact, a further identification by an equivalence relation is already implicit on it. Such final process of quotienting by an equivalence relation has been discussed in the context of the Riccati equation in the previous section.

Of course reparametrization and reduction via invariant surfaces and invariant equivalence relations can be used together to get nonlinear systems out of linear ones that will be called reduced dynamical systems. To be more precise, given a dynamical system, that will be supposed to be linear, the restriction to an invariant surface and the quotienting by a compatible equivalence leads to a 'reduced' system. It is obvious from what we said that the reduced dynamical system will be integrable if the initial system is.

Now we will restrict ourselves to show how it works using a couple of examples. In addition we will show how to obtain a family of the so called completely integrable systems, Calogero–Moser, Toda, etc., by reduction from simple systems.

1.2.12.1 Reducing the Free System on $\mathbb{R}^3$: The Calogero–Moser System

We consider again the free system on $\mathbb{R}^3$, namely the equations of motion

$$\ddot{\mathbf{r}} = 0. \tag{1.99}$$

This equation has a very large symmetry group and a large number of constants of motion. In fact, the Lie symmetry group of linear transformations is of dimension 11, the semidirect product $\mathbb{R}^3 \rtimes GL(3, \mathbb{R})$ (see later Sect. 4.6.1 for definitions). Therefore we have a large choice of invariant surfaces and symmetry groups.

We can use for instance an equivalence relation defined by the vector field

$$Y = x_2 \frac{\partial}{\partial x_1} - x_1 \frac{\partial}{\partial x_2} + x_3 \frac{\partial}{\partial x_2} - x_2 \frac{\partial}{\partial x_3} + v_2 \frac{\partial}{\partial v_1} - v_1 \frac{\partial}{\partial v_2} + v_3 \frac{\partial}{\partial v_2} - v_2 \frac{\partial}{\partial v_3} . \quad (1.100)$$

By using rectifying coordinates (q_1, q_2, φ) and the corresponding velocities (or momenta) in such a way that $Y = \partial/\partial\varphi$, say

$$\begin{aligned} x_1 &= q_1 \cos^2\varphi + q_2 \sin^2\varphi \\ x_2 &= \tfrac{\sin 2\varphi}{\sqrt{2}}(q_2 - q_1) \\ x_3 &= q_1 \sin^2\varphi + q_2 \cos^2\varphi \end{aligned} \quad (1.101)$$

and 'momenta'

$$\begin{aligned} p_1 &= v_1 \cos^2\varphi + v_3 \sin^2\varphi - v_2 \tfrac{\sin 2\varphi}{\sqrt{2}} \\ p_2 &= v_1 \cos^2\varphi + v_3 \sin^2\varphi + v_2 \tfrac{\sin 2\varphi}{\sqrt{2}} \\ p_\varphi &= (q_2 - q_1)(v_1 - v_3)\sin 2\varphi + \sqrt{2}(q_2 - q_1)\cos 2\varphi \end{aligned} \quad (1.102)$$

we get

$$\frac{1}{2}\mathbf{p}^2 = \frac{1}{2}(p_1^2 + p_2^2) + \frac{p_\varphi^2}{4(q_2 - q_1)^2} . \quad (1.103)$$

Therefore, on the invariant surface $p_\varphi = 2k$ associated with $\partial/\partial\varphi$ we find a quotient manifold diffeomorphic to $T^*\mathbb{R}^2$ with dynamics described by the Hamiltonian function

$$H = \frac{1}{2}(p_1^2 + p_2^2) + \frac{k^2}{(q_2 - q_1)^2} , \quad (1.104)$$

i.e., we have obtained the so-called Calogero–Moser potential, and k behaves like a coupling constant.

Now we shall continue the discussion of the example in Eq. (7.47) out of which we may find various interesting interactions.

1.2.12.2 Interactions from a Simple Cubic System on $\mathbb{R}^3$

Consider again the cubic system defined by Eq. (7.47). We shall consider the cases when it is a sphere and the case when it is an hyperboloid. For that we can consider now the invariant surface

$$\Sigma = \left\{ (\mathbf{q}, \mathbf{p}) \in T^*\mathbb{R}^3 \mid \langle \mathbf{q}, \mathbf{p} \rangle = 0,\ \langle \mathbf{q}, \mathbf{q} \rangle = 1 \right\} , \quad (1.105)$$

where $\langle\cdot,\cdot\rangle$ can denote either the pseudometric on $\mathbb{R}^3$ of signature $-++$ or the Euclidian metric. The second constraint defines either the two-sphere S^2 or an hyperboloid. By virtue of the first one, instead, at each point the momenta $\mathbf{p}$ span in both cases the Euclidean plane $\mathbb{R}^2$. All in all, the phase space of the system turns out to be in the Euclidean case TS^2 which locally looks like $S^2 \times \mathbb{R}^2$, and a similar four-dimensional surface in the pseudo-Euclidean case. On this surface the equations of motion acquire the form

$$\frac{d}{dt}\begin{pmatrix}\mathbf{q}\\ \mathbf{p}\end{pmatrix} = \begin{pmatrix}0 & 1\\ -\langle\mathbf{p},\mathbf{p}\rangle & 0\end{pmatrix}\begin{pmatrix}\mathbf{q}\\ \mathbf{p}\end{pmatrix} \tag{1.106}$$

Let us examine then separately the two cases.

Euclidean Metric: The Sphere

In this case we have

$$\langle\mathbf{q},\mathbf{q}\rangle = q_0^2 + q_1^2 + q_2^2 = 1\,, \tag{1.107}$$

and then Σ is given by

$$\Sigma = \{(\mathbf{q},\mathbf{p}) \mid \langle\mathbf{q},\mathbf{p}\rangle = q_0p_0 + q_1p_1 + q_2p_2 = 0\,. \tag{1.108}$$

The equations of motion on Σ give the geodesical equations of motion on phase space. We have

$$\dot{\mathbf{q}} = \mathbf{p}\,, \qquad \dot{\mathbf{p}} = -p^2\mathbf{q} \tag{1.109}$$

with the constraints on $\mathbf{q}$ and $\mathbf{p}$ being understood.

Now we introduce polar spherical coordinates on the unit sphere. We have

$$q_0 = \cos\theta\,, \quad q_1 = \sin\theta\cos\varphi\,, \quad q_2 = \sin\theta\sin\varphi\,. \tag{1.110}$$

In these coordinates we obtain

$$p_\theta = p_1\cos\theta\cos\varphi + p_2\cos\theta\sin\varphi - p_0\sin\theta, \qquad p_\varphi = p_2\sin\theta\cos\varphi - p_1\sin\theta\sin\varphi \tag{1.111}$$

and the Hamiltonian becomes

$$H = \frac{1}{2}\left(p_\theta^2 + \frac{p_\varphi^2}{\sin^2\theta}\right). \tag{1.112}$$

Therefore on the surface $\Sigma = \left\{(\mathbf{q},\mathbf{p})\,|\,p_\varphi = \sqrt{2}g\right\}$, by using the equivalence relation associated with $\partial/\partial\varphi$, i.e., by the curves $\dot{\varphi} = 1$, we find

$$H = \frac{1}{2}p_x^2 + \frac{g^2}{\sin^2 x} \tag{1.113}$$

where we have denoted by x the angle $0 < \theta < \pi$. The equations of motion on the sphere are easily integrated to

$$\mathbf{q}(t) = \mathbf{a}\cos kt + \mathbf{b}\sin kt \tag{1.114}$$

with $(\mathbf{a}, \mathbf{a}) = 1$, $(\mathbf{a}, \mathbf{b}) = 0$, $(\mathbf{b}, \mathbf{b}) = 1$. Hence we find:

$$x(t) = \cos^{-1}(\cos q_0 \cos kt) \tag{1.115}$$

where $k = p_\varphi / \sin q_0$.

Hyperboloid Metric: The Pseudo–Sphere

We notice that Σ can be identified with the phase space of the pseudo–sphere $(\mathbf{q}, \mathbf{q}) = 1$.

When we are on a hyperboloid of one sheet, say

$$(\mathbf{q}, \mathbf{q}) = q_0^2 - q_1^2 - q_2^2 = -1\,, \tag{1.116}$$

we introduce pseudo–spherical polar coordinates

$$q_0 = \sinh x\,, \quad q_1 = \cosh x \cos\varphi\,, \quad q_2 = \cosh x \sin\varphi\,, \tag{1.117}$$

and proceed in a similar way to find the following reduced system

$$H = \frac{1}{2}p_x^2 - \frac{g^2}{\cosh^2 x}\,. \tag{1.118}$$

However now there are three different types of geodetical orbits. We will therefore have three types of orbits for $x(t)$ depending on the initial conditions.

(a) Initial conditions such that $\langle \mathbf{a}, \mathbf{a}\rangle = -1$, $\langle \mathbf{a}, \mathbf{b}\rangle = 0$, $\langle \mathbf{b}, \mathbf{b}\rangle = 1$. Then

$$\mathbf{q}\,(t) = \mathbf{a}\cosh kt + \mathbf{b}\sinh kt \tag{1.119}$$

and

$$x(t) = \sinh^{-1}(\alpha \sinh kt)\,, \quad k = \sqrt{2}g/\sqrt{\alpha^2 - 1}\,. \tag{1.120}$$

(b) Initial conditions such that $\langle \mathbf{a}, \mathbf{a}\rangle = -1$, $\langle \mathbf{a}, \mathbf{b}\rangle = 0$, $\langle \mathbf{b}, \mathbf{b}\rangle = -1$. We obtain

$$\mathbf{q}(t) = \mathbf{a}\cos kt + \mathbf{b}\sin kt \tag{1.121}$$

$$x(t) = \sinh^{-1}(\alpha \cos kt) , \quad k = \sqrt{2}g/\sqrt{\alpha^2 + 1} . \tag{1.122}$$

(c) Initial conditions such that $\langle \mathbf{a}, \mathbf{a} \rangle = -1$, $\langle \mathbf{a}, \mathbf{b} \rangle = 0$, $\langle \mathbf{b}, \mathbf{b} \rangle = 0$.

$$\mathbf{q}(t) = \mathbf{a} + \mathbf{b}t \tag{1.123}$$

$$x(t) = \sinh^{-1}(\alpha t) , \quad k = \sqrt{2}g/\alpha . \tag{1.124}$$

1.2.12.3 Exponential Interaction

As indicated before, we can obtain different interactions, i.e., nonlinear mechanical systems with a non-trivial potential term, out of the geodesical motion on the hyperboloid, by using a different surface Σ and a different equivalence relation. We consider the above defined pseudoscalar product. We recall that the equations of motion are

$$\ddot{\mathbf{q}} = -2(\mathbf{p} \times \mathbf{q})^2 \mathbf{q} . \tag{1.125}$$

Introducing new coordinates

$$q_0 - q_1 = e^x , \quad q_0 + q_1 = e^y , \quad e^{2z} = q_2^2 , \tag{1.126}$$

it is not difficult to find

$$\ddot{x} = -g^2 e^{-2x} \tag{1.127}$$

where $g^2 = (\dot{x}^2 + (\mathbf{p} \times \mathbf{q})^2) e^{2x}$ evaluated on Σ is a constant of motion and defines an invariant surface Σ_1. The Hamiltonian function has the expression

$$H = \frac{1}{2} p_x^2 + g^2 e^{-2x} \tag{1.128}$$

which is equivalent to the Toda interaction of two particles.

For the equations of motion one has Eq. (1.119). If for simplicity we use the following initial conditions $\mathbf{a} = (a_0, a_1, 0)$, $\mathbf{b} = (0, 0, 1)$, we get

$$x(t) = \log(\alpha \cosh kt) ,$$

where $k\alpha = \sqrt{2}g$.

These various systems can be extended to higher dimensions. The main idea is to use spaces of matrices and use the group $SO(p, q)$ to introduce pseudo–spherical polar coordinates and $SU(n)$ and the Hermitian matrices. We refer to the literature for further details [Pe90]. Here we limit ourselves to derive the Calogero–Moser potential from this different viewpoint.

1.2.12.4 Free Motion on Matrix Spaces: The Calogero–Moser System

We consider a parametrization of $\mathbb{R}^3$ in terms of symmetric 2×2 matrices,

$$X = \frac{1}{\sqrt{2}}(x_1\, I + x_2\, \sigma_1 + x_3\, \sigma_3) = \frac{1}{\sqrt{2}} \begin{pmatrix} x_1 + x_3 & x_2 \\ x_2 & x_1 - x_3 \end{pmatrix} . \tag{1.129}$$

The free motion equations become in this notation

$$\ddot{X} = 0 \tag{1.130}$$

and then we see that:

$$M = [X, \dot{X}] = -(x_2\, \dot{x}_3 - x_2\, \dot{x}_3)\, \sigma \ ,$$

where σ denotes

$$\sigma = i\, \sigma_2 = \begin{bmatrix} 0 & 1 \\ -1 & 0 \end{bmatrix}$$

for which $\sigma^2 = -I$, is a matrix constant of motion whose nonzero elements are proportional to the third component ℓ_3 of the angular momentum. In fact,

$$\dot{M} = [\dot{X}, \dot{X}] + [X, \ddot{X}] = 0 \, .$$

We can introduce some new coordinates by using the rotation group: any symmetric matrix X can be diagonalized by means of an orthogonal transformation G, thus, X can be written as

$$X = G\, Q\, G^{-1} , \tag{1.131}$$

with

$$Q = \begin{bmatrix} q_1 & 0 \\ 0 & q_2 \end{bmatrix} , \qquad G = \begin{bmatrix} \cos\varphi & \sin\varphi \\ -\sin\varphi & \cos\varphi \end{bmatrix}$$

and therefore, as

$$G\, Q\, G^{-1} = \begin{bmatrix} q_1 \cos^2\varphi + q_2 \sin^2\varphi & (q_2 - q_1)\sin\varphi\cos\varphi \\ (q_2 - q_1)\sin\varphi\cos\varphi & q_1 \sin^2\varphi + q_2 \cos^2\varphi \end{bmatrix}$$

we get the relation

$$x_1 = \frac{1}{\sqrt{2}}(q_1 + q_2) , \quad x_2 = \frac{1}{\sqrt{2}}(q_2 - q_1)\sin 2\varphi , \quad x_3 = \frac{1}{\sqrt{2}}(q_1 - q_2)\cos 2\varphi \, .$$

We also note that $G\, \sigma = \sigma\, G$.

Then, using:

$$\frac{d}{dt}G^{-1} = -G^{-1}\dot{G}G^{-1},$$

we see that

$$\dot{X} = \dot{G}\,Q\,G^{-1} - G\,Q\,G^{-1}\,\dot{G}G^{-1} + G\,\dot{Q}\,G^{-1} = G\left([G^{-1}\dot{G}, Q] + \dot{Q}\right)G^{-1},$$

with

$$G^{-1}\,\dot{G} = \dot{G}\,G^{-1} = \dot{\varphi}\begin{bmatrix} 0 & 1 \\ -1 & 0 \end{bmatrix} = \dot{\varphi}\,\sigma\ ,$$

i.e.

$$\dot{X} = G(\dot{Q} + \dot{\varphi}\,[\sigma, Q])\,G^{-1}.$$

Notice that $[\sigma, Q] = (q_2 - q_1)\,\sigma$ and $[Q, \dot{Q}] = 0$. Consequently,

$$M = [X, \dot{X}] = G\,[Q, \dot{\varphi}\,[\sigma, Q] + \dot{Q}]\,G^{-1} = \dot{\varphi}\,(q_2 - q_1)G\,[Q, \sigma]\,G^{-1} = -\dot{\varphi}\,(q_2 - q_1)^2\,\sigma\ ,$$

and then ℓ_3 is given by

$$\ell_3 = \dot{\varphi}\,(q_2 - q_1)^2.$$

The equations of motion in the new coordinates are given by:

$$\frac{d}{dt}\mathrm{Tr}\,M\sigma = 0\,, \qquad \frac{d}{dt}Q = \dot{\varphi}^2[\sigma, [\sigma, Q]]\,. \tag{1.132}$$

By setting:

$$-\frac{1}{2}\mathrm{Tr}\,M\sigma = \dot{\varphi}(q_2 - q_1)^2 = g \tag{1.133}$$

we find the equations of motion:

$$\ddot{q}_1 = -\frac{2g^2}{(q_2 - q_1)^3}\,, \qquad \ddot{q}_2 = \frac{2g^2}{(q_2 - q_1)^3}\,, \tag{1.134}$$

which are the (nonlinear) Euler–Lagrange equations associated with the Lagrangian function

$$L = \frac{1}{2}\left(\dot{q}_1^2 + \dot{q}_2^2\right) - \frac{g^2}{(q_2 - q_1)^2}\,. \tag{1.135}$$

and called the Calogero-Moser equations.

1.3 Linear Systems with Infinite Degrees of Freedom

Although the main body of this book deals with the simpler case of dynamical systems modelled on finite-dimensional linear spaces, there are many instances where we have to face the description of dynamical systems whose state spaces are infinite-dimensional. That is the case, for instance, of continuum mechanics, field theory, etc., and many other cases where the framework for the description of the system is that of systems of (linear) partial differential equation.

Examples of this situation are the wave equation, the Schrödinger equation, the Klein–Gordon equation, the Pauli-Dirac equation and the Proca and Maxwell equations. In all these cases the state space of the system consists of a set of functions ϕ, called the fields of the theory, defined on an Euclidean space,[3] and therefore, it is an infinite-dimensional linear space that in addition supports a representation of the corresponding symmetry group of the theory.

From an historical viewpoint these equations arose first on phenomenological grounds and later on, their Lagrangian and Hamiltonian descriptions were investigated, mainly because of the quantization problem and the description of conservation laws.

These equations can be given the form of an ordinary (linear) differential equation in some infinite-dimensional linear space $\mathcal{F} = \{\phi\}$ once a time-coordinate t has been fixed and a splitting of the space–time of the theory has been chosen. Thus the initial data of the equation is fixed by choosing a Cauchy surface $\{t = 0\}$, and the values of the fields on it:

$$\frac{d}{dt}\phi = A\phi\,, \qquad \phi(t = 0) = \phi_0\,. \tag{1.136}$$

Very often the linear operator A is a differential operator (see Chap. 10), the simplest cases being multiplication by a function, $V(x)$, the derivative operator d/dx (the gradient in the three-dimensional case), or functions of them.

1.3.1 The Klein-Gordon Equation and the Wave Equation

The Klein-Gordon equation (KG) is usually written in the form

$$u_{tt}(x,t) - \Delta u(x,t) + m^2 u(x,t) = 0; \qquad x \in \mathbb{R}^n,$$

with initial conditions (Cauchy data): $u(x,0) = f(x),\ u_t(x,0) = g(x)$.

This equation corresponds to the relationship between linear momentum and energy:

$$E^2 = \mathbf{p}^2 + m^2,$$

[3] More generally, on manifolds.

(in natural units $\hbar = c = 1$), with the formal substitution of E by $i\,\partial/\partial t$ and $\mathbf{p}$ by $-i\,\nabla$, as operators acting on wave functions.

In order to treat this equation in the general setting of first-order systems we may use different approaches. The first one is very much similar to our procedure to associate first-order equations with second-order Newtonian ones. We set: $v(x,t) = u_t(x,t)$ and write the KG equation as the system

$$\begin{aligned} u_t - v &= 0 \\ v_t - \Delta u + m^2 u &= 0 \end{aligned} \tag{1.137}$$

with the initial conditions: $u(x,0) = f(x),\ v(x,0) = g(x)$.

We may introduce now a column vector:

$$\phi(x,t) = \begin{pmatrix} u(x,t) \\ v(x,t) \end{pmatrix},$$

and rewrite the KG equation as a first-order system:

$$\frac{d}{dt}\phi = \begin{pmatrix} 0 & \mathbb{I} \\ \Delta - m^2 & 0 \end{pmatrix}\phi, \quad \phi(x,0) = \begin{pmatrix} f(x) \\ g(x) \end{pmatrix}.$$

Remark 1.2 The operator: $H = -\Delta + m^2$ is a positive self-adjoint operator on $L^2(\mathbb{R}^n)$, the space of square integrable functions on $\mathbb{R}^n$, whose domain are the functions $f \in L^2(\mathbb{R}^n)$ such that $(k^2 + m^2)\widehat{f}(k) \in L^2(\mathbb{R}^n)$, where $\widehat{f}(\cdot)$ denotes the Fourier transform of f.

Let us denote now by B the positive, self-adjoint square root of $H = -\Delta + m^2$, $B = H^{1/2} \geq 0$, defined as it is usual in functional analysis. Since B is strictly positive and closed, its domain $D(B)$, is a Hilbert space under the inner product $(Bu, Bv)_{L^2}$. Then,we may introduce a Hilbert space: $\mathcal{H}_B = D(B) \oplus L^2(\mathbb{R}^n)$ with the inner product

$$\langle \begin{pmatrix} u \\ v \end{pmatrix}, \begin{pmatrix} u' \\ v' \end{pmatrix} \rangle = (Bu, Bu')_{\mathcal{L}^2} + (v, v')_{L^2}.$$

In analogy with what we did in the case of the harmonic oscillator (Sect. 1.2.7), we set

$$\mathbb{A} = i\begin{pmatrix} 0 & \mathbb{I} \\ -B^2 & 0 \end{pmatrix},$$

and check that $\mathbb{A}$ is a self-adjoint operator on $\mathcal{H}_B$ with domain: $D(\mathbb{A}) = D(B^2) \oplus D(B)$. Moreover, $\mathbb{A}$ is closed since B and B^2 are closed.

We may now 'integrate' the equation:

$$\frac{d}{dt}\phi = -i\mathbb{A}\phi, \quad \phi(0) = \phi_0$$

by considering, for each t,

$$\mathbb{W}(t) = \begin{pmatrix} \cos(tB) & B^{-1}\sin(tB) \\ -B\sin(tB) & \cos(tB) \end{pmatrix}$$

where each entry is defined by the usual procedure with respect to B on $L^2(\mathbb{R}^n)$.

One may check directly that $\{\mathbb{W}(t)\}_{t\in\mathbb{R}}$ is a strongly continuous one-parameter group, that is,

$$\mathbb{W}(t) \circ \mathbb{W}(s) = \mathbb{W}(t+s), \qquad \forall t, s \in \mathbb{R},$$

and the map $t \mapsto \mathbb{W}(t)f$ is continuous for any $f \in L^2(\mathbb{R}^n)$. Moreover for any $\psi \in D(\mathbb{A})$ the strong derivative of $\mathbb{W}(t)\psi$ exists and is equal to $-i\mathbb{A}\psi$. Therefore, by Stone's theorem [Re80] the operator $\mathbb{A}$ is the infinitesimal generator of a one-parameter unitary group, and the equations of motion will integrate by exponentiation as

$$\phi(t) = \exp(-i\mathbb{A}t)\,\phi_0 = \mathbb{W}(t)\phi_0\,. \tag{1.138}$$

Alternatively we may write the KG equation as a first-order differential equation by considering the change of variables:

$$\xi = Bu$$
$$\eta = v = u_t$$

and we find that the KG equation can be recasted in the form

$$\frac{d}{dt}\begin{pmatrix} \xi \\ \eta \end{pmatrix} = \begin{pmatrix} 0 & B \\ -B & 0 \end{pmatrix}\begin{pmatrix} \xi \\ \eta \end{pmatrix}.$$

This form is very similar to the way one can write the Maxwell equations (see below) and the harmonic oscillator Sect. 1.2.7. The unitary time-evolution operator is given now by

$$\widetilde{\mathbb{W}}(t) = \exp\left(-i\widetilde{\mathbb{A}}t\right),$$

where

$$\widetilde{\mathbb{A}} = i\begin{pmatrix} 0 & B \\ -B & 0 \end{pmatrix},$$

and

$$\begin{pmatrix} \xi(t) \\ \eta(t) \end{pmatrix} = \exp\left(-i\widetilde{\mathbb{A}}t\right)\begin{pmatrix} \xi(0) \\ \eta(0) \end{pmatrix}.$$

A more elementary approach to integrate the linear system above consists in using the elementary system of 'solutions' (the plane waves):

$$u_{k,\omega}(x,t) = \exp\{i\,(kx - \omega t)\}\,,$$

where k and ω have to satisfy the dispersion relation:

$$-\,\omega^2 + k^2 + m^2 = 0\,. \tag{1.139}$$

Then, the $u_{k,\omega}$'s will be locally integrable solutions and the linear superposition principle can be used to build up wave packets satisfying the assigned initial conditions.

Setting $m = 0$ we get the scalar wave equation from the KG equation. For example, in spherical polar coordinates (r, θ, ψ) the wave equation becomes

$$\frac{\partial^2 u}{\partial r^2} + \frac{2}{r}\frac{\partial u}{\partial r} + \frac{1}{r^2 \sin\theta}\frac{\partial}{\partial\theta}\left(\sin\theta\frac{\partial u}{\partial\theta}\right) + \frac{1}{r^2\sin^2\theta}\frac{\partial^2 u}{\partial\psi^2} = \frac{1}{c^2}\frac{\partial^2 u}{\partial t^2}\,.$$

If we are interested in solutions possessing spherical symmetry we have to solve the simpler equation

$$\frac{\partial^2 u}{\partial r^2} + \frac{2}{r}\frac{\partial u}{\partial r} = \frac{1}{c^2}\frac{\partial^2 u}{\partial t^2}\,,$$

that can be rewritten as

$$\frac{\partial^2}{\partial r^2}(ru) = \frac{1}{c^2}\frac{\partial^2}{\partial t^2}(ru),$$

showing that the solutions acquire the form

$$u = \frac{1}{r}f(r - ct) + \frac{1}{r}g(r + ct)\,. \tag{1.140}$$

If we look instead for solutions that, apart from time, depend only on one coordinate, x say, then they are of the form $u = f(x - ct) + g(x + ct)$. By using the generators of rotations ($x\partial/\partial y - y\partial/\partial x$, etc.), or the Lorentz boosts it is possible to generate new solutions out the given ones. This gives an example of the use of symmetries to generate new solutions, but more on this later on.

1.3.2 The Maxwell Equations

The well-known Maxwell equations describing the electric and magnetic fields, supposed to be measurable at any point of space-time $(\,x, t) \in \mathbb{R}^4$, are:

$$\frac{1}{c}\frac{\partial \mathbf{B}}{\partial t} = -\,\nabla\times\mathbf{E} = -\mathrm{rot}\,\mathbf{E}$$

$$\frac{1}{c}\frac{\partial \mathbf{D}}{\partial t} + \frac{4\pi}{c}\mathbf{j} = \nabla\times\mathbf{H} = \mathrm{rot}\,\mathbf{H}$$

$$\nabla \cdot \mathbf{B} = \operatorname{div} \mathbf{B} = 0$$
$$\nabla \cdot \mathbf{D} = \operatorname{div} \mathbf{D} = \rho$$

where **E** and **H** are the electric and magnetic fields, respectively, **D** and **B** the electric and magnetic inductions and ρ and $\mathbf{j}$ the macroscopic charge and current densities, respectively. The fields **D** and **H** are not independent of **E** and **B**, but are related by the constitutive equations.

In the absence of dielectrics and diamagnetics we may avoid dealing with all four fields **D**, **E**, **B** and **H**, and consider only the electric field **E** and the magnetic induction **B**. The Maxwell equations, in Gaussian units, are then:

$$\nabla \cdot \mathbf{E} = \operatorname{div} \mathbf{E} = 4\pi\rho\,, \qquad \text{Coulomb's law} \tag{1.141}$$
$$\nabla \cdot \mathbf{B} = \operatorname{div} \mathbf{B} = 0\,, \qquad \text{Absence of monopoles} \tag{1.142}$$
$$\frac{1}{c}\frac{\partial \mathbf{B}}{\partial t} = -\nabla \times \mathbf{E} = -\operatorname{rot} \mathbf{E}\,, \quad \text{Faraday's law} \tag{1.143}$$
$$\frac{1}{c}\frac{\partial \mathbf{E}}{\partial t} + \frac{4\pi}{c}\mathbf{j} = \nabla \times \mathbf{B} = \operatorname{rot} \mathbf{B}\,, \text{ Ampère's law} \tag{1.144}$$

The first three equations are the expression in differential form of experimental facts. The first equation is the Coulomb law for electric charges, while the second one is the Coulomb law for magnetic poles, supplemented by the fact that no free magnetic poles do seem to exist in nature. The third equation is Faraday's induction law while the fourth is the generalization of Ampère's law

$$\nabla \times \mathbf{B} = \operatorname{rot} \mathbf{B} = \frac{4\pi}{c}\mathbf{j} \tag{1.145}$$

by the addition of the term $\frac{1}{c}\,\partial\mathbf{E}/\partial t$, which is called the Maxwell displacement term. Maxwell added this term because otherwise the equations lead to $\nabla \cdot \mathbf{j} = 0$, and this would contradict the experience for a non-steady flow of current. When this term is included we get instead

$$\nabla \cdot \mathbf{j} = -\frac{1}{c}\frac{\partial \rho}{\partial t} \tag{1.146}$$

which by putting

$$\mathbf{j} = \frac{1}{c}\,\rho\mathbf{v}$$

gives raise to

$$\nabla \cdot (\rho\mathbf{v}) + \frac{\partial \rho}{\partial t} = 0. \tag{1.147}$$

Maxwell's additional term is also necessary for **E** and **B** to satisfy the wave equation in free space. If the term is not included we would get:

$$\nabla^2\mathbf{E} = 0 \qquad \nabla^2\mathbf{B} = 0,$$

which do not admit of plane wave solutions and again would contradict the experience.

These equations have to be supplemented with the Lorentz force law:

$$\mathbf{f} = \rho \left[\mathbf{E} + \frac{1}{c} \mathbf{v} \times \mathbf{B} \right] = \rho\, \mathbf{E} + \frac{1}{c} \mathbf{j} \times \mathbf{B}\,, \tag{1.148}$$

when considering also interactions of fields with charged particles.

In this section we shall concentrate on Maxwell equations in free space. Putting $\rho = 0$ and $\mathbf{j} = 0$ we find

$$\begin{aligned} \nabla \cdot \mathbf{E} &= 0, & \frac{1}{c}\frac{\partial \mathbf{B}}{\partial t} &= -\nabla \times \mathbf{E}\,, \\ \nabla \cdot \mathbf{B} &= 0, & \frac{1}{c}\frac{\partial \mathbf{E}}{\partial t} &= \nabla \times \mathbf{B}\,. \end{aligned}$$

These equations can be written in the form

$$\frac{1}{c}\frac{\partial}{\partial t}\begin{pmatrix} \mathbf{E} \\ \mathbf{B} \end{pmatrix} = \begin{pmatrix} 0 & \mathrm{rot} \\ -\mathrm{rot} & 0 \end{pmatrix}\begin{pmatrix} \mathbf{E} \\ \mathbf{B} \end{pmatrix} = \begin{pmatrix} 0 & 1 \\ -1 & 0 \end{pmatrix}\begin{pmatrix} \mathrm{rot}\,\mathbf{E} \\ \mathrm{rot}\,\mathbf{B} \end{pmatrix} \tag{1.149}$$

$$\nabla \cdot \begin{pmatrix} \mathbf{E} \\ \mathbf{B} \end{pmatrix} = 0. \tag{1.150}$$

The first equation appears as a first-order (in time) evolution equation. It can be formally integrated to yield (as in the KG equation):

$$\begin{pmatrix} \mathbf{E}(t) \\ \mathbf{B}(t) \end{pmatrix} = \mathbb{W}(t) \begin{pmatrix} \mathbf{E}(0) \\ \mathbf{B}(0) \end{pmatrix} \tag{1.151}$$

with

$$\mathbb{W}(t) = \begin{pmatrix} \cos(ct(\mathrm{rot}\,)) & \sin(ct(\mathrm{rot}\,)) \\ -\sin(ct(\mathrm{rot}\,)) & \cos(ct(\mathrm{rot}\,)) \end{pmatrix} = \exp\left(ict(\mathrm{rot}\,)\sigma_2\right) \tag{1.152}$$

where the differential operators (see Appendix G) in the entries are defined through their series expansions, and σ_2 is the Pauli sigma matrix $\begin{pmatrix} 0 & -i \\ i & 0 \end{pmatrix}$.

The second equation (1.150) appears instead as a constraint. It is immediate to show that, in view of the known fact that $\nabla \cdot \mathrm{rot}\, \equiv 0$, it will be satisfied at all times if it is at the initial time, i.e., evolution is compatible with constraints.

It will be shown (Sect. 4.2.3) that the appearance of the skew-symmetric matrix Λ,

$$\Lambda = \begin{pmatrix} 0 & \mathrm{rot} \\ -\mathrm{rot} & 0 \end{pmatrix},$$

means that the Maxwell equations can be written in Hamiltonian form.

To close this section, let us recall how the Maxwell equations can be rewritten in a manifestly covariant form. We shall assume as known that if we 'lump' together the electric and magnetic fields **E** and **B** to form the Faraday tensor:

$$F = (F_{\mu\nu}) = \begin{pmatrix} 0 & E_x & E_y & E_z \\ -E_x & 0 & -B_z & B_y \\ -E_y & B_z & 0 & -B_x \\ -E_z & -B_y & B_x & 0 \end{pmatrix}, \tag{1.153}$$

the latter transforms as a rank-two (skew-symmetric) covariant tensor.

Using the standard Minkowski metric $g_{\mu\nu}$ on $\mathbb{R}^4$ whose diagonal entries are: diag$(1, -1, -1, -1)$, the contravariant components of the Faraday tensor will be given by

$$*F = (*F^{\mu\nu}) = g^{\mu\eta} g^{\xi\nu} F_{\eta\xi}.$$

Explicitly,

$$(F^{\mu\nu}) = \begin{pmatrix} 0 & -E_x & -E_y & -E_z \\ E_x & 0 & -B_z & B_y \\ E_y & B_z & 0 & -B_x \\ E_z & -B_y & B_x & 0 \end{pmatrix}.$$

Exercise 1.1 Show that under rotations and/or inversions $\Lambda^0_0 = 1$, $\Lambda^0_\mu = 0$, $\mu = 1, 2, 3$ the F^{0k}'s and the F^{ij}'s $i, j, \neq 0$ transform separately as the components of a vector and of a pseudo-vector respectively. Show also that under a Lorentz boost with velocity **v** (and setting $c = 1$) the fields transform as:

$$\mathbf{E}' = (\mathbf{E} \cdot n)\, n + \frac{(n \times \mathbf{E}) \times n + \mathbf{v} \times \mathbf{B}}{\sqrt{1 - v^2}}$$

and,

$$\mathbf{B}' = (\mathbf{B} \cdot n) + \frac{(n \times \mathbf{B}) \times n - \mathbf{v} \times \mathbf{E}}{\sqrt{1 - v^2}},$$

where: $n = \mathbf{v}/|\mathbf{v}|$.

Introducing now the totally antisymmetric tensor $\epsilon_{\mu\nu\rho\eta}$,[4] we can define the *dual*tensor

$$\widetilde{F}_{\mu\nu} = \frac{1}{2} \epsilon_{\mu\nu\rho\eta} F^{\rho\eta}$$

or, explicitly

[4] Defined as: $\epsilon^{\mu\nu\rho\eta} = 1$ if (μ, ν, ρ, η) is an even permutation of $(0, 1, 2, 3)$, $\epsilon^{\mu\nu\rho\eta} = -1$ if (μ, ν, ρ, η) is an odd permutation, and 0 otherwise. It is left as an exercise to show that $\epsilon^{\mu\nu\rho\eta}$ is invariant under the transformations of the Poincaré group and that: $\epsilon_{\mu\nu\rho\eta} = -\epsilon^{\mu\nu\rho\eta}$.

$$(\widetilde{F}_{\mu\nu}) = \begin{pmatrix} 0 & B_x & B_y & B_z \\ -B_x & 0 & E_z & -E_y \\ -B_y & -E_z & 0 & E_x \\ -B_z & E_y & -E_x & 0 \end{pmatrix},$$

the Maxwell equations can be rewritten (again setting $c = 1$) as

$$\partial_\mu \widetilde{F}^{\mu\nu} = 0 \qquad \partial_\mu F^{\mu\nu} = 4\pi j^\nu \tag{1.154}$$

where: $(j^\nu) = (\rho, \mathbf{j})$ is the four-current. Antisymmetry of $F^{\mu\nu}$ ensures at once that j^μ is a conserved current, i.e., that it satisfies the continuity equation

$$\partial_\mu j^\mu = 0 \tag{1.155}$$

Remark 1.3 The way it is written, the continuity equation is manifestly Poincaré-invariant if j^μ does indeed transform as a four-vector. It is actually the requirement that the total charge, being a measurable quantity, be an invariant that leads to this requirement, and this entails in turn that $F_{\mu\nu}$ should transform as a rank-two tensor.

1.3.2.1 The Electromagnetic Potentials

Expressing the dual tensor in terms of the Faraday tensor the homogeneous Maxwell equations can be rewritten as:

$$\partial_\mu F_{\nu\eta} + \partial_\eta F_{\mu\nu} + \partial_\nu F_{\eta\mu} = 0.$$

Let us show in an elementary way that this implies (and is implied by) the existence of a four-vector A_μ , the *four-potential*, such that

$$F_{\mu\nu} = \partial_\mu A_\nu - \partial_\nu A_\mu\,. \tag{1.156}$$

That if A_μ exists, then the derived Faraday tensor satisfies the homogeneous Maxwell equations is obvious. Viceversa, given $F_{\mu\nu}$, let us define in the neighborhood of any point, that we will take as the origin ($x = 0$), where the homogeneous equations hold:

$$A_\mu(x) = -\int_0^1 d\lambda\,\lambda F_{\mu\nu}(\lambda x) x^\nu|.$$

Then, using again the homogeneous equations, one can prove with some algebraic effort that:

$$\partial_\mu A_\nu(x) - \partial_\nu A_\mu(x) = \int_0^1 d\lambda \frac{d}{d\lambda}\left(\lambda^2 F_{\mu\nu}(\lambda x)\right) = F_{\mu\nu}(x)\,, \tag{1.157}$$

and this completes the proof.

Spelling out explicitly the dependence of the fields from the potentials we find the more familiar relations

$$\mathbf{B} = \nabla \times \mathbf{A}, \qquad \mathbf{E} = -\frac{\partial \mathbf{A}}{\partial t} - \nabla \phi$$

where: $(\phi, \mathbf{A}) = (A^\mu)$ and $A^\mu = g^{\mu\nu} A_\nu$.

In terms of the four-potential, the inhomogeneous Maxwell equations become

$$\partial_\mu \partial^\mu A^\nu - \partial^\nu (\partial_\mu A^\mu) = 4\pi j^\nu . \tag{1.158}$$

It is known that gauge transformations of the form:

$$A_\mu \mapsto A'^\mu = A_\mu + \partial_\mu \chi \tag{1.159}$$

with χ an arbitrary (C^2) function will leave the Faraday tensor invariant. This 'gauge freedom' can be used to choose, e.g., the *Lorentz gauge*, that is defined by the covariant condition

$$\partial_\mu A^\mu = 0 \tag{1.160}$$

whereby the equation(s) for the potentials (1.158) acquire the form

$$\partial_\mu \partial^\mu A^\nu = 4\pi j^\nu , \tag{1.161}$$

i.e., in the Lorentz gauge, the potentials satisfy the inhomogeneous wave equation with the four-current j^μ as a source term.

Remark 1.4 Fixing the Lorentz gauge leaves a residual gauge freedom in which the 'gauge function' χ is obliged to satisfy the homogeneous wave equation: $\partial_\mu \partial^\mu \chi = 0$.

Remark 1.5 The explicitly covariant form of the equation for the potentials can be, to some extent, misleading. Indeed, spelling out explicitly the equations, that for the scalar potential ϕ reads:

$$\Delta \phi + \frac{\partial}{\partial t} (\nabla \cdot \mathbf{A}) = -4\pi \rho , \tag{1.162}$$

and choosing then the (non covariant) Coulomb gauge:

$$\nabla \cdot \mathbf{A} = 0 \tag{1.163}$$

(with the residual gauge freedom dictated by: $\Delta \chi = 0$) makes it explicit that the scalar potential obeys the Poisson equation

$$\Delta \phi = -4\pi \rho , \tag{1.164}$$

which does not contain time derivatives. Therefore, it is not an evolution, but rather a constraint equation (see also the previous discussion).

1.3.3 The Schrödinger Equation

The Schrödinger equation arising in one-dimensional problems in Quantum Mechanics for a quantum mechanical particle moving under the action of a potential $V(x)$, is

$$i\hbar\frac{d\psi}{dt} = -\frac{\hbar^2}{2m}\frac{d^2\psi}{dx^2} + V(x)\psi\,. \tag{1.165}$$

It corresponds to the 'Newtonian' relationship between linear momentum and energy:

$$E = \frac{p^2}{2m} + V(x)\;, \tag{1.166}$$

with the formal substitution of E by $i\,\hbar\,\partial/\partial t$ and p by $-i\,\hbar\,\partial/\partial x$, as operators acting on wave functions ψ. A general discussion on the phenomenological origins of this equation can be found for instance in [Es04].

In this case the linear space of the theory is the set of square-integrable functions $L^2(\mathbb{R})$ and the linear operator is (in natural units $\hbar = 2m = 1$):

$$A = -\frac{d^2}{dx^2} + V(x)\,.$$

The corresponding time-dependent Schrödinger equation describing the dynamical evolution of a quantum system in a three-dimensional space will be:

$$i\frac{d\psi}{dt} = -\Delta\psi + V(x)\psi\,, \tag{1.167}$$

where Δ denotes the Laplace operator on $\mathbb{R}^3$. Equation (1.167) is a complex equation that can be rewritten in terms of the real and imaginary parts of the wave function ψ [Ma96]:

$$q(x,t) = \mathrm{Re}\,\psi\,, \quad p(x,t) = \mathrm{Im}\,\psi\,, \tag{1.168}$$

as follows:

$$\frac{d}{dt}\begin{pmatrix} q \\ p \end{pmatrix} = \begin{pmatrix} 0 & -d^2/dx^2 + V(x) \\ d^2/dx^2 - V(x) & 0 \end{pmatrix}\begin{pmatrix} q \\ p \end{pmatrix} \tag{1.169}$$

i.e.,

$$\frac{d}{dt}\begin{pmatrix} q \\ p \end{pmatrix} = \begin{pmatrix} 0 & 1 \\ -1 & 0 \end{pmatrix}\left(-d^2/dx^2 + V(x)\right)\begin{pmatrix} q \\ p \end{pmatrix} = \Lambda\begin{pmatrix} q \\ p \end{pmatrix} \tag{1.170}$$

which will be shown to be of a Hamiltonian form because of the skew symmetry of the operator Λ.

Remark 1.6 The space of square-integrable complex wave functions is endowed with the standard scalar product that we will denote as $\langle\cdot,\cdot\rangle$. Then, if $\psi = q + ip$ and $\psi' = q' + ip'$:

$$\langle\psi,\psi'\rangle = \langle q,q'\rangle + \langle p,p'\rangle + i(\langle q,p'\rangle - \langle p,q'\rangle).$$

On the other hand, the real linear space of the two-dimensional real vectors $\zeta = \begin{pmatrix} q \\ p \end{pmatrix}$ has the natural scalar product:

$$\langle\zeta,\zeta'\rangle = \langle q,q'\rangle + \langle p,p'\rangle,$$

but this product captures only the real part of the complex scalar product. Notice however that the norms of the vectors will coincide in the two cases.

Let us consider now the case of a free particle (i.e., $V(x) \equiv 0$) in an arbitrary number of space dimensions. Then the realified Schrödinger equation can be written as

$$\frac{\partial}{\partial t}\begin{pmatrix} q(x,t) \\ p(x,t) \end{pmatrix} = -iK\begin{pmatrix} q(x,t) \\ p(x,t) \end{pmatrix}$$

where

$$K = i\frac{\hbar\Delta}{2m}\begin{pmatrix} 0 & -1 \\ 1 & 0 \end{pmatrix} = \frac{\hbar\Delta}{2m}\sigma_2 \,.$$

Then the evolution of initial Cauchy data is provided by:

$$\begin{pmatrix} q(x,t) \\ p(x,t) \end{pmatrix} = \mathbb{W}(t)\begin{pmatrix} q(x,0) \\ p(x,0) \end{pmatrix}$$

where

$$\mathbb{W}(t) = \exp[-iKt] = \begin{pmatrix} \cos(\hbar t\Delta/2m) & -\sin(\hbar t\Delta/2m) \\ \sin(\hbar t\Delta/2m) & \cos(\hbar t\Delta/2m) \end{pmatrix}.$$

The action of $\mathbb{W}(t)$ on the (realified) wave functions is best seen by going to Fourier space. If we define the Fourier transforms of q and p as

$$\widehat{q}(k,t) = \frac{1}{(2\pi)^{r/2}}\int d^r x e^{i\langle k,x\rangle} q(x,t)$$

(with r being the space dimension) and similarly for p, then $\mathbb{W}$ will act on the Fourier transform as the matrix (a numerical matrix for every k)

$$\widehat{\mathbb{W}}(t) = \begin{pmatrix} \cos(\hbar t k^2/2m) & \sin(\hbar t k^2/2m) \\ -\sin(\hbar t k^2/2m) & \cos(\hbar t k^2/2m) \end{pmatrix}.$$

Each entry of $\widehat{\mathbb{W}}(t)$ will act on square-integrable functions $\hat{f}(k)$ ($\hat{f} = \hat{q}$ or $\hat{p}$) and the domain of $\widehat{\mathbb{W}}(t)$ will be that of the square-integrable $\widehat{f}$'s such that $\widehat{f}(k)$ $\cos(\hbar t k^2/2m)$ ($\widehat{f}(k)\sin(\hbar t k^2/2m)$) is also square-integrable.[5] The same will be true then for $\mathbb{W}(t)$ and this, together with $\mathbb{W}\,\mathbb{W}^\dagger = \mathbb{W}^\dagger\,\mathbb{W} = \mathbb{I}$, shows in an elementary way that $\mathbb{W}$ is actually a unitary operator.

In the more general case of a nonvanishing potential, $\mathbb{W}(t)$ will have the same expression at the price of the replacement

$$\Delta \mapsto \Delta - \frac{2m}{\hbar^2} V(x) \tag{1.171}$$

and, by Stone's theorem, $\mathbb{W}(t)$ will be unitary to the extent that the potential is such that $\Delta - 2mV(x)/\hbar^2$ is self-adjoint.

1.3.3.1 Transformation Properties of the Schrödinger Equation

Consider the Galilei transformation:

$$x' = x - x_0 - vt, \qquad t' = t\,, \tag{1.172}$$

which implies that $\dot{x}' = \dot{x} - v$.

Let us consider the effect of the transformation (1.172) on the differential operators (see Appendix G) occurring in the Schrödinger equation. The elementary rules of differentiation of composite functions yield

$$\frac{\partial}{\partial t'} = \frac{\partial}{\partial t} + v\frac{\partial}{\partial x} \qquad \text{and} \qquad \frac{\partial}{\partial x} = \frac{\partial}{\partial x'}$$

which implies

$$i\hbar\frac{\partial}{\partial t'} + \frac{\hbar^2}{2m}\frac{\partial^2}{\partial x'^2} = i\hbar\left(\frac{\partial}{\partial t} + v\frac{\partial}{\partial x}\right) + \frac{\hbar^2}{2m}\frac{\partial^2}{\partial x^2}.$$

If

$$\psi'(x', t') = e^{\frac{i}{\hbar}\left(\frac{1}{2}mv^2 t - mv\cdot x\right)}\,\psi(x, t) \tag{1.173}$$

[5] But, as: $|\widehat{f}(k)\cos(\hbar t k^2/2m)|^2$ ($|\widehat{f}(k)\sin(\hbar t k^2/2m)|^2$) $\le |\widehat{f}(k)|^2$, the domain will be the whole space of square-integrable functions.

in the frame (t', x'), one finds that $\psi'(x', t')$ solves the equation:

$$\left(i\hbar\frac{\partial}{\partial t'} + \frac{\hbar^2}{2m}\frac{\partial^2}{\partial x'^2}\right)\psi' = 0 \tag{1.174}$$

if ψ solves the equation

$$\left(i\hbar\frac{\partial}{\partial t} + \frac{\hbar^2}{2m}\frac{\partial^2}{\partial x^2}\right)\psi = 0, \tag{1.175}$$

in the frame (t, x). This happens because the left-hand sides of Eqs. (1.174) and (1.175) differ by the function:

$$e^{if/\hbar}\left[-\frac{\partial f}{\partial t}\psi - v\cdot\frac{\partial f}{\partial x}\psi + i\hbar v\cdot\frac{\partial \psi}{\partial x} + \frac{\hbar^2}{2m}\sum_{l=1}^{3}\left(\frac{2i}{\hbar}\frac{\partial f}{\partial x_l}\frac{\partial \psi}{\partial x_l} + \frac{i}{\hbar}\frac{\partial^2 f}{\partial x_l^2}\psi - \frac{1}{\hbar^2}\left(\frac{\partial f}{\partial x_l}\right)^2\psi\right)\right]$$

which is found to vanish if $f = \frac{1}{2}mv^2t - mv\cdot x$.

Remark 1.7 Note that the phase of a plane wave: $\frac{i}{\hbar}(\mathbf{p}\cdot\mathbf{x} - Et)$ is not invariant under Galilei transformations, because:

$$\mathbf{p}'\cdot\mathbf{x}' - E't' = (\mathbf{p}\cdot\mathbf{x} - Et) + \frac{1}{2}m\mathbf{v}^2t - m\mathbf{v}\cdot\mathbf{x}. \tag{1.176}$$

From the relation $\mathbf{p}\cdot d\mathbf{x} - Hdt = Ldt$, it follows that the phase changes exactly by the quantity occurring in the variation of the Lagrangian. The transformation (1.173) is called a gauge transformation on the wave function, and the physical quantities which remain invariant under such transformations are called gauge-invariant. The analysis performed here shows therefore that the wave function of Quantum Mechanics does not change as a (scalar) function under transformations of reference frames. Its deeper meaning can only become clear after a thorough investigation of the geometric formulation of modern physical theories, but this goes somehow beyond the scope of this introductory chapter.

1.3.4 Symmetries and Infinite-Dimensional Systems

All field equations that we have considered so far can be written in the general form:

$$D\phi = 0$$

where ϕ is the field (scalar, real or complex, vector and so on) and D a suitable differential operator, see Appendix G, (e.g., $D = \partial_{tt} - \Delta + m^2$ for the KG equation, $D = i\hbar\partial_t - \left(\hbar^2/2m\right)\Delta + V$ for the Schrödinger equation, etc.).

A symmetry for the field equations will be, just as in the finite-dimensional case of Sect. 1.2.6, any transformation of the space-time coordinates and (if it is the case) of the fields that 'sends solutions into solutions', i.e., such that the transformed fields satisfy the same wave equation. In general one considers transformations belonging to a group. These will be either discrete transformations (like, e.g., parity and time-reversal) or transformations belonging to the connected component of the identity of the group. There can be also transformations that act on the fields alone, like the gauge transformations that we will consider later on. For example, under translations

$$x^{\mu} \mapsto x'^{\mu} = x^{\mu} + a^{\mu}, \qquad \text{then} \quad \phi(x) \mapsto \phi'(x') = \phi(x+a) \tag{1.177}$$

and that is true for *all* kinds of fields, i.e., the fields do not transform (except for their dependence on the coordinates) under space-time translations. As a second example, under a homogeneous Lorentz transformation

$$x^{\mu} \mapsto x'^{\mu} = \Lambda^{\mu}{}_{\nu} x^{\nu} \tag{1.178}$$

we will have

$$\phi(x) \mapsto \phi'(x') = \phi(\Lambda x) \tag{1.179}$$

for a scalar field while, e.g.

$$A^{\mu}(x) \mapsto A'^{\mu}(x') = \Lambda^{\mu}{}_{\nu} A^{\nu}(\Lambda x) \tag{1.180}$$

for a vector field, and so on. Actually, these transformation laws *define* for us what we mean for a field being a 'scalar', a 'vector' and so on.

Let then g be a transformation belonging to a group G. Requiring that g be a symmetry amounts to

$$D\phi = 0 \Longrightarrow D(g(\phi)) = 0, \tag{1.181}$$

where, in a rather succinct notation, we denote by $g(\cdot)$ the action of g both on the coordinates and on the form of the field.

Notice that all we are requiring here is that $g(\phi)$ be a solution whenever ϕ is (and vice versa, since g is invertible by assumption). That is a less stringent requirement than the requirement that g should commute with D, i.e., that: $gD = Dg$, this last requirement defining the invariance group of the differential operator itself, indeed a transformation which changes D into $e^{\lambda}D$ would still satisfy Eq. (1.181). This invariance group will of course be a group of symmetries, but it may well be that there are additional symmetries that do not necessarily leave D invariant. Of course it is the former that can be often identified by simple inspection, while finding the latter can be a considerably more complicated task.

For example, the full Poincaré group will be an invariance group for the KG operator and hence a symmetry group for the KG equation (and, as a special case, for the scalar wave equation) as well as of Maxwell's equations once it is recognized that the electric field **E** and the magnetic induction **B** can be lumped together to form the contravariant Faraday rank-two tensor $F^{\mu\nu}$. Translations in time will be symmetries for the Schrödinger equation for time-independent potentials. Space translations will be also symmetries in the absence of a potential, or some of them will again be symmetries if the potential is invariant under translations in some space directions. As the Schrödinger equation is manifestly nonrelativistic, the relevant group of space-time transformations will be the Galilei group. In the absence of a potential, we expect the Galilei group to be a symmetry group. Rotations and (spacetime) translations will be obvious symmetries in this case. We will digress now briefly on the covariance properties of the Schrödinger equation under the remaining part of the group, namely under the Galilei boosts.

1.3.5 Constants of Motion

A *constant of motion* will be a function or a functional of the fields that remains constant in time when the solutions of the field equation are substituted for the fields, i.e., a function (or a functional) that is 'constant along the trajectories' of the field equation. Normally, in Classical Field Theory, there appear *conserved currents*, i.e. four-vectors j^μ (but they could be higher-rank tensors as well) that are built up using the fields and such that they satisfy a *continuity equation* of the form

$$\partial_\mu j^\mu = 0. \tag{1.182}$$

Remark 1.8 The formally 'covariant' notation that we are using here has to be intended only as a convenient notation. We will employ the same notation in non-relativistic cases as well, like when it will be the case of the Schrödinger equation. So, the previous equation stands simply for (we will set $c = 1$ everywhere in what follows whenever the velocity of light enters the equations):

$$\frac{\partial \rho}{\partial t} + \nabla \cdot \mathbf{j} = 0 \tag{1.183}$$

where $\rho = j^0$ *and* $\mathbf{j} = (j^1, j^2, j^3)$.

If a continuity equation holds, integrating it on a constant-time surface (or, for that matter,on any space-like surface) and defining

$$Q = \int_{\mathbb{R}^3} \rho(\mathbf{x}, t)\, d^3\mathbf{x} \tag{1.184}$$

we find

$$\frac{dQ}{dt} = \int_{\mathbb{R}^3} \frac{\partial \rho(\mathbf{x}, t)}{\partial t}\, d^3\mathbf{x} = -\int_{\mathbb{R}^3} \nabla \cdot \mathbf{j}(\mathbf{x}, t)\, d^3\mathbf{x} \tag{1.185}$$

and, if the fields are 'well behaved', i.e., they vanish fast enough at space infinity, Gauss's theorem shows that we can drop the resulting surface term and end up with

$$\frac{dQ}{dt} = 0. \tag{1.186}$$

i.e., Q is the required constant of motion.

We give below a few examples of conserved currents and of the associated constants of motion, warning that is far from being an exhaustive list.

1. **The real KG field.**

Consider the four-vector $(h, \mathbf{j}_H)$, where

$$h(x,t) = \frac{1}{2}\left\{(\partial_0\phi)^2 + (\nabla\phi)^2 + m^2\phi^2\right\} \tag{1.187}$$

and

$$\mathbf{j}_H = -(\partial_0\phi)\nabla\phi\,. \tag{1.188}$$

It is not hard to see that, by virtue of the equation of motion obeyed by the KG field, it satisfies the continuity equation

$$\frac{\partial h}{\partial t} + \nabla\cdot\mathbf{j}_h = 0\,. \tag{1.189}$$

The quantity

$$E = \int_{\mathbb{R}^3} h(\mathbf{x},t)\, d^3\mathbf{x} \tag{1.190}$$

is therefore a constant of motion and, indeed, as will be seen in the next Subsection, it can be identified with the total Hamiltonian, and hence with the total energy, for the KG field.

Remark 1.9 The conditions of 'good behavior at infinity' that allows us to deduce the conservation law from the continuity equation are just the same that ensure that the total energy of the field is finite.

2. **The complex KG field.**

It is left as an exercise to show that if ϕ is complex-valued the definitions of h and $\mathbf{j}_H$ modify into

$$h(\mathbf{x},t) = \frac{1}{2}\left\{|\partial_0\phi|^2 + \nabla\phi\cdot\nabla\overline{\phi} + m^2|\phi|^2\right\} \tag{1.191}$$

and

$$\mathbf{j}_H(\mathbf{x}, t) = -\frac{1}{2}\left\{(\partial_0\overline{\phi})\nabla\phi + \partial_0\phi\nabla\overline{\phi}\right\} \tag{1.192}$$

and the same equation of continuity holds.

Again using the equations of motion it can be proved immediately that the four-vector

$$j^\mu = -\frac{i}{2}\phi \overleftrightarrow{\partial}^\mu \overline{\phi} \tag{1.193}$$

where

$$\phi \overleftrightarrow{\partial}^\mu \overline{\phi} = \phi\partial^\mu\overline{\phi} - \overline{\phi}\partial^\mu\phi \tag{1.194}$$

satisfies a continuity equation. The associated constant of motion will be

$$Q = -\frac{i}{2}\int_{\mathbb{R}^3} d^3\mathbf{x}\, \phi \overleftrightarrow{\partial}^0 \overline{\phi} \tag{1.195}$$

It will be useful to express the conservation laws that we have just found for the KG equation in terms of the Fourier transform of the field.

For reasons that will become clear shortly, the Fourier expansion of the field is done customarily with a slightly different normalization, namely as

$$\phi(x) = \frac{\sqrt{2}}{(2\pi)^{3/2}}\int d^4k \exp(-i\langle k, x\rangle)\widehat{\phi}(k) \tag{1.196}$$

where: $\langle k, x\rangle = k_\mu x^\mu$. The KG equation becomes then simply

$$(k_\mu k^\mu - m^2)\widehat{\phi}(k) = 0 \tag{1.197}$$

The solutions will be therefore distributions concentrated on the 'mass shell': $k_\mu k^\mu - m^2 = 0$. Explicitly, with: $(k^\mu) = (k_0, \mathbf{k})$, $(k_\mu) = (k_0, -\mathbf{k})$, the mass shell will be defined by the equation: $k_0^2 = \mathbf{k}^2 + m^2$. It will be then the union of the two disjoint hyperboloids M_+ and M_- defined by

$$M_\pm = \{(k_0, \mathbf{k}) \in \mathbb{R}^4 |\ k_0 = \pm E(\mathbf{k}),\quad E(\mathbf{k}) = \sqrt{\mathbf{k}^2 + m^2}\,. \tag{1.198}$$

The solution(s) of the KG equation in Fourier space will be of the form:

$$\widehat{\phi}(k) \propto \delta(k_\mu k^\mu - m^2) \equiv \delta(k_0^2 - \mathbf{k}^2 - m^2) \tag{1.199}$$

Using some standard results in the manipulation of δ functions

$$\delta(k_0^2 - \mathbf{k}^2 - m^2) = \frac{1}{2E(\mathbf{k})}\left\{\delta(k_0 - E(\mathbf{k})) + \delta(k_0 + E(\mathbf{k}))\right\}. \tag{1.200}$$

As the mass shell is the union of two disjoint components, $\widehat{\phi}$ will be fixed by the independent values $\widehat{\phi}_+(\mathbf{k})$ and $\widehat{\phi}_-(\mathbf{k})$ it assumes on M_+ and M_- respectively, and the most general solution of the KG equation will be of the form

$$\widehat{\phi}(k) = \frac{1}{2E(\mathbf{k})}\left\{\widehat{\phi}_+(\mathbf{k})\delta(k_0 - E(\mathbf{k})) + \widehat{\phi}_-(\mathbf{k})\delta(k_0 + E(\mathbf{k}))\right\}. \tag{1.201}$$

Eventually, the most general solution of the KG equation in space-time can be written as

$$\phi(x) = \phi_+(x) + \phi_-(x) \tag{1.202}$$

with

$$\phi_\pm(x) = \int \frac{d^3\mathbf{k}}{(2\pi)^{3/2}\sqrt{2}E(\mathbf{k})} \exp[i(\mathbf{k}\cdot\mathbf{x} \mp E(\mathbf{k})x^0)]\widehat{\phi}_\pm(\mathbf{k}). \tag{1.203}$$

It will be convenient also to introduce the amplitudes

$$a(\mathbf{k}) = [E(\mathbf{k})]^{-1/2}\widehat{\phi}_+(\mathbf{k}) \tag{1.204}$$

and

$$\overline{b(\mathbf{k})} = [E(\mathbf{k})]^{-1/2}\widehat{\phi}_-(-\mathbf{k}). \tag{1.205}$$

Then

$$\phi(x) = \int \frac{d^3k}{(2\pi)^{3/2}E(\mathbf{k})}\left[a(\mathbf{k})\exp(-i\langle k, x\rangle + \overline{b(\mathbf{k})}\exp(i\langle k, x\rangle\right] \tag{1.206}$$

where: $\langle k, x\rangle = E(k)x^0 - \mathbf{k}\cdot\mathbf{x}$.

In terms of the new amplitudes that we have just introduced it is not hard to check that the two constants of motion that we have just found take the form

$$E = \frac{1}{2}\int d^3k E(\mathbf{k})\left\{[a(\mathbf{k})]^2 + [b(\mathbf{k})]^2\right\} \tag{1.207}$$

and

$$Q = \frac{1}{2}\int d^3k\left\{[a(\mathbf{k})]^2 - [b(\mathbf{k})]^2\right\} \tag{1.208}$$

and are both additive in Fourier space. This leads to interpreting $(1/2)\ |a(\mathbf{k})|^2$ $((1/2)\ |b(\mathbf{k})|^2)$ as a density of positive (negative) charges with energy $E(k)$ associated with the plane wave of momentum $\mathbf{k}$.

Remark 1.10 Of course the same construction applies to a real KG field, in which case reality of the field implies: $a(\mathbf{k}) = b(\mathbf{k})$, and hence $Q = 0$ for any solution of the KG equation. With the interpretation outlined above, a real KG field is a neutral field, while a complex KG field is a charged field.

3. **The electromagnetic field.** Let us reconsider now the Maxwell equations in vacuum, namely the inhomogeneous equations

$$\nabla \cdot \mathbf{E} = 4\pi\rho, \quad \frac{1}{c}\frac{\partial \mathbf{E}}{\partial t} + \frac{4\pi}{c}\mathbf{j} = \nabla \times \mathbf{B} \tag{1.209}$$

and the homogeneous ones

$$\nabla \cdot \mathbf{B} = 0, \quad \frac{1}{c}\frac{\partial \mathbf{B}}{\partial t} = -\nabla \times \mathbf{E}. \tag{1.210}$$

As a first example of a locally conserved current associated with the Maxwell equations, it is easy to deduce from the first pair of equations that the charge and current densities obey the continuity equation

$$\frac{\partial \rho}{\partial t} + \nabla \cdot \mathbf{j} = 0 \tag{1.211}$$

that is a local conservation law which implies that, under rather mild assumptions on the distribution of charges and currents (e.g., if they are localized) the *total charge*

$$Q = \int d^3\mathbf{x}\rho(\mathbf{x}, t) \tag{1.212}$$

will be a conserved quantity

$$\frac{dQ}{dt} = 0. \tag{1.213}$$

In Electrostatics the energy density associated with the electric field is (in Gaussian units) $E^2/8\pi$. Quite similarly, the energy density of the field is $B^2/8\pi$ in Magnetostatics. It makes therefore sense to consider

$$h(\mathbf{x}, t) = \frac{E^2 + B^2}{8\pi} \tag{1.214}$$

as the energy density associated with a general electromagnetic field. Taking then time derivatives and using again Maxwell's equations one finds easily that

$$\frac{\partial h}{\partial t} = -\frac{c}{4\pi}\left[\mathbf{B} \cdot \nabla \times \mathbf{E} - \mathbf{E} \cdot \nabla \times \mathbf{B}\right] - \mathbf{E} \cdot \mathbf{j} \tag{1.215}$$

But

$$\mathbf{B} \cdot \nabla \times \mathbf{E} - \mathbf{E} \cdot \nabla \times \mathbf{B} \equiv \nabla \cdot (\mathbf{E} \times \mathbf{B}) \tag{1.216}$$

and therefore, defining the *Poynting vector*

$$\mathbf{S} = \frac{c}{4\pi} \mathbf{E} \times \mathbf{B} \tag{1.217}$$

we obtain

$$\frac{\partial h}{\partial t} + \nabla \cdot \mathbf{S} = -\mathbf{E} \cdot \mathbf{j} \tag{1.218}$$

In the absence of charges and currents this becomes the local conservation law

$$\frac{\partial h}{\partial t} + \nabla \cdot \mathbf{S} = 0 \tag{1.219}$$

and this implies that, under suitable assumptions on the distribution of the fields

$$H = \int d^3\mathbf{x}\, \frac{E(\mathbf{x}, t)^2 + B(\mathbf{x}, t)^2}{8\pi} \tag{1.220}$$

will be a conserved quantity, and will represent the total energy stored in the electromagnetic field. Whenever $\mathbf{j} \neq 0$, the term on the right-hand side of Eq. (1.218) can be interpreted as the rate at which energy is being exchanged with the system of sources, and a more general conservation law can be established in that context as well.

4. **The Schrödinger equation.**

Now we consider the Schrödinger equation and its complex conjugate equation which is automatically satisfied:

$$-i\hbar \frac{\partial \psi^*}{\partial t} = \left(-\frac{\hbar^2}{2m} \triangle + V \right) \psi^*(\mathbf{x}, t) \tag{1.221}$$

and we multiply the Schrödinger equation by $\psi^*(x, t)$, and its complex conjugate by $\psi(x, t)$. On subtracting one equation from the other one then finds

$$-\frac{\hbar^2}{2m} (\psi^* \triangle \psi - \psi \triangle \psi^*) = i\hbar \left(\psi^* \frac{\partial \psi}{\partial t} + \psi \frac{\partial \psi^*}{\partial t} \right). \tag{1.222}$$

Notice that this equation does not depend on the potential V as long as the potential is real. Thus the conclusions of this paragraph are valid for all Schrödinger operators. Interestingly, on defining

$$\mathbf{j} \equiv \frac{\hbar}{2mi}(\psi^* \Delta(\psi) - \psi \Delta(\psi^*)) \tag{1.223}$$

and $\rho \equiv \psi^* \psi$, Eq. (1.222) takes the form of a *continuity equation* for the current

$$\frac{\partial \rho}{\partial t} + \mathrm{div}\mathbf{j} = 0 \tag{1.224}$$

that is a *local conservation law*. If one integrates the continuity equation on a volume V, the divergence theorem yields

$$\int\limits_V \frac{\partial \psi^*}{\partial t} \psi d^3\mathbf{x} + \int\limits_V \psi \frac{\partial \psi^*}{\partial t} d^3\mathbf{x} = \frac{d}{dt} \int\limits_V \psi^* \psi d^3\mathbf{x} = -\frac{\hbar}{2mi} \int\limits_\Sigma \left(\psi^* \frac{\partial \psi}{\partial n} - \psi \frac{\partial \psi^*}{\partial n} \right) d\sigma \tag{1.225}$$

where $\Sigma = \partial V$ is the boundary surface of V and $\frac{\partial}{\partial n}$ denotes differentiation along the direction normal to Σ. Thus, if ψ vanishes in a sufficiently rapid way as $\|\mathbf{x}\| \to \infty$, and if its first derivatives remain bounded in that limit, the integral on the right-hand side of (1.225) vanishes when the surface Σ is pushed off to infinity. The volume $\mathcal{V}$ extends then to the whole of $\mathbf{R}^3$, and one finds the *global conservation property*

$$\frac{d}{dt} \int\limits_{\mathbf{R}^3} \psi^* \psi \, d^3\mathbf{x} = 0 \tag{1.226}$$

which shows that the integral of $|\psi|^2$ over the whole space is independent of t. We have thus found a conservation law for the dynamics defined by Eq. (1.221). By virtue of the linearity of the Schrödinger equation, if ψ is square-integrable, one can then rescale the wave function so that the integral is set to 1

$$\int\limits_{\mathbf{R}^3} \psi^* \psi \, d^3\mathbf{x} = 1 \tag{1.227}$$

The interpretation of $|\psi|^2$ as a probability density will be discussed later on (see Sect. 6.4.1).

References

[Mm85] Marmo, G., Saletan, E.J., Simoni, A., Vitale, B.: Dynamical Systems: A Differential Geometric Approach to Symmetry and Reduction. John Wiley, Chichester (1985)

[La16] Lagrange, J.L.: Mécanique Analytique, 1816. A. Blanchard, Paris (1965)

[Ja62] Jackson, J.D.: Classical Electrodynamics. Wiley, New York (1962)

[Ga59] Gantmacher, F.R.: The Theory of Matrices, vol. II. Chelsea, New York (1959)

[Pe90] Perelomov, A.M.: Integrable Systems of Classical Mechanics and Lie Algebras. Birkhauser, Boston (1990)

[Re80] Reed, M., Simon, B.: Functional Analysis, vol. I. Academic Press, New York (1980)
[Es04] Esposito, G., Marmo, G., Sudarshan, G.: From Classical to Quantum Mechanics: An Introduction to the Formalism. Cambridge University Press, Cambridge (2004)
[Ma96] Marmo, G., Vilasi, G.: Symplectic structures and quantum mechanics. Modern Phys. Lett. B **10**, 545–553 (1996)

Chapter 2
The Language of Geometry and Dynamical Systems: The Linearity Paradigm

La filosofia è scritta in questo grandissimo libro che continuamente ci sta aperto innanzi agli occhi (io dico l'universo), ma non si può intendere, se prima non s'impara a intender la lingua, e conoscer i caratteri ne'quali è scritto. Egli è scritto in lingua matematica, e i caratteri son triangoli, cerchi ed altre figure geometriche, senza i quali mezzi è impossibile a intenderne umanamente parola; senza questi è un aggirarsi vanamente per un oscuro laberinto.
Philosophy is written in this grand book, the universe, which stands continually open to our gaze. But the book cannot be understood unless one first learns to comprehend the language and read the letters in which it is composed. It is written in the language of mathematics, and its characters are triangles, circles, and other geometric figures without which it is humanly impossible to understand a single word of it; without these, one wanders about in a dark labyrinth.

Galileo Galilei Il Saggiatore

2.1 Introduction

We can infer from the examples given in Chap. 1 that linear dynamical systems are interesting on their own. Moreover they can be explicitly integrated providing therefore a laboratory to explore new ideas and methods. We will use them systematically to illustrate all new notions and ideas to be introduced in this book.

We begin by elaborating more systematically the elementary, i.e., algebraic theory, for finite-dimensional linear dynamical systems, whose discussion was only initiated in the previous chapter. Later on, we will see how these algebraic ideas can be better expressed using a combination of geometry and analysis, that is, differential geometry.

We will use our experience with linear systems to build the foundations of differential geometry on vector spaces and from there to move to more general carrier spaces. This simple relation between linear algebra and elementary differential geometry is highlighted at the end of the chapter under the name of the 'easy' tensorialization principle, a simple idea that will prove to be very useful throughout the book.

J.F. Cariñena et al., *Geometry from Dynamics, Classical and Quantum*,
DOI 10.1007/978-94-017-9220-2_2

Thus in building a geometrical language for the theory of dynamical evolution we will start by discussing the notion of vector fields versus dynamical systems and the accompanying notions of tangent space, flows, forms, exterior algebra and the more abstract notion of derivations of the algebra of smooth functions on a linear space. This will be the content of Sects. 2.3 and 2.4.

Finally we will address in Sect. 2.5 the integration problem not just for a single vector field, but for a family of them, stating a simple form of the Frobenius theorem that provides necessary and sufficient conditions for the simultaneous integration of a family of vector fields. Then we will use a variation of this idea to solve the integration problem for Lie algebras, offering in this way a proof of Lie's third theorem and introducing in a more systematic way the theory of Lie groups.

2.2 Linear Dynamical Systems: The Algebraic Viewpoint

2.2.1 Linear Systems and Linear Spaces

A mathematical setting that embraces most examples in Chap. 1 is provided by a linear space E (also sometimes called a vector space) and a linear map $A\colon E \to E$ that helps us in defining the dynamics we are interested in. Later on we will discuss the extent to which these assumptions are reasonable so that these considerations will become an argument throughout all the book.

Thus we will consider a real linear space E, i.e. there are two binary operations defined on it: addition, denoted by $u + v$ for any $u, v \in E$, and multiplication by real numbers (scalars), denoted by λu for any $\lambda \in \mathbb{R}$. The linear map A satisfies $A(u + v) = Au + Av$ and $A(\lambda u) = \lambda Au$.

The interpretation of these objects regarding a given dynamics is that the vectors of the linear space E characterize our knowledge of the system at a given time t, thus we may think of vectors of E as describing partially the 'state' of the system we are studying (if the knowledge we can obtain from the system under scrutiny is maximal, i.e., no further information on the system can be obtained besides the one provided by the vectors u, then we would say that description provided by the vectors u is complete and they actually characterize the 'states' of the system). This physical interpretation is the reason that leads in many occasions to consider that there is a given linear structure in E, even though that is at this moment just an assumption we are introducing. In this setting, the trajectories $u(t)$ of our system, the data we can observe, are functions $t \mapsto u(t)$, i.e., curves on E. The parameter t used to describe the evolution of the system is associated to the 'observer' of the system, that is, to the people who are actually performing the experiments that allow description of the change on the vectors u, thus the parameter t has the meaning of a 'time' provided by an observer's clock.

We will assume first that the linear space is finite-dimensional with finite dimension n, thus the choice of any linear basis $\mathcal{B} = \{e_i \mid i = 1, \ldots, n\}$ allows us to identify it with $\mathbb{R}^n$ by means of $u \mapsto (u^i)$, $u = u^i e_i$, thus all notions of differential calculus of several variables can be transported to E by means of the previous identification. Hence if the trajectories are differentiable functions, they are characterized by the tangent vectors du/dt.

The expression for du/dt has to be determined by the experimental data, i.e., the observation of the actual trajectories of the system. If exhaustive experiments could be performed that will give us the value $F(u, t)$ of du/dt for all values of u at all possible times t, then we will have an expression of the form

$$\frac{du}{dt} = F(u, t),$$

that will be the mathematical law describing the dynamics of our system. However that is not the case, because performing such exhaustive measurements is impossible. Thus the determination of du/dt must be done combining experimental data and the ingenuity of the theoretician (at this point we should recall again Einstein's quote mentioned in the introduction (p. 12) of this book). On how to construct a vector field out of experimental data see [MS85].

The simplest situation we may envisage happens when the system has the form:

$$\frac{du}{dt} = A \cdot u. \tag{2.1}$$

(Let us recall that in Sect. 1.2.3 it was shown how it is possible by a simple manipulation to transform a system possessing an inhomogeneous term into a linear one, thus we will omit in what follows possible inhomogeneous terms in Eq. (2.1)). In other words, we assume that the tangent vector to the curve $u(t)$ depends just on the vector $u(t)$ at any time and does it linearly. Any description of a dynamics in the form given by Eq. (2.1) will be called a linear dynamical system or a linear system for short. Notice that the full description of a linear system involves the prescription of a space E with its linear structure $+, \cdot$ and a linear map A on it.

It is interesting to observe that linearity appears at two different levels here. On one side, we are assuming that the vectors describing states of the system can be composed and that the composition law we use for them satisfies the axioms of a linear space. On the other hand, we are considering that the infinitesimal variations to the trajectories of the states of the system, which are vectors by themselves, are tied to the linear structure of the states themselves by means of a linear map. Thus we are identifying a natural mathematical linear structure, that possessed by tangent vectors to curves, with a linear structure on the space of states which depends on the experimental setting we are preparing to describe the system under study. The nature of the exact relation between both structures will be the substance of much discussion in what follows. It will be enough to point out here that it cannot be assumed that the linear structure on the space of states will be uniquely determined in general (what can be thought to be a trivial statement) and that among all possible linear structures on such space, there could be some of them compatible with the dynamics we are trying to describe, in other words, a single dynamics can have various different descriptions as a linear system.

We must also notice here that most of the finite-dimensional examples discussed in the previous chapter were not linear systems. Actually none of the systems for which we have had a direct experience, like the free falling system (Sect. 1.2.4), the motion of celestial bodies, the Kepler system (Sect. 7.4.3), or other systems like the motion of charged particles in constant magnetic fields like the ones exhibited in picture 0.1 (Sect. 1.2.5) are linear systems (notice that we describe them using very specific parameters). Among these simple systems only the harmonic oscillator (Sect. 1.2.7) is a linear system. However all of them can be related to linear systems in different ways, either by direct manipulation or by the general theory of reduction as it will be done exhaustively in Chap. 7.

Contrary to systems in finite dimensions, all infinite-dimensional systems we discussed before, the Klein-Gordon field (Sect. 1.3.1), Maxwell equations (Sect. 1.3.2), and on top of all them, the Schrödinger equation (Sect. 1.3.3), were linear systems. The paramount role played by linearity in the analysis of infinite-dimensional systems has overcast its dynamical character and has taken over its physical interpretation. Without trying to elaborate a full physical analysis of them from a different perspective where linearity will not play a primitive role, we will emphasize later on some of the geometrical structures attached to them not depending on linearity.

2.2.2 Integrating Linear Systems: Linear Flows

Linear systems can be easily integrated. A linear system (2.1) defines a linear homogeneous (and autonomous) differential equation on E. Thus the problem we want to solve is to find differentiable curves $u(t)$ with t, defined on some interval I in $\mathbb{R}$, $t_0 \in I$ with values on E, such that

$$\frac{d}{dt}u(t) = A \cdot u(t), \quad \forall t \in I.$$

Such a curve will be called an integral curve or a (parametrized) solution of the linear system (2.1). We may think of this equation as an initial value problem if we select a vector $u_0 \in E$ and we look for solutions $u(t)$ that at time t_0 satisfies $u(t_0) = u_0$. The pair t_0, u_0 are called the Cauchy data of the initial value problem. The general existence and uniqueness theorem on the theory of differential equations guarantees that if E is a finite-dimensional linear space there exists a unique smooth curve $u(t)$, $t \in \mathbb{R}$ solving such a problem. However we will not need to use such a theorem as we will provide a direct constructive proof of this fact in Sect. 2.2.4. For the moment it will be enough for us to assume that such solutions exist and are unique.

Without loss of generality, see later, we can set $t_0 = 0$ in the present case. A natural way of thinking about solving this equation is to find a family of linear maps

$$\phi_t : E \to E, \quad t \in \mathbb{R}, \tag{2.2}$$

differentiable[1] on the variable t such that for any initial condition $u_0 \in E$ the curve $u(t) = \phi_t(u_0)$ is a solution of Eq. (2.1) with the given Cauchy data, namely,

$$\frac{d}{dt}\phi_t(u_0) = A(\phi_t(u_0)), \quad \phi_0(u_0) = u_0. \tag{2.3}$$

The family of maps $\{\phi_t \mid t \in \mathbb{R}\}$ will be called the flow of the linear system (2.1). Characteristic properties of a flow are:

$$\phi_t \circ \phi_s = \phi_{t+s}, \quad \phi_0 = I. \tag{2.4}$$

(with I the identity map on E). As an example, it is easy to check that the flow of the differential equation associated to the identity map $A = I$ is given by $\phi_t(x) = e^t x$. From the additive properties of the family of maps φ_t it is immediate to get that all of them are invertible and

$$\varphi_t^{-1} = \varphi_{-t}.$$

Both properties in (2.4) are immediate consequences of the uniqueness of solutions for the differential equation (2.1) (see for instance [Ar73] and [HS74] for a general discussion on the subject).

Exercise 2.1 Prove properties (a), (b) of Eq. (2.4).

Definition 2.1 Given a vector space E, a one-parameter flow (or just a flow) is a family of linear maps φ_t, $t \in \mathbb{R}$ depending smoothly on t and such that they satisfy properties (2.4) above.

In the case of E being finite-dimensional, it is easy to show from (2.4) that the smooth dependence on t of the family φ_t of linear maps defining a flow is equivalent to the much weaker property of the curve $t \mapsto \varphi_t(u)$ from $\mathbb{R}$ into E is continuous for all $u \in E$. In the later case we will say that the family φ_t is a strongly continuous one-parameter family of linear maps. In infinite dimensions the situation is much more subtle leading eventually to Stone's theorem that will be discussed in Chap. 6.

It is also evident from the linearity of Eq. (2.1) that the space of its solutions is a linear space. Thus if $u(t) = \phi_t \cdot u_0$ is the solution with initial condition u_0 and $v(t) = \phi_t \cdot v_0$ is the solution with initial data v_0, then, for any real number λ, $\lambda\, u(t)$ is the solution with initial data $\lambda\, u_0$ and $u(t) + v(t)$ is the solution with initial data

[1] 'Differentiable' here could be understood simply as the statement that the maps $t \mapsto \phi_t e_i$, with $\{e_i\}$ a basis in E, are differentiable.

$u_0 + v_0$. It follows from Eq. (2.4) that $\phi_{-t} = \phi_t^{-1}$, and it is also easy to show that the transformation sending $t \mapsto -t$ sends solutions of the linear equation given by A into solutions of the linear equation given by $-A$, because

$$\frac{d}{dt}(\phi_{-t} \cdot u_0) = -\frac{d}{ds}(\phi_s \cdot x_0)|_{s=-t} = -A \cdot (\phi_s u_0)|_{s=-t} = -A \cdot (\phi_{-t} \cdot u_0).$$

We have just proved that

Proposition 2.2 *Let E be a finite-dimensional space and A a linear map on it. Then the space of solutions $\mathcal{S}$ of the linear system $du/dt = A \cdot u$ is a linear space isomorphic to E. Such isomorphism is not canonical and a linear basis of the space of solutions is called a system of fundamental solutions.*

Proof A way to establish an isomorphism between $\mathcal{S}$ and E is by mapping $u(t)$ into $u(0)$. Clearly this map is invertible with inverse $u_0 \mapsto u(t) = \phi_t(u_0)$. □

For any initial condition $u_0 \in E$, Eq. (2.3) can be written as

$$\left(\frac{d\phi_t}{dt}\right)(u_0) = (A \circ \phi_t)(u_0),$$

and therefore we can also set[2]

$$\frac{d\phi_t}{dt} = A \circ \phi_t, \tag{2.5}$$

or, equivalently,

$$\left(\frac{d\phi_t}{dt}\right) \cdot \phi_t^{-1} = \left(\frac{d\phi_t}{dt}\right) \cdot \phi_{-t} = A. \tag{2.6}$$

If we think of $t \mapsto \phi_t$ as a curve on the space of linear maps on E, fancily written as $\mathrm{End}(E)$, then Eq. (2.5) can be thought as an equation on $\mathrm{End}(E)$ whose solutions are the flow of the linear system we are looking for. In order to put such a problem as a genuine initial value problem we should provide it with the initial data ϕ_0, which we can take without losing generality as being the identity matrix I. Notice that because of the existence and uniqueness theorem for solutions of differential equations, the solution of this initial value problem exists and must coincide with the flow of the linear system $du/dt = A(u)$ described in (2.4). This means that the evolution described by Eq. (2.5) takes place actually in the subset $GL(E) \subset \mathrm{End}(E)$ of invertible linear maps on E. It is clear that the set $GL(E)$ has the algebraic structure

[2] The derivative of ϕ_t could be easily understood as thinking of ϕ_t as a curve of matrices, once we have selected any basis on E, then the space of $n \times n$ matrices can be identified with $\mathbb{R}^{n^2}$.

of a group, that is it carries its binary associative composition law given by the standard composition of linear maps, the identity linear map is the unit element and, finally, any element has an inverse (see later on Sect. 2.6.1 for a detailed description of the class of groups that will be of interest in this book). We call the initial value problem

$$d\phi_t/dt = A \circ \phi_t, \quad \phi_0 = I,$$

the group theoretical picture of the linear system (2.1). It is remarkable that the space of linear maps of E denoted before by $\mathrm{End}(E)$ carries a natural Lie algebra structure given by the standard commutator $[\cdot, \cdot]$ of linear maps, that is:

$$[A, B] = A \circ B - B \circ A. \tag{2.7}$$

(Let us recall that a Lie algebra structure on a linear space L is a skew symmetric bilinear composition law $[\cdot, \cdot] \colon L \times L \to L$ such that it satisfies Jacobi identity, i.e., $[a, [b, c]] + [c, [a, b]] + [b, [c, a]] = 0, \forall a, b, c \in L$, see 10 for more details on the basic algebraic notions concerning Lie algebras and groups). In what follows we will start exploring the relation between the Lie algebra structure in the space $\mathrm{End}(E)$ and the group structure in the group of automorphisms of E, $GL(E)$.

From the form of our equation on $\mathrm{Aut}\,(E)$ it is clear that ϕ_t can be found easily by exponentiation, namely,[3]

$$\phi_t = \exp(tA) = e^{tA} = \sum_{k=0}^{\infty} \frac{t^k}{k!} A^k. \tag{2.8}$$

Indeed, since

$$\frac{d}{dt}\left(\sum_{n=0}^{\infty} \frac{t^n}{n!} A^n\right) = \sum_{n=1}^{\infty} \frac{t^{n-1}}{(n-1)!} A^n = A \sum_{m=0}^{\infty} \frac{t^m}{m!} A^m = A \exp(tA)\,, \tag{2.9}$$

we see that

$$\left(\frac{d}{dt} e^{tA}\right) \cdot e^{-tA} = A. \tag{2.10}$$

The operator A is called the infinitesimal generator of the one-parameter group $\phi_t = \exp\,(tA)$, $t \in \mathbb{R}$. We recall some useful properties of the exponential map exp defined above, Eq. (2.8):

[3] Notice that the operator-valued power series $\sum_{k=0}^{\infty} \frac{t^k}{k!} A^k$ is convergent because it can be bounded by the numerical series $\sum_{k=0}^{\infty} \frac{t^k}{k!} ||A||^k$ with $|| \cdot ||$ any norm in the finite-dimensional linear space of linear maps on E.

1. If B is an isomorphism of E and A is in End (E), we have $B^{-1}A^nB = (B^{-1}AB)^n$, for any integer number n, and therefore,

$$e^{B^{-1}AB} = B^{-1}e^A B. \tag{2.11}$$

2. If $A, B \in$ End (E), and $A \circ B = B \circ A$, then, $e^{A+B} = e^A e^B$.
3. Every exponential is a square, for $e^A = (e^{A/2})^2$. If E is a real, finite-dimensional vector space, then $\det e^A > 0$. It follows that the map $\exp\colon \mathrm{End}(E) \to GL(E)$ is not surjective.
4. For a finite-dimensional vector space E,

$$\det e^A = e^{\mathrm{Tr}\, A},$$

where det stays for the determinant and Tr for the trace.

The flow ϕ_t is a symmetry of the system for any value of t according with the notion of symmetry presented in Sect. 1.2.6. If $\phi_t(u_0)$ is a solution of Eq. (2.1) with initial condition u_0, then $(\phi_s \circ \phi_t)(u_0)$ is again a solution starting at $\phi_s(u_0)$. On the other hand is obvious that:

$$[\phi_t, A] = 0\,. \tag{2.12}$$

Having written $u(t) = e^{tA} \cdot u(0)$ as a solution of our equation $\dot{u} = A \cdot u$, we might believe that the problem of analyzing the linear system is solved. However that is not so. Indeed the series Eq. (2.8) defining e^{tA} may be unsuitable for computations or we could have questions on the dynamics that are not easy to answer from the expression before. For instance we may be interested in knowing if the solution $e^{tA}u(0)$ is periodic or, if $e^{tA}u(0)$ is bounded when $t \to \pm\infty$, and so on.

There are however some situations where most of the questions we may raise about the properties of the solutions have an easy answer:

1. A is diagonalizable. In this generic case E has a basis $\{e_j\}$ of eigenvectors of A with corresponding eigenvalues λ_j, i.e., $Ae_j = \lambda_j e_j$. It follows that $A^n e_j = (\lambda_j)^n e_j$ and therefore the set of curves

$$e_k(t) = e^{tA}\,(e_k) = e^{t\lambda_k}\,(e_k) \tag{2.13}$$

forms a fundamental system of solutions. Hence any solution $u(t)$ will have the form $u(t) = c^k e_k(t)$ (sum over k) and c^k determined by the initial conditions. However we must point it out that even if we know in advance that the operator A is diagonalizable, for instance if it is symmetric, solving the eigenvalue problem is often a difficult problem.
2. The endomorphism A is nilpotent. We recall that $A \in$ End (E) is nilpotent with index n if n is the smallest positive integer for which $A^n = 0$. In this case e^{tA} reduces to a polynomial and the general solution will have the form:

$$u(t) = \left(1 + tA + \cdots + \frac{(tA)^{n-1}}{(n-1)!}\right) u(0). \tag{2.14}$$

The general solution $\exp(tA)u(0)$ will be a vector with components which are polynomials in t of degree less than n.
As a particular instance, if we consider E to be the set of real polynomials in x of degree less than n, then E is a vector space of dimension n on which the operator d/dx is nilpotent of index n. On the basis $\{1, x, x^2, \ldots, x^{n-1}\}$, the linear operator d/dx is represented by the matrix:

$$\left(\frac{d}{dx}\right) = \begin{pmatrix} 0\,1\,0 & & 0 & 0 \\ 0\,0\,2 & \ddots & 0 & 0 \\ 0\,0\,0 & \ddots & 0 & 0 \\ \vdots\,\vdots\,\vdots & \ddots & \vdots & \vdots \\ 0\,0\,0 & & n-2 & 0 \\ 0\,0\,0 & \cdots & 0 & n-1 \\ 0\,0\,0 & \cdots & 0 & 0 \end{pmatrix},$$

and

$$\exp(td/dx) = 1 + t\frac{d}{dx} + \cdots + \frac{t^{n-1}}{(n-1)!}\frac{d^{(n-1)}}{dx^{(n-1)}}.$$

If $p \in E$ we find that

$$\left[\exp(td/dx)p\right](x) = p(x) + tp'(x) + \cdots + \frac{t^{n-1}}{(n-1)!}p^{(n-1)}(x) = p(x+t).$$

Indeed the operator $\exp(td/dx)$ is the translation operator by $-t$. It is clear that the series expansion above would still make sense for any real analytic function on the real line, while the translation operator makes sense for any function, even if it is not differentiable.

3. The case of a general endomorphism A. A general endomorphism A can be decomposed in an essentially unique way as $A = D + N$ with D diagonalizable and N nilpotent (see later on the discussion of Jordan canonical forms) such that $[D, N] = 0$. Therefore the general situation will reduce to the preceding cases, but again a similar warning as in the diagonalizable case should be raised: finding D and N is often unpracticable.

2.2.2.1 Linear Changes of Coordinates

If we perform a linear change of coordinates $x \mapsto x' = Px$ in the vector space E where the linear system (2.1) is defined, clearly we obtain a new equation of motion with: $A \mapsto A' = PAP^{-1}$. That is clear because: $\dot{x}' = P\dot{x} = PA \cdot x =$

$PAP^{-1}Px = PAP^{-1} \cdot x'$. Of course all the elements of the family of equations of motion obtained in this way are equivalent under all respects and the corresponding flows are then related by $\phi'_t = P\phi_t P^{-1}$. In fact: $d\left(P\phi_t P^{-1}\right)/dt = P\left(d\phi_t/dt\right)P^{-1} = PA\phi_t P^{-1} = A'P\phi_t P^{-1}$.

From the active viewpoint, if $\Psi\colon E \to E$ is a linear isomorphism and the curve $x(t)$ is a solution of (2.1), then using the chain rule it is easy to see that the curve $(\Psi \circ x)(t)$ is a solution of,

$$\frac{d}{dt}(\Psi \cdot x) = (\Psi \circ A \circ \Psi^{-1})(\Psi \circ x) \tag{2.15}$$

A linear isomorphism Ψ will be a linear symmetry for our linear dynamical system if it maps solutions of the linear system into solutions of the same system. That means that if $x(t) = \exp(tA) \cdot x(0)$ is a solution, then also $(\Psi \circ x)(t) = (\Psi \circ \exp(tA)) \cdot x(0)$ is a solution of $\dot{x} = A(x)$. This clearly implies that $\Psi \circ A \circ \Psi^{-1} = A$. Conversely, if $\Psi \circ A \circ \Psi^{-1} = A$, then the linear map Ψ sends solutions into solutions. The set of symmetries obtained in this way is thus a group characterized as the subgroup G_A of $GL(E)$ of those isomorphisms Ψ such that $[A, \Psi] = 0$.

2.2.2.2 Symmetries

Given the evolution equation: $dx/dt = A \cdot x$, it is clear that we can construct new evolution equations as

$$\frac{dx}{ds_{(k)}} = A^k \cdot x, \tag{2.16}$$

for any positive integer k (we emphasize here that the evolution parameters are all taken to be independent. Of course: $s_{(1)} \equiv t$). It follows then quite easily that the flows $\Psi_k\left(s_{(k)}\right) = \exp\left\{s_{(k)}A^k\right\}$ are symmetries for the original equation and that, moreover, all these one-parameter groups pairwise commute.

Whenever the characteristic polynomial of A coincides with the minimal polynomial the powers of A yield all the infinitesimal symmetries of our equation, i.e., they generate the algebra of symmetries of A, which turns out then to be Abelian. That is the case when the eigenvalues of A are non-degenerate. If instead A has degenerate eigenvalues, then additional symmetries will be provided by all the elements of the general linear group acting on the corresponding eigenspace on which A acts as a multiple of the identity, and the symmetry group will be in general larger and no longer Abelian (see, however, below Sect. 3.6 for more details).

2.2.3 Linear Systems and Complex Vector Spaces

2.2.3.1 The Canonical Form of the Flow of a Linear Complex System

When the linear map A defining a linear system on the real vector space V is not diagonalizable (i.e., the representative matrix associated to it is not diagonalizable over the reals) it is convenient to use complex numbers so that the fundamental theorem of algebra is available. Let then E be a complex n-dimensional vector space (we will discuss its structure immediately after these remarks) and A a complex linear map on E. Then the characteristic polynomial $p(\lambda) = \det(A - \lambda I)$ of $A \in \mathrm{End}\,(E)$ factorizes as

$$p(\lambda) = (\lambda - \lambda_1)^{r_1} \cdots (\lambda - \lambda_m)^{r_m}, \tag{2.17}$$

where the complex numbers $\lambda_1, \ldots, \lambda_m$, $m \leq n$ are the different eigenvalues of A, r_i is the algebraic multiplicity of the eigenvalue λ_i, $i = 1, \ldots, m$, and $r_1 + \cdots + r_m = n$. We consider the operators $(A - \lambda_k I)^{r_k}$ along with their kernels E_k. Each E_k has dimension r_k and $E = \bigoplus_k E_k$. This splitting leads to a decomposition of the identity by means of a family of projections P_i on each subspace E_i in such a way that $E_k = P_k E$, $\sum_{i=1}^m P_i = \mathrm{id}_E$ and $P_i P_j = \delta_{ij} P_i$. Then:

1. A leaves each E_k invariant, that is $A(E_k) \subset E_k$, therefore it makes sense to define $A_k = AP_k$. Note that A-invariance of E_k implies that $P_k A P_k = A P_k$. As the supplementary subspace $E_k' = \bigoplus_{l \neq k} E_l$ is also A-invariant, $P_k A = A P_k$.
2. $A = \sum_k A_k$, with $[A_k, A_j] = 0$, $\forall k, j$. that is a consequence of $A_k A_j = AP_k AP_j = A^2 P_k P_j = \delta_{kj} A^2 P_k$. Also $(A_k - \lambda_k I)^{r_k} = 0$.

We now find the flow of A decomposes as:

$$e^{tA} = e^{tA_1} \cdots e^{tA_m}. \tag{2.18}$$

If a vector $x \in E$ is decomposed as $x = \sum_k P_k x = \sum_k x_k$, we get $\exp(tA)x = \sum_k \exp(tA_k) x_k$. If we write e^{tA_k} as $e^{tA_k} = e^{t(A_k - \lambda_k I) + t\lambda_k I}$, and since the identity I commutes with everything, we have also

$$e^{tA_k} = e^{t(A_k - \lambda_k I)} e^{t\lambda_k I};$$

but now, $(A_k - \lambda_k I)$ is nilpotent of index not higher than r_k and we get

$$e^{tA} x = \sum_k e^{t\lambda_k} \mathcal{P}_k(t) x_k, \tag{2.19}$$

where $\mathcal{P}_k(t)$ is a polynomial in t with coefficients in $\mathrm{End}\,(E)$ and of degree less than r_k.

We have shown that the general solution of $\dot{x} = A \cdot x$ is the sum of vectors of the form $e^{t\lambda_k} x_k(t)$, with $x_k(t)$ a vector whose components are polynomials in t of degree less than the multiplicity r_k of the eigenvalues λ_k.

Remark 2.1 This splitting of A into the sum of commuting operators is the prototype of separability of differential equations. We will take back this theme when we discuss the notion of integrability of dynamical systems and we will relate this notion to the existence of normal forms (see Chap. 8).

Remark 2.2 We will come back to this discussion again when analyzing systems with compatible generic Hermitean structures (see Sect. 6.2.5).[4]

2.2.3.2 Complexification of Linear Dynamical Systems

It is time now to discuss with some care a structure that is going to play a relevant role in what follows: complex linear structures.

We recall that a complex linear space E is a vector space over the field of complex numbers $\mathbb{C}$. Any complex vector space has a basis. We will assume in what follows that E is finite-dimensional with complex dimension $\dim_{\mathbb{C}} E = n$. Let $\mathcal{B} = \{u_1, \ldots, u_n\}$ be a basis for E. Thus, any vector $v \in E$ can be written as $v = \sum_k z^k u_k$, with $z^k \in \mathbb{C}$. The map $\varphi_{\mathcal{B}} \colon E \to \mathbb{C}^n$ given by $\varphi_{\mathcal{B}}(v) = (z^1, \ldots, z^n)$ is an isomorphism of complex vector spaces where $\mathbb{C}^n$ is a complex vector space with the natural action of $\mathbb{C}$.

We can also think of E as a real vector space by considering the action of $\mathbb{R}$ on E given by $\lambda \cdot v = (\lambda + i\,0)v$, $\lambda \in \mathbb{R}$. We shall denote this real vector space as $E_{\mathbb{R}}$ and we will call it the realification of E. It is clear that the vectors u and $i\,u$ are linearly independent on $E_{\mathbb{R}}$. Hence the set $\mathcal{B}_{\mathbb{R}} = \{u_1, \ldots, u_n, i\,u_1, \ldots, i\,u_n\}$ is a linear basis for $E_{\mathbb{R}}$. We will call this a real basis adapted to the complex structure on E. It is clear that $\dim E_{\mathbb{R}} = 2 \dim_{\mathbb{C}} E$.

The realification $E_{\mathbb{R}}$ of a complex vector space E carries a natural endomorphism J verifying $J^2 = -I$ defined by $J(u) = i\,u$. Conversely if the real vector space V is equipped with a real linear map J such that $J^2 = -I$ it becomes a complex linear space with the action of $\mathbb{C}$ on V defined by

$$z \cdot v = xv + yJ(v)$$

for all $z = x + iy \in \mathbb{C}$ and $v \in V$. We shall denote by $V(J)$ the complex vector space defined in this way. The realification of the complex space $V(J)$ is the original real vector space V and the endomorphism induced on it coincides with J. We see

[4] In Quantum Mechanics a similar decomposition of the total space can be achieved by using a compact group of symmetries for A. The irreducible invariant subspaces of our group will be finite-dimensional and the restriction of the Hamiltonian operator A to each invariant subspace gives raise to a finite-dimensional problem. Motions in central potentials are often studied in this way by using the rotation group and spherical harmonics. The radial part is then studied as a one-dimensional problem.

in this way that a complex structure on a real vector space is defined by a linear map whose minimal polynomial is the simplest possible $p(\lambda) = \lambda^2 + 1$.

Definition 2.3 A linear complex structure on the real vector space E is a real linear map $J : E \to E$ such that $J^2 = -I$.

Exercise 2.2 Prove that E must be even-dimensional as a real space.

Remark 2.3 The canonical model for a complex structure J is provided by the real space $\mathbb{R}^{2n}$ and the endomorphism J_0 given by

$$J_0 = \left(\begin{array}{c|c} 0 & -I_n \\ \hline I_n & 0 \end{array}\right). \tag{2.20}$$

Thus, if we denote by $(x^1, \ldots, x^n, y^1, \ldots, y^n)$ a generic point in $\mathbb{R}^{2n}$, multiplication by J_0 gives, $(-y^1, \ldots, -y^n, x^1, \ldots, x^n)$, hence that is equivalent to multiplication by i if we identify $\mathbb{R}^{2n}$ with $\mathbb{C}^n$ by $(x^1, \ldots, x^n, y^1, \ldots, y^n) \mapsto (x^1 + iy^1, \ldots, x^n + iy^n)$.

We consider now V to be an n-dimensional real vector space. It is possible however to exploit the previous discussions leading to the structure of the flow of a complex linear system by constructing a complex vector space out of V. Such construction is called the complexification of V and it proceeds as follows:

Consider the set $E = V \times V$. We can endow E with the structure of a complex space by defining [Ar73]:

1. $(v_1, w_1) + (v_2, w_2) = (v_1 + v_2, w_1 + w_2)$,
2. $(a + ib)(v, w) = (av - bw, bv + aw)$.

Exercise 2.3 Check that $V \times V$ with the binary composition law + defined by (1) above and the action of $\mathbb{C}$ defined by (2) satisfies the axioms of a complex linear space.

The set $V \times V$ when endowed with the new structure of complex space will be denoted $V^{\mathbb{C}}$ and it is said to be the complexification of V. The space V is embedded as a real linear subspace in $V^{\mathbb{C}}$ by means of the map $j : V \to V \times V$, $j(v) = (v, 0)$. Moreover, every element $(v, w) \in V^{\mathbb{C}}$ can be written as $(v, w) = (v, 0) + i(w, 0) = j(v) + i\, j(w)$, because $i(w, 0) = (0, w)$.

A vector of $V^{\mathbb{C}}$ is said to be real if it is of the form $j(v) = (v, 0)$, and then it will be denoted as v instead of $j(v)$. An arbitrary element $(v, w) \in V^{\mathbb{C}}$ can be written as a sum $v + iw$, with v and w being real vectors.

Notice that if $\mathcal{B} = \{e_i \mid i \in I\}$ is a basis of the real space V, then, $\overline{\mathcal{B}} = \{(e_i, 0) \mid i \in I\}$ will be a basis of the complex space $V^{\mathbb{C}}$. Therefore,

$$\dim_{\mathbb{R}} V = \dim_{\mathbb{C}} V^{\mathbb{C}} \tag{2.21}$$

Another important remark is that if the real linear space V was endowed with a real inner product $(\cdot,\cdot)$, that is a positive definite symmetric bilinear form on E, then $V^{\mathbb{C}}$ becomes a pre-Hilbert space (actually a Hilbert space in the finite-dimensional case) by means of the Hermitean product:

$$\langle v_1 + i v_2, w_1 + i w_2 \rangle = (v_1, w_1) + (v_2, w_2) + i(v_1, w_2) - i(v_2, w_1) \tag{2.22}$$

which satisfies $\langle v, w\rangle = (v, w)$. We will discuss Hermitean products in depth in Sect. 6.2.

We introduce now the important notion of complex linear map.

Definition 2.4 A complex linear map φ between two complex linear spaces E_1 and E_2, is a map $\varphi\colon E_1 \to E_2$ such that

$$\varphi(z_1 \cdot u_1 + z_2 \cdot u_2) = z_1 \cdot \varphi(u_1) + z_2 \cdot \varphi(u_2), \quad \forall z_1, z_2 \in \mathbb{C}, \quad u_1, u_2 \in E_1.$$

Equivalently, we have:

Proposition 2.5 *A complex linear map φ between two complex linear spaces (E_1, J_1) and (E_2, J_2), is a real linear map $\varphi\colon E_1 \to E_2$ such that*

$$\varphi \circ J_1 = J_2 \circ \varphi.$$

The first example of complex linear maps is provided by the complexification of a real linear map. Given a linear map $A\colon V \to V$ of the real linear space V, It is possible to complexify A as

$$A^{\mathbb{C}}(v + iw) = (Av) + i(Aw) \tag{2.23}$$

and this correspondence satisfies:

1. $(\lambda A)^{\mathbb{C}} = \lambda A^{\mathbb{C}}$.
2. $(A + B)^{\mathbb{C}} = A^{\mathbb{C}} + B^{\mathbb{C}}$.
3. $(AB)^{\mathbb{C}} = A^{\mathbb{C}} B^{\mathbb{C}}$.
4. $(A^T)^{\mathbb{C}} = (A^{\mathbb{C}})^\dagger$.

Moreover, if $A^{\mathbb{C}}$ has a real eigenvalue λ, then λ is also an eigenvalue of A. More specifically, if $v + iw$ is an eigenvector of $A^{\mathbb{C}}$ corresponding to the eigenvalue $\lambda + i\mu$, then,

$$Av + i\,Aw = A^{\mathbb{C}}(v + iw) = (\lambda + i\mu)(v + iw) = (\lambda v - \mu w) + i(\lambda w + \mu v),$$

and as a consequence of the uniqueness of the splitting

$$Av = \lambda v - \mu w, \quad Aw = \lambda w + \mu v$$

In particular, when $\mu = 0$, we get:

$$Av = \lambda v, \quad Aw = \lambda w \tag{2.24}$$

But $v + iw$ was an eigenvector of $A^{\mathbb{C}}$ and therefore v and w cannot vanish simultaneously.

Notice that if A is a real linear operator, the complexification $A^{\mathbb{C}}$ will define a complex liner dynamical system:

$$\frac{dz}{dt} = A^{\mathbb{C}} \cdot z; \quad z \in V^{\mathbb{C}} \tag{2.25}$$

We notice that a solution of our complexified equation stays real if the initial condition is real. In fact the operator that associates to any vector $z = x + iy$ in $V^{\mathbb{C}}$ its complex conjugate $\bar{z} = x - iy$, commutes with $A^{\mathbb{C}}$. That means that if $z(t)$ is a solution of Eq. (2.25), then $\bar{z}(t)$ will be a solution too. Then by uniqueness of the existence of a solution for given Cauchy data, if the Cauchy data are real the solution must be real for all values of t. This solution is also a solution of the real differential equation. It is also clear that a curve $t \mapsto z(t) = x(t) + iy(t)$ is a solution of the complexified equation iff the real and imaginary parts are solutions of the real equation. All this follows from the fact that $A^{\mathbb{C}}$ commutes with the multiplication by i.

Now we can look for solutions of $\dot{x} = A \cdot x$ with initial conditions $x(0)$ by solving $\dot{z} = A^{\mathbb{C}} \cdot z$ with initial condition $z(0) = x(0)$. Now a diagonalization is possible, all our previous considerations apply and we get the desired solution. It remains to express it into real form for it is the sum of vectors like $e^{t\lambda_k} x_k(t)$ where $x_k(t)$ is a polynomial of degree at most r_k. The eigenvalue λ_k is not necessarily real. What we need is to use combinations of the real and imaginary part of the solution. If $\lambda = a + ib$, $a, b \in \mathbb{R}$ we shall consider terms of the form $e^{at}\cos(bt)x(t)$ and $e^{at}\sin(bt)x(t)$ where $x(t)$ is a polynomial in t with real coefficients.

The following proposition shows that all complex structures in a given linear space are isomorphic, and isomorphic to the canonical model $(\mathbb{R}^{2n}, J_0)$.

Proposition 2.6 *Let (E, J) be a complex linear space. Then, there exists a linear isomorphism $\varphi\colon E \to \mathbb{R}^{2n}$ such that $J_0 \circ \varphi = \varphi \circ J$, i.e., (E, J) is complex isomorphic to $(\mathbb{R}^{2n}, J_0)$.*

Proof We give two proofs. Because E is a complex linear space of dimension n, let $u_1, \ldots, u_n$ denote a basis for it. The real linear space $E_{\mathbb{R}}$ is the realification of the complex linear space (E, J). A linear basis of it is provided by $u_1, \ldots, u_n, Ju_1, \ldots, Ju_n$. Then, we identify E with $\mathbb{R}^{2n}$ by means of the linear isomorphism $\psi\colon \mathbb{R}^{2n} \to E$ given by:

$$\psi(x^1, \ldots, x^n, y^1, \ldots, y^n) = x^k u_k + y^k J(u_k).$$

Then notice that

$$\psi(J_0(x^1, \ldots, x^n, y^1, \ldots, y^n)) = \psi(-y^1, \ldots, -y^n, x^1, \ldots, x^n) = -y^k u_k + x^k J(u_k)$$

and then,

$$\psi(J_0(x^1, \dots, x^n, y^1, \dots, y^n)) = J(y^k J(u_k) + x^k u_k) = J(\psi(x^1, \dots, x^n, y^1, \dots, y^n)),$$

and the result is established.

An alternative proof that will be useful for us later on, works as follows.

Consider the complexification $E^{\mathbb{C}}$ of the real space E and complexity J to $E^{\mathbb{C}}$. Thus if we denote by $u^{\mathbb{C}} = u_1 + i\, u_2$ a vector on $E^{\mathbb{C}}$, the complexified map $J^{\mathbb{C}}$ acts as $J^{\mathbb{C}}(u^{\mathbb{C}}) = J(u_1) + i J(u_2)$. Moreover $(J^{\mathbb{C}})^2 = -I$, hence $E^{\mathbb{C}}$ carries two alternative complex structures. We can diagonalize $J^{\mathbb{C}}$ with respect to the complex structure on $E^{\mathbb{C}}$ defined by multiplication by i, then denote by $K_{\pm} = \ker(J^{\mathbb{C}} \mp i)$, the eigenspaces of $\pm i$ respectively. Denote by $R_{\pm} = \operatorname{Im}(J^{\mathbb{C}} \pm i)$. Then, because $(J^{\mathbb{C}} + i)(J^{\mathbb{C}} - i) = 0$ it is clear that $R_{\pm} = K_{\pm}$. But $E^{\mathbb{C}}/K_{\pm} \cong R_{\mp}$. Hence,

$$\dim R_+ = \dim R_- = \dim K_+ = \dim K_-,$$

and $\dim_{\mathbb{C}} E^{\mathbb{C}}$ is even (hence the real dimension of E is even). Let us call this dimension $2n$. The dimension of K_+ is n and let $w_1^{\mathbb{C}}, \dots, w_n^{\mathbb{C}}$ a complex basis of it. The vectors $w_k^{\mathbb{C}}$ have the form $w_k^{\mathbb{C}} = u_k + i v_k$, $k = 1, \dots, n$, and they satisfy

$$J(u_k) = v_k, \quad J(v_k) = -u_k.$$

Thus we have found a basis such that the natural identification with $\mathbb{R}^{2n}$ provided by it gives the desired map. □

The set of complex linear isomorphisms $\varphi \colon E \to E$ defines a subgroup of the real general linear group $GL(E)$. We shall denote such group by $GL(E, J)$.

As it was shown in Proposition 2.6 we can identify (E, J) with $(\mathbb{R}^{2n}, J_0)$, hence the group $GL(E, J)$ is isomorphic with a subgroup of the group $GL(2n, \mathbb{R})$. Such subgroup will be denoted by $GL(n, \mathbb{C})$ and is characterized as the set of matrices $A \in GL(2n, \mathbb{R})$ such that

$$AJ_0 = J_0 A$$

or, equivalently that A has the block form

$$A = \begin{pmatrix} X & -Y \\ Y & X \end{pmatrix},$$

with X, Y $n \times n$ real matrices. Notice that if we identify E with $\mathbb{C}^n$, then $GL(E, J)$ becomes simply the group of invertible complex $n \times n$ matrices, i.e., the group $GL(n, \mathbb{C})$. The identification between these two representations of the same group (the fundamental one and the $2n$-dimensional) is given by the map

$$Z = X + iY \mapsto \begin{pmatrix} X & -Y \\ Y & X \end{pmatrix}.$$

The (real) dimension of the group $GL(n, \mathbb{C})$ is $2n^2$.

Notice that if we consider the realification of $V^{\mathbb{C}}$, then $V^{\mathbb{C}}_{\mathbb{R}}$ is isomorphic to $V \oplus V$. A basis for $V^{\mathbb{C}}_{\mathbb{R}}$ is obtained from a basis $B = \{e_i\}_{i\in I}$ of V by $\bar{B} = \{(e_i, 0), (0, e_i)\}_{i\in I}$. In such a basis, the $\mathbb{R}$-linear map J of $V^{\mathbb{C}}_{\mathbb{R}}$ corresponding to multiplication for the imaginary unit i is represented by the standard model matrix J_0 Eq. (2.20). Notice that

$$V \cap JV = \{\mathbf{0}\}, \qquad V^{\mathbb{C}}_{\mathbb{R}} = V \oplus JV. \tag{2.26}$$

2.2.4 *Integrating Time-Dependent Linear Systems: Dyson's Formula*

We can consider again the problem of integrating a linear dynamical system. In the previous section we found the flow of the system by integrating the equation on the group and checking the convergence of the series found in that way. A similar idea can be used to integrate and find explicitly the flow of a time-dependent (i.e., non-autonomous) linear system on E like

$$\frac{dx}{dt} = A(t) \cdot x. \tag{2.27}$$

To discuss the solution of this equation we reconsider first the time-independent case and put it into a slightly different perspective.

We assume that the Eq. (2.27) has a solution,

$$x(t) = \phi(t, t_0)x(t_0) \tag{2.28}$$

i.e., $\phi(t, t_0)$ is the evolution matrix of the Cauchy datum $x_0 = x(t_0)$. Of course, due to the previous results we know that if A is constant $\phi(t, t_0) = e^{A(t-t_0)}$, however for the time being we are arguing independently of it. Taking the time derivative of $x(t)$ we get,

$$\dot{x}(t) = \dot{\phi}(t, t_0)x(t_0) = \dot{\phi}(t, t_0)\phi^{-1}(t, t_0)x(t) \tag{2.29}$$

therefore we have $A = \dot{\phi}(t, t_0)\phi^{-1}(t, t_0)$. Because by assumption A is independent of time, the right-hand side can be computed for any time t. Thus we find that our initial equation on E can be replaced by the equation on $GL(n, \mathbb{R})$,

$$\frac{d}{dt}\phi = A\phi \tag{2.30}$$

We use now our knowledge that $\phi(t, t_0) = e^{A(t-t_0)}$ is a solution to consider the series expansion,

$$\phi(t, t_0) = I + A\,(t - t_0) + \cdots + A^n\,\frac{(t - t_0)^n}{n!} + \cdots \tag{2.31}$$

We denote the term $A^n\,\frac{(t-t_0)^n}{n!}$ in the previous expansion by $R_n(t, t_0)$ and we notice that,

$$R_{n+1}(t, t_0) = \int_{t_0}^{t} A R_n(s, t_0)\, ds \tag{2.32}$$

Therefore, we can consider the matrix,

$$S_n(t, t_0) = I + \sum_{k=0}^{n} R_k(t, t_0) \tag{2.33}$$

as providing us with an approximate solution of our initial equation. It is clear that because of the independence of A on t we have,

$$R_{n+1}(t, t_0) = A^{n+1} \int_{t_0}^{t} \frac{(s - t_0)^n}{n!}\, ds \tag{2.34}$$

Therefore, the sequence: $R_0(t, t_0), \ldots, R_n(t, t_0)$ converges uniformly to the limit,

$$S(t, t_0) = I + \int_{t_0}^{t} A\, S(s, t_0)\, ds \tag{2.35}$$

Now, $S(t, t_0)$ is differentiable and satisfies:

$$\frac{d}{dt} S = A\, S \tag{2.36}$$

and we have found our solution in terms of an integral.

This new way of finding $S = e^{A(t-t_0)}$ holds true in the time dependent case as long as $A\colon I \subset \mathbb{R} \to \mathfrak{gl}(n, \mathbb{R})$ is a continuous map with M an upper bound for $||A(t)||$ in I. Then we define again,

$$R_{n+1}(t, t_0) = \int_{t_0}^{t} A(s) R_n(s, t_0)\, ds, \qquad R_0(t, t_0) = I \tag{2.37}$$

and notice that $||R_n(t, t_0)|| \leq |t - t_0|^n \frac{M^n}{n!}$. Therefore the sequence $R_0(t, t_0) + \cdots + R_n(t, t_0)$ converges uniformly on I to the limit,

$$S(t, t_0) = I + \int_{t_0}^{t} A(s)S(s, t_0)\, ds \tag{2.38}$$

with S satisfying the differential equation,

$$\frac{d}{dt} S(t, t_0) = A(t)S(t, t_0) \tag{2.39}$$

with initial condition: $S(t_0, t_0) = I$.

The matrix we have found is called the resolvent or the *resolvent kernel* of the equation $\dot{x}(t) = A(t)x(t)$.

In general one can write for $R_n(t, t_0)$ the formula,

$$R_n(t, t_0) = \int_{t_0 \leq s_1 \leq \ldots \leq s_n \leq t} A(s_n) \cdots A(s_1)\, ds_1 \cdots ds_n \tag{2.40}$$

where due care is required for the order of factors because in general $[A(s_i), A(s_j] \neq 0$.

Defining a time-ordering operation T as,

$$T\left\{A(s)\, A\left(s'\right)\right\} = \begin{cases} A(s)\, A\left(s'\right), & s \geq s' \\ A\left(s'\right) A(s), & s < s' \end{cases} \tag{2.41}$$

and similarly for products with more than two factors (i.e., $T\{\ \}$ will order factors in decreasing (or non-increasing) order of the time arguments from left to right), it is not hard to convince oneself that,

$$R_n(t, t_0) = \frac{1}{n!} \int_{t_0}^{t} ds_1 \cdots ds_n T\left\{A(s_1) \cdots A(s_n)\right\} \tag{2.42}$$

and hence that $S(t, t_0)$ can be expressed a 'time-ordered exponential',[5] quoted as Dyson's formula:

$$S(t, t_0) = T\left\{\exp \int_{t_0}^{t} ds\, A(s)\right\} \tag{2.43}$$

[5] Also-called a 'product integral' (see, e.g.: [DF79]).

The matrix $S(t, t_0)$ can be given a simple interpretation. Starting (in the time-independent case, for the time being) from the linear equation, $dx/dt = A \cdot x$, we can consider a fundamental system of solutions, say $\{x_\alpha(t)\}$, $\alpha = 1, \ldots, n$, such that any other solution can be written as: $y(t) = c^\alpha x_\alpha(t)$, $y(0) = c^\alpha x_\alpha(t = 0)$. We can then construct an $n \times n$ matrix $X(t)$ whose α-th column ($\alpha = 1, \ldots, n$) is given by $x_\alpha(t)$, i.e., $X_{j\alpha}(t) = x_\alpha^j(t)$. Then, for this 'matrix of solutions' we have

$$X(t) = e^{tA} X(0)$$

i.e.,

$$e^{tA} = X(t) \circ X^{-1}(0) .$$

Therefore, from a fundamental set of solutions we can construct

$$S(t, t_0) = X(t) \circ X^{-1}(t_0) . \tag{2.44}$$

This relation holds true also for time-dependent equations and any solution can be written as

$$x(t) = S(t, t_0)\, x(t_0) . \tag{2.45}$$

2.2.5 From a Vector Space to Its Dual: Induced Evolution Equations

From the equations of motion on the vector space E it is possible to induce equations of motion on any other vector space that we may build canonically out of E. We consider first the induced motion on the dual space E^*, the vector space of linear functions on E.

Starting with the linear system $dx/dt = A \cdot x$, we consider its linear flow φ_t, then we define the linear flow $\widetilde{\varphi}_t$ on E^* defined as:

$$(\widetilde{\varphi}_t \alpha)(x) = \alpha(\varphi_{-t}(x)), \quad \forall \alpha \in E^*, \quad x \in E, \tag{2.46}$$

It is a simple matter to check that:

$$\widetilde{\varphi}_{t+s} = \widetilde{\varphi}_t \circ \widetilde{\varphi}_s, \quad \widetilde{\varphi}_0 = I,$$

and that the dependence on t of $\widetilde{\varphi}_t$ is the same as that of φ_t, i.e., the family of linear maps $\widetilde{\varphi}_t$ defines a flow on E^*. Moreover, a simple computation shows:

$$\left(\frac{d}{dt}\widetilde{\varphi}_t\alpha\right)(x) = \frac{d}{dt}\alpha(\varphi_t^{-1}x) = \alpha\left(\frac{d}{dt}\varphi_t^{-1}x\right) = -\alpha(\varphi_t^{-1}A \cdot x)$$

i.e.,

$$\frac{d}{dt}\left(\widetilde{\varphi}_t\alpha\right) = -A^*\alpha$$

where $A^* \colon E^* \to E^*$ is the linear map defined as $A^*\alpha = \alpha \circ A$ and usually called the dual map to A. A simple computation shows us that the matrix representing the linear map A^* in the dual basis of a given basis in E is the transpose of the matrix representing the linear operator A. Thus we conclude stating that the dynamics on the dual vector space is induced by $-A^*$, the *opposite* of the dual map to A, or in linear coordinates the opposite of the transpose matrix which defines the dynamics on E. If we consider now the curves $x(t) = \varphi_t(x_0)$ and $\alpha(t) = \widetilde{\varphi}_t\alpha_0$ on E and E^* respectively, it is trivial to check that:

$$\frac{d}{dt}[\alpha(x)] = 0 \tag{2.47}$$

that is the quantity $\alpha(x) = \alpha_0(x_0)$ is constant under evolution.

The particular requirement we have considered here to induce a dynamics on E^* is instrumental to inducing an isospectral dynamics on linear maps or, equivalently, a dynamics that is compatible with the product of linear maps. The same can be stated for more complicated tensorial objects.

Once we have defined the induced flow on E^* it is easy to extend it to any other tensor space. For instance the induced flow on linear maps $B \colon E \to E$, that is on tensors of order $(1, 1)$, we require the following diagram,

$$\begin{array}{ccc} E & \xrightarrow{B} & E \\ {\scriptstyle \widetilde{\varphi}_t}\downarrow & & \downarrow{\scriptstyle \widetilde{\varphi}_t} \\ E & \xrightarrow{B_t} & E \end{array}$$

to be commutative, i.e., $B(t) = \widetilde{\varphi}_t \circ B \circ \widetilde{\varphi}_t^{-1}$.

From this we get,

$$\frac{d}{dt}B(t) = [A, B(t)], \tag{2.48}$$

and the evolution is isospectral. This last equation is the analog of Heisenberg equations in a classical context.

Returning to dual dynamics on the dual vector space E^*, let us consider a linear transformation of the most elementary type, namely: $B = x \otimes \alpha$, with $x \in E$ and $\alpha \in E^*$. By using the derivation property on the tensor product we find: $dB/dt = \dot{x} \otimes \alpha - x \otimes \dot{\alpha} = A(x \otimes \alpha) - (x \otimes \alpha)A$, i.e. once again Eq. (2.48).

Remark 2.4

1. If we require the evolution on linear maps Φ_t to satisfy

$$\Phi_t (M \circ N) = \Phi_t (M) \circ \Phi_t (N) , \tag{2.49}$$

we find that there exists a one-parameter group $\{\phi_t \mid t \in \mathbb{R}\}$ of automorphisms of the vector space such that

$$\Phi_t (M) = \phi_t \circ M \circ \phi_t^{-1} . \tag{2.50}$$

In one direction the statement is easily verified. In the opposite direction it relies on preservation of the 'row-by-column' product $\alpha(x) = \Phi_t [\alpha(x)] = \left(\Phi_t^* (\alpha)\right) \circ (\Phi_t (x)) = \alpha \circ \phi_t^{-1} \circ \phi_t (x)$. Then, by writing only elementary blocks in M and N, say $y \otimes \beta$, $x \otimes \alpha$, we find

$$(y \otimes \beta) \cdot (x \otimes \alpha) = \beta (x) \, y \otimes \alpha \tag{2.51}$$

and,

$$\begin{aligned} \Phi_t (\beta (x) \, y \otimes \alpha) &= \Phi_t (y) \cdot \Phi_t (\beta (x)) \otimes \Phi_t (\alpha) \\ &= \Phi_t (y) \left[(\Phi_t (\beta)) (\Phi_t (x)) \right] \otimes \Phi_t (\alpha) \\ &= (\Phi_t (y) \otimes \Phi_t (\beta)) \cdot (\Phi_t (x) \otimes \Phi_t (\alpha)) \\ &= [\Phi_t (y \otimes \beta)] \cdot [\Phi_t (x \otimes \alpha)] \end{aligned}$$

and thus, Eq. (2.49) is satisfied.

2. In describing the evolution of the so-called open quantum systems one has to give up the requirement of the preservation of the product structure on linear maps (in which they represent density states) and we get a more general dynamics which cannot be described in terms of a commutator bracket.

2.3 From Linear Dynamical Systems to Vector Fields

2.3.1 Flows in the Algebra of Smooth Functions

In the previous sections we have been discussing some aspects concerning the structure and properties of linear systems using elementary notions from linear algebra.

Now we are going to present the basic tools from calculus and geometry needed to get a deeper understanding of them. For this purpose we are going to depart from the presentation of these subjects found in most elementary textbooks. We will emphasize the fact that the linear structure of the carrier space is not relevant for construction of a differential calculus in the sense that any other linear structure will define the same differential calculus.

This will be made explicit by using systematically a description in the space of (smooth) functions, where the particular choice of a linear structure will not play any role. This will have also far reaching consequences when we discuss linearization of vector fields and vice versa.

Together with the development of differential calculus we will find immediately the basic notions of differential geometry, vector fields, differential forms, etc. Then we will construct the exterior differential calculus and we will discuss its appealing algebraic structure. Finally we will use all these ideas to present the elementary geometry of dynamical systems, thus setting the basis for later developments.

We assume a basic acquaintance of the reader with the elementary notions of differential calculus and linear algebra in $\mathbb{R}^n$ or in finite-dimensional linear spaces E, such as they are discussed in standard textbooks such as [HS74], etc. We would like to discuss them again briefly here, to orient the reader in building what are going to be the notions on which this book is founded. A systematic use of concepts like those of rings, modules, algebras and the such will done throughout the text, thus the basic definitions are collected for the benefit of the reader in Appendix A. Related concepts, such as that of graded algebras, graded Lie algebras and graded derivations are also discussed in the same Appendix.

We denote by E a real linear space, finite-dimensional for most of the present discussion. In a finite-dimensional linear space all norms are equivalent, thus we will not specify which one we are using in the underlying topological notions we will introduce. Whenever we deal instead with infinite-dimensional spaces we will assume them to be Banach spaces with a given fixed norm $\|\cdot\|$, and often, more specifically Hilbert spaces.

We know from elementary analysis that functions with good regularity properties do not have to be polynomial; let us think of trigonometric or exponential functions. However these functions do share the property that they can be well approximated pointwise by polynomial functions of arbitrary degree. In fact, we may say that a function f is differentiable of class C^r at x if there exists a polynomial function P_r of degree r such that $f(y) - P_r(y)$ is continuous at the point x, and goes to zero faster than $||y - x||^r$ when $y \to x$. The class of smooth or C^∞ functions is defined as the family of functions which are of class C^r for every r at any point x in E. Thus, smooth functions are approximated by polynomial functions of arbitrary degree in the neighborhood of any point and it is the class of functions that extends most naturally the properties of the algebra of polynomials.[6] In this sense we can say that the algebra $\mathcal{F}(E)$ (or $\mathcal{F}$ for short if there is no risk of confusion) of the smooth functions on E extends the algebra $\mathcal{P}$ of the polynomials.[7]

Exercise 2.4 Prove that if f is a function on E homogeneous of degree 1 and differentiable at 0, then it must be linear.

---

[6] See however below, Remark 2.5.

[7] To be quite honest, the class of functions that extend more naturally the algebra of polynomials is the algebra of real analytic functions. However in this book we will restrict our attention to the algebra of smooth functions.

Exercise 2.5 Find examples of functions which are homogeneous of degree k and which are not k-tic forms, $k \geq 1$.

Because linear functions generate the algebra of polynomials and polynomials approximate arbitrarily well smooth functions in the neighborhood of any point, we will say that linear functions will 'generate' the algebra of smooth functions (we will make precise the notion of 'approximation' when introducing the notion of differentiable algebras in Sect. 3.2.2 and the meaning of 'generating' used here in Sect. 3.2.3).

As we will see in the chapters to follow, the algebra $\mathcal{F}(E)$ will play a central role in our exposition and we will come back to its structure and properties when needed.

Remark 2.5 All that has been said up to now (what we mean by linear functions and so on) depends of course in a crucial way on the linear structure that is assumed to have been assigned on E, and a different linear structure (see below, Sect. 3.5.3 for a discussion of this point) will lead to different notions of 'linear functions', 'polynomials' and so on. On the other hand, being smooth is a property of a function that is independent of the linear structure. We can conclude therefore that $\mathcal{F}(E)$ is obtained anyway as the closure (in the sense specified above) of the polynomial algebras associated with the different linear structures.

2.3.2 Transformations and Flows

Invertible linear maps L from E to E are the natural transformations of E preserving its linear structure. They form a group, the group $GL(E)$ of automorphisms of E. If we fix a linear basis $\{e_i \mid i = 1, \ldots, n\}$ on E, then E is identified with $\mathbb{R}^n$, $n = \dim E$, and linear maps from E to E are represented by square matrices. Invertible linear maps correspond in this representation to regular matrices and the group $GL(E)$ becomes the general linear group $GL(n, \mathbb{R})$. Later on, we will discuss in more detail this and other related groups of matrices (see Sect. 2.6.1). Using the previous identification of E with $\mathbb{R}^n$, any map $\phi\colon E \to E$ can be written as a n-tuple of component functions $\phi = (\phi^1, \ldots, \phi^n)$. Thus the notion of smoothness translates directly to ϕ via the components ϕ^i.

A smooth diffeomorphism of E is a smooth invertible map $\phi\colon E \to E$ whose inverse is also smooth. Clearly, if f is a smooth function, $f \circ \phi$ is also smooth and we obtain in this way a map $\phi^*\colon \mathcal{F} \to \mathcal{F}$, called the pull-back map along ϕ, as

$$\phi^* (f) (x) = (f \circ \phi) (x) \tag{2.52}$$

which preserves the product structure and, as $(\phi_2 \circ \phi_1)^* = \phi_1^* \circ \phi_2^*$, it is invertible if ϕ is a diffeomorphism, $(\phi^*)^{-1} = (\phi^{-1})^*$. Thus the set of diffeomorphisms transforms smooth functions into smooth functions and they are the natural set of maps preserving the differentiability properties of functions, as we shall see shortly.

The composition of two diffeomorphisms is again a diffeomorphism, hence they constitute a group, denoted as Diff (E), because of the associativity property of the composition of maps, and clearly contains the group $GL(E)$. The group of diffeomorphisms of E will also be called the group of transformations of the space E. We must remark that a transformation ϕ will destroy in general the linear structure on E but will leave invariant the class of smooth functions $\mathcal{F}$. This means that the notion of smoothness is not related to any particular linear structure on the space E (see Remark 2.5) and depends only on what is called the differential structure of E.

We can also consider local transformations, i.e., smooth maps defined only on open sets of E and which are invertible on their domains. In what follows we will not pay attention to the distinction between local transformations and transformations in the sense discussed above, because, given a transformation, we can always restrict our attention to an arbitrary open set on E.

2.3.3 The Dual Point of View of Dynamical Evolution

One of the aspects we would like to stress in this chapter is that all notions of differential calculus can be rephrased completely in terms of functions and their higher order analogues, differential forms, i.e., we can take the dual viewpoint and use instead of points in a linear space E, functions on it. In fact, let us consider again the algebra $\mathcal{F}(E)$ of smooth functions on E. It is clear that $\mathcal{F}(E)$ contains the same information as the set of points E itself. In fact, as we discussed earlier, we can reconstruct E from $\mathcal{F}(E)$ by considering the set of homogeneous functions of degree one.

This attitude, one of the main aspects of this book, is very common in the physical construction of theories where, implicitly, states and observables are used interchangeably to describe physical systems. The states are usually identified with points making up the space E and the observables with smooth functions (or more general objects) in E. This means that they define the value taken by the observable on any possible state. If the description of a system in terms of observables is complete, we can reconstruct the states from the observables by taking appropriate measures on them. This is the essence of Gelfand-Naimark theorem.

Evolution can then be described not in terms of how points (states) actually evolve, but in terms of the evolution of the observables themselves. This approach is often taken in physical theories where the actual description of states can be very complicated (if possible at all) but we know instead a generating set of observables. In fact, that is what it is usually done in elementary textbooks, where points are described by their coordinates x^i, thus if we say, for instance, that the position at time t of the point is given by $x^i(t)$, what we mean is that the basic position observables x^i have evolved and at time t they are given by the new functions $x^i(t)$ that turn out to be a new set of 'basic' position observables. Because the position observables x^i generate the full algebra $\mathcal{F}(E)$, describing how they evolve gives us the evolution of all other observables. An interesting observation is that the evolution described by means of observables or functions is always linear.

If we are given two observables f, g and their sum $f+g$ at a given time t_0, then their 'evolved' functions $f(t), g(t)$ and $(f+g)(t)$ satisfy $f(t)+g(t) = (f+g)(t)$. It is also clear that evolution must be invertible, thus if we denote by Φ_t the evolution operator, then $\Phi_t : \mathcal{F}(E) \to \mathcal{F}(E)$ is a linear isomorphism of the (infinite-dimensional) vector spaces $\mathcal{F}(E)$. Using this notation we will write $\Phi_t(f)$ for the observable $f(t)$ and the previous equation would be written as,

$$\Phi_t(f) + \Phi_t(g) = \Phi_t(f+g).$$

It is also clear that, when considering the (pointwise) product $f \cdot g$, we have,

$$\Phi_t(f \cdot g) = \Phi_t(f) \cdot \Phi_t(g) \tag{2.53}$$

i.e., that Φ_t preserves products as well. The family $\{\Phi_t\}_{t\in\mathbb{R}}$ appears therefore as a one-parameter family of automorphisms of the algebra $\mathcal{F}(E)$ of functions on E.

Later on, see Chap. 6, if E is a Hilbert space, we would describe the dynamics on expectation value functions, a subspace of $\mathcal{F}(E)$.

Remark 2.6 Of course not all linear automorphisms of $\mathcal{F}(E)$ are also algebra automorphisms. For example the mapping: $f \mapsto \exp(\lambda k)\, f$ with k a fixed function is a linear automorphism which is not an algebra automorphism.

Another example borrowed from Quantum Mechanics is given by the linear map $\Phi_t(f) = \exp(-i\Delta^2 t)(f)$ (Δ now denotes the Laplace operator). The family Φ_t is a one-parameter group on square integrable functions but it does not define a group of automorphisms for the product because the infinitesimal generator is a second-order differential operator, which is not a derivation.

Equation (2.53) or, otherwise stated, the requirement that time evolution of 'observables' (i.e., functions in $\mathcal{F}(E)$) should preserve both the linear structure and the algebra structure, has some interesting consequences. The most relevant among them is that we can characterize the evolution operator Φ_t in more mundane terms, and precisely as a transformation on E. With reference again to Eq. (2.53) we can go further and think of the effects of the iteration of evolution, i.e., of the result of applying Φ_t and Φ_s successively. In an autonomous world, i.e., the system has no memory of the previous history of the state it is acting upon, then necessarily we must have,

$$\Phi_t(\Phi_s(f)) = \Phi_{t+s}(f). \tag{2.54}$$

Obviously,

$$\Phi_0(f) = f, \tag{2.55}$$

and we find again (cfr. Eq. (2.50)) a one-parameter group. These properties may be satisfied even without the existence of ϕ_t^{-1}; in this ocurrence we would have a semi-group. Thus, evolution will be given by a one-parameter group of isomorphisms of the algebra $\mathcal{F}(E)$. Then, we can conclude this paragraph by postulating the axioms of autonomous evolution as given by a smooth one-parameter group of automorphisms of the algebra $\mathcal{F}(E)$.

Notice that the axioms for autonomous evolution are satisfied for a smooth one-parameter group of diffeomorphisms φ_t on E:

$$\Phi_t(f) = f \circ \varphi_{-t} = \varphi_{-t}^*(f). \tag{2.56}$$

(In Sect. 3.2.2 it will be shown that this is the most general situation).

2.3.4 *Differentials and Vector Fields: Locality*

From their transformation properties, smooth functions are scalar quantities, i.e., they verify a transformation rule that (cfr. Eq. (2.52)) can be stated as 'the transformed function at a transformed point takes the value of the untransformed function at the given point'.

The usual definition of the differential of a given function f as

$$df = \frac{\partial f}{\partial x^i} dx^i$$

requires however the explicit introduction of a coordinate system, i.e., an identification of E with $\mathbb{R}^n$. However, partial derivatives and differentials transform under changes of coordinates (local transformations) in a contragradient manner. In other words, under a change of coordinates

$$x^i \mapsto y^i = \phi^i(x), \tag{2.57}$$

we have

$$df = \frac{\partial f}{\partial x^i} dx^i = \frac{\partial f}{\partial y^i} dy^i,$$

or

$$dx^i \frac{\partial}{\partial x^i} = dy^i \frac{\partial}{\partial y^i}.$$

All that is rather elementary, but proves that the association of differentials to functions is an invariant operation. In more intrinsic terms, we can rewrite the above invariance property as the commutation of the operator d and the pull-back along the map ϕ (see later Sect. 2.4.1):

$$\phi^* \circ d = d \circ \phi^*,$$

and will refer to this property by saying that "d" is a "scalar" operator, or that it is "natural" with respect to the group Diff (E) of diffeomorphisms of E.

It is well-known from elementary courses in calculus that the differential of the function $f\colon E \to \mathbb{R}$ at the point x defines a linear map $df(x)\colon E \to \mathbb{R}^n$ via $df(x)(v) = (\partial f(x)/\partial x^i)v^i$, the v^i's being the components of $v \in E$ in the given coordinate system. It is an easy exercise to prove that $df(x)(v)$ is actually a coordinate-independent expression. The differential $df(x)$ belongs therefore to the dual space E^*, i.e., to the space of (real) linear functionals on E, whose elements are called *covectors*. Thus, the differential of f at x is a covector. A basis of covectors at x will be provided by the dx^i's, which denote consistently the differentials of the linear maps $x^i\colon E \to \mathbb{R}$, $x^i(v) = v^i$, and a covector at x will be given, in local coordinates, by an expression of the form $\alpha = \alpha_i\, dx^i$, with the α_i's transforming in the appropriate way.

The operator d does actually something more. Acting on the algebra $\mathcal{F}(E)$ of smooth functions it will produce a smooth field of covectors. Smooth fields of covectors are called 1-forms and their space is denoted by $\Omega^1(E)$. So, d is actually a map $d : \mathcal{F}(E) \to \Omega^1(E)$, but more on this later in this chapter.

Another class of objects to be defined starting from a coordinate representation but that has actually an invariant nature is that of vector fields. Any such an object will be defined, in a given system of coordinates, x^i say, as a first-order differential operator of the form $X = f^i(x)\partial/\partial x^i$. If we require the set f^i to transform under a change of coordinates like (2.57), as

$$f^i(x) \mapsto g^i(y) = f^k(x)\,\frac{\partial y^i}{\partial x^k}, \tag{2.58}$$

(i.e., just as the dx^i's do in df), then X will acquire an intrinsic character as well.

By using the chain rule we see immediately that there is associated, in a natural way, to a vector field a first-order ordinary differential equation, namely,

$$\frac{dx^i}{dt} = f^i\,(x)\,. \tag{2.59}$$

In this way we are 'reading' the components of the vector field as those of a velocity field on E. We obtain therefore a definition of the action of X on functions as,

$$X(h) = \frac{dh}{dt} = \frac{\partial h}{\partial x^i}\frac{dx^i}{dt} = \frac{\partial h}{\partial x^i}\,f^i(x).$$

It is precisely the requirement that the evolution of a function be expressed in an invariant manner (or equivalently, that the previous system of ordinary differential equations be covariant with respect to changes of coordinates) that fixes the transformation laws for the components of a vector field. As $X(f)$ contains only the basic ingredients defining both X and df, we may read it also as an action of df on X itself,

$$X(f) = df(X). \tag{2.60}$$

Notice that, in this sense, we are identifying the value of X at a given point x with a vector dual to $df(x)$, thus $X(x)$ is a vector on E. Then a vector field is simply what its name indicates, a field of vectors, i.e., a map $X : E \to E$. More specifically, $X : E \to E \times E, x \mapsto (x, X(x))$; i.e., they are vectors along with their point of application.

2.3.5 *Vector Fields and Derivations on the Algebra of Smooth Functions*

As we have already pointed out, from an algebraic point of view, the set of smooth functions $\mathcal{F}(E)$ on a vector space is an algebra (over the field $\mathbb{R}$ in the present context). Now, a derivation over a ring $\mathcal{A}$ is a map $D\colon \mathcal{A} \to \mathcal{A}$, such that, $D(f+g) = D(f) + D(g)$ and

$$D(fg) = D(f)g + f D(g), \tag{2.61}$$

for every $f, g \in \mathcal{A}$. This equation is known as Leibniz's rule. If, instead of being simply a ring, $\mathcal{A}$ is an algebra over a field $\mathbb{K}$ (we will consider always only $\mathbb{K} = \mathbb{R}$ or $\mathbb{K} = \mathbb{C}$) we can extend the requirement of linearity in an obvious manner as,

$$D(\alpha f + \beta g) = \alpha D(f) + \beta D(g)\,, \quad \alpha, \beta \in \mathbb{K}\,, f, g, \in \mathcal{A}. \tag{2.62}$$

We can turn the set of derivations into an $\mathcal{A}$-module by defining,

$$(D_1 + D_2)(f) = D_1(f) + D_2(f)$$

and

$$(f\,D)(g) = f\,D(g)\,, \quad \forall f, g \in \mathcal{A}\,.$$

Furthermore, we can define a product of derivations as the commutator,

$$[D_1, D_2](f) = D_1(D_2(f)) - D_2(D_1(f))\,, \qquad \forall f \in \mathcal{A}\,. \tag{2.63}$$

One can check that if the algebra is associative, then $[D_1, D_2]$ is again a derivation. Notice that, however, $D_1 \circ D_2$ and $D_2 \circ D_1$ separately are not derivations. It is also easy to check that the Lie bracket $[\cdot, \cdot]$ above, Eq. (2.63), satisfies the *Jacobi identity*,

$$[D_1, D_2], D_3] + [[D_3, D_1], D_2] + [[D_2, D_3], D_1] = 0\,, \quad \forall D_1, D_2, D_3\,. \tag{2.64}$$

In this way, the set of derivations over a ring becomes actually a Lie algebra.

Now we have the following:

Proposition 2.7 *The derivations over $\mathcal{F}(E)$ are the vector fields on E. Explicitly the Lie bracket of two vector fields X, Y is given by*

$$[X, Y](f) = X(Y(f)) - Y(X(f)).$$

Proof It is clear that vector fields are derivations. We shall consider now an arbitrary derivation D and prove that it defines a vector field. Let $D(x^i) = X^i$ be the images of a coordinate set of functions x^i. Then let us consider the first-order Taylor expansion of f around a given point x_0. Then,

$$f(x) = f(x_0) + \sum_{i=1}^{n} \left.\frac{\partial f}{\partial x^i}\right|_{x'} (x^i - x_0^i), \tag{2.65}$$

with x' lying in the segment joining x_0 and x. Then,

$$(Df)(x) = \sum_{i=1}^{n} \left(D\left(\left.\frac{\partial f}{\partial x^i}\right|_{x'} \right) (x^i - x_0^i) + \left.\frac{\partial f}{\partial x^i}\right|_{x'} X^i(x) \right). \tag{2.66}$$

If we take the limit $x \to x_0$ in the previous equation we get,

$$(Df)(x_0) = \sum_{i=1}^{n} X^i(x_0) \left.\frac{\partial f}{\partial x^i}\right|_{x_0}, \tag{2.67}$$

namely, $D(f) = X(f)$ for the vector field defined by the local components X^i. □

As the set $\partial/\partial x^i$, for i running from 1 to n, form a local basis for the module $\mathcal{F}(E)$, we can easily compute the commutator of two derivations,

$$X = X^i \frac{\partial}{\partial x^i}, \quad Y = Y^j \frac{\partial}{\partial x^j} \tag{2.68}$$

as follows:

$$[X, Y] = \left(X^i \frac{\partial Y^j}{\partial x^i} - Y^i \frac{\partial X^j}{\partial x^i} \right) \frac{\partial}{\partial x^j}. \tag{2.69}$$

We will denote as $\mathfrak{X}(E)$ the set of vector fields on E.

Definition 2.8 The action of $X \in \mathfrak{X}(E)$ on $f \in \mathcal{F}(E)$ defines the Lie derivative $\mathcal{L}_X$ of f along X, i.e.,

$$X(f) = \mathcal{L}_X f. \tag{2.70}$$

From the very definition of Lie brackets we obtain (always on functions),

$$\mathcal{L}_X \mathcal{L}_Y - \mathcal{L}_Y \mathcal{L}_X = \mathcal{L}_{[X,Y]}. \tag{2.71}$$

2.3.6 The 'Heisenberg' Representation of Evolution

The discussion in the previous sections has led us very close to what is known as the Heisenberg representation of evolution in Quantum Mechanics. This approach consists in translating the attention of the observer of the evolution of a given system from the states to the observables and to consider that the states do not evolve but that the observables change. In this sense, the states should be considered not as actual descriptions of the system at a given time, but as abstract descriptions of 'all' possible descriptions of the system. Then the actual evolution takes place by changing the observables while we measure the position, for instance, of the system. These considerations will sound familiar to those readers familiar with Quantum Mechanics but they are not really related with a 'quantum' description of the world but only with the duality between states and observables sketched in the previous section.

The postulate that evolution must also preserve the product of observables necessarily implies that infinitesimal evolution will be given by a derivation of the algebra. We notice immediately by differentiating Eq. (2.53) that,

$$\frac{d}{dt}\Phi_t(f \cdot g)\bigg|_{t=s} = \frac{d}{dt}\Phi_t(f)\bigg|_{t=s} \cdot \Phi_s(g) + \Phi_s(f) \cdot \frac{d}{dt}\Phi_t(f)\bigg|_{t=s}$$

Thus the infinitesimal evolution operator $\Gamma = d\Phi_t/dt|_{t=0}$ is a derivation in the algebra $\mathcal{F}(E)$. But from the previous discussions we know that derivations of the algebra of smooth functions on E are in one-to-one correspondence with vector fields on E, thus we can conclude that the axioms of evolution for observables discussed above imply that the evolution is described by a vector field Γ on E whose flow is given by the (local) one-parameter group of diffeomorphisms φ_t. Then, we will have

$$\Gamma(f) = \frac{d}{dt}\Phi_{-t}(f) = \frac{d}{dt} f \circ \varphi_t, \tag{2.72}$$

and,

$$\Gamma(x) = \frac{d}{dt}\varphi_t(x)\mid_{t=0} .$$

This equation, relating a one-parameter group of diffeomorphisms to its infinitesimal generator, is strongly reminiscent, for those who are familiar with Quantum Mechanics, of the Stone-von Neumann theorem, and can be taken actually to constitute its 'classical' version. Note however that the above is valid under the assumption that f

be a smooth (or at least a C^1) function, while the one-parameter group Φ_t, per se, can act on more general classes of functions like, e.g., continuous or simply measurable functions. In the latter cases we cannot revert from the group to the infinitesimal generator acting on functions, and again that is the classical counterpart of the problems with domains (of the infinitesimal generator) that are well known to occur in the transition from the unitary evolution to its self-adjoint generator. A prototypical example of an operator for which this kind of problems arises is provided by the operator d/dx, the infinitesimal generator of translations on the real line. A nice discussion of the relation between completeness and self-adjointness may be found in [ZK93].

Remark 2.7 A few remarks are in order here. First, not all vector fields on E arise in this form. The existence theorem for ordinary differential equations, states only the local existence of the flow φ_t, thus in general for an arbitrary vector field we will not be able to extend the local solutions $x(t)$ for all values of t, hence we will not be able to define a one-parameter family of diffeomorphisms φ_t but only a local one-parameter group (We will discuss this issue in the next section).

An elementary example is provided, in one dimension, by the vector field: $\Gamma(x) = \alpha x^2 \partial/\partial x$, $\alpha = \text{const.}$, $x \in \mathbb{R}$, whose integral curves are of the form: $x(t) = x_0/(1 - \alpha x_0 t)$, $x_0 = x(0)$, that, for every $x_0 \neq 0$, will 'explode' to infinity in a finite time: $t^* = 1/\alpha x_0$ and therefore is not complete.

A less elementary example is provided by the vector fields[8]: $X_{(i)} = \epsilon_{ijk} x^j \partial/\partial x^k$, $i = 1, 2, 3$ on $\mathbb{R}^3$, where ϵ_{ijk} is the totally antisymmetric (Ricci or Levi-Civita) tensor ($\epsilon_{123} = 1$), which close on the Lie algebra of $SO(3)$ (or of $SU(2)$), i.e.: $\left[X_{(i)}, X_{(j)}\right] = \epsilon_{ijk} X_{(k)}$ and generate the rotations in $\mathbb{R}^3$. They can be restricted to the unit sphere S^2, where they become, in spherical polar coordinates (ϕ, θ):

$$
\begin{aligned}
X_1 &= \sin\phi \frac{\partial}{\partial\theta} + \cot\theta \cos\phi \frac{\partial}{\partial\phi}, \\
X_2 &= \cos\phi \frac{\partial}{\partial\theta} - \cot\theta \sin\phi \frac{\partial}{\partial\phi}, \\
X_3 &= \frac{\partial}{\partial\phi}.
\end{aligned}
\tag{2.73}
$$

Spherical polar coordinates are, of course, a system of coordinates only for the sphere without the poles and one meridian passing through them, i.e. for: $\theta \in (0, \pi)$ and $\phi \in (0, 2\pi)$. So, the vector fields (2.73) are actually globally defined on the cylinder: $(0, \pi) \times [0, 2\pi]$, and it is not difficult to convince oneself that, out of these three vector fields, only $X_{(3)}$ is complete.

An even more startling example is provided by the realization of the pseudo-rotation group on the real line;

[8] These fields are not independent. In fact, denoting them collectively as: $\mathbf{X} = \left(X_{(1)}, X_{(2)}, X_{(3)}\right)$, with: $\mathbf{x} = \left(x^1, x^2, x^3\right)$, it is obvious that: $\mathbf{x} \cdot \mathbf{X} = 0$.

$$X_1 = \frac{\partial}{\partial x}, \quad X_2 = \sin x \frac{\partial}{\partial x}, \quad X_3 = \cos x \frac{\partial}{\partial x}.$$

The above notions can be carried over to the level of the algebra of smooth functions modifying in an appropriate way the axioms above but we will not do it here in order not to create an unnecessary complication in the definitions. Thus in what follows by evolution we will understand the (local) smooth one-parameter group of transformations defined by a derivation Γ of the algebra of smooth functions $\mathcal{F}(E)$ or equivalently by the vector field Γ on E.

Secondly, the autonomous condition on evolution introduced on the axioms above can be removed because systems can have at a given point, and in fact they often do, memory of their previous history. Then, evolution will be given simply by a (local) smooth one-parameter family of diffeomorphisms on E. That is equivalent to giving (local) smooth one-parameter family of derivations Γ_t on E, which is usually called a time-dependent vector field.

2.3.7 The Integration Problem for Vector Fields

Thus we have seen that a one-parameter group of automorphisms Φ_t of $\mathcal{F}$ defines, at least formally, a derivation Γ. Derivations are identified with vector fields, thus we have a way to recover the group of automorphisms by integrating the differential equation defined by the vector field. Because we know the action of a vector field on functions, $f \mapsto X(f)$ we could try to compute the flow φ of X on a function f by 'integrating' the previous formula and we will get:

$$\varphi_t^*(f) = \sum_{k \geq 0} \frac{t^k}{k!} \mathcal{L}_X^k(f). \tag{2.74}$$

However the previous formula could raise a few eyebrows. When is the series on the right-hand side of equation (2.74) convergent? Even if it is convergent with respect to some reasonable topology on some class of functions, would the family φ_t of maps thus obtained be the 'integral flow' of X as in the linear case?

This is the first nonlinear integrability problem we are facing and its solution provides the key to predict the evolution of a given system. The answer to this problem is, of course, well known and it constitutes the main theorem in the theory of ordinary differential equations. Before discussing it, we would like to elaborate further on the notion of tangent vectors and vector fields, both for the sake of the stament of the solution to this problem and for further use in the construction of geometrical structures associated to given dynamics.

2.3.7.1 The Tangent and the Cotangent Bundle

The interpretation of vector fields and 1-forms as fields of vectors and covectors respectively captures only a partial aspect of their use. The ordinary differential equation (2.59), associated with a given vector field X, shows us an alternative interpretation of vector fields which is at the basis of the present geometrical construction. The components $f^i(x)$ of the vector field X are interpreted according to Eq. (2.59) as the components of the velocity of a curve $\gamma(t)$ which is a solution of the ordinary differential equation defined by X. Thus, the value of the vector field X at a given point x can be geometrically thought of as a tangent vector to a curve $\gamma(t)$ passing through x at $t = 0$. Notice that two curves $\gamma_1(t)$ and $\gamma_2(t)$ having a contact of order 1 at x define the same tangent vector, i.e.

$$\left.\frac{d\gamma_1(t)}{dt}\right|_{t=0} = \left.\frac{d\gamma_2(t)}{dt}\right|_{t=0} . \tag{2.75}$$

Therefore a tangent vector at the point x can be thought as an equivalence class of curves[9] passing through x with respect to the equivalence relation of having a contact of order 1 at x. We shall denote by T_xE the collection of all tangent vectors to E at x. If v_x denotes one of these tangent vectors we can define the variation of a function f at the point x in the direction of v_x as

$$v_x(f) = \left.\frac{d}{dt} f \circ \gamma(t)\right|_{t=0} , \tag{2.76}$$

where $\gamma(t)$ is any representative in the equivalence class defining v_x. Notice that the definition of the numerical value $v_x(f)$ does not depend on the choice of the representative γ we make. Thus v_x is a linear map from the space of differentiable functions defined in a given neighborhood of the considered point x into $\mathbb{R}$ with the additional property that

$$v_x(f_1 f_2) = f_1(x)v_x(f_2) + f_2(x)v_x(f_1)$$

This additional property is what characterizes vectors v_x at a point x among all linear maps. We can define an addition on the space of tangent vectors as

$$(v_x + u_x)(f) = v_x(f) + u_x(f) \tag{2.77}$$

Of course we need to guarantee that the object thus defined, $u_x + v_x$, corresponds again to a tangent vector, i.e., we need to find a curve passing through x such that its tangent vector will be $u_x + v_x$. Because we are in a linear space E, that is actually very easy. Given a vector $v \in E$ there is a natural map identification $v \mapsto v_x$ with a tangent vector at x, which is the equivalence class corresponding to the curve

[9] See however below, Appendix C, Sect. C.2 for a similar discussion in a more general context.

$\gamma(t) = x + tv$. Such identification is clearly one-to-one and the vector corresponding to $v_x + u_x$ is the vector corresponding to the curve $\gamma(t) = x + t(u + v)$.[10] From this perspective, we see that Eq. (2.77) actually defines the addition on the tangent space $T_x E$ and it shows that this addition does not depend on any linear structure of E. However, in this particular setting it is also true that we have a natural isomorphism between $T_x E$ and E as linear spaces. Hence, we can think that the tangent space at the point x is a copy of the background space E put at the point x. The tangent vectors corresponding to the curves $\gamma_i(t) = x + t\, e_i$, where $\{e_i \mid i = 1, \ldots, n\}$ is a given basis in E, are denoted by $(\partial/\partial x^i)_{|x}$. The notation is consistent with the operator defined on functions, because,

$$\left.\frac{\partial}{\partial x^i}\right|_x (f) = \left.\frac{d}{dt}(f \circ \gamma_i)(t)\right|_{t=0} = \left.\frac{d}{dt} f(x + te_i)\right|_{t=0} = \left.\frac{\partial f}{\partial x^i}\right|_x . \tag{2.78}$$

The union of all tangent spaces $T_x E$ is collectively denoted by,

$$TE = \bigcup_{x \in E} T_x E \tag{2.79}$$

and it is clearly isomorphic as a linear space to $E \oplus E$. An element of TE is thus a pair (x, v) where x is a point (vector) in E and v is a tangent vector at x. There is a natural projection $\tau_E \colon TE \to E$, defined as $\tau_E(x, v) = x$. Such a structure, the triple (TE, τ_E, E), is called the tangent bundle over E.

If $\phi \colon E \to E$ denotes a smooth map and v_x is a tangent vector at x, then $\phi_*(v_x)$ is a tangent vector at $\phi(x)$ defined as:

$$(\phi_*(v_x))\,(f) = v_x(f \circ \phi)\,,$$

for any f defined in an open neighborhood of $\phi(x)$. Thus we have defined a map $\phi_* \colon TE \to TE$ called the tangent map to ϕ or, sometimes, the differential of ϕ. It is clearly satisfied that $(\phi \circ \psi)_* = \phi_* \circ \psi_*$ which is just a consequence of the chain rule. Thus if ϕ is a diffeomorphism, then $\phi_*^{-1} = (\phi^{-1})_*$.

Turning back to the notion of a vector field, we see that a vector field consists of a smooth selection of a tangent vector at each point of x, the values of the vector field X being the vectors $X(x)$ tangent to E at x. Thus, a vector field is a smooth map $X \colon E \to TE$ such that it maps x into $(x, X(x))$ where $X(x) \in T_x E$.

The above maps satisfy $\tau_E \circ X = \mathrm{id}_E$ and are called *cross sections* of the tangent bundle. Therefore, in this terminology a vector field is just a cross section of the tangent bundle. Moreover, if ϕ is a diffeomorphism, then we may define the push-forward $\phi_* X$ of X along ϕ as follows: $(\phi_* X)(x) = \phi_*(X(x))$.

---

[10] However this idea will also work in a more abstract setting in the sense that it is possible to show that there is a one-to-one correspondence between equivalence classes of curves possessing a contact of order 1 at x and linear first-order differential operators v acting locally on functions at x.

Because E is already a vector space, there is a distinguished vector field called the dilation (or Liouville) vector field Δ[11]:

$$\Delta : E \to TE; \quad x \mapsto \Delta(x) = (x, x), \quad x \in E. \tag{2.80}$$

The Liouville vector field Δ allows us to identify the vector space structure on the tangent space with the vector space structure on E (see later on Sect. 3.3.1), i.e., Δ can be defined only if E itself is a vector space. The graph of Δ in $TE \approx E \times E$ is a subspace of $TE \oplus TE$, the diagonal vector subspace.

Together with the tangent bundle we do have its dual, the cotangent bundle. We describe it briefly here. Again, as in the case of tangent vectors being identified with equivalence classes of curves, we have a natural identification of functions at a given point by possessing the same differential (the actual value of the function is irrelevant). Any such equivalence class actually defines a covector $\alpha_x = df(x) \in E^*$, for some f. Thus the space of covectors at x (differentials of functions at x) defines the cotangent space denoted by $T_x^* E$. Such a space is obviously naturally isomorphic to E^* and is dual to $T_x E$. The set $\{dx^i(p) \mid i = 1, \ldots n\}$ is the dual basis of $\{(\partial/\partial x^i)_p \mid i = 1, \ldots n\}$ at each point $p \in E$.

The pairing between both is given as follows: If γ is a curve representing the tangent vector v_x and f is a function representing the cotangent vector α_x, then,

$$\langle \alpha_x, v_x \rangle = \frac{d}{dt}(f \circ \gamma)(t)\Big|_{t=0} . \tag{2.81}$$

The union of all cotangent spaces $T_x^* E$ is denoted by T^*E. Clearly, T^*E is naturally isomorphic to $E \oplus E^*$ and it carries a natural projection $\pi_E : T^*E \to E$, defined as $\pi_E(x, \alpha_x) = x$. The triple (T^*E, π_E, E) is called the cotangent bundle of E.

A smooth assignment of a covector at x to any point $x \in E$ is called a 1-form. Thus a 1-form α is a smooth map $\alpha \colon E \to T^*E$ such that $x \mapsto (x, \alpha(x))$, i.e., again a 1-form is a cross section of the cotangent bundle, and therefore such that $\pi_E \circ \alpha = \mathrm{id}_E$.

2.3.7.2 Vector Fields and Local Flows

Thus given a vector field X on E, we want to determine the existence of a flow φ_t for X. In general, as it was pointed out before, this cannot be done globally, however it is always possible to do it locally (see [Ar73] by a masterly exposition of the subject).

Theorem 2.9 (Fundamental theorem of ordinary differential equations) *Given a smooth vector field X on E, for every $x \in E$ there exists an open neighborhood U of x and a number $\epsilon > 0$ such that given any point $y \in U$ and any t with $|t| < \epsilon$, the solution $\varphi_t(y)$ of the equation $du/dt = X(u)$ satisfying the initial condition y at $t = 0$ exists, is unique, depends smoothly on y and t and satisfies:*

---

[11] Also-called the *Euler differential operator.*

$$\varphi_{t+s}(y) = \varphi_t \circ \varphi_s(y), \quad |t| < \epsilon, \quad |s| < \epsilon, \quad |t+s| < \epsilon.$$

A vector field whose solutions can be extended from $-\infty$ to $+\infty$ so as to give raise to a one-parameter group will be said to be a 'complete' vector field. As already said, generic vector fields need not be complete. However, if the vector field is defined on a compact set, or better, is different from zero only on a compact set, it is complete. This result is also true for smooth manifolds ([Ar73], Theorem 35.1), but we will state it here just in the case of linear spaces.

Theorem 2.10 *Let X be a smooth vector field different from zero only in a compact subset K of E. Then there exists a one-parameter group of diffeomorphisms $\varphi_t : E \to E$ for which X is the velocity field:*

$$\frac{d}{dt}\varphi_t(x) = X(\varphi_t(x)), \quad \forall x \in E.$$

Thus if Γ is a vector field on E, then we may find a one-parameter group of diffeomorphisms φ_t that describes the trajectories of it on a compact set, or in other words, there is a one-parameter group of automorphisms of $\mathcal{F}(E)$ that restricted to $\mathcal{F}(K)$ for K a compact set satisfies:

$$\frac{d}{dt}\varphi_t = \Gamma \circ \varphi_t\,. \tag{2.82}$$

The picture we can get from this situation is that given a vector field and choosing a compact set K neighborhood of a given point, there is a complete flow φ_t that acting on points on K will produce the trajectories of Γ but that a little bit after exiting K the flow will 'freeze' leaving the points fixed.

The idea to prove this is simple. Given a compact set K, we may construct (taking for instance the a closed ball containing K) a smooth 'bump' function ρ adapted to K, that is a function such that $\rho = 1$ on K and $\rho = 0$ in the complementary of the closure of a ball containing K. Thus multiplying X by ρ we have a vector field to which we may apply Theorem 2.10 and whose complete flow is a one-parameter group of diffeomorphims satisfying the previous equation (2.82) on K.

The previous formula (2.82) provides a rigorous setting for Eq. (2.72) and makes sense of the formal integration idea expressed by Eq. (2.74).

Thus, using the previous observation, we may assume that we have a one-parameter group of diffeomorphisms integrating a given dynamics (that is describing the trajectories of our system on compact sets) and we will use this in what follows without explicit mention to it.

We will close this collection of ideas by noticing that the Lie derivative of a function f along a vector field Γ that was derived before represents the infinitesimal variation of the function in the direction of the vector field and can be defined (by using Eq. (2.82)) as:

$$\mathcal{L}_\Gamma f = \Gamma(f) = \frac{d}{dt}\varphi_{-t}^* f \mid_{t=0} = \frac{d}{dt} f \circ \varphi_{-t} \mid_{t=0} .$$

2.4 Exterior Differential Calculus on Linear Spaces

2.4.1 Differential Forms

Having defined the differentials as objects that behave as 'scalars' one may also say that they are 'natural' under arbitrary (smooth) changes of coordinates. Thus we may form $\mathcal{F}(E)$-linear combinations of differentials of functions, i.e., sums of monomials of the form $f\,dg$, $f, g \in \mathcal{F}(E)$. Then if we transform f and g by using a diffeomorphisms ϕ of E, then $f\,dg$ transforms accordingly. The monomials $f\,dg$ will generate an $\mathcal{F}(E)$-module $\Omega^1(E)$ whose elements are differential 1-forms over E. Notice that $\Omega^1(E)$ is just the space of sections of the cotangent bundle T^*E because any 1-form α can be written as $\alpha = \alpha_i\,dx^i$, once a linear coordinate system x^i on E has been chosen, i.e., a linear basis e_i has been selected. Then it is clear that the dx^i's are a basis of $\Omega^1(E)$, i.e., any 1-form can be given uniquely as an $\mathcal{F}(E)$-linear combination of them. Equivalently, the dx^i's are linearly independent. We may choose also any other basis, i.e., any other set df^i, provided they are also independent. Of course, the df^i's are linearly independent iff the f^i's are functionally independent, i.e., no nonconstant function $\Phi = \Phi(f^1, \ldots, f^n)$ exists such that $\Phi(f^1, \ldots, f^n) = \text{const}$. Later on, see Eq. (2.89), we will give a compact characterization of this condition). A 1-form α which is the differential of a function, i.e., $\alpha = df$ will be called exact.

We may also wish to extend the action of d from functions, i.e., zero-forms, to 1-forms. Let us start with a monomial like $f\,dg$, and let $d(f\,dg)$ be defined by acting on the coefficients of the dx^i's (which are functions, after all). We are immediately faced with the problem of defining products of differentials. If we choose, e.g., tensor products, we obtain for $d(f\,dg)$ the expression,

$$d(f\,dg) = \left(\frac{\partial f}{\partial x^i}\frac{\partial g}{\partial x^j} + f\frac{\partial^2 g}{\partial x^i \partial x^j}\right) dx^i \otimes dx^j .$$

Changing coordinates $x \mapsto x'$, a tedious but straightforward calculation yields,

$$d\,(f dg) = \left(\frac{\partial f}{\partial x'^i}\frac{\partial g}{\partial x'^j} + f\frac{\partial g}{\partial x^k}\frac{\partial^2 x^k}{\partial x'^i \partial x'^j} + f\frac{\partial^2 g}{\partial x'^i \partial x'^j}\right) dx'^i \otimes dx'^j .$$

So, naturality of the operator d gets lost (except for linear changes of coordinates) unless we redefine the product of differentials in such a way as to eliminate symmetric parts. This leads us to define the *wedge* (or *exterior*) product $dx^i \wedge dx^j$ as the antisymmetrized product,

$$dx^i \wedge dx^j = \frac{1}{2}(dx^i \otimes dx^j - dx^j \otimes dx^i), \tag{2.83}$$

and, by extension,

$$df \wedge dg = \frac{\partial f}{\partial x^i}\frac{\partial g}{\partial x^j} dx^i \wedge dx^j = \frac{1}{2}\left(\frac{\partial f}{\partial x^i}\frac{\partial g}{\partial x^j} - \frac{\partial g}{\partial x^i}\frac{\partial f}{\partial x^j}\right) dx^i \wedge dx^j,$$

and to extend (by definition now) the action of d on monomials of the form $f\,dg$ as

$$d(f\,dg) = df \wedge dg. \tag{2.84}$$

This definition has the advantage of retaining the naturality of the *exterior differential*, as d will be called from now on. It is remarkable that d is the only derivation which is a 'scalar operator' with respect to the full diffeormorphism group [Pa59].

Remark 2.8 With the Definition (2.83), the wedge product differs by a normalization factor $1/2$ in the case of the product of two one-forms as in Eq. (2.83), $1/n!$ for the product of n one-forms, from the antisymmetrized product one would obtain using Eq. (10.6). This normalization turns out however to be more convenient, and we will use it throughout whenever we will deal with differential forms.

The $\mathcal{F}(E)$-linear space spanned by the monomials $dx^i \wedge dx^j$ (or by the independent wedge products in any other basis) will be called the space of smooth two-forms, or just 2-forms for short, and will be denoted by $\Omega^2(E)$. A general 2-form will have the expression,

$$\alpha = \alpha_{ij}\, dx^i \wedge dx^j, \quad \alpha_{ij} = -\alpha_{ji}, \ \alpha_{ij} \in \mathcal{F}(E) \tag{2.85}$$

It is left to the reader to work out the transformation law of the coefficients α_{ij}'s under arbitrary (smooth) changes of coordinates.

Because E is a linear space then the $\mathcal{F}(E)$-module of 1-forms is finitely generated (a system of generators is provided by the differentials dx^i of a linear system of coordinates x^i for instance). Not only that, $\Omega^1(E)$ is a free module over $\mathcal{F}(E)$ and the 1-forms dx^i provide a basis for it, then $\dim_{\mathcal{F}(E)} \Omega^1(E) = \dim E$.

Similarly the $\mathcal{F}(E)$-module $\Omega^2(E)$ is free and the 2-forms $dx^i \wedge dx^j$ provide a basis for it. Then clearly if $\dim E = n$, $\dim_{\mathcal{F}(E)} \Omega^2(E) = \binom{n}{2}$. Here too, if $\alpha = d\theta$ for some $\theta \in \Omega^1(E)$, α will be called an exact two-form.

As we discussed before, starting with the tangent bundle TE we can form the cotangent bundle or the bundle of covectors or linear 1-forms over E, but we could also form the bundle of linear 2-forms (or skew symmetric linear (0, 2) tensors) over E. We shall denote such bundle as $\Omega^2(T^*E)$, and it is just the union of all spaces $\Omega^2(T_xE)$, $x \in E$. We also have as before that $\Omega^2(T^*E) \cong E \oplus \Omega^2(E)$. Notice that $\Omega^1(T^*E) = T^*E$. Cross sections of $\Omega^2(T^*E)$ are smooth 2-forms $\omega \in \Omega^2(E)$. It is also customary to denote the space of cross sections of the bundles TE, T^*E, $\Omega^2(T^*E)$ by $\Gamma(TE)$, $\Gamma(T^*E)$, and $\Gamma(\Omega^2(T^*E))$, etc., thus $\Gamma(T^*E) = \Omega^1(E)$ and so on.

Let us recall that a linear 1-form α acts on a vector producing a number. Thus we may also think that $\Omega^1(E)$ is the dual with respect to the algebra $\mathcal{F}(E)$ of the module of vector fields $\mathfrak{X}(E)$, i..e, a 1-form α is a $\mathcal{F}(E)$-linear map $\alpha\colon \mathfrak{X}(E) \to \mathcal{F}(E)$. $\alpha(X) \in \mathcal{F}(E)$, for all $X \in \mathfrak{X}(E)$ and being defined as $\alpha(X)(x) = \langle \alpha(x), X(x)\rangle$ where $\langle \cdot, \cdot \rangle$ denotes as usual the natural pairing between a linear space and its dual.

A similar argument applies to 2-forms. A 2-form $\omega \in \Omega^2(E)$ can be considered as defining a skew symmetric $\mathcal{F}(E)$-bilinear map on the module $\mathfrak{X}(E)$, that is $\omega(X, Y)$ is a smooth function on E for any $X, Y \in \mathfrak{X}(E)$.

Now from the definition of the wedge product it follows at once that,

$$(dx^i \wedge dx^j)(X, Y) = dx^i(X)dx^j(Y) - dx^i(Y)dx^j(X)$$

for any pair of vector fields X and Y.

In particular, having in mind the definition of the Lie derivative on functions, it is not hard to see that, if $\alpha = d\theta$ is exact, then,

$$d\theta(X, Y) = \mathcal{L}_X(\theta(Y)) - \mathcal{L}_Y(\theta(X)) - \theta([X, Y]), \qquad \forall X, Y \in \mathfrak{X}(E)\,. \tag{2.86}$$

It is left as an exercise to prove that, notwithstanding the differential nature of the Lie derivative, the right-hand side of Eq. (2.86) is actually $\mathcal{F}(E)$-linear in both X and Y (besides being manifestly skew-symmetric). Together with Eq. (2.71), this tells us also that, if $\theta = df$ is an exact 1-form, then, $d\theta = d^2 f = 0$, i.e., that $d^2 = d \circ d = 0$ (on functions only, for the time being).

2.4.2 Exterior Differential Calculus: Cartan Calculus

If we extend the wedge product into an associative product, we can generate forms of higher degree by taking wedge products of forms of lower order. In a systematic manner, if $\alpha_1, \ldots, \alpha_n$ is a basis of 1-forms (e.g., $\alpha_i = df_i$ for a set of functionally independent functions), then, monomials of the form,

$$\alpha_{i_1 \ldots i_k} = \alpha_{i_1} \wedge \cdots \wedge \alpha_{i_k}, \quad i_1, \ldots, i_k = 1, \ldots, n, \tag{2.87}$$

will generate the $\mathcal{F}(E)$-linear space of k-forms on E, denoted as $\Omega^k(E)$. Alternatively, we may think that a k-form α is an $\mathcal{F}(E)$-multilinear map

$$\alpha\colon \mathfrak{X}(E) \times \overset{k}{\cdots} \times \mathfrak{X}(E) \to \mathcal{F}(E),$$

such that $\alpha(X_1, \ldots, X_i, \ldots, X_j, \ldots, X_k) = -\alpha(X_1, \ldots, X_j, \ldots, X_i, \ldots, X_k)$ for all i, j. Then, given any smooth map $\phi\colon E \to E$, we may define the pull-back of any k-form along the map ϕ as

$$\phi^*\alpha(X_1, \ldots, X_k) = \alpha(\phi_* X_1, \ldots, \phi_* X_k), \qquad \forall X_i \in \mathfrak{X}(E). \tag{2.88}$$

Remark 2.9 The previous formula should be understood pointwise, that is considering the k-form α as a section of the bundle $\Omega^k(T^*E) \to E$. Then we will write:

$$(\phi^*\alpha)_x(v_1, \ldots, v_k) = \alpha_x(\phi_* v_1, \ldots, \phi_* v_k), \quad \forall v_1, \ldots, v_k \in T_x E.$$

However the formula above (2.88) makes perfect sense if ϕ is a diffeomorphism.

Remark 2.10 It is not hard to prove that the wedge product of two 1-forms (and hence of any number) vanishes iff the forms are linearly dependent. The condition for the linear independence for k-monomials will be then,

$$\alpha_{i_1} \wedge \cdots \wedge \alpha_{i_k} \neq 0.$$

Hence, if $k \leq n$, $\dim_{\mathcal{F}(E)} \Omega^k(E) = \binom{n}{k}$ and there will be no room for forms of degree higher than the dimension n of E, actually $\Omega^n(E) = \mathcal{F}(E)$. We note parenthetically that functional independence of a set of k functions $f_1, \ldots, f_k \in \mathcal{F}(E)$ will be expressed by the condition

$$df_1 \wedge \cdots \wedge df_k \neq 0. \tag{2.89}$$

Note that $\Omega^\bullet(E) = \bigoplus_{k \geq 0} \Omega^k(E)$ is an associative graded algebra (see Appendix A) and the elements of $\Omega^k(E)$ are said to be homogenous of degree k.

We wish now to extend the action of the exterior differential d to forms of arbitrary rank. Let us start again with monomials in a basis generated by exact 1-forms. To make things simpler, let $\alpha = f\, dx^{i_1} \wedge \cdots \wedge dx^{i_k}$ be a monomial of rank k (for a fixed set of i_l's, $l = 1, \ldots, k$). Then, we define $d\alpha$ as the monomial of rank $k + 1$,

$$d\alpha = df \wedge dx^{i_1} \wedge \ldots \wedge dx^{i_k}. \tag{2.90}$$

The exterior differential, with this obvious extension, is then defined on forms of arbitrary rank and is an $\mathbb{R}$-linear map $d\colon \Omega^k(E) \to \Omega^{k+1}(E)$, with $\Omega^0(E) = \mathcal{F}(E)$, and $\Omega^{n+1}(E) = \mathbf{0}$.

If a set of coordinates $\{x^1, \ldots, x^n\}$ has been chosen for E, then, $\varpi = dx^1 \wedge \cdots \wedge dx^n$ will be a basis for the ($\mathcal{F}(E)$-one-dimensional) module $\Omega^n(E)$, i.e., any $n-$form will be expressible as $f\,\varpi$ for some $f \in \mathcal{F}(E)$. In view of the fact that coordinates are globally defined for a vector space, ϖ can be thought of as well as a basis for the space $\Omega^n(T_x^*E)$ of the n-forms based at any point $x \in E$. As such, it will be better denoted as $\varpi(x)$ although the notation may appear somewhat redundant in this particular case. It enjoys the features that:

1. $\varpi(x) \neq 0, \forall x \in E$,
2. If we perform a permutation of coordinates, $x^i \mapsto y^i = x^{\pi(i)}$, $\pi \in S_n$ the group of permutations of n elements, then: $\varpi' = dy^1 \wedge \cdots \wedge dy^n = \mathrm{sign}(\pi)\varpi$, where $\mathrm{sign}(\pi)$ stands for the signature (the parity) of the permutation π. So, $\varpi' = \pm\varpi$ according to the parity of the permutation.

In general, a nowhere vanishing form of maximal rank will be called a *volume form*, and we have just seen that a volume form always exists on a vector space. This may not be so in more general situations in which we deal with spaces that can be modeled on vector (and hence Euclidean) spaces only locally, in which case it may well be that volume forms exist only locally, but this more general case is, for the time being, outside our scopes. We have also seen that $-\varpi$ is an equally acceptable volume-form if ϖ is. Each choice will be said to define an *orientation* on E. Again, that is a globally defined notion as long as E is a vector space, but need not be so in more general situations.

Let now ϕ be a linear map from E to E. In a given system of coordinates, ϕ : $x^i \mapsto y^i = A^i{}_j x^j$, i.e., ϕ will be represented by the matrix $A = (A^i{}_j)$. Then, by using the properties of the wedge product, it is not difficult to show that

$$(\phi^*\varpi)(x) = \det(A)\,\varpi(x).$$

More generally, if ϕ is a smooth map (not necessarily a linear one), the pull-back $\phi^*\varpi$ will be again an n-form, and hence proportional to ϖ itself, and this motivates the following:

Definition 2.11 Let $\phi\colon E \to E$ be a smooth map and let ϖ be a volume-form. Then the determinant of ϕ, $\det(\phi)$, is defined by

$$\phi^*\varpi = \det(\phi)\,\varpi\,. \tag{2.91}$$

A straightforward calculation leads then to the result that, if ϕ_1, ϕ_2, are smooth maps, then the determinant function enjoys the property we are familiar with in the linear case, i.e., that,

$$\det(\phi_1 \circ \phi_2) = \det(\phi_2 \circ \phi_1) = \det(\phi_1)\,\det(\phi_2). \tag{2.92}$$

Remark 2.11 If the volume-form is realized as: $\varpi = dx^1 \wedge \cdots \wedge dx^n$ in a given system of coordinates, then $\det(\phi)$ at point x is, of course, nothing but the familiar Jacobian determinant of ϕ at $x \in E$.

Equation (2.90) defines also the action of d on a wedge product of 1-forms. For example, let $\alpha = g\,df$ and $\beta = h\,dk$, $f, g, h, k \in \mathcal{F}(E)$, then, $\alpha \wedge \beta = (gh)\,df \wedge dk$ and one proves immediately that,

$$d(\alpha\wedge\beta) = d(gh)\wedge df\wedge dk = (g\,dh+h\,dg)\wedge df\wedge dk = d\alpha\wedge\beta-\alpha\wedge d\beta. \quad (2.93)$$

Using bilinearity, if α were a 2-form (actually a monomial of rank 2) we would get instead,

$$d(\alpha\wedge\beta) = d\alpha\wedge\beta - \alpha\wedge d\beta. \quad (2.94)$$

Extending these results in an obvious way from monomials to forms we obtain eventually:

Proposition 2.12 *If* $\alpha\in\Omega^p(E)$ *and* $\beta\in\Omega^q(E)$*, then* $\alpha\wedge\beta\in\Omega^{p+q}(E)$ *and the graded Leibniz rule is satisfied:*

$$d(\alpha\wedge\beta) = d\alpha\wedge\beta + (-1)^p\alpha\wedge d\beta. \quad (2.95)$$

Morevoer, we have that $d^2=0$.

Finally, if we consider a vector field, $X = X^i\frac{\partial}{\partial x^i}$, the Lie derivative of a volume form ϖ is proportional to the volume form, $\mathcal{L}_X\varpi = f\varpi$. As

$$\mathcal{L}_X(dx^1\wedge\cdots\wedge dx^n) = \left(\frac{\partial X^1}{\partial x^1}+\cdots+\frac{\partial X^n}{\partial x^n}\right)dx^1\wedge\cdots\wedge dx^n$$

when $\varpi = dx^1\wedge\cdots\wedge dx^n$, the proportionality factor is called the *divergence* of X, $f = \mathrm{div}\,(X)$, because using such global coordinate system,

$$\mathcal{L}_X(dx^1\wedge\cdots\wedge dx^n) = \mathrm{div}\,(X)\,dx^1\wedge\cdots\wedge dx^n.$$

It is also possible to associate with any element $X\in\mathfrak{X}(E)$ a derivation of degree -1 in the graded algebra of forms $\Omega^\bullet(E)$, called an *inner derivation* (or a *contraction*). We set $i_X\colon\Omega^p(E)\to\Omega^{p-1}(E)$, where $\alpha\mapsto i_X\alpha$ with $i_Xf=0$, for f a function (a 0-form) and,

$$(i_X\alpha)(X_1,\ldots,X_{p-1}) = \alpha(X,X_1,\ldots,X_{p-1}). \quad (2.96)$$

One finds that, as before, if α is homogeneous of degree $|\alpha|$, then,

$$i_X(\alpha\wedge\beta) = (i_X\alpha)\wedge\beta + (-1)^{|\alpha|}\alpha\wedge i_X\beta), \quad (2.97)$$

and that, for any vector fields X and Y, the graded commutator of the associated inner derivations vanishes,

$$[i_X, i_Y] = i_X\circ i_Y + i_Y\circ i_X = 0. \quad (2.98)$$

Recall that the graded commutator of two derivations is defined by

$$[D_1, D_2] = D_1 \circ D_2 - (-1)^{|D_1||D_2|} D_2 \circ D_1 .$$

As d and i_X are (graded) derivations, of degree $+1$ and -1, respectively, their graded commutator: $[d, i_X] = d \circ i_X + i_X \circ d$ is a derivation of degree zero. We denote it by $\mathcal{L}_X$ and it will be called the *Lie derivative* with respect to X. From its definition we have

$$\mathcal{L}_X = d \circ i_X + i_X \circ d \,. \tag{2.99}$$

On functions, the Lie derivative coincides with the action of X on $\mathcal{F}(E)$, and it extends to general forms the action of derivations on $\mathcal{F}(E)$, i.e. $\mathcal{L}_X f = X(f)$, and

$$\mathcal{L}_X(\alpha \wedge \beta) = (\mathcal{L}_X \alpha) \wedge \beta + \alpha \wedge (\mathcal{L}_X \beta) \tag{2.100}$$

Together with $d \circ d = 0$, this has the consequence that the exterior differential and the Lie derivative commute,

$$d \circ \mathcal{L}_X = d \circ i_X \circ d = \mathcal{L}_X \circ d \tag{2.101}$$

Moreover,

$$[\mathcal{L}_X, \mathcal{L}_Y] = \mathcal{L}_{[X,Y]} \tag{2.102}$$

(the graded commutator of two derivations of degree 0 is again a derivation of degree 0) and finally,

$$\mathcal{L}_X \circ i_Y - i_Y \circ \mathcal{L}_X = i_{[X,Y]} . \tag{2.103}$$

In particular, when $X = Y$, then $\mathcal{L}_X \circ i_X = i_X \circ \mathcal{L}_X$.

With these ingredients at hand, one can prove (see e.g., [AM78] (Prop. 2.4.15) and [Ne67] for details, [KN63] uses a slightly different normalization) that, with $X \in \mathfrak{X}(E)$ and $\beta \in \Omega^p(E)$,

$$\begin{aligned} d\beta(X_1, \ldots, X_{p+1}) = & \sum_{i=1}^{p+1} (-1)^{i+1} X_i(\beta(X_1, \ldots, \widehat{X}_i, \ldots, X_{p+1})) \\ & + \sum_{i<j} (-1)^{i+j} \beta([X_i, X_j], X_1, \ldots, X_i, \ldots, X_j, \ldots, X_{p+1}) . \end{aligned} \tag{2.104}$$

$$(\mathcal{L}_X \beta)(X_1, \ldots, X_p) = \mathcal{L}_X(\beta(X_1, \ldots, X_p) - \sum_{i=1}^{p} \beta(X_1, \ldots, [X, X_i], \ldots, X_p) . \tag{2.105}$$

where the symbol $\widehat{}$ means that the corresponding vector field should be omitted. This formula was used by R. Palais to provide an intrinsic definition of the exterior derivative [Pa54].

In this way we have defined a sequence of maps $d\colon \Omega^p(E) \to \Omega^{p+1}(E)$, all of them denoted by d such that $d^2 = 0$. The pair $(\Omega^\bullet(E), d)$ is called a graded differential algebra.

Remark 2.12 More generally, the definition of d can be generalized to antisymmetric multilinear maps $\phi\colon \mathfrak{X}(E) \times \cdots \times \mathfrak{X}(E) \to M$ with M any vector space carrying an action of $\mathfrak{X}(E)$, i.e., a linear map $\rho\colon \mathfrak{X}(E) \to \mathrm{End}\,(M)$. In this case we would have,

$$\begin{aligned} d_\rho \phi(X_1, \ldots, X_{p+1}) &= \sum_{i=1}^{p} (-1)^{i+1} \rho(X_i)(\phi(X_1, \ldots, \widehat{X}_i, \ldots, X_{p+1})) \\ &\quad + \sum_{i<j} (-1)^{i+j} \phi([X_i, X_j], X_1, \ldots, \widehat{X}_i, \ldots, \widehat{X}_j, \ldots, X_{p+1}). \end{aligned} \tag{2.106}$$

We find in an analogous way that $d_\rho \circ d_\rho = 0$ iff ρ is a Lie algebra homomorphism. The exterior differential defined by equation (2.104) will be recovered when $M = \mathcal{F}(E)$ and $\rho(X)(f) = \mathcal{L}_X(f)$, the Lie derivative.

We wish to stress again that all our constructions rely only on the commutative algebra structure of $\mathcal{F}(E)$. The linearity of E never played any role, therefore our calculus will be 'insensitive' to the kind of transformations we might perform on E.

Finally let us point out that the Lie derivative can be extended to the set of vector fields, dual space of that of 1-forms, by requiring that, if $X, Y \in \mathfrak{X}(E)$ and $\alpha \in \Omega^1(E)$,

$$\mathcal{L}_X \langle \alpha, Y \rangle = \langle \mathcal{L}_X \alpha, Y \rangle + \langle \alpha, \mathcal{L}_X Y \rangle$$

and then we obtain the following definition for $\mathcal{L}_X Y$,

$$\mathcal{L}_X Y = [X, Y]$$

In fact (cfr. Eq. (2.105)), $\mathcal{L}_X \alpha$ was defined in such a way that

$$\langle \mathcal{L}_X \alpha, Y \rangle = \mathcal{L}_X \langle \alpha, Y \rangle - \langle \alpha, [X, Y] \rangle$$

from where we find that $\mathcal{L}_X Y$ is given by $\mathcal{L}_X Y = [X, Y]$.

Once that $\mathcal{L}_X$ has been defined on functions, on vector fields and on 1-forms, an extension to the space of all tensors can be obtained requiring that $\mathcal{L}_X$ be a derivation of degree zero.

2.4.3 The 'Easy' Tensorialization Principle

It should be clear now that for any linear object associated with the abstract vector space E we may think of it as realized in terms of 'applied' vectors at x, i.e., of tangent vectors, or covectors, or any other linear tensor constructed in the tangent space to E at x. Then we can transform them into tensor fields and operations depending on x. This simple statement is what we call the 'easy' tensorialization principle. We will provide now various examples of the effective use of this principle that will be of use along the rest of the text.

2.4.3.1 Linear Algebra and Tensor Calculus

We will start by geometrizing a linear map $A \in \mathrm{End}(E)$. First, we can geometrize A by considering the associated (1, 1) tensor $T_A \colon TE \to TE$, defined as:

$$T_A \colon (x, v) \mapsto (x, Av); \qquad x \in E, \quad v \in T_xE \cong E, \tag{2.107}$$

or, dually $T_A^* \colon T^*E \to T^*E$,

$$\langle T_A^* \alpha, v \rangle = \langle \alpha, Tv \rangle; \quad x \in E, \; v \in T_xE \cong E, \; \alpha \in T_xE^* \cong E^* . \tag{2.108}$$

Using linear coordinates x^k induced by a given base $\{e_k\}$,

$$T_A = A^j{}_i dx^i \otimes \frac{\partial}{\partial x^j} \tag{2.109}$$

Then, $A^i{}_k$ is given by

$$T_A(\partial/\partial x^k) = A^i{}_k \, \partial/\partial x^i .$$

The correspondence $A \mapsto T_A$ is an algebra homomorphism, i.e.,

$$T_{A \cdot B} = T_A \circ T_B \tag{2.110}$$

because,

$$(T_A \circ T_B)\left(\frac{\partial}{\partial x^k}\right) = T_A\left(B^i{}_k \frac{\partial}{\partial x^i}\right) = B^i{}_k A^j{}_i \frac{\partial}{\partial x^j} = (AB)^j{}_k \frac{\partial}{\partial x^j} \tag{2.111}$$

Notice that once we have geometrized the linear map A on the vector space and promoted it to a (1, 1) tensor on the space E, then we are not restricted to consider linear coordinates or linear transformations. We can use any system of coordinates to describe it (even if the underlying linear structure becomes blurred) and use the exterior differential calculus as was discussed in the previous section.

The tensorialization T_A of A we have constructed retains all the algebraic properties of A which are hidden in the fact that when we express T_A in linear coordinates, the tensor is constant. That is we can de-geometrize it by choosing the appropriate set of coordinates, returning to the algebraic ground. This property will be instrumental in the general notion of tensorialization that will be discussed later on.

It is also possible, as it was shown before, to tensorialize a linear map A by associating to it a vector field X_A defined as:

$$X_A(x) = (x, Ax), \quad x \in E, \ Ax \in T_x E \cong E. \tag{2.112}$$

Now the geometrized object is not constant anymore and in linear coordinates x^k takes the form:

$$X_A = A^j{}_i x^i \frac{\partial}{\partial x^j}.$$

In this case we have been using the additional feature offered by linear spaces that $T_x E$ can be identified naturally with the base space E and with the linear space where A is defined. Notice that in the definition of T_A we were simply using the fact that $T_x E$ can be identified with the linear space where A is defined.

However on this occasion, the association $A \mapsto X_A$ fails to be a homomorphism of algebras and is only a Lie algebra homomorphism because $X_{AB}(f) \neq X_A(X_B(f))$, while,

$$X_{(AB-BA)}(f) = X_A(X_B(f)) - X_B(X_A(f)) \tag{2.113}$$

i.e.,

$$X_{[A,B]} = [X_A, X_B]. \tag{2.114}$$

Another interesting formula that transforms an algebraic identity into a geometrical operation is obtained by computing:

$$\mathcal{L}_{X_A} T_B = T_{AB},$$

We will see other formulae similar to the preceding one as we apply the tensorialization principle to other objects. The 'easy' tensorialization principle can also be stated as the thumb rule that transforms a tensor in a linear space E replacing e_k (the vectors of a given base) by $\partial/\partial x^k$ and the dual base elements e^j by dx^j. We can write it more precisely as follows:

The 'Easy' Tensorialization Principle

Given a tensorial object $\mathbf{t}$ in a linear space E, and given a linear base $\{e_k\}$, by replacing e_k by $\partial/\partial x^k$ and the dual base elements e^j by dx^j in the expression of $\mathbf{t}$ we will define a geometrical tensor $T_{\mathbf{t}}$ with the same structure as $\mathbf{t}$.

The 'easy' tensorialization principle as stated before could seem to depend on the choice of a system of linear basis or on a system of linear coordinates. However that is not so. The choice of a base $\{e_1, , \ldots, e_n\}$ on E (and $\{e^1, \ldots, e^n\}$ its dual basis in E^*), provides an injection of the linear space E in the set of constant vector fields (i.e., homogeneous vector fields of degree -1) given by: $v \mapsto v^i\, \partial/\partial x^i$ (with v^i being the coordinates of the vector v with respect to the given basis). We also have the injection: $\alpha = \alpha_i\, e^i \mapsto \alpha_i\, dx^i$ for the dual E^*.

This injection does not depend on the choice of the basis. Actually, we can originate the previous association by the following construction: For a fixed $x \in E$ we have a linear map $\xi_x \colon E \to E_x = T_x E$ defined associating to $v \in E$ the tangent vector to the curve $\gamma \colon \mathbb{R} \to E$ given by $t \mapsto x + vt$, that is:

$$\xi_x\,(v) = \frac{d}{dt}\,(x + vt)\,|_{t=o} \tag{2.115}$$

In other words, when acting on functions,

$$\xi_x(v) f = \frac{d}{dt} f(x + vt)\bigg|_{t=0} = (df)_x(v) = v^i\, \frac{\partial f}{\partial x^i}\bigg|_x , \tag{2.116}$$

i.e.

$$v \mapsto v^i\, \frac{\partial}{\partial x^i}\bigg|_x .$$

Then, given a bilinear pairing $b \colon E \otimes E \to \mathbb{R}$, say $B = B_{ij}\, e^i \otimes e^j$, with e^i a basis for E^*, we can apply the previous principle and associate to B the tensor,

$$T_B = B_{ij}\, dx^i \otimes dx^j . \tag{2.117}$$

Of course we can evaluate it pointwise on vector fields like the dilation field $\Delta = X_I$ whose expression in local coordinates is $\Delta = x^i \partial/\partial x^i$ and get $f_B(x) = \tau_B(\Delta, \Delta)\,(x) = B_{ij} x^i x^j$. By using T_A we obtain also,

$$T_A(\Delta) = X_A \tag{2.118}$$

This shows that, while X_A depends on the linear structure, T_A depends only on the affine structure of E, i.e., the tangent bundle structure.

Similarly we can geometrize an algebraic bivector. If Λ is an algebraic bivector, that is an algebraic skew symmetric contravariant tensor on E. Selecting a base $\{e_i\}$, Λ will take the form $\Lambda = \Lambda^{ij} e_i \wedge e_j$, and we may construct a tensor field on E by means of:

$$\Lambda = \Lambda^{ij} \frac{\partial}{\partial x^i} \wedge \frac{\partial}{\partial x^j} \tag{2.119}$$

In other words, bi-vectors like $\Lambda^{ij} e_i \wedge e_j$, with $\Lambda^{ij} \in \mathbb{R}$, are to be identified with the corresponding constant bivector fields.

We will discuss further extensions of this principle in Sect. 3.4.

2.4.4 Closed and Exact Forms

We say that a form $\beta \in \Omega^p(E)$ is *closed* if $d\beta = 0$, and *exact* if $\beta = d\theta$ for some $\theta \in \Omega^{p-1}(E)$. Quite obviously, by virtue of the fact that $d \circ d = 0$, every exact form is closed. The converse is the content of the following:

Proposition 2.13 (Poincare's Lemma for vector spaces) *If E is a linear vector space, then every closed form in E is exact.*

Proof For the proof, let us note preliminarily that, in view of the fact that d is'natural with respect to diffeomorphisms' i.e., as already stressed previously, $\phi^* \circ d = d \circ \phi^*$, then: (i) Once a coordinate system has been fixed, $E \approx \mathbb{R}^n$, and we might as well conduct the proof directly in $\mathbb{R}^n$, and (ii) for the same reasons, the proof will hold for any open set that is diffeomorphic to an open ball in $\mathbb{R}^n$ (and hence to $\mathbb{R}^n$ itself).

The theorem will be proved if we can construct a mapping, $T\colon \Omega^p(E) \to \Omega^{p-1}(E)$ such that, $T \circ d + d \circ T = Id$, for then, $\beta = (T \circ d + d \circ T)\beta$, and, in view of the closure of β, we see that $\beta = d\theta$ with $\theta = T\beta$.

We claim then that the required mapping is provided by,

$$(T\beta)(x) = \int_0^1 t^{p-1}(i_\Delta \beta)(tx)\, dt, \quad x \in \mathbb{R}^n, \tag{2.120}$$

where Δ denotes the dilation field in E. If β has the expression: $\beta(x) = \frac{1}{p!}\beta_{i_1 \dots i_p}(x)\, dx^{i_1} \wedge \dots \wedge dx^{i_p}$, the integrand has to be understood as, $\beta(tx) = \frac{1}{p!}\beta_{i_1 \dots i_p}(tx)\, dx^{i_1} \wedge \dots \wedge dx^{i_p}$.

Also, for any vector field X, we have that

$$\mathcal{L}_X \beta = \frac{1}{p!}\left(X^k \frac{\partial}{\partial x^k}\beta_{i_1 \dots i_p} + p \frac{\partial X^k}{\partial x^{i_1}}\beta_{k, i_2 \dots i_p}\right) dx^{i_1} \wedge \dots \wedge dx^{i_p},$$

and

$$\begin{aligned}\frac{d}{dt}\beta(tx) &= \frac{d}{dt}\frac{1}{p!}\beta_{i_1\dots i_p}(tx)\,dx^{i_1}\wedge\dots\wedge dx^{i_p}\\ &= \frac{1}{t}\left(\frac{1}{p!}y^i\frac{\partial}{\partial y^i}\beta_{i_1\dots i_p}(y)\right)_{y=tx} dx^{i_1}\wedge\dots\wedge dx^{i_p}\\ &= \frac{1}{t}\left(y^i\frac{\partial}{\partial y^i}\beta(y)\right)_{y=tx}.\end{aligned}$$

Therefore,

$$\frac{d}{dt}\left(t^p\beta(tx)\right) = t^{p-1}\left(p\,\beta(tx) + \left[y^i\frac{\partial}{\partial y^i}\beta(y)\right]_{y=tx}\right),$$

and hence,

$$\frac{d}{dt}\left(t^p\beta(tx)\right) = t^{p-1}(\mathcal{L}_\Delta\beta)(tx).$$

But then,

$$\begin{aligned}(T\circ d + d\circ T)\beta(x) &= \int_0^1 t^{p-1}(i_\Delta\circ d + d\circ i_\Delta)\beta(tx)dt = \int_0^1 t^{p-1}(\mathcal{L}_\Delta\beta)(tx)dt\\ &= \int_0^1 \frac{d}{dt}\{t^p\beta(tx)\}dt = \beta(x).\end{aligned}$$

□

Remark 2.13 (i) From the way the proof of Poincare's Lemma has been constructed, it is quite clear that, if we consider a differential form defined on an arbitrary open set U, the relevant condition for the validity of the Lemma is that there is a point (taken above as the origin in $\mathbb{R}^n$) such that any other point in U can be joined to it by a straight line segment lying entirely in U. That's why, in an equivalent way, Poincare's Lemma is often stated with reference to 'star-shaped' open sets.

(ii) As long as we consider forms that are differentiable over the whole of E (a vector space) there will be no distinction between closed and exact forms. The situation will change however as soon as we consider topologically less trivial spaces. As a simple example, consider $\mathbb{R}^2 - \{0\}$ with (Cartesian) coordinates (x, y) and the 1-form,

$$\beta = \frac{x\,dy - y\,dx}{x^2 + y^2}. \tag{2.121}$$

It is a simple computation to show that: $\beta = d\theta$, with $\theta = \tan^{-1}(y/x)$, the polar angle. Hence: $d\beta = 0$, i.e., β is closed, but it fails to be exact, because θ is not a globally defined function.

2.5 The General 'Integration' Problem for Vector Fields

2.5.1 The Integration Problem for Vector Fields: Frobenius Theorem

After the discussion in the last few sections we have arrived to the understanding that our modeling of dynamical systems is done in terms of vector fields, i.e., first-order differential equations, or equivalently, in the algebraic setting we have started to develop, as derivations in an algebra of smooth functions. We have also emphasized that in the case of linear systems the system is described completely by means of a flow of linear maps, and in general—nonlinear—case, the uniqueness and existence theorem of solutions of initial value problems for ordinary differential equations, guarantees the existence of local flows describing the dynamical behaviour of our system. The existence of such flows, globally defined in the case of linear systems and only locally defined in the general case, does not imply that we have a simple way of computing it.

We will say that the 'integration' problem for a given dynamics Γ consists in determining its (local) flow φ_t explicitly.

Again a few remarks are in order here regarding what do we mean by the 'explicit' determination of the flow. For instance, an 'explicit' solution of the dynamics could be an approximate numerical determination of the solution given for some initial condition x_0 and a time interval $[0, T]$. Varying the initial condition x_0 using some discrete approximation on a given domain U would provide an approximate explicit description of the dynamics. Unfortunately that is the most that can be done in many occasions when dealing with arbitrary nonlinear equations, and often only after devising very clever numerical algorithms and solving a number of hard problems regarding the stability and convergence of them. Even solving the integration problem for linear systems could be a hard problem. As we know the flow of the system is given by $\varphi_t = \exp tA$, so it seems that we have a closed expression for it, hence an 'explicit' description of the dynamics. However that is not so. Of course the infinite-dimensional situation, like the ones we face when dealing with Maxwell, Schrödinger, and other systems of interest, could be very hard to analyze because the structure of the linear operator A could be difficult to grasp, but also the finite-dimensional case could have interesting features which are not displayed in the simple minded expression for the flow above. Thus we have seen that looking for constants of motion and symmetries is often quite helpful in discussing the structure of linear systems offering new and deep insights into its properties, recall for instance the discussion of the harmonic oscillator in Chap. 1. That is the existence of structures compatible with the dynamics provides useful leads to analyze it even in the linear situation.

Even more, it is a fact that many dynamical systems arising from physical theories have a rich structure (for instance symmetries, constants of motion, Hamiltonian and/or Lagrangian descriptions, etc.) and exploring such intricacies has been proved to be the best way to approach their study. Actually it happens that in some particular instances, a judiciously and in some cases, extremely clever, use of such structures leads to an 'explicit' description of the dynamics, where now 'explicit' means that the actual solutions to the dynamical equations can be obtained by direct manipulation of some algebraic quantities. In such cases the systems are usually called 'integrable' with various adjectives depending on the context. For instance if the system is Hamiltonian and it possesses a maximal number of independent commuting constants of motion, the system is called completely integrable, etc.

As it has already been stated, our interest in this work is focused in unveiling the simplest and most significative structures compatible with a given dynamics but as it will be shown along these pages, anytime that there is a new structure compatible with the dynamics, we learn something on its integration problem. To the point that in some cases we can actually integrate it. This process will be spread out along the book and it will culminate in the last two chapter where the solution of the integration problem for various classes of dynamics will be discussed at length.

We would like to close this digression pointing out another twist of the 'integration' problem that permeates some parts of this work and that has played a relevant role in the development of modern geometry. Clearly if instead of having a single dynamics, let us say now, a vector field, we had two or more, we may ask again for the determination of the solutions of all of them. To be precise, suppose that we are given vector fields $X_1, \ldots, X_r$ in some vector space E (or on an open set in it). We can integrate them locally and obtain solutions $x^{(1)}(t_1), \ldots, x^{(r)}(t_r)$ for a given common initial data x_0. Nothing fancy so far, but we may ask, can be combine the r functions above in a single one, i.e., does there exist a function $x(t_1, \ldots, t_r)$ such that it reproduces the integral curves of the vector field X_1 when we fix the parameters $t_2, \ldots, t_r$ and so on? Or, in other words, if we change the initial data x_0 moving it for instance along the solution of the flow of the first vector field the solutions we would obtain now will be compatible with the ones obtained if we move the initial data in the direction of any other vector field?

The answer to this question is the content of the so-called Frobenius theorem and provides the backbone for the theory of foliations, but again we could raise the same questions as in the case of a single vector field: can be describe 'explicitly' such collective solutions?

Theorem 2.14 (Frobenius theorem: local form) *Let $X_1, \ldots, X_r$ be a family of vector fields on an open set U of a linear space E such that the rank of the linear subspace spanned by them at each point is constant. Then, this family can be integrated in the sense before, i.e., for each point $x \in U$ there exist an open set $V \in \mathbb{R}^r$ and a smooth injective function $\varphi\colon U \subset \mathbb{R}^r \to U \subset E$ such that the local flows of the vector fields X_i are given by the curves $\varphi(c_1, \ldots, c_{i-1}, t, c_{i+1}, \ldots, c_r)$ iff $[X_i, X_j]$ can be expressed as a superposition of the vector fields X_k.*

2.5.2 Foliations and Distributions

We can get a better grasp on the meaning of the local form of Frobenius theorem, Theorem 2.14, by 'geometrizing' the notion of 'collective solutions' used before. First we will define a smooth submanifold M of $\mathbb{R}^n$ of dimension r as a subset of $\mathbb{R}^n$ such that at each point $x \in S$ there is a neighborhood U of it which is the graph of a smooth function $\Psi \colon V \subset \mathbb{R}^r \to \mathbb{R}^{n-r}$, and open set $V \subset \mathbb{R}^r$, that is, $x = (u, \psi(u))$ for all $x \in U \subset M$, $u \in V$ (see a detailed account of the notions of manifolds and submanifolds in Appendix C),

In particular a submanifold M of dimension r of $\mathbb{R}^n$ can be defined in virtue of the Implicit Function Theorem (10.44), as a level set of a regular value c of a map $F \colon \mathbb{R}^n \to \mathbb{R}^{n-r}$. If we allow c to vary in a small neighborhood of $\mathbb{R}^{n-r}$ over the range of the map F, we will generate a family of submanifolds such that one and only one such submanifold will pass through each point of $\mathbb{R}^n$ (or of the domain of F that, in case it is not the whole of $\mathbb{R}^n$, will be always assumed to be an open subset thereof), and we obtain what might be called a 'slicing' of $\mathbb{R}^n$ (or, again, of the domain of F) into a 'stack' of closely packed submanifolds. That is the basic idea of what 'foliating a manifold' will mean for us, an idea that we will try to make slightly more precise here, together with the associated notions of 'distributions' and of 'integrable distributions'.

So, let us begin with a more precise definition of what we mean by a 'foliation' of a manifold.

Definition 2.15 Let U be an open set of a linear space of dimension n. A foliation $\mathcal{L}$ of U (of codimension m) is a family $\{\mathcal{L}_\alpha\}$ of disjoint connected subsets $\mathcal{L}_\alpha$ of U (to be called from now on the leaves of the foliation), one passing through each point of M, such that the identification mapping, $i \colon \mathcal{L}_\alpha \to U$ is injective, and for each point $x \in U$, there exists a neighborhood V of x such that V is diffeomorphic to $\mathcal{L}_x \cap V \times B$ where $\mathcal{L}_x$ is the leaf passing through x and B is an open ball in $\mathbb{R}^m$.

Then, $\mathcal{L}$ will consist of a family of connected submanifolds, each of dimension $n - m$ which stack together to fill up U.

A generalized notion of foliation including the possibility of having leaves of different dimension could be introduced. Sometimes such foliations are called singular foliations. In what follows we are going to discuss mainly non-singular foliations, even though throughout the text, examples of singular foliations will show up (and will be discussed in its context). We will give now some simple examples of foliations, referring to the literature [AM78, AM88, BC70, MS85] for further details:

1. The simplest example of a foliation is provided by the set of the integral curves of a non-vanishing vector field. The leaves of the foliation are its integral curves. They are all one-dimensional for non-vanishing vector fields, but could be also zero-dimensional if we allow the vector field to vanish somewhere.
   For example, in: $M = \mathbb{R}^2 - \{\mathbf{0}\}$ with the usual Cartesian coordinates, the (images of the) integral curves of the Liouville field:

$$\Delta = x\frac{\partial}{\partial x} + y\frac{\partial}{\partial y} \tag{2.122}$$

(or: $\Delta = r\partial/\partial r$ in polar coordinates: $x = r\cos\theta$, $y = r\sin\theta$) are the rays from the origin: ($r > 0$, $\theta = const.$). They are all one-dimensional and diffeomorphic to each other. If the origin were included, then $\mathbf{0}$ itself would be an integral curve, thus providing us with an example of a singular foliation, i.e., one whose leaves are not all of the same dimension.

2. A similar example is provided by the following construction. Consider the map,

$$\varphi\colon U = \mathbb{R}^3\setminus\{\mathbf{0}\} \to S^2 \tag{2.123}$$

which maps all the points (in spherical polar coordinates) $(r, \theta, \phi) \in U$ to the point $n = (\theta, \phi)$ in S^2. Then: $\mathcal{L} = \left\{\varphi^{-1}(n)\right\}_{n\in S^2}$ will foliate U with leaves that are rays through the origin.

3. Consider next in:

$$M = \mathbb{R}^3 - \{(-1, 0, 0) \cup (1, 0, 0)\} \tag{2.124}$$

again with the usual Cartesian coordinates, the foliation: $\Phi = \{l_b\}_{b\in\mathbb{R}}$ whose leaves l_b are given by,[12]

$$\left[(x-1)^2 + y^2 + z^2\right]^{-1/2} - \left[(x+1)^2 + y^2 + z^2\right]^{-1/2} = b \tag{2.125}$$

This foliation is depicted in the figure below:
Being level sets, the leaves are regular submanifolds in the sense of Appendix C. They are all one-dimensional, compact and diffeomorphic to each other for $b \neq 0$. However, the leaf corresponding to $b = 0$ is the y-axis, which is again one-dimensional but non-compact.

4. If we change the relative sign in the left-hand side of Eq. (2.125) and consider instead,

$$\left[(x-1)^2 + y^2 + z^2\right]^{-1/2} + \left[(x+1)^2 + y^2 + z^2\right]^{-1/2} = b,\ b > 0 \tag{2.126}$$

we obtain the foliation that is depicted in the figure below:

Now, the leave corresponding to $b = 2$, which contains the origin, is the 'bubble-eight' (8), which is not even a submanifold of $\mathbb{R}^3$ (with the induced topology). On top of that, the leaves are not connected for $b > 2$.

[12] The leaves are essentially the equipotential surfaces of two opposite electric charges (a dipole) located at $(-1, 0, 0)$ and $(1, 0, 0)$ respectively.

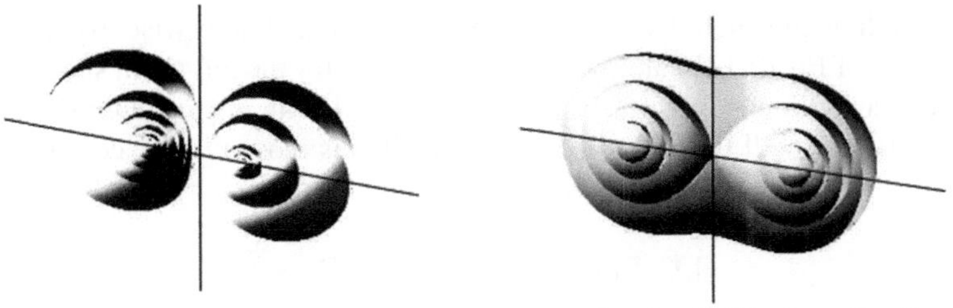

$[(x-1)^2+y^2+z^2]^{-1/2}-[(x+1)^2+y^2+z^2]^{-1/2}=b$ $[(x-1)^2+y^2+z^2]^{-1/2}+[(x+1)^2+y^2+z^2]^{-1/2}=b$

Fig. 2.1 Leaves of the foliation defined by equipotential surfaces of two charges of the same sign (*left*), and two charges of opposite sign defining a singular foliation (*right*)

The examples given above show that foliations can exhibit various kinds of pathologies. In order to avoid them, we will always consider what are called *regular* foliations [MS85], that, among other desirable features, are characterized by the leaves being all of the same dimension and all diffeomorphic to each other (Fig. 2.1).

2.5.2.1 Distributions and Integrability

Let Φ be a nonsingular foliation of M of dimension n. Then, a leaf l_α of the foliation passes through each $m \in M$. The tangent space $T_m l_\alpha$ will be a vector space of dimension n and will be spanned by a set of n vectors in $T_m M$. At least locally, i.e., in a neighborhood U of m, it will be possible to single out a set of n vector fields: $X_1, \ldots, X_n \in \mathfrak{X}(U)$ that span $T_m l_\alpha$ for all $m \in U$.

A *distribution* $\mathcal{D}$ will be the assignment of a similar set of vector subspaces of $T_m M$ at each point $m \in M$, all of the same dimension n, spanned in each neighborhood U by a set of smooth independent local vector fields $X_1, \ldots, X_n \in \mathfrak{X}(U)$. The $X_j(m)$'s, $j = 1, \ldots, n$ will be called a *basis* for the distribution at the point m, that will be denoted as $\mathcal{D}(m)$.

In the specific case of the distribution associated with a foliation Φ, the distribution will be denoted as $\mathcal{D}_\Phi$, and,

$$T_m l_\alpha = \mathcal{D}_\Phi(m) . \tag{2.127}$$

As a simple example, we may consider the foliation of M determined by the integral curves (actually the *images* in M of the integral curves) of a vector field X having no zeros, so that all the leaves of the foliation will be one-dimensional. Denoting as $\mathcal{D}_X$ the one-dimensional distribution associated with this foliation, we will have,

$$\mathcal{D}_X(m) = \operatorname{span}(X(m)) \equiv \{aX(m) \mid a \in \mathbb{R}\} . \tag{2.128}$$

It is clear that every foliation Φ defines a distribution, one that, moreover, satisfies the property expressed by Eq. (2.127). Whenever, vice versa, a distribution $\mathcal{D}$ is given satisfying the same property, i.e., we can find at every point m a submanifold l

passing through m and such that $\mathcal{D}(m)$ spans $T_m l$ or, stated otherwise, we know the right-hand side of Eq. (2.127) and we are able to solve for the left-hand side, we will say that the distribution is *integrable*, and l will be called an *integral manifold* of $\mathcal{D}$.

A distribution $\mathcal{D}$ will be said to be *involutive* if it is closed under commutation, i.e., if,

$$[X, Y] \in \mathcal{D}, \qquad \forall X, Y \in \mathcal{D}. \tag{2.129}$$

In the case of the distribution $\mathcal{D}_\Phi$ associated with the foliation Φ, the involutivity property of Eq. (2.129) is granted by the fact that (cfr. Eq. (2.127)) l_α is a (sub)manifold, and the tangent vectors to a (sub)manifold are obviously closed under commutation. Therefore integrable distributions are involutive.

The converse of this constitutes the main content of *Frobenius' theorem* [BC70, Wa71], which we will not prove here but simply state as:

Theorem 2.16 (Frobenius integrability theorem) *A distribution is integrable if and only if is involutive.*

Not all distributions need to be involutive, as the following example shows.

Consider, on $\mathbb{R}^3$, the two-dimensional distribution $\mathcal{D}$ defined (globally) by the vector fields: $X = \partial/\partial x$ and: $Y = \partial/\partial y + x\partial/\partial z$. As: $[X, Y] = \partial/\partial z \notin \mathcal{D}$, the distribution is not integrable. In fact, if it were, we could find a surface defined as a level function of a function, i.e., as: $f(x, y, z) = b$ for some $f \in \mathcal{F}(\mathbb{R}^3)$ and $b \in \mathbb{R}$ such that X and Y span the tangent space at every point, i.e., such that: $\mathcal{L}_X f = \mathcal{L}_Y f = 0$. But it is immediate to see that the only solution to these equations is: $f = const.$, i.e., no level surfaces at all.

To complete this rather long digression, we state now the conditions of the Frobenius theorem in a dual way, i.e., in terms of one-forms. Let then $\theta^1, \ldots, \theta^{m-n}$ be a set of linearly independent one-forms, and let: $\omega = \theta^1 \wedge \ldots \wedge \theta^{m-n}$. The intersection of the kernels of the θ^j's is a distribution that will be involutive if one of the following equivalent conditions holds:

1. $\theta^i \wedge d\omega = 0, \forall i = 1, \ldots, m - n$.
2. There exists a 1-form α such that: $d\omega = \alpha \wedge \omega$.
3. There exist local one-forms α^i_j, $i, j = 1, \ldots, m - n$ such that: $d\theta^i = \alpha^i_j \wedge \theta^j$.
4. There exist functions f^i and g^i_j such that: $d\theta^i = g^i_j df^j$.

2.6 The Integration Problem for Lie Algebras

In this section we will solve the problem of integrating the Lie algebra of symmetries of a given dynamical system, rephrasing in this way the so-called Lie's third theorem. In doing so we will move forward towards the notion of Lie group.

Let us recapitulate a situation we have found in the previous sections. Let Γ be a vector field. We had seen that the collection of infinitesimal symmetries of a dynamical system is a real Lie algebra (see Sect. 3.6.3). Let us suppose that the Lie algebra of infinitesimal symmetries of Γ is finite-dimensional generated by a family of vector fields $X_1, \ldots, X_r$ such that;

$$[X_i, X_j] = c_{ij}{}^k X_k ,$$

where $c_{ij}{}^k$ are the structure constants of the Lie algebra. The analysis of such dynamics will be done in full depth in Chap. 9. Here we will try to understand the structure of the flows of the family of vectors X_i. Because the vector fields X_i do not commute we cannot pretend to integrate their flows by using a single function $\varphi(t_1, \ldots, t_r)$ (remember the discussion on Sect. 2.5 about the simultaneous integration of a family of vector fields). However it may happen that there is another space such that the flows of the vector fields are just curves on it. It happens that such space is what we call a Lie group and we will devote the next few sections to this idea.

2.6.1 Introduction to the Theory of Lie Groups: Matrix Lie Groups

We have already found in the preceding sections some examples of groups, $GL(n, \mathbb{R})$, $U(n)$, etc. Throughout the book many other groups are going to have a relevant role, like $SU(2)$, $SO(3)$, $SL(2, \mathbb{C})$, $HW(n)$, etc. A rigorous discussion of their properties and representations would require a detailed development of the theory of Lie groups (i.e., groups that are equipped with the structure of a smooth manifold). We will not attempt to do that in this book, referring the reader to the abundant literature on the subject (see for instance [Wa71], etc.) even if we will provide an intrinsic definition of the class of Lie groups in the next chapter.

However, most of the previous examples (and many more) are groups of matrices, or as we will call them, matrix Lie groups, and contrary to the general class of Lie groups, only elementary calculus is required to discuss some of their properties and structure. This section will constitute an approximation to the theory of Lie groups by means of the study of an important family of them, closed subgroups of the general linear group.

Immediately after we will address the problem of integrating Lie algebras, the infinitesimal trace of a Lie group, arriving to the main theorem in Lie's theory that establishes a one-to-one correspondence between Lie algebras and connected and simply connected (Lie) groups.

2.6.1.1 The General Linear Group and the Orthogonal Group

Consider the general linear group in $\mathbb{R}^n$, that is the set of all invertible $n \times n$ real matrices $GL(n, \mathbb{R})$. It can be considered as an open subset in R^{n^2} by means of the map:

$$GL(n, \mathbb{R}) \to \mathbb{R}^{n^2}; \qquad A = (a_{ij}) \mapsto (a_{11}, \ldots, a_{1n}, \ldots, a_{n1} \ldots, a_{nn}).$$

Obviously, the set $GL(n, \mathbb{R})$ is a group, because if $A, B \in GL(n, \mathbb{R})$, then $AB \in GL(n, \mathbb{R})$ and $A^{-1} \in GL(n, \mathbb{R})$. The multiplication function is differentiable, because

$$(AB)_{ij} = \sum_{k=1}^{n} a_{ik} b_{kj},$$

say, the elements $(AB)_{ij}$ are quadratic polynomial functions of the elements of A and B, respectively.

In all that follows we will assume that $GL/n, \mathbb{R})$ is a subset of $\mathbb{R}^{n^2}$ because of the previous identification. Because the determinant map det is continuous (is a polynomial of degree n), we get that the group $GL(n, \mathbb{R}) = \det^{-1}(\mathbb{R} - \{0\})$ is an open subset of $\mathbb{R}^{n^2}$.

If we consider now the group $O(n)$ of orthogonal matrices, $O(n) = \{R \in GL(n, \mathbb{R}) \mid R^T R = R R^T = \mathbb{I}\}$ from the orthogonality condition $R^t R = \mathbb{I}$, we get:

$$\sum_j R_{ij} R_{jk} = 0, \quad i \neq k; \quad \sum_j R_{ij}^2 = 1; \qquad i = k, \tag{2.130}$$

showing that $|R_{ij}| \leq 1$, for all i, j. The subset $O(n) \subset \mathbb{R}^{n^2}$ is closed because it is defined by a set of algebraic equation (2.130). Notice that $O(n)$ is $F^{-1}(\mathbb{I})$ where $F \colon GL(n, \mathbb{R}) \to GL(n, \mathbb{R})$ is the smooth map $F(R) = R^T R$. Moreover $O(n)$ is bounded because $\sum_{i,j} R_{ij}^2 = n$, then $O(n)$ is compact. Notice, however that $O(n)$ is not connected because in general we just have $(\det R)^2 = 1$ and $O(n)$ has two connected components characterized by the sign of $\det R$. The connected component containing the neutral element is a normal subgroup:

$$SO(m, \mathbb{R}) = \{X \in GL(m, \mathbb{R}) \mid X^T X = I_m, \ \det A = 1\}.$$

We may now compute its tangent space as a subset of $\mathbb{R}^{n^2}$ (it is actually a submanifold, see Appendix C). Let $\gamma \colon (-\epsilon, \epsilon) \to O(n)$ be a smooth curve passing through the identity matrix, i.e., $\gamma(0) = \mathbb{I}$. Then $\gamma(t)^t \gamma(t) = \mathbb{I}$ for all t and computing the derivative at $t = 0$ we get: $\dot{\gamma}(0)T + \dot{\gamma}(0)$. Then the tangent vector $\dot{\gamma}(0)$ is a skew symmetric matrix. Conversely any skew symmetric matrix A is the tangent vector to a smooth curve in $O(n)$. It is enough to consider the curve $\gamma(t) = \exp tA$. We conclude that the tangent space to $O(n)$ at the identity can be identified with the set of skew symmetric matrices:

$$T_I O(n) = \{A \in M_n(\mathbb{R}) | A^T = -A\}.$$

In a similar way we can compute the tangent space to $O(n)$ at a given orthogonal matrix R. It suffices to consider a curve $\gamma(t)$ as before passing through the identity and multiply it by R on the right. Then the tangent space $T_R O(n)$ will be identified with matrices of the form AR with A skew symmetric.

The set of skew symmetric matrices $n \times n$ is a linear space of dimension $n(n-1)/2$ and we will say that the orthogonal group $O(n)$ is a manifold of dimension $n(n-1)/2$.

Definition 2.17 We will say that G is a matrix Lie group if it is an algebraic subgroup of $GL(n, \mathbb{R})$ and it is closed as a subset of $\mathbb{R}^{n^2}$.

Groups of $n \times n$ matrices with complex coefficients will be considered in a natural way as subgroups of $GL(2n, \mathbb{R})$ identifying $\mathbb{C}$ with $\mathbb{R}^2$. Thus the complex entry z_{jk} will be replaced by the 2×2 real matrix

$$\begin{pmatrix} x_{jk} & -y_{jk} \\ y_{jk} & x_{jk} \end{pmatrix}$$

with $z_{jk} = x_{jk} + iy_{jk}$.

It can be shown that because of the group law any closed subgroup of $GL(n, \mathbb{R})$ has a well defined tangent space at any point (that is, a smooth submanifold of $\mathbb{R}^{n^2}$) [MZ55, Gl52].[13] Thus the considerations we have made for $O(n)$ can be extended to any matrix Lie group.

Definition 2.18 The tangent space at the identity $T_I G$ of a matrix Lie group G will be called the Lie algebra of the group and will be denoted as $\mathfrak{g}$.

Example 2.6 Not every subgroup of $GL(n, \mathbb{R})$ is a matrix Lie group. Consider for instance the subgroup of $GL(2, \mathbb{C}) \cong GL(4, \mathbb{R})$ of matrices:

$$\begin{pmatrix} e^{it} & 0 \\ 0 & e^{i\lambda t} \end{pmatrix}$$

where $t \in \mathbb{R}$ and λ is an irrational number. It is easy to check that such subgroup is not closed in $\mathbb{R}^4$.

2.6.1.2 The Lie Algebra of a Matrix Lie Group

Definition 2.19 A Lie algebra L is a linear space with a skew symmetric bilinear map $[\cdot, \cdot]\colon L \times L \to L$ such that it satisfies Jacobi's identity:

$$[[\xi, \zeta], \chi] + [[\zeta, \chi], \xi] + [[\chi, \xi], \zeta] = 0,$$

[13] This is an application of a deep result in the theory of Lie groups, also-called Hilbert's fifth problem that shows that any finite-dimensional locally compact topological group without "small subgroup" is a Lie group [MZ55].

for all $\xi, \zeta, \chi \in L$.

Let L be a Lie algebra and $\mathcal{B} = \{E_i\}$ a linear basis, then we get:

$$[E_i, E_j] = c_{ij}^k E_k. \tag{2.131}$$

The constants c_{ij}^k are called the structure constants of the Lie algebra L with respect to the basis $\mathcal{B}$. It is immediate to check that the structure constants c_{ij}^k satisfy $c_{ij}^k = -c_{ji}^k$ and

$$c_{ij}^l c_{lk}^m + c_{jk}^l c_{li}^m + c_{ki}^l c_{lj}^m = 0, \qquad \forall i, j, k, m. \tag{2.132}$$

Conversely, given a family of numbers c_{ij}^k satisfying the previous conditions they are the structure constants of a unique Lie algebra with respect to some linear basis on it.

Example 2.7

1. the associative algebra $M_n(\mathbb{R})$ of $n \times n$ square matrices with real coefficients can be endowed with the Lie product $[A, B]$ given by the commutator of matrices $[A, B] = AB - BA$, and then $M_n(\mathbb{R})$ is endowed with a Lie algebra structure of dimension m^2. A basis is given by n^2 matrices E_{ij} with elements given by $(E_{ij})_{kl} = \delta_{ik}\delta_{jl}$. Each matrix $A = (a_{ij}) \in \mathcal{M}_m(\mathbb{R})$ can be written in a unique way as a linear combination

$$A = \sum_{i,j=1}^{n} a_{ij} E_{ij}.$$

   The structure constants in such a basis are

$$[E_{ij}, E_{kl}] = \delta_{jk} E_{il} - \delta_{il} E_{kj},$$

   because $E_{ij} E_{kl} = \delta_{jk} E_{il}$.
2. The cross product of vectors $x \times y$ defines a Lie algebra structure on $\mathbb{R}^3$.
3. Let $\mathcal{F}$ be the linear space of smooth functions on $\mathbb{R}^{2n}$. The bilinear map given by the standard Poisson bracket:

$$\{f, g\} = \sum_{i=1}^{n} \frac{\partial f}{\partial x^i}\frac{\partial g}{\partial x^{i+n}} - \frac{\partial g}{\partial x^i}\frac{\partial f}{\partial x^{i+n}}$$

   defines a Lie algebra structure on $\mathcal{F}$.

Let now G be a matrix Lie group and consider the map $\Phi\colon G \times \mathfrak{g} \to \mathfrak{g}$, defined as

$$\Phi(g, \xi) = \frac{d}{dt} g \cdot \gamma(t) \cdot g^{-1} \mid_{t=0} = g \cdot \xi \cdot g^{-1}, \tag{2.133}$$

where $\dot{\gamma}(0) = \xi$ and $g \in G$. This map defines an action of G on $\mathfrak{g}$, that is, it satisfies: $\Phi(g, \Phi(h, \xi)) = \Phi(gh, \xi)$, $\Phi(\mathbb{I}, \xi) = \xi$, called the adjoint action of G on its Lie algebra $\mathfrak{g}$.

We will denote by $\mathrm{Ad}_g \colon \mathfrak{g} \to \mathfrak{g}$ the linear map $\mathrm{Ad}_g(\xi) = \Phi(g, \xi) = g \cdot \xi \cdot g^{-1}$. Moreover $\mathrm{Ad}_g \circ \mathrm{Ad}_h = \mathrm{Ad}_{gh}$. Then the adjoint action defines a linear representation of G as linear maps on its Lie algebra.

We have that the tangent space at the identity of a matrix Lie group is a Lie algebra (hence the name).

Proposition 2.20 *Let G be a matrix Lie group. The tangent space at the identity $\mathfrak{g} = T_{\mathbb{I}} G$ is a Lie algebra with respect to the commutator of matrices.*

Proof Let $\xi, \zeta \in \mathfrak{g}$ and $g(t) \colon (-\epsilon, \epsilon) \to G$ be a smooth curve such that $\dot{g}(0) = \xi$. Then the curve $\sigma(t) = \mathrm{Ad}_{g(t)} \zeta$ is in $\mathfrak{g}$ and $\sigma(0) = \zeta$. Computing the derivative of $\sigma(t)$ at 0, we get:

$$\dot{\sigma}(0) = \frac{d}{dt}(g(t).\zeta.g(t)^{-1}) \mid_{t=0} = \xi \cdot \zeta - \zeta \cdot \xi = [\xi, \zeta],$$

where we have used that $d(g(t)^{-1})/dt = -g(t)^{-1} \cdot (dg(t)/dt) \cdot g(t)^{-1}$. □

Example 2.8 The following list describes some relevant groups of matrices:

1. $SO(n) = \{R \in GL(n, \mathbb{R}) | RR^t = I, \det R = 1\}$.
2. $U(n) = \{U \in GL(n, \mathbb{C}) | UU^\dagger = I\}$.
3. $SU(n) = \{ U \in GL(n, \mathbb{C}) | UU^\dagger = I, \det U = 1 \}$.
4. $SL(n, \mathbb{R}) = \{ S \in GL(n, \mathbb{R}) | \det S = 1 \}$.

and their Lie algebras:

1. $\mathfrak{so}(n) = \{ A \in M_n(\mathbb{R}) | A^t = -A, \mathrm{Tr}\, A = 0 \}$.
2. $\mathfrak{u}(n) = \{ V \in M_n(\mathbb{C}) | V^\dagger = -V \}$.
3. $\mathfrak{su}(n) = \{ V \in M_n(\mathbb{C}) | V^\dagger = -V, \mathrm{Tr}\, V = 0 \}$.
4. $\mathfrak{sl}(n) = \{ A \in M_n(\mathbb{R}) | \mathrm{Tr}\, A = 0 \}$.

As a consequence we obtain that their dimensions are:

1. $\dim SO(n) = n(n-1)/2$.
2. $\dim U(n) = n^2$.
3. $\dim SU(n) = n^2 - 1$.
4. $\dim SL(n, \mathbb{R}) = n^2 - 1$.

We will devote the next few sections to work out in detail the Lie algebras and other properties of some groups that are of capital importance.

2.6.1.3 The Lie Algebra of $SO(3)$ and $SU(2)$

The Lie algebra of $SO(3)$ will be obtained by computing the tangent vectors to smooth curves passing through $\mathbb{I}$. We consider the rotations around the axis:

$$R(e_1,\theta)=\begin{pmatrix}1 & 0 & 0\\ 0 & \cos\theta & -\sin\theta\\ 0 & \sin\theta & \cos\theta\end{pmatrix},\quad R(e_2,\theta)=\begin{pmatrix}\cos\theta & 0 & \sin\theta\\ 0 & 1 & 0\\ -\sin\theta & 0 & \cos\theta\end{pmatrix},$$

and

$$R(e_3,\theta)=\begin{pmatrix}\cos\theta & -\sin\theta & 0\\ \sin\theta & \cos\theta & 0\\ 0 & 0 & 1\end{pmatrix}.$$

Denoting by M_i the tangent vector to $R(e_i,\theta)$ at $\theta=0$, we get:

$$M_1=\begin{pmatrix}0 & 0 & 0\\ 0 & 0 & -1\\ 0 & 1 & 0\end{pmatrix},\quad M_2=\begin{pmatrix}0 & 0 & 1\\ 0 & 0 & 0\\ -1 & 0 & 0\end{pmatrix},\quad M_3=\begin{pmatrix}0 & -1 & 0\\ 1 & 0 & 0\\ 0 & 0 & 0\end{pmatrix}, \tag{2.134}$$

and we check immediately:

$$[M_1,M_2]=M_3,\quad [M_2,M_3]=M_1,\quad [M_3,M_1]=M_2$$

that will be written as:

$$[M_i,M_j]=\epsilon_{ijk}M_k. \tag{2.135}$$

Conversely, if A is in the Lie algebra, i.e., is a skew-symmetric 3×3 matrix, we get:

$$A=\begin{pmatrix}0 & -\zeta_3 & \zeta_2\\ \zeta_3 & 0 & -\zeta_1\\ -\zeta_2 & \zeta_1 & 0\end{pmatrix} \tag{2.136}$$

that can be written as: $A=\zeta_1M_1+\zeta_2M_2+\zeta_3M_3$. It is clear that this construction generalizes immediately to $SO(n,\mathbb{R})$ with $n\geq 3$.

To obtain the Lie algebra of $SU(2)$ it is sufficient to consider the curves:

$$U(e_k,\varphi)=\cos\varphi-i\sigma_k\sin\varphi \tag{2.137}$$

where σ_k denotes Pauli's sigma matrices:

$$\sigma_0 = \mathbb{I}_2 = \begin{pmatrix} 1 & 0 \\ 0 & 1 \end{pmatrix}, \quad \sigma_1 = \begin{pmatrix} 0 & 1 \\ 1 & 0 \end{pmatrix}, \quad \sigma_2 = \begin{pmatrix} 0 & i \\ -i & 0 \end{pmatrix}, \quad \sigma_3 = \begin{pmatrix} 1 & 0 \\ 0 & -1 \end{pmatrix}. \tag{2.138}$$

Then we get:

$$U(e_1,\varphi) = \begin{pmatrix} \cos\varphi & -i\sin\varphi \\ -i\sin\varphi & \cos\varphi \end{pmatrix}, \quad \frac{dU(e_1,\varphi)}{d\varphi}\,|_{\varphi=0} = \begin{pmatrix} 0 & -i \\ -i & 0 \end{pmatrix} = N_1 = -i\sigma_1 \tag{2.139}$$

$$U(e_2,\varphi) = \begin{pmatrix} \cos\varphi & -\sin\varphi \\ \sin\varphi & \cos\varphi \end{pmatrix}, \quad \frac{dU(e_2,\varphi)}{d\varphi}\,|_{\varphi=0} = \begin{pmatrix} 0 & -1 \\ 1 & 0 \end{pmatrix} = N_2 = -i\sigma_2 \tag{2.140}$$

$$U(e_3,\varphi) = \begin{pmatrix} e^{-i\varphi} & 0 \\ 0 & e^{i\varphi} \end{pmatrix} \quad \frac{dU(e_3,\varphi)}{d\varphi}\,|_{\varphi=0} = \begin{pmatrix} -i & 0 \\ 0 & i \end{pmatrix} = N_3 = -i\sigma_3 \tag{2.141}$$

and the Lie algebra $\mathfrak{su}(2)$ of $SU(2)$ is given by:

$$[N_1, N_2] = 2N_3, \quad [N_2, N_3] = 2N_1, \quad [N_3, N_1] = 2N_2\,. \tag{2.142}$$

A natural basis for the Lie algebra $\mathfrak{su}(2)$ consists of the matrices $-i\sigma_k$ tangent to the curves (2.139). We realize immediately that the Lie algebra $\mathfrak{su}(2)$ is isomorphic to the Lie algebra $\mathfrak{so}(3)$ of $SO(3)$; the isomorphism $\eta\colon \mathfrak{so}(3) \to \mathfrak{su}(2)$, is given by:

$$\eta(M_i) = \frac{1}{2}N_i = -\frac{i}{2}\sigma_i. \tag{2.143}$$

However the Lie groups $SU(2)$ and $SO(3)$ are not isomorphic because their topological properties are different and if they were they should be homeomorphic, but it is easy to see that $SU(2)$ is simply connected by identifying it with the 3-dimensional sphere while this is not the case of $SO(3)$ (see below). However they are locally isomorphic, they have the same Lie algebra.

The general theory of Lie groups shows that $SU(2)$ is the universal covering (see Sect. 2.6, Theorem. 2.25) of all Lie groups with Lie algebra isomorphic to $\mathfrak{su}(2)$. Any other group possessing the same Lie algebra can be obtained as a quotient group of $SU(2)$ by a central discrete subgroup.

In our case because the center of $SU(2)$ is $\mathbb{Z}_2$, the only two groups with the same Lie algebra are $SO(3)$ and $SU(2)$.

Exercise 2.9 Compute the center of $SU(2)$ and $SO(3)$. Prove that a central subgroup is a subgroup of the center of the group.

The covering map $\pi\colon SU(2) \to SO(3)$ is defined as follows: let x be a vector in $\mathbb{R}^3$ and $x \cdot \boldsymbol{\sigma}$ the 2×2 Hermitean matrix:

$$x \cdot \boldsymbol{\sigma} = x_1\sigma_1 + x_2\sigma_2 + x_3\sigma_3 = \begin{pmatrix} x_3 & x_1 + ix_2 \\ x_1 - ix_2 & -x_3 \end{pmatrix}.$$

The map $x \mapsto x \cdot \boldsymbol{\sigma}$ defines a one-to-one correspondence between $\mathbb{R}^3$ and the linear space of traceless 2×2 Hermitean matrices. Then we define:

$$(\pi(U)\,x) \cdot \boldsymbol{\sigma} = U(\,x \cdot \boldsymbol{\sigma})U^\dagger, \quad \forall\, x \in \mathbb{R}^3. \tag{2.144}$$

Exercise 2.10 Check that $||\pi(U)\,x|| = ||\,x||$, and $\det \pi(U) = 1$, hence $\pi(U) \in SO(3)$.

2.6.1.4 More Examples: The Euclidean Group in 2-Dimensions and the Galilei Group

If we consider the Euclidean group of transformations in two dimensions $E(2)$,

$$\begin{aligned} x_1' &= x_1 \cos\varphi - x_2 \sin\varphi + a_1, \\ x_2' &= x_1 \sin\varphi + x_2 \cos\varphi + a_2, \end{aligned} \tag{2.145}$$

it is a Lie group of dimension three for which the composition law is

$$(\mathbf{a}\,', \varphi') \cdot (\mathbf{a}, \varphi) = (\mathbf{a}\,' + R(\varphi')\mathbf{a}, \varphi' + \varphi).$$

These transformations can be written in a matrix form as

$$\begin{pmatrix} x_1' \\ x_2' \\ 1 \end{pmatrix} = \begin{pmatrix} \cos\varphi & -\sin\varphi & a_1 \\ \sin\varphi & \cos\varphi & a_2 \\ 0 & 0 & 1 \end{pmatrix} \begin{pmatrix} x_1 \\ x_2 \\ 1 \end{pmatrix}.$$

Hence, the infinitesimal generators are just the matrices

$$J = \begin{pmatrix} 0 & -1 & 0 \\ 1 & 0 & 0 \\ 0 & 0 & 0 \end{pmatrix}, \quad P_1 = \begin{pmatrix} 0 & 0 & 1 \\ 0 & 0 & 0 \\ 0 & 0 & 0 \end{pmatrix}, \quad P_2 = \begin{pmatrix} 0 & 0 & 0 \\ 0 & 0 & 1 \\ 0 & 0 & 0 \end{pmatrix},$$

with commutation defining relations for the Lie algebra:

$$[J, P_1] = P_2, \quad [J, P_2] = -P_1, \quad [P_1, P_2] = 0.$$

Another interesting example is the Galilei group. We can identify it with a subgroup of $GL(5, \mathbb{R})$ (or the corresponding affine group in four dimensions)

$$\begin{pmatrix} \mathbf{x}' \\ t' \\ 1 \end{pmatrix} = \begin{pmatrix} R & \mathbf{v} & \mathbf{a} \\ 0 & 1 & b \\ 0 & 0 & 1 \end{pmatrix} \begin{pmatrix} \mathbf{x} \\ t \\ 1 \end{pmatrix}.$$

The commutation relations defining the Lie algebra of Galilei group are then,

$$\begin{aligned}
&[\mathbf{J}, \mathbf{J}] = \mathbf{J}, \quad [\mathbf{J}, \mathbf{K}] = \mathbf{K}, \quad [\mathbf{J}, \mathbf{P}] = \mathbf{P}, \quad [\mathbf{J}, H] = 0\,, \\
&[\mathbf{K}, \mathbf{K}] = 0, \quad [\mathbf{K}, \mathbf{P}] = 0, \quad [\mathbf{K}, H] = \mathbf{P}\,, \\
&[\mathbf{P}, \mathbf{P}] = 0, \quad [\mathbf{P}, H] = 0\,.
\end{aligned} \tag{2.146}$$

Here **P** are the generators of the one-parameter groups of space translations, H is that of time translations, **J** are the generators of proper rotations and **K** those of pure Galilei transformations.

Finally, when considering one-parameter groups of transformations of an affine space M, for instance e^{tA}, each point $x \in M$, transforms into $x' = \Phi(e^{tA}, x)$, and for small values of the parameter t, which we will denote by ϵ,

$$x'^i = x^i + \epsilon\,\xi^i(x) + O(\epsilon^2)\,,$$

and

$$\xi(x) = \left(\frac{d\Phi(e^{\epsilon A}, x)}{d\epsilon}\right)\Big|_{\epsilon=0}\,.$$

For instance, for the one-parameter group of translations in the x_1 direction for the case of the Euclidean group in two dimensions, $\xi^1 = 1$, $\xi^2 = 0$, while for the one-parameter group of translations in the other direction, $\xi^1 = 0$, $\xi^2 = 1$. For the proper rotation subgroup, $\xi^1 = -x_2$, $\xi^2 = x_1$.

In an analogous way, in the group of proper rotations in three dimensions, for the subgroup of rotations around the axis determined by the vector $\mathbf{n}$, $\xi^i = \varepsilon^{ijk} n_j x_k$.

2.6.1.5 Group Homomorphisms and Lie Algebras

Definition 2.21 A homomorphism between the Lie algebras $(\mathfrak{g}_1, [,]_1)$ and $(\mathfrak{g}_2, [,]_2)$, is a linear map $\phi\colon \mathfrak{g}_1 \to \mathfrak{g}_2$ such that: $\phi([\xi, \zeta]_1) = [\phi(\xi), \phi(\zeta)]_2$, for all $\xi, \zeta \in \mathfrak{g}_1$. If the homomorphism is bijective we will call it an isomorphism.

Example 2.11

1. The Lie algebras $\mathfrak{so}(3)$ and $\mathfrak{su}(2)$ are isomorphic, the isomorphism given by equation (2.143).
2. The Lie algebra $\mathfrak{sl}(2, \mathbb{C})$ is isomorphic to the complexification $\mathfrak{sl}(2, \mathbb{R})^{\mathbb{C}}$ of the Lie algebra $\mathfrak{sl}(2, \mathbb{R})$. (The complexification $\mathfrak{g}^{\mathbb{C}}$ of a real Lie algebra $\mathfrak{g}$ is the natural Lie algebra structure obtained on the complexification of the linear space $\mathfrak{g}$ by extending the bilinear map $[\cdot, \cdot]$ to a complex bilinear map.)
3. The Lie algebra $\mathfrak{sl}(2, \mathbb{C})$ is isomorphic to the complexification of the Lie algebras $\mathfrak{so}(3)$ and $\mathfrak{so}(1, 2)$ (where $\mathfrak{so}(1, 2)$ is the Lie group of the linear isomorphisms preserving the metric with signature $(- + +)$).

Definition 2.22 Given two matrix Lie groups G_1 and G_2, a Lie group homomorphism between them is a smooth group homomorphism $\psi \colon G_1 \to G_2$ (i.e., $\psi(gh) = \psi(g)\psi(h)$ for all $g, h \in G$).

Notice that if $\psi \colon G_1 \to G_2$ is a differentiable map, the differential of this map at $\mathbb{I}$ is a linear map, $d\psi(\mathbb{I}) \colon T_{\mathbb{I}}G_1 \to T_{\mathbb{I}}G_2$, that is, a linear map between the corresponding Lie algebras. We will denote in what follows the map $d\psi(\mathbb{I})$ as ψ_* and we will check that it is a Lie algebra homomorphism.

Actually, one-parameter Lie subgroups, described by curves $\gamma \colon \mathbb{R} \to G$ which are a group homomorphism, namely, such that

$$\gamma(t_1)\gamma(t_2) = \gamma(t_1 + t_2) ,$$

play a relevant role. Indeed, this last property means that $\gamma(t)$ is determined by the tangent vector to the curve in the neutral element, $\gamma(0) = e \in G$. When G is a subgroup of $GL(m, \mathbb{R})$, if A is the matrix

$$A = \frac{d}{dt}\gamma(t)_{|t=0} ,$$

then $\gamma(t) = e^{tA}$. In fact, it suffices to take into account the relation $\gamma(t_1)\gamma(t_2) = \gamma(t_1 + t_2)$, and to take derivative with respect to t_1 at $t_1 = 0$, and then we find $A\gamma(t) = \dot{\gamma}(t)$, and as $\gamma(0) = I$, we obtain $\gamma(t) = e^{tA}$.

Thus, the matrices A obtained as tangent vectors to one-parameter subgroups of $GL(m, \mathbb{R})$ at the identity matrix, close on the Lie algebra $\mathfrak{gl}(n, \mathbb{R})$, and those corresponding to Lie subgroups of $GL(m, \mathbb{R})$ are Lie subalgebras, i.e. they are linear subspaces stable under the Lie product.

By using exponentiation we can obtain the elements in a neighbourhood of $I \in G$, and these are generators of G when it is connected. For instance, the set of all traceless matrices is a linear space and the commutator of two traceless matrices is also traceless. They determine a Lie subalgebra, usually denoted $\mathfrak{sl}(n, \mathbb{R})$ and by exponentiation of these matrices we obtain the elements in the subgroup $SL(m, \mathbb{R})$.

The exponential map for an arbitrary matrix Lie group G is defined as the map $\exp \colon \mathfrak{g} \to G$ given by the standard exponential function of matrices. In more abstract terms, we would use the correspondence above between one-parameter subgroups and elements in the Lie algebra to define the exponential, that is, if $\xi \in \mathfrak{g}$ and $\gamma_\xi(t)$ is the corresponding one-parameter subgroup, then $\exp t\xi = \gamma_\xi(t)$ for all $t \in \mathbb{R}$.

It is not hard to see that the exponential map is surjective in any compact group but in non-compact groups is usually not surjective.

Exercise 2.12 Prove that the exponential map is not surjective for $SL(n, \mathbb{R})$ but it is surjective for $GL(n, \mathbb{C})$.

Because $\mathrm{Ad}_{\exp t\xi}\zeta = e^{t\xi}\zeta e^{-t\xi}$, computing the derivative with respect to t, we get:

$$\frac{d}{dt} |_{t=0} (\mathrm{Ad}_{\exp t\xi}\zeta) = [\xi, \zeta] = \mathrm{ad}\,(\xi)\zeta ,$$

with ad $(\xi)\zeta = [\xi, \zeta]$. Thus we get:

$$\text{Ad}_{\exp \xi} = \exp(\text{ad}\, \xi), \qquad \forall \xi \in \mathfrak{g}\,.$$

Let us compute now the differential of the exponential map. Let $\xi\colon (-\epsilon, \epsilon) \to \mathfrak{g}$ a smooth curve and we denote by $\delta\xi(t) = d\xi(t)/dt$. The differential of exp on the tangent vector $\delta\xi(t)$ is by definition $\exp_*(\delta\xi(t)) = d(\exp \xi(t))/dt$, thus $\exp_*\colon T_\xi \mathfrak{g} \to T_{\exp \xi} G$. It is a simple exercise to check that:

$$\frac{d}{dt} e^{\xi} = \int_0^1 e^{s\xi} \frac{d\xi}{dt} e^{(1-s)\xi} ds.$$

Then

$$\begin{aligned}
\exp_*(\delta\xi) &= \frac{d \exp \xi}{dt} = \int_0^1 e^{s\xi} \frac{d\xi}{dt} e^{(1-s)\xi} ds = \left(\int_0^1 e^{s\xi} \frac{d\xi}{dt} e^{-s\xi} \right) e^{\xi} \\
&= \left(\int_0^1 e^{s\,\text{ad}\,\xi} ds \right) \frac{d\xi}{dt} e^{\xi} = \sum_{k=0}^{\infty} \frac{1}{(k+1)!} (\text{ad}\,\xi)^k \frac{d\xi}{dt} e^{\xi} = F(\text{ad}\,\xi)(\delta\xi) e^{\xi}
\end{aligned}$$

Then:

$$\exp_*(\delta\xi) e^{-\xi} = F(\text{ad}\,\xi)(\delta\xi). \tag{2.147}$$

with $F(x) = (e^x - 1)/x$. Notice that if $\xi(0) = 0$ and we evaluate the previous formula at $t = 0$, we get $\exp_*(0) = \text{Id}$ which shows that the exponential map is a local diffeomorphism.

The next two propositions will provide more information on the relation between matrix Lie groups and their Lie algebras.

Proposition 2.23 *Let $\psi\colon G_1 \to G_2$ be a homomorphisms of matrix Lie groups, then the differential at the identity $\psi_*\colon \mathfrak{g}_1 \to \mathfrak{g}_2$ at the identity is a homomorphism of Lie algebras.*

Proof The proof is simple. Consider two vectors ξ and ζ in $\mathfrak{g}_1$ with integral curves $g(t)$ and $h(t)$ respectively. Then:

$$\frac{d}{dt} \psi_*(\text{Ad}_{g(t)}\zeta) \mid_{t=0} = [\psi_*(\xi), \psi_*(\zeta)]_2.$$

On the other hand, because the differential is linear, we will get that the previous expression is equal to:

$$\psi_* \frac{d}{dt}(Ad_{g(t)}\zeta) \mid_{t=0}$$

and computing it again we get: $\psi_*([\xi, \zeta]_1)$. □

Thus, associated to any group homomorphism there is a homomorphism between the corresponding Lie algebras. This relation can be qualified further because of the following theorem that we establish without proof (see [Wa71]):

Theorem 2.24 *Let $\psi\colon G \to H$ be a homomorphism of matrix Lie groups and let $\psi_*\colon \mathfrak{g} \to \mathfrak{h}$, be the corresponding Lie algebras homomorphism. Then:*

i. If ψ_* is onto, then ψ is an onto on H_0 (the connected component of H containing $\mathbb{I}$).
ii. If ψ_* is mono, then ψ is mono in a neighborhood of $\mathbb{I}$ in G.
iii. If ψ_* is bijective, then ψ is a local isomorphism between G_0 and H_0.

*2.6.2 The Integration Problem for Lie Algebras**

Now we are ready to prove Lie's third theorem that provides the solution for the integration of a finite-dimensional Lie algebra of vector fields. The global object that integrates a Lie algebra is a Lie group.

Theorem 2.25 (Lie's integration theorem) *Let $\mathfrak{g}$ be a finite-dimensional Lie algebra, then there exists a unique, up to isomorphisms, connected and simply connected Lie group G whose Lie algebra is $\mathfrak{g}$.*

It is interesting to notice that after more than one hundred years since Lie's construction there is not an 'easy' proof of this theorem. The simplest way to address it (and its meaning) is to use Ado's theorem first [Ja79]. Ado's theorem establishes that any finite-dimensional Lie algebra can be seen as a subalgebra of the Lie algebra $M_n(\mathbb{R})$ of $n \times n$ matrices for some n. Then we can try to work inside the general linear group $GL(n, \mathbb{R})$. It is important to notice that Ado's theorem does not extend to Lie groups, in other words, not every Lie group is a subgroup of the general linear group (recall Example 2.6). At the end of this section we will comment a bit on an 'intrinsic' proof of Lie's theorem without recurring to Ado's theorem that will be significant later on.

2.6.2.1 Proving Lie's Theorem I: Using Ado's Theorem

To address the proof of Lie's theorem without using more sophisticated tools, we may rely on Ado's theorem stating that any (finite-dimensional) Lie algebra is isomorphic to a subalgebra of the Lie algebra of $n \times n$ real matrices for some n. We will denote such Lie algebra as $\mathfrak{gl}(n)$, and Ado's theorem can be restated saying that given a finite-dimensional Lie algebra $\mathfrak{g}$ there exist n and an injective homomorphism of Lie algebras $i: \mathfrak{g} \to \mathfrak{gl}(n)$. In what follows we will identify $\mathfrak{g}$ with its image $i(\mathfrak{g}) \subset \mathfrak{gl}(n)$ and we will not distinguish between the element $\xi \in \mathfrak{g}$ and the $n \times n$ matrix $i(\xi)$.

We will follow now the arguments in [Mi83b]. We consider the collection of all smooth maps:

$$\xi: [0, 1] \to \mathfrak{g} \subset \mathfrak{gl}(n)$$

such that $\xi(0) = \xi(1) = \xi'(0) = \xi'(1) = 0$. Given any such map $\xi(t)$ we may integrate the time-dependent linear dynamical system defined on the space of $n \times n$ matrices:

$$\frac{d\varphi}{dt} = \xi(t)\varphi(t), \qquad \varphi(0) = 1. \tag{2.148}$$

As we know from the general analysis of linear systems, such an initial value problem has a unique solution $\varphi^\xi: [0, 1] \to GL(n, \mathbb{R})$. The invertible matrices $\varphi^\xi(t)$ are constructed from a family of fundamental solutions $x^{(i)}$ of the system $dx/dt = \xi(t)x$ with initial conditions $x^{(i)}(0) = e_i$ (see Sect. 2.2.2). Then $\varphi^\xi(t)$ is just the matrix whose columns are the vectors $x^{(i)}(t)$ solutions of the previous initial value problem.

We notice now that the elements $\varphi^\xi(1) \in GL(n, \mathbb{R})$ satisfy:

$$\varphi^\xi(1)\varphi^\zeta(1) = \varphi^{\xi\star\zeta}(1), \tag{2.149}$$

where $\xi \star \zeta$ is the concatenation of the paths ξ and ζ on $\mathfrak{g}$, that is:

$$\xi \star \zeta(t) = \begin{cases} 2\zeta(2t) & \text{if } 0 \leq t \leq 1/2 \\ 2\xi(2t-1) & \text{if } 1/2 \leq t \leq 1 \end{cases}.$$

To prove that, check by direct substitution that the curve:

$$\varphi(t) = \begin{cases} \varphi^\zeta(2t) & \text{if } 0 \leq t \leq 1/2 \\ \varphi^\xi(2t-1)\varphi^\zeta(1) & \text{if } 1/2 \leq t \leq 1 \end{cases}$$

satisfies that $d\varphi/dt = (\xi \star \zeta)\varphi$.

Now consider the space of equivalence classes of smooth maps as before $\xi: [0, 1] \to \mathfrak{g}$ where

$$\xi \sim \zeta \quad \text{iff} \quad \varphi^\xi(1) = \varphi^\zeta(1). \tag{2.150}$$

Let us denote by G such space. We will show now that G is the object integrating $\mathfrak{g}$. First we check that we can define a composition law in G as follows: if g^ξ, g^ζ denote two equivalence classes of paths on $\mathfrak{g}$ with representatives ξ and ζ respectively, then we define:

$$g^{\xi} \cdot g^{\zeta} = g^{\xi \star \zeta}.$$

Notice that this composition law is well defined because of Eq. (2.149). This composition law is associative. The proof requires some work because $\xi \star (\zeta \star \eta)$ is not equal to $(\xi \star \zeta) \star \eta$, however we know that they are homotopic which is enough to guarantee that their equivalence classes with respect to $\sim$ are the same. There is a neutral element $e = g^0$ corresponding to the trivial curve 0 on $\mathfrak{g}$ and each element g^{ξ} has an inverse element $g^{\xi^{-1}}$ where ξ^{-1} is the opposite to the path ξ. Thus the set G becomes a group.

The set G inherits a natural topology from the topology of the space of paths ξ (for instance that induced by the supremum norm $||\xi||_\infty = \{||\xi(t)|| \mid 0 \leq t \leq 1\}$ and $||\cdot||$ any norm in $\mathfrak{g}$), and the composition law as well as taking the inverse are continuous with respect to this topology. Notice that G is trivially simply connected because $\mathfrak{g}$ is. In this way G becomes a topological group, however we are not interested in this approach as we want to construct directly a local identification of G with $\mathfrak{g}$.

Given a matrix A we have defined $\exp A$. Similarly we can define $\ln A$ (that will be uniquely determined provided that A is close enough to the identity matrix).[14] If the map $\xi(t)$ is small enough (that is $||\xi(t)|| < \epsilon$ for some $\epsilon > 0$ and a norm $||\cdot||$ in $\mathfrak{g}$), then $\varphi^{\xi}(t)$ will be close enough to I for all $t \in [0, 1]$. We want to check that now $\ln \varphi^{\xi}(t)$ is in $\mathfrak{g}$ for all t. Once we do that we have identified a neighborhood of the identity element e in G with a neighborhood of $\mathbf{0}$ in $\mathfrak{g}$.

To check that $\ln \varphi^{\xi}$ is in $\mathfrak{g}$ we will compute its derivative. Taking $A = \ln \varphi^{\xi}$ in Eq. (2.147) we obtain:

$$\xi(t) = F(\operatorname{ad} A - I)(\delta A) \in \mathfrak{g}$$

as we wanted to show.

2.6.2.2 Proving Lie's Theorem II: Extending Lie Algebra Homomorphisms

Once we have constructed G out of $\mathfrak{g}$ we would like to understand how we can construct a homomorphism $f: G_1 \to G_2$ of the groups G_1 and G_2 obtained by the procedure above from the Lie algebras $\mathfrak{g}_1, \mathfrak{g}_2$, that 'integrates' a homomorphism $\alpha: \mathfrak{g}_1 \to \mathfrak{g}_2$,

If we consider the path $g_1(t): [0, 1] \to G_1$, we define 'tangent' path $t \mapsto \xi_1(t) = \gamma_1'(t) g_1^{-1}(t) \in \mathfrak{g}_1$. Now we take its image under α, i.e., $t \mapsto \xi_2(t) = \alpha(\xi_1(t)) \in \mathfrak{g}_2$. Then we solve the differential equation on G_2:

$$\frac{dg_2}{dt} = \xi_2(t)\varphi_2, \qquad g_2(0) = I.$$

[14] The map $\exp : \mathfrak{gl}(n) \to GL(n)$ is differentiable with differential the identity at I, hence by the inverse theorem there is local inverse of exp which is differentiable, that is the map ln we are using.

Then we define $f(g(1)) = g_2(1)$ and the proof finishes if we show that $g_2(1)$ does not depend on the path $g_1(t)$ (we could have worked similar formulae and conclusions using a representative ξ_1 for g_1 in the space of paths in $\mathfrak{g}_1$).

Now if we have two different paths $g_1(t)$ and $g_1'(t)$ ending in the same point $g \in G_1$, then because G_1 is simply connected g_1 and g_2 are homotopic, that is there exists a family of paths $g(t,s)$ all from e to a fixed element $g \in G_1$ such that $g(t,0) = g_1(t)$ and $g(t,1) = g_2(t)$. Then we have the tangent vectors:

$$X_1(t,s) = \frac{\partial g(t,s)}{\partial t} g(t,s)^{-1}, \quad Y_1(t,s) = \frac{\partial g(t,s)}{\partial s} g(t,s)^{-1},$$

and after a simple computation we get:

$$\frac{\partial X_1}{\partial s} - \frac{\partial Y_1}{\partial t} = [X_1, Y_1]. \tag{2.151}$$

Then define $X_2 = \alpha(X_1)$ and $Y_2 = \alpha(Y_1)$. Then because α is linear and is a Lie algebras homomorphism we get for X_2 and Y_2:

$$\frac{\partial X_2}{\partial s} - \frac{\partial Y_2}{\partial t} = [X_2, Y_2].$$

But these equations are just the compatibility conditions for the system of linear equations:

$$\frac{\partial g'(t,s)}{\partial t} = X_2(t,s) g'(t,s), \quad \frac{\partial g'(t,s)}{\partial s} = Y_2(t,s) g'(t,s)$$

hence this system has a solution $g' \colon [0,1] \times [0,1] \to G_2$ that shows that $f(g)$ is well defined (notice that $Y_1(t,1) = 0$, then $Y_2(t,0) = 0$ for all t, then $g'(t,1)$ is constant.

Thus we may state:

Theorem 2.26 *Let $\alpha \colon \mathfrak{g}_1 \to \mathfrak{g}_2$ be a homomorphism of Lie algebras, then if G_1 and G_2 are two Lie groups with Lie algebras $\mathfrak{g}_1$ and $\mathfrak{g}_2$ respectively, there exists a homomorphism of Lie groups $\psi \colon G_1 \to G_2$ such that $\alpha = \psi_*$.*

2.6.2.3 Proving Lie's Theorem III: The Hard Proof, Not Using Ado's Theorem

Now it is easy to devise how we can avoid Ado's theorem in the proof of Lie's theorem. We were using the realization of the Lie algebra $\mathfrak{g}$ as a subalgebra of the algebra of matrices $\mathfrak{gl}(n)$ to define the equivalence relation (2.150) via the explicit integration of the linear system Eq. (2.148). owever this can be replaced by simply asking that the two paths $\xi(t)$ and $\zeta(t)$ on $\mathfrak{g}$ are equivalent if they can be joined by a curve of paths $X(s,t)$ such that it satisfies the compatibility equations above Eq. (2.151). Then the quotient space of paths module this equivalence relation will give us the group G as before. Again the hardest step in finishing the proof is to show that locally G is like $\mathfrak{g}$. Again we have to compute $\ln \varphi^\xi$ and we can proceed along similar lines as we did before (for that we need to show that the formula for the differential of the exponential still makes sense but we will not insist on this here).

Remark 2.14 It is pertinent to notice here that this way of approaching the integration of a Lie algebra has been continued in proving a much harder integration problem, that of integrating a Poisson structure solved by Crainic and Fernandez [CF04]. Then the compatibility condition is substituted by a more involved condition but the spirit is the same.

References

[MS85] Marmo, G., Saletan, E.J., Simoni, A., Vitale, B.: Dynamical Systems: A Differential Geometric Approach to Symmetry and Reduction. John Wiley, Chichester (1985)

[Ar73] Arnol'd, V.I.: Ordinary Differential Equations. MIT Press, Cambridge, 4th printing (1985).

[HS74] Hirsch, M., Smale, S.: Differential Equations, Dynamical Systems, and Linear Algebra. Academic Press Inc., New York (1974)

[DF79] Dollard, J.C., Friedman, C.N.: Product Integrals. Addison-Wesley, Cambridge (1979)

[ZK93] Zhu, J.C., Klauder, J.R.: Classical symptoms of quantum illness. Amer. J. Phys. **61**, 605–611 (1993)

[Pa59] Palais, R.: Natural operations on differential forms. Trans. Amer. Math. Soc. **92**, 125–141 (1959)

[AM78] Abraham, R., Marsden, J.E.: Foundations of Mechanics, 2nd edn. Benjamin, Reading (1978).

[Ne67] Nelson, E.: Tensor Analysis. Princeton University Press, New Jersey (1967)

[KN63] Kobayashi, S., Nomizu K.: Foundations of Differential Geometry, vol. 2. Interscience, New York (1963, 1969).

[Pa54] Palais, R.: A definition of the exterior derivative in terms of Lie derivatives. Proc. Amer. Math. Soc. **5**, 902–908 (1954)

[AM88] Abraham, R., Marsden, J.E., Ratiu, T.: Manifolds, Tensor Analysis and Applications, 2nd edn. Springer, New York (1988)

[BC70] Brickell, F., Clarke, R.S.: Foundations of Differentiable Manifolds (An Introduction). Van Nostrand, Reinhold (1970)

[Wa71] Warner, F.: Foundation of Differentiable Manifolds and Lie Groups. Scott Foresman, New York (1971)

[MZ55] Montgomery, D., Zippin, L.: Topological Transformation Groups, vol. 1. Interscience Publishers, New York (1955)

[Gl52] Gleason, A.M.: Groups without small subgroups. Ann. Math. **56**, 193–212 (1952)

[Ja79] Jacobson, N.: Lie Algebras (No. 10). Dover Publications, New York (1979).

[Mi83b] Milnor, J.: Remarks on infinite-dimensional Lie groups. In: DeWitt B. (ed.) Proceedings of Summer School on Quantum Gravity, Les Houches (1983).

[CF04] Crainic, M., Fernandes, R.L.: Integrability of Poisson brackets. J. Diff. Geom. **66**, 71–137 (2004)

Chapter 3
The Geometrization of Dynamical Systems

ΑΓΕΩΜΕΤΡΕΤΟΣ ΜΗΔΕΙΣ ΕΙΣΙΤΟ
Let none but geometers enter here.
Frontispiece of Plato's Academy

3.1 Introduction

In this chapter we would like to move a step forward and discuss the notions discussed in the previous chapter in such a way that they will not depend on being defined on a linear space. Such carrier space, as it was suggested before would be closely related to the notion of the space of 'states' of our system, and it is not always true that there is a linear structure on it compatible with the given dynamics.

It is true however that in most dynamical problems related to physical theories, the measurable quantities that characterize the state of the system are related among them, at least locally, by regular transformations on some real linear space, the quantities themselves not determining a linear structure though.

The standard way to model mathematically such situation is by means of the notion of smooth manifolds. A smooth manifold looks locally like a linear space, so tensorial objects can be defined and constructed locally, while globally it can differ from a linear space and exhibit a non trivial topology. The consistency of the local constructions is guaranteed by the transformations between different local pictures that are required to be as regular as needed. Such transformations correspond to local changes of coordinates in an operational description of our system and are implicit in the description of any dynamical system.

The algebra of smooth functions on a manifold provides an alternative way of describing the carrier space of a given dynamics, this time focusing the attention on the quantities that determine it. This point of view, equivalent to the previous one, is somehow more appealing from an algebraic point of view as some of the structural properties of the dynamics are reflected directly in terms of the algebraic structure of

J.F. Cariñena et al., *Geometry from Dynamics, Classical and Quantum*,
DOI 10.1007/978-94-017-9220-2_3

this algebra and it is gaining more weight in describing the foundations of physical theories.

In this chapter we will take the point of view of developing the general notion of a dynamical system from the point of view of the algebra of their measurable quantities or observables. Of course the class of algebras that, at least in principle, could constitute the observables of a dynamical system, is extremely large, so some assumptions would have to be made on their structure. We will introduce the notion of differentiable spaces in the first part of the chapter by abstracting some of the fundamental properties of the algebra of smooth functions on a linear space discussed in the previous chapter, essentially assuming that they provide a good 'local' description of the observables of dynamical systems.

Differentiable spaces actually embrace the notion of smooth manifolds providing the natural setting for a geometrical-algebraic description of their properties. Examples of both circumstances will be exhibited, that is examples that just amount to the standard description of smooth manifolds, and situations where spaces other than smooth manifolds can be used. For the most part of this book however, either smooth manifolds or just linear spaces are required.

We call the process of extending the definition and description of dynamical systems from the linear realm to the much broader class of dynamical systems on differentiable spaces the 'geometrization' of dynamical systems that provides the title for this chapter.

One of the first applications of such geometrization process will be to provide a tensorial characterization of the linear structures, i.e., we will dissociate the carrier space considered as a set of points equipped with a differentiable structure and the algebraic structures making it into a linear space, i.e., addition and multiplication by scalars. We will show that such algebraic structure is completely characterized by a vector field satisfying a few properties. In this way we can formulate the first compatibility condition between a dynamical system and a geometrical structure leading to the notion of alternative linear structures for a given dynamical system. An exhaustive discussion on this subject, elementary but subtle, will be provided in Sect. 3.3.1.

The geometrical viewpoint emphasized here will payback immediately as it will show how the tensorial characterization of linear structures will lead right away to the notion of vector bundles. We will discuss this in Sects. 3.3.2 and 3.3.3. The exterior differential calculus on manifolds introduced earlier will be extended to the new setting as it will be discussed in Appendix E.

Of course it is very difficult to know in advance if given a differentiable space or a smooth manifold, there exists on it a tensorial object on it satisfying some given properties. In a linear space, all algebraic tensors are at our disposal, however in differentiable spaces, even if locally they are like linear spaces, in the large that is not necessarily so, thus it could happen that there is no a geometrical tensor with the required properties, e.g., in general there will not exist a vector field with the properties required to define a linear structure. However locally it is always possible to construct tensorial quantities associated to algebraic tensors, extending the 'easy' tensorialization principle stated in the previous chapter to the class of differentiable spaces. A detailed account of this idea will be given in Sect. 3.4.1.

In general we would ask for more. We would like to find out if it is possible to construct (even locally) geometrical tensors that behaves like algebraic ones. The existence or not of such geometrical tensors, for a given class of algebraic ones, will be called the holonomic tensorialization principle. We will discuss a few instances when the holonomic tensorialization principle could be applied, e.g., symplectic structures. It is clear that in most cases such principle does not apply, e.g., Riemannian metrics. This matters will be discussed in Sect. 3.4.

We will conclude this chapter reviewing the basic notions about dynamical systems in the geometrical language provided by this chapter. This will be the content of Sect. 3.5, thus the arena to study further geometrical structures determined by them will be set.

3.2 Differentiable Spaces*

As we have mentioned in the introduction, once we have devoted Chap. 2 to develop a comprehensive discussion on differential calculus on linear spaces, we would like to extend it to make it suitable for spaces that are not linear. A first choice for that would be to introduce the notion of smooth manifold that captures exactly that.

The main idea behind the notion of a smooth manifold is that a smooth manifold locally it looks like an open set in a linear space. We can define all notions from exterior differential calculus as we did in the previous chapter, using these open sets then, in order to get a global consistent picture, we have to impose that whenever we are considering two different local pictures to describe the same object, the two pictures must be consistent, that is, it must behave accordingly transforming smoothly with the same specific algebraic rules determined by its tensorial nature.

The standard intrinsic approach to build the mathematical description of this idea is to introduce the notion of 'local charts' and 'atlases'. An alternative way to proceed, simpler in a sense, to build up the basic notions of differential calculus on manifolds is provided by considering that our spaces are lying inside some finite-dimensional linear space. Then introduce the objects we are interested in by taking advantage of that. Both approaches are well developed in the literature, even if the former one has gained weight along the years because, even if more abstract, is less cumbersome in its development.

As it was commented in the introduction, we will depart from these approaches because, among other reasons, we want to emphasize the algebraic approach to the main notions discussed in this book. Thus we prefer to concentrate on the algebra of functions (and their associated structures) rather than the space itself. Then many of the arguments become more transparent and they provide with the right intuition when trying to use them in other circumstances, for instance when we try to study dynamical systems whose algebras of observables are non commutative.

The algebraic approach to describe smooth manifolds has a long history and can be traced back to the work on singularity theory developed in the 50's.[1] Thus it turned out to that the algebraic picture provided the tools needed to understand the analog of factorization properties of polynomials in the setting of smooth manifolds.

The notion of differentiable algebras was coined by Malgrange [Ma66] to indicate algebras $\mathcal{F}$ with the properties of algebras of smooth functions of smooth manifolds and the properties of such algebras were characterized intrinsically. It turns out that such algebras are always quotients of the algebra $C^\infty(\mathbb{R}^N)$ by closed (with respect to the appropriate topology) ideals $\mathcal{J}$, thus in order to construct a differential calculus all that we have to do is to consider the differential calculus on $\mathbb{R}^N$ and restrict it to structures that are compatible with the ideal $\mathcal{J}$.

We will devote the next few sections to make these notions precise and develop the basic notions of exterior calculus on manifolds from this perspective. It is remarkable that this approach allows us to extend easily the well-known notions of exterior calculus in manifolds to more general spaces, for instance possessing some 'mild' singularities, like orbifolds.

The family of spaces described in this way are called differentiable spaces, in contrast to differentiable or smooth manifolds, and there exists a substantial body of work around them. Because one of the main themes of this book is the systematic description of nonlinear systems obtained by reduction from simpler ones and taking into account that in the reduction process singularities could arise (and they really do), the framework provided by differentiable spaces is particularly suitable to discuss their properties. In any case, for the sake of the reader, Appendix C contains a concise self-contained introduction to smooth manifolds using the standard language of local charts, atlases, local coordinates, etc.

3.2.1 Ideals and Subsets

The main idea behind the notion of differentiable algebra is the description of a subset Y of a given set X by means of an ideal in the algebra of functions of the later. Thus, consider X be a set, i.e., a well specified collection of objects, and $\mathcal{F}(X)$ its algebra of real-valued functions $f\colon X \to \mathbb{R}$, that is, $\mathcal{F}(X)$ carries a natural structure of associative algebra by means of the operations,

$$\begin{aligned} (f+g)(x) &= f(x)+g(x), \quad (\lambda f)(x) = \lambda(f(x)) \\ (f\cdot g)(x) &= f(x)g(x), \quad \forall x\in X, \quad f,g\in\mathcal{F}(X), \quad \lambda\in\mathbb{R}\,. \end{aligned} \tag{3.1}$$

Notice that both, the pointwise addition and product, are induced from the addition and multiplication in $\mathbb{R}$. Moreover the algebra $\mathcal{F}(X)$ has a unit given by the constant function $\mathbf{1}$ ($\mathbf{1}(x) = 1, \forall x \in X$) and is obviously commutative or Abelian.

[1] In a wider perspective this approach is foundational in what is called today Algebraic Geometry.

If $Y \subset X$ is a subset of X, then the set of functions which vanish on Y define an ideal of $\mathcal{F}(X)$ (bilateral because $\mathcal{F}(X)$ is Abelian). In fact, denoting by $\mathcal{J}_Y$ such set, i.e.,

$$\mathcal{J}_Y = \{f \in \mathcal{F}(X) \,|\, f(y) = 0, \forall y \in Y\}, \tag{3.2}$$

we have that if $f \in \mathcal{F}(X)$ and $g \in \mathcal{J}_Y$, then clearly $f \cdot g \in \mathcal{J}_Y$. However such ideal is not maximal if Y contains more than just one point. It is a simple exercise to check that maximal ideals are given precisely by the sets of functions of the form $\mathcal{J}_Y$ when $Y = \{y\}$ consists ofly of one point y of X.

We may try to introduce a notion of distance on the space of functions $\mathcal{F}(X)$ by using a norm, for instance, we may use the supremum norm given by: $||f||_\infty = \sup\{|f(x)| \mid x \in X\}$. However such norm is not defined in general for all possible functions on X. If we restrict our attention to the subalgebra $\mathcal{B}(X)$ of $\mathcal{F}(X)$ of bounded functions on X, the norm above is well defined and $\mathcal{B}(X)$ becomes a Banach space.[2] It also becomes a Banach algebra in the sense that:

$$||f \cdot g||_\infty \leq ||f||_\infty ||g||_\infty, \quad \forall f, g \in \mathcal{B}(X).$$

If the set X carries some additional structure we would like to restrict the algebra of functions $\mathcal{F}(X)$ and consider only those that are compatible with the given structure. For instance, if X is a topological space, then we would like to consider the subalgebra $C(X)$ of $\mathcal{F}(X)$ of continuous functions on X. Again by considering the subalgebra of bounded continuous functions on X we can equip this space with a natural norm, the supremum norm as before. We just want to mention here that these algebras have constituted one of the main sources of inspiration for the theory of operator algebras developed along the XX century.

If $Y \subset X$ now is a subset of the topological space X and we consider it with the induced topology from X, sometimes called the trace topology on Y, then the ideal $\mathcal{J}_Y$ of bounded continuous functions on X vanishing at Y is closed with respect to the topology induced by the norm $|| \cdot ||_\infty$ above. Even more, it is easy to convince oneself that the spaces $\mathcal{B}(Y)$ of bounded continuous functions on Y and $\mathcal{B}(X)/\mathcal{J}_Y$ are homeomorphic. The quotient space $\mathcal{B}(X)/\mathcal{J}_Y$ carries a natural topology defined by the quotient norm $||f + \mathcal{J}_Y||_\infty = \sup\{||f + g||_\infty \mid g \mid_Y = 0\}$.

The proof is simple, we consider the map $\alpha \colon f + \mathcal{J}_Y \mapsto f \mid_Y$ from $\mathcal{B}(X)/\mathcal{J}_Y$ to $\mathcal{B}(Y)$. This map is well defined and has an obvious inverse, $\alpha^{-1} \colon f_Y \mapsto f + \mathcal{J}_Y$ where f is any continuous extension of the continuous function f_Y defined on Y.[3] The map α is not only bijective but it also preserves the topologies. But there is more then that. If we consider an arbitrary closed ideal $\mathcal{J}$, then it is clear that the subset

---

[2] It is not hard to show that the space $\mathcal{B}(X)$ is complete, i.e., any Cauchy sequence of functions is convergent.

[3] We must impose some separability properties on our topological space X to guarantee that such extension always exist but we do not insist on this here.

$Y = \{x \in X \mid g(x) = 0, \forall g \in \mathcal{J}\}$ is such that $\mathcal{J}_Y = \mathcal{J}$ and the correspondence between subspaces of X and closed ideals is one-to-one.

However as we keep introducing additional structures on X things become more insidious. Thus if we consider now a finite-dimensional linear space E and we consider the algebra of smooth functions $\mathcal{F}(E) = C^\infty(E)$ that we have been using so far, there is no a single norm on such space that captures the "smoothness" of them, i.e., there will be always Cauchy sequences of smooth functions that will not be convergent (that is whose limit would loose some of the regularity properties of the elements of the sequence).

Thus we will need stronger topologies to capture "smoothness" and, even if we are not going to need this in the future, we just want to point out that there is a natural topology on the algebra of differentiable functions that actually characterizes it. This topology is called the strong Whitney topology and then the algebra $C^\infty(E)$ (E finite-dimensional) becomes a Fréchet space (that is a separable metrizable space), such that the algebra multiplication is continuous, sometimes called a Fréchet algebra. It suffices to say that a basis of open neighborhoods in this topology are provided by considering for each function $f \in C^\infty(E)$ all smooth functions h such that h and all its derivatives a "close" to f and their derivatives; more precisely, consider a covering of E by compact sets K_i and a family of positive numbers ϵ_i, then h will be in the open neighborhood defined by (f, K_i, ϵ_i) iff $||\partial^k(f - h)/\partial x^k||_{\infty, K_i} < \epsilon_i$ for all i, where $||\cdot||_{\infty, K_i}$ denotes the supremum norm restricted to functions defined on the compact set K_i.

Now again, if we have a subset $Y \in X$, we may consider the ideal $\mathcal{J}_Y$ of smooth functions vanishing at Y. Such ideal in general will not be closed with respect to the strong Whitney topology described above. In other words we may ask what additional structure should be added to Y that guarantee that $\mathcal{J}_Y$ is closed and that the quotient space $C^\infty(E)/\mathcal{J}_Y$ would be isomorphic to that structure. A partial answer is easy to provide. Suppose that Y is a smooth regular submanifold of E, that is suppose that there is a map $\Phi \colon E \to \mathbb{R}^r$ which is smooth, $Y = \Phi^{-1}(\mathbf{0})$ and $\mathbf{0}$ is a regular value of Φ (the rank of the differential map Φ_* is maximal at all points $x \in \Phi^{-1}(\mathbf{0})$), then it is possible to show that the ideal of functions vanishing at Y is closed.

Exercise 3.1 Prove that the ideal of functions vanishing on a $Y = \Phi^{-1}(\mathbf{0})$ with $\mathbf{0}$ a regular value for $\Phi \colon \mathbb{R}^m \to \mathbb{R}^n$ is closed in the strong Whitney topology.

It can be shown that such ideal $\mathcal{J}_Y$ is the ideal in $C^\infty(E)$á generated by the function Φ or, using the customary notation from algebra, $\mathcal{J}_Y = (\Phi)$. Now, it is clear that the quotient space $C^\infty(E)/\mathcal{J}_Y$ is isomorphic to $C^\infty(Y)$ where we will say that a function on Y is differentiable if it is the restriction of a smooth function on E (or conversely if it can be extended to a smooth function on E).

For instance, if we consider $\mathbb{R}$ and $\Phi \colon \mathbb{R} \to \mathbb{R}$, $\Phi(x) = x$, then $\Phi^{-1}(0) = \{0\}$. Because 0 is a regular value of Φ, then $\{0\}$ is a regular submanifold (of dimension 0) and the algebra of smooth functions on it $C^\infty(\{0\}) = \mathbb{R}$ is just $C^\infty(\mathbb{R})/(x)$. However if we consider the map $\Phi \colon \mathbb{R} \to \mathbb{R}$, $\Phi(x) = x^2$, then 0 is not a regular value and the space defined by the algebra $\mathcal{F}_2 := C^\infty(\mathbb{R})/(x^2)$ is the space of "two infinitesimally closed points". Notice that the algebra $\mathcal{F}_2$ is the two-dimensional real algebra generated by elements $1, \theta$ satisfying the relations $1^2 = 1$, $1 \cdot \theta = \theta \cdot 1 = \theta$ and $\theta^2 = 0$.

Thus the strategy to describe smooth manifolds will be to define their algebras of smooth functions as quotient spaces of $C^\infty(E)$ by closed ideals. We must remark here that because of Whitney's embedding theorem [Wh44] this definition actually includes all (separable paracompact) smooth manifolds (see Appendix C). Any such manifold can be embedded on $\mathbb{R}^N$ with N large enough, then its algebra of functions has the form above. Taking into account all these comments we will proceed now to establish the appropriate definitions.

3.2.2 Algebras of Functions and Differentiable Algebras

Definition 3.1 A differentiable algebra is a unital commutative associative Fréchet topological[4] algebra isomorphic to $C^\infty(\mathbb{R}^N)/\mathcal{J}$ with $\mathcal{J}$ a closed ideal (with respect to the strong Whitney's topology).

Remark 3.1 As it was indicated above, because of the Whitney's embedding theorem any smooth separable paracompact manifold M can be embedded into $\mathbb{R}^N$ for some N, hence $\mathcal{F}(M) \cong C^\infty(\mathbb{R}^N)/\mathcal{J}$, where $\mathcal{J}$ is the closed ideal of smooth function on $\mathbb{R}^N$ vanishing on M (as a closed submanifold of it). Hence $\mathcal{F}(M) := C^\infty(M)$ is a differentiable algebra too. Notice however that the class of differentiable algebras is strictly larger than the class of algebras of differentiable functions on a manifold (for instance the example above, $C^\infty(\mathbb{R})/(x^2)$ is not an algebra of differentiable functions over a smooth manifold).

Given a real topological algebra $\mathcal{F}$ we define its spectrum $\mathrm{Spec}_{\mathbb{R}}(\mathcal{F})$ as the space of continuous $\mathbb{R}$-morphisms $\varphi \colon \mathcal{F} \to \mathbb{R}$, i.e., $\varphi(f \cdot g) = \varphi(f)\varphi(g)$, $f, g \in \mathcal{F}$. It is a simple exercise to show such space can be naturally identified with the space of real maximal ideals of $\mathcal{F}$.

In the particular instance that $\mathcal{F} = C^\infty(E)$, E finite-dimensional linear space, then it is obvious that $\mathrm{Spec}_{\mathbb{R}}(\mathcal{F}) \cong E$, where each point $x \in E$ defines a maximal ideal $\mathcal{J}_{\{x\}}$ (that according with the uses of algebra is denoted as $\mathfrak{m}_x$). Conversely if we choose a linear basis e_i of E and the corresponding linear coordinate functions x^i, any morphism $\varphi \colon \mathcal{F} \to \mathbb{R}$ defines a collection of coordinates $\phi^i = \varphi(x^i)$. Then the point $x = \phi^i e_i$ is such that $\mathfrak{m}_x = \ker \varphi$ (actually notice that $\varphi(f) = \varphi(f(0) + f_i(x')x^i) = f(0) + f_i(x')\phi^i = f(x)$).[5]

Notice that given an element $f \in \mathcal{F}$ we can define the function $\hat{f} \colon \mathrm{Spec}_{\mathbb{R}}\mathcal{F} \to \mathbb{R}$ (sometimes called the Gel'fand transform)

---

[4] A real algebra with a topology such that the operations of the algebra are continuous and such that it is separable and metrizable.

[5] It can be shown that if M a separable smooth manifold satisfying the second countability axiom, then $\mathrm{Spec}_{\mathbb{R}}(\mathcal{F}(M))$ inherits a differentiable structure and it becomes diffeomorphic to M.

$$\hat{f}(x) := x(f), \qquad \forall x \in \mathrm{Spec}_{\mathbb{R}}\mathcal{F}. \tag{3.3}$$

In particular if we consider $\mathcal{F} = C^\infty(\mathbb{R}^N)$, then $\hat{f}$ coincides with f.[6]

The spectrum of a differentiable algebra $\mathcal{F}$ provides the local model for a class of spaces called $\mathcal{C}^\infty$-spaces (or, sometimes differentiable spaces). These spaces extend in a natural way the notion of smooth manifolds and allow us to work in a large class of topological spaces that share most of the natural notions of smooth manifolds and the differential calculus attached to them.

Remark 3.2 We will not need the full theory of differentiable spaces to present the geometrical notions discussed in this book but just the local model for them, that is, differentiable algebras. In any case we can just say that a differentiable space is obtained by 'patching together' the spectra of differentiable algebras (technically it is a ringed space over a sheaf of differentiable algebras) thus, it is clear that once a geometrical notion has been discussed for differentiable algebras it can be extended without pain to the category of differentiable spaces.

Differentiable spaces given by the real spectrum of a differentiable algebra are called affine differentiable spaces and again it can be proved that a differentiable space is affine if and only if it is Haussdorff with bounded dimension and its topology has a countable basis (see for instance [NS03], p. 67). Thus all spaces arising in our examples and applications are going to be affine differentiable spaces, that is, obtained from differentiable algebras.

Given a differentiable algebra $\mathcal{F}$ and a point $x \in \mathrm{Spec}_{\mathbb{R}}\mathcal{F}$, we can consider the localization $\mathcal{F}_x := \mathcal{F}_{\mathfrak{m}_x}$ of $\mathcal{F}$, or the field of fractions of $\mathcal{F}$ at x, i.e., elements on $\mathcal{F}_x$ are equivalence classes of fractions f/g where $g(x) \neq 0$; i.e., g belongs to the multiplicative system $S_x = \mathcal{F} - \mathfrak{m}_x$, and $f/g \sim f'/g'$ iff $fg' = f'g$. Elements on $\mathcal{F}_x$ are called germs of elements f of $\mathcal{F}$ at x and they can be also be considered as residue classes of elements on $\mathcal{F}$ with respect to the ideal of elements of $\mathcal{F}$ vanishing on open neighborhoods of x and will be denoted as $[f]_x$.

Given a differentiable algebra $\mathcal{F}$ we can consider its space of derivations equipped with its canonical Lie algebra structure $[\cdot,\cdot]$. That is, a derivation X of $\mathcal{F}$ is a linear map such that $X(f \cdot g) = X(f) \cdot g + f \cdot X(g), \forall f, g \in \mathcal{F}$ and the Lie bracket

$$[X, Y](f) = X(Y(f)) - Y(X(f)), \quad \forall X, Y \in \mathfrak{X}(\mathcal{F}),$$

is a derivation again (which is not true anymore for the anticommutator $[X, Y]_+(f) = X(Y(f)) + Y(X(f))$). It is clear that if $\mathcal{F} = C^\infty(E)/\mathcal{J}$ and $\tilde{X}$ is a derivation, i.e., a vector field in E, such that $X(\mathcal{J}) \subset \mathcal{J}$, then $\tilde{X}$ induces a derivation on $\mathcal{F}$ by the simple formula:

$$X(f + \mathcal{J}) = \tilde{X}(f) + \mathcal{J}.$$

[6] And similarly for $C^\infty(M)$, M a separable smooth manifold.

It is again true, even if proving it requires some non trivial arguments, that all derivations on $\mathcal{F}$ have this form. Thus we may define the space of derivations of $\mathcal{F}$, which will be denoted by $\mathfrak{X}(\mathcal{F})$ or $\mathrm{Der}(\mathcal{F})$, as the space of derivations of $C^\infty(E)$ leaving $\mathcal{J}$ invariant.

Notice that if X is a derivation then $X(1) = 0$. Moreover if f/g is a representative of a germ $[f]_x$ at x, then the derivation X defines a derivation in the space of germs at x. The space of all these derivations is a linear space called the tangent space at x to the affine differentiable space $M = \mathrm{Spec}_{\mathbb{F}}(\mathcal{F})$ and denoted as $T_x M$.

The space of derivations $\mathfrak{X}(\mathcal{F})$ is a $\mathcal{F}$-module (recall the properties of vector fields on E) and we can consider in analogy as we did in the case of linear spaces E when we introduced 1-forms, the space $\mathrm{Der}(\mathcal{F})^*$ of $\mathcal{F}$-valued $\mathcal{F}$-linear maps on $\mathrm{Der}(\mathcal{F})$, i.e., the space of 1-forms of the differentiable algebra $\mathcal{F}$. The space $\mathrm{Der}(\mathcal{F})^*$ is a $\mathcal{F}$-module again and we usually denote it as $\Omega^1(\mathcal{F})$ (or simply Ω^1 if there is no risk of confusion).

Given an open set U in M all objects considered so far can be localized to it by standard constructions, thus we can consider the differentiable algebra $\mathcal{F}_U$, $\mathfrak{X}_U$, Ω^1_U, etc. Moreover differentiable spaces have the fundamental property of the existence of partitions of unity (see [NS03], p. 52).

The differential map $d\colon \mathcal{F} \to \Omega^1$ is defined as in Sect. 2.3.4, Eq. (2.60),

$$df(Y) = Y(f),$$

for any $Y \in \mathfrak{X}(\mathcal{F})$. In the case of $\mathcal{F} = C^\infty(M)$, the space Ω^1 is just the module of smooth 1-forms $\Omega^1(M)$ on the manifold M. Alternatively, we can consider for any point $x \in \mathrm{Spec}_{\mathbb{R}}\mathcal{F}$ the linear space $\mathfrak{m}_x/\mathfrak{m}_x^2$ denoted by T_x^*X where $X = \mathrm{Spec}_{\mathbb{R}}\mathcal{F}$. Again, the space T_x^*X is the localization of the module Ω^1 with respect to the multiplicative system defined by $\mathfrak{m}_x$. Then given $f \in \mathfrak{m}_x$, we have that $df(x) := [df]_x$ can be identified with the class of f in $\mathfrak{m}_x/\mathfrak{m}_x^2$.

In what follows we will call the elements f of a differentiable algebra $\mathcal{F}$ "functions" unless there is risk of confusion.

3.2.3 Generating Sets

The main separability properties regarding 'points' and 'functions' on differentiable algebras can be stated as follows:

Definition 3.2 Given a differentiable algebra $\mathcal{F}$, we will say that a family of functions $g_1, \ldots, g_n$ separate points if they separate elements v_x in $T_x X$, i.e., for any $v_x \neq w_x$ there is g_k such that $v_x(g_k) \neq w_x(g_k)$.

A family of functions $g_1, \ldots, g_n$ of the differentiable algebra $\mathcal{F}$ will be said to separate derivations if $X(g_k) = 0$, for all k, implies that $X = 0$. We will say that a subset $\mathcal{S} \subset \mathcal{F}$ generates the algebra $\mathcal{F}$ if it separates derivations.

Notice that if $\mathcal{S}$ separates derivations, two derivations X, Y are equal iff $X(f) = Y(f)$ for all $f \in \mathcal{S}$.

Definition 3.3 We say that the family of functions $\mathcal{G} = \{g_\alpha\}$ of the differentiable algebra $\mathcal{F}$ is a generating set if they separate points in $X = \mathrm{Spec}_{\mathbb{R}}\mathcal{F}$ and derivations.

It is not hard to see that smooth manifolds (paracompact second countable) possess generating sets (all it is needed is to embed them in $\mathbb{R}^n$ and choose enough coordinate functions there). Again, it is not difficult to show that if $\mathcal{G}$ is a generating set for the algebra $\mathcal{F}$ as an associative algebra then it is a differential generating set for $\mathcal{F}$.

Notice that since derivations can be localized, generating sets also separate local derivations. In fact that is used to prove the following important property of differential generating set that justifies its name.

Lemma 3.4 *A set of functions $\mathcal{G} = \{g_\alpha\}$ of $\mathcal{F}(M)$ is a generating set iff the set of 1-forms $\mathrm{d}\mathcal{G} = \{\mathrm{d}g_\alpha\}$ generates the algebraic dual of the space of derivations $\mathrm{Der}(\mathcal{F})^*$ (that coincides with the space $\Omega^1(M)$ of smooth 1-forms on M if $\mathcal{F} = \mathcal{F}(M)$ with M a smooth manifold).*

Proof We will write the proof for a smooth manifold, but the general case is similar. Consider first an open set U contained in the domain of a local chart of M. If there were a local 1-form σ_U that could not be written as $\sigma_U = \sigma_{U,\alpha} dg_\alpha$, then the span of dg_α would be a proper subspace of T^*U and there would exist a vector field X lying in the annihilator of such subspace. Hence $dg_\alpha(X) = 0$ for all α. This argument can be made global by using partitions of unity: for any 1-form σ on M there must exist a family of functions σ_α with compact support on M such that $\sigma = \sigma_\alpha \mathrm{d}g_\alpha$. □

In the subsequent analysis, we will assume the existence of a generating set $\mathcal{G}$ for $\mathcal{F}$. Let us mention another direct consequence of the properties of a generating set $\mathcal{G}$ in the case of smooth manifolds. Since locally the differentials dg_α generate T^*M, we can extract a subset dg_{α_i} that is locally independent, i.e., such that $dg_{\alpha_1} \wedge \cdots \wedge dg_{\alpha_m} \neq 0$ on a open neighborhood of any given point. Therefore, the differentials dg_{α_i} will define a local coordinate system. Later on we will exploit this fact to write explicit formulae in terms of subsets of generating sets.

A differentiable algebra $\mathcal{F}(M)$ is said to be differentiable finite if it admits a finite differential generating set $\mathcal{G} = \{g_1, \ldots, g_N\}$, $N \in \mathbb{N}$. This more restrictive condition is satisfied for instance if the manifold is compact or of finite type because in such case it can be embedded in a finite-dimensional euclidean space whose coordinate functions restricted to the embedded manifold would provide a differental generating set. As a consequence of the previous discussion, it can be shown that a differential generating set of a differentiable finite algebra $\mathcal{F}(M)$ provides, by restriction to small enough open sets, and sieving out dependent functions, a set of local coordinate systems, i.e., an atlas for the manifold M.

3.2.3.1 Derivations and Their Flows on Differentiable Algebras

Let Γ be a derivation of the differentiable algebra $\mathcal{F}$, hence $\mathcal{F} \cong C^\infty(\mathbb{R}^n)/\mathcal{J}$. Denoting as before by X the real spectrum of $\mathcal{F}$, the derivation Γ defines an element $\Gamma_x \in T_x X$ for each $x \in X$. Hence because X is a closed differentiable subspace of

$\mathbb{R}^n$, the canonical injection $i\colon X \to \mathbb{R}^n$ maps Γ_x to a tangent vector $i_*\Gamma_x \in T_{i(x)}\mathbb{R}^n$, moreover we can extend the vector field $i_*\Gamma$ along $i(X)$ to a vector field $\widetilde{\Gamma}$ in $\mathbb{R}^n$. Let $\tilde{\varphi}_t$ be the flow of $\widetilde{\Gamma}$, hence by construction $\tilde{\varphi}_t$ leaves $i(X)$ invariant. Let us denote by φ_t the restriction of $\tilde{\varphi}_t$ to X (which always exists because of the universal property of closed differentiable subspaces [NS03], p. 60). We will call φ_t the flow of the derivation Γ. The flow φ_t will act on elements $f \in \mathcal{F}$ as:

$$\varphi_t^*(f) = \tilde{\varphi}_t^*(\tilde{f}) + \mathcal{J}, \quad \tilde{f} + \mathcal{J} = f,$$

and the flow φ_t can be integrated formally by using a close analogue to formula (2.74):

$$\varphi_t^*(f) = \sum_{n=0}^{\infty} \frac{t^n}{n!} \widetilde{\Gamma}^n(\tilde{f}) + \mathcal{J}\,. \tag{3.4}$$

Notice that we have proved not only that derivations on differentiable algebras can be extended to derivations on $\mathcal{C}^\infty(\mathbb{R}^n)$, but that they are continuous maps with respect to the canonical topology on $\mathcal{F}$ because they are the quotient of derivations on $\mathcal{C}^\infty(\mathbb{R}^n)$ preserving the closed ideal $\mathcal{J}$ and they are continuous.

Finally we will observe that formulae similar to those defining the local flow for a vector field hold for derivations on differentiable algebras:

$$\frac{df}{dt} = \Gamma(f), \quad f \in \mathcal{F}\,. \tag{3.5}$$

and the local flow φ_t satisfies:

$$\frac{d\varphi_t}{dt} = \Gamma \circ \varphi_t\,. \tag{3.6}$$

3.2.4 Infinitesimal Symmetries and Constants of Motion

3.2.4.1 Infinitesimal Symmetries

Given a derivation Γ of the differentiable algebra $\mathcal{F}$, we will define the space of its infinitesimal symmetries as the Lie subalgebra of $\mathfrak{X}(\mathcal{F})$ of derivations Z such that $[\Gamma, Z] = 0$. More generally, if $\mathcal{S}$ is any subset of a Lie algebra $\mathcal{L}$ the space of infinitesimal symmetries of $\mathcal{S}$ is the commutant $\mathcal{S}'$ of $\mathcal{S}$ in $\mathcal{L}$. We will define the nth commutant of $\mathcal{S}$ recursively as:

$$\mathcal{S}' = \{\xi \in \mathcal{L} \mid [\xi, x] = 0, \text{forall } x \in \mathcal{L}\}, \quad \mathcal{S}^{(k+1)} = (\mathcal{S}^{(k)})', \quad k \geq 1. \tag{3.7}$$

Notice that by definition $\mathcal{S} \subset \mathcal{S}''$ where $\mathcal{S}''$ is called the bicommutant of $\mathcal{S}$. We will have:

Lemma 3.5 *Let $\mathcal{L}$ be a Lie algebra and $\mathcal{S} \subset \mathcal{L}$ a subset. Then:*

$$\mathcal{S} \subset \mathcal{S}'' = \mathcal{S}^{(4)} = \mathcal{S}^{(6)} = \cdots, \quad \mathcal{S}' = \mathcal{S}''' = \mathcal{S}^{(5)} = \cdots. \tag{3.8}$$

Moreover, if $\mathcal{S}$ is abelian, we have $\mathcal{S} \subset \mathcal{S}'' \subset \mathcal{S}'$ and the bicommutant $\mathcal{S}''$ is an abelian Lie subalgebra of $\mathcal{L}$.

Proof Notice that if $\mathcal{S}_1 \subset \mathcal{S}_2$, then $\mathcal{S}_2'' \subset \mathcal{S}_1'$. Because $\mathcal{S} \subset \mathcal{S}'$, we have $\mathcal{S}''' \subset \mathcal{S}'$. On the other hand because for any set $\mathcal{S} \subset \mathcal{S}''$ we have that $\mathcal{S}' \subset (\mathcal{S}')''$, we conclude $\mathcal{S}' = \mathcal{S}'''$ and (3.8) follows.

If $\mathcal{S}$ is abelian, i.e., $[x, y] = 0$ for all $x, y \in \mathcal{S}$, then $\mathcal{S} \subset \mathcal{S}'$, hence $\mathcal{S}'' \subset \mathcal{S}'$. Moreover if $\xi, \zeta \in \mathcal{S}'' \subset \mathcal{S}'$, then $\zeta \in \mathcal{S}'$, and $[\xi, \zeta] = 0$, which shows that $\mathcal{S}''$ is abelian.

The bicommutant $\mathcal{S}''$ is the minimal abelian Lie subalgebra of $\mathcal{L}$ containing $\mathcal{S}$. More can be said when the Lie algebra is represented as an algebra of derivations. Let $\rho\colon \mathcal{L} \to \mathfrak{X}(\mathcal{F})$ be a morphism of Lie algebras, i.e., the Lie algebra $\mathcal{L}$ is represented as derivations of the algebra $\mathcal{F}$. To each element $\xi \in \mathcal{L}$ we associate the derivation $\rho(\xi)$, satisfying $\rho([\xi, \zeta]) = [\rho(\xi), \rho(\zeta)]$.

Remark 3.3 General derivations of a topological algebra $\mathcal{F}$ will not be in general continuous, hence there is no a canonical topology on $\mathfrak{X}(\mathcal{F})$.

3.2.4.2 Constants of Motion

Definition 3.6 Given the dynamics Γ on $\mathcal{F}$, the subalgebra of constants of motion of Γ, denoted by $\mathcal{C}(\Gamma)$ (or simply by $\mathcal{C}$) is defined by

$$\mathcal{C}(\Gamma) = \{f \in \mathcal{F}(M) \mid \Gamma(f) = 0\}. \tag{3.9}$$

More generally, we can consider as in the previous section a Lie algebra $\mathcal{L}$ which is represented as derivations of the algebra $\mathcal{F}$ (in particular we can consider the Lie algebra $\mathfrak{X}(\mathcal{F})$ itself with the tautological representation). Then given any subset $\mathcal{S} \in \mathcal{L}$ we can define the subalgebra of its constants of motion $\mathcal{C}(\mathcal{S})$ defined as:

$$\mathcal{C}(\mathcal{S}) = \{f \in \mathcal{F} \mid \rho(x)(f) = 0, \text{ for all } x \in \mathcal{S}\}. \tag{3.10}$$

We could use a notation reminiscent of the one used in the previous section by denoting the subalgebra of constants of motion by $\mathcal{S}'_{\mathcal{F}} \subset \mathcal{F}$ and calling it the "commutant" of $\mathcal{S}$ in $\mathcal{F}$ with respect to the representation ρ. Similarly we can define the bicommutant of $\mathcal{S}$ in $\mathcal{F}$ as the Lie subalgebra of $\mathcal{L}$, denoted as $\mathcal{S}''_{\mathcal{F}}$, of elements $\xi \in \mathcal{L}$ such that $\rho(\xi)(f) = 0$ for all $f \in \mathcal{C}(\mathcal{S}) = \mathcal{S}'_{\mathcal{F}} \subset \mathcal{F}$. We will define recursively $\mathcal{S}_{\mathcal{F}}^{(2k+1)} \subset \mathcal{F}$ as the subalgebra of elements $f \in \mathcal{F}$ such that $\rho(\xi)(f) = 0$ for all $\xi \in \mathcal{S}_{\mathcal{F}}^{(2k)}\acute{a}$ and $\mathcal{S}_{\mathcal{F}}^{(2k)} \subset \mathcal{L}$ as the Lie subalgebra of elements of $\mathcal{L}$ such that $\rho(\xi)(f) = 0$ for all $f \in \mathcal{S}_{\mathcal{F}}^{(2k-1)}$, with $\mathcal{S}_{\mathcal{F}}^{(1)} = \mathcal{S}'_{\mathcal{F}} = \mathcal{C}(\mathcal{S})$, $k \geq 1$.

Lemma 3.7 *With the notations above,*

$$\mathcal{S} \subset \mathcal{S}''_{\mathcal{F}} = \mathcal{S}^{(4)}_{\mathcal{F}} = \cdots, \quad \mathcal{C}(\mathcal{S}) = \mathcal{S}'_{\mathcal{F}} = \mathcal{S}'''_{\mathcal{F}} = \cdots. \tag{3.11}$$

Moreover, if $\mathcal{S}$ is an abelian subset of $\mathcal{L}$, then $\mathcal{S}''_{\mathcal{F}}$ is the minimal abelian subalgebra of $\mathcal{L}$ containing $\mathcal{S}$.

Notice that the kernel of any derivation of an algebra is a subalgebra. Since $\mathcal{C}$ is a subalgebra of $\mathcal{F}$, if $\mathcal{M}$ is a $\mathcal{F}$-module, $\mathcal{M}$ will be a $\mathcal{C}$-module too. In particular $\mathcal{F}$ is a $\mathcal{C}$-module. The space of $\mathcal{C}$-linear maps from $\mathcal{F}$ to $\mathcal{C}$ is called the space of sections of $\mathcal{F}$ over $\mathcal{C}$ (see later for the set theoretical interpretation of such set). The algebra of constants of motion for the derivation Γ is closed in $\mathcal{F}$ with respect to its canonical topology because Γ is a continuous map. However it is not true in general that a closed subalgebra of a differentiable algebra is a differentiable algebra. We will assume in what follows that the algebra of constants of motion for Γ is a differentiable algebra. As we will see in the various examples discussed later on, this assumption is actually satisfied by a large class of derivations.

We will denote the real spectrum of the differentiable algebra $\mathcal{C}$ by C, i.e., $C = \mathrm{Spec}_{\mathbb{R}}(\mathcal{C})$. Hece the canonical inclusion morphism $i : \mathcal{C} \to \mathcal{F}$ induces a continuous map $\pi : X \to C$, such that $i = \pi^*$. The injectivity of i implies that the map π is a surjective projection map. Let $c \in C$ be a point in the real spectrum of $\mathcal{C}$, then $\mathcal{F}/\mathcal{F}i(\mathfrak{m}_c)$ is the differentiable algebra of the fibre $\pi^{-1}(c)$ of π over the point c, Notice that the flow of Γ leaves invariant the fibres of π and projects to the trivial derivation on C.

3.2.5 Actions of Lie Groups and Cohomology

Lie groups have appeared so far from two different perspectives: on one side they have appeared, both matrix Lie groups and their Lie algebras, as symmetries of dynamical systems and, on the other hand Lie groups are the global objects integrating Lie algebras. The first perspective offers a 'dynamical' interpretation of Lie groups because it presents them as 'acting' on the elements describing a dynamical system.

We want to elaborate further on this idea. We would like to consider groups as transformations on a given space. Thus if we are given an abstract group G and we provide for each element of the group $g \in G$ a transformation $\rho(g)$ on some space such that the composition law of the group becomes the composition of transformations we will say that ρ is a representation of the group. Let us make this definition precise in the setting of differentiable algebras.

3.2.5.1 Lie Groups

As we have anticipated before, Lie groups are smooth manifolds G with a smooth group structure. We will write first the formal definition in the standard setting of smooth manifolds and make some commentaries from the differentiable algebras viewpoint afterwards.

Definition 3.8 A Lie group G is both a group and a smooth manifold such that the product map $m: G \times G \to G$, $m(g, h) = gh$ for all $g, h \in G$, and the inverse map $s: G \to G$, $s(g) = g^{-1}$, $g \in G$ are diffeomorphisms.

From the point of view of their algebra of functions $\mathcal{F}(G)$ the characterization of such properties is natural but requires the dual notion of product, called a coproduct that will lead to the notion of a Hopf algebra (see for instance [BM10] for a physicist's introduction to the subject).

Consider the map $\Delta: \mathcal{F}(G) \to \mathcal{F}(G \times G)$ such that $\Delta(f)(g, h) = f(gh)$ for all $f \in \mathcal{F}(G)$ and $g, h \in G$. The algebraic tensor product $\mathcal{F}(G) \otimes \mathcal{F}(G)$ is dense in $\mathcal{F}(G \times G)$ and we write its completion as $\mathcal{F}(G)\widehat{\otimes}\mathcal{F}(G)$. The map Δ is called a coproduct and satisfies the dual properties of the group product m:

i. Coassociativity: $(\Delta \otimes \mathbb{I}) \circ \Delta = (\mathbb{I} \otimes \Delta) \circ \Delta$.
ii. Counit: The map $\epsilon: \mathcal{F}(G) \to \mathbb{R}$, given by $\epsilon(f) = f(e)$ satisfies $(\epsilon \otimes \mathbb{I})(\Delta f) = f$ for all $f \in \mathcal{F}(G)$.
iii. Antipodal map: The map $S: \mathcal{F}(G) \to \mathcal{F}(G)$ given by $S(f)(g) = f(g^{-1})$, satisfies that $(\mathbb{I} \otimes S)\Delta = (S \otimes \mathbb{I})\Delta = \epsilon$.

It is clear that from the point of view of algebras of functions, the natural way to consider Lie groups is by means of the Hopf algebra structure on $\mathcal{F}(G)$. This approach has proved to be extremely fruitful but we will not discuss the subject further, only invite the reader to dwell on this approach.

Matrix Lie groups together with discrete groups, constitute the simplest families of Lie groups. Their smooth structure is inherited directly from the space of matrices on which are embedded (for matrix Lie groups).

We will denote by L_g the natural action of an element $g \in G$ in the group G by left translations, that is $L_g: G \to G$, $Lg(h) = gh$ and, similarly, R_g is the action of g on G by right translations, $R_g(h) = hg$. In the case of matrix groups, these operations correspond to the standard multiplication of matrices on the left or on the right respectively. On Lie groups, both operations L_g, R_g commute, i.e., $L_g \circ R_h = R_h \circ L_g$ for all $g, h \in G$, and they define diffemorphisms of G that may be interpreted as the group acting on itself by transformations. If the group G is not Abelian, the effect of both could be different. These natural actions of G on itself inspire a general notion of groups acting on other spaces, which is the natural way of modeling the notion of group of transformations. We will devote the next section to discussing this idea.

3.2.5.2 Representations and Actions of Lie Groups

Definition 3.9 Let G be a group and $\mathcal{F}$ be an associative algebra. A representation of G on $\mathcal{F}$ is a map $\rho: G \to \text{Aut}(\mathcal{F})$ such that:

$$\rho(gg') = \rho(g) \circ \rho(g'); \qquad \rho(e) = \mathrm{id}_{\mathcal{F}}; \quad \forall g, g' \in G\,. \tag{3.12}$$

Recall that the group of automorphisms of the differential algebra $\mathcal{F}$ consists of all linear maps $\Phi\colon \mathcal{F} \to \mathcal{F}$ which are invertible and such that $\Phi(f \cdot f') = \Phi(f) \cdot \Phi(f')$, $f, f' \in \mathcal{F}$. If ρ is a representation of G on $\mathcal{F}$, then $\rho(g)(f \cdot f') = \rho(g)(f) \cdot \rho(g)(f')$.

If $\mathcal{F}$ is the algebra of smooth functions on a manifold M, then it can be shown immediately that any automomorphism Φ of the algebra $\mathcal{F}$ is induced from a diffeomorphism $\varphi\colon M \to M$. In fact if $\mathfrak{m}_x$ is the ideal of smooth functions vanishing at x then $\Phi(\mathfrak{m}_x)$ is another maximal ideal, hence we define $\varphi(x)$ as the point such that $\mathfrak{m}_{\varphi(x)} = \Phi(\mathfrak{m}_x)$. The fact that Φ is bijective implies that φ must be bijective too and, finally, we check that

$$\Phi(f) = f \circ \varphi^{-1}\,,$$

using Gel'fand's formula Eq. (3.3).

Thus if we have a representation ρ of a Lie group G on the algebra of functions $\mathcal{F}$ of a manifold (the same arguments apply for a differentiable algebra), then $\rho(g)$ are automorphisms of the algebra, hence there will exist $\varphi_g\colon M \to M$ such that

$$\rho(g)(f) = f \circ \varphi_g^{-1}, \qquad \forall g \in G\,.$$

The conditions defining the representation imply that $\varphi_g = \mathrm{id}_M$ and that $\varphi_g \circ \varphi_h = \varphi_{gh}$. This allows as to define the notion of the action of a group G on a manifold M as follows:

Definition 3.10 Let G be a Lie group and M a smooth manifold. An action of G on M (on the left) is a smooth map $\varphi\colon G \times M \to M$, such that:

$$\varphi(e, x) = x, \qquad \varphi(gh, x) = \varphi(g, \varphi(h, x)), \qquad \forall g, h \in G, \quad x \in M\,. \tag{3.13}$$

Given $g \in G$, we define $\varphi_g\colon M \to M$ by $\varphi_g(x) := \varphi(g, x)$. Then properties Eq. (3.13) imply that $\varphi_g \circ \varphi_h = \varphi_{gh}$ and φ_g defines a representation of G on the algebra $\mathcal{F}(M)$. If there is no risk of confusion we will often write $g \cdot x$ (or simply gx) instead of $\varphi(g, x)$ or $\varphi_g(x)$, and we will call the element gx "the action of g on x". We will say that the action of G on M is free if $gx = x$ for some $x \in M$ implies $g = e$ and the action will be called transitive if for any $x, y \in M$ there exists at least one element $g \in G$ such that $y = gx$. There are completely analogous definitions for actions of groups on the right on a given manifold M. In this case, the action of g on x will be written as xg instead.

The family of left translations L_g introduced above, define a free and transitive action of the Lie group G on itself (in the same way the family of right translation R_g above define a right action of G on itself).

We have found actions of groups in various places, either as the definition of the groups themselves (think for instance of matrix Lie groups acting on the linear space that we use to define them) or, for instance, when defining the adjoint action of a matrix Lie group on its Lie algebra, Eq. (2.133). Exactly in the same terms we can extend the definition of the adjoint action of a Lie group on its Lie algebra. We will denote it again as Ad.

Given a point $x \in M$, the subset of elements $G_x = \{g \in G \mid gx = x\}$ are called the isot(r)opy group of x.[7] The set G_x is actually a closed subgroup of G. Again given a point $x \in M$ the collection of points $\Gamma \cdot x = \{g \cdot x \mid g \in G\}$ is called the orbit through x of G. The set $G \cdot x$ is an immersed submanifold of M. It is easy to prove that G/G_x is diffeomorphic to $G \cdot x$:

$$G \cdot x \cong G/G_x\,. \tag{3.14}$$

This diffeomorphism is defined by $gx \mapsto gG_x$ and the inverse is given by $gG_x \mapsto gx$.

Exercise 3.2 Given a differentiable algebra $\mathcal{F}$, and ρ a representation of G on $\mathcal{F}$, establish the notions of isotopy group G_x an orbit $G \cdot x$ of a given $x \in X = \mathrm{Spec}_{\mathbb{R}}(\mathcal{F})$ and show that:

$$\mathcal{F}(G/G_x) = \mathcal{F}(G)^{G_x}, \qquad \mathcal{F}(G \cdot x) = \mathcal{F}/\mathfrak{m}_{G\cdot x}\,,$$

where $\mathcal{F}(G)^{G_x}$ denotes the subalgebra of smooth functions in $\mathcal{F}(G)$ invariant under G_x, i.e., $f(gh) = f(h)$ for all $g \in G_x$, and $\mathfrak{m}_{Gx}$ is the intersection of all ideals $\rho(g)(\mathfrak{m}_x)$, $g \in G$. Finally prove that:

$$\mathcal{F}(G)^G \cong \mathcal{F}/\mathfrak{m}_{G\cdot x}\,.$$

We would like to point out at this moment that the group of symmetries G of a given dynamical system Γ defined on the manifold M acts in M. It is also clear that because Γ in invariant under the action of G, then Γ induces a dynamical system in the quotient space M/G.

3.2.5.3 Actions of Lie Groups and Cohomology

If we have a Lie group G acting on a manifold we can start to understand properties of this action by observing how G changes elements on the exterior algebra of M. Thus let us suppose that α is a k-form on M and consider the action $g \cdot \alpha := (g^{-1}) * \alpha$ of g on α (that is the pull-back of α with respect to the diffeomorphism defined by φ_g^{-1}). Then we define the k-form $\beta(g)$ as:

$$g \cdot \alpha = \alpha + \beta(g), \qquad g \in G\,. \tag{3.15}$$

Now if we compute $(gh) \cdot \alpha$ and $g \cdot (h \cdot \alpha)$ we will obtain:

$$\beta(gh) = \beta(g) + g \cdot \beta(h)\,,$$

[7] In the past it was also-called the "little group" of x.

that is, $\beta\colon G \to \Omega^k(M)$ is a 1-cocycle in G with values in the G-module $\Omega^k(M)$. Recall (see Appendix F for more details) that a p-cochain on G with values in a G-module Ω is just a map $\beta\colon G \times \overset{p}{\cdots} \times G \to \Omega$. The cohomology operator δ maps 1-cochains into 2-cochains by $\delta\beta(g,h) = \beta(gh) - \beta(g) - g\cdot\beta(h)$ (see Eq. (10.282)). Thus the 1-cochain β defined in Eq. (3.15) satisfies $\delta\beta = 0$, that is, β is a 1-cocycle. Moreover, if we change α adding an arbitrary k-form Ψ, say $\alpha' = \alpha + \psi$, then

$$g\cdot\alpha' = \alpha' + \beta'(g) = \alpha + \psi + \beta'(g) = g\cdot\alpha - \beta(g) + \psi + \beta'(g) = g\cdot\alpha' - g\cdot\psi + \psi - \beta(g) + \beta'(g)\,,$$

that is

$$\beta'(g) = \beta(g) + g\cdot\psi - \psi,$$

but if ψ is a 0-cochain (i.e., just an element in Ω), then $\delta\psi(g) = g\cdot\psi - \psi$. Then

$$\beta' = \beta + \delta\psi\,,$$

and the family of k-forms $\Omega^k(M)$ defines a cohomology class $[\beta] \in H^1(G, \Omega^k(M))$.

Now, let us assume that $d\alpha = \omega$ is such that $g\cdot\omega = \omega$ for any $g \in G$, that is ω is G-invariant, then clearly $d(g\cdot\alpha - \alpha) = 0$, hence $d\beta(g) = 0$, thus β is a 1-cocycle on G with values in closed k-forms. Taking now ψ to be closed, we get that $[\beta] \in H^1(G, Z^k(M))$.

In a similar way we may construct the Chevalley cohomology on the Lie algebra $\mathfrak{g}$ with values in a $\mathfrak{g}$-module Ω. We define p-cochains in $\mathfrak{g}$ with values in Ω, $C^p(\mathfrak{g},\Omega)$, as maps $c\colon \mathfrak{g} \times \overset{p}{\cdots} \times \mathfrak{g} \to \Omega$, and the cohomology operator $d_{\hat\rho}$ where $\hat\rho\colon \mathfrak{g} \to \mathrm{Aut}(\Omega)$ describes the representation of $\mathfrak{g}$ as automorphisms of Ω. Then $d_{\hat\rho}\colon C^p(\mathfrak{g},\Omega) \to C^{p+1}(\mathfrak{g},\Omega)$ defined as (eq. (10.286)):

$$\begin{aligned} d_{\hat\rho}c(\xi_1,\dots,\xi_{p+1}) = &\sum_{i=1}^{p+1}(-1)^i\hat\rho(\xi_i)c(\xi_1,\dots,\hat\xi_i,\dots,\xi_{p+1}) \\ &+\sum_{i\neq j}(-1)^{i+j}c([\xi_i,\xi_j]\xi_1,\dots,\hat\xi_i,\dots,\hat\xi_j,\dots,\xi_{p+1})\,. \end{aligned} \tag{3.16}$$

Then a 1-cocycle c_1 satisfies

$$c_1([\xi,\zeta]) = \hat\rho(\xi)c_1(\zeta) - \hat\rho(\zeta)c_1(\xi)\,. \tag{3.17}$$

Similarly, if $\Omega = \mathbb{R}$ and the action is trivial, we get that a 2-cocycle c_2 satisfies:

$$c_2([\xi_1,\xi_2],\xi_3) + \mathrm{cyclic}(1,2,3) = 0\,.$$

It is clear (and the notation suggests it) that if ρ is an action of G on Ω, and we denote by $\hat{\rho}$ the induced action of $\mathfrak{g}$ on Ω, that is:

$$\hat{\rho}(\xi)(w) = \frac{d}{ds}\rho(\exp(s\xi)(w) \mid_{s=0} ,$$

each cochain β on G defines a cochain on $\mathfrak{g}$ by derivation, that is, for instance for 1- and 2-cochains we get:

$$c_1(\xi) = \frac{d}{ds}\beta_1(\exp(s\xi)) \mid_{s=0}, \qquad c_2(\xi,\zeta) = \frac{\partial^2}{\partial s \partial t}\beta_2(\exp(s\xi), \exp(t\zeta)) \mid_{s=0,t=0} .$$

Then if $\beta(g)$ is the 1-cocycle defined by a k-form α, then $c_1(\xi) = \mathcal{L}_{\xi_M}\alpha$.

Example 3.3 (Cohomology class defined by a 1-form) A particular instance that will be of some use later on is provided by the following setting: let θ be a 1-form defined on a manifold M and let G be a Lie group acting on M, and such that $\omega = d\theta$ is G-invariant. Consider now the 1-cocycle defined by θ, that is the cohomology class $[\beta] \in H^1(G, Z^1(M))$, that is $g \cdot \theta = \theta + \beta_g$. The corresponding 1-cocycle in $\mathfrak{g}$ with values in $Z^1(M)$ will be defined as:

$$c_1(\xi) = \mathcal{L}_{\xi_M}\theta, \qquad \forall \xi \in \mathfrak{g} .$$

We will assume now that either M is connected and simply connected, that is $H^1(M) = 0$ (which in turn implies that $Z^1(M) = B^1(M)$ or that β_g is exact, that is $\beta_g = dJ_g$ for all $g \in G$ and $J_g \colon M \to \mathbb{R}$ is a real function. Now we have the exact sequence of maps:

$$0 \to \mathbb{R} \to \mathcal{F}(M) \stackrel{d}{\to} B^1(M) \to 0 .$$

Then a general theorem (see Appendix F) permits us to construct what is called the long exact cohomology sequence for the cohomology groups of $\mathfrak{g}$ with values in the $\mathfrak{g}$-modules $\mathbb{R}$, $\mathcal{F}(M)$ and $B^1(M)$:

$$\cdots \to H^1(\mathfrak{g}, \mathbb{R}) \to H^1(\mathfrak{g}, \mathcal{F}(M)) \to H^1(\mathfrak{g}, B^1(M)) \stackrel{\Delta}{\to} H^2(\mathfrak{g}, \mathbb{R}) \to H^2(\mathfrak{g}, \mathcal{F}(M)) \to \cdots$$

The connecting homomorphism $\Delta \colon H^1(\mathfrak{g}, B^1(M)) \to H^2(\mathfrak{g}, \mathbb{R})$ is defined as:

$$\Delta c_1(\xi,\zeta) = \xi(J_\zeta) - \zeta(J_\xi) - 2J_{[\xi,\zeta]},$$

where $c_1(\xi) = dJ_\xi$.

3.3 The Tensorial Characterization of Linear Structures and Vector Bundles

In the previous sections we have introduced a formalism that allows us to deal with spaces that are not necessarily linear. In particular, we have discussed the geometrization of the notion of dynamical system as a vector field on a differentiable space (or in a smooth manifold).

In what follows we are going to follow this pattern. First we will analyze the structures compatible with a given dynamics at the linear level and afterwards we will geometrize them, that is, we will study how these structures and their properties can be formulated in a purely geometrical context, or in other words, in general differentiable spaces or smooth manifolds.

Thus each structure that we can describe as a linear algebraic object will be 'promoted' to a geometrical structure. The first, and most fundamental example is the geometrization of the linear space structure itself. In other terms we may ask, is there a natural tensorial object that characterizes geometrically a linear space structure on a smooth manifold?

The answer to this question is, of course, positive but subtle and we will devote this section to answer it. Along the discussion we will see that a slight weakening on their properties leads immediately to the notion of vector bundle, another of the fundamental building bricks of the language of differential geometry.

3.3.1 A Tensorial Characterization of Linear Structures

The 'easy' tensorialization of linear maps as linear vector fields, Eq. (2.112), sending $A \mapsto X_A$ relies, as it was already pointed it out then, on the vector space structure of E. Indeed, as we have seen, a vector space E carries a dilation operator, or Liouville vector field Δ_E, which is defined by the flow,

$$\phi_t\,(v) = \exp\,(t)\, v\,, \tag{3.18}$$

and that in linear coordinates x^i has the simple form,

$$\Delta = x^i \frac{\partial}{\partial x^i} \tag{3.19}$$

because,

$$\Delta(x^i)(v) = \frac{dx^i}{dt}(\phi_t(v))\bigg|_{t=0} = \frac{dx^i}{dt}(e^t v)\bigg|_{t=0} = \frac{d\left(e^t x^i\,(v)\right)}{dt}\bigg|_{t=0} = x^i\,(v)\,. \tag{3.20}$$

The operator Δ creates a gradation on the ring $\mathcal{F}(E)$ of smooth functions on E as follows:

Let $\mathcal{F}^{(k)}(E)$ (or $\mathcal{F}^{(k)}$ for short) be the subspace of homogeneous functions of degree $k \in \mathbb{Z}, \geq 0$, i.e.,

$$\mathcal{F}^{(k)} = \{f \in \mathcal{F} | \mathcal{L}_\Delta f = kf\} \ . \tag{3.21}$$

Notice that in linear coordinates x^i, the subspace $\mathcal{F}^{(k)}$ is a finite dimensional space generated by homogeneous polynomials of degree k on the variables $x^1, \ldots, x^n$. This fact is an easy consequence of the following exercises:

Exercise 3.4 Prove that $\mathcal{F}^{(0)} = \mathbb{R}$.

The argument in this exercise can generalized easily to show that:

Exercise 3.5 Prove that $f \in \mathcal{F}^{(k)}$ if $f(tv) = t^k f(v)$ for all $t \in \mathbb{R}$ and $v \in E$.

Then we can easily prove:

Proposition 3.11 (Euler's Theorem) *The space $\mathcal{F}^{(k)}$ is a linear space of dimension $\binom{n+k-1}{k}$ of homogeneous polynomials of degree k denoted by $S^k[x]$.*

Proof Let $f \in \mathcal{F}^k$. Consider the function $\phi(t)$ as before for a fixed $v \neq 0$. Then $\phi(t) = t^k f(v)$ and $\phi(t)$ is an homogeneous polynomial of degree k. Taylor's formula for $\phi(t)$ around $t = 0$, will give,

$$\phi(t) = \sum_{k \geq 0} \frac{1}{k!} \phi^{(k)}(0)\, t^k \tag{3.22}$$

and the computation of the k-th derivative of $\phi(t)$ gives,

$$\phi^{(k)}(0) = \frac{\partial^k f}{\partial x^{i_1} \ldots \partial x^{i_k}}(0)\, x^{i_1} \ldots x^{i_k} \tag{3.23}$$

Then, taking $t = 1$, we obtain,

$$f(x) = \frac{1}{k!} \frac{\partial^k f}{\partial x^{i_1} \ldots \partial x^{i_k}}(0)\, x^{i_1} \ldots x^{i_k} \tag{3.24}$$

thus $f \in S^k[x]$. □

Proposition 3.12 *The space $\mathcal{F}^{(1)}$ of homogenous functions of degree one separates derivations.*

Proof In fact, using that the derivations $\partial/\partial x^i$ form a basis of $\mathfrak{X}(E)$, where x^i are linear coordinates on E defined by the choice of a basis of E, it follows immediately that if $X = X^i \partial/\partial x^i$, then $0 = X(x^i) = X^i$. □

We have seen to what extent the Liouville vector field Δ_E determines the structure of the ring of functions of a vector space. Conversely, the existence of a vector field on a manifold possessing similar properties with respect to the ring of functions is enough to characterize a vector space structure on the manifold. Indeed:

Theorem 3.13 (Tensorial characterization of linear spaces) *Let M be a smooth finite-dimensional manifold of dimension m possessing a complete vector field Δ with only one non-degenerate critical point such that:*

i. $\mathcal{F}^0 = \mathbb{R}$,

ii. The family of functions $\mathcal{F}^1$ separates derivations, where,

$$\mathcal{F}^k = \{f \in \mathcal{F} | \Delta(f) = kf\}, \qquad k \in \mathbb{Z}. \tag{3.25}$$

Then M can be given a unique vector space structure such that Δ is the vector field generated by dilations.

Proof Notice first that $\mathcal{F}^{(1)}$ is a finite-dimensional linear space. Actually, $\mathcal{F}^{(1)}$ is a linear subspace of the algebra $\mathcal{F}$ of smooth functions on M. Then two functions f, g which are linearly independent define independent covectors df and dg, but M is finite-dimensional, thus there can be at most a finite number n of linearly independent functions in $\mathcal{F}^{(1))}$ and then $n \leq m$ Then we can choose a linear basis for $\mathcal{F}^{(1)}$ made up of the functions $f^1, \ldots, f^n$. Then they are functionally independent, i.e., $df^1 \wedge \ldots \wedge df^n \neq 0$ everywhere (we use that they form a basis and condition i).

On the other hand, because $\mathcal{F}^{(1)}$ separates derivations, then $n \geq m$, thus $n = m$ and we may use them as a global coordinate chart. It is better to define a family of vector fields X_i by,

$$X_i(f^j) = \delta_i^j \tag{3.26}$$

i.e.,

$$\frac{df^j}{ds^i} = \delta_i^j \tag{3.27}$$

These vector fields are well defined. If X_i and X_i' satisfy the same definition property, then for an arbitrary f, $(X_i - X_i')f = 0$, then $X_i = X_i'$. If we compute the commutator of Δ and X_i we obtain,

$$[\Delta, X_i](f^j) = \Delta(X_i(f^j)) - X_i(\Delta(f^j)) = \Delta(\delta_i^j) - X_i(f^j) = -\delta_i^j \tag{3.28}$$

Then, $[\Delta, X_i] = -X_i$ for all $i = 1, \ldots n$. Besides,

$$[X_i, X_j](f_k) = X_i(X_j(f^k)) - X_j(X_i(f^k)) = 0 \tag{3.29}$$

for all k, then $[X_i, X_j] = 0$. The vector fields X_i are complete, because each one of them has the following flow,

$$\begin{aligned} \phi_{s_j}\left(f^k\right) &= f^k,\ k \neq j \\ \phi_{s_j}\left(f^j\right) &= f^j + s^j \end{aligned} \tag{3.30}$$

Then Eq. (3.30) define an action of $\mathbb{R}^n$ on M which is free and transitive.

Using the critical point x_0 we can identify now $\mathbb{R}^n$ with M by means of the map $(s_1, \ldots, s_n) \mapsto \phi_{s_1} \cdots \phi_{s_n}(x_0)$. □

Definition 3.14 A vector field Δ on a smooth manifold M satisfying conditions (i) an (ii) in Theorem 3.13 will be called a Liouville vector field.

It is clear then that if a manifold carries a Liouville vector field it must be connected and contractible (because it is a linear space). In fact that is the only topological obstruction to the existence of a Liouville vector field on a manifold, because if M is a contractible finite-dimensional manifold, then it is diffeomorphic to a ball, hence to $\mathbb{R}^n$ and it can carry a Liouville vector field.

It is now clear that a space carrying a Liouville vector field Δ allows us to define the set of linear vector fields X as the set of all vector fields which commute with Δ, $[X, \Delta] = 0$. From this definition and the Jacobi identity it follows that linear vector fields close in a Lie algebra. If we denote by $\mathcal{F}^{(1)}(M)$ as before, the set of linear functions on M, defined by $\mathcal{L}_\Delta f = f$, we also find that for any linear vector field X:

$$\mathcal{L}_X \mathcal{F}^{(1)} \subset \mathcal{F}^{(1)} \tag{3.31}$$

Indeed: $\mathcal{L}_\Delta \mathcal{L}_X f = \mathcal{L}_X \mathcal{L}_\Delta f = \mathcal{L}_X f$, then $\mathcal{L}_X f$ is a linear function if it is not identically vanishing. If $\mathcal{F}^{(1)}$ is finite-dimensional and we select a basis $f^1, f^2, \ldots, f^n$, we have,

$$\mathcal{L}_X f^i = A^i{}_j(X) f^j \tag{3.32}$$

and we recover the matrix representation of X. It is clear that $A^i{}_j(X)$ depends on the basis we have chosen.

It is possible to use this matrix representation for X to show that its flow exists and in the given basis is represented by e^{tA}. Of course it is possible to write the flow in a way that is independent of the choice of the basis. Indeed, we can write $e^{t\mathcal{L}_X}$ and we could write,

$$f(x(t)) = (e^{t\mathcal{L}_X} f)(x(0)) = f(x(0)) + t(\mathcal{L}_X f)(x(0)) + \frac{t^2}{2!}\left((\mathcal{L}_X)^2 f\right)(x(0)) + \cdots \tag{3.33}$$

When f is linear, let us say $f = a_i f^i$, it is clear that $\mathcal{L}_X f = a_i A^i{}_j f^j$ and we simply recover $f(x(t)) = f(e^{tA} x(0))$, i.e.,

$$x(t) = e^{tA} x(0) . \tag{3.34}$$

One has to be cautious however because the right-hand side in terms of Lie series only makes sense if f is real analytic. It should be noticed that as far as the flow is concerned it is enough to restrict to C^ω functions because they are enough to separate points and therefore a cancellation holds true. In other terms, from $f(x(t)) = f(e^{tA}x(0))$, we can conclude that $x(t) = e^{tA}x(0)$ if the set of functions for which the result is true are enough to separate points on the space on which they are defined.

Remark 3.4 The formula we have written for the flow of a linear vector field X, $\varphi_t = e^{t\mathcal{L}_X}$, might have additional limitations when X is not linear. In this case we also need X to be a complete vector field.

3.3.2 Partial Linear Structures

Let us consider a derivation Δ in the algebra $\mathcal{F}(E)$ of differentiable functions in the linear space E, i.e., a vector field on E. Then, the operator Δ may create a gradation on the ring $\mathcal{F}(E)$ of smooth functions on E as follows:

Let $\mathcal{F}^{(k)}(E)$ be the subspace of homogeneous functions, with respect to Δ, of degree $k \in \mathbb{Z}$, $k \geq 0$, i.e.,

$$\mathcal{F}^{(k)}(E) = \{f \in \mathcal{F}(E) \,|\, \mathcal{L}_\Delta f = kf\} \tag{3.35}$$

assumed to be not empty for $k \geq 0$.

We should denote better $\mathcal{F}^{(k)}(E)$ as $\mathcal{F}^{(k)}_\Delta(E)$ to indicate the dependence on the derivation Δ, however we will not do it unless risk of confusion. We will often omit the reference to the underlying space E in what follows.

We first remark that $\mathcal{F}^{(0)}(E)$ is a subalgebra, because if $f_1, f_2 \subset \mathcal{F}^{(0)}$, then $\mathcal{L}_\Delta f_1 = \mathcal{L}_\Delta f_2 = 0$, and thus, for each real number λ, $\mathcal{L}_\Delta(f_1 + \lambda f_2) = \mathcal{L}_\Delta f_1 + \lambda \mathcal{L}_\Delta f_2 = 0$, and $\mathcal{L}_\Delta(f_1 f_2) = f_2 \mathcal{L}_\Delta f_1 + f_1 \mathcal{L}_\Delta f_2 = 0$. The subalgebra $\mathcal{F}^{(0)}$ will possibly be identified with the algebra of functions defined on a quotient space out of E. This quotient space is the collection of "linear fibres" identified by Δ. On the other side, $\mathcal{F}^{(1)}(E)$ is not a subalgebra, but a $\mathcal{F}^{(0)}(E)$-module, because if $f_0 \in \mathcal{F}^{(0)}(E)$ and $f_1 \in \mathcal{F}^{(1)}(E)$, the conditions $\mathcal{L}_\Delta f_0 = 0$ and $\mathcal{L}_\Delta f_1 = f_1$ imply $\mathcal{L}_\Delta(f_0 f_1) = f_0 \mathcal{L}_\Delta f_1 = f_0 f_1$.

The subalgebra $\mathcal{F}^{(0)}(E)$ determines a subalgebra of vector fields, spanned by the vector fields $X \in \mathfrak{X}(E)$ such that,

$$\mathcal{L}_X \mathcal{F}^{(0)}(E) = 0 \tag{3.36}$$

Obviously if $X, Y \in \mathcal{D}$, then $\mathcal{L}_X f = \mathcal{L}_Y f = 0$, $\forall f \in \mathcal{F}^{(0)}(E)$, and therefore, $\mathcal{L}_{[X,Y]} f = \mathcal{L}_X(\mathcal{L}_Y f) - \mathcal{L}_Y(\mathcal{L}_X f) = 0$, $\forall f \in \mathcal{F}^{(0)}(E)$, and therefore,

$$[X, Y] \in \mathcal{D} \tag{3.37}$$

Before proceeding, let us digress briefly on some additional notions that are needed here.

According then to what has been discussed in Sect. 2.5.2, the subalgebra $\mathcal{D}$ is (cfr. Eq. (3.37)) an involutive distribution, and it is integrable according to Frobenius' theorem [AM78].

Let us assume now that the differentials of functions of $\mathcal{F}^{(0)}(E)$ and $\mathcal{F}^{(1)}(E)$ span the set of 1-forms $\Omega^1(E)$ as a $\mathcal{F}^{(0)}(E)$-module.

Remark 3.5 If we consider $\Delta\colon E \to TE$, we may consider $I\colon E \to TE$, $I(x) = (x, x)$. The difference $I - \Delta\colon E \to TE$, defines another partial structure which gives a "representative" of the quotient space associated with $\mathcal{F}^{(0)}(E)$. This would be a "tensorial version" of the decomposition of the vector space E as a direct sum $E = E_1 \oplus E_2$.

In the present situation, it is actually possible to define a basis for $d\mathcal{F}^{(0)}(E)$. Moreover we will assume that the set $\mathcal{F}^{(0)}(E)$ is finitely generated by $f_i^{(0)}$, $i = 1, \ldots, r$, and the restriction of Δ to the level sets of the functions $f_i^{(0)}$ defines linear structures. Similarly, we will assume that $\mathcal{F}^{(1)}(E)$ is finitely generated and we will denote one such a basis of functions by $f_\alpha^{(1)}$, $\alpha = 1, \ldots, k$. Then we will say that Δ is a *partial linear structure*.

Definition 3.15 Let E be a linear space and let Δ be a complete vector field which is a derivation on it. We will say that Δ is a partial linear structure if both $\mathcal{F}^{(0)}(E)$ and $\mathcal{F}^{(1)}(E)$ are finitely generated and moreover $d\mathcal{F}^{(0)}(E)$ and $d\mathcal{F}^{(1)}(E)$ span the set of 1-forms $\Omega^1(E)$ as an $\mathcal{F}^{(0)}(E)$-module. Moreover we require that the critical set of Δ be a submanifold whose algebra of functions is isomorphic with $\mathcal{F}^{(0)}(E)$.

In order to compare with the usual coordinate approach and what we know already of linear structures, we will denote the functions $f_i^{(0)}$ as x^i, while we denote the functions $f_\alpha^{(1)}$ by y^α Then the coordinate expression for Δ will be given by,

$$\Delta = y^\alpha \frac{\partial}{\partial y^\alpha} \tag{3.38}$$

Definition 3.16 If Δ is a partial linear structure we will say that a vector field X is partially linear (with respect to Δ) if,

$$[X, \Delta] = 0 \tag{3.39}$$

Its coordinate expression will be then:

$$X = f^i\left(x^1, \ldots, x^k\right) \frac{\partial}{\partial x^i} + a^\alpha{}_\beta\,(x)\, y^\beta \frac{\partial}{\partial y^\alpha} \tag{3.40}$$

3.3.3 Vector Bundles

Given a partial linear structure Δ on E, we define a projection map $\pi_\Delta \colon E \to B$, with the property $\pi_\Delta^*(\mathcal{F}(B)) = \mathcal{F}_\Delta^{(0)}(E)$. In the particular case we are considering, we may think of E as a product of vector spaces $B \times V$, where V is a typical fibre representing the linear subspace of E identified by Δ.

Given coordinates $(x, y) \in B \times V$, we find that $F \colon E \to \mathbb{R} \in \mathcal{F}^{(1)}$, i.e., satisfying $\mathcal{L}_\Delta F = F$, can be written in the chosen coordinates as $F(x, y) = f_r(x) y^r$. Thus we can represent the map $F \colon B \times V \to \mathbb{R}$ as $\tilde{F} \colon B \to V^*$. In this way, functions which are 'partially linear' on E are in one-to-one correspondence with vector valued functions $\tilde{F} \colon B \to V^*$, which leads us to consider, together with the vector bundle $B \times V$, the dual bundle $B \times V^*$.

It is appropriate to introduce here a notion of 'projectability' of vector fields with respect to the projection π_Δ.

Definition 3.17 A vector field $Y \in \mathfrak{X}(E)$ will be said to be projectable with respect to the projection π_Δ if there exists a vector field $X_Y \in \mathfrak{X}(B)$ such that

$$T\pi_\Delta \circ Y = X_Y \circ \pi_\Delta \,. \tag{3.41}$$

Equivalently, for every function $f \in \mathcal{F}(B)$,

$$\mathcal{L}_Y \left(\pi_\Delta^* f\right) = \pi_\Delta^* \left(\mathcal{L}_{X_Y} f\right) \,. \tag{3.42}$$

We will call X_Y the projected vector field associated with Y.

It is now a natural question to inquire to what extent the differential calculus we have elaborated on E can be put to work to give a differential calculus on vector-valued functions $\tilde{F} \colon B \to V^*$.

Let us consider first the Lie derivative and exterior derivative of functions $F \colon E \to \mathbb{R}$ satisfying the requirement $\mathcal{L}_\Delta F = 0$, that is, of functions in the subalgebra $\mathcal{F}^{(0)}(E)$. It is clear that to remain within the same vector space of functions, we should consider vector fields Y which commute with Δ, in this way $\mathcal{L}_\Delta \mathcal{L}_Y F = \mathcal{L}_Y \mathcal{L}_\Delta F = 0$. In the chosen coordinates this amounts to requiring, see Eq. (3.40),

$$Y = h^i(x) \frac{\partial}{\partial x^i} + A^\alpha{}_\beta(x)\, y^\beta \frac{\partial}{\partial y^\alpha} \,. \tag{3.43}$$

Thus, the projected vector field is $X_Y = h\,(x)\, \partial/\partial x^i$. On the other hand, if we want to associate a vector field on B, say,

$$X = h\,(x) \frac{\partial}{\partial x^i} \tag{3.44}$$

with a vector field on E commuting with Δ and respecting linearity with projected vector field X itself, we would set,

$$Y_X = h^i(x)\frac{\partial}{\partial x^i} + h^i A^\alpha{}_{\beta,i}(x)\, y^\beta \frac{\partial}{\partial y^\alpha} \tag{3.45}$$

and it is clear that, in this association, we are using a chosen (1, 1) tensor field, namely,

$$T_A = A^\alpha{}_{\beta,i}(x) y^\beta dx^i \frac{\partial}{\partial y^\alpha} \tag{3.46}$$

so that, with some abuse of notation,

$$Y_X = X + T_A(X)\ . \tag{3.47}$$

Now we consider $\mathcal{L}_Y F$ in the chosen coordinate system. We find,

$$\mathcal{L}_Y(f_\alpha(x)y^\alpha) = h^j\left(\frac{\partial f_\alpha}{\partial x^j} + A^\beta_{\alpha j} f_\beta\right) y^\alpha\ . \tag{3.48}$$

Therefore, we may conclude that for our function $\tilde{F}\colon B \to V^*$, we have a derivative induced from $X = h^j(x)\frac{\partial}{\partial x^j}$ given by,

$$D_X\tilde{F} = h^j\left(\frac{\partial f_\alpha}{\partial x^j} + A^\beta_{\alpha j} f_\beta\right) y^\alpha \tag{3.49}$$

or equivalently by,

$$D_X f_\alpha = h^j\left(\frac{\partial f_\alpha}{\partial x^j} + A^\beta_{\alpha j} f_\beta\right) \tag{3.50}$$

which is usually known as the covariant derivative (see below, Appendix E) of the vector-valued function $\tilde{F}$ associated with the endomorphism defined by $A^\alpha_{i\beta}\, y^\beta\, dx^i \otimes (\partial/\partial y^\alpha)$. Along similar lines we can define now a covariant exterior differential of f_α by setting,

$$Df_\alpha = df_\alpha + A^\beta_{\alpha j} f_\beta dx^j \tag{3.51}$$

providing us with a differential 1-form associated with the vector-valued function f_α.

The expression:

$$\frac{\partial f_\alpha}{\partial x^j} + A^\beta_{\alpha j} f_\beta$$

in the right-hand side of Eq. (3.51) shows that we are writing the derivative in terms of first-order differential operators (see Appendix G) which are not derivations anymore (i.e., they do not satisfy Leibnitz's rule). For this reason we should not expect the covariant exterior derivative D to be a derivation. On the other hand, if we go back to our starting point where we have considered the Lie derivative of F along Y we remark the possibility of describing covariant derivatives in terms of derivations, i.e., by going backwards, from $\widetilde{F}$ we define a function on the vector bundle, and we replace the vector field X with the associated vector field Y_X.

Let us give now two simple examples.

Example 3.6 Consider first: $E = \mathbb{R}^5 = \mathbb{R}^4 \times \mathbb{R}$, with coordinates (x, y), $B = \mathbb{R}^4$. The partial linear structure is given by the vector field $\Delta = y\partial/\partial y$. Thus $\mathcal{F}^{(0)} = \{f(x) \,|x \in \mathbb{R}^4\}$. Moreover $\mathcal{F}^{(1)} = \{yf(x) \,|\, f \in \mathcal{F}^{(0)}\}$. Consider now a basis of vector fields along B, $\partial/\partial x^\mu$. A vector field Y, commuting with Δ and satisfying the linearity requirement with respect to the vector fields along the base manifold will be,

$$Y_\mu = \frac{\partial}{\partial x^\mu} + A_\mu(x) y \frac{\partial}{\partial y} \tag{3.52}$$

associated with $\partial/\partial x^\mu$ by means of the (1, 1) tensor field,

$$A = yA_\mu(x)\, dx^\mu \otimes \frac{\partial}{\partial y} \tag{3.53}$$

The covariant derivative of $\widetilde{f}\colon \mathbb{R}^4 \to \mathbb{R}^*$ will be expressed in terms of the associated function $y\widetilde{f}$ on $\mathbb{R}^5$ as,

$$\mathcal{L}_{Y_\mu}(y\widetilde{f}) = y\frac{\partial \widetilde{f}}{\partial x^\mu} + A_\mu(x)\widetilde{f} y = y\left(\frac{\partial \widetilde{f}}{\partial x^\mu} + A_\mu(x)\widetilde{f}\right) \tag{3.54}$$

It is now interesting to compute the commutator of two vector fields of the selected type,

$$[Y_\mu, Y_\nu] = \left(\frac{\partial A_\nu}{\partial x^\mu} - \frac{\partial A_\mu}{\partial x^\nu}\right) y\frac{\partial}{\partial y} = F_{\mu\nu} y \frac{\partial}{\partial y} \tag{3.55}$$

i.e., if A_μ is interpreted as a vector potential on space-time, the commutator provides us with the electromagnetic field tensor $F_{\mu\nu}$.

Example 3.7 Another example along the same lines would be provided by $E = \mathbb{R}^{4+k} = \{(x, y) \,|x \in \mathbb{R}^4, y \in \mathbb{R}^k\}$. $E = B \times V$, $B = \mathbb{R}^4$, $V = \mathbb{R}^k$. The partial linear structure is $\Delta = \sum_{\alpha=1}^k y^\alpha \partial/\partial y^\alpha$, the spaces of base functions $\mathcal{F}^{(0)} = \{f(x) \,|x \in \mathbb{R}^4\}$ and sections $\mathcal{F}^{(1)} = \{y^\alpha \widetilde{f}_\alpha(x)\}$.

Our selected fields Y_μ to consider the Lie derivative of functions: $F = y^\alpha \widetilde{f}_\alpha$ satisfying: $\mathcal{L}_\Delta F = 0$, will be,

$$Y_\mu = \frac{\partial}{\partial x^\mu} + A^\beta_{\mu\alpha} y^\alpha \frac{\partial}{\partial y^\beta} \tag{3.56}$$

with,

$$T_A = A^\beta_{\mu\alpha} y^\alpha dx^\mu \otimes \frac{\partial}{\partial y^\beta} \tag{3.57}$$

and,

$$\mathcal{L}_{Y_\mu} F = y^\alpha \frac{\partial f_\alpha}{\partial x^\mu} + A^\beta_{\mu\alpha} f_\beta y^\alpha = y^\alpha \left(\frac{\partial f_\alpha}{\partial x^\mu} + A^\beta_{\mu\alpha} f_\beta \right) \tag{3.58}$$

therefore,

$$D_\mu f_\alpha = \frac{\partial f_\alpha}{\partial x^\mu} + A^\beta_{\mu\alpha} f_\beta \tag{3.59}$$

and, again, the commutator,

$$[Y_\mu, Y_\nu] = (F_{\mu\nu})^\beta_\alpha y^\alpha \frac{\partial}{\partial y^\beta} \tag{3.60}$$

defines the Yang-Mills field strength.

It is now appropriate to give a 'coordinate free' presentation of our covariant derivatives. On E we consider a $(1, 1)$-tensor field T with the properties:

$$T(\Delta) = \Delta, \quad \mathcal{L}_\Delta T = 0, \quad T \circ T = T.$$

We will assume in addition that,

$$T \cdot d\mathcal{F}^{(0)}_\Delta = 0\,. \tag{3.61}$$

Then for any vector field commuting with Δ, say $[Y, \Delta] = 0$, we consider the associated vector field $Y_T = (I - T)(Y) = A(Y)$. The Lie derivative along the special family of vector fields we have considered earlier is given by,

$$\mathcal{L}_{Y_T} F = dF(Y_T) = dF(A(Y)) \tag{3.62}$$

which gives immediately $DF = A(dF)$. By replacing F with $\tilde{F}$, the "vector valued" function, we obtain the usual exterior covariant differential. Thus the role of "vector potential" is played by the $(1, 1)$ tensor field $A = I - T$ as a "vector valued" one-form. Again, the curvature as a "vector valued" 2-form is obtained by considering,

$$[A(X_1), A(X_2)] - A([X_1, X_2]) = F(X_1, X_2) \tag{3.63}$$

Then, to make contact with more traditional presentations of the covariant derivative, we notice that the algebra of vector fields commuting with Δ is a subalgebra of the algebra of projectable vector fields. Therefore, starting with a projectable Y, after projection we get a vector field X, the usual lifting of X to a horizontal vector field on E is provided by $(I - T)(Y) = A(Y)$. Thus the choice of the $(1, 1)$-tensor field T is equivalent to the choice of a 'connection one-form'.

Example 3.8 The tangent bundle TE. Here the partial linear structure is provided by $\Delta = v^a \partial/\partial v^a$. $\mathcal{F}^{(0)}(TE) = \mathcal{F}(E)$ and $\mathcal{F}^{(1)}(TE) = \{v^\alpha f_\alpha\}$.

A choice for the $(1, 1)$ tensor field T could be,

$$T = \frac{\partial}{\partial v^i} \otimes (dv^i - v^k \Gamma^i_{kj} dx^j) \tag{3.64}$$

and,

$$I - T = dx^j \otimes \left(\frac{\partial}{\partial x^j} + v^k \Gamma^i_{kj} \frac{\partial}{\partial v^i} \right) \tag{3.65}$$

It is now possible to consider vector fields on E as vector-valued functions. They may be obtained by projecting on E vector fields on TE which commute with Δ.

On the cotangent bundle T^*E, we have the partial linear structure $\Delta = p_a \frac{\partial}{\partial p_a}$ and we may proceed exactly as for the tangent bundle.

In Appendix G we present the covariant derivative in more traditional terms. We hope that in this way the reader will make contact with the traditional way of presenting the covariant differential calculus.

3.4 The Holonomic Tensorialization Principle*

3.4.1 The Natural Tensorialization of Algebraic Structures

We are in a situation now of establishing on a firm ground some of the intuition developed previously about the relationship between linear algebra and geometry. In Sect. 2.4.3 we discussed the 'easy' tensorialization principle that states that any algebraic tensorial object gives raise to a geometrical tensor possessing the same structure as the original tensor. Even as simple as this observation is it provides a useful hint in the geometrical structures of interest in analyzing the properties of dynamical systems. This point of view will be exploited systematically in what follows. There is however another aspect of this relation that we want to study now. If we are given a geometrical tensor in a given smooth manifold, say a $(1, 1)$ tensor L for instance, then at each point $x \in M$, $L_x \colon T_x M \to T_x M$ defines a linear endomorphism of the linear space $E = T_x M$. The linear map L_x will change from point to point, however under some circumstances, such change is only apparent in the sense that, at least locally, that is in an open neighborhood U of x, all $L_{x'}, x' \in U$, are "equivalent".

The simpler notion of equivalence for the $(1, 1)$ tensor at different points of U, would be that there exists a local frame $\{e_i(x)\}$ defined on U (that is a smooth choice of a base for each tangent space $T_x M$) such that the matrices $(L_i^j(x))$ associated to L_x with respect to the base $\{e_i(x)\}$ are equivalent, or in other worlds, if we select a point $x_0 \in U$, then there exist a change of base $P_i^j(x)$ such that $L(x) = P(x)^{-1} L(x_0) P(x)$. Equivalently, we can find a local frame $\{e_i(x)\}$ such that the matrix representation of L in this frame is constant (and equal to the matrix $L(x_0)$). The specific form of the matrix $L(x_0)$ at a point x_0 would be the algebraic model for the tensor L. We will say in such case that the tensor L is locally constant.

In some cases the family of algebraic models or normal forms for a given algebraic tensor **t** are known and we have nice structure theorems form them (see for instance the structure theorem for a bivector or a linear symplectic form in Chaps. 4 and 5 respectively). Then given a smooth manifold M of the same dimension as the linear space E where the algebraic tensor **t** is defined, a geometrization of the tensor **t** is a locally constant tensor T on M such that, restricted to each tangent space, $T_x M$ is equivalent to **t**. In most cases the tensorialization of a given algebraic tensor amounts to solving a topological problem in the sense that the existence of such tensorialization amounts to determining if there is a topological obstruction to the existence of such a locally constant geometrical tensor.

We may consider for instance the simple tensorialization problem posed by a given non-zero vector $v \in E$. Notice that given a manifold M of dimension dim E, a tensorialization of the $(1, 0)$ tensor v is just a vector field X in M which is non-zero everywhere. Notice that a non-zero vector field is locally constant. Actually a vector field which is non-zero at a given point x_0 can be straightened out in a neighborhood of that point. Even more it is possible to find a local system of coordinates in an open neighborhood of x_0 such that X takes the simple form $\partial/\partial x^1$ in such neighborhood. So every non-zero vector field is locally constant and performing a change of base it can be made into any vector v. Hence the existence of a tensorialization of a vector v in a manifold M amounts to proving the existence of a nowhere zero vector field on M, however it is well-known that there are topological obstructions to that determined by the Euler characteristic $\chi(M)$ of the manifold. For instance in the sphere S^2 there exists no non zero vector fields as $\chi(S^2) = -2$.

Typically, the first step to study the global properties of a geometrical structure on a given manifold M is to determined its existence, that is, to study the existence of a tensorialization on an algebraic model for it. For instance, if we wish to study the global properties of a Riemannian metric on a manifold M, i.e., of a symmetric non-degenerate $(0, 2)$ tensor on M, we will start studying the existence of a tensorialization of an algebraic model for such structure. As it happens there is a unique model for such algebraic structures in any linear space equipped with a linear symmetric non-degenerate bilinear form, i.e., a scalar product, has an orthonormal base. Then a tensorialization of a scalar product always exists provided that M is orientable (and second countable and paracompact as we will assume all the time).

Such a tensorialization problem is not always 'easy' in the sense that there is no guarantee of the existence of such tensorialization when the manifold M is not the linear space E itself; however as we were pointing out, commonly the obstruction to the existence of such tensorialization is topological and can be solved by using "soft" techniques. Because this proposed geometrization seeks to construct a tensorial object along the fibres of the tangent bundle (or of any natural tensorial bundle) over a manifold M, we will call it a natural tensorialization of the given algebraic structure.

The example of the tensorialization of a scalar product brings to the fore a fundamental issue regarding the structure of the geometrized object. By definition a geometrized object is locally constant but we use a general local frame to express such constancy. For instance in the case of a metric, what we do is to construct a locally constant metric $g = g_{ij}e^i(x) \otimes e^j(x)$, thus the non-constant character of g is absorbed in the local frame components $e^i(x)$. We may ask ourselves if it is possible to find a local frame of the natural form $e^i(x) = dx^i$ for some local set of coordinates x^i or, in other worlds, if the "non-holonomic" local frame $e^i(x)$ can be integrated and be described by a function $\varphi\colon U \to \mathbb{R}^n, \varphi = (x^1, \ldots, x^n)$ of maximal rank, i.e., by a local chart. Notice that if this were possible then our metric g will have locally the form

$$g = g_{ij}dx^i \otimes dx^j \,. \tag{3.66}$$

However as we will see in the next section, if this were the case, the curvature of g would be zero and again for some manifolds this would be impossible because there are tight relations between the topology of the manifold and the curvature tensor of any metric on it. For instance if M is an orientable closed[8] connected Riemann surface Σ, then the Gauss-Bonet theorem establishes that $\int_\Sigma \kappa = 2\pi\chi(\Sigma)$ with κ the scalar curvature of the metric. Then if $\chi(\Sigma) \neq 0$, then κ cannot vanish and there are no metrics that can be written locally as in Eq. (3.66).

The existence of such integrated or holonomic form for a geometrized algebraic structure will be the subject of the forthcoming section. The statement of the existence (or not) of such geometrized structure will be called the holonomic tensorialization principle even if we may call it properly the holonomic tensorialization principle.

3.4.2 The Holonomic Tensorialization Principle

Before stating with precision the main idea of this section, a holonomic version of the natural tensorialization principle, we will explore two more examples.

Consider again a linear map $A\colon V \to V$ and a natural tensorialization of it $L\colon TM \to TM$ with M a smooth manifold of dimension $\dim V$. Notice that the $(1, 1)$ tensor L has the constant local form $L = A_i^j\, \xi_j \otimes \theta^i$ where $\{\xi_i\}$ is a local frame for M and $\{\theta^j\}$ its dual frame, i.e., $\langle \theta^j, \xi_i\rangle = \delta_i^j$. This situation can be achieved for

[8] That is compact without boundary.

instance if M is parallelizable. In such case there exists a global frame ξ_i and L is defined as T_A with respect to the frame ξ_i and its coframe θ^j.

Now if A is generic and has real spectrum, we can write $A = \sum_a \lambda_a P_a$, with $\sigma(A) = \{\lambda_a\} \subset \mathbb{R}$ and P_a the projectors on the corrresponding spaces of eigenvectors V_a (of rank one if A is generic), i.e., $V = \bigoplus V_a$, $AP_a = \lambda_a P_a$. Then, the geometrized object will decompose as $L = \sum_a \lambda_a T_{P_a}$, and $TM = \bigoplus_a D_a$ where D_a is the subbundle of TM defined by the range of T_{P_a}. In general the subbundles D_a will not define an integrable distribution and it will not be possible to find a holonomic expression for L, i.e., there could not exist a local system of coordinates μ^a such that:

$$L = \sum_a \lambda_a \frac{\partial}{\partial \mu^a} \otimes d\mu^a.$$

If the linear map A is a projector, $\sigma(A) = \{0, 1\}$, then $L^2 = L$ and $L(TM) = D$ is just a distribution. In such case D is integrable iff the Nijenhuis torsion tensor $N_L = [L, L]$ vanishes and L can be written as its local model in a local coordinates.

Thus the existence of an holonomic natural tensorialization for a given projector P of rank r is equivalent to the existence of an integrable distribution D of rank r (the dimension of the linear subspace $D_x \subset T_x M$). It actually happens that if there exists a distribution (not necessarily integrable) D of rank r in the manifold M, then if the manifold is open we can always deform D to an integrable distribution that will provide the holonomic natural tensorialization of P. that is essentially the content of Haefliger's theorem [Ha70]:

Theorem 3.18 *Let M be an open manifold, then any distribution D can be deformed homotopically to an integrable distribution (i.e., into a foliation) of the same dimension.*

This theorem can be completed by stating the topological conditions that guarantee the existence of a distribution of rank r and that consists of the vanishing of the cohomology groups $H^{k+1}(M), \ldots, H^n(M)$, $n = \dim M$ (notice that $H^n(M) = 0$ imply that M must be open). Thus we may state now:

Corollary 3.19 *A rank k projector P on a n-dimensional linear space has a holonomic tensorialization on an open manifold M iff $H^{k+1}(M) = \cdots = H^n(M) = 0$.*

3.4.2.1 Gromov's *h*-Principle

Let us consider now as another example of the subtleties involved in finding a holonomic tensorialization of an algebraic structure, the problem of geometrizing constant Poisson structures (we are anticipating here material that will be discussed at length in Sect. 4.3).

Let Λ be a skew symmetric bivector on a linear space V. Choosing a linear basis $\{e_i\}$, Λ can be written as: $A = A^{ij} e_i \wedge e_j$. We will say that Λ is regular if the induced linear map (denoted with the same symbol $\Lambda \colon V^* \to V$ is invertible. In such case, there will exists a skew symmetric bilinear form ω such that the natural map (again, denoted with the same symbol) $\omega \colon V \to V^*$ is such that $\Lambda \circ \omega = I$. Then $\omega = \omega_{ij} e^i \wedge e^j$, with $\{e^i\}$ the dual basis to the previous basis $\{e_i\}$. In other words, $\Lambda^{ij}\omega_{jk} = \delta^i_k$.

The 'easy' tensorialization of Λ produces a geometric bisector, a section of $TV \wedge TV$, that will be denoted with the same symbol, and that will read as:

$$\Lambda = \Lambda^{ij} \frac{\partial}{\partial x^i} \wedge \frac{\partial}{\partial x^j} \,.$$

The range of the natural map $\Lambda \colon T^*V \to TV$ defines a distribution $S = \Lambda(TV)$, that by definition is integrable (Λ is constant). In general, we may try to find a natural tensorialization of Λ in a manifold M. The existence of a natural tensorialization has the same obstructions as geometrizing a projector whose rank is the same as the rank of the range of Λ, because if we find a distribution whose rank is the rank of Λ all we have to do is to construct an arbitrary non-degenerate bivector there and then bring it to its normal form (which is unique as in the case of a positive definite metric). Then we may conclude stating that provided that M is an open manifold and $H^{r+1}(M) = \cdots = H^n(M) = 0$ with $r = \mathrm{rank}(\Lambda)$, then there exists a skew symmetric bivectdor Λ on M such that locally it can be written as:

$$\Lambda = \Lambda^{ij} \xi_i \wedge \xi_j \,.$$

where ξ_i is a local frame of TM.

We must notice that the distribution $S = \Lambda(TM)$ defined by the an arbitrary biector Λ doesn't have to be integrable. The integrability conditions for S are, applying Frobenius theorem, that $[X, Y] \in S$ for any $X, Y \in S$, but it is easily checked that vector fields in S are spanned by vector fields of the form $X = \Lambda(df)$, hence if we compute $[\Lambda(dx^i), \Lambda(dx^j)]$ we get (taking into account that $\Lambda(dx^I) = \Lambda^{ij}(x)\partial/\partial x^j$) that $\mathrm{Ran}(\Lambda) = S$ is integrable iff

$$\Lambda^{il} \frac{\partial \Lambda^{ik}}{\partial x^l} + \mathrm{cyclic}(i, j, k) = 0 \,. \qquad (3.67)$$

The previous set of equations must be satisfied to find a holonomic tensorialization of Λ. As it will be discussed later on, a bisector Λ on M satisfying Eq. (3.67) is called a Poisson tensor, and it provides the geometrical definition of a Poisson structure on M.

The existence of such holonomic tensorialization has been elucidated recently by Crainic and Fernandes [LF12]. Let us analyze first the regular case, i.e., Λ non-degenerate.

Then a non-degenerate skew symmetric tensor Λ on a linear space can be thought of as a section of the bundle $TE \wedge TE$ over E, and thus the holonomic condition Eq. (3.67) can be thought as a condition on the space of fields $(x^i, \Lambda^{ij}, \Sigma^{ij}_k)$, where for a specific section $\Lambda(x)$, then $\Sigma^[_k ij] = \partial \Lambda^{ij}/\partial x^k$. The bundle before is called the first jet bundle of E and the algebraic scripture of the holonomic condition above is:

$$\sum_{(i,j,k)\text{ cyclic}} \Lambda^{il}\Sigma_l^{jk} = 0\,.$$

Thus, the holonomic condition can be written as a submanifold $\mathcal{R}$ of J^1E, i.e.,

$$\mathcal{R} = \{(x,\Lambda,\Sigma) \mid \Lambda^{ij} = -\Lambda^{ji}, \det\Lambda^{ij} \neq 0, \sum_{(i,j,k)\text{cyclic}} \Lambda^{il}\Sigma_l^{jk} = 0\} \subset J^1E\,.$$

In the case of symplectic structures, we have a positive solution to the holonomic tensorialization problem due to M. Gromov:

Theorem 3.20 (Gromov's theorem) [Gr86a] *Let M be an even-dimensional open manifold. Then for each $[\alpha] \in H^2(M)$ there exists a symplectic form ω cohomologous to α. Moreover if ω_0, ω_1 are two cohomologous symplectic forms, then there exists a symplectic homotopy ω_t, $t \in [0,1]$ joining ω_0 and ω_1 and $\omega_t = \omega_0 + d\psi_t$.*

Therefore in the language we were developing before, we can state:

Corollary 3.21 *A linear symplectic form on a vector space of dimension $2n$ has a holonomic tensorialization on any open manifold M of dimension $2n$.*

Proof Let M be an open manifold of dimension $2n$. Then a simple transversality argument shows that there is no obstruction to the existence of a non-degenerate 2-form on M, i.e., a section of $\Lambda^2(T^*M)$ of maximal rank. Gromov's theorem states that there exists, not only the natural tensorialization of a linear symplectic form, but a closed one. But then because of Darboux's theorem, such form is holonomic because there are local charts such that the form is constant and equal to its algebraic model (see Sect. 5.3.1 for a proof of Darboux's theorem). □

The proof of Gromov's theorem is based on Gromov's h-principle [Gr86a].

Given a bundle $E \to M$, its first jet bundle $J^1E \to M$ and a first-order partial differential relation $\mathcal{R} \subset J^1E$. A section $\sigma: M \to J^1E$ of the first jet bundle is a solution of the partial differential relation $\mathcal{R}$ if $\sigma(x) \in \mathcal{R}$ for all $x \in M$. We will say that σ is a holonomic solution if there exists $\phi: M \to E$ such that $\sigma = j^1\phi$, where $j^1\phi(x) = (x^i, \phi(x), \partial\phi/\partial x^i)$.

As it was shown before, solutions of partial differential relations often exist (the obstructions for their existence is just topological), however, as it was also shown in the previous examples, holonomic solutions are hard to find (and often do not exist). Thus $\mathcal{R}$ is said to satisfy the h-principle if for every solution σ there exists σ_t, a homotopy of solutions such that σ_1 is holonomic, i.e., $\sigma_0 = \sigma$ and $\sigma_1 = j^1\phi$.

There is a parametric version of the h-principle. Given σ_1, σ_2 solutions of $\mathcal{R}$, the parametric h-principle is satisfied if there exists σ_t joining σ_1 and σ_2. Moreover if $\sigma_1 = j^1\phi_1, \sigma_2 = j^1\phi_2$, then $\sigma_t = j^1\phi_t$.

Then the main question regarding the existence of holonomic tensorialization for a given algebraic structure is when the partial differential relation $\mathcal{R}$ defined by it and its derived 'integrability conditions' satisfies the h-principle. There are various techniques to prove (if possible) the h-principle, e.g., removal of singularities), flexible continuous sheaves and convex integration. A discussion of the ideas involved in these methods fails completely out of the scope of this book, so we will refer the reader to the existing literature [Gr86b].

3.4.3 Geometric Structures Associated to Algebras

Let V be a vector space and V^* its dual. To any element $v \in V$, there is a corresponding element in the bi-dual $\hat{v} \in (V^*)^*$ given by:

$$\hat{v}(\alpha) = \alpha(v)\ ,\ \ \forall \alpha \in V^* \tag{3.68}$$

Thus any multilinear function on V^*, $f\colon V^* \times \cdots \times V^* \to \mathbb{R}$ defines, by restricting it to the diagonal, a polynomial function $\tilde{f} \in \mathcal{F}(V^*)$, $\tilde{f}(\alpha) = f(\alpha, \ldots, \alpha)$, which can be obtained from the "monomials of degree one", $\hat{v} \in (V^*)^*$, on which one has defined the (commutative) product:

$$(\hat{v}_1 \cdot \hat{v}_2)(\alpha) := \hat{v}_1(\alpha)\, \hat{v}_2(\alpha) \tag{3.69}$$

Suppose now that on V there is defined an additional bilinear operation:

$$B : V \times V \to V \tag{3.70}$$

which induces a (in general noncommutative) product $\times_B$ on $V \subset \mathcal{F}(V^*)$ by:

$$\hat{v}_1 \times_B \hat{v}_2 = \widehat{B(v_1, v_2)} \tag{3.71}$$

Then we can define a 2-tensor τ_B in $\mathcal{F}(V^*)$, at the point α, by the relation:

$$\tau_B(d\hat{v}_1, d\hat{v}_2)(\alpha) := \alpha(B(v_1, v_2)) \tag{3.72}$$

which satisfies Leibniz' rule:

$$\tau_B(d\hat{v}, d(\hat{v}_1 \cdot \hat{v}_2)) = \tau_B(d\hat{v}, \hat{v}_1 \cdot d\hat{v}_2 + d\hat{v}_1 \cdot \hat{v}_2) = \hat{v}_1 \cdot \tau_B(d\hat{v}, \hat{v}_2) + \tau_B(d\hat{v}, \hat{v}_1) \cdot \hat{v}_2 \tag{3.73}$$

Thus, $\tau_B(d\hat{v}, \cdot)$ defines a derivation on $V \subset \mathcal{F}(V^*)$ with respect to the commutative product (3.69).

In particular, suppose that B is a skew-symmetric bilinear operation which satisfies the Jacobi identity, so that $g = (V, B)$ is a Lie algebra. The corresponding 2-tensor $\Lambda := \tau_B$:

$$\Lambda(d\hat{v}_1, d\hat{v}_2) = \widehat{B(v_1, v_2)} \tag{3.74}$$

is a Poisson tensor in $\mathcal{F}(V^*)$ and $\Lambda(d\hat{v}, \cdot)$ is a derivation with respect to the commutative product (3.69). Moreover, $\Lambda(d\hat{v}, \cdot)$ is a derivation also with respect to the product (3.71). Indeed, by using the fact that B is antisymmetric and satisfies the Jacobi identity, one has

$$\begin{aligned} \Lambda(d\hat{v}, d(\hat{v}_1 \cdot \hat{v}_2)) &= \widehat{B(v, B(v_1, v_2))} \\ &= \widehat{B(v_1, B(v, v_2))} + \widehat{B(B(v, v_1), v_2)} \\ &= \hat{v}_1 \cdot \Lambda(d\hat{v}, d\hat{v}_2) + \Lambda(d\hat{v}, d\hat{v}_1) \cdot \hat{v}_2 \end{aligned} \tag{3.75}$$

Similarly, if on V one has a Jordan product B', the corresponding 2-tensor $G := \tau_{B'}$ is a metric tensor and $G(d\hat{v}, \cdot)$ is a derivation with respect to the commutative product (3.69), but not with respect to the product (3.71).

If now $V = \mathcal{A}$ is a $\mathbb{C}^*$-algebra, where we have defined both a Lie product and a Jordan product as:

$$B(a_1, a_2) := [a_1, a_2] = \frac{1}{2i}(a_1 a_2 - a_2 a_1) \ , \ \ \forall a_1, a_2 \in \mathcal{A} \tag{3.76}$$

and a Jordan product

$$B'(a_1, a_2) := a_1 \circ a_2 = \frac{1}{2}(a_1 a_2 + a_2 a_1) \ , \ \ \forall a_1, a_2 \in \mathcal{A} \tag{3.77}$$

in $\mathcal{F}(\mathcal{A}^*)$ we have defined both a Poisson tensor Λ and a metric tensor G such that $\Lambda(d\hat{a}, \cdot)$ and $G(d\hat{a}, \cdot)$ are both derivations with respect to the pointwise commutative product, with the former being also a derivation with respect to the Lie product. It is also not difficult to check that the subalgebra $\mathcal{B} \subset \mathcal{A}$ composed of all real elements, when embedded in $\mathcal{F}(\mathcal{A}^*)$, comes equipped with an antisymmetric and a symmetric product, denoted by $[\cdot, \cdot]$ and $\circ$ respectively, such that:

1. The Leibniz rule is satisfied: $[a, b \circ c] = [a, b] \circ c + b \circ [a, c]$,
2. The Jacobi identity is satisfied: $[a, [b, c]] = [[a, b], c] + [b, [a, c]]$, and
3. The identity: $(a \circ b) \circ c - a \circ (b \circ c) = [[a, c], b]$ holds.

This means that $(\mathcal{B}, [\cdot, \cdot], \circ)$ is a Lie-Jordan algebra [Em84, FF13]. Finally, we notice that the Hamiltonian vector fields:

$$X_{\hat{a}} := \Lambda(\cdot, d\hat{a}) = -[\hat{a}, \cdot] \tag{3.78}$$

are derivations with respect to the Jordan product, since, by using the properties above:

$$\begin{aligned} X_{\hat{a}}(d(\hat{a}_1 \circ \hat{a}_2)) &= -[\hat{a}, \hat{a}_1 \circ \hat{a}_2] = -[\hat{a}, \hat{a}_1] \circ \hat{a}_2 + -\hat{a}_1 \circ [\hat{a}, \hat{a}_2] \\ &= X_{\hat{a}}(d\hat{a}_1) \circ \hat{a}_2 + \hat{a}_1 \circ X_{\hat{a}}(d\hat{a}_2) \end{aligned} \tag{3.79}$$

3.5 Vector Fields and Linear Structures

3.5.1 Linearity and Evolution

In the previous chapter it was shown that the dynamics of a given system can be described in various equivalent forms: for instance as a local flow φ_t on a carrier space (Sects. 2.2.2 and 2.3.2), i.e., a vector field Γ, or as a one-parameter family of automorphisms Φ_t on the algebra of observables (Sects. 2.3.3 and 2.3.6). This makes it clear that we must be careful about any statement of the type 'linear evolution' of a given system because evolution from the point of view of observables is always linear with respect to the linear structure on the algebra of observables. Something completely different happens when we ask the evolution of the system to preserve some linear structure on the carrier space E. This will be the case if the algebraic operations done on states will be carried out onto themselves by the evolution defined by the diffeomorphisms φ_t,

$$\varphi_t(u + v) = \varphi_t(u) + \varphi_t(v), \quad \varphi_t(\lambda u) = \lambda \varphi_t(u) \tag{3.80}$$

for all $u, v \in E$, $\lambda \in \mathbb{R}$. We will say then that the the dynamics preserves the linear structure, that the dynamics and the linear structure given on E are compatible, or that the linear structure on E is admissible for the given dynamics.

As we learned in Sect. 3.3.1 a linear structure on E is characterized by an Euler or Liouville vector field Δ, thus it can be easily checked that:

Proposition 3.22 *A linear structure characterized by the Euler vector field Δ is admissible for the dynamics Γ if and only if $[\Delta, \Gamma] = 0$.*

Proof Let φ_t the local flow defined by the vector field Γ. Then if the linear structure Δ is admissible for Γ, then because of Eq. (3.80) we have $\varphi_t(e^s u) = e^s(\varphi_t(u))$ for all $s \in \mathbb{R}$, but multiplication by e^s is the flow of Δ, hence both flows commute and this implies that $[\Delta, \Gamma] = 0$.

Conversely, if $[\Delta, \Gamma] = 0$, then both flows, the local flow φ_t of Γ and the flow $\varphi_s^\Delta = e^s$ commute. Let us recall that a linear function f on E satisfies that $\Delta(f) = f$. Moreover if f is a linear function we have $f(u + v) = f(u) + f(v)$ (that is in fact the definition of the first symbol '+' in the previous equation). Then it is trivial to check that if f is linear $f \circ \varphi_t$ is linear too. In fact we get:

$$\Delta(f\circ\varphi_t)(u) = \frac{d}{ds}(f\circ\varphi_t)(e^s(u)) \mid_{s=0} = \frac{d}{ds} f(e^s\varphi_t(u)) \mid_{s=0} = \frac{d}{ds} e^s(f\circ\varphi_t)(u) \mid_{s=0} = (f\circ\varphi_t)(u),$$

but then $f(\varphi_t(u+v)) = (f\circ\varphi_t)(u+v) = (f\circ\varphi_t)(u) + (f\circ\varphi_t)(v) = f(\varphi_t(u)) + f(\varphi_t(v))$ for all linear f and the conclusion follows.

The reasons why one would like the evolution to preserve a given tensor, a linear structure for instance, must come from the experiments performed on the system. We may be able to compose states of the system additively, then we will be extremely interested to see if the dynamics of the system is compatible with such composition. This will actually happen with the states of a harmonic oscillator described as points on phase space or with the states of a medium described as the position of the elements of it. A different example is provided by pure states of a quantum system described as vectors on a Hilbert space. In all these cases we are able to compose the states of the system in a linear way.

3.5.2 Linearizable Vector Fields

In the previous chapter we saw that if the carrier state space of the system whose dynamics we want to describe has a differentiable structure then, vector fields constitute an appropriate way of describing such dynamics.

Thus we will consider a system whose dynamics is determined by a vector field Γ on a smooth finite-dimensional manifold M. The vector field Γ defines a differential equation on M, that is for each local chart x^i on M we will have:

$$\frac{dx^i}{dt} = \Gamma^i(x), \qquad i = 1, \ldots, n, \tag{3.81}$$

where $\Gamma^i(x)$ are the local components of the vector field Γ in the given chart.

Because of the theorem of existence and uniqueness of solutions for ordinary differential equations, the previous equation provides, locally, a description of the evolution of the states of the system associated with the vector field X.

If M carries a Liouville vector field Δ, then there is a linear structure on M whose dilation vector field is Δ. We will denote by E the linear space (M, Δ) to emphasize the choice of a linear structure on M defined by Δ. It is clear that if Γ is a linear vector field, then

$$[\Gamma, \Delta] = 0\,. \tag{3.82}$$

Conversely, if Γ is a vector field on E commuting with Δ, then it is linear, because using the expression of Δ in linear coordinates we have that if $\Gamma = \Gamma^j\,\partial/\partial x^j$, Eq. (3.82) becomes

$$x^i\,\frac{\partial\Gamma^j}{\partial x^i} = \Gamma^j\,,$$

i.e., the Γ^j are homogeneous smooth functions of degree 1, and they are assumed not to have singularities, hence they are linear.

A vector field will be called globally linearizable if there exists a global diffeomorphism $\phi\colon E \to E$ such that $\phi_*\Gamma$ is a linear vector field, i.e., $[\Delta, \phi_*\Gamma] = 0$. Notice that in such case $\phi(0) = 0$.

A globally linearizable vector field can be easily integrated by transforming it into the linear form $\phi_*\Gamma$, obtaining the linear flow for it $\psi_t = \exp tA$, and changing back to old coordinates, $\phi^{-1} \circ \psi_t \circ \phi$, i.e.,

$$\psi_t^X = \phi^{-1} \circ \exp tA \circ \phi.$$

In other words we can think that we were looking at Γ in a wrong set of coordinates and that after rectifiyng them by using the diffeomorphisms ϕ, the vector field Γ looks linear.

It is important to notice that $[\Delta, \Gamma] \neq 0$, but $[\phi_*^{-1}\Delta, \Gamma] = 0$. We can think of the vector field $\Delta_\phi = \phi_*^{-1}\Delta$ as defining an alternative linear structure on E such that Γ is a linear vector field with respect to it.

So far the previous observations are rather trivial as the reader may think immediately that if we have a linear vector field on a linear space and we perform a nonlinear change of coordinates, the vector field in the new coordinates will not look linear at all, but that is only because are looking at it in wrong coordinates. But what would happen if the vector field in the new coordinates will look linear! This situation can actually occur.

For that to happen it is only necessary to find a diffeormorphism ϕ such that $\phi_*\Gamma = \Gamma$ and $\phi_*\Delta \neq \Delta$. In such case, what we will have in our hands would be a system that possesses an alternative linear structure description, one given by Δ and the other by $\Delta_\phi = \phi_*\Delta$. Consider the following example in two dimensions:

$$\Gamma = y\frac{\partial}{\partial x} - x\frac{\partial}{\partial y},$$

and consider the change of coordinates ϕ:

$$X = \phi^1(x, y) = x(1 + x^2 + y^2), \quad Y = \phi^2(x, y) = y(1 + x^2 + y^2),$$

which is obviously nonlinear. If we compute the vector field in the new coordinates we find

$$\phi_*\Gamma = Y\frac{\partial}{\partial X} - X\frac{\partial}{\partial Y}.$$

Thus it is linear with respect to the linear structure $\phi_*\Delta$.

As another simple example, consider in $\mathbb{R}^2$, with coordinates (x, y), the dynamical system described by

$$\begin{cases} \dot{x} = \dfrac{y}{\cosh x} \\ \dot{y} = -\sinh x \end{cases} . \tag{3.83}$$

By setting: $\xi = \sinh x$, $\eta = y$, we find:

$$\begin{cases} \dot{\xi} = \eta, \\ \dot{\eta} = -\xi \end{cases} , \tag{3.84}$$

i.e., the harmonic oscillator. Therefore, our original system, Eq. (3.83), is diffeomorphic to a linear one by using the diffeomorphism

$$\phi : \mathbb{R}^2 \longrightarrow \mathbb{R}^2, \quad \phi(x, y) = (\sinh x, y) = (\xi, \eta) \tag{3.85}$$

Description of the linear structure $\phi_*\Delta$: In this particular example it is relatively easy to describe the new linear structure because of the symmetry of the transformation. First notice that the map $\phi\colon \mathbb{R}^2 \to \mathbb{R}^2$ maps circles of radius r into circles of radius $R = r(1 + r^2)$ and sends the rays lines l passing trough the origin into themselves, $\phi(l) = l$. This would imply that if we denote by $+_\phi$ the addition and by $\cdot_\phi$ the multiplication by scalars defined respectively by Δ_ϕ we will have:

$$u +_\phi v = u + v, \quad \lambda \cdot_\phi u = \frac{\lambda(1 + \lambda^2)}{2} u,$$

for any $u, v \in \mathbb{R}^2$ and $\lambda \in \mathbb{R}$.

This observation raises a number of interesting questions that have an importance beyond the purely mathematical level. For instance, if a vector field is linear with respect to two alternative linear structures, then it will have different eigenvalues in both of them because the matrices defined by it using the two linear structures will not be equivalent. Thus, if the eigenvalues of a given vector field play any relevant role in the (physical, engineering, etc.) interpretation of such vector field, which one should we choose? In the particular instance of quantum-like systems that is a crucial question that demands a clear answer.

Given $x_0 \in E$, we will say that Γ is linearizable (or locally linearizable) around x_0 if there exists a local chart (U, ψ), $x_0 \in U$, such that in the local coordinates y^i, the vector field Γ is linear, i.e.,

$$\Gamma_U = A^j{}_i \, y^i \, \frac{\partial}{\partial y^j} .$$

This notion is equivalent to claiming the existence of a local diffeomorphism $\phi_U\colon U \to \phi(U) = V$, where V is an open set of E, such that $\phi_*\Gamma$ commutes with Δ.

We must point out that this notion is quite different from the straightening out property. In fact, a vector field could be (locally) straightened out if there exists a (local) diffeomorphism ϕ such that $\phi_* \Gamma = \Gamma_0$, where Γ_0 is a constant vector field. In other words, the vector field could be straightened out if there are local coordinates y^i (not necessarily linear ones) such that in these coordinates the vector field X takes the form

$$\Gamma = \frac{\partial}{\partial y^1} .$$

It is well known (we will come back to this question later on) that any vector field can be straightened out around a noncritical point x_0. Thus, a vector field which is linearizable around x_0 will always have a critical point at x_0. Thus, the first normal form theorem will concern the linearizability of a vector field around a critical point. This problem has been discussed in a series of papers by a number of people since Poincaré (Poincaré, Hermann, Guillemin, Sternberg, etc.)

3.5.3 Alternative Linear Structures: Some Examples

It is known that all finite-dimensional linear spaces of the same dimension are linearly isomorphic.[9] However, alternative (i.e., not linearly related) linear structures can be constructed easily on a given set. For instance consider a linear space E with addition $+$ and multiplication by scalars $\cdot$, and a nonlinear diffeomorphism $\phi \colon E \to E$. Now we can define a new addition $+_{(\phi)}$ and a new multiplication by scalar $\cdot_{(\phi)}$ by setting

$$u +_{(\phi)} v = \phi(\phi^{-1}(u) + \phi^{-1}(v)) \tag{3.86}$$

and

$$\lambda \cdot_{(\phi)} u = \phi\left(\lambda \phi^{-1}(u)\right) . \tag{3.87}$$

These operations have all the usual properties of addition and multiplication by a scalar. In particular,

$$\left(\lambda\lambda'\right) \cdot_{(\phi)} u = \lambda \cdot_{(\phi)} \left(\lambda' \cdot_{(\phi)} u\right) \tag{3.88}$$

and

$$\left(u +_{(\phi)} v\right) +_{(\phi)} w = u +_{(\phi)} \left(v +_{(\phi)} w\right) . \tag{3.89}$$

[9] The same is true for infinite-dimensional separable Hilbert spaces (even more, the isomorphism can be chosen to be an isometry).

Indeed,

$$\lambda \cdot_{(\phi)} \left(\lambda' \cdot_{(\phi)} u\right) = \phi\left(\lambda\phi^{-1}\left(\lambda' \cdot_{(\phi)} u\right)\right) = \phi\left(\lambda\lambda'\phi^{-1}(u)\right) = \left(\lambda\lambda'\right) \cdot_{(\phi)} u, \quad (3.90)$$

which proves (3.88), and similarly for (3.89).

Obviously, the two linear spaces $(E, +, \cdot)$ and $(E, +_{(\phi)}, \cdot_{(\phi)})$ are finite-dimensional vector spaces of the same dimension and hence are isomorphic. However, the change of coordinates defined by ϕ that we are using to 'deform' the linear structure is a nonlinear diffeomorphism. In other words, we are using two different (diffeomorphic but not linearly related) global charts to describe the same manifold space E.

Using then the two different linear structures defined in this way, one obtains also two different (and, again, not linearly related) realizations of the translation group $\mathbb{R}^n$ ($n = \dim E$).

As a simple (but significant) example of this idea consider the linear space $\mathbb{R}^2$. This can also be viewed as a Hilbert space of complex dimension 1 that can be identified with $\mathbb{C}$. We denote its coordinates as (q, p) and choose the nonlinear transformation

$$\begin{aligned} q &= Q(1 + \lambda R^2), \\ p &= P(1 + \lambda R^2), \end{aligned} \quad (3.91)$$

with $R^2 = P^2 + Q^2$, which can be inverted as

$$\begin{aligned} Q &= qK(r) \\ P &= pK(r), \end{aligned} \quad (3.92)$$

where $r^2 = p^2 + q^2$, and the positive function $K(r)$ is given by the relation $R = rK(r)$, satisfying the equation

$$\lambda r^2 K^3 + K - 1 = 0 \quad (3.93)$$

(hence, actually, $K = K\left(r^2\right)$ and when $\lambda = 0$ then $K \equiv 1$). Using this transformation we can construct an alternative linear structure on $\mathbb{C}$ by using formulae (3.86) and (3.87). Let us denote by $+_K$ and $\cdot_K$ the new addition and multiplication by scalars. Then, with

$$\begin{aligned} \phi &: (Q, P) \mapsto (q, p) = \left(Q\left(1 + \lambda R^2\right), P\left(1 + \lambda R^2\right)\right), \\ \phi^{-1} &: (q, p) \mapsto (Q, P) = \left(qK(r), pK(r)\right), \end{aligned}$$

one finds

$$(q,p) +_{(K)} (q',p') = \phi\left(\phi^{-1}(q,p) + \phi^{-1}(q',p')\right) = \\ = \phi\left((Q+Q', P+P')\right) = \phi\left(qK + q'K', pK + p'K'\right), \tag{3.94}$$

where $K = K(r)$, $K' = K(r')$, i.e.

$$(q,p) +_{(K)} (q',p') = S(r,r')\left((qK + q'K'), (pK + p'K')\right), \tag{3.95}$$

where

$$S(r,r') = 1 + \lambda\left((qK + q'K')^2 + (pK + p'K')^2\right). \tag{3.96}$$

Quite similarly,

$$a \cdot_{(K)} (q,p) = \phi\left(a\phi^{-1}(q,p)\right) = \phi\left((aqK(r), apK(r))\right) \\ = S'(r)\left(aK(r)q, aK(r)p\right) \tag{3.97}$$

where

$$S'(r) = 1 + \lambda a^2 r^2 K^2(r). \tag{3.98}$$

The two different realizations of the translation group in $\mathbb{R}^2$ are associated with the vector fields $(\partial/\partial q, \partial/\partial p)$ and $(\partial/\partial Q, \partial/\partial P)$ respectively. The two are connected by

$$\begin{pmatrix} \frac{\partial}{\partial Q} \\ \frac{\partial}{\partial P} \end{pmatrix} = A \begin{pmatrix} \frac{\partial}{\partial q} \\ \frac{\partial}{\partial p} \end{pmatrix}, \tag{3.99}$$

where A is the Jacobian matrix:

$$A = \frac{\partial(q,p)}{\partial(Q,P)} = \begin{pmatrix} 1+\lambda(3Q^2+P^2) & 2\lambda PQ \\ 2\lambda PQ & 1+\lambda(Q^2+3P^2) \end{pmatrix} \\ = \begin{pmatrix} 1+\lambda K(r)^2(3q^2+p^2) & 2\lambda K(r)^2 pq \\ 2\lambda K(r)^2 pq & 1+\lambda K(r)^2(q^2+3p^2) \end{pmatrix}.$$

In the sequel we will write simply A as

$$A = \begin{pmatrix} a & b \\ d & c \end{pmatrix}, \tag{3.100}$$

with an obvious identification of the entries. Then,

$$A^{-1} = \frac{\partial (Q, P)}{\partial (q, p)} = D^{-1} \begin{pmatrix} c & -b \\ -d & a \end{pmatrix}, \quad D = ac - bd. \tag{3.101}$$

Thus the $2D$ translation group $\mathbb{R}^2$ is realized in two different ways.

Going back to the general case, we have just proved that to every linear structure there is associated in a canonical way a dilation (or Liouville) field Δ which is the infinitesimal generator of dilations. Therefore, in the framework of the new linear structure, it makes sense to consider the mapping

$$\Psi : \mathbb{R} \times E \to E \tag{3.102}$$

via,

$$\Psi(t, u) = e^t \cdot_{(\phi)} u = u(t), \tag{3.103}$$

where again, we are considering a transformation $\phi \colon E \to E$. The transformed flow takes the explicit form

$$u(t) = \phi\left(e^t \phi^{-1}(u)\right). \tag{3.104}$$

Property 3.88 ensures that

$$\Psi\left(t, u\left(t'\right)\right) = \Psi\left(t + t', u\right), \tag{3.105}$$

i.e., that (2.80) is indeed a one-parameter group of transformations of E. Then, the infinitesimal generator of the group is defined as

$$\Delta(u) = \left[\frac{d}{dt} u(t)\right]_{t=0} = \left[\frac{d}{dt}\phi\left(e^t \phi^{-1}(u)\right)\right]_{t=0}, \tag{3.106}$$

or, explicitly, in components,

$$\Delta = \Delta^i \frac{\partial}{\partial u^i}, \tag{3.107}$$

$$\Delta^i = \left[\frac{\partial \phi^i(w)}{\partial w^j} w^j\right]_{w=\phi^{-1}(u)}. \tag{3.108}$$

In other words, if we denote by $\Delta_0 = w^i \partial / \partial w^i$ the Liouville field associated with the linear structure $(+, \cdot)$ on E,

$$\Delta = \phi_* \Delta_0, \tag{3.109}$$

where ϕ_* denotes, as usual, the push-forward.
It is clear that, if ϕ is a linear (and invertible) map, $\phi^i(w) = a^i{}_j\, w^j$, then (3.108) yields: $\Delta^i = u^i$, i.e.,

$$\phi_* \Delta_0 = \Delta_0. \tag{3.110}$$

Conversely it is simple to see that if a map ϕ satisfies (3.110) then, $\partial\phi^i/\partial w^j = a^i{}_j$ and ϕ is linear with respect to the linear structure defined by Δ_0.

The above scheme can be generalized to the case of a diffeomorphism $\phi : E \to M$ between a vector space E and a manifold M possessing 'a priori' no linear structures whatsoever. This will require, of course, that M be such that it can be equipped with a one-chart atlas. Then it is immediate to see that Eqns. (3.86) and (3.87) (with $u, v \in M$, now) apply to this slightly more general case as well.

As a simple example of this sort, let us consider: $E = \mathbb{R}$, $M = (-1, 1)$ and

$$\phi : E \to M;\ \ x \to X = \tanh x. \tag{3.111}$$

Then,

$$\lambda \cdot_{(\phi)} X = \tanh\left(\lambda \tanh^{-1}(X)\right) \tag{3.112}$$

and

$$\begin{aligned} \lambda \cdot_{(\phi)} \left(\lambda' \cdot_{(\phi)} X\right) &= \lambda \cdot_{(\phi)} \tanh\left(\lambda' \tanh^{-1}(X)\right) = \\ &= \tanh\left(\lambda\lambda' \tanh^{-1}(X)\right) = \left(\lambda\lambda'\right) \cdot_{(\phi)} X, \end{aligned} \tag{3.113}$$

while

$$X +_{(\phi)} Y = \tanh\left(\tanh^{-1}(X) + \tanh^{-1}(Y)\right) = \frac{X+Y}{1+XY}, \tag{3.114}$$

which is nothing but the elementary one-dimensional relativistic law (in appropriate units) for the addition of velocities. It is also simple to prove that

$$\begin{aligned} \left(X +_{(\phi)} Y\right) +_{(\phi)} Z &= = \tanh\left(\tanh^{-1}\left(X +_{(\phi)} Y\right) + \tanh^{-1}(Z)\right) = \\ &= \tanh\left(\tanh^{-1} X + \tanh^{-1}(Y) + \tanh^{-1}(Z)\right) \end{aligned} \tag{3.115}$$

i.e., that

$$\left(X +_{(\phi)} Y\right) +_{(\phi)} Z = X +_{(\phi)} \left(Y +_{(\phi)} Z\right). \tag{3.116}$$

Explicitly,

$$X +_{(\phi)} Y +_{(\phi)} Z = \frac{X + Y + Z + XYZ}{1 + XY + XZ + YZ}. \tag{3.117}$$

The mapping (3.104) is now

$$X(t) = \tanh\left(e^t \tanh^{-1}(X)\right) \tag{3.118}$$

and we obtain, for the Liouville field on $(-1, 1)$,

$$\Delta(X) = \left(1 - X^2\right)\tanh^{-1}(X)\frac{\partial}{\partial X}, \tag{3.119}$$

and $\Delta(X) = 0$ for $X = 0$, while, as $\tanh^{-1} x = (1/2)\ln((1+x)/(1-x))$, $\Delta(X)$ tends to 0 for X going to ± 1.

3.6 Normal Forms and Symmetries

3.6.1 The Conjugacy Problem

If we are given a vector field X_A, we can try to find the simplest possible form it takes in any linear structure we can construct on E, i.e., we can look for the form that X_A takes with respect to an arbitrary diffeomorphism ϕ. In other words, we will be looking for the orbits of linear vector fields of the Lie algebra $\mathfrak{X}(E)$ with respect to the group Diff$(E, 0)$ of diffeomorphisms of E fixing 0. An elementary argument shows that it is not possible to simplify more a linear vector field by using smooth transformations. First we disregard the possibility of converting the vector field X_A into a constant vector field because of the completely different nature of their flows. Then, we can consider whether we can transform the linear structure of the vector field into another simpler linear one, but that is impossible because a smooth transformation will preserve the spectrum of the matrix A. In fact denoting by P the Jacobian matrix of the transformation defined by ϕ at $\mathbf{0}$ we obtain after an easy computation that

$$AP = PB\,,$$

where B will denote the matrix defining the transformed linear vector field. Hence, $B = P^{-1}AP$ and the spectrum of A is unchanged. Thus, if we are looking for transformations simplifying our vector field, they must transform our linear vector field into another linear vector field without adding spurious higher order terms to X_A. Such a family of maps will contain linear transformations $x \mapsto P \cdot x$ and eventually more transformations (that can be called 'linearoid' with respect to X_A), but the change they can produce on X_A is at most the change that would be produced just by linear transformations. Thus we are lead to look for the simplest form that X_A takes under linear transformations, i.e., under diffeomorphisms preserving the geometry of the problem, in this case the vector field Δ (in further studies we will address the same questions with respect to geometrical structures different from Δ). It is clear now that such a subgroup of the group of diffeomorphisms is precisely the group of linear automorphisms of E. Thus looking for normal forms of linear vector fields amounts to describing the orbits of linear vector fields under linear changes of coordinates. Under a linear change of coordinates $\phi(x) = P \cdot x$ the vector field X'_A transforms into $\phi_* X_A = X_{PAP^{-1}}$, thus classifying orbits of linear vector fields under linear changes of coordinates is equivalent to understanding the structure of the set of matrices under conjugation $A \mapsto PAP^{-1}$, i.e., the adjoint representation of $GL(n, \mathbb{R})$. It is well-known from elementary courses in linear algebra that the classification of conjugacy problem of real matrices is solved by Jordan's Normal Form Theorem that we rephrase succinctly here.

Theorem 3.23 *[Normal Form Theorem] For any real $n \times n$ matrix A there exists a (essentially unique) representative on each adjoint orbit under the group $GL(n, \mathbb{R})$ which has the block diagonal form*

$$J = \begin{pmatrix} J_1 & 0 & & \\ 0 & J_2 & & \\ & & \ddots & \\ & & & J_k \end{pmatrix}, \tag{3.120}$$

where the matrices J_i correspond to the i-th eigenvalue λ_i of the matrix A and are made up of elementary blocks whose structure depends on λ_i and have the form

$$J_i = \begin{pmatrix} J_i^{(1)} & 0 & & \\ 0 & J_i^{(2)} & & \\ & & \ddots & \\ & & & J_i^{(r_i)} \end{pmatrix}. \tag{3.121}$$

Here the $J_i^{(l)}$ are elementary Jordan blocks corresponding to the same eigenvalue which are listed below, such that the order of $J_i^{(1)}$ is precisely the multiplicity of λ_i in the minimal polynomial if λ_i is real, and twice the multiplicity of λ_i in the minimum polynomial if λ_i is complex. The number r_i of blocks is the geometric multiplicity of λ_i, i.e., the dimension of its eigenspace. The sum of all orders is the algebraic multiplicity of λ_i, and the order of $J_i^{(l)}$ is decreasing with l.

The elementary real Jordan blocks are the following:

I. A single element (λ) (corresponding to a non-degenerate real eigenvalue).

II. A 2×2 matrix $\begin{pmatrix} 0 & \nu \\ -\nu & 0 \end{pmatrix}$, $\nu \in \mathbb{R}$ (corresponding to a non-degenerate purely imaginary eigenvalue $\lambda = i\nu$, $\nu \neq 0$).

III. A 2×2 matrix $\begin{pmatrix} a & b \\ -b & a \end{pmatrix}$, $ab \neq 0$, $a, b \in \mathbb{R}$ (corresponding to a non-degenerate complex eigenvalue $\lambda = a + ib$).

IVa. If λ is a degenerate real eigenvalue,

$$J_\lambda = \begin{pmatrix} \lambda & 1 & 0 & \dots \\ 0 & \lambda & 1 & \dots \\ 0 & 0 & \lambda & \dots \\ \vdots & \vdots & \vdots & \ddots \end{pmatrix} .$$

IVb. If $\lambda = i\nu$ is a degenerate purely imaginary eigenvalue,

$$\left(\begin{array}{cc|cc|cc|c} 0 & \nu & 1 & 0 & \multicolumn{2}{c|}{\multirow{2}{*}{0}} & \multirow{2}{*}{\dots} \\ -\nu & 0 & 0 & 1 & & & \\ \hline \multicolumn{2}{c|}{\multirow{2}{*}{0}} & 0 & \nu & 1 & 0 & \multirow{2}{*}{\dots} \\ & & -\nu & 0 & 0 & 1 & \\ \hline \multicolumn{2}{c|}{\multirow{2}{*}{0}} & \multicolumn{2}{c|}{\multirow{2}{*}{0}} & 0 & \nu & \multirow{2}{*}{\dots} \\ & & & & -\nu & 0 & \\ \hline \multicolumn{2}{c|}{\vdots} & \multicolumn{2}{c|}{\vdots} & \multicolumn{2}{c|}{\vdots} & \ddots \end{array}\right) .$$

IVc. If $\lambda = a + i\,b$, $ab \neq 0$, is degenerate complex eigenvalue,

$$\left(\begin{array}{cc|cc|cc|c} a & b & 1 & 0 & \multicolumn{2}{c|}{\multirow{2}{*}{0}} & \multirow{2}{*}{\dots} \\ -b & a & 0 & 1 & & & \\ \hline \multicolumn{2}{c|}{\multirow{2}{*}{0}} & a & b & 1 & 0 & \multirow{2}{*}{\dots} \\ & & -b & a & 0 & 1 & \\ \hline \multicolumn{2}{c|}{\multirow{2}{*}{0}} & \multicolumn{2}{c|}{\multirow{2}{*}{0}} & a & b & \multirow{2}{*}{\dots} \\ & & & & -b & a & \\ \hline \multicolumn{2}{c|}{\vdots} & \multicolumn{2}{c|}{\vdots} & \multicolumn{2}{c|}{\vdots} & \ddots \end{array}\right) .$$

The linear vector fields associated with any of the elementary blocks listed above are given by:

1. $X_{\mathrm{I}} = \lambda x\, \partial/\partial x$.
2. $X_{\mathrm{II}} = \nu\,(-x\,\partial/\partial y + y\,\partial/\partial x)$.
3. $X_{\mathrm{III}} = (ax + by)\,\partial/\partial x - (bx - ay)\,\partial/\partial y$.

4.
$$X_{\text{IVa}} = \lambda\left(x_1\frac{\partial}{\partial x_1} + \cdots + x_n\frac{\partial}{\partial x_n}\right) + x_2\frac{\partial}{\partial x_1} + \cdots + x_n\frac{\partial}{\partial x_{n-1}}$$
$$X_{\text{IVb}} = \nu\left(-x_1\frac{\partial}{\partial x_2} + x_2\frac{\partial}{\partial x_1}\right) + \cdots + \nu\left(-x_{2n-1}\frac{\partial}{\partial x_{2n}} + x_{2n}\frac{\partial}{\partial x_{2n-1}}\right) +$$
$$+ x_1\frac{\partial}{\partial x_3} + x_2\frac{\partial}{\partial x_4} + \cdots + x_{2n-3}\frac{\partial}{\partial x_{2n-1}} + x_{2n-2}\frac{\partial}{\partial x_{2n}}$$
$$X_{\text{IVc}} = (ax_1 + bx_2)\frac{\partial}{\partial x_1} + (ax_2 - bx_1)\frac{\partial}{\partial x_2} + \cdots +$$
$$+ (ax_{2n-1} + bx_{2n})\frac{\partial}{\partial x_{2n-1}} + (ax_{2n} - bx_{2n-1})\frac{\partial}{\partial x_{2n}} + x_1\frac{\partial}{\partial x_3} +$$
$$+ x_2\frac{\partial}{\partial x_4} + \cdots x_{2n-3}\frac{\partial}{\partial x_{2n-1}} + x_{2n-2}\frac{\partial}{\partial x_{2n}}$$

Thus any linear vector field can be brought, by a linear change of coordinates, to a sum of elementary vector fields as listed. In this sense the conjugacy problem is completely solved. As each one of these vector fields belongs to a different subspace, they all commute among themselves, so that in finding the flow associated with any direct sum we simply take the product of the component flows.

We might comment very briefly on these elementary vector fields.

1. X_{I} is an infinitesimal generator of dilations, it has a flow given by $\phi_t = e^{\lambda t}$.
2. X_{II} is the generator of rotations in the plane, it has a flow given by

$$\phi_t = \begin{pmatrix} \cos\nu t & \sin\nu t \\ -\sin\nu t & \cos\nu t \end{pmatrix}.$$

3. For X_{III} the two vector fields $a(x\,\partial/\partial x + y\,\partial/\partial y)$ and $b(x\,\partial/\partial y - y\,\partial/\partial x)$ commute, so we can write

$$\phi_t = \begin{pmatrix} e^{at} & 0 \\ 0 & e^{at} \end{pmatrix}\begin{pmatrix} \cos bt & \sin bt \\ -\sin bt & \cos bt \end{pmatrix}.$$

4. For X_{IV}:

   a. The vector field X_{IVa} is again the sum of two commuting vector fields, that in dimension two are given by, $\lambda\,(x\,\partial/\partial x + y\,\partial/\partial y)$ and $y\,\partial/\partial x$, therefore we obtain:

$$\phi_t = \begin{pmatrix} 1 & t \\ 0 & 1 \end{pmatrix}\begin{pmatrix} e^{\lambda t} & 0 \\ 0 & e^{\lambda t} \end{pmatrix}.$$

   b. The vector field X_{IVb} is again the sum of two commuting vector fields, that in dimension four are given by, $\lambda(x_1\,\partial/\partial x_2 - x_2\,\partial/\partial x_1) + \lambda(x_3\,\partial/\partial x_4 - x_4\,\partial/\partial x_3)$ and $x_1\,\partial/\partial x_3 + x_2\,\partial/\partial x_4$.

$$\left(\begin{array}{cc|cc} e^{at} & 0 & 0 & 0 \\ 0 & e^{at} & 0 & 0 \\ \hline 0 & 0 & e^{at} & 0 \\ 0 & 0 & 0 & e^{at} \end{array}\right)\left(\begin{array}{cc|cc} 1 & 0 & t & 0 \\ 0 & 1 & 0 & t \\ \hline 0 & 0 & 1 & 0 \\ 0 & 0 & 0 & 1 \end{array}\right).$$

c. Finally, the vector field X_{IVa} is again the sum of two commuting vector fields, that in dimension four are given by, $(ax_1 + bx_2)\,\partial/\partial x_1 + (ax_2 - bx_1)\,\partial/\partial x_2 + (ax_3 + bx_4)\,\partial/\partial x_3 + (ax_4 - bx_3)\,\partial/\partial x_4$ and $x_1\,\partial/\partial x_3 + x_2\,\partial/\partial x_4$.

$$\left(\begin{array}{cc|cc} e^{at} & 0 & 0 & 0 \\ 0 & e^{at} & 0 & 0 \\ \hline 0 & 0 & e^{at} & 0 \\ 0 & 0 & 0 & e^{at} \end{array}\right)\left(\begin{array}{cc|cc} \cos bt & \sin bt & 0 & 0 \\ -\sin bt & \cos bt & 0 & 0 \\ \hline 0 & 0 & \cos bt & \sin bt \\ 0 & 0 & -\sin bt & \cos bt \end{array}\right)\left(\begin{array}{cc|cc} 1 & 0 & t & 0 \\ 0 & 1 & 0 & t \\ \hline 0 & 0 & 1 & 0 \\ 0 & 0 & 0 & 1 \end{array}\right).$$

3.6.2 Separation of Vector Fields

The conjugation problem discussed in the previous section can be cast in a more general scheme which is suitable for discussing other situations of big interest. We have already noticed that E can be embedded in $\mathcal{F}(E^*)$ by the canonical identification $E \cong E^{**}$. In fact, if $\{ e_i \mid i = 1, \ldots, n \}$ is a linear basis for E, we will denote by $x_i \in \mathcal{F}(E^*)$ the associated linear functions defined by

$$x_i(\mu) = \langle \mu, e_i \rangle\,, \quad \mu \in E^*.$$

We notice that a generic vector $u = u^i\, e_i \in E$ will define a linear function f_u on E^* by

$$f_u(\mu) = \langle \mu, u \rangle\,,$$

i.e., $f_u = u^i\, x_i$. A linear vector field X on E can be thought as a vector field X^* on E^* by means of

$$X^*(f)(\mu) = f(X(\mu))\,,$$

where $X(\mu)$ is the action of X in the linear function $\mu \in \mathcal{F}(E)$. Notice that if X is linear, $X(\mu)$ is linear too, thus $X(\mu)$ is an element of E^*. Hence, if $X = A^j{}_i\, x^i\, \partial/\partial x^j$, then $X^* = A^{\,j}{}_i\, x_j\, \partial/\partial x_i$. Notice that

$$\mathcal{L}_{X^*} f_u = A^{\,j}{}_i\, x_j\, \frac{\partial}{\partial x_i}(u^k\, x_k) = A^{\,j}{}_i\, x_j\, u^i\,,$$

and if u is an eigenvector of A with eigenvalue λ, we have

$$\mathcal{L}_{X^*} f_u = \lambda\, f_u\,, \tag{3.122}$$

i.e., the linear function f_u on E^* is an eigenvector of the linear operator X^* acting on functions on E^* with the same eigenvalue λ, and conversely.

Up to here there is nothing new except for a change of language. However, because E is a subspace of the associative and commutative algebra $\mathcal{F}(E^*)$, we can consider the eigenvalue problem Eq. (3.122) as an equation directly on $\mathcal{F}(E^*)$. Therefore from now on we can perform any coordinate transformation and the eigenvalue problem will be unchanged. Notice that

$$\Delta^*(f_u) = x_i \frac{\partial}{\partial x_i}(u^k x_k) = x_i u^i = f_u ,$$

thus, we can reformulate our eigenvalue problem in the form

$$\mathcal{L}_{X^* - \lambda \Delta^*} f = 0 .$$

This implies that we are looking for constants of the motion of the linear vector field $X_\lambda = X^* - \lambda \Delta^*$. Notice that the family of vector fields $X_{\lambda_1}, X_{\lambda_2}, \ldots, X_{\lambda_k}$, commute, because $[X, \Delta] = 0$. The space of constants of the motion for a vector field X is a subalgebra $\mathcal{F}_\lambda(E^*)$ of $\mathcal{F}(E^*)$. The first step in the conjugacy problem is equivalent to finding a decomposition of the algebra $\mathcal{F}(E^*)$ as generated by a direct product of subalgebras: $\mathcal{F}_{\lambda_1}(E^*)$, …, $\mathcal{F}_{\lambda_k}(E^*)$, i.e.

$$\mathcal{F}(E^*) = \mathcal{F}_{\lambda_1}(E^*) \otimes \cdots \otimes \mathcal{F}_{\lambda_k}(E^*) .$$

The vector field results $X^* = X_1^* + \cdots X_r^*$ where X_i^* is the restriction of X^* to $\mathcal{F}_{\lambda_i}(E^*)$. Then $[X_i^*, X_j^*] = 0$, and

$$\mathcal{L}_{X_i^*} \mathcal{F}_{\lambda_j}(E^*) = 0, \quad i \neq j ,$$

$$\mathcal{L}_{X_i^*} \mathcal{F}_{\lambda_i}(E^*) \subset \mathcal{F}_{\lambda_i}(E^*) .$$

We can find the flow ϕ_t of X^* by determining the flow of each X_k^* on the associated eigenalgebra $\mathcal{F}_{\lambda_i}(E^*)$, which in terms of the exponential map can be written as

$$\phi_t^*(f) = \exp(t\,\mathcal{L}_X) f = \sum_{n=0}^{\infty} \frac{t^n}{n!} (\mathcal{L}_X)^n f = f + t\,\mathcal{L}_X f + \frac{t^2}{2} (\mathcal{L}_X)^2 f + \cdots ,$$

where the right-hand side makes sense for instance if f is real analytic and X is a complete vector field. that is the case when both X and f are linear.

It is clear now that by replacing our eigenvalue problem on the vector space with an eigenvalue problem on the algebra of functions, Eq. (3.122), we can try to decompose a vector field on any manifold. This decomposition is usually known as a separation of the vector field or it is said that the vector field X is separable. The notion of separability of a dynamics will be used later on to analize the intregrability properties of a dynamics (see Chapter 8) and in the foundations of reduction theory (Sect. 7.2).

3.6.3 Symmetries for Linear Vector Fields

Let us investigate now the existence of symmetries for a linear vector field. We must notice first that the algebra of symmetries of a vector field can be strictly larger than the algebra of linear symmetries. In fact, a linear vector field X_A will be a symmetry of the dynamical vector field Γ, if

$$[\Gamma, X_A] = 0.$$

But that is equivalent to the matrix equation

$$[M, A] = 0.$$

As it was discussed in Sect. 3.6.1, every matrix A can be written as a direct sum

$$A = \begin{pmatrix} A_1 & 0 & \cdots & 0 \\ 0 & A_2 & \dots & 0 \\ \vdots & \vdots & \ddots & \vdots \\ 0 & 0 & \cdots & A_k \end{pmatrix},$$

where each A_j has the form

$$A_j = \begin{pmatrix} J_1^{(j)} & 0 & \cdots & 0 \\ 0 & J_2^{(j)} & \cdots & 0 \\ \vdots & \vdots & \ddots & \vdots \\ 0 & 0 & \cdots & J_{n_j}^{(j)} \end{pmatrix},$$

with $J_r^{(j)}$ an elementary Jordan block corresponding to the eigenvalue λ_j. The sizes of the matrices $J_m^{(j)}$ do not increase as m increases. The matrix A_j is an $r_j \times r_j$ matrix, with r_j the algebraic multiplicity of the eigenvalue λ_j. The number n_j is the number of independent eigenvectors associated with the eigenvalue λ_j, i.e., its geometric multiplicity. The first Jordan block in the matrix A_j is a $d_j \times d_j$ matrix, where d_j is the multiplicity of λ_j as a root of the minimal polynomial. We recall here that if we denote by $\mathcal{P}[x]$ the algebra of polynomials in the indeterminate x, with any polynomial $p = a_0 + a_1 x + \cdots + a_n x^n$, we can associate a matrix $p(A) = a_0 I + a_1 A + \cdots + a_n A^n$. This association is an algebra homomorphism and its image is the algebra associated to A. Thus, any polynomial p defines a matrix $p(A)$ commuting with A. The kernel of the homomorphism $p \mapsto p(A)$ is an ideal in $\mathcal{P}$. The minimal polynomial for A is the (unique) monic generator of the ideal associated with A, and we denote it by p_A. If p_A has degree s, we have $a_0+a_1A+\cdots+a_sA^s=0$. Therefore, there is no linear combination of $A^0, A, A^2, \ldots, A^{s-1}$ which is zero. It follows that it is possible to generate, by taking powers of A, s independent matrices commuting with A. By the previous decomposition we can restrict our considerations to A_j. If A_j contains only one elementary Jordan block, the minimal polynomial is of maximum degree (i.e., coincides with the characteristic polynomial) and we find a number of independent matrices equal to the algebraic multiplicity of the eigenvalue λ_j. If A_j contains an elementary Jordan block two or more times, let say three times, we have

$$A_j = \begin{pmatrix} J^{(j)} & 0 & 0 \\ 0 & J^{(j)} & 0 \\ 0 & 0 & J^{(j)} \end{pmatrix},$$

and, in addition to the powers of each elementary Jordan block taken separately, we have additional commuting matrices of the form

$$C_j^{(1)} = \begin{pmatrix} 0 & I & 0 \\ 0 & 0 & 0 \\ 0 & 0 & 0 \end{pmatrix}, \quad C_j^{(2)} = \begin{pmatrix} 0 & 0 & 0 \\ 0 & 0 & I \\ 0 & 0 & 0 \end{pmatrix}, \quad C_j^{(3)} = \begin{pmatrix} 0 & 0 & I \\ 0 & 0 & 0 \\ 0 & 0 & 0 \end{pmatrix},$$

along with transposed matrices $(C_j^{(a)})^t$. For instance, for

$$A_\lambda = \begin{pmatrix} \lambda & 0 & 0 \\ 0 & \lambda & 0 \\ 0 & 0 & \lambda \end{pmatrix},$$

the algebra of matrices we construct with the previous method is $\mathfrak{gl}(3, \mathbb{R})$.

We conclude therefore that for any matrix with minimal polynomial of maximum degree the algebra of commuting matrices is Abelian and generated by powers of the matrix itself. For other matrices the algebra of symmetries is bigger and not commutative.

We list now a few symmetries for vector fields in normal forms. For a single element Jordan block, $\Gamma = \lambda\, x\, \partial/\partial x$ we have $X_A = x\, \partial/\partial x$.

For a 2×2 matrix we have

(1) $\Gamma = \nu \left(y\, \partial/\partial x - x \frac{\partial}{\partial y} \right)$, $X_A = y\, \partial/\partial x - x\, \partial/\partial y$, $X_{A^2} = x\, \partial/\partial x + y\, \partial/\partial y$.
(2) $\Gamma = (ax + by)\, \partial/\partial x - (bx - ay)\, \partial/\partial y$, $X_A = \Gamma$, $X_{A^2} = ((a^2 - b^2)x + 2aby)\, \partial/\partial x + ((a^2 - b^2)y - 2abx)\, \partial/\partial y$.
(3) $\Gamma = (x\, \partial/\partial x + y\, \partial/\partial y) + y\, \partial/\partial x$, $X_A = \Gamma$, $X_{A^2} = x\, \partial/\partial x + y\, \partial/\partial y + 2y\, \partial/\partial x$.

We conclude this section by saying that, apart from a more thorough analysis of the use of symmetry algebras in coming sections, we can use symmetries to generate new solutions out of a given one. For the sake of illustration let us consider a simple example on $\mathbb{R}^2$. Let $\Gamma = \lambda\, (x\partial/\partial x + y\partial/\partial y)$. Symmetries for Γ are given by $\mathfrak{gl}(2,\mathbb{R})$, generated by $x\partial/\partial x$, $y\partial/\partial y$, $y\partial/\partial x$ and $x\partial/\partial y$. A particular element is $X_R = x\partial/\partial y - y\partial/\partial x$ with associated one parameter group

$$\psi_s(x,y) = \begin{pmatrix} \cos s & -\sin s \\ \sin s & \cos s \end{pmatrix} \begin{pmatrix} x \\ y \end{pmatrix}.$$

A particular solution of the system

$$\dot{x} = \lambda\, x\,, \quad \dot{y} = \lambda\, y\,,$$

with initial conditions $x = x(0)$, $y = y_0 = 0$, is given by

$$x(t) = e^{\lambda\, t} x(0)\,, \quad y(t) = 0\,.$$

By using our symmetry we have new solutions

$$\begin{pmatrix} \tilde{x}(t) \\ \tilde{y}(t) \end{pmatrix} = \begin{pmatrix} \cos s & -\sin s \\ \sin s & \cos s \end{pmatrix} \begin{pmatrix} e^{\lambda t} x(0) \\ 0 \end{pmatrix} = \begin{pmatrix} e^{\lambda t} x(0) \cos s \\ e^{\lambda t} x(0) \sin s \end{pmatrix}.$$

Proposition 3.24 *A generic linear vector field on an n-dimensional linear space possesses an n-dimensional Abelian symmetry algebra.*

Non-generic vector fields define bigger symmetry algebras which are not necessarily Abelian. The space of linear vector fields supports a bundle of symmetry algebras over the generic open dense submanifold. In fact, consider the subset of $M_n(\mathbb{K}) \times M_n(\mathbb{K})$ given by

$$\{\, (M, A) \mid M \in M_n(\mathbb{K}), [M, A] = 0 \,\}\,.$$

The bundle over the generic part is trivial because we can choose the family of sections $I, A, A^2, \ldots, A^{n-1}$. But the singular set has an involved structure. We will discuss some of these sets in the forthcoming sections.

3.6.4 Constants of Motion for Linear Dynamical Systems

As it was stated before a constant of motion for the vector field Γ is any function f such that $f(\phi_t(x)) = f(x)$, where ϕ_t is the flow of the vector field Γ. If f is differentiable we get

$$\Gamma(f) = 0\,. \tag{3.123}$$

Let us first search for a linear constant of motion, i.e., $f_a = a_i\, x^i$. We find that Eq. (3.123) implies, $A^i{}_j\, x^j\, a_i = 0$ for all x, therefore, $a = (a_1, \ldots, a_n)$ must be an eigenvector of A^t with zero eigenvalue, or in other words, $a \in \ker A^t$.

If $f_1, f_2, \ldots, f_k$, are a maximal set of independent linear constants of motion for Γ we can use them to define a map $F\colon \mathbb{R}^n \to \mathbb{R}^k$ by means of $F(x) = (f_1(x), f_2(x), \ldots, f_k(x))$. By taking $F^{-1}(c) \subset \mathbb{R}^n$ for any $c \in \mathbb{R}^k$ we find a family of $n-k$ dimensional vector spaces (affine spaces) in $\mathbb{R}^n$ parametrized by points in $\mathbb{R}^k$.

Remark 3.6 In fact the map F defines a (trivial) principal fibration of $\mathbb{R}^n$ over $\mathbb{R}^k$, the group $\mathbb{R}^{n-k}$ acting on the fibres.

Solutions of our dynamics are contained in $V_c = F^{-1}(c)$ if they started with initial data $x(0)$ such that $F(x(0)) = c$. It is clear that there exists a linear change of coordinates such that the dynamical system X_A can be brought into the form X_B with

$$B = \begin{pmatrix} B_0 & 0 \\ 0 & 0 \end{pmatrix}$$

and B_0 being a $(n-k) \times (n-k)$ matrix on the space V_c. The change of coordinates is constructed by fixing an invariant supplementary subspace to V_c and choosing and adapted linear basis. In other words, let $(y^1, \ldots, y^n)$ be coordinates which are linear combinations of x^i coordinates and such that $y^{n-k+j} = f_j$, for $j = 1, \ldots, k$. The expression of the vector field Γ in these new coordinates $\Gamma = B^i{}_j\, y^j\, \partial/\partial y^i$ is such that for $\Gamma(y^{n-k+j}) = 0$ implies $B^{n-k+j}{}_i\, y^i = 0$, for $j = 1, \ldots, k$.

It is clear that linear constants of motion are few because generically matrices do not have zero eigenvalues. In fact, a linear dynamical system will have as many linear constants of motion as the dimension of the kernel of its associated linear operator. In such situation the null degrees of freedom can be thrown away by an appropriate linear change of coordinates and restricting the system to a subspace.

If we allow for quadratic constants of motion, the situation improves. However, a similar map, that we will call “momentum map” for reasons that will become clear later on,

$$F\colon \mathbb{R}^n \to \mathbb{R}^k\,,$$

can be constructed, where $F = (f_1, \ldots, f_k)$, and the set $\{f_i \mid i = 1, \ldots, k\}$ constitutes a maximal set of quadratic constants of motion. Now, however, there is no, in general, regular slicing for the map F. The solutions of Γ are forced to sit on the level set $F^{-1}(c)$ if the initial condition $x(0)$ lies on it, but we cannot decompose Γ linearly as we did before for the case of linear constants of motion.

A more general analysis of this situation, i.e., foliations compatible with dynamical systems, will be taken up later. Here we limit ourselves to a simple example showing us some of the difficulties we will encounter in the analysis of this problem.

3.6.4.1 A Few Examples

On $\mathbb{R}^3$ we consider the linear system

$$y\frac{\partial}{\partial z} + (x+z)\frac{\partial}{\partial y} - y\frac{\partial}{\partial x},$$

corresponding to the matrix

$$A = \begin{pmatrix} 0 & -1 & 0 \\ 1 & 0 & 1 \\ 0 & 1 & 0 \end{pmatrix}.$$

Three commuting symmetries are provided by Γ itself,

$$\Delta = x\frac{\partial}{\partial x} + y\frac{\partial}{\partial y} + z\frac{\partial}{\partial z},$$

described by the identity matrix, and

$$X_2 = x\frac{\partial}{\partial z} - z\frac{\partial}{\partial x} + z\frac{\partial}{\partial z} - x\frac{\partial}{\partial x},$$

corresponding to

$$B = A^2 = \begin{pmatrix} -1 & 0 & -1 \\ 0 & 0 & 0 \\ 1 & 0 & 1 \end{pmatrix}.$$

There are two constants of motion. One is linear, given by $f_1 = x+z$, corresponding to the null eigenvector $(1,0,1)^t$ of the matrix A^t. It is easy to check that another one is $f_2 = x^2 + y^2 - z^2 = y^2 + (x-z)f_1$. Of course, these constants of motion can be found directly from the adjoint system

$$\frac{dx}{-y} = \frac{dy}{x+z} = \frac{dz}{y}.$$

Only in the points $(t, 0, -t)$ are the differentials df_1 and df_2 linearly dependent. This corresponds to the choice $c_1 = c_2 = 0$.

The map $F\colon \mathbb{R}^3 \to \mathbb{R}^2$ has one-dimensional slices given by $f_1 = c_1$, $f_2 = y^2 + c_1(x-z) = c_2$, except for $c_1 = 0$, $c_2 = 0$ when $F^{-1}(0,0)$ is a straight line $(t, 0, -t)$ and a point. For f_2 we have a "slicing" by hyperboloids and f_1 gives a slicing by planes.

Solutions for our vector fields are given by the intersection of these two slicings. It is clear that the "slicing" associated with f_2 is not "regular". For $c_2 > 0$ we get connected hyperboloids, for $c_2 = 0$ we get a cone, for $c_2 < 0$ we get two disconnected hyperboloids (space-like, light-like and time-like).

The existence of constants of motion may suggest us to use them as part of a new coordinate system. In our example we can take f_1, f_2, $\eta = x - z$, as new coordinates, which define a compatible change of coordinates except for $y = 0$. Our dynamical system has now the form $\dot{f}_1 = 0$, $\dot{f}_2 = 0$, $\dot{\eta}^2 = 2(f_2 - f_1\eta)$, i.e., it is an implicit differential equation. For fixed values of f_1 and f_2 it looks like a parametric implicit differential equation. In the plane $(\dot{\eta}, \eta)$ we have the following situation:

Another example in the same spirit is provided on $\mathbb{R}^3$ by

$$\Gamma = (x + z)\frac{\partial}{\partial y} - y\left(\frac{\partial}{\partial x} + \frac{\partial}{\partial z}\right).$$

This time we have constants of motion $f_2 = x^2 + y^2 + z^2$ and $f_1 = \frac{1}{2}(x - z)$. By using coordinates $\eta = \frac{1}{2}(x + z)$, f_1 and f_2 we get a compatible coordinate system except for $y = 0$. In the new coordinates our dynamical system is given by

$$\dot{f}_1 = 0\,, \qquad \dot{f}_2 = 0\,, \qquad \dot{\eta}^2 = f_2 - 2(f_1^2 + \eta^2)\,,$$

or $\dot{\eta}^2 + \eta^2 = f_2 - f_1^2$.

Trajectories of our dynamical system are provided by circles on concentric spheres lying in planes parallel to the $x = z$ plane. In the implicit equation, of course, f_2 is the square of the radius of the sphere and $|f_1|$ is the distance of the plane from the $x = z$ plane. If $f_1^2 = f_2$ the plane is tangent to the sphere and the trajectory reduces to a point. For $f_1^2 > f_2$ there is no intersection. The two equations that we get in normal form are due to the fact that there are two planes at distance $|f_1|$ from the $x = z$ plane.

Let us investigate now how many constants of motion we might find for a generic linear vector field. Let us first discuss an example. On $\mathbb{R}^2$ we consider

$$\Gamma = x\frac{\partial}{\partial x} + y\frac{\partial}{\partial y}\,.$$

It is easy to show that there are no global constants of the motion. Indeed any trajectory contains the fixed point in its closure, therefore if a continuous constant of motion exists it has the same value on a given trajectory and on the "accumulation point" constituted by the fixed point. As this point is in the closure of any trajectory it follows that the constant of motion is identically constant.

Now it is clear that if $\mathcal{O}(m_1)$ and $\mathcal{O}(m_2)$ are two orbits with initial conditions m_1 and m_2 respectively, a continuous constant of motion has the same value on $\mathcal{O}(m_1)$ and its closure $\overline{\mathcal{O}(m_1)}$, therefore if $\overline{\mathcal{O}(m_1)} \cap \overline{\mathcal{O}(m_2)}$ is not empty, then any constant of the motion has the same constant value on both orbits. For a generic linear dynamical system the representative matrix will have simple eigenvalues, therefore we find in $\mathbb{R}^n$ a situation similar to the one we have discussed in $\mathbb{R}^2$. We conclude that a generic linear dynamical system has no globally defined constants of motion. We notice that various eigenvalues will only change the component of the speed along the "star" phase portrait. To have constants of motion for linear dynamical systems we shall consider non generic families.

Proposition 3.25 *A generic linear vector field has no constants of the motion.*

We may conclude this section summarizing the most important properties we have found for linear dynamical systems.

1. Every linear dynamical system Γ on a vector space E can be decomposed into a direct sum $\Gamma = \Gamma_0 \oplus \Gamma_1 \oplus \cdots \oplus \Gamma_k$. of irreducible commuting dynamical systems Γ_k. The vector space E decomposes into Γ-invariant subspaces $E_1 \oplus \cdots \oplus E_k$ such that $\Gamma_j = \Gamma_{|E_j}$ and $[\Gamma_i, \Gamma_j] = 0$. If A_j is the representative matrix of Γ_j we get additional commuting symmetries by considering linear vector fields associated with powers of A_j (there are as many independent ones as the degree of the minimal polynomial).
2. Linear constants of motion correspond to zero eigenvalues, i.e., there existence implies that there is an invariant subspace on which Γ is identically zero.
3. Nonlinear constants of motion allow to "factor out" an identically zero part of Γ, the price to pay is that we have to introduce nonlinear coordinates.
4. By using symmetries we are able to generate new solutions out of a given one.
5. Generic linear dynamical systems do not have global constants of the motion. Non generic families can be selected by requiring that they preserve some non-degenerate $(0, 2)$ tensor. This tensor can be used to raise and lower indices, allowing us to associate invariant covector fields with various infinitesimal symmetries we have found.

References

[Ma66] Malgrange, B.: Ideals of differentiable functions. Oxford University Press, Published by the Tata Institute of Fundamental Research, Bombay (1966)

[Wh44] Whitney, H.: The self intersections of a smooth n-manifold in $2n$-space. Annals Math. **45**, 220–246 (1944)

[NS03] Navarro González, J., Sancho de Salas, J.: $^{\infty}$-Differentiable Spaces. Lecture Notes in Maths, 1824. Springer, Berlin (2003)

[BM10] Balachandran, A.P., Marmo, G.: Group Theory and Hopf Algebras: Lectures for Physicists. World Scientific, Hackensack (2010)

[AM78] Abraham, R., Marsden, J.E.: Foundations of Mechanics, 2nd edn. Benjamin, Reading (1978)

[Ha70] Haefliger, A.: Homotopy and Integrability. Lecture Notes in Maths, vol. 197, pp. 133–164, Springer, Berlin (1970)

[LF12] Loja Fernandes, R., Frejlich, P.: A h-principle for symplectic foliations. Int. Math. Res. Not. **2012**, 1505–1518 (2012)

[Gr86a] Gromov, M.: Soft and hard symplectic geometry. In: Proceedings of the International Congress of Mathematicians at Berkeley, vol. 1, pp. 81–98 (1986)

[Gr86b] Gromov, M.: Partial Differential Relations. Springer, Berlin (1986)

[Em84] Emch, G.G.: Mathematical and Conceptual Foundations of 20th Century Physics. North-Holland, Amsterdam (1984)

[FF13] Falceto, F., Ferro, L., Ibort, A., Marmo, G.: Reduction of Lie-Jordan Banach algebras and quantum states. J. Phys. A Math. Theor. **46**, 015201 (2013)

Chapter 4
Invariant Structures for Dynamical Systems: Poisson Dynamics

The miracle of the appropriateness of the language of mathematics for the formulation of the laws of physics is a wonderful gift which we neither understand nor deserve. We should be grateful for it and hope that it will remain valid in future research and that it will extend, for better or for worse, to our pleasure, even though perhaps also to our bafflement, to wide branches of learning.

Eugene Wigner, *The Unreasonable Effectiveness of Mathematics in the Natural Sciences,* 1960.

4.1 Introduction

In this chapter we will start developing systematically one of the inspiring principles of this book: all geometrical structures should be dynamically determined. In other words, given a dynamical system Γ we try to determine the geometrical structures determined by Γ. The exact nature of the geometrical structure determined by Γ that we will be interested in will depend on the problem we are facing, however the simplest ones will always be of interest: symmetries and constants of motion as it was discussed in the previous chapter. Higher order objects like contravariant o covariant tensors of order 2 tensorial will be discussed now. This problem will lead us in particular to the study of Poisson and symplectic structures compatible with our given dynamical system Γ.

Because of the special place that linear systems occupy in this work, we will complete the discussion on linearity started in the previous chapter and we will determine all linear structures compatible with a given dynamics. We will relate this with the problem of linearizability of a given dynamics and the study of normal forms. This will merge with another of the main themes of the book, the study of linear systems and their geometrical structures. In particular we will study the factorization problem for a linear dynamical vector field. This problem will lead us immediately to the study of Poisson structures compatible with a given vector field. We will discuss the properties and structure of Poisson tensors and the particular instance of symplectic structures that will be fully developed in the next chapter.

J.F. Cariñena et al., *Geometry from Dynamics, Classical and Quantum*,
DOI 10.1007/978-94-017-9220-2_4

We will solve the factorization problem for linear dynamics, that is under what conditions a linear dynamical system has a Poisson structure. After solving the inverse problem we will come back to the study of infinitesimal symmetries of a dynamical system. When considering not one alone, but all the alternative ones, we will obtain the notion of Lie algebra of symmetries of a dynamical system and their integrated form, a Lie group of symmetries of a dynamical system. We will establish the basic ideas of this problem whose full description will be postponed to Chaps. 6, 7 and 8.

The situation where the system has a compatible linear structure and we ask for the existence not just of a Hamiltonian factorization but a gradient factorization will be left to Chap. 6 where Hermitean structures compatible with a given dynamics and the geometry of quantum systems will be thoroughly discussed.

4.2 The Factorization Problem for Vector Fields

4.2.1 The Geometry of Noether's Theorem

We have explored the basic properties of dynamical systems starting from the basic assumption that "dynamics is first" i.e., any description of a dynamical system has to consider the dynamical evolution as the main conceptual ingredient and all additional geometrical structures used on its description should be 'dynamically determined'.

In order to substantiate this program some additional mathematical ingredients are needed. We have provided them with a minimum of additional assumptions, the main one, that the space of observables carries the structure of an algebra.

We have isolated linear structures as natural ingredients of the description of a large family of dynamical systems emphasizing its non-uniqueness and the special properties derived of the compatibility of a given dynamical system with one particular linear structure, i.e., its linearity and the explicit integrability of the system.

However, the nonexistence of linear structures compatible with a given dynamics makes necessary to use the full algebra of smooth observables, and thus we have opened the door for the use of differential calculus 'in the large' in our exploration of dynamical systems.

A whole family of structures compatible with a given dynamics have already emerged: smooth structures, linear structures, constants of the motion and symmetry algebras. This is the first generation of compatible structures determined by dynamical systems and constitute the basic vocabulary of the theory. We would like to go beyond this and discover additional structures compatible with a given dynamics.

In this book we will follow an "Occam's razor" principle in the economy of additional structures used in the description of systems, thus we will always try to search for structures that will emerge in simple situations, e.g., the existence of regularities in the trajectories (like periodic behavior, trajectories easily mapped into other trajectories, etc.). Then we have shown that we should explore the existence of constants of the motion and symmetry properties. The examples discussed in Chap. 1

show that in many occasions both things come together: the system has constants of motion and has a rich symmetry algebra.

The existence of a link between constants of the motion and symmetries for a given dynamical system will be called generically "Noether's theorem". We will analyze first some implications of this situation. All these observations will lead us naturally to the notion of Poisson structures, the first of the second generation of structures compatible with dynamical systems that are going to be discussed in this book.

4.2.1.1 Noether's Theorem and 2-Tensors

The link between infinitesimal symmetries and constants of the motion is given by additional invariant structures, and therefore this fact motivates the search for such invariant structures.

We have discussed some basic properties of infinitesimal symmetries and of constants of motion for a given dynamical system Γ. We are also used to the fact that these two notions are related. In fact, in Chap. 1 we were able to show in concrete examples how this link is established (see for instance Sect. 1.2.8 for a discussion of some examples).

Thus, we are used to think that symmetries are associated to invariants that can be measured, i.e., to observables. However, it is not obvious how this connection between symmetries and constants of motion is actually established.

Infinitesimal symmetries are described by vector fields X and constants of motion by functions f, thus, we will need an object that maps vector fields into functions. On the other hand, the particular value of a constant of motion at a given point is irrelevant because if f is a constant of motion for Γ, so it is $f + c$, where c is an arbitrary constant. Then as we discussed in previous sections, this implies that the relevant object is the 1-form df rather than f and hence we will consider the space of closed (locally exact) invariant 1-forms rather than the space of constants of motion.

The fact of looking for, instead of constants of the motion, their differentials, i.e., 1-forms, leads us to consider the natural geometrical objects mapping vector fields into 1-forms, $(0, 2)$-tensors or, 1-forms into vector fields, that is $(2, 0)$ tensors.

This is the road we are going to follow to understand the link between symmetries and constants of the motion, looking for compatible $(0, 2)$-tensors or $(2, 0)$-tensors such that they transform infinitesimal symmetries into constants of motion or vice versa for a given dynamical system Γ.

4.2.2 Invariant 2-Tensors

As we have already discussed before, constants of motion for a given dynamical system Γ close a subalgebra $\mathcal{C}(\Gamma)$ of the algebra of observables. But, as indicated above, we should consider the space of closed invariant 1-forms, rather than the space of constants of the motion. On the other side, the elements of the symmetry algebra of Γ close a Lie subalgebra $\mathfrak{X}(\Gamma)$ of the Lie algebra of derivations.

A 1-form α will be said to be Γ-invariant if $\mathcal{L}_\Gamma \alpha = 0$. The set of all such 1-forms will be denoted by $\Omega^1(\Gamma)$. If $f \in \mathcal{C}(\Gamma)$ is a constant of motion for Γ, then df is a Γ-invariant 1-form $df \in \Omega^1(\Gamma)$, because

$$\mathcal{L}_\Gamma df = d\mathcal{L}_\Gamma f = 0\,,$$

and hence, $d\mathcal{C}(\Omega) \subset \Omega^1(\Gamma)$. Conversely, if the 1-form α is exact, $\alpha = df$, the Γ-invariance of α implies that $\Gamma(f)$ is a constant, because $d(\mathcal{L}_\Gamma(f)) = \mathcal{L}_\Gamma(df) = 0$. Notice that this implies that such a function f can always locally be written at a neighbourhood of a regular point x of Γ ($\Gamma(x) \neq 0$) as $f(x^1, x^2, \ldots, x^n) = c\,x^1$, where x^1 is the natural flow parameter of Γ (it suffices to choose local coordinates straightening out the vector field Γ).

Let us consider now a Γ-invariant 1-form α and a rule to assign to it a symmetry vector field X_α, $\alpha \mapsto X_\alpha$. We want such a map to be $\mathbb{R}$-linear. Moreover, if we multiply a Γ-invariant 1-form by a function which is zero in the neighborhood of a given point then, when restricting to such a neighborhood, the corresponding symmetry has to vanish identically. Then, as we discussed in Sect. 2.3.4, the correspondence has to be local. If we assume, and that is the only assumption which could be relaxed, that such correspondence is established not only for invariant 1-forms, but for all 1-forms, we conclude that it has to be given by a (2, 0)-tensor T on the state space E of the system, and then, there will exist a map $T\colon T^*E \to TE$, linear along the fibers of the space, defined by contraction with T, such that

$$X_\alpha = \widehat{T}(\alpha)\,.$$

The same argument could be reproduced in the converse direction. Thus, starting with infinitesimal symmetries we could have argued in the same way, concluding that there should exists a (0, 2)-tensor L such that

$$\alpha_X = L(X)\,,$$

where now α_X denotes the 1-form which is associated to the infinitesimal symmetry X. If one of the two tensors T or L is non-degenerate, then clearly they are the inverse one of each other, $T = L^{-1}$, and that is what we will assume for the purposes of the discussion now. Degenerate tensors can actually appear and they will do, but their discussion will be postponed until we have advanced in our discussion.

The mere existence of such tensor T imposes some conditions on it. Thus, because X_α is a symmetry of Γ, we have

$$0 = \mathcal{L}_\Gamma(X_\alpha) = (\mathcal{L}_\Gamma \widehat{T})(\alpha) + \widehat{T}(\mathcal{L}_\Gamma \alpha) = (\mathcal{L}_\Gamma \widehat{T})(\alpha)\,,$$

hence the tensor T itself must be invariant under Γ along the Γ-invariant 1-forms. Thus, in order to obtain such tensors, we can look, in particular, for tensors that are invariant under Γ, that is tensors which are compatible with the given dynamics.

On the other side, we should remark that $\mathcal{L}_\Gamma T = 0$ if and only if $\mathcal{L}_\Gamma \widehat{T} = 0$, because for any pair of forms α and β,

$$(\mathcal{L}_\Gamma T)(\alpha, \beta) = \langle \beta, (\mathcal{L}_\Gamma \widehat{T})(\alpha) \rangle \,.$$

We shall consider now the Γ-invariant (0, 2) tensor L. We can decompose such a tensor into its symmetric and skew-symmetric parts, $L = S + A$. Let us first consider the case in which L is a (0, 2)-symmetric tensor, and we shall study later on the case in which L is skew-symmetric.

4.2.2.1 Invariant Symmetric Tensors

We study first the problem for symmetric (0, 2)-tensors. Following the linear paradigm we will start first with a linear vector field X_A and we would like to find symmetric tensors $g = g_{ij}\, dx^i \otimes dx^j$, with g_{ij} constants such that $g_{ij} = g_{ji}$, satisfying

$$\mathcal{L}_{X_A} g = 0 \,. \tag{4.1}$$

We will only consider the simplest case in which $\det G \neq 0$, where G is the symmetric square matrix with elements g_{ij}. We first remark that the invariance condition is equivalent to:

$$A^T\, G + G\, A = 0 \,, \tag{4.2}$$

because

$$\mathcal{L}_X g = g_{ij} \left[d(A^i{}_k x^k) \otimes dx^j + dx^i \otimes d(A^j{}_l x^l) \right] = \left[(A^T G)_{kl} + (GA)_{kl} \right] dx^k \otimes dx^l \,.$$

The above condition is equivalent to

$$(GA)^T = -GA \,, \tag{4.3}$$

then the matrix GA must be skew-symmetric. Moreover, as we assume that $\det G \neq 0$, we see that

$$A^T = -GAG^{-1} \,,$$

and this implies that $\mathrm{Tr}\, A^T = -\mathrm{Tr}\, A = 0$. By taking powers of our relation we get

$$(A^T)^k = (-1)^k\, G\, A^k\, G^{-1} \,, \tag{4.4}$$

and, therefore, all odd powers of A should be traceless.

It is to be remarked that (4.4) means that the vector fields associated with odd powers of A will also preserve g and the functions $g(X_{A^k}, X_{A^j})$, for any pair of values k, j, are constants of the motion because

$$\mathcal{L}_{X_{A^i}}[g(X_{A^k}, X_{A^j})] = (\mathcal{L}_{X_{A^i}} g)(X_{A^k}, X_{A^j}) + g(\mathcal{L}_{X_{A^i}}(X_{A^k}), X_{A^j}) + g(X_{A^k}, \mathcal{L}_{X_{A^i}}(X_{A^j})) = 0\,.$$

To have a better understanding of this situation, let us consider the two dimensional case with

$$X = (ax + by)\frac{\partial}{\partial x} + (mx + ny)\frac{\partial}{\partial y}\,,$$

and

$$g = \alpha\, dx \otimes dx + \beta\, dy \otimes dy + \gamma\, (dx \otimes dy + dy \otimes dx)\,,$$

i.e.,

$$A = \begin{pmatrix} a & b \\ m & n \end{pmatrix}\,, \qquad G = \begin{pmatrix} \alpha & \gamma \\ \gamma & \beta \end{pmatrix}\,.$$

Then,

$$A^T G + GA = \begin{pmatrix} 2\,(a\,\alpha + m\,\gamma) & \gamma(a+n) + \alpha\, b + m\,\beta \\ \gamma(n+a) + b\,\alpha + \beta\, m & 2\,(b\,\gamma + n\,\beta) \end{pmatrix}\,,$$

and the invariance conditions for g are

$$\begin{cases} a\,\alpha + m\,\gamma = 0\,, \\ \gamma(a+n) + \alpha\, b + m\,\beta = 0\,, \\ b\,\gamma + n\,\beta = 0\,. \end{cases}$$

which is a homogeneous linear system for the unknown α, β, γ. The determinant of the matrix of coefficients is $(a+n)(b\,m - a\,n) = -\mathrm{Tr}\, A \cdot \det A$ and, therefore, if $\det A \neq 0$, the compatibility condition is $\mathrm{Tr}\, A = a + n = 0$. Then, a solution for these equations is provided by $m = \alpha$, $n = \gamma = -a$, $\beta = -b$. Consequently, the most general constant non-degenerate (0, 2)-tensor invariant under X_A is given by:

$$g = m\, dx \otimes dx - b\, dy \otimes dy - a\, (dx \otimes dy + dy \otimes dx)\,.$$

We see that if $\mathrm{Tr}\, A = a + n \neq 0$, there is no solution. Actually, if $\det A = 0$ as well, there is a second solution with a degenerate tensor.

Exercise 4.1 Complete the previous computation finding out the constants of motion of the system and the algebra of symmetries of X_A.

4.2.2.2 Invariant Skew-Symmetric Tensors

We next study the characterization of those skew-symmetric $(2, 0)$-tensors in the linear space E which are invariant under a linear dynamics. We will start with a fixed linear dynamics given by the vector field $X_A = A^i{}_j x^j \frac{\partial}{\partial x^i}$. If the invariant bi-vector field Λ we are looking for is a constant, i.e., $\Lambda^{ij} =$ constant, then because of the relations

$$\mathcal{L}_{X_A} \frac{\partial}{\partial x^i} = -A^j{}_i \frac{\partial}{\partial x^j}, \qquad i = 1, \ldots, n,$$

which imply that

$$\begin{aligned}\mathcal{L}_{X_A} \Lambda &= \Lambda^{ij} \left[-A^k{}_i \frac{\partial}{\partial x^k} \wedge \frac{\partial}{\partial x^j} - A^l{}_j \frac{\partial}{\partial x^i} \wedge \frac{\partial}{\partial x^l} \right] \\ &= -(A\Lambda)^{kj} \frac{\partial}{\partial x^k} \wedge \frac{\partial}{\partial x^j} - (\Lambda A^T)^{il} \frac{\partial}{\partial x^i} \wedge \frac{\partial}{\partial x^l} .\end{aligned}$$

and because the matrix Λ is skew-symmetric, the condition for $\mathcal{L}_{X_A} \Lambda = 0$ is

$$A\,\Lambda + \Lambda\,A^T = 0\,, \tag{4.5}$$

which can be rewritten $(A\Lambda)^T = A\Lambda$ (compare with Eq. (4.3)). It is easy to show by iteration of (4.5) that $A^{2k+1}\Lambda = -\Lambda(A^T)^{2k+1}$ while $A^{2k}\Lambda = \Lambda(A^T)^{2k}$. Therefore, $A^{2k}\Lambda$ (and $\Lambda(A^T)^{2k}$) is a skew-symmetric matrix, and $A^{2k+1}\Lambda$ is a symmetric one, for any positive integer number k. Therefore, $\Lambda(A^T)^{2k}$ provide new solutions of admissible skew-symmetric matrices, because

$$A[\Lambda\,(A^T)^{2k}] = -\Lambda\,(A^T)^{2k+1}\,.$$

Note, however, that when A is not invertible the rank of each power may change and we would get a sequence of admissible skew-symmetric tensors with different rank.

When considering a general dynamics on a smooth manifold M, we have as it happens in the symmetric case, much less control on the situation, however we can still obtain a theorem connecting the constant of motion and infinitesimal symmetries.

Theorem 4.1 *Let Γ be a dynamical system possessing a $(2, 0)$ Γ-invariant tensor Λ. Then,*

i. *If f is a constant of motion for Γ, then the vector field $X_f = -\widehat{\Lambda}(df)$ is an infinitesimal symmetry for Γ.*
ii. *If X is a symmetry of Γ such that there exists a function f with $\widehat{\Lambda}(df) = -X$, and Λ is regular, then $\Gamma(f)$ is a constant.*

Proof 1. Having in mind that

$$[X_f, \Gamma] = \mathcal{L}_\Gamma(\widehat{\Lambda}(df)) = (\mathcal{L}_\Gamma \widehat{\Lambda})(df) + \widehat{\Lambda}(\mathcal{L}_\Gamma(df))\,,$$

as we have assumed that Λ is invariant under Γ, $\mathcal{L}_\Gamma \widehat{\Lambda} = 0$, and as f is a constant of motion, we also have $\mathcal{L}_\Gamma(df) = d(\mathcal{L}_\Gamma f) = 0$. Therefore, $[\Gamma, X_f] = 0$.

2. If X is a symmetry of Γ such that there exists a function f with $\widehat{\Lambda}(df) = -X$, then,

$$\mathcal{L}_\Gamma(\widehat{\Lambda}(df)) = 0 = (\mathcal{L}_\Gamma \widehat{\Lambda})(df) + \widehat{\Lambda}(d(\mathcal{L}_\Gamma f)),$$

and therefore, when $\widehat{\Lambda}$ is regular,

$$d(\mathcal{L}_\Gamma f) = 0,$$

i.e., $\mathcal{L}_\Gamma f$ is constant. □

4.2.3 Factorizing Linear Dynamics: Linear Poisson Factorization

After the previous discussion, given a dynamical system Γ we observe that Γ itself is a symmetry of Γ, hence we would like to know if there exists a constant of motion H corresponding to it. For that we should find a skew-symmetric $(2, 0)$-tensor such that $-\widehat{\Lambda}(dH) = \Gamma$. Thus, we are led to solve the equation

$$\Gamma = \widehat{\Lambda}(\alpha),$$

where Γ is given and Λ, and the exact form α are unknown. Notice that all vector fields Γ can be written in this way. We will analyze this problem by thinking first in the linear case.

Thus, let Γ be a linear vector field determined by the linear map $A : E \to E$, i.e., $\Gamma = X_A$. Then, Γ is homogeneous of degree 0, because it is linear, $\mathcal{L}_\Delta \Gamma = 0$. If we denote the degree of homogeneity of a (homogeneous) tensor T by $h(T)$, i.e., $\mathcal{L}_\Delta T = h(t)T$, then we have that if $\Gamma = \Lambda(\alpha)$, with Λ and α homogeneous, then $h(\Gamma) = h(\Lambda) + h(\alpha)$. Thus, $h(\Lambda) = -h(\alpha)$ if we want Γ to be linear. It happens that $h(\alpha) \geq 1$ and $h(\Lambda) \geq -2$ if we want α and Λ to be smooth on E. Hence, $-2 \leq h(\Lambda) = -h(\alpha) \leq -1$. Thus the only solutions for these conditions are given by $h(\Lambda) = -2 = -h(\alpha)$ and $h(\Lambda) = -1 = -h(\alpha)$.

4.2.3.1 First Case: Λ Constant

The first case, means that the bi-vector Λ has the form

$$\Lambda = \Lambda^{ij} \frac{\partial}{\partial x^i} \wedge \frac{\partial}{\partial x^j}, \quad \Lambda^{ij} = -\Lambda^{ji} \in \mathbb{R},$$

with Λ^{ij} constants, and α has the form

$$\alpha = -H_{ij}\, x^i\, dx^j,$$

with H_{ij} constants. Moreover if α is closed, then $H_{ij}\, dx^i \wedge dx^j = 0$ and therefore $H_{ij} = H_{ji}$, i.e., H is a symmetric matrix. Then $\alpha = -dH$ with H being the quadratic function

$$H = \frac{1}{2} H_{ij}\, x^i\, x^j \, .$$

We are going to see that finding a constant of motion for a linear dynamics amounts in the first case to solve the factorization problem of finding a skew-symmetric matrix Λ and a symmetric matrix H, which will provide us the quadratic constant of motion, such that

$$A = \Lambda\, H \, , \tag{4.6}$$

because

$$-\widehat{\Lambda}(dH) = \Lambda^{il}\, H_{kl}\, x^k \, \frac{\partial}{\partial x^i} \, ,$$

and therefore,

$$A^j{}_k\, x^k = x^k\, H_{ki}\, \Lambda^{ji} \, .$$

Conversely, if we suppose that the matrix $A = (A^i{}_j)$ can be factorized as a product of matrices $A = \Lambda\, H$, i.e.

$$A^i{}_j = \Lambda^{ik}\, H_{kj} \, ,$$

with Λ being a skew-symmetric matrix and H being a symmetric one,

$$\Lambda^{ij} = -\Lambda^{ji} \, , \qquad H_{ij} = H_{ji} \, ,$$

then the vector field X_A is Hamiltonian with respect to the Poisson structure defined by the matrix Λ and the Hamiltonian function is the quadratic function defined by the matrix H.

Example 4.2 (The 1-dimensional harmonic oscillator) As an example we can consider the 1-dimensional harmonic oscillator, described by the system:

$$\dot{q} = p \, , \qquad \dot{p} = -q \, ,$$

whose associated vector field is $\Gamma = X_A$ with A being the matrix

$$A = \begin{pmatrix} 0 & 1 \\ -1 & 0 \end{pmatrix} \, ,$$

which can be factorized as a product $A = \Lambda\, H$, where

$$\Lambda = \begin{pmatrix} 0 & 1 \\ -1 & 0 \end{pmatrix} \, , \qquad H = \begin{pmatrix} 1 & 0 \\ 0 & 1 \end{pmatrix} \, ,$$

which correspond to

$$\Lambda = \frac{\partial}{\partial q} \wedge \frac{\partial}{\partial p}, \qquad H = \frac{1}{2}(q^2 + p^2) .$$

The function f is a constant of motion if, and only if,

$$p \frac{\partial f}{\partial q} - q \frac{\partial f}{\partial p} = 0 ,$$

whose general solution is given by an arbitrary function of $q^2 + p^2$. The symmetry of X_A corresponding to the quadratic function $H = 1/2\,(q^2 + p^2)$ is X_A itself.

A vector field X_B commutes with X_A if it takes the form

$$B = \begin{pmatrix} a & b \\ -b & a \end{pmatrix}, \qquad a, b \in \mathbb{R} ,$$

because that is the general form of a matrix commuting with the given matrix A.

Then X_B can be written as $X_B = a\,X_I + b\,X_{i\,\sigma_2}$, where σ_2 is the second Pauli matrix. We can see that X_I is not a Hamiltonian vector field with respect to Λ, because there is no symmetric matrix such that

$$\begin{pmatrix} 1 & 0 \\ 0 & 1 \end{pmatrix} = \begin{pmatrix} 0 & 1 \\ -1 & 0 \end{pmatrix} \begin{pmatrix} \alpha & \beta \\ \beta & \gamma \end{pmatrix} .$$

However, as we have seen above, $X_{i\,\sigma_2}$ is the given dynamical system, which is Hamiltonian.

Another remarkable property is that if the vector field X_A is Hamiltonian, then the bivector field Λ is invariant under X_A, because if A admits a factorization like in (4.6), then

$$\Lambda\, A^T + A\,\Lambda = \Lambda\, H\, \Lambda^T + \Lambda\, H\, \Lambda = \Lambda\, H\, (\Lambda^T + \Lambda) = 0 ,$$

and therefore condition (4.5) holds.

In the second case we mentioned before, when $h(\Lambda) = -1$, the bi-vector Λ should have the form

$$\Lambda = c^{ij}{}_k\, x^k \frac{\partial}{\partial x^i} \wedge \frac{\partial}{\partial x^j},$$

where $c^{ij}{}_k$ are constants. The 1-form α (with $h(\alpha) = 1$) looks like

$$\alpha = a_i\, dx^i ,$$

with each a_i being a constant. The inverse problem now for Γ is equivalent to determining

$$A^j{}_k = -c^{ji}{}_k a_i ,$$

because

$$-i(a_l\, dx^l)\left(c^{ij}{}_k\, x^k \frac{\partial}{\partial x^i} \wedge \frac{\partial}{\partial x^j}\right) = a_i\, c^{ji}{}_k\, x^k \frac{\partial}{\partial x^j} ,$$

where c is a (2, 1) tensor, skew-symmetric in the superscripts and a is a constant vector. From now on we will concentrate on the first situation, leaving the second case for a further discussion that will bring us into the general theory of Lie algebras and Lie groups (see Sect. 4.3.3).

4.2.3.2 The Factorization Theorem

Let assume that the vector field X_A is Hamiltonian with respect to the constant Poisson structure Λ and with Hamiltonian the quadratic function defined by the matrix H. Therefore, A admits the factorization $A = \Lambda H$, with $H^T = H$ and $\Lambda^T = -\Lambda$. Then, $A^T = H^T \Lambda^T = -H \Lambda$, and consequently,

$$A \Lambda = \Lambda H \Lambda = -\Lambda A^T . \tag{4.7}$$

A few simple consequences of these facts are:

1. $A^T = -H \Lambda$,
2. $H A = H \Lambda H = -A^T H$.
3. Having in mind that the trace of a matrix coincides with that of its transpose, and that the trace is invariant under circular permutation, we see that $\mathrm{Tr}\,(A) = 0$.
4. For any integer number k (if A is invertible, for k positive in general), a similar argument works for A^{2k+1}, and therefore,

$$\mathrm{Tr}\,(A^{2k+1}) = 0 . \tag{4.8}$$

5. From the previous property it follows that the minimal polynomial and the characteristic polynomial of the matrix A only contains even degree powers. As a consequence, both real and imaginary eigenvalues will appear in pairs; real eigenvalues $\pm\lambda$ with the same multiplicity, while other complex eigenvalues λ are such that $-\lambda$ and their conjugates $\pm\lambda^*$ are also eigenvalues with the same multiplicity.
6. On the other side, $A^3 = \Lambda H \Lambda H \Lambda H = \Lambda\,(-A^T H A)$, shows that the vector field X_{A^3} is Hamiltonian with respect to the same Poisson structure. A simple iteration of the reasoning will lead to

$$A^{2k+1} = (-1)^k\, \Lambda\, ((A^T)^k\, H\, A^k) ,$$

and therefore, all odd powers of A give raise to Hamiltonian vector fields with respect to the same Poisson structure. Of course, all of them pairwise commute, but however they are not independent, because Cayley theorem establishes a relation among these powers: the matrix A satisfies its characteristic equation.

Moreover, if we recall that the correspondence between the set of linear homogeneous vector fields and associated Hamiltonians is a Lie algebra homomorphism, we can conclude that any linear Hamiltonian vector field admits constants of motion in involution provided by

$$f_k(x) = (-1)^k((A^T)^k\, H\, A^k)_{ij}\, x^i\, x^j\ .$$

We will study in another section what happens in the particular case of a regular Λ, i.e. when we require Λ to be invertible.

Summing up these properties we have that a necessary condition for A to be factorizable in the product of an skew-symmetric and a symmetric matrix is that the trace of all odd powers is zero. This implies that the characteristic and minimal polynomials of A have only even degree terms. Thus the real and purely imaginary eigenvalues occur always in pairs $(\lambda, -\lambda)$, the complex eigenvalues occurs in quadruples $(\lambda, \bar{\lambda}, -\lambda, -\bar{\lambda})$ and the zero eigenvalue has even degeneracy. If Λ is a skew-symmetric factor for A, it is also a skew-symmetric factor for any odd power of A. Even powers of A (A invertible) will permute admissible skew-symmetric factors.

The previous conditions are also sufficient to characterize linear systems that can be factorized. We would like to end this section by stating it in the form of a theorem [Gi93].

Theorem 4.2 (Factorization theorem) *Let Γ be a linear dynamical system on the real vector space E with associated linear operator A. A necessary and sufficient condition for the vector field Γ to be factorizable, i.e., that there exists a skew-symmetric* $(2,0)$ *tensor Λ and a symmetric* $(0,2)$ *tensor H such that $\Gamma = \Lambda \circ H$, is that all odd powers of A are traceless* ($\mathrm{Tr}\, A^{2k+1} = 0$, $k \geq 0$) *and:*

i. *No further condition if the eigenvalues of A are non-degenerate or purely imaginary,*
ii. *For degenerate real or complex eigenvalues λ, the Jordan block belonging to λ has the same structure as the Jordan block belonging to $-\lambda$,*
iii. *Zero eigenvalues have even multiplicity.*

In such case the flows corresponding to odd powers of A will leave invariant both the skew-symmetric and the symmetric tensors Λ and H.

Moreover the linear flows $\varphi_t^{(k)} = \exp(tA^{2k})$, $k = 1, 2, \ldots$ corresponding to even powers of A will transform the previous factorization of Γ in an alternative one $\Gamma = \Lambda_t^{(k)} H_t^{(k)}$ with $\Lambda_t^{(k)} = \exp(tA^{2k})\Lambda \exp(tA^{2k})$ and $H_t^{(k)} = \exp(tA^{2k})H \exp(tA^{2k})$.

Proof The necessity follows from the discussion before. We show now that the previous conditions are sufficient.

The canonical real Jordan blocks for A are assume to be of the following types (see the classification after Theorem 3.23 and Eq. (3.121)): (simple Jordan blocks) type I, type II, type III, (non-simple Jordan blocks) type IVa, type IVb and type IVc.

There are three cases to consider (of course in all them we assume Tr $A^{2k+1} = 0$):

1. In the first case A has non-degenerate eigenvalues, real or complex. We consider first the simple case when the eigenvalues are all real and distinct. In this case an invertible matrix P will bring A to the canonical form:

$$\widetilde{A} = P^{-1}AP = \begin{pmatrix} \lambda_1 & \cdots & 0 & 0 & \cdots & 0 \\ \vdots & \ddots & \vdots & \vdots & \ddots & \vdots \\ 0 & & \lambda_n & 0 & & 0 \\ 0 & & 0 & -\lambda_1 & \cdots & 0 \\ \vdots & & & \vdots & \ddots & \vdots \\ 0 & & & 0 & \cdots & -\lambda_n \end{pmatrix}$$

   which is obviously Hamiltonian with respect to the canonical Poisson structure (the canonical form of a non-degenerate Λ) that in linear coordinates $x_1, \ldots, x_n, x_{n+1}, \ldots, x_{2n}$ is given by $\Lambda = \sum_k \frac{\partial}{\partial x\,k} \wedge \frac{\partial}{\partial x\,k+n}$. Then we use P^{-1} to transform Λ into a possible Poisson tensor for A.
   It is also possible to bring A to the form:

$$\widetilde{A}' = P'^{-1}AP' = \begin{pmatrix} \lambda_1 & 0 & & & & & 0 \\ 0 & -\lambda_1 & 0 & & & & 0 \\ 0 & & \lambda_2 & 0 & & & 0 \\ 0 & & 0 & -\lambda_2 & 0 & & 0 \\ \vdots & & & & \ddots & & \vdots \\ 0 & & & & & \lambda_n & 0 \\ 0 & & & & & 0 & -\lambda_n \end{pmatrix},$$

   which is Hamiltonian with respect to the Poisson tensor:

$$\widetilde{\Lambda} = \begin{pmatrix} \widetilde{\Lambda}_1 & 0 & & 0 \\ 0 & \widetilde{\Lambda}_2 & & 0 \\ \vdots & & \ddots & \vdots \\ 0 & & & \widetilde{\Lambda}_n \end{pmatrix}$$

   with $\widetilde{\Lambda}_k = \begin{pmatrix} 0 & \lambda_k \\ -\lambda_k & 0 \end{pmatrix}$. This Poisson structure shows that our space $\mathbb{R}^{2n}$ decomposes into 2-dimensional spaces W_k such that our starting vector field Γ decomposes accordingly into the direct sum of Hamiltonian vector fields $\Gamma_k = \Gamma\,|_{W_k}$ for the subspace corresponding to the pair $(\lambda_k, -\lambda_k)$.
2. We consider now the situation of complex non-degenerate eigenvalues. We can construct a non-degenerate Poisson structure for the subspace W_k corresponding to the quadruple $\{\lambda\} = (\lambda, \bar{\lambda}, -\lambda, -\bar{\lambda})$ and then we consider the direct sum of the Poisson tensors, the direct sum of Hamiltonian forms and vector fields. We

can write the restriction of A to $W_{\{\lambda\}}$ in the form:

$$\begin{pmatrix} J_\lambda & 0 \\ 0 & -J_\lambda^T \end{pmatrix}$$

with J_λ of type III. This matrix is Hamiltonian with respect to the canonical Poisson structure.
Therefore when A contains only simple Jordan blocks and real or complex non-degenerate eigenvalues, the system can be factorized by requiring that the trace of odd powers vanishes.

3. As a second case we consider the situation when there is degeneration in the eigenvalues of A, but eigenvalues are not purely imaginary or zero.
For a Hamiltonian vector field Γ the Jordan block belonging to a real or complex eigenvalue λ has the same structure as the Jordan block belonging to $lambda$, that is, if in the almost diagonal form of A there is a certain Jordan block J, there must be also J' conjugate to $-J$ (see [Ar76], Appendices 6 and 7). Then when this is the case $A_\lambda = A\mid_{W_\lambda}$ can be brought to the form:

$$\widetilde{A}_\lambda = \begin{pmatrix} J & 0 \\ 0 & -J^T \end{pmatrix},$$

which again is Hamiltonian with respect to the canonical Poisson structure. Thus for degenerate eigenvalues (excluding imaginary eigenvalues and zero eigenvalues) the requirement that the Jordan blocks belonging to λ and $-\lambda$ have the same structure gives a necessary and sufficient condition (along with the zero trace of all odd powers of A) for A to be Hamiltonian.

4. We consider now imaginary eigenvalues. It is clearly sufficient to consider a matrix A with only one pair $\pm i\lambda$ n times degenerate with one non-simple Jordan block. The almost diagonal representation of A is:

$$\widetilde{A} = \begin{pmatrix} J & \mathbb{I} & & \\ \vdots & J & & \\ \vdots & & \ddots & \mathbb{I} \\ 0 & \cdots & 0 & J \end{pmatrix} \tag{4.9}$$

with $J = \begin{pmatrix} 0 & \lambda \\ -\lambda & 0 \end{pmatrix}$ repeated n times. Let us first suppose that n is odd. For $n = 1$ we have, of course, $A = J$ and a possible Λ is given by the canonical one:

$$\Lambda_0 = \begin{pmatrix} 0 & 1 \\ -1 & 0 \end{pmatrix}.$$

For arbitrary (but odd) n, we may set:

$$\widetilde{\Lambda} = \begin{pmatrix} 0 & 0 & \cdots & 0 & \Lambda_0 \\ 0 & \vdots & & -\Lambda_0 & 0 \\ \vdots & & \ddots & \vdots & \vdots \\ 0 & -\Lambda_0 & & & 0 \\ \Lambda_0 & 0 & \cdots & 0 & 0 \end{pmatrix}$$

that is, the Λ_0 are repeated on the counterdiagonal with alternating sign. Then:

$$\widetilde{\Lambda}^{-1}\widetilde{A} = \begin{pmatrix} 0 & 0 & \cdots & 0 & -\Lambda_0 J \\ 0 & \vdots & & \Lambda_0 J & \Lambda_0 \\ \vdots & & \ddots & \vdots & \vdots \\ 0 & \Lambda_0 J & -\Lambda_0 & & 0 \\ -\Lambda_0 J & \Lambda_0 & \cdots & 0 & 0 \end{pmatrix}$$

Now the terms in the principal counterdiagonal, which are in even place, after transposition of $\widetilde{\Lambda}^{-1}\widetilde{A}$ go in terms of even place too, and the sign is unchanged. Moreover $\Lambda_0 J$ is symmetrical, thus as for these terms $\widetilde{\Lambda}^{-1}\widetilde{A}$ is symmetrical. Let us now consider the second non-null counterdiagonal. There is an even number of them and the transposition sends the transpose of a term Λ_0 into the place occupied by a former $-\Lambda_0$ but this is just Λ_0^T and also for these terms $\widetilde{\Lambda}^{-1}\widetilde{A}$ turns out to be symmetric.
For even n, a possible Λ is:

$$\widetilde{\Lambda} = \begin{pmatrix} 0 & 0 & \cdots & 0 & \mathbb{I} \\ 0 & \vdots & & -\mathbb{I} & 0 \\ \vdots & & \ddots & \vdots & \vdots \\ 0 & \mathbb{I} & & & 0 \\ -\mathbb{I} & 0 & \cdots & 0 & 0 \end{pmatrix} \tag{4.10}$$

and the proof goes along the same lines as before.

5. Finally, we consider zero eigenvalues. When $\lambda = 0$ the expression Eq. (4.9) becomes:

$$\widetilde{A} = \begin{pmatrix} 0 & \mathbb{I} & & \\ \vdots & 0 & & \\ \vdots & & \ddots & \mathbb{I} \\ 0 & \cdots & 0 & 0 \end{pmatrix}$$

with n even, and then the $\widetilde{\Lambda}$ given in Eq. (4.10) gives a Poisson structure for Γ. □

4.2.3.3 The Symmetric Case Again

At this point we compare the condition for X_A to preserve a symmetric (0, 2)-tensor and a symplectic structure. We have found $A^T\Omega+\Omega A=0$ for the case, $A^TG+GA=0$ for the symmetric case. By using $\Omega A=H$ symmetric and $(GA^T)^T=B$ skew-symmetric, we have $A=\Omega^{-1}H$ or $A=G^{-1}B$ respectively. Therefore if det $A\neq 0$, any solution of the first problem also provides a solution for the second problem.

Actually, if $A^T\Omega+\Omega A=0$, then G defined by $G=\Omega A$ is such that

$$A^TG+GA=A^T\Omega A+\Omega AA=(A^T\Omega+\Omega A)A=0,$$

and conversely, if $A^TG+GA=0$, then $\Omega=GA^{-1}$ is such that

$$A^T\Omega+\Omega A=A^TGA^{-1}+GA^{-1}A=(A^TG+GA)A^{-1}=0.$$

To put it differently, if the symmetric matrix associated with the Hamiltonian is not degenerate we can use it to define a (pseudo) metric tensor. Vice versa, the skew symmetric matrix GA can be used to construct a symplectic structure and G will be used to define the Hamiltonian function. Thus our problem amounts to find necessary and sufficient conditions for A to be decomposed in the product of a symmetric matrix by a skew-symmetric one with one or both factors non degenerate.

From $A^T\Omega+\Omega A=0$ or $A^TG+GA=0$, we find

$$(A^T)^3\Omega=-(A^T)^2\Omega A=A^T\Omega A^2=-\Omega A^3$$
$$(A^T)^3g=-(A^T)^2GA=A^TGA^2=-GA^3$$

Therefore all odd powers share the same property. In either one of the two problems we have

$$A=S(-A^T)S^{-1},$$

i.e., a necessary and sufficient condition for A to be decomposed into the product of a symmetric matrix by a skew-symmetric one is that the matrix is conjugated to the opposite of its transpose. This has the consequence that the characteristic polynomial and minimal polynomial only contain even powers. Both polynomials can be considered as polynomials in the variable $y=x^2$ and degree $\frac{1}{2}$ of the starting ones.

It is possible now to make a few comments on the uniqueness of the decomposition. We start with $A^T\Omega+\Omega A=0$ and consider any symmetry transformation for A, i.e., an invertible matrix S such that $SAS^{-1}=A$. We notice that also $S^TA^T=A^TS^T$. Now, from $A^T\Omega+\Omega A=0$ we get by left multiplication by S^T and right multiplication by S,

$$S^TA^T\Omega S+S^T\Omega AS=0$$

and using that $SA=AS$,

$$A^T S^T \Omega S + S^T \Omega S A = 0,$$

i.e., $S^T \Omega S$ is another alternative structure if S is not a canonical transformation. Therefore with any noncanonical transformation that is also a symmetry for A we find a new structure. Another way to approach the same problem is to start with the decomposition $A = \Omega^{-1} H$. If S is an invertible linear transformation we have

$$\begin{aligned} A &\mapsto S^{-1} A S \\ \Omega &\mapsto S^T \Omega S \\ H &\mapsto S^T H S \\ \Omega^{-1} &\mapsto S^{-1} \Omega^{-1} (S^T)^{-1}. \end{aligned}$$

Now from $A = \Omega^{-1} H$ we find

$$S^{-1} A S = S^{-1} \Omega^{-1} (S^T)^{-1} S^T H S$$

and for $S^{-1} A S = A$ we get the new decomposition.

Notice that not all possible different decompositions of A can always be obtained by the previous method. However notice that if A admits two different decompositions if

$$A = \Omega_1^{-1} H_1 = \Omega_2^{-1} H_2$$

there exists a linear coordinate transformation taking Ω_1 into Ω_2 (the existence of Darboux charts implies that $Gl(n, \mathbb{R})$ acts transitively on linear symplectic structures, see Sect. 5.2.4). Let us say that

$$\Omega_1 = S^T \Omega_2 S,$$

then $\Omega_1^{-1} = S^{-1} \Omega_2^{-1} (S^T)^{-1}$ and we have $\tilde{H}_1 = S^T H_2 S$ therefore

$$\Omega_1^{-1} \tilde{H}_1 = S^{-1} \Omega_2 (S^T) H_2 S = S^{-1} \Omega_2 H_2 S = S^{-1} A S.$$

Then if $H_1 = \tilde{H}_1$ we get

$$A = \Omega_1^{-1} H_1 = S^{-1} A S,$$

i.e., S is a symmetry for A. As for noncanonical symmetries we know that all even powers of A will be infinitesimal generators of non canonical symmetries. For a generic A, i.e., all eigenvalues are simple, these symmetries exhaust all noncanonical symmetries. Therefore a linear Hamiltonian system has at least an n-parameter family of alternative Hamiltonian descriptions. As a matter of fact, because odd powers, in the generic case, generate a locally free action of $\mathbb{R}^n$ which is a canonical action we also find n independent quadratic functions which are pairwise in involution.

Therefore a generic linear Hamiltonian system is completely integrable and admits an n-parameter family of alternative Hamiltonian descriptions.

4.3 Poisson Structures

4.3.1 Poisson Algebras and Poisson Tensors

We will start the discussion of Poisson structures by considering first the simplest possible case: constant Poisson structures on linear spaces. We have found such structures already in the discussion of the factorization and inverse problem for linear dynamics Sect. 4.2.3, and in the analysis of the geometry of Noether's theorem Sects. 4.2.1 and 4.2.2. We will summarize here some of the properties that we have already found and we will prepare to move to the general situation.

Let E be a finite-dimensional linear space of dimension n and E^* will denote its dual space. Consider a skew-symmetric bilinear form Λ on E^*. Remember that because of the easy principle of tensorialization of algebraic structures Sect. 2.4.3, there exists a unique bivector field on E associated to Λ denoted by the same symbol. Let us recall that if we choose a linear basis $\{e_k\}$, and we denote by $\{e^l\}$ its dual basis, and Λ^{kl} denotes the matrix representing Λ in such basis, i.e., $\Lambda(e^k, e^l) = \Lambda^{kl}$, then the bivector Λ takes the form:

$$\Lambda = \Lambda^{ij} \frac{\partial}{\partial x^i} \wedge \frac{\partial}{\partial x^j} . \tag{4.11}$$

This allows us to define a Poisson algebra structure structure on the linear space $\mathcal{F}(E)$ of real functions in E, by setting

$$\{f, g\} = \Lambda(df, dg) = \Lambda^{ij} \frac{\partial f}{\partial x^i} \frac{\partial g}{\partial x^j} , \tag{4.12}$$

and therefore the coefficients Λ^{ij} are just

$$\Lambda^{ij} = \Lambda(dx^i, dx^j) ,$$

i.e.,

$$\{f, g\} = \Lambda(df, dg) = \frac{\partial f}{\partial x^i} \Lambda(dx^i, dx^j) \frac{\partial g}{\partial x^j} = \frac{\partial f}{\partial x^i} \{x^i, x^j\} \frac{\partial g}{\partial x^j} . \tag{4.13}$$

The function $\{f, g\}$ is called the Poisson bracket of f and g and the bilinear form on the space of smooth real-valued functions on E given by $(f, g) \mapsto \{f, g\}$ is called the Poisson bracket associated with Λ. The Poisson bracket (4.12) is skew-symmetric and bilinear by construction, and it is easy to check that this bracket will also satisfy for any triplet of functions f, g, h the property:

$$\{f, \{g, h\}\} + \{h, \{f, g\}\} + \{g, \{h, f\}\} = 0, \tag{4.14}$$

which is called the Jacobi identity. This means that $\{\cdot, \cdot\}$ endows $\mathcal{F}(E)$ with a Lie algebra structure. Moreover, it is easy to check that it satisfies Leibniz identity too:

$$\{gh, f\} = \{g, f\}h + \{h, f\}g\,, \tag{4.15}$$

and therefore, for every function f the map $g \mapsto \{g, f\}$ is a derivation of the associative algebra $\mathcal{F}(E)$ with the usual product of functions as composition law. The properties above define a Poisson algebra structure on the space of smooth functions on E. To be precise we have the definition:

Definition 4.3 A Poisson algebra is a real linear space $\mathcal{F}$ (finite or infinite-dimensional) which is an associative algebra with respect to a bilinear product $\cdot$, a Lie algebra with respect to a skew-symmetric bilinear product $\{\cdot, \cdot\}$ and both operations are compatible in the sense of Eq. (4.15), i.e., the Lie bracket satisfies the Leibniz identity with respect to the associative product.

Notice that because of the previous discussion, any constant bivector Λ on a linear space E defines a Poisson algebra structure on the space of smooth functions on E. Conversely, if we have a Poisson structure on the algebra of smooth functions on a linear space E, we may define the bivector:

$$\Lambda = \{x^i, x^j\} \frac{\partial f}{\partial x^i} \frac{\partial g}{\partial x^j}\,, \tag{4.16}$$

where x^i are linear coordinates on E.

Exercise 4.3 Prove that the object Λ defined by Eq. (4.16) actually defines a bivector field on E.

Notice that in general the bivector field Λ associated to a Poisson algebra structure will not be constant. In fact we can extend the construction of Λ not just considering the algebra of smooth functions on a linear space E but to the algebra of smooth functions on a smooth manifold M. Thus if we have a Poisson algebra structure on the associative algebra of smooth functions on M, we may define a bivector field Λ by Eq. (4.16) with x^i any local chart on M. Because of Exercise 4.3 the local expression provided by (4.16) actually provides a globally well-defined bivector field on M.

Definition 4.4 Given a Poisson algebra structure $(\mathcal{F}(M), \cdot, \{\cdot, \cdot\})$ on the associative algebra of smooth functions on a manifold M, we will call the bivector field Λ defined locally by Eq. (4.16) the Poisson tensor associated to the given Poisson algebra structure on $\mathcal{F}(M)$.

If Λ is the Poisson tensor of a Poisson algebra structure on $\mathcal{F}(M)$ it is an immediate consequence of Leibnitz's rule, Eq. (4.15), that the assignment:

$$X_f(g) = \{g, f\} \tag{4.17}$$

defines a vector field on M because it actually defines a derivation on $\mathcal{F}(M)$. We will denote this vector field by X_f and it can also be written as:

$$X_f = -\widehat{\Lambda}(df)$$

where $\widehat{\Lambda}$ denotes as before the natural bundle map $\widehat{\Lambda}\colon T^*M \to TM$ defined as $\langle\widehat{\Lambda}(\alpha), \beta\rangle = \Lambda(\alpha, \beta)$. The vector field X_f will be called the Hamiltonian vector field associated to f.

When the bi-vector Λ is such that $\ker \widehat{\Lambda} = 0$, we will say that it is non-degenerate. In this case the bundle map $\widehat{\Lambda}$ will be invertible and we can construct a 2-form ω on M associated to Λ by means of:

$$\omega(X, Y) = \Lambda(\widehat{\Lambda}^{-1}(X), \widehat{\Lambda}^{-1}(Y)) .$$

We will show later that the 2-form ω defined in this way is closed (and non-degenerate). Such forms will be called symplectic (see Sect. 5.2.1, Eq. (5.4)).

Taking into account the explicit expression of the Poisson bracket given above, we see that the Hamiltonian vector field X_f can be written as:

$$X_f = \{x^i, x^j\} \frac{\partial f}{\partial x^j} \frac{\partial}{\partial x^i} . \tag{4.18}$$

It is easy to check that:

Proposition 4.5 *Let $\mathcal{F}$ be Poisson differentiable algebra (that is, $\mathcal{F}$ is a differentiable algebra), the assignment $f \mapsto X_f$, with $X_f(g) = \{g, f\}$ is a Lie algebra anti-homomorphism:*

$$[X_f, X_g] = -X_{\{g,f\}} \qquad f, g \in \mathcal{F}$$

Proof Obviously, the correspondence is $\mathbb{R}$-linear. Let f, g and h be arbitrary functions in $\mathcal{F}$. Thus,

$$[X_f, X_g]h = X_f(X_g h) - X_g(X_f h) = \{X_g h, f\} - \{X_f h, g\} = \{\{h, g\}, f\} - \{\{h, f\}, g\} .$$

Using Jacobi identity the preceding relation can be written

$$[X_f, X_g]h = \{\{h, g\}, f\} + \{\{f, h\}, g\} = -\{\{g, f\}, h\} = -\{h, \{f, g\}\} = -X_{\{f,g\}}h ,$$

from which we obtain $[X_f, X_g] = -X_{\{f,g\}}$. □

Consequently, the distribution defined by the Hamiltonian vector fields is involutive.

A function h is a constant of the motion for X_f if and only if $\{f, h\} = 0$. Due to the Jacobi identity, the Poisson bracket of two constants of motion is still a constant of the motion, because if $\{f, g\} = 0$ and $\{f, h\} = 0$, then

$$\{f, \{g, h\}\} = -\{g, \{h, f\}\} - \{h, \{f, g\}\} = 0 .$$

On the other side, when the Poisson structure is degenerate there can exist functions f such that $\{f, h\} = 0$ for any function h, i.e., $df \in \ker \Lambda$, and therefore they will be constant for any possible motion. These functions are called Casimir functions. The associated vector field vanishes identically.

The Jacobi identity (4.14) can be expressed in terms of the components Λ^{ij} of the bivector field Λ. This condition is equivalent to the quadratic relation:

$$\Lambda^{il}\frac{\partial \Lambda^{jk}}{\partial x^l} + \Lambda^{kl}\frac{\partial \Lambda^{ij}}{\partial x^l} + \Lambda^{jl}\frac{\partial \Lambda^{ki}}{\partial x^l} = 0 . \tag{4.19}$$

Of course this condition is automatically satisfied when the components Λ^{ij} are constant. It is clear that the non degeneracy of Λ played no role in the stated properties, therefore we can also allow Λ to be degenerate and eventually of a non constant rank. Therefore we have that a bivector field Λ, degenerate or not, defines a Poisson bracket if and only if,

$$0 = \Lambda(df, \Lambda(dg, dh)) + \Lambda(dh, \Lambda(df, dg)) + \Lambda(dg, \Lambda(dh, df)) . \tag{4.20}$$

An important remark is that this relation is not additive, and then, even if Λ_1 and Λ_2 satisfy this condition, it may be false for the sum $\Lambda_1 + \Lambda_2$. When $\Lambda_1 + \Lambda_2$ is also a Poisson structure we will say that the Poisson structures Λ_1 and Λ_2 are compatible, and then any linear combination $\lambda_1 \Lambda_1 + \lambda_2 \Lambda_2$ is a Poisson structure too.

Pairs of compatible bivectors are characterized by:

$$\Lambda_1^{il}\frac{\partial \Lambda_2^{jk}}{\partial x^l} + \Lambda_1^{kl}\frac{\partial \Lambda_2^{ij}}{\partial x^l} + \Lambda_1^{jl}\frac{\partial \Lambda_2^{ki}}{\partial x^l} + \Lambda_2^{il}\frac{\partial \Lambda_1^{jk}}{\partial x^l} + \Lambda_2^{kl}\frac{\partial \Lambda_1^{ij}}{\partial x^l} + \Lambda_2^{jl}\frac{\partial \Lambda_1^{ki}}{\partial x^l} = 0 . \tag{4.21}$$

An interesting way of expressing this property is by means of a commutator $[\,\cdot\,,\cdot\,]$ introduced by Schouten [Sc53] and used by Lichnerowicz in this context [Li77]. In fact it was Lichnerowicz who showed that Jacobi's identity is given by

$$[\Lambda, \Lambda] = 0 . \tag{4.22}$$

We recall that the Schouten bracket (see Appendix D for other related brackets) is the unique extension of the Lie bracket of vector fields to the exterior algebra of multivector fields, making it into a graded Lie algebra (the grading in this algebra given by the ordinary degree of the multivectors minus one). In fact, given a multivector V, the linear operator $[V, \cdot\,]$ defines a derivation on the exterior algebra of multivector fields on M. The degree of $[V, \cdot\,]$ equals the ordinary degree of V. So, for any triplet (U, V, W) of multivectors,

$$[U, V \wedge W] = [U, V] \wedge W + (-1)^{u(v+1)} V \wedge [U, W]\,,$$

u and v being the ordinary degree of U and V respectively. In the particular case of the wedge product of vector fields,

$$[X\wedge Y, U\wedge W] = [X, U]\wedge Y\wedge W + X\wedge[Y, U]\wedge V + Y\wedge[X, W]\wedge U + X\wedge U\wedge[Y, W]\,. \tag{4.23}$$

Therefore if $V = X \wedge Y$ is a monomial bivector, then

$$[V, V] = 2X \wedge Y \wedge [X, Y]\,. \tag{4.24}$$

Now, it is clear that if Λ_1 and Λ_2 are Poisson structures, then $\Lambda_1 + \Lambda_2$ is a Poisson structure if and only if $[\Lambda_1 + \Lambda_2, \Lambda_1 + \Lambda_2] = 0$, i.e., iff $[\Lambda_1, \Lambda_2] = 0$. Provided this condition is met, any linear combination of Λ_1 and Λ_2 with real coefficients is also a Poisson structure.

4.3.2 The Canonical 'Distribution' of a Poisson Structure

A Poisson structure provides a linear correspondence, $f \mapsto X_f$, which is equivalent to giving for any point $x \in M$ an $\mathbb{R}$-vector bundle map $\widehat{\Lambda}(x)\colon T_x^*M \to T_xM$ which preserves the base manifold. The rank of $\widehat{\Lambda}(x)$ at the point $x \in E$ was called the rank of the Poisson structure at the point x. In order to simplify notation we will not distinguish Λ from $\widehat{\Lambda}$.

Now, for each $x \in M$ the image of T_xM under $\widehat{\Lambda}_x$ is a linear subspace $\Lambda_x(T_x^*M) \subset T_xM$. In general the family of subspaces $\Lambda_x(T_x^*M) \subset T_xM$ will not define a distribution on TM because its rank can change.

However because of Proposition 4.5, the family of Hamiltonian vector fields X_f are involutive. Then it is possible to extend Frobenius theorem to this setting to prove that there is a singular foliation of the manifold such that the tangent spaces to the leaves at each point are actually the subspace $\Lambda_x(T_x^*M)$. This result is sometimes referred to as Kirillov's theorem and the singular foliation of a Poisson structure called Kirillov's foliation [Ki62].

Actually it can be proved by using transversality arguments that the set of points where the rank of Λ is maximal (hence constant) is an open dense set in M called the regular part of the Poisson tensor. If we consider now the restriction of Λ to the regular part of M, it will define an involutive distribution and we can apply the Frobenius theorem, Theorem 2.16.

We have started with Poisson algebra $\mathcal{F}$ and considered the set of Casimir functions

$$\mathcal{C}(\mathcal{F}) = \{h \in \mathcal{F} \mid \{f, h\} = 0, \forall f \in \mathcal{F}\} .$$

Obviously the Casimir functions define a Lie subalgebra of $\mathcal{F}$. On the other side we recall that every such subalgebra has associated an equivalence relation, denoted $\approx$, in the state space M by saying that two points of M, x and y, are equivalent if $f(x) = f(y)$, for each function $f \in \mathcal{C}(\mathcal{F})$. Correspondingly we have a foliation whose leaves are the level set of the map $\pi\colon M \to M/\approx$. Notice that in general $\mathcal{C}(\mathcal{F})$ will not be a differentiable algebra thus the space of leaves $M/\approx$ will not be a manifold.

The level sets of the Casimir functions describe the regular leaves of the Kirillov's foliation. However the singular part of M, that is, the points where the rank of Λ is not maximal, is the union of smooth submanifolds whose tangent spaces are spanned by Hamiltonian vector fields, but in order to define them we need extra functions which are not Casimirs (we will see that when studying spaces of coadjoint orbits of groups Sect. 4.3.5).

Moreover, these leaves are invariant for any Hamiltonian dynamics because the vector field X_H corresponding to a function H is by definition tangent to the foliation. This allows us to define a new Poisson structure on the leaves of the foliation which is non-degenerate, that is a symplectic form ω, by considering:

$$\omega_x(v, u) = \Lambda_x(df(x), dg(x)) ,$$

with $v = X_f(x)$ and $u = X_g(x)$. Notice that the definition of ω does not depend on the choice of the functions f such that $X_f(x) = v$. All these remarks together lead to the following theorem:

Theorem 4.6 (Kirillov's symplectic foliation theorem) *Let Λ be a Poisson structure on a smooth manifold M. Then there exists a singular foliation of M whose leaves are such that their tangent spaces are spanned by the Hamiltonian vector fields. Each leaf carries a canonical symplectic structure induced by the Poisson brackets on M.*

4.3.3 Poisson Structures and Lie Algebras

When looking for a solution of the inverse problem we can fix our attention in a concrete family of Poisson tensors. The factorization problem of dynamics discussed in Sect. 4.2.3 targets constant Poisson tensors, i.e., Poisson tensors homogeneous of degree -2. There the only condition on the components of the tensor is the skew-symmetry of them, $\Lambda^{ij} = -\Lambda^{ji}$. An extremely interesting case is when Λ^{ij} are linear functions in the coordinates,

$$\Lambda^{ij}(x) = c^{ij}{}_k\, x^k . \tag{4.25}$$

Then conditions for Λ to define a Poisson structure becomes the Jacobi identity for the structure constants of a Lie algebra Eq. (2.132):

$$\sum_{\text{cyclic } i,j,k} \sum_{l} c^{ij}{}_l \, c^{lk}{}_k = 0 \, .$$

This shows that for any Lie algebra there should be a related Poisson structure. We will discuss this correspondence in detail.

Let $\mathfrak{g}$ be a finite-dimensional Lie algebra (Definition 2.19), and let $\mathfrak{g}^*$ be its dual space. The choice of a basis $\{\xi_k\}$ for the linear pace $\mathfrak{g}$ provides us a dual basis $\{\theta^k\}$, $\langle \theta^k, \xi_l \rangle = \delta^k_l$, for $\mathfrak{g}^*$ and the corresponding set of linear coordinates x_k, $\mu = x_k \theta^k$, for all $\mu \in \mathfrak{g}^*$. Let $c_{ij}{}^k$ denote the structure constants relative to the given basis $[\xi_i, \xi_j] = c_{ij}{}^k \xi_k$. The family of linear functions

$$\Lambda^{ij}(x) = c_{ij}{}^k \, x_k$$

define the components of a Poisson tensor in $\mathfrak{g}^*$. The corresponding Poisson algebra structure on $\mathcal{F}(\mathfrak{g}^*)$ can be defined intrinsically as follows.

If $f \in C^\infty(\mathfrak{g}^*)$, the differential $(\mathrm{d}f)_\mu$ at $\mu \in \mathfrak{g}^*$ is a linear map $(\mathrm{d}f)_\mu \colon T_\mu \mathfrak{g}^* \to \mathbb{R}$. There is a natural identification of $\mathfrak{g}^*$ with $T_\mu \mathfrak{g}^*$, and in this sense $(\mathrm{d}f)_\mu$ can be thought of as an element of $(\mathfrak{g}^*)^*$ and therefore of $\mathfrak{g}$. Let us assume that $\delta_u f$ denotes such an element, i.e.,

$$\langle \nu, \delta_\mu f \rangle = (\mathrm{d}f)_\mu(\nu) = \frac{d}{dt} f(\mu + t\nu) \bigg|_{t=0} .$$

The canonical Poisson structure on $\mathfrak{g}^*$, called in what follows the Lie–Poisson structure on $\mathfrak{g}^*$, is then defined by

$$\{f, g\}(\mu) = \langle \mu, [\delta_\mu f, \delta_\mu g] \rangle \, . \tag{4.26}$$

Exercise 4.4 Check that Eq. (4.26) defines a Poisson structure on $\mathcal{F}(\mathfrak{g}^*)$.

The linear subspace $\mathcal{F}^{(1)}(\mathfrak{g}^*)$ of linear functions on $\mathcal{F}(\mathfrak{g}^*)$ is closed with respect to the Lie–Poisson algebra structure. In fact, for any $\xi \in \mathfrak{g}$ we define the linear function $x_\xi \in \mathcal{F}^{(1)}(\mathfrak{g}^*)$:

$$x_\xi(\mu) = \langle \mu, \xi \rangle \, . \tag{4.27}$$

It is quite easy to see that,

$$\delta_\mu x_\xi = \xi \, ,$$

because,

$$(\mathrm{d}x_\xi)_\mu(\nu) = \frac{d}{dt} x_\xi(\mu + t\nu) \bigg|_{t=0} = \langle \nu, \xi \rangle ,$$

and consequently, if ξ and ζ are elements of $\mathfrak{g}$,

$$\{x_\xi, x_\zeta\}(\mu) = \langle \mu, [\xi, \zeta] \rangle = x_{[\xi,\zeta]}(\mu) \,. \tag{4.28}$$

Therefore, the Poisson bracket of two linear functions is again a linear function. Thus we have proved:

Theorem 4.7 *Any Poisson algebra structure on the space of smooth functions of a finite-dimensional linear space E whose associated Poisson tensor is homogeneous of degree -1 defines a Lie algebra structure on E^*. Conversely, given a finite-dimensional Lie algebra $\mathfrak{g}$, there is a canonical Poisson structure on the space of smooth functions on $\mathfrak{g}^*$ called the Lie–Poisson structure on $\mathfrak{g}^*$. Moreover the original Lie algebra $\mathfrak{g}$ is canonically isomorphic to the Lie subalgebra of linear functions on $\mathfrak{g}^*$.*

Proof Regarding the first assertion we have already seen that if Λ is a Poisson tensor associated to a given Poisson algebra structure on $\mathcal{F}(E)$ which is homogeneous of degree -1, then its component functions with respect to a chart of linear coordinates x^i have the linear form Eq. (4.25). Then we define the Lie algebra structure on E^* by:

$$[\alpha, \beta] = \alpha_i \beta_j c^{ij}{}_k \theta^k ,$$

where θ^k is the dual basis to the basis e_i used to define the linear chart x^i, i.e., $u = x^i e_i$, $u \in E$.

Conversely, we have already seen that the Lie–Poisson bracket Eq. (4.26) defines a Poisson algebra structure on $\mathcal{F}(\mathfrak{g}^*)$. Moreover formula (4.28) shows that the linear subspace $\mathcal{F}^1(\mathfrak{g}^*)$ is closed with respect to the Lie bracket defined by the Lie–Poisson bracket. Then if we consider the map $x \colon \mathfrak{g} \to \mathcal{F}^1(\mathfrak{g}^*)$, given by $\xi \mapsto x_\xi$, again formula (4.28) says that such a map is a Lie algebra homomorphism and trivially it is an isomorphism too. □

The Poisson tensor associated to the Lie–Poisson structure on $\mathfrak{g}^*$ is easily computed from (4.28) because if we denote by x_i the linear coordinate functions in $\mathfrak{g}^*$ associated to the linear basis θ^k then with the notations above we have:

$$\{f, g\} = \frac{\partial f}{\partial x_i}\frac{\partial g}{\partial x_j}\{x_i, x_j\} = \frac{\partial f}{\partial x_i}\frac{\partial g}{\partial x_j} x_{[\xi_i,\xi_j]} = \frac{\partial f}{\partial x_i}\frac{\partial g}{\partial x_j} c_{ij}{}^k x_k \,.$$

and the Poisson tensor associated to the Lie–Poisson structure, sometimes called the Lie–Poisson tensor, has components given by (4.25). The previous construction and theorem is another instance of the tensorialization principle stated in Sect. 2.4.3, showing that an algebraic structure, in this case a Lie algebra structure, can be geometrized, in this case by defining a Poisson tensor and all the algebraic properties are captured by it.

Recall that if A is an endomorphism of a vector space E of dimension n, the determinant of A is defined by choosing a basis $\{e_1, \dots, e_n\}$ and then $Ae_1 \wedge \cdots \wedge$

$Ae_n = (\det A)\, e_1 \wedge \cdots \wedge e_n$. The choice of the basis allows us to introduce linear coordinates in E and we can consider a volume element $\Omega = dx^1 \wedge dx^2 \wedge \cdots \wedge dx^n$. Given any automorphism $A \in \mathrm{Aut}(E)$, $\det A$ is then the factor in $dA(x^1) \wedge dA(x^2) \wedge \cdots \wedge dA(x^n) = (\det A)\, \Omega$. The set of transformations with $\det A = 1$, corresponding to those linear transformations preserving the volume form, is the subgroup $SL(n) \subset GL(n)$.

In terms of geometrical objects, the previous notions become:

$$\mathcal{L}_{X_A} \Omega = (\mathrm{Tr}\, A)\, \Omega\ ,$$

and then, if $\mathrm{Tr}\, A = 0$, we get $\mathcal{L}_{X_A} \Omega = 0$.

In fact, if we define the $n-1$-form $\Theta_A = i_{X_A} \Omega$, its local expression is:

$$\Theta_A = \sum_{j=1}^{n} (-1)^{j+1} A^j{}_i x^i \, dx^1 \wedge \cdots \wedge \widehat{dx^j} \wedge \cdots \wedge dx^n\ ,$$

where $\widehat{dx^j}$ means that this term is omitted from the expression. Clearly,

$$\mathcal{L}_{X_A} \Omega = d\Theta_A = (\mathrm{Tr}\, A)\, dx^1 \wedge \cdots \wedge dx^n\ .$$

Therefore, we see that traceless endomorphisms A defining the Lie subalgebra $\mathfrak{sl}(n)$ are those such that Θ_A is exact.

Since $\mathcal{L}_\Delta \Omega = n\Omega$, i.e., Ω is homogeneous of degree n, given any linear vector field X_A, we can decompose it in the following way

$$X_A = \left(X_A - \frac{\mathrm{Tr}\, A}{n} \Delta \right) + \frac{\mathrm{Tr}\, A}{n} \Delta, \tag{4.29}$$

which means that X_A is decomposed into a traceless part and a part proportional to the Liouville vector field Δ. This corresponds to the splitting

$$A = \left(A - \frac{\mathrm{Tr}\, A}{n} I \right) + \frac{\mathrm{Tr}\, A}{n} \mathbb{I}\ .$$

Because with any vector v we can associate the constant vector field $Y_v = v^i \frac{\partial}{\partial x^i}$ and $\mathcal{L}_{Y_v} \Omega = 0$, we can construct the inhomogeneous special linear algebra $\mathfrak{isl}(n)$ by putting together traceless linear vector fields X_A and constant ones. Since

$$[X_A, Y_v] = Y_{Av}\ ,$$

the vector fields Y_v generate an ideal, and the set of vector fields X_A define a subalgebra whose intersection with the previous ideal reduces to the zero vector field. Therefore the set of vector fields $\{X_A, Y_v\}$ span the Lie algebra $\mathfrak{isl}(n)$ which is a semi-direct sum of both the subalgebra and the ideal.

Consider now an arbitrary vector field $X = X^i(x)\frac{\partial}{\partial x^i}$. We have

$$\mathcal{L}_X\Omega = dX^1(x)\wedge dx^2\wedge\cdots\wedge dx^n+\cdots+dx^1\wedge\cdots\wedge dX^n(x) = \\ = \left(\sum_{i=1}^{n}\frac{\partial X^i}{\partial x^i}\right)dx^1\wedge dx^2\wedge\cdots\wedge dx^n = (\text{div}\, X)\ \Omega .$$

We can consider the Lie algebra $\mathfrak{g}$ of vector fields $\mathfrak{X}(M)$ and the cohomology corresponding to its action on the set of functions in M (see Sect. 3.2.5 and Appendix F). In other words, we define $\Psi(X)f := \mathcal{L}_X f$, which is a linear representation of $\mathfrak{X}(M)$ because of

$$[\Psi(X)\circ\Psi(Y)-\Psi(Y)\circ\Psi(X)]f = (\mathcal{L}_X\mathcal{L}_Y-\mathcal{L}_Y\mathcal{L}_X)f = \mathcal{L}_{[X,Y]}f = \Psi([X,Y])f.$$

The elements of $Z^1(\mathfrak{g}, \mathcal{F}(M))$ are linear maps (not $\mathcal{F}(M)$-linear) $\alpha\colon \mathfrak{g}\to\mathcal{F}(M)$ satisfying (see Eq. (3.17)):

$$\mathcal{L}_X\alpha(Y)-\mathcal{L}_Y\alpha(X) = \alpha([X,Y])\ .$$

We have seen that the divergence of a vector field is the generalization of the trace and that the set of all divergence-less vector fields is an infinite-dimensional Lie algebra. We want to remark that the association div : $\mathfrak{X}(E)\to\mathcal{F}(E)$, given by $X\mapsto \text{div}X$ is a 1-cocycle, i.e.,

$$\mathcal{L}_X\text{divY}-\mathcal{L}_Y\text{div}X = \text{div}[X,Y]\ , \tag{4.30}$$

and therefore, the set ker div $\subset\mathfrak{X}(E)$ is the Lie subalgebra which generalizes $\mathfrak{isl}(n)$. We notice that in general it is not possible to decompose a vector field like in (4.29). However, it is possible to consider a new vector space $E\times\mathbb{R}$, and a new volume

$$\widetilde{\Omega} = dx^1\wedge dx^2\wedge\cdots\wedge dx^n\wedge ds = \Omega\wedge ds$$

such that we can associate to any vector field X a new vector field $\widetilde{X} = X+(\text{div}X)s\frac{\partial}{\partial s}$ which is divergenceless with respect to $\widetilde{\Omega}$. In fact,

$$\mathcal{L}_{\widetilde{X}}\widetilde{\Omega} = (\mathcal{L}_X\Omega)\wedge ds-(\text{div}X)\Omega\wedge ds = 0\,.$$

4.3.4 The Coadjoint Action and Coadjoint Orbits

We recall that if G is a Lie group and $\mathfrak{g}$ its Lie algebra, for every $g\in G$ the inner automorphism $i_g = L_g\circ R_g^{-1}\colon G\to G$, given by $i_g\colon g'\mapsto gg'g^{-1}$ induces a Lie algebra automorphism $i_{g*}:\mathfrak{g}\to\mathfrak{g}$, which gives raise to the adjoint representation of G on $\mathfrak{g}$, given by $\mathrm{Ad}_g = i_{g*}$. Let $\mathfrak{g}^*$ denote the dual space of $\mathfrak{g}$, the coadjoint representation of G on $\mathfrak{g}^*$ is defined naturally by:

$$\langle \mathrm{Coad}\,_g(\mu), \xi\rangle = \langle \mu, \mathrm{Ad}_{g^{-1}}(\xi)\rangle, \qquad \mu \in \mathfrak{g}^*, \quad \xi \in \mathfrak{g}\,.$$

We will call coadjoint orbits the orbits of the coadjoint action and we denote by $\mathcal{O}_\mu$ the coadjoint orbit passing through the element $\mu \in \mathfrak{g}^*$, that is $\mathcal{O}_\mu = \{\mathrm{Coad}\,_g\mu \mid g \in G\}$. Notice that $\mathfrak{g}^*$ is foliated (the foliation would be singular in general) by the coadjoint orbits of G. We want to explore the properties and structure of them.

The fundamental vector fields X_ξ corresponding to the coadjoint action of G on $\mathfrak{g}^*$, are given by

$$(X_\xi f)(\mu) = \frac{d}{dt} f[\mathrm{Coad}\,(\exp(-t\xi))\mu] \mid_{t=0},$$

where $f \in C^\infty(\mathfrak{g}^*)$ and $\mathrm{ad} \equiv \mathrm{Ad}_*$ is the adjoint representation of $\mathfrak{g}$, that is $\mathrm{ad}\,\xi(\zeta) = [\xi, \zeta]$. Then it follows, $\exp(-t\,\mathrm{ad}\,\xi) = \mathrm{Ad}\,(\exp(-t\,\xi))$. Moreover we get that for every $\zeta \in \mathfrak{g}$,

$$\langle \mathrm{Coad}\,(\exp(-t\xi))\mu, \zeta\rangle = \langle \mu, \mathrm{Ad}\,(\exp t\xi)\zeta\rangle = \langle \mu, (\exp(t\mathrm{ad}\,\xi))\zeta\rangle\,,$$

then

$$\langle \mathrm{Coad}\,(\exp(-t\xi))\mu, \zeta\rangle = \langle (\exp t\mathrm{ad}\,(\xi))^*\mu, \zeta\rangle = \langle \exp t\,(\mathrm{ad}\,(\xi))^*\mu, \zeta\rangle\,,$$

and hence

$$\mathrm{Coad}\,(\exp(-t\xi))\mu = \exp t\,(\mathrm{ad}\,(\xi))^*\mu\,.$$

Taking into account the preceding relation, the expression of the fundamental vector field X_ξ reduces to

$$(X_\xi f)(\mu) = \frac{d}{dt} f[\exp(\mathrm{ad}\,t\xi)^*\mu]|_{t=0} = df[(\mathrm{ad}\,\xi)^*\mu]\,.$$

When the function f is linear it coincides with its differential and then

$$(X_\xi f)(\mu) = f((\mathrm{ad}\,\xi)^*\mu)\,.$$

If we consider now the linear functions x_ξ (see Eq. (4.27)) we get

$$X_\xi x_\zeta(\mu) = x_\zeta[(\mathrm{ad}\,\xi)^*\mu] = \langle (\mathrm{ad}\,\xi)^*\mu, \zeta\rangle = \langle \mu, (\mathrm{ad}\,\xi)\zeta\rangle = \langle \mu, [\xi, \zeta]\rangle = x_{[\xi,\zeta]}(\mu)\,.$$

More specifically, if $\{\xi_k\}$ and $\{\theta^k\}$, for k running from 1 to dim G, are dual basis of $\mathfrak{g}$ and $\mathfrak{g}^*$ respectively, the expression in coordinates for the fundamental vector field X_{ξ_i}, now written X_i is

$$X_i = c_{ij}{}^k x_k \frac{\partial}{\partial x_j} ,$$

where $c_{ij}{}^k$ are the structure constants of $\mathfrak{g}$ with respect to the basis $\{\xi_i\}$.

Let us remark that the vector field X_i is but the Hamiltonian vector field on $\mathfrak{g}^*$ with x^i as Hamiltonian function. Moreover this means that coadjoint orbits are level sets of Casimir functions for the Lie–Poisson structure on $\mathfrak{g}^*$. In fact, if f is a Casimir function, then $\{x^i, f\} = 0$ is equivalent to $X_i f = 0$. The level sets of the Casimir functions determine submanifolds in which the Poisson structure is non-degenerate and the Hamiltonian vector fields are tangent to such submanifolds, that is the orbits of the coadjoint action.

All these remarks together, lead to the following remarkable theorem by Kostant, Kirillov and Souriau on the structure of the symplectic foliation of the Lie–Poisson structure on $\mathfrak{g}^*$.

Theorem 4.8 (Kostant-Kirillov-Souriau) *Let G be a connected Lie group. Then the symplectic leaves of the canonical Lie–Poisson structure on the dual of the Lie algebra $\mathfrak{g}^*$ of a Lie group G are the coadjoint orbits of the coadjoint action of G. Moreover the symplectic structures induced on them are given by*

$$\omega_\mu(X_f, X_h) = \langle \mu, [df(\mu), dh(\mu)] \rangle, \qquad \mu \in \mathfrak{g}^*, \quad f, h \in \mathcal{F}(\mathfrak{g}^*) .$$

Proof The proof is a direct consequence of the previous remarks. The symplectic foliation (singular in general) defined by a Poisson structure is the integral foliation corresponding to the distribution defined by the range of the Poisson tensor. But now the range of the Poisson tensor is spanned by the fundamental vector fields X_i before, hence the integral leaves are just the orbits of the connected Lie group. □

Notice that the symplectic structure ω_μ defined on the coadjoint orbit $\mathcal{O}_\mu$ takes the following simple explicit form when evaluated in fundamental fields:

$$\omega_\mu(X_i, X_j) = c_{ij}{}^k x_k ,$$

with x_k the linear coordinates of μ induced by the basis used to define the fundamental fields X_i.

We will discuss now a few examples of these results.

4.3.5 The Heisenberg–Weyl, Rotation and Euclidean Groups

4.3.5.1 The Heisenberg–Weyl Group

As a particularly interesting example we will study next the so called Heisenberg or Heisenberg–Weyl group. It is the group arising naturally in Quantum Mechanics because position and momenta have the same commutation rules as the Lie algebra of this group.

Let $HW(3)$ denote the 3-dimensional Heisenberg group,[1] namely,

$$HW(3) := \left\{ \begin{pmatrix} 1 & x & z \\ 0 & 1 & y \\ 0 & 0 & 1 \end{pmatrix} : x, y, z \in \mathbb{R} \right\}.$$

A (global) chart (or parametrization) ϕ for the group is given by:

$$\phi \begin{pmatrix} 1 & x & z \\ 0 & 1 & y \\ 0 & 0 & 1 \end{pmatrix} = (x, y, z) .$$

The composition law is:

$$(x_1, y_1, z_1) \cdot (x_2, y_2, z_2) = (x_1 + x_2, y_1 + y_2, z_1 + z_2 + x_1 y_2) .$$

The neutral element e is such that $\phi(e) = (0, 0, 0)$ and if $g \in HW(3)$ is such that $\phi(g) = (x, y, z)$ the differential of the left translation L_g at the point $e \in HW(3)$ is given by

$$(L_g)_*(e) = \begin{pmatrix} 1 & 0 & 0 \\ 0 & 1 & 0 \\ 0 & x & 1 \end{pmatrix} .$$

Thus the left-invariant vector fields determined by $\partial/\partial x|_0, \partial/\partial y|_0, \partial/\partial z|_0$ are

$$X = \frac{\partial}{\partial x}, \quad Y = \frac{\partial}{\partial y} + x\frac{\partial}{\partial z}, \quad Z = \frac{\partial}{\partial z},$$

which satisfy the commutation relations

$$[X, Y] = Z, \quad [X, Z] = [Y, Z] = 0 ,$$

from which we obtain

$$\operatorname{ad} X = \begin{pmatrix} 0 & 0 & 0 \\ 0 & 0 & 0 \\ 0 & 1 & 0 \end{pmatrix}, \quad \operatorname{ad} Y = \begin{pmatrix} 0 & 0 & 0 \\ 0 & 0 & 0 \\ -1 & 0 & 0 \end{pmatrix}, \quad \operatorname{ad} Z = 0 .$$

Let (α, β, γ) be the corresponding coordinates of a vector $\mu \in \mathfrak{hw}(3)^* \cong \mathbb{R}^3$. Then the Poisson structure in $\mathfrak{hw}(3)^*$ is given by

[1] The extension to the $2n + 1$-dimensional Heisenberg–Weyl group is straightforward.

$$\{\alpha, \beta\} = \gamma\,, \quad \{\alpha, \gamma\} = \{\beta, \gamma\} = 0\,,$$

and the Hamiltonian vector fields for the coordinate functions are

$$X_\alpha = -\gamma\frac{\partial}{\partial\beta}\,, \quad X_\beta = \gamma\frac{\partial}{\partial\alpha}\,, \quad X_\gamma = 0\,.$$

The vector fields X_α, X_β and X_γ span a distribution of dimension two (except at the points of the plane $\gamma = 0$ where all fields vanish). Therefore the integral submanifold of the distribution passing through a point (α, β, γ) with $\gamma \neq 0$ is given by the level sets of a function f solution of the system of partial differential equations $X_\alpha f = X_\beta f = 0$, i.e., the generic integral submanifolds are planes $\gamma =$ const. On the other hand, the orbits of points like $(\alpha, \beta, 0)$ reduce to one point.

To compute explicitly the fundamental vector fields of the coadjoint representation of the Heisenberg group, we note that

$$\mathrm{Ad}\,\exp aX = e^{\mathrm{ad}\,aX} = \begin{pmatrix} 1 & 0 & 0 \\ 0 & 1 & 0 \\ 0 & a & 1 \end{pmatrix}, \quad \mathrm{Ad}\,\exp bY = \begin{pmatrix} 1 & 0 & 0 \\ 0 & 1 & 0 \\ -b & 0 & 1 \end{pmatrix}, \quad \mathrm{Ad}\,\exp cZ = I\,,$$

and,

$$\mathrm{Coad}\,\exp aX = \begin{pmatrix} 1 & 0 & 0 \\ 0 & 1 & -a \\ 0 & 0 & 1 \end{pmatrix}, \quad \mathrm{Coad}\,\exp bY = \begin{pmatrix} 1 & 0 & b \\ 0 & 1 & 0 \\ 0 & 0 & 1 \end{pmatrix}, \quad \mathrm{Coad}\,\exp cZ = I\,.$$

Using the above-mentioned coordinates (α, β, γ) in $\mathfrak{g}^* \cong \mathbb{R}^3$, we have

$$\mathrm{Coad}\,\exp aX(\alpha) = \alpha\,, \quad \mathrm{Coad}\,\exp aX(\beta) = \beta - a\gamma\,, \quad \mathrm{Coad}\,\exp aX(\gamma) = \gamma\,,$$

$$\mathrm{Coad}\,\exp bY(\alpha) = \alpha + b\gamma\,, \quad \mathrm{Coad}\,\exp bY(\beta) = \beta\,, \quad \mathrm{Coad}\,\exp bY(\gamma) = \gamma\,,$$

and, finally,

$$\mathrm{Coad}\,\exp cZ = \mathrm{Id}\,.$$

Thus, the fundamental fields are:

$$\tilde{X} = \gamma\frac{\partial}{\partial\beta}\,, \quad \tilde{Y} = -\gamma\frac{\partial}{\partial\alpha}\,, \quad \tilde{Z} = 0\,,$$

namely, the opposite vector fields of X_α, X_β and X_γ.

The coadjoint orbits may also be seen directly from the explicit expression of $\mathrm{Coad}\,\exp aX$, $\mathrm{Coad}\,\exp bY$ and $\mathrm{Coad}\,\exp cZ$. If $\gamma \neq 0$, then $\mathcal{O}_\gamma = \{(\alpha, \beta, \gamma) : \alpha, \beta \in \mathbb{R}\}$, while for $\gamma = 0$ every point $(\alpha, \beta, 0)$ is an orbit.

Notice that the tensor Λ defining the Poisson structure is given by $\Lambda = \gamma\frac{\partial}{\partial\alpha}\wedge\frac{\partial}{\partial\beta}$. Here Casimir functions are arbitrary functions of γ, on the orbit $\mathcal{O}_\gamma$, parameterized by (α, β), and the non-degenerate two form corresponding to the Poisson tensor is $\omega = \frac{1}{\gamma}\,d\alpha\wedge d\beta$.

4.3.5.2 The Coadjoint Orbits of the Group of Plane Motions E_2

Consider now again the Euclidean group of plane motions $E(2)$ (Sect. 2.6.1). The Lie algebra of the group E_2 is generated by $\{\, J, P_1, P_2 \,\}$ with the relations:

$$[J, P_1] = P_2\,, \quad [J, P_2] = -P_1\,, \quad [P_1, P_2] = 0\,.$$

The Lie–Poisson structure in $\mathfrak{e}(2)^*$ is determined in the corresponding coordinates (s, p_1, p_2) by

$$\{s, p_1\} = p_2\,, \quad \{s, p_2\} = -p_1\,, \quad \{p_1, p_2\} = 0\,,$$

the Hamiltonian vector fields being

$$X_s = -p_2\frac{\partial}{\partial p_1} + p_1\frac{\partial}{\partial p_2}\,, \quad X_{p_1} = p_2\frac{\partial}{\partial s}\,, \quad X_{p_2} = -p_1\frac{\partial}{\partial s}\,.$$

The rank of the distribution generated by these vector fields is the rank of the matrix:

$$\begin{pmatrix} 0 & -p_2 & p_1 \\ p_2 & 0 & 0 \\ -p_1 & 0 & 0 \end{pmatrix},$$

and therefore the rank is two except at the points $(s, 0, 0)$. The generic orbits are therefore determined by functions f solutions of the system $X_s f = X_{p_1} f = X_{p_2} f = 0$, namely, the orbits are cylinders $p_1^2 + p_2^2 = \text{const}$. Coordinates adapted to the distribution will be 'cylindrical' coordinates (s, ρ, θ) given by $p_1 = \rho\cos\theta$, $p_2 = \rho\sin\theta$.

The tensor Λ defining the Poisson structure is given by

$$\Lambda = p_2\frac{\partial}{\partial s}\wedge\frac{\partial}{\partial p_1} - p_1\frac{\partial}{\partial s}\wedge\frac{\partial}{\partial p_2}\,.$$

In such 'cylindrical' coordinates Λ is expressed as

$$\Lambda = -\frac{\partial}{\partial s}\wedge\frac{\partial}{\partial\theta}\,,$$

because $\{s, \rho\} = 0$, and from

$$d\theta = \frac{p_1 dp_2 - p_2 dp_1}{p_1^2 + p_2^2}$$

and $d\rho = 2p_1 dp_1 + 2p_2 dp_2$ we obtain that $\{\rho, \theta\} = 0$ and $\{s, \theta\} = -1$. Note that in this example the Casimir functions are arbitrary functions of ρ.

The adjoint representation is given by

$$\operatorname{ad} J = \begin{pmatrix} 0 & 0 & 0 \\ 0 & 0 & -1 \\ 0 & 1 & 0 \end{pmatrix}, \quad \operatorname{ad} P_1 = \begin{pmatrix} 0 & 0 & 0 \\ 0 & 0 & 0 \\ -1 & 0 & 0 \end{pmatrix}, \quad \operatorname{ad} P_2 = \begin{pmatrix} 0 & 0 & 0 \\ 1 & 0 & 0 \\ 0 & 0 & 0 \end{pmatrix},$$

$$\operatorname{Ad} \exp \phi J = e^{\operatorname{ad} \phi J} = \begin{pmatrix} 1 & 0 & 0 \\ 0 & \cos\phi & -\sin\phi \\ 0 & \sin\phi & \cos\phi \end{pmatrix},$$

$$\operatorname{Ad} \exp a_1 P_1 = e^{\operatorname{ad} a_1 P_1} = \begin{pmatrix} 1 & 0 & 0 \\ 0 & 1 & 0 \\ -a_1 & 0 & 1 \end{pmatrix}, \quad \operatorname{Ad} \exp a_2 P_2 = e^{\operatorname{ad} a_2 P_2} = \begin{pmatrix} 1 & 0 & 0 \\ a_2 & 1 & 0 \\ 0 & 0 & 1 \end{pmatrix}.$$

Thus

$$\operatorname{Coad} \exp \phi J = \begin{pmatrix} 1 & 0 & 0 \\ 0 & \cos\phi & -\sin\phi \\ 0 & \sin\phi & \cos\phi \end{pmatrix},$$

and

$$\operatorname{Coad} \exp a_1 P_1 = \begin{pmatrix} 1 & 0 & a_1 \\ 0 & 1 & 0 \\ 0 & 0 & 1 \end{pmatrix}, \quad \operatorname{Coad} \exp a_2 P_2 = \begin{pmatrix} 1 & -a_2 & 0 \\ 0 & 1 & 0 \\ 0 & 0 & 1 \end{pmatrix}.$$

Using coordinates (s, p_1, p_2) in $\mathfrak{e}(2)^* \cong \mathbb{R}^3$, we find that if $(0, p_1, p_2)$ lies in an orbit $\mathcal{O}$ and $p_1^2 + p_2^2 \neq 0$, then $(s, p_1', p_2') \in \mathcal{O}$ for $s \in \mathbb{R}$, ${p_1'}^2 + {p_2'}^2 = p_1^2 + p_2^2$. Hence these orbits are cylinders. If $p_1 = p_2 = 0$, the orbit of $(s, 0, 0)$ reduces to a point.

Since,

$$\begin{aligned}
(\operatorname{Coad} \exp \phi J)(j, p_1, p_2) &= (j, p_1 \cos\phi - p_2 \sin\phi, p_1 \sin\phi + p_2 \cos\phi), \\
(\operatorname{Coad} \exp a_1 p_1)(j, p_1, p_2) &= (j + a_1 p_2, p_1, p_2), \\
(\operatorname{Coad} \exp a_2 p_2)(j, p_1, p_2) &= (j - a_2 p_1, p_1, p_2),
\end{aligned}$$

we obtain the fundamental vector fields:

$$X_J = p_2 \frac{\partial}{\partial p_1} - p_1 \frac{\partial}{\partial p_2}, \quad X_{P_1} = -p_2 \frac{\partial}{\partial s}, \quad X_{P_2} = p_1 \frac{\partial}{\partial s}.$$

Notice that

$$[X_J, X_{P_1}] = \left[p_2 \frac{\partial}{\partial p_1} - p_1 \frac{\partial}{\partial p_2}, -p_2 \frac{\partial}{\partial s} \right] = p_1 \frac{\partial}{\partial s} = X_{P_2} \,,$$

and

$$[X_J, X_{P_2}] = \left[p_2 \frac{\partial}{\partial p_1} - p_1 \frac{\partial}{\partial p_2}, p_1 \frac{\partial}{\partial s} \right] = p_2 \frac{\partial}{\partial s} = -X_{P_1} \,.$$

In ‘cylindrical’ coordinates (s, ρ, θ) the fundamental vector fields become:

$$X_J = -\frac{\partial}{\partial \theta}, \qquad X_{P_1} = -\rho \sin\theta \frac{\partial}{\partial s}, \qquad X_{P_2} = \rho \cos\theta \frac{\partial}{\partial s} \,.$$

On the orbit $\mathcal{O}_\rho := \{\, (s, \rho, \theta) \mid s \in \mathbb{R}, -\pi < \theta \leq \pi \,\}$ the non-degenerate 2-form ω corresponding to the Poisson bivector field Λ will be

$$\omega = -ds \wedge d\theta = d\theta \wedge ds \,.$$

4.3.5.3 The Rotation Group $SO(3)$

The Lie algebra of the rotation group (see Sect. 2.6.1) is:

$$\mathfrak{so}(3) = \left\{ B = \begin{pmatrix} 0 & x & -y \\ -x & 0 & z \\ y & -z & 0 \end{pmatrix} \right\} \cong \mathbb{R}^3 \,,$$

where we identify these matrices with the corresponding vectors $\mathbf{x} = (x, y, z) \in \mathbb{R}^3$. With this identification, the adjoint action is the natural action of $SO(3)$ on $\mathbb{R}^3$. The Lie algebra structure of $\mathfrak{so}(3)$ is determined by $[J_i, J_k] = \epsilon_{ikl} J_l$. Furthermore $\mathfrak{so}(3)$ is identified with $\mathfrak{so}(3)^*$ using the Killing–Cartan form $\langle B_1, B_2 \rangle = -\mathrm{Tr}\, B_1 B_2$.

The tensor Λ defining the Lie–Poisson structure is:

$$\Lambda = \epsilon_{ikl}\, x_l \,\frac{\partial}{\partial x_i} \wedge \frac{\partial}{\partial x_k} .$$

The Hamiltonian vector fields corresponding to the coordinate functions are

$$X_i = -\epsilon_{ijk} x_k \,\frac{\partial}{\partial x_j} \,,$$

which correspond, up to sign, to the fundamental vector fields. The coadjoint orbits are spheres, except for $(0, 0, 0)$, which is a point. Using coordinates in $\mathfrak{so}(3)^* \cong \mathbb{R}^3$ adapted to the characteristic distribution, namely spherical coordinates (r, θ, ϕ), we obtain the fundamental vector fields on the orbit $\mathcal{O}_r$:

$$X_1 = \sin\phi\frac{\partial}{\partial\theta} + \cot\theta\cos\phi\frac{\partial}{\partial\phi}, \quad X_2 = -\cos\phi\frac{\partial}{\partial\theta} + \cot\theta\sin\phi\frac{\partial}{\partial\phi}, \quad X_3 = -\frac{\partial}{\partial\phi}.$$

To compute the non-degenerate 2-form $\omega = f(\theta, \phi)\, d\theta \wedge d\phi$ on $\mathcal{O}_r$ we observe that the Casimir functions are arbitrary functions of r and that $\{\theta, \phi\} = 1/\sin\theta$ because

$$d\theta = \frac{1}{(x_1^2 + x_2^2 + x_3^2)\sqrt{x_1^2 + x_2^2}} \left(x_3 x_1\, dx_1 + x_3 x_2\, dx_1 - (x_1^2 + x_2^2)\, dx_3 \right) ,$$

$$d\phi = \frac{1}{x_1^2 + x_2^2}(x_1\, dx_2 - x_2\, dx_1) ,$$

and therefore

$$\{\theta, \phi\} = \Lambda(d\theta, d\phi) = \frac{1}{\sqrt{x_1^2 + x_2^2}} = \frac{1}{r\sin\theta} ,$$

from which we see that the restriction on every orbit is $\omega = r\sin\theta\, d\theta \wedge d\phi$.

4.4 Hamiltonian Systems and Poisson Structures

4.4.1 Poisson Tensors Invariant Under Linear Dynamics

Given a vector field X_A, we will be interested in looking for bivector fields invariant under X_A and which are not of the constant type. Here we are much better off, indeed we can construct quite a few invariant bivector fields which define Poisson structures. Let us start with the vector field X_A associated with the matrix A. We construct the sequence of vector fields associated with $A^0, A^1, \ldots, A^{s-1}$, with s being the degree of the minimal polynomial associated with A, and let X_j denote the vector field $X_j = X_{A^j}$. We claim that

$$\Lambda_B = B^{ij}\, X_i \wedge X_j ,$$

with $B^{ij} = -B^{ji} \in \mathbb{R}$, provides us with a Poisson bracket on $\mathcal{F}(\mathbb{R}^n)$ whose components are quadratic in coordinates and which is invariant under X_A. The invariance of Λ_B under X_A is obvious, because the B^{ij} are constant and $\mathcal{L}_{X_A} X_i = 0$. We need only to prove the Jacobi identity for the bracket defined by B. We have to show that given three arbitrary functions f_1, f_2, f_3

$$\{\{f_2, f_3\}, f_1\} = \{\{f_2, f_1\}, f_3\} + \{f_2, \{f_3, f_1\}\} ,$$

and having in mind that

$$\mathcal{L}_{X_{f_1}}(\Lambda_B(df_2, df_3)) = \{\{f_2, f_3\}, f_1\},$$

and

$$\Lambda_B(\mathcal{L}_{X_{f_1}} df_2, df_3) = \{\{f_2, f_1\}, f_3\}, \quad \Lambda_B(df_2, \mathcal{L}_{X_{f_1}} df_3) = \{f_2, \{f_3, f_1\}\},$$

we have to prove that $X_{f_1} = -\widehat{\Lambda}_B(df_1)$ satisfies

$$\mathcal{L}_{X_{f_1}}(\Lambda_B(df_2, df_3)) = \Lambda_B(\mathcal{L}_{X_{f_1}} df_2, df_3) + \Lambda_B(df_2, \mathcal{L}_{X_{f_1}} df_3),$$

which is equivalent to

$$\mathcal{L}_{X_{f_1}} \Lambda_B = 0 .$$

Now

$$\begin{aligned} \mathcal{L}_{X_{f_1}} \Lambda_B &= B^{ij}[X_{f_1}, X_i] \wedge X_j + B^{ij} X_i \wedge [X_{f_1}, X_j] \\ &= -B^{ij} \mathcal{L}_{X_i} X_{f_1} \wedge X_j - B^{ij} X_i \wedge \mathcal{L}_{X_j} X_{f_1}, \end{aligned}$$

and taking into account that

$$\mathcal{L}_{X_i} X_{f_1} = -\mathcal{L}_{X_i}(\Lambda_B(df_1)) = -(\mathcal{L}_{X_i} \Lambda_B)(df_1) + \Lambda_B(\mathcal{L}_{X_i} df_1),$$

we can write

$$\mathcal{L}_{X_{f_1}} \Lambda_B = B^{ij} \Lambda_B(\mathcal{L}_{X_i} df_1) \wedge X_j - B^{ij} X_i \wedge \Lambda_B(\mathcal{L}_{X_j} df_1),$$

and using that

$$\Lambda_B(\mathcal{L}_{X_i} df_1) = \Lambda_B(d(X_i f_1)) = B^{kl}[\mathcal{L}_{X_k} \mathcal{L}_{X_i} f_1] X_l,$$

we obtain

$$\begin{aligned} \mathcal{L}_{X_{f_1}} \Lambda_B = &-B^{ij}[B^{kl}(\mathcal{L}_{X_k} \mathcal{L}_{X_i} f_1) X_l - (\mathcal{L}_{X_l} \mathcal{L}_{X_i} f_1) X_k] \wedge X_j \\ &-B^{ij} X_i \wedge B^{kl}(\mathcal{L}_{X_k} \mathcal{L}_{X_j} f_1) X_l - (\mathcal{L}_{X_l} \mathcal{L}_{X_i} f_1) X_k], \end{aligned}$$

and by relabeling some indexes and using skew-symmetry it reduces to $\mathcal{L}_{X_{f_1}} \Lambda_B = 0$.

In a more general case in which we start with a given dynamics, i.e., a vector field X_A, we will be interested in those bivector fields for which time evolution satisfies

$$\frac{d}{dt}\{f, g\} = \{\dot{f}, g\} + \{f, \dot{g}\},$$

or in more geometric terms, the vector field X_A giving the dynamics satisfies

$$X_A\{f, g\} = \{X_A f, g\} + \{f, X_A g\} . \tag{4.31}$$

This corresponds in physical terminology to the fact that the evolution is given by canonical transformations.

The preceding relation can be written as

$$\mathcal{L}_{X_A}[\Lambda(df, dg)] - \Lambda(d\mathcal{L}_{X_A} f, dg) - \Lambda(df, d\mathcal{L}_{X_A} g) = 0 ,$$

i.e.

$$(\mathcal{L}_{X_A}\Lambda)(df, dg) = \mathcal{L}_{X_A}[\Lambda(df, dg)] - \Lambda(\mathcal{L}_{X_A} df, dg) - \Lambda(df, \mathcal{L}_{X_A} dg) = 0 ,$$

for any pair of functions f and g, and therefore, the relation (4.31) is equivalent to $\mathcal{L}_{X_A}\Lambda = 0$. Vector fields satisfying this property will be called infinitesimal Poisson symmetries.

An interesting property of Λ is the following statement: The Jacobi identity for the Poisson bracket defined by the tensor field Λ is equivalent to the fact that the bivector field Λ is preserved by $X_f = -\widehat{\Lambda}(df)$ for any function f, i.e.,

$$\mathcal{L}_{X_f}\Lambda = 0 .$$

In fact,

$$(\mathcal{L}_{X_f}\Lambda)(dg, dh) = \mathcal{L}_{X_f}[\Lambda(dg, dh)] - \Lambda(\mathcal{L}_{X_f} dg, dh) - \Lambda(dg, \mathcal{L}_{X_f} dh) ,$$

and then,

$$(\mathcal{L}_{X_f}\Lambda)(dg, dh) = \{\{g, h\}, f\} - \{\{g, f\}, h\} - \{g, \{h, f\}\} ,$$

from which we can see that $\mathcal{L}_{X_f}\Lambda = 0$, for any function f, if and only if Jacobi identity holds.

Of course, Jacobi identity suffices to prove that a Hamiltonian vector field, X_f, is such that

$$X_f\{g, h\} = \{\{g, h\}, f\} = -\{\{h, f\}, g\} - \{\{f, g\}, h\} = \{g, X_f h\} + \{X_f g, h\} .$$

It may turn out to be quite surprising that we are able to construct many independent Poisson structures for any linear dynamical system.

To help in visualizing the situation let us consider a few examples.

We consider

$$A = \begin{pmatrix} 1 & 0 \\ -1 & 1 \end{pmatrix},$$

i.e.,

$$X_A = x\,\frac{\partial}{\partial x} + y\,\frac{\partial}{\partial y} - x\,\frac{\partial}{\partial y}\,.$$

We have commuting vector fields provided by

$$X_1 = X_I = x\,\frac{\partial}{\partial x} + y\,\frac{\partial}{\partial y}\,, \quad X_2 = X_A = -x\,\frac{\partial}{\partial y} + X_1\,,$$

and we construct

$$\Lambda = X_1 \wedge X_2 = -x^2 \frac{\partial}{\partial x} \wedge \frac{\partial}{\partial y},$$

i.e.

$$\widehat{\Lambda} = \begin{pmatrix} 0 & x^2 \\ -x^2 & 0 \end{pmatrix}\,,$$

or, in other words,

$$\{x, y\} = -x^2\,.$$

We look for a 1-form $\alpha = \alpha_x\,dx + \alpha_y\,dy$ such that

$$\widehat{\Lambda}(\alpha) = X_A\,,$$

and we get $\widehat{\Lambda}(\alpha) = -x^2\,\alpha_x\,\partial/\partial y + x^2\,\alpha_y\,\partial/\partial x$, and then, if $x_2 \neq 0$,

$$\alpha(x, y) = -\frac{y-x}{x^2}\,dx + \frac{1}{x}\,dy = +d\left(\log|x| + \frac{y}{x}\right)\,.$$

In this way we have found a non-globally defined constant of the motion as it was to be expected because our vector field does not allow for globally defined constants of motion.

Another example in the same spirit is provided by

$$A = \begin{pmatrix} 1 & 0 & 0 & 0 \\ 0 & 1 & 0 & 0 \\ 0 & 0 & -1/2 & 1/2 \\ 0 & 0 & -1/2 & -3/2 \end{pmatrix} \Longrightarrow A^2 = \begin{pmatrix} 1 & 0 & 0 & 0 \\ 0 & 1 & 0 & 0 \\ 0 & 0 & 0 & -1 \\ 0 & 0 & 1 & 2 \end{pmatrix}\,,$$

then

$$X_A = x\,\frac{\partial}{\partial x} + y\,\frac{\partial}{\partial y} - \frac{1}{2}\,(-w + z)\,\frac{\partial}{\partial z} - \frac{1}{2}\,(z + 3w)\,\frac{\partial}{\partial w}\,,$$

and

$$X_{A^2} = x\,\frac{\partial}{\partial x} + y\,\frac{\partial}{\partial y} + 2\,w\,\frac{\partial}{\partial w} + \left(z\,\frac{\partial}{\partial w} - w\,\frac{\partial}{\partial z}\right),$$

we construct one of the many bivector fields

$$\Lambda = \left(x\frac{\partial}{\partial x} + y\frac{\partial}{\partial y}\right) \wedge \left(y\frac{\partial}{\partial x} - x\frac{\partial}{\partial y}\right) + \left(-\frac{1}{2}(w+z)\frac{\partial}{\partial z} + \frac{1}{2}(z-3w)\frac{\partial}{\partial w}\right)$$
$$\wedge \left(w\frac{\partial}{\partial z} + (2w-z)\frac{\partial}{\partial w}\right).$$

We point out that X_A is not compatible with any constant maximal rank Poisson structure.

Remark 4.1 From our construction of Λ it is clear that all that matters for the Jacobi identity is that it is constructed out of vector fields commuting between them and with X_A. We can show that if $Y_1, Y_2, \ldots, Y_m$ are any algebra of commuting vector fields, then

$$\Lambda_B = B^{ij}\,Y_i \wedge Y_j\,, \quad B^{ij} = -B^{ji} \in \mathbb{R}\,,$$

will define a Poisson structure which is invariant under the action of each one of the Y_k (and of any linear combination of them with coefficients Casimir elements for Λ_B).

4.4.2 Poisson Maps

Definition 4.9 A map $\phi\colon M_1 \to M_2$ between Poisson manifolds $(M_1, \{\cdot,\cdot\}_1)$ and $(M_2, \{\cdot,\cdot\}_2)$ is said to be a Poisson map when

$$\{F \circ \phi, G \circ \phi\}_1 = \{F, G\}_2 \circ \phi$$

for any couple of functions F, G in M_2.

In particular it is easy to see that if $\phi\colon M_1 \to M_2$ is a diffeomorphism, then ϕ is a Poisson map if and only if $\phi_*\Lambda_1 = \Lambda_2$. The flow ϕ_t of a vector field X in M is made of Poisson maps iff $\mathcal{L}_X\Lambda = 0$, because of

$$\phi_{t*}\mathcal{L}_X\Lambda = \frac{d}{dt}\phi_{t*}\Lambda\,.$$

A vector field X will be said to be canonical if

$$\mathcal{L}_X\Lambda = 0\,. \tag{4.32}$$

Canonical vector fields are characterized as follows: A vector field X in M is canonical iff

$$X\{F, G\} = \{XF, G\} + \{F, XG\} , \tag{4.33}$$

for any pair of functions F, G in M. Notice that in this case the vector field X is a derivation of the Lie algebra structure defined on M by $\{\cdot, \cdot\}$. Equation (4.33) follows from

$$(\mathcal{L}_X\Lambda)(dF \wedge dG) = X(\Lambda(dF \wedge dG)) - \Lambda(\mathcal{L}_X dF \wedge dG) + \Lambda(dF \wedge \mathcal{L}_X dG),$$

namely,

$$(\mathcal{L}_X\Lambda)(dF \wedge dG) = X\{F, G\} - \{XF, G\} - \{F, XG\} .$$

In particular, when X_H is a Hamiltonian vector field, it is canonical, because of the relation

$$\begin{aligned}(\mathcal{L}_{X_H}\Lambda)(dF \wedge dG) &= \mathcal{L}_{X_H}\{F, G\} - \Lambda(d(X_H F) \wedge dG) - \Lambda(dF \wedge d(X_H G)) \\ &= \{\{F, G\}, H\} - \{\{F, H\}, G\} - \{F, \{G, H\}\} , \end{aligned} \tag{4.34}$$

and then $\mathcal{L}_{X_H}\Lambda = 0$ as a consequence of the Jacobi identity. Obviously the same proof works for a closed 1-form θ, for which $\mathcal{L}_{X_\theta}\Lambda = 0$ with $X_\theta = -\Lambda(\theta)$. However, not every canonical vector field is a Hamiltonian one. As an example, let us consider the Poisson structure in $\mathbb{R}^3$, parametrized by x^1, x^2, x^3, defined by

$$\{F, G\} = \frac{\partial F}{\partial x^2}\frac{\partial G}{\partial x^1} - \frac{\partial F}{\partial x^1}\frac{\partial G}{\partial x^2}.$$

The tensor Λ defining the Poisson structure is $\Lambda = \partial/\partial x^2 \wedge \partial/\partial x^1$. The vector field $Y = \partial/\partial x^3$ is canonical, but it cannot be Hamiltonian, for if a corresponding Hamiltonian H did exist, then $Y(F) = \{F, H\}$, for any function F. Taking $F = x^3$ we will find that x^3 being a Casimir function, leads to a contradiction: $1 = Y(x^3) = \{x^3, H\} = 0$.

We can also consider the more general case of an arbitrary 1-form α in M. If we denote by X_α the vector field: $-\Lambda(\alpha)$. Then taking into account that for any pair of functions $F, G \in \mathcal{F}(M)$,

$$(\mathcal{L}_{X_\alpha}\Lambda)(dF \wedge dG) = X_\alpha\{F, G\} - \{X_\alpha F, G\} - \{F, X_\alpha G\} ,$$

and

$$\Lambda(dF \wedge \alpha) = -X_\alpha F = -\alpha(X_F) .$$

we see that

$$(\mathcal{L}_{X_\alpha}\Lambda)(dF \wedge dG) = X_\alpha\{F, G\} - X_G(X_\alpha F) + X_F(X_\alpha G) .$$

Also if we recall that

$$\alpha([X_F, X_G]) = -\alpha(X_{\{F,G\}}) = -X_\alpha\{F, G\} ,$$

then

$$(\mathcal{L}_{X_\alpha}\Lambda)(dF \wedge dG) = -\alpha([X_F, X_G]) + X_F(\alpha(X_G)) - X_G(\alpha(X_F)) = (d\alpha)(X_F \wedge X_G) .$$

Then we see that X_α is a derivation iff $d\alpha$ is closed along Hamiltonian vector fields. Thus for generic Poisson manifolds, there will be derivations that will not be locally Hamiltonian vector fields.

Definition 4.10 A submanifold S of M endowed with a Poisson structure is called a Poisson submanifold of M when $i: S \to M$ is a Poisson map.

If S is a Poisson submanifold, the Hamiltonian vector fields in M are tangent to S. In fact, let ϕ be a constraint function for S, i.e., it vanishes onto S, $\phi \circ i = 0$. Then, for every $F \in \mathcal{F}(M)$,

$$\{F, \phi\} \circ i = \{F \circ i, \phi \circ i\} = 0 ,$$

and hence $0 = \{F, \phi\} \circ i = -X_F(\phi) \circ i$, which means that the restriction of X_F onto S is tangent to S.

Conversely, if $i: S \to M$ is a submanifold such that the Hamiltonian vector fields in M are tangent to S, then we can define a Poisson structure in S in such a way that it is a Poisson submanifold of M. It suffices to define the Poisson bracket of $F, G \in \mathcal{F}(S)$, by

$$\{F, G\} = \{\bar{F}, \bar{G}\} \circ i ,$$

where $\bar{F}, \bar{G}$ are arbitrary extensions of F and G respectively, i.e., $F = \bar{F} \circ i$, $G = \bar{G} \circ i$.

We remark that in the computation of $\{F, G\} = X_{\bar{G}} F$ only the values of F along the integral curves of $X_{\bar{G}}$, which lie in S by the tangency condition, are relevant. The Poisson bracket is then well defined, as it does not depend on the choice of the representative $\bar{G}$, nor does it depend on the representative of $\bar{F}$.

4.4.3 Symmetries and Constants of Motion

In Sect. 1.5 we mentioned that contraction of invariant tensors are still invariant and Lie algebraic products of invariant objects generate new invariants.

In this section we would like to consider a dynamical system Γ which is Hamiltonian with respect to a Poisson structure Λ with Hamiltonian function H.

We will first make some general comments and then we further analyze the particular cases in which Γ is linear and the components of Λ are either constant, or linear, or quadratic. Our special interest is concentrated in the relationship between symmetries and constants of motion on one side and on alternative Poisson descriptions and their relevance for the correspondence between constants of motion and symmetries, on the other side.

We start with general aspects. If $\Gamma = -\widehat{\Lambda}(dH)$, then $\mathcal{L}_\Gamma \Lambda = 0$. When $\alpha_1, \ldots, \alpha_n$, are invariant 1-forms, say $\mathcal{L}_\Gamma \alpha_j = 0$, $j = 1, \ldots, n$, by contracting with Λ we will find vector fields $X_j = \widehat{\Lambda}(\alpha_j)$ and obviously $[\Gamma, X_j] = 0$, for any index $j = 1, \ldots, n$, i.e., the vector fields X_j are infinitesimal symmetries of Γ. We also have that X_j, $j = 1, \ldots, n$, preserve Λ, $\mathcal{L}_{X_j} \Lambda = 0$.

As we already remarked, the commutator of two symmetries, X_i and X_j, $[X_i, X_j]$, is again a symmetry. Therefore, by taking all possible commutators of our symmetries we generate a (possibly infinite-dimensional) Lie algebra. One may ask the natural question of a possible counterpart at the level of invariant 1-forms, i.e., is it possible to construct a binary operation which associates an invariant 1-form out of an invariant pair?

As a matter of fact that is possible with the following binary operation[2]:

$$\{\alpha, \beta\} = \mathcal{L}_{X_\alpha} \beta - \mathcal{L}_{X_\beta} \alpha_i - d[\Lambda(\alpha, \beta)] \,.$$

When $\alpha = df$ and $\beta = dg$ we see that

$$\{df, dg\} = d\{f, g\} \,.$$

Therefore, a Poisson structure allows us to define a Lie algebra product on 1-forms and if all ingredients are invariant under the vector field Γ, the result is still invariant. The following particular situations are of great interest.

1. Let us assume that the invariant 1-forms are associated with constants of the motion, i.e., $\alpha_i = df_i$, with $\Gamma f_i = 0$. In this case we find that $\{f_i, f_j\}$ is again a constant of motion and we have a realization with the bracket defined on 1-forms
$$d\{f_i, f_j\} = \{df_i, df_j\} \,.$$
2. Let assume that the vector fields X_i are such that there exist real constants ${c_{ij}}^k$ satisfying
$$[X_i, X_j] = \sum_{k=1}^{n} {c_{ij}}^k X_k \,,$$
or in other words, our symmetries close on a finite-dimensional real Lie algebra and will define (at least locally) an action of a Lie group G which preserves Γ and Λ.
3. The 1-forms $\alpha_1, \ldots, \alpha_n$, satisfy Maurer–Cartan relationships,

[2] This composition law is just the Lie bracket arising in the algebroid structure in the cotangent bundle of a Poisson manifold.

$$d\alpha_j + \frac{1}{2}\sum_{i,k=1}^{n} b_j{}^{ik}\alpha_i \wedge \alpha_k = 0\,,$$

where $b_j{}^{ik}$ are real numbers that are the structure constants of some finite-dimensional real Lie algebra.
In this case there will be a coaction of some Lie group G', i.e., a map

$$\Phi : E \to G'$$

such that our invariant 1-forms are generated by $\Phi^*(g^{-1}\,dg)$, $g \in G'$. Our dynamics Γ is related to a one parameter subgroup on G'.
We remark that this relation is not related to

$$\{\alpha_i, \alpha_j\} = \sum_{k=1}^{n} c_{ij}{}^{k}\,\alpha_k\,.$$

4. Let us now suppose that we are in a situation in which the conditions of (2) and (3) are satisfied, i.e., invariant vector fields closing on a Lie algebra are associated with invariant 1-forms that satisfy Maurer–Cartan relations. In general $c_{ij}{}^{k}$ and $b_j{}^{ik}$ are different structure constants and are related by a compatibility condition.
5. We can assume now that the invariant 1-forms α_j are closed along Hamiltonian vector fields,

$$d\alpha_j(X_f, X_g) = 0\,, \quad \forall f, g \in \mathcal{F}(E)\,.$$

In this case, we have

$$\mathcal{L}_{X_j}\Lambda = 0\,.$$

If

$$[X_i, X_j] = \sum_{k=1}^{n} c_{ij}{}^{k}\,X_k\,,$$

we still have a local action of a Lie group G. This situation generalizes the most familiar case we are going to discuss now as a next item.

Let us suppose that the invariant 1-forms are exact, $\alpha_j = df_j$, and that the corresponding vector fields satisfy the relation

$$[X_i, X_j] = \sum_{k=1}^{n} c_{ij}{}^{k}\,X_k \qquad c_{ij}{}^{k} \in \mathbb{R}\,.$$

Here f_j is only defined up to addition of a constant function a_j. Notice that because of the relation $d\{f_i, f_j\} = \{df_i, df_j\}$ we have, in general,

$$\{f_i, f_j\} = \sum_{k=1}^{n} c_{ij}{}^k f_k + d_{ij} ,$$

which is a Lie algebra extension of that defined by the vector fields X_i.

The 'integration constants' d_{ij} can be eliminated if and only if there exists a vector with components a_k such that

$$d_{ij} = \sum_{k=1}^{n} c_{ij}{}^k a_k ,$$

because, since $\{f, a_i\} = 0$, for any function f, we see that it suffices to choose $\tilde{f}_j = f_j + a_j$ and then the corresponding relations for the functions $\tilde{f}_j$ become

$$\{\tilde{f}_i, \tilde{f}_j\} = \sum_{k=1}^{n} c_{ij}{}^k f_k + d_{ij} = \sum_{k=1}^{n} c_{ij}{}^k \tilde{f}_k .$$

The correspondence $\phi : E \to \mathbb{R}^n$ associating with any point x on our carrier space $\phi(x) = (\tilde{f}_1(x), \ldots, \tilde{f}_n(x))$ is called the momentum map.

Remark 4.2 This situation considered from the point of view of (c) gives $b_j{}^{ik} = 0$, i.e., we are dealing with the coaction of an Abelian group $\mathbb{R}^n$ and the structure constants $c_{ij}{}^k$ arise because there is an action of the group G, associated with structure constants $c_{ij}{}^k$, induced via the Poisson tensor.

We are now ready to consider situation (e) again. We consider the algebra of Casimir functions on E, $\mathcal{C} \subset \mathcal{F}(E)$. Our invariant 1-forms will have (locally) the structure

$$\alpha_j = \sum_{i=1}^{n} c^i{}_j \, df_i + a^k{}_j \, dc_k ,$$

where $c^i{}_j$ and c_k are Casimir functions and f_i are constants of motion for Γ.

If the associated vector fields close on a finite-dimensional real Lie algebra, the "structure constants" may change from one to another leaf. The momentum map would also depend on the particular orbit (of Hamiltonian vector fields).

Let us now study particular cases. We will always assume that the components of the vector field Γ are linear, i.e., the coordinate expression is

$$\Gamma = \sum_{i,j=1}^{N} A^j{}_i x^i \frac{\partial}{\partial x^j} .$$

Let us first suppose that the components of the Poisson tensor Λ are constant. Its coordinate expression will be

$$\Lambda = \sum_{i,j=1}^{N} \Lambda^{ij} \frac{\partial}{\partial x^i} \wedge \frac{\partial}{\partial x^j} .$$

Therefore, if H is the Hamiltonian of the vector field Γ, it should be a quadratic function that we will write

$$H = \frac{1}{2} \sum_{i,j=1}^{N} M_{ij}\, x^i\, x^j ,$$

where M_{ij} is a symmetric matrix. The condition $\Gamma = -\widehat{\Lambda}(dH)$, implies that

$$\sum_{j=1}^{N} \Lambda^{ij}\, M_{jk} = -A^i{}_k .$$

In this case, the search for linear Casimir functions $C(x) = a_i\, x^i$ amounts to determinating the vectors a such that

$$\sum_{i=1}^{N} a_i\, \Lambda^{ij} = 0, \qquad i = 1, \ldots, N .$$

We can factorize the kernel of Λ and induce a dynamics on the quotient $E^* / \ker \Lambda \approx V$, where Λ is not degenerate anymore.

However, we continue to discuss the situation on E.

As for symmetries, as already discussed in Sects. 1.3 and 3.5, the Lie algebra is Abelian and its dimension is equal to the codimension of $\ker \Lambda$.

In this situation quadratic constants of motion will be associated with a symmetric matrix N given by the equation

$$\sum_{k=1}^{N} \Lambda^{ik}\, N_{kj} = (A^{2m+1})^i{}_j .$$

This equation may be solved on V and then associated quadratic functions may be pulled-back to E.

Let us now investigate the existence of quadratic invariant 1-forms. We set

$$\alpha = \sum_{i,j=1}^{N} a_{ij}\, x^i\, dx^j .$$

By writing it as a symmetric part and a skew-symmetric part we have

$$\alpha = \sum_{i,j=1}^{N} d[b_{ij}\, x^i\, x^j] + \sum_{i,j=1}^{N} c_{ij}(x^i\, dx^j - x^j\, dx^i)\,.$$

Imposing the condition $\mathcal{L}_\Gamma \alpha = 0$ gives that $\sum_{i,j=1}^{N} b_{ij}\, x^i\, x^j$ should be a constant of motion and the additional matrix condition

$$M^T C + CM = 0\,,$$

which is equivalent to the invariance of the 2 form

$$d\alpha = \sum_{i,j=1}^{N} c_{ij}\, dx^i \wedge dx^j = \sum_{i,j=1}^{N} a_{ij}\, dx^i \wedge dx^j\,.$$

Therefore, the analysis of invariant quadratic 1-forms is encompassed by the search for quadratic constants of motion and invariant 2-forms.

If we want to associate infinitesimal canonical symmetries with these quadratic 1-forms we see immediately that the composition of Λ with C should be vanishing. The skew-symmetric part of the quadratic 1-form must be in $\ker \Lambda$.

The situation becomes more interesting if we drop the assumption of quadratic invariant 1-forms and allow them to be arbitrary.

Example 4.5 Let be $E = \mathbb{R}^4$ with coordinates (x^a, y^a), $a = 1, 2$, and consider the following dynamics

$$\dot{x}^a = y^a$$
$$\dot{y}^a = -x^a$$

which corresponds to the vector field

$$\Gamma = \sum_{a=1}^{2} \left(y^a \frac{\partial}{\partial x^a} - x^a \frac{\partial}{\partial y^a} \right)\,.$$

A compatible Poisson bracket is provided by

$$\{y^a, x^b\} = \delta^{ab}\,, \quad \{x^a, x^b\} = 0 = \{y^a, y^b\} = \delta^{ab}\,,$$

with a Hamiltonian

$$H = \frac{1}{2} \sum_{a=1}^{2} \left((x^a)^2 + (y^a)^2 \right)\,.$$

The following vector fields are infinitesimal symmetries

$$Z_1 = \frac{1}{(x^2)^2+(y^2)^2}\left(y^1\frac{\partial}{\partial x^1} - x^1\frac{\partial}{\partial y^1}\right), \qquad Z_2 = \frac{1}{(x^2)^2+(y^2)^2}\left(y^2\frac{\partial}{\partial x^2} - x^2\frac{\partial}{\partial y^2}\right).$$

Corresponding invariant 1-forms are given by

$$\alpha_1 = \frac{1}{(x^2)^2+(y^2)^2}(y^1\,dy^1 + x^1\,dx^1), \qquad \alpha_2 = \frac{1}{(x^2)^2+(y^2)^2}(y^2\,dy^2 + x^2\,dx^2).$$

These two 1-forms satisfy the Maurer–Cartan relations

$$d\alpha_1 + \alpha_2 \wedge \alpha_1 = 0, \qquad d\alpha_2 = 0,$$

corresponding to the Lie algebra of $SB(2, \mathbb{R})$. The coaction is provided by

$$\Phi\colon (x^1, y^1, x^2, y^2) \mapsto \begin{pmatrix} (x^2)^2+(y^2)^2 & (x^1)^2+(y^1)^2 \\ 0 & 1 \end{pmatrix}.$$

It is not difficult to show that

$$\begin{pmatrix} (x^2)^2+(y^2)^2 & (x^1)^2+(y^1)^2 \\ 0 & 1 \end{pmatrix}^{-1} d\begin{pmatrix} (x^2)^2+(y^2)^2 & (x^1)^2+(y^1)^2 \\ 0 & 1 \end{pmatrix}$$
$$= \alpha_1 \begin{pmatrix} 0 & 1 \\ 0 & 0 \end{pmatrix} + \alpha_2 \begin{pmatrix} 1 & 0 \\ 0 & 0 \end{pmatrix}.$$

Vector fields Z_1 and Z_2 act trivially on α_1 and α_2, i.e., $\{\alpha_1, \alpha_2\} = 0$.

As for alternative invariant Poisson tensors, it is clear form our previous analysis that all even powers of the matrix associated with Γ will give raise to alternative Poisson tensors according to Sect. 2.1. Let us now consider the case of Poisson tensor with linear coordinates. We have again

$$\Gamma = \sum_{i,j=1}^{N} A^j{}_i\, x^i \frac{\partial}{\partial x^j},$$

and

$$\Lambda = \sum_{i,j,k=1}^{N} C^{ij}{}_k\, x^k \frac{\partial}{\partial x^i} \wedge \frac{\partial}{\partial x^j}.$$

and this time the Hamiltonian function must be linear, i.e.,

$$H = \sum_{i=1}^{N} a_i\, x^i, \qquad a_i \in \mathbb{R}.$$

We find

$$A^j{}_i = \sum_{k=1}^{N} a_k\, C^{ik}{}_j \,.$$

Of course, it is still true that powers of A will generate symmetries for Γ; this time, however, we cannot conclude easily that with these powers (if they are not canonical) we generate new Poisson tensors.

It is clear that if

$$\Lambda_1 = \sum_{i,j,k=1}^{N} C^{ij}{}_k\, x^k \frac{\partial}{\partial x^i} \wedge \frac{\partial}{\partial x^j}\,, \quad \Lambda_2 = \sum_{i,j,k=1}^{N} B^{ij}{}_k\, x^k \frac{\partial}{\partial x^i} \wedge \frac{\partial}{\partial x^j}\,.$$

provide two Poisson descriptions for Γ we have

$$A^j{}_i = \sum_{k=1}^{N} a_k C^{ik}{}_j = \sum_{k=1}^{N} b_k B^{ik}{}_j$$

where $C^{ik}{}_j$ and $B^{ik}{}_j$ are the structure constants of N-dimensional Lie algebras, i.e., they satisfy some quadratic algebraic relations.

As we have already remarked, a Poisson tensor with linear components defines a Lie algebra structure on the vector space. The fact that Γ is Hamiltonian implies that the one-parameter subgroup it generates can be considered as a subgroup of different N-dimensional Lie groups. Linear Hamiltonian symmetries will be given by the centralizer of Γ within the corresponding Lie algebra.

Before starting a general analysis let us consider an example:

Let be $E = \mathbb{R}^3 = \{(x^1, x^2, x^3)\}$ and Γ be the vector field

$$\Gamma = x^1 \frac{\partial}{\partial x^2} - x^2 \frac{\partial}{\partial x^1}\,.$$

It is clear from our group theoretical considerations that Γ can be considered as a subgroup of $E(2)$, $SL(2,\mathbb{R})$, $SO(3,\mathbb{R})$ and $SO(2,1)$. It follows that the corresponding Lie algebra, realized in terms of Poisson brackets will provide an alternative Poisson description for Γ. We have:

1. $E(2)\,,\quad \Lambda_1 = x^1 \frac{\partial}{\partial x^3} \wedge \frac{\partial}{\partial x^2} + x^2 \frac{\partial}{\partial x^1} \wedge \frac{\partial}{\partial x^3}\,,\quad H = x^3.$
2. $SO(3,\mathbb{R})\,,\quad \Lambda_2 = x^1 \frac{\partial}{\partial x^3} \wedge \frac{\partial}{\partial x^2} + x^3 \frac{\partial}{\partial x^2} \wedge \frac{\partial}{\partial x^1} + x^2 \frac{\partial}{\partial x^1} \wedge \frac{\partial}{\partial x^3}\,,\quad H = x^3.$
3. $SL(2,\mathbb{R})\,,\quad \Lambda_3 = x^1 \frac{\partial}{\partial x^3} \wedge \frac{\partial}{\partial x^2} - x^3 \frac{\partial}{\partial x^2} \wedge \frac{\partial}{\partial x^1} + +x^2 \frac{\partial}{\partial x^1} \wedge \frac{\partial}{\partial x^3}.$

It is reasonable to try to use the previous idea of getting one Poisson tensor from the other via an invariant matrix. Because Λ_1 and Λ_2 are both homogeneous of degree -1, a connecting tensor must be homogeneous of degree zero. We consider

$$T = T^i{}_j dx^j \otimes \frac{\partial}{\partial x^i} \,.$$

If $\Lambda_1(X, Y)$ and $\Lambda_2(X, Y)$ stay for the commutator of vectors X and Y, respectively, we need

$$\Lambda_1(TX, TY) = T\Lambda_2(X, Y)$$

with

$$\mathcal{L}_\Gamma T = 0.$$

This last condition says that we are allowed to experiment with

$$T = \sum_{i,j,k=1}^{N} a_k C^{ik}{}_j dx^j \otimes \frac{\partial}{\partial x^i}$$

and try to solve for B the equation

$$\sum_{l,m=1}^{N} C^{lm}{}_n T_l{}^r T_m{}^s = \sum_{p=1}^{N} T_n{}^p B^{rs}{}_p \,,$$

with the additional algebraic quadratic constraint on the components of B. Another possibility would be to start with

$$T = e^{\lambda M}$$

and try to solve for M, this time by taking the λ-derivative at the origin of

$$T_\lambda^{-1} \Lambda_1(T_\lambda X, T_\lambda Y) = \Lambda_\lambda(X, Y) \,.$$

It is convenient to define the Poisson bracket on 1-forms by setting

$$\Lambda_1 = \sum_{i,j,k=1}^{n} C^{ij}{}_k dx^k \otimes \frac{\partial}{\partial x^i} \wedge \frac{\partial}{\partial x^j} \,.$$

From $e^{\lambda M}$ with

$$M = M_i{}^j \, dx^i \otimes \frac{\partial}{\partial x^j} \,,$$

the condition for

$$\frac{d}{d\lambda} \Lambda_\lambda$$

to be a Poisson bracket is given by

$$\Lambda_1(M\,df, M\,dg) + M^2\Lambda_1(df, df) - M\Lambda_1(M\,df, dg) - M\Lambda_1(df, M\,dg) = 0\,,$$

i.e., M has to satisfy the zero Nijenhuis condition.

When we apply this condition to the previous example it is not difficult to show that all previous Poisson tensors are mutually compatible (the sum of structure constants still satisfies the quadratic condition corresponding to Jacobi identity).

We may conclude that if a vector field Γ is an element of a Cartan subalgebra of some simple algebra, all real forms of the complex algebras to which Γ belongs will provide alternative Poisson descriptions. Moreover, any combination of the structure constants of the real form gives Lie algebra structures.

When going to the case of quadratic brackets, as we remarked before, the situation is simpler. A general result is the following: If Γ is any vector field and T any (1,1)-tensor field satisfying the Nijenhuis condition $\mathbb{N}_T = 0$ then setting

$$T^k(\Gamma) = X_k$$

we find that

$$\Lambda = \sum_{k,l=1}^{n} B^{kl} X_k \wedge X_l$$

is a Poisson tensor.

When the components of Γ are linear and T is homogeneous of degree zero, previous construction provides linear vector fields X_k and compatible quadratic brackets. Let us consider the following example

$$\Gamma = \sum_a \omega^a \left(x^a \frac{\partial}{\partial y^a} - y^a \frac{\partial}{\partial x^a} \right)$$

and

$$T = \sum_a \omega^a \left(dx^a \otimes \frac{\partial}{\partial y^a} - dy^a \otimes \frac{\partial}{\partial x^a} \right)\,.$$

We find

$$\begin{aligned}\Lambda_0 &= \sum_{a,b} \left[\omega^a \left(x^a \frac{\partial}{\partial y^a} - y^a \frac{\partial}{\partial x^a} \right) \wedge \left((\omega^b)^2 (x^b \frac{\partial}{\partial x^b} + y^b \frac{\partial}{\partial y^b} \right) \right] \delta^{ab} \\ &= \sum_a (\omega^a)^3 ((x^a)^2 + (y^a)^2) \frac{\partial}{\partial y^a} \wedge \frac{\partial}{\partial x^a}\end{aligned}$$

It should be noticed that even if Λ_0 is invertible on some open dense submanifold, linear canonical symmetries are not associated with smooth Hamiltonian functions.

The reason being that Hamiltonian functions would be homogeneous of degree zero and therefore could not be smooth on a vector space.

4.5 The Inverse Problem for Poisson Structures: Feynman's Problem

As it was pointed out before, it is natural to ask when a dynamical system Γ can be given a description in terms of some Poisson bracket and some Hamiltonian function. That is, given a vector field $\Gamma = \Gamma^k(x)\partial/\partial x^k$ defined on a manifold M with local coordinates x^k, the hamiltonian inverse problem, or general factorization problem, for Γ consists in finding a Poisson bracket $\{\cdot, \cdot\}$ and a smooth function H such that

$$\dot{x}^k = \Gamma^k(x) = \{H, x^k\}. \tag{4.35}$$

In other words, the hamiltonian inverse problem for Γ is equivalent to determining whether or not Γ is a Hamiltonian vector field with respect to some Poisson brackets.

In this formulation, neither the Poisson brackets nor the Hamiltonian are determined a priori, that is, if they carry physical information it has to be determined by further information not contained in the vector field Γ alone. Needless to say, this problem in its full generality has not been solved.

The Hamiltonian inverse problem can be considered locally or globally, i.e., we can discuss the existence of a Poisson structure Λ in the neighborhood of an ordinary point for the vector field Γ such that $\Gamma = \Lambda(dH)$, and in a second step, we can discuss the existence of a globally defined Poisson structure. We will not pursue this aspect of the problem here (see [Ib90] for a discussion of the related issue of global aspects of the inverse problem in the Lagrangian formalism).

The inverse problem of Poisson dynamics stated in such general terms it is hopeless, difficult, or useless because of the lack of uniqueness of solutions and the lack of understanding of their physical meaning.

One way to improve the problem is to make some further assumptions on the Poisson manifold M. For instance, we can assume that $M = \mathbb{R}^{2n}$ and that the classical system is described by a second order differential equation. In this way we obtain new problems that could be called restricted inverse problems for Poisson dynamics. Obviously attached to these problems is the question of uniqueness of the Poisson description for the given classical system. This was already raised by Wigner in [Wi50].

If the inverse problem of Poisson dynamics is solved for a given dynamical system Γ on M and a Poisson tensor Λ is found, we can ask if there exists a symplectic realization for it, that is, a symplectic manifold (P, Ω) and a Poisson map $\Phi \colon P \to M$. Such a symplectic realization exists locally, but in general there are obstructions to obtaining globally defined ones [We83, We87].

Together with the problem of finding symplectic realizations for Poisson manifolds, arises the problem of finding Lagrangian realizations. That is, does there exists a Lagrangian for a given dynamical system? Once a symplectic realization has been found, and if the symplectic manifold is a tangent bundle, we will apply the inverse problem of the calculus of variations to look for a Lagrangian describing the system. The conditions assuring the existence of such Lagrangians, both local and global [Mo91, Ib90], are well known and can be systematically applied to the problems at hand.

We will address some of these questions in Sect. 5.6.

4.5.1 Alternative Poisson Descriptions

Thus given Γ, we should look for a bivector field $\Lambda = \Lambda^{ij}(x)\frac{\partial}{\partial x^i}\wedge\frac{\partial}{\partial x^j}$, and a function H such that

$$\Gamma = X_f = \widehat{\Lambda}(dH)\,,$$

i.e., such that the coordinate expression of Γ is,

$$\Gamma = X_f = \Lambda^{ij}\frac{\partial f}{\partial x^j}\frac{\partial}{\partial x^i}\,.$$

Note, however, that there are vector fields X satisfying the derivation property (4.31) and which are not Hamiltonian vector fields. In other words, not every vector field X such that $\mathcal{L}_X\Lambda = 0$ is a Hamiltonian one.

As an instance, let us consider the Poisson structure in $\mathbb{R}^3$ defined by

$$\{f, g\} = \frac{\partial f}{\partial x_2}\frac{\partial g}{\partial x_1} - \frac{\partial f}{\partial x_1}\frac{\partial g}{\partial x_2}\,.$$

The tensor Λ defining the Poisson structure is

$$\Lambda = \frac{\partial}{\partial x_2}\wedge\frac{\partial}{\partial x_1}\,.$$

In this case the vector field $Y = \partial/\partial x_3$ is an infinitesimal Poisson symmetry, but it cannot be Hamiltonian; if a Hamiltonian H does exist, then $Yf = \{f, H\}$, for any function f. It is remarkable that the obstruction for the existence of the function f is not only local, but global, due to the degeneracy of the Poisson bivector.

Taking $f = x_3$ we will find that x_3 being a Casimir function, it would be

$$1 = Y(x_3) = \{x_3, H\} = 0\,,$$

which is impossible.

Of course, in case of existence of a Poisson bracket such that Γ is Hamiltonian, we can then ask whether such a Poisson bracket is unique and, in the contary case of existence of different Poisson brackets for which Γ is Hamiltonain what additional information we have. In fact, we shall see that the knowledge of alternative Poisson structures will provide us some constants of motion.

Let us recall that if $(\mathcal{L}, [\cdot, \cdot])$ is a Lie algebra, then for each $x \in \mathcal{L}$ the map $\mathrm{ad}\, x : \mathcal{L} \to \mathcal{L}$ defined by $\mathrm{ad}\, x(y) = [x, y]$, is a derivation. These derivations $\mathrm{ad}\, x$ are said to be inner derivations. However there can be derivations of $(\mathcal{L}, [\cdot, \cdot])$ that are not inner. For the case of Poisson manifolds, for any function f, the inner derivation $\mathrm{ad}\, f$ is X_f.

We can also consider the more general case of an arbitrary 1-form α in E. If we denote by X_α the vector field: $X_\alpha = -\widehat{\Lambda}(\alpha)$, then taking into account that for any pair of functions $f, g \in \mathcal{F}(E)$,

$$(\mathcal{L}_{X_\alpha}\Lambda)(df \wedge dg) = X_\alpha\{f, g\} - \{X_\alpha f, g\} - \{f, X_\alpha g\} ,$$

and

$$\Lambda(df \wedge \alpha) = -X_\alpha f = -\alpha(X_f) ,$$

we see that

$$(\mathcal{L}_{X_\alpha}\Lambda)(df \wedge dg) = X_\alpha\{f, g\} - X_g(X_\alpha f) + X_f(X_\alpha g) .$$

Also if we recall that

$$\alpha([X_f, X_g]) = -\alpha(X_{\{f,g\}}) = -X_\alpha\{f, g\} ,$$

then

$$(\mathcal{L}_{X_\alpha}\Lambda)(df \wedge dg) = -\alpha([X_f, X_g]) + X_f(\alpha(X_g)) - X_g(\alpha(X_f)) = (d\alpha)(X_f \wedge X_g) .$$

Consequently, we see that X_α is a derivation of the Poisson algebra iff $d\alpha$ is closed along Hamiltonian vector fields. Thus, for generic Poisson manifolds, there will be derivations that are not locally Hamiltonian vector fields.

Given a linear vector field X_A we have analyzed the question of whether or not there exists a constant bivector field invariant under X_A. We will also be interested in the uniqueness problem or in finding alternative bivectors. We should remark that every infinitesimal symmetry Y for X_A that is not canonical, will take from one admissible Poisson structure to another.

In fact, from $[Y, X_A] = 0$ and $\mathcal{L}_Y\Lambda \neq 0$, but $\mathcal{L}_{X_A}\Lambda = 0$, we see that

$$\mathcal{L}_{X_A}(\mathcal{L}_Y\Lambda) = 0 .$$

We can say more however. As the Jacobi identity is equivalent to some tensor field being identically zero, we have that this property is preserved under a general change of coordinates. Therefore if Λ is a compatible Poisson structure for X_A, any symmetry for X_A, if it is non canonical for Λ will define a new compatible Poisson bracket. It is clear however, that the rank of Λ may change.

We are also interested in vector fields X_A which are bi-Hamiltonian, what means that there exist two Poisson structures Λ_1 and Λ_2 and two Hamiltonian functions H_1 and H_2 such that

$$X_A = \widehat{\Lambda}_1(dH_1) = \widehat{\Lambda}_2(dH_2) .$$

This corresponds to different factorizations of the matrix A as products of a skew-symmetric matrix times a symmetric one. A constructive method for finding such bi-Hamiltonian structures is the following: If S is any invertible matrix and A can be factorized as before, $A = \Lambda\, H$, then

$$S^{-1}\, A\, S = S^{-1}\, \Lambda\, H\, S = (S^{-1}\, \Lambda\, (S^{-1})^T)\, (S^T\, H\, S) ,$$

and then the vector field corresponding to $S^{-1}\, A\, S$ is Hamiltonian with respect to the Poisson structure defined by the skew-symmetric matrix $S^{-1}\, \Lambda\, (S^{-1})^T$ with a quadratic Hamiltonian function defined by the symmetric matrix $S^T\, H\, S$. This shows that if the linear map S commutes with A but is not canonical, i.e., $S^{-1}\, \Lambda \neq \Lambda\, S^T$, it will carry from one Hamiltonian description to another one.

Remark 4.3 As our Λ is given by

$$\Lambda = \Lambda^{ij}\, \frac{\partial}{\partial x^i} \wedge \frac{\partial}{\partial x^j} ,$$

from $[\partial/\partial x^i, \partial/\partial x^j] = 0$ we find that we can add any two compatible Poisson structures without spoiling the Jacobi identity. Therefore if f is any even function we can consider $\Lambda \cdot f(A)^t$ and obtain a new compatible Poisson structure.

If B is any matrix commuting with A, degenerate or not, we get another admissible Poisson structure by considering $\Lambda^{(1)} = B^t \Lambda + \Lambda B$. Indeed,

$$(\Lambda^{(1)})^t = \Lambda^t B + B^t \Lambda^t = -(B^t \Lambda + \Lambda B)$$

and in the basis $\frac{\partial}{\partial x^i}$ it will satisfy the Jacobi identity because of an earlier remark. In the vector field language the result is quite trivial, from $\mathcal{L}_\Gamma \Lambda = 0$, and $[X, \Gamma] = 0$ we get $\mathcal{L}_\Gamma(\mathcal{L}_X \Lambda) = 0$. What is non trivial in general, and in our case depends on the linearity of X and the constancy of Λ, is the assertion that $\mathcal{L}_X \Lambda$ satisfies the Jacobi identity.

4.5.2 Feynman's Problem

An alternative to the inverse problem for Poisson dynamics as discussed above, consists in asking as Feynman did, which dynamical laws are compatible with certain basic physical assumptions. One of Feynman's aims was to obtain dynamical laws beyond those in the Lagrangian formalism. Starting with just a few postulates for the Quantum Mechanics of a single particle, Feynman was able to recover the standard interaction with an electromagnetic field. The postulates which we will here review include assuming the existence of position and velocity operators with commutation relations consistent with particle localizability. (In order to avoid unnecessary complication due to operator ordering we shall be concerned only with the classical analogue of Feynman's procedure.) This program failed, as described by Dyson [Dy90], because the conditions imposed on the Poisson brackets were restrictive enough to show that only electromagnetic forces were compatible with them.

Feynman's problem deals with a full family of dynamical systems, those being second-order differential equations (SODE's) on a fixed velocity phase space TE (we will come back to the general discussion of Feynman's problem for Lagrangian systems in Sect. 5.6). The classical Feynman problem can then be stated in a more general setting as the problem of finding all Poisson tensors on TE such that they have Hamiltonian vector fields which correspond to SODE's and such that the systems are localizable.

Let us now recall Feynman's derivation of the Lorentz force for a point particle. For the choice of phase space variables we take $(\xi^a) = (x^i, v^i)$ for $i = 1, 2, 3$, x^i denoting position and v^i velocity.

The analogue of Feynman's postulates gives some information about the Poisson tensor Λ and the equations of motion. The assumptions on the Poisson brackets are that

$$\{x^i, x^j\} = 0 \tag{4.36}$$

and

$$m\{x^i, v^j\} = \delta_{ij} \ . \tag{4.37}$$

In terms of Poisson tensors Λ, we are postulating that Λ has the following form on $\mathbb{R}^6$:

$$\Lambda = \frac{1}{m}\frac{\partial}{\partial x^i} \wedge \frac{\partial}{\partial v^i} + f^{ij}(\mathbf{x}, \mathbf{v})\frac{\partial}{\partial v^i} \wedge \frac{\partial}{\partial v^j} \ , \tag{4.38}$$

where the form of f^{ij} is to be determined. Equation (4.36) is the classical analogue of the condition of localizability, while m in (4.37) refers to the particle's mass. The equations of motion were postulated to have the form

$$\frac{d}{dt}x^i = \{x^i, H\} = v^i \tag{4.39}$$

$$\frac{d}{dt}v^i = \{v^i, H\} = \frac{1}{m}F^i(\mathbf{x}, \mathbf{v}). \tag{4.40}$$

Notice that Eq. (4.39) implies that the Hamiltonian dynamical system is a SODE. For convenience we will at first assume that F^i has no explicit time dependence.

The problem which was then posed was to find dynamics compatible with assumptions (4.36), (4.37), (4.39) and (4.40). Alternatively this means that one should determine the Hamilton function H and the remaining Poisson brackets defining Λ_{ab}, which here are just $\{v^i, v^j\}$. A unique solution to this problem gives the coupling to electromagnetic fields, as we now review.

One first defines the magnetic field B^i according to $f^{ij} = \frac{1}{m^2}\epsilon^{ijk}B^k$, or

$$m^2\{v^i, v^j\} = \epsilon^{ijk}B^k. \tag{4.41}$$

It is then easy to show that B^i is a function of only the spatial coordinates x^i. For this, one writes down the Jacobi identity

$$\{x^i, \{v^j, v^k\}\} + \{v^j, \{v^k, x^i\}\} + \{v^k, \{x^i, v^j\}\} = 0\ , \tag{4.42}$$

and applies (4.37) and (4.41) to find that $\{x^i, B^j\} = 0$. Hence B^i must be independent of the particle velocity v^i. It also has zero divergence, which follows from the Jacobi identity involving v^i, v^j and v^k, and thus we recover one of the Maxwell equations.

Next one takes the time derivatives of Poisson brackets (4.37) and (4.41). The former gives

$$\{v^i, v^j\} + \{x^i, \dot{v}^j\} = 0\ ,$$

or

$$\{x^i, F^j\} = -\frac{1}{m}\epsilon^{ijk}B^k(\mathbf{x})\ , \tag{4.43}$$

from which it follows that F^i is at most linear in the velocities

$$F^i = E^i(\mathbf{x}) + \epsilon^{ijk}\, v^j\, B^k(\mathbf{x})\ . \tag{4.44}$$

Equation (4.44) is the Lorentz force law (where the electric charge is set equal to one) and it here defines the electric field $E^i(\mathbf{x})$.

From the time derivative of (4.41) one gets

$$m\ \epsilon^{ijk}\{F^i, v^j\} = \dot{B}^k\ . \tag{4.45}$$

Next apply (4.44) and the divergenceless condition on B^i to simplify (4.45) to

$$m\ \epsilon^{ijk}\{E^i, v^j\} = \dot{B}^k - m\{B^k, v^i\}v^i\ . \tag{4.46}$$

But the right-hand side of (4.46) vanishes because we have assumed B^i to have no explicit time dependence, and so we are just left with

$$\epsilon^{ijk}\{E^i, v^j\} = 0\ , \tag{4.47}$$

implying that E^i has zero curl, as is the case for time independent magnetic fields.

Upon repeating this calculation for time dependent fields, one instead recovers the Faraday law and thus both of the homogeneous Maxwell equations. In this case both the velocity v^i and the force F^i are allowed to have explicit dependence on t, so that Eq. (4.40) is generalized to

$$\frac{d}{dt}v^i = \{v^i, H\} + \frac{\partial v^i}{\partial t} = \frac{1}{m}F^i(\mathbf{x}, \mathbf{v}, t) \ . \tag{4.48}$$

The analysis then proceeds as in the above with the static fields $E^i(\mathbf{x})$ and $B^i(\mathbf{x})$ replaced by $E^i(\mathbf{x}, t)$ and $B^i(\mathbf{x}, t)$. Now the right hand side of (4.46) is $\partial B^i/\partial t$ and hence the Faraday law is recovered.

The Hamilton function for the above system is

$$H = \frac{m}{2}\mathbf{v}^2 + \phi(\mathbf{x}, t) \ ,$$

where $\phi(\mathbf{x}, t)$ is the standard scalar potential. From it we construct the electric field according to,

$$\mathbf{E} = -\nabla\phi + m\frac{\partial \mathbf{v}}{\partial t} \ .$$

4.5.3 Poisson Description of Internal Dynamics

4.5.3.1 A Spinning Particle Interacting with a Magnetic Field

Let us go back to discuss the inverse problem for Poisson dynamics. For this we concentrate on one particular system—that of a spinning particle interacting with a magnetic field. In addition to giving the standard Hamiltonian formulation of this system, we will exhibit examples of nonstandard formulations.

In this section we shall ignore the spatial degrees of freedom of the particle and rather only study the spin variables. The spatial coordinates will be included in the following section.

Let us denote the spin variables by $S_i, i = 1, 2, 3$. They span some 3-manifold M. (We initially impose no constraints on S_i.) Since S_i correspond to *internal* degrees of freedom (as opposed to the spatial coordinates discussed in the previous section) we should not make the same assumptions as Feynman used for getting dynamics on Minkowski space. That is, we should not in general assume localizability (i.e., that the Poisson brackets of S_i with themselves are zero), nor should we assume that the equations of motion are second order in time derivatives. So let us instead start with the most general Poisson tensor

$$\Lambda = \epsilon_{ijk} F^k(\mathbf{S}) \frac{\partial}{\partial S_i} \wedge \frac{\partial}{\partial S_j} \ . \tag{4.49}$$

Then, the Jacobi identity requires that $\mathbf{F} \cdot (\nabla \times \mathbf{F}) = 0$, where the gradient ∇ is with respect to $\mathbf{S}$. If we introduce the 1-form $\alpha = F^i \, dS_i$, the condition reads: $\alpha \wedge d\alpha = 0$. Locally, α can then be written $\alpha = f_0 \, df_1$, f_0 and f_1 being functions of S_i and we have

$$\{S_i, S_j\} = f_0(\mathbf{S}) \, \epsilon_{ijk} \, \frac{\partial f_1}{\partial S_k} \, .$$

As a result f_1 has zero Poisson bracket with all S_i and it therefore defines a classical Casimir function. If we now require that the dynamics be such that there is a Hamilton function $H = H(S_i)$, then f_1 is also a constant of motion and therefore

$$\dot{S}_i \frac{\partial f_1}{\partial S_i} = 0 \, . \tag{4.50}$$

For dynamics, which is non-trivial (i.e., $\dot{S}_i \neq 0$), Eq. (4.50) implies a condition on f_1. To see what that is, let us now specialize to the dynamical system of a spin interacting with a magnetic field $\mathbf{B}$. The standard equation of motion for such a system gives the precession of the spin S_i,

$$\dot{S}_i = \mu \, \epsilon_{ijk} \, S_j B^k \, , \tag{4.51}$$

where μ denotes the magnetic moment. From (4.50) and (4.51) one finds that

$$\frac{\partial f_1}{\partial S_k} = \rho_1 S_k + \sigma_1 B^k \, , \tag{4.52}$$

and so f_1 must be of the form

$$f_1 = f_1(\mathbf{S}^2, \mathbf{S} \cdot \mathbf{B}) \, . \tag{4.53}$$

The Hamilton function $H(S_i)$ must have an analogous form. To see this, substitute (4.51) into $\dot{S}_i = \{S_i, H\}$ to get:

$$f_0(\mathbf{S}) \, \epsilon_{ijk} \, \frac{\partial H}{\partial S_j} \frac{\partial f_1}{\partial S_k} = \mu \, \epsilon_{ijk} S_j B^k \, . \tag{4.54}$$

It follows that

$$\frac{\partial H}{\partial S_k} = \rho_H S_k + \sigma_H B^k,$$

and thus that the most general solution for H is of the form

$$H = H(\mathbf{S}^2, \mathbf{S} \cdot \mathbf{B}). \tag{4.55}$$

From (4.54) we get the following restriction on the derivatives of f_1 and H:

$$f_0(\rho_H\sigma_1 - \rho_1\sigma_H) = \mu \ . \tag{4.56}$$

The functions f_1 and H given by (4.53) and (4.55) subject to the constraint (4.56) define a family of Poisson brackets and Hamiltonian functions all of which lead to the same equations of motion (4.51) for the spin. They correspond in general to canonically inequivalent descriptions for the dynamics. We first review the standard Hamiltonian description.

4.5.3.2 Standard and Alternative Formalisms

The standard canonical formalism for a spinning particle is recovered when we write

$$f_0 = \frac{1}{2} \ , \quad f_1 = \mathbf{S}^2 \ , \quad H = -\mu\mathbf{S}\cdot\mathbf{B} \ .$$

For this choice the Poisson bracket algebra (4.49) corresponds to the $\mathfrak{su}(2)$ Lie algebra. Reducible representations of $\mathfrak{su}(2)$ then arise upon performing a canonical quantization of this system. Alternatively, to obtain irreducible representations for $\mathfrak{su}(2)$ one must impose a constraint on the variables S_i. Namely, one must require that the classical Casimir function f_1 takes on certain constant values (and as a result S_i will then span S^2) [Ba82].

There exist alternative Hamiltonian formulations associated which yield the classical equation of motion (4.51), which are canonically inequivalent to the standard one. One such formulation results from the choice [So92, St93],

$$f_0 = \frac{1}{2} \ , \quad f_1 = S_1^2 + S_2^2 + \frac{1}{2\lambda}\left[\frac{\cosh 2\lambda S_3}{\sinh\lambda} - \frac{1}{\lambda}\right] \tag{4.57}$$

and

$$H = -\mu\lambda S_3 \ , \tag{4.58}$$

where with no loss of generality we have taken the magnetic field to be in the third direction. Here λ is a 'deformation' parameter as the standard formalism is recovered when $\lambda \to 0$. For non zero λ, (4.58) leads to

$$\{S_2, S_3\} = S_1 \ , \quad \{S_3, S_1\} = S_2 \ , \tag{4.59}$$

$$\{S_1, S_2\} = \frac{1}{2}\,\frac{\sinh 2\lambda S_3}{\sinh\lambda} \ . \tag{4.60}$$

Unlike in the standard formulation, the Poisson bracket algebra of observables S_i does not correspond to $\mathfrak{su}(2)$ or any Lie algebra.[3]

---

[3] These brackets are a classical realization of the quantum commutation relations for generators of the $U_q(\mathfrak{sl}(2))$ Hopf algebra [Ta89, Ma90, Tj92].

Let us denote the latter generators by $\mathbf{I}_+$, $\mathbf{I}_-$ and $\mathbf{I}_0$. Their commutation relations are

$$[\mathbf{I}_0, \mathbf{I}_\pm] = \pm \mathbf{I}_\pm \ , \qquad [\mathbf{I}_+, \mathbf{I}_-] = \frac{q^{2\mathbf{I}_0} - q^{-2\mathbf{I}_0}}{q - q^{-1}} \ . \tag{4.61}$$

These commutation relations reduce to the $\mathfrak{su}(2)$ Lie algebra relations in the limit $q \to 1$. To obtain the above classical system, one can replace the quantum operators $\mathbf{I}_+$, $\mathbf{I}_-$ and $\mathbf{I}_0$ by $S_1 + iS_2$, $S_1 - iS_2$ and S_3, respectively, q by $\exp \lambda$, and the commutation relations (4.61) by i times Poisson brackets.

There is another choice for f_0 and f_1 (which is known to correspond to the classical limit of the $U_q(\mathfrak{sl}(2))$ Hopf algebra):

$$f_0 = \frac{\lambda}{4} S_3 \ , \quad f_1 = (S_1)^2 + (S_2)^2 + (S_3)^2 + (S_3)^{-2} \ . \tag{4.62}$$

It leads to the following brackets for S_i:

$$\{S_2, S_3\} = \frac{\lambda}{2} \ S_1 S_3 \ , \quad \{S_3, S_1\} = \frac{\lambda}{2} \ S_2 S_3 \ ,$$

$$\{S_1, S_2\} = \frac{\lambda}{2} \left((S_3)^2 - (S_3)^{-2} \right) , \tag{4.63}$$

or if we define $S_\pm = S_1 \pm i S_2$, these relations can be expressed by

$$\{S_\pm, S_3\} = \pm \frac{i\lambda}{2} \ S_\pm S_3 \ , \quad \{S_+, S_-\} = -i\lambda \left((S_3)^2 - (S_3)^{-2} \right) \ . \tag{4.64}$$

Now in order to obtain the equations of motion for the spin (4.51) one can choose the Hamiltonian

$$H = -\frac{2\mu B}{\lambda} \ln S_3 \ ,$$

where again we have chosen the magnetic field to be along the third direction.

To show that the Poisson structure defined in (4.64) has something to do with the classical limit of the algebra generated by $\mathbf{I}_+$, $\mathbf{I}_-$ and $\mathbf{I}_0$, let us define new operators $\mathbf{S}_+$, $\mathbf{S}_-$ and $\mathbf{S}_0$ according to

$$\mathbf{S}_\pm = -\sqrt{q}(q - q^{-1})\mathbf{I}_\pm \ , \quad \mathbf{S}_3 = q^{I_0} \ .$$

Then the commutation relations (4.61) can be expressed according to

$$\mathbf{S}_3\mathbf{S}_\pm = q^{\pm 1}\mathbf{S}_\pm\mathbf{S}_3 \ , \quad [\mathbf{S}_+, \mathbf{S}_-] = q(q - q^{-1}) \left((\mathbf{S}_3)^2 - (\mathbf{S}_3)^{-2} \right) \ . \tag{4.65}$$

Next we introduce Planck's constant $\hbar$ by setting $q = \exp \frac{\lambda\hbar}{2}$ and expand (4.65) in powers of $\hbar$. The result is:

$$[\mathbf{S}_3, \mathbf{S}_\pm] = \pm\frac{\lambda\hbar}{2}\mathbf{S}_\pm\mathbf{S}_3 + O(\hbar^2)\,, \quad [\mathbf{S}_+, \mathbf{S}_-] = \lambda\hbar\left((\mathbf{S}_3)^2 - (\mathbf{S}_3)^{-2}\right) + O(\hbar^2)\,. \tag{4.66}$$

The lowest order terms in the commutation relations (4.66) for $\mathbf{S}_+$, $\mathbf{S}_-$ and $\mathbf{S}_0$ correspond with $i\hbar$ times the Poisson brackets (4.64) of S_+, S_- and S_0.

4.5.4 Poisson Structures for Internal and External Dynamics

We adapt Feynman's procedure to examine the dynamics of particles which trace out trajectories in space-time, as well as in some internal space M. Physically, coordinates of the latter space can be associated with an extrinsic spin of the particle (as discussed in the preceding section) or with some internal degrees of freedom like isospin or color. In the following general treatment we let the manifold M be of arbitrary dimension. Further, the algebra of observables spanning M need not be associated with any Lie algebra (as was the case in the alternative formulations of spin given in the previous section). The treatment given here thus allows for the possibility of finding new kinds of particle dynamics. We recover some familiar systems upon specializing to the case where the Poisson bracket algebra for the internal variables corresponds to a Lie algebra; for instance, the Wong equations [Wo70] describing the coupling of particles to a Yang–Mills field. We can also recover the magnetic dipole coupling of a spinning particle with a magnetic field when we take the Poisson bracket algebra on M to be $\mathfrak{su}(2)$. We will postpone the discussion of the general setting for these questions as well as the general results to Sect. 5.6.2. Of course the general treatment will include particles on curved space-time [Ta92]. Again in Sect. 5.6.4 the discussion will again be specialized to the case of M being a Lie group and a cotangent bundle of a group.

4.5.4.1 Free System

We first consider free motion on $\mathbb{R}^3 \times M$. The manifold M now denotes a d-dimensional internal space which we parametrize by I_a, $a = 1, \ldots, d$. Free motion on $\mathbb{R}^3$ corresponds to

$$\ddot{x}^i = 0\,, \quad i = 1, 2, 3\,, \tag{4.67}$$

while on M one standardly writes

$$\dot{I}_a = 0\,. \tag{4.68}$$

We parametrize the corresponding phase space of the system by x^i, v^i and I_a. In writing down a Poisson description on $T\mathbb{R}^3 \times M$ we let

$$m\{x^i, v^j\} = \delta_{ij}; \qquad \{I_a, I_b\} = C_{ab} \tag{4.69}$$

define the non zero Poisson brackets for the system, where C_{ab} may be any general function on $\mathbb{R}^3 \times M$. Notice that the zero brackets $\{x^i, x^j\} = \{x^i, I_a\} = 0$ reflect the localizability of the system and the simultaneous measurability of inner variables with external coordinates. From Jacobi identity involving I_a, I_b and x^i, we get that $\{x^i, C_{ab}\} = 0$ and consequently C_{ab} can have no dependence on v^i.

The Poisson tensor defined by (4.69) reads as:

$$\Lambda = \frac{1}{m}\frac{\partial}{\partial x^i} \wedge \frac{\partial}{\partial v^i} + \frac{1}{2}C_{ab}\frac{\partial}{\partial I_a} \wedge \frac{\partial}{\partial I_b}\,. \tag{4.70}$$

The Jacobi identity $[\Lambda, \Lambda] = 0$, implies that $[\Lambda_I, \Lambda_I] = 0$, Λ_I being the second term in (4.70). As a result C_{ab} satisfies the following condition:

$$0 = C_{af}\frac{\partial C_{bc}}{\partial I_f} + C_{bf}\frac{\partial C_{ca}}{\partial I_f} + C_{cf}\frac{\partial C_{ab}}{\partial I_f}\,, \tag{4.71}$$

and Λ_I defines a Poisson tensor itself on $\mathbb{R}^3 \times M$. For the case where the Poisson brackets of I_a define a Lie algebra, we have $C_{ab} = c_{ab}{}^f I_f$, $c_{ab}{}^f$ being the corresponding structure constants. Then the Jacobi identity is satisfied due to

$$0 = c_{af}{}^d c_{bc}{}^f + c_{bf}{}^d c_{ca}{}^f + c_{cf}{}^d c_{ab}{}^f\,.$$

Finally, the free motion (4.67) and (4.68) results from (4.69) along with the Hamilton function

$$H = \frac{m}{2} > \mathbf{v}^2\,.$$

4.5.4.2 Interactions

What kind of interaction can be introduced into the system by generalizing the previous Poisson brackets and Hamilton function? In order to better define this question we generalize Feynman's assumptions (4.36–4.40). We assume equations of motion of the form

$$\begin{aligned} \dot{x}^i &= v^i\,, \\ \dot{v}^i &= \frac{1}{m}F^i(\mathbf{x}, \mathbf{v}, I)\,, \\ \dot{I}_a &= f_a(\mathbf{x}, \mathbf{v}, I)\,, \end{aligned} \tag{4.72}$$

while for the Poisson brackets we take

$$\{x^i, x^j\} = 0\,, \quad \{x^i, I_a\} = 0\,, \quad m\{x^i, v^j\} = \delta^{ij}\,, \tag{4.73}$$

and

$$\{I_a, I_b\} = C_{ab}\ , \quad m^2\{v^i, v^j\} = \mathcal{F}^{ij}\ , \quad m\{v^i, I_a\} = \mathcal{A}^i_a\ , \tag{4.74}$$

where the functions F^i, f_a, $\mathcal{F}^{ij}$, $\mathcal{A}^i_a$ and C_{ab} are initially undetermined. This corresponds to defining a Poisson tensor which is not a direct sum, since it will contain cross terms. After the brackets have been determined, we once again demand that the equations of motion are Hamiltonian. Free dynamics is recovered if all functions except for C_{ab} vanish.

It is clear that we have two problems here. The first one is to find conditions on the functions C_{ab}, $\mathcal{F}^{ij}$ and $\mathcal{A}^i_a$ such that the Poisson brackets satisfy the Jacobi identity. The second one is to ensure that the functions are compatible with the requirement that the dynamics defined by (4.72) be Hamiltonian.

We begin by taking the time derivative of (4.73). This can be done under the assumption that the evolution equation is Hamiltonian with respect to the Poisson tensor defined by the Poisson brackets (4.73) and (4.74). Upon using (4.72) we get

$$\begin{aligned}
&\{x^i, v^j\} + \{v^i, x^j\} = 0\ ,\\
&\{v^i, I_a\} + \{x^i, f_a\} = 0\ ,\\
&m\{v^i, v^j\} + \{x^i, F^j\} = 0\ .
\end{aligned} \tag{4.75}$$

Of course the first equation is identically satisfied. From the second equation we notice that if f_a depends non trivially on velocity we need $\mathcal{A}^i_a \neq 0$. Similarly, from the third equation, if F^i depends non trivially on velocity we need $\mathcal{F}^{ij} \neq 0$. Now using the Jacobi identity one finds that the coordinates x^i have zero brackets with the functions C_{ab}, $\mathcal{A}^i_a$ and $\mathcal{F}^{ij}$, and consequently none of these functions can have a dependence on velocity. The Poisson tensor will then be said to be velocity independent. (This will be discussed further in Sect. 5.51.)

For $\mathcal{F}^{ij} \neq 0$ or $\mathcal{A}^i_a \neq 0$, we must modify the free Poisson tensor Λ (4.70) to

$$\Lambda = \frac{1}{m}\mathcal{D}_i \wedge \frac{\partial}{\partial v^i} + \frac{1}{2}C_{ab}\frac{\partial}{\partial I_a} \wedge \frac{\partial}{\partial I_b} + \frac{1}{2m^2}\mathcal{F}^{ij}\frac{\partial}{\partial v^i} \wedge \frac{\partial}{\partial v^j}\ , \tag{4.76}$$

where

$$\mathcal{D}_i = \frac{\partial}{\partial x^i} - \mathcal{A}^i_a(\mathbf{x}, I)\,\frac{\partial}{\partial I_a}$$

defines the covariant derivative associated to the connection $\mathcal{A}$ (see Eq. (3.51) and Appendix E).

The Hamiltonian vector field associated with dx^i is $\Lambda(dx^i) = \partial/\partial v^i$. Being Hamiltonian then implies that $\mathcal{L}_{\partial/\partial v^i}\Lambda = 0$, and we once again have the result that C_{ab}, $\mathcal{A}^i_a$ and $\mathcal{F}^{ij}$ can have no v^i dependence.

This result further implies that functions F^i and f_a appearing in the equations of motion (4.72) are at most first-order in the velocity. From (4.75)

$$F^i = \mathcal{F}^{ij}(\mathbf{x}, I)v^j + \mathcal{F}^{i0}(\mathbf{x}, I) , \quad f_a = -\mathcal{A}^i_a(\mathbf{x}, I)v^i - \mathcal{A}^0_a(\mathbf{x}, I) . \tag{4.77}$$

Here we have introduced new functions $\mathcal{F}^{i0}$ and $\mathcal{A}^0_a$ which are by definition independent of velocity.

By expanding the Jacobi identity

$$[\Lambda, \Lambda] = 0 , \tag{4.78}$$

we obtain additional conditions on C_{ab}, $\mathcal{A}^i_a$ and $\mathcal{F}^{ij}$ and consequently additional conditions on the functions F^i and f_a. From (4.78) we obtain

$$\frac{1}{m^2}[\mathcal{D}_i \wedge \frac{\partial}{\partial v^i}, \mathcal{D}_j \wedge \frac{\partial}{\partial v^j}] + [\Lambda_I, \Lambda_I] + 2[\mathcal{D}_i \wedge \frac{\partial}{\partial v^i}, \Lambda_F] + \frac{2}{m}[\mathcal{D}_i \wedge \frac{\partial}{\partial v^i}, \Lambda_I] \\ + 2[\Lambda_I, \Lambda_F] = 0 . \tag{4.79}$$

Λ_I and Λ_F denote the second and third terms, respectively, in Λ (4.76). Using

$$[\mathcal{D}_i, \frac{\partial}{\partial v^i}] = 0 ,$$

we can reduce (4.79) to

$$\frac{1}{m^2}[\mathcal{D}_i, \mathcal{D}_j] \wedge \frac{\partial}{\partial v^j} \wedge \frac{\partial}{\partial v^i} + [\Lambda_I, \Lambda_I] + 2[\mathcal{D}_i, \Lambda_F] \wedge \frac{\partial}{\partial v^i} + \frac{2}{m}[\mathcal{D}_i, \Lambda_I] \wedge \frac{\partial}{\partial v^i} \\ + 2[\Lambda_I, \Lambda_F] = 0 . \tag{4.80}$$

Collecting our results we find the following conditions on the functions $\mathcal{F}_{ij}$, $\mathcal{A}_{ia}$ and C_{ab}:

$$\begin{aligned} \mathcal{D}_i C_{ab} &= -C_{af}\frac{\partial \mathcal{A}_{ib}}{\partial I_f} + C_{bf}\frac{\partial \mathcal{A}_{ia}}{\partial I_f} , \\ \mathcal{D}_{[i}\mathcal{A}_{j]a} &= C_{ab}\frac{\partial \mathcal{F}^{ij}}{\partial I_b} , \\ \mathcal{D}_{[i}\mathcal{F}_{jk]} &= 0 , \end{aligned} \tag{4.81}$$

in addition to the condition (4.71) on C_{ab}. Of the three conditions in (4.81), the third is the most familiar as it resembles the Bianchi identity. Its connection to the Bianchi identity will be clarified in the examples which follow.

The conditions (4.81) more generally apply if we replace the spatial indices $i, j, \ldots,$ going from 1 to 3 by space-time indices $\mu, \nu, \ldots,$ going from 0 to 3. As usual 0 denotes time, with $\mathcal{F}^{i0} = -\mathcal{F}^{0i}$ and $\mathcal{A}^0_a$ defined in (4.77). The more general result follows upon taking the time derivative of Poisson brackets (4.74). For example, the time derivative of $\{I_a, I_b\} = C_{ab}$ yields

$$\{\dot{I}_a, I_b\} + \{I_a, \dot{I}_b\} = \frac{\partial C_{ab}}{\partial I_f}\dot{I}_f + \frac{\partial C_{ab}}{\partial x^i}\dot{x}^i . \tag{4.82}$$

Then after substituting (4.72) and (4.74) into (4.82) one obtains an equation which is at most linear in velocity v^i. The linear terms however cancel due to the first equation in (4.81) and we are left with

$$\mathcal{D}_0 C_{ab} = -C_{af}\frac{\partial \mathcal{A}_{0b}}{\partial I_f} + C_{bf}\frac{\partial \mathcal{A}_{0a}}{\partial I_f} . \tag{4.83}$$

Equation (4.83) is the first equation in (4.81) with i replaced by 0. Similarly, by taking the time derivative of $m\{v^i, I_a\} = \mathcal{A}^i_a$ and $m^2\{v^i, v^j\} = \mathcal{F}^{ij}$ one obtains

$$\mathcal{D}_{[0}\mathcal{A}_{i]a} = C_{ab}\frac{\partial \mathcal{F}^{0i}}{\partial I_b} \tag{4.84}$$

and

$$\mathcal{D}_{[0}\mathcal{F}_{ij]} = 0 , \tag{4.85}$$

respectively, thus proving the result.

In the above system the particle interacts with the "fields" $C_{ab}(x, I)$, $\mathcal{A}^{\mu}_a(x, I)$ and $\mathcal{F}^{\mu\nu}(x, I)$. Since these fields are functions on the internal space as well as on space-time, they resemble Kaluza–Klein fields. To reduce the theory to one where the coupling is to fields defined on space-time only, it will be necessary to make certain *ansätze* for the form of the functions $C_{ab}(x, I)$, $\mathcal{A}^{\mu}_a(x, I)$ and $\mathcal{F}^{\mu\nu}(x, I)$. These ansätze must be consistent with conditions (4.71) and (4.81). In the following we give examples which lead to some standard particle interactions.

4.5.4.3 Wong Equations

Here we let the Poisson bracket algebra for I_a correspond to a Lie algebra; i.e., $C_{ab} = c_{ab}{}^f I_f$, $c_{ab}{}^f$ being structure constants. As stated earlier, condition (4.71) is then identically satisfied. The standard coupling of a particle to a Yang–Mills field is recovered upon choosing

$$\mathcal{A}^{\mu}_a(x, I) = c_{ab}{}^f I_f A^{b\mu}(x) , \quad \mu = 0, 1, 2, 3, \tag{4.86}$$

where $A^{b\mu}$ is independent of the internal variable I_a and is to be identified with the Yang–Mills potential. This choice identically satisfies the first equation in (4.81). Upon substituting (4.86) into the second equation of (4.81), we get

$$\frac{\partial \mathcal{F}^{\mu\nu}}{\partial I_a} = \frac{\partial A^{a\nu}}{\partial x_\mu} - \frac{\partial A^{a\mu}}{\partial x_\nu} + c_{bf}{}^a A^{b\mu} A^{f\nu} . \tag{4.87}$$

The right-hand side of (4.87) is independent of the internal variables I_a and is to be identified with the Yang–Mills field strength $F^{a\mu\nu}(x)$. Therefore,

$$\mathcal{F}^{\mu\nu}(x, I) = F^{a\mu\nu}(x) I_a \ , \tag{4.88}$$

plus terms which are independent of the internal variables. Finally, the third equation in (4.81) corresponds to the usual Bianchi identities, while the equations of motion following from (4.77) are the usual Wong particle equations,

$$\begin{aligned} m\ddot{x}^\mu &= F^{a\mu\nu}(x) I_a \dot{x}_\nu \ , \\ \dot{I}_a &= -c_{ab}{}^f A^{b\mu}(x) I_f \dot{x}_\mu \ . \end{aligned} \tag{4.89}$$

These equations of motion are obtainable using the Hamiltonian

$$H = \frac{m}{2} > \mathbf{v}^2 - A^{a0}(x) I_a \ . \tag{4.90}$$

The above system of equations can be generalized to the case where I_a does not span a Lie algebra. In this case (4.86) is replaced by

$$\mathcal{A}^\mu_a(x, I) = C_{ab}\ A^{b\mu}(x) \ , \tag{4.91}$$

Equation (4.91) satisfies the first condition in (4.81) provided C_{ab} satisfies (4.71) and it is independent of x. Upon substituting (4.91) into the second equation of (4.81), we get

$$C_{ab}(I)\left(\frac{\partial \mathcal{F}^{\mu\nu}}{\partial I_a} - \frac{\partial A^{a\nu}}{\partial x_\mu} + \frac{\partial A^{a\mu}}{\partial x_\nu} - \frac{\partial C_{bf}}{\partial I_a} A^{b\mu} A^{f\nu}\right) = 0 \ . \tag{4.92}$$

Ignoring the I_a-independent terms, (4.92) is solved by

$$\mathcal{F}^{\mu\nu}(x, I) = \left(\frac{\partial A^{a\nu}}{\partial x_\mu} - \frac{\partial A^{a\mu}}{\partial x_\nu}\right) I_a + C_{bf}(I) A^{b\mu}(x) A^{f\nu}(x) \ . \tag{4.93}$$

Thus $\mathcal{F}^{\mu\nu}$ is not linear in I if C_{ab} is not. The third equation in (4.81) is identically satisfied for $\mathcal{A}^\mu_a$ and $\mathcal{F}^{\mu\nu}$ of the form (4.91) and (4.93), and thus all conditions (4.81) hold for this case. The equations of motion which follow from (4.91) and (4.93) are a generalization of the Wong particle equations (4.89) (although their physical meaning is not clear when I_a does not span a Lie algebra). These equations of motion are also obtainable from the Hamiltonian (4.90).

4.5.4.4 Magnetic Moment Coupling

We next show how the standard Hamiltonian formulation for a spinning particle interacting with a magnetic field can be obtained from the above treatment. For this we take to I_a denote spin. Its Poisson brackets should define an $\mathfrak{su}(2)$ Lie algebra on $\mathbb{R}^3$,

$$\{F, G\} = \epsilon^{abc} I_c \frac{\partial F}{\partial I_a} \frac{\partial G}{\partial I_b} , \tag{4.94}$$

and thus

$$C_{ab} = \epsilon_{abf} I^f . \tag{4.95}$$

Further we set $\mathcal{A}_a^i = 0$, from which it follows using (4.75) that f_a has no velocity dependence. Since $\mathcal{D}_i C_{ab} = 0$, the first condition in (4.81) is satisfied, while the second one implies that $\mathcal{F}^{ij}$ is independent of I_a. If we then set

$$\mathcal{F}^{ij} = \epsilon^{ijk} B^k , \tag{4.96}$$

the third equation in (4.81) tells us that $\mathbf{B}$ has zero divergence and hence can be identified with a magnetic field. It remains to specify $\mathcal{A}_a^0$ and $\mathcal{F}^{0i}$. They must be chosen to be consistent with conditions (4.83–4.85). We do not identify $\mathcal{A}_a^0$ with an electromagnetic potential but rather take

$$\mathcal{A}_a^0 = -\mu \, \epsilon_{abc} I_b B_c , \tag{4.97}$$

where μ represents the magnetic moment. Then (4.83) is identically satisfied while (4.84) gives

$$\mu \, \epsilon_{abc} I_b \partial_i B_c = -\epsilon_{abc} I_b \frac{\partial \mathcal{F}^{0i}}{\partial I_c} \tag{4.98}$$

and thus

$$\mathcal{F}^{0i} = -\mu \, I_c \partial_i B_c , \tag{4.99}$$

plus some function of $I_a I_a$. Here we see that $\mathcal{F}^{0i}$ is not interpreted as an electric field. Condition (4.85) is satisfied only for static magnetic fields.

Upon substituting the above results into the equations of motion (4.72) and (4.77) we recover the standard dynamics for a spinning particle,

$$m\ddot{x}^i = \epsilon^{ijk} \dot{x}^j B^k + \mu \, \partial_i B^j \, I_j , \tag{4.100}$$

$$\dot{I}_a = \mu \, \epsilon_{abc} I_b B_c . \tag{4.101}$$

These equations of motion are obtainable from the Hamilton function

$$H = \frac{m}{2} \mathbf{v}^2 - \mu B_a I_a . \tag{4.102}$$

The above dynamical system can be generalized by replacing the $\mathfrak{su}(2)$ Poisson bracket algebra for the spin by an arbitrary algebra. Thus instead of (4.95) we can take $C_{ab} = C_{ab}(I)$, where we assume that condition (4.71) holds. Here we still take

$\mathcal{A}^i_a = 0$, from which it again follows that the first equation in (4.81) is satisfied, the second one implies that $\mathcal{F}^{ij}$ is independent of I_a, and the third equation gives that **B** has zero divergence. For $\mathcal{A}^0_a$, we can generalize (4.97) to

$$\mathcal{A}^0_a = \mu\, C_{ab}(I) B_b \,, \tag{4.103}$$

which satisfies (4.83). Conditions (4.84) and (4.85) are once again fulfilled by (4.99) with static magnetic fields **B**. Now the equation of motion (4.100) for the spatial coordinates is still valid,

$$\dot{I}_a = -\mu\, C_{ab}(I) B_b \,.$$

These equations are obtainable from the Hamilton function (4.102).

4.6 The Poincaré Group and Massless Systems

Massless systems must be dealt with repeatedly. In order to understand them better we should discuss first their underlying symmetry group which is none other than the Poincaré group.

4.6.1 The Poincaré Group

We will devote this section to the study of the Poincaré group that constitutes the mathematical model of Special Relativity.

4.6.1.1 Minkowski Space-Time

The collection of physical events could be modeled approximately by an affine space,[4] that we call Minkowski space-time and will be denoted by $\mathcal{M}$, over a four-dimensional linear space V carrying a metric $\langle \cdot, \cdot \rangle$ of signature $(- + ++)$.

The points A of $\mathcal{M}$ are called events, and the interval s_{AB} between two events $A, B \in \mathcal{M}$ is defined as $s^2_{AB} = \langle v(A, B), v(A, B) \rangle$, where $v(A, B)$ denotes the unique vector in V such that $B = A + v(A, B)$. The vectors in V can be classified as:

i. Temporal: vectors $x \in M$ such that $\langle x, x \rangle < 0$.
ii. Spatial: vectors $x \in M$ such that $\langle x, x \rangle > 0$.
iii. Light or isotropic: vectors $x \in M$ such that $\langle x, x \rangle = 0$, $x \neq 0$.
iv. $x = 0$.

[4] Recall that an affine space is a pair $(\mathcal{M}, V)$ where V is a linear space that as an abelian group acts freely and transitively on $\mathcal{M}$.

Two events are said to be temporal, spatial or light related if the associated vector in V is temporal, spatial or light respectively.

The collection of affine invertible maps of $\mathcal{M}$[5] form a group know as the affine linear group $AL(\mathcal{M})$. The translation of vector $u \in V$ is defined as the transformation that maps any point A into $B = A + u$. The associated linear map in V is the identity and, conversely, any affine linear map whose associated linear map is the identity is a translation. The set of all translations defines a normal subgroup and $AL(\mathcal{A}) = T_4 \rtimes GL(V)$, where $\ltimes$ denotes the semidirect product of the Abelian group of translations $T_4 \cong \mathbb{R}^4$ and the general linear group $GL(V)$ (see later Sect. 4.6.1 for definitions).

4.6.1.2 The Poincaré and Lorentz Groups

Observers describe events. In our extremely simplified picture, an observer would be a point O in Minkowski space-time $\mathcal{M}$ and the choice of a pseudo-orthonormal basis in V, i.e., a basis e_μ of V such that $\langle e_\mu, e_\nu \rangle = g_{\mu\nu}$, with $g_{\mu\nu} = (-1)^{\delta_{0\nu}} \delta_{\mu\nu}$.[6] In other words, the only non zero elements are: $g_{11} = g_{22} = g_{33} = -g_{00} = 1$.

The point $O \in \mathcal{M}$ allows us to identify $\mathcal{M}$ with V, and the basis of V provides an identification of V with $\mathbb{R}^4$. Thus an observer would associate to each event coordinates x^μ, $\mu = 0, 1, 2, 3$. Another observer would associate to the same event different coordinates x'^μ and the two observers would be (inertially) equivalent if both sets of coordinates are related as:

$$x'^\mu = a^\mu + \Lambda^\mu{}_\nu x^\nu \,. \tag{4.104}$$

where Λ^μ_ν satisfy:

$$g_{\mu\nu} \Lambda^\mu{}_\alpha \Lambda^\nu{}_\beta = g_{\alpha\beta}. \tag{4.105}$$

The set of transformations of the form (4.104) satisfying (4.105) define a group called the Poincaré group that will be denoted in what follow as $\mathcal{P}$.

Transformations such that $a^\mu = 0$ are called (pure) Lorentz transformations and define a subgroup of the Poincaré group called the homogeneous Lorentz group $\mathcal{L}$. Lorentz transformations can be written in matrix form as:

$$\begin{pmatrix} x'^0 \\ x'^1 \\ x'^2 \\ x'^3 \end{pmatrix} = \begin{pmatrix} \Lambda^0{}_0 & \Lambda^0{}_1 & \Lambda^0{}_2 & \Lambda^0{}_3 \\ \Lambda^1{}_0 & \Lambda^1{}_1 & \Lambda^1{}_2 & \Lambda^1{}_3 \\ \Lambda^2{}_0 & \Lambda^2{}_1 & \Lambda^2{}_2 & \Lambda^2{}_3 \\ \Lambda^3{}_0 & \Lambda^3{}_1 & \Lambda^3{}_2 & \Lambda^3{}_3 \end{pmatrix} \begin{pmatrix} x^0 \\ x^1 \\ x^2 \\ x^3 \end{pmatrix} . \tag{4.106}$$

Then condition (4.105) is written as:

---

[5] A map $g \colon \mathcal{M} \to \mathcal{M}$ is called affine if there exists a linear map $\bar{g} \colon V \to V$ such that $v(g(A), g(B)) = \bar{g}(v(A, B))$.

[6] We will always assume Einstein's convention of sum over repeated indices.

$$\Lambda^T G \Lambda = G, \tag{4.107}$$

and we get immediately that:

$$|\det \Lambda| = 1 .$$

Lorentz transformations such that $\det \Lambda = +1$ define a unimodular subgroup in $\mathcal{L}$ denoted by $\mathcal{L}_+$.

Because Lorentz transformations preserve the metric then, in particular $|\Lambda^0{}_0| \geq 1$, and as a topological space $\mathcal{L}$ is not connected, but it has four connected components characterized as follows:

$$\mathcal{L}_0 = \mathcal{L}_+^\uparrow \quad \text{Proper orthocronous,} \qquad \det \Lambda = 1, \quad \Lambda^0{}_0 \geq 1 , \tag{4.108}$$

$$\mathcal{L}_-^\uparrow \quad \text{Improper orthocronous,} \qquad \det \Lambda = -1, \quad \Lambda^0{}_0 \geq 1 , \tag{4.109}$$

$$\mathcal{L}_+^\downarrow \quad \text{Proper antiorthocronous,} \qquad \det \Lambda = 1, \quad \Lambda^0{}_0 \leq -1 , \tag{4.110}$$

$$\mathcal{L}_-^\uparrow \quad \text{Improper antiorthocronous,} \qquad \det \Lambda = -1, \quad \Lambda^0{}_0 \leq -1 . \tag{4.111}$$

Among all of them only the component that contains the identity $\mathcal{L}_0$, is a subgroup. Other important subgroups of $\mathcal{L}$, are:

$$\mathcal{L}^\uparrow = \mathcal{L}_+^\uparrow \cup \mathcal{L}_-^\uparrow, \quad \mathcal{L}_+ = \mathcal{L}_+^\uparrow \cup \mathcal{L}_+^\downarrow, \quad \mathcal{L}_{+-}^{\uparrow\downarrow} = \mathcal{L}_+^\uparrow \cup \mathcal{L}_-^\downarrow .$$

The matrix form of a pure Lorentz transformation in the Z direction is:

$$\begin{pmatrix} x'^0 \\ x'^1 \\ x'^2 \\ x'^3 \end{pmatrix} = \begin{pmatrix} c & 0 & 0 & d \\ 0 & 1 & 0 & 0 \\ 0 & 0 & 1 & 0 \\ a & 0 & 0 & b \end{pmatrix} \begin{pmatrix} x^0 \\ x^1 \\ x^2 \\ x^3 \end{pmatrix} . \tag{4.112}$$

and it must satisfy:

$$a^2 - c^2 = 1, \quad d^2 - b^2 = 1, \qquad ab - cd = 1,$$

from which we get $a^2 = d^2$ and $b^2 = c^2$. We can write them so that the transformation belongs to $\mathcal{L}_0$, as

$$a = d = \cosh \beta, \qquad b = c = \sinh \beta . \tag{4.113}$$

Let us recall that the physical meaning of β is given by:

$$\tanh \beta = \frac{v}{c} . \tag{4.114}$$

where v is the relative velocity of the two systems.

4.6.1.3 Semidirect Products: $\mathcal{P}_0$ as a Semidirect Product

Given a group H and an action of H on an Abelian group N by automorphisms, that is a group homomorphism $\rho\colon H \to \mathrm{Aut}(N)$, where $\rho(h)\colon N \to N$ is a group isomorphism, we can define a group structure on the product space $G = N \times H$ in a natural way and the resulting group will be called the semidirect product of N and H. The semidirect product will be denoted as $H \rtimes N$ and the group law is given by:

$$(n, h) \cdot (n', h') = (n + \rho(h)n, hh'), \qquad \forall n, n' \in N, \quad h, h' \in H .$$

We will denote, as usual $\rho(h)n$ by hn. We check immediately that the composition law is associative and that $(0, e)$ is the identity element. Finally $(-h^{-1}n, h^{-1})$ is the inverse element of (n, h).

Notice that: $(0, h) \cdot (n, e) \cdot (0, h)^{-1} = (hn, h) \cdot (0, h^{-1}) = (hn, e)$, which shows that N is a normal subgroup of $N \rtimes H$, and $N \rtimes H/N \cong H$.

If $\mathfrak{h}$ and $\mathfrak{n}$ denote the Lie algebras of H and N respectively, then the Lie algebra of $N \rtimes H$ is, as a linear space $\mathfrak{n} \oplus \mathfrak{h}$, and the Lie bracket is given by:

$$[(\alpha, \xi), (\beta, \zeta)] = (\hat{\rho}(\xi)\beta - \hat{\rho}(\zeta)\alpha, [\xi, \zeta]), \qquad (\alpha, \xi), (\beta, \zeta) \in \mathfrak{n} \oplus \mathfrak{h} ,$$

where $\hat{\rho}(\xi)$ is the representation of the Lie algebra $\mathfrak{h}$ induced by ρ, this is $\hat{\rho}(\xi)n = \frac{d}{dt}\rho(\exp(t\xi)n \mid_{t=0}$.

4.6.1.4 The Poincaré Group as a Semidirect Product

The iterated application of Eq. (4.104) allows us to obtain the group law of Poincaré's group. Thus if,

$$x_1{}^\mu = a_1^\mu + \Lambda_1{}^\mu{}_\nu x^\nu, \quad x_2{}^\mu = a_2^\mu + \Lambda_2{}^\mu{}_\nu x_1{}^\nu,$$

then

$$x_2{}^\mu = (\Lambda_2\Lambda_1)^\mu{}_\alpha x^\alpha + \Lambda_2{}^\mu{}_\nu a_1^\nu + a_2^\mu$$

and in condensed form:

$$(a_2, \Lambda_2) \cdot (a_1, \Lambda_1) = (a_2 + \Lambda_2 a_1, \Lambda_2\Lambda_1)$$

that is the composition law of the semidirect product of the Lorentz group $\mathcal{L}$ and the Abelian group of translations T_4 with the natural action of $\mathcal{L}$ on T_4 given by matrix multiplication. Thus,

$$\mathcal{P} = T_4 \rtimes \mathcal{L} .$$

In particular, if we restrict ourselves to the connected components we have $\mathcal{P}_0 = T_4 \rtimes \mathcal{L}_0$. Notice the normal subgroup relation for T_4:

$$(0, \Lambda)(a, \mathbb{I})(0, \Lambda^{-1}) = (\Lambda a, \mathbb{I})\,. \tag{4.115}$$

4.6.1.5 The Lie Algebra of the Poincaré Group

An infinitesimal Lorentz transformation will be written as

$$x'^{\mu} = (g^{\mu}{}_{\nu} + \epsilon^{\mu}{}_{\nu})x^{\nu}\,, \tag{4.116}$$

and must satisfy the isometry condition:

$$\begin{aligned} g_{\mu\nu}x'^{\mu}x'^{\nu} &= g_{\mu\nu}(g^{\mu}{}_{\alpha} + \epsilon^{\mu}{}_{\alpha})(g^{\nu}{}_{\beta} + \epsilon^{\nu}{}_{\beta})x^{\alpha}x^{\beta} \\ &= g_{\alpha\beta}x^{\alpha}x^{\beta} + g_{\mu\nu}(g^{\mu}{}_{\alpha}\epsilon^{\nu}{}_{\beta} + g^{\nu}{}_{\beta}\epsilon^{\mu}{}_{\alpha}) + O(\epsilon^2), \end{aligned}$$

that is,

$$\epsilon_{\alpha\beta} + \epsilon_{\beta\alpha} = 0\,.$$

We will denote by $M^{\alpha\beta}$ the infinitesimal generators of $\mathcal{L}_0$, then we will rewrite (4.116) as:

$$x'^{\mu} = (g^{\mu}{}_{\nu} + \frac{1}{2}\epsilon_{\alpha\beta}[M^{\alpha\beta}]^{\mu}{}_{\nu})x^{\nu}\,.$$

and the generators $M^{\alpha\beta}$ can be defined as:

$$(M^{\alpha\beta})^{\mu}{}_{\nu} = g^{\mu\alpha}g_{\nu}{}^{\beta} - g_{\nu}{}^{\alpha}g^{\mu\beta}\,, \tag{4.117}$$

Notice that, as a consequence of equation (4.117), we get:

$$M^{\alpha\beta} + M^{\beta\alpha} = 0\,.$$

The commutation relations that define the Lie algebra structure of $\mathcal{L}_0$ are obtained easily from (4.117):

$$[M^{\alpha\beta}, M^{\gamma\delta}] = g^{\beta\gamma}M^{\alpha\delta} + g^{\alpha\delta}M^{\beta\gamma} + g^{\alpha\gamma}M^{\delta\beta} + g^{\delta\beta}M^{\gamma\alpha}, \tag{4.118}$$

Notice that if we define $\mathbf{M} = (M^1, M^2, M^3)$ by

$$M^1 = M^{23}, \qquad M^2 = M^{31}, \qquad M^3 = M^{12}\,,$$

from Eq. (4.118) we find the commutation rules:

$$[M^1, M^2] = [M^{23}, M^{31}] = g^{33} M^{21} = -M^3 \,, \tag{4.119}$$

and in a similar way:

$$[M^2, M^3] = -M^1, \qquad [M^3, M^1] = -M^2 \,. \tag{4.120}$$

If we define now $\mathbf{N} = (N^1, N^2, N^3)$ as:

$$N^i = M^{0i}, \qquad i = 1, 2, 3 \,, \tag{4.121}$$

again, using Eq. (4.118), we find the corresponding commutation rules:

$$[N^1, N^2] = [M^{01}, M^{02}] = g^{00} M^{21} = M^3 \,, \tag{4.122}$$

together with

$$[N^2, N^3] = M^1, \qquad [N^3, N^1] = M^2 \,. \tag{4.123}$$

Moreover M^i satisfy the commutation rules with N^j

$$[M^1, N^2] = [M^{23}, M^{02}] = M^{30} = -N^3 \,, \tag{4.124}$$

and

$$[M^2, N^3] = -N^1, \quad [M^3, N^1] = -N^2 \,,$$

$$[M^2, N^1] = N^3, \quad [M^1, N^3] = N^2 \,, \qquad [M^3, N^2] = N^1 \,. \tag{4.125}$$

Notice that if we define $\mathbf{K}_{\pm} = \frac{1}{2}(\mathbf{M} \pm i\mathbf{N})$, from Eqs. (4.120)–(4.125), we get the commutation rules:

$$[K_+^i, K_+^j] = -\epsilon^{ijl} K_+^l \qquad [K_-^i, K_-^j] = -\epsilon^{ijl} K_-^l \qquad [K_+^i, K_-^j] = 0 \,, \tag{4.126}$$

that shows that the complexification of the Lie algebra of $\mathcal{L}_0$ is the Lie algebra of $SO(3) \times SO(3)$.

The Lie algebra of the Poincaré group includes the generators of translations P^μ, satisfying $[P^\mu, P^\nu] = 0$. Recall that because of Eq. (4.115), we get:

$$[M^{\alpha\beta}, P^\mu] = [M^{\alpha\beta}]^\mu{}_\nu P^\nu \tag{4.127}$$

and taking into account the explicit form of $M^{\alpha\beta}$, we finally get:

$$[M^{\alpha\beta}, P^\mu] = (g^{\mu\alpha} g_\nu{}^\beta - g_\nu{}^a g^{\mu\beta}) P^\nu = g^{\mu\alpha} P^\beta - g^{\mu\beta} P^\alpha \,. \tag{4.128}$$

4.6.1.6 The Universal Covering $SL(2, \mathbb{C})$ of $\mathcal{L}_0$

In a similar way as we did with the unitary group $SU(2)$ and the group of rotations (see Sect. 2.6.1, Eq. (2.144)) we want to show that the group $SL(2, \mathbb{C})$ is the universal covering of the group $\mathcal{L}_0$. We denote by $\mathcal{H}(2)$ the linear space of 2×2 Hermitean matrices.

Let us consider the map $\chi : \mathbb{R}^4 \to \mathcal{A}$ given by:

$$\chi(x) = x^0 \mathbb{I} + \mathbf{x} \cdot \boldsymbol{\sigma} , \tag{4.129}$$

Clearly χ is an isomorphism of linear spaces. Moreover the determinant of $\chi(x)$ is $-\langle x, x \rangle$.

The image of the standard basis in $\mathbb{R}^4$ is the basis $\{\sigma^0 = \mathbb{I}, \boldsymbol{\sigma}\}$ of $\mathcal{H}(2)$. The pre-image of a matrix $X \in \mathcal{H}(2)$ is the vector x with coordinates:

$$x^\mu = -\frac{1}{2} \mathrm{Tr}\,(X \hat{\sigma}^\mu) , \tag{4.130}$$

where

$$\hat{\sigma}_0 = \mathbb{I}, \qquad \hat{\sigma}_k = -\sigma_k, \quad k = 1, 2, 3 .$$

Then, because:

$$\chi(x) = \begin{pmatrix} x^0 + x^3 & x^1 - ix^2 \\ x^1 + ix^2 & x^0 - x^3 \end{pmatrix} ,$$

we get:

$$\det \chi(x) = (x^0)^2 - \mathbf{x} \cdot \mathbf{x} = -\langle x, x \rangle .$$

Proposition 4.11 i. *For any $A \in SL(2, \mathbb{C})$ and $X \in \mathcal{H}(2)$, then $AXA^\dagger \in \mathcal{H}(2)$.*
ii. *The transformation $\Lambda_A = \chi^{-1} \circ p(A) \circ \chi$ is a Lorentz transformation, where $p : \mathcal{H}(2) \to \mathcal{H}(2)$ is the linear map given by $p(A) : X \mapsto XAX^\dagger$, for all $A \in SL(2, \mathbb{C})$.*

Proof i. It suffices to check $(AXA^\dagger)^\dagger = AX^\dagger A^\dagger = AXA^\dagger$.
ii. Notice that $\langle \Lambda_A x, \Lambda_A x \rangle = -\det p(A)\chi(x) = -\det AXA^\dagger = -\det X = \langle x, x \rangle$, thus
the transformation Λ_A preserves the Minkowski metric. □

Proposition 4.12 *The map $\Lambda : SL(2, \mathbb{C}) \to \mathcal{L}_0$, defined by $A \mapsto \Lambda_A = \chi^{-1} \circ p(A) \circ \chi$ is a group homomorphism with kernel $\mathbb{Z}_2$.*

Proof It suffices to check that $p(AB)X = ABX(AB)^\dagger = ABXB^\dagger A^\dagger = p(A)p(B)X$. Then $\Lambda_{AB} = \chi^{-1} \circ p(AB) \circ \chi = \chi^{-1} \circ p(A) \circ \chi \circ \chi^{-1} \circ p(B) \circ \chi^{-1} = \Lambda_A \Lambda_B$.

The kernel is given by matrices A such that $p(A)X = X$, $\forall X$, but this implies that $[A, \sigma_i] = 0$, and then $A = \pm \mathbb{I}$. □

Notice that the elements of the matrix Λ_A can be written as:

$$(\Lambda_A)^\mu{}_\nu = -\frac{1}{2}\mathrm{Tr}\,(\hat{\sigma}^\mu A \sigma_\nu A^\dagger)\,.$$

4.6.2 A Classical Description for Free Massless Particles

The description of massless elementary systems with non-zero spin and/or helicity, like gluons, gravitons, neutrinos or photons, when thought of as classical particles should have some features inherited from our quantum description of them. In particular:

a. If $\mathbf{p}$ denotes the particle 3-momentum, the value $\mathbf{p} = 0$ should not be part of our carrier space for obvious reasons.
b. Denoting with $\mathbf{x}$ the position, a description in terms of Poisson brackets should be such that

$$\{x_i, x_j\} \neq 0, \qquad i \neq j,$$

to take into account that massless particles are not localizable.
c. The angular momentum of a massless particle should be non-zero along the direction of the momentum of the particle, if its spin is nonzero.
d. The translation group should state that the momentum $\mathbf{p}$ and the total angular momentum $\mathbf{J}$ are conserved, i.e.,

$$\frac{d\mathbf{p}}{dt} = 0; \qquad \frac{d\mathbf{J}}{dt} = 0. \tag{4.131}$$

Starting from these requirements, we may wish to build a Hamiltonian description, i.e., to define a carrier space, endow it with a Poisson structure and specify the Hamiltonian function.

Let us make a few considerations on our requirements. They will consists of,

1. The carrier space should be $\mathbb{R}^3 \times (\mathbb{R}^3 - \{0\})$.
2. The dynamical vector field on this space is $\Gamma = p_i \partial/\partial x_i$.
3. The rotation group generated by

$$R_i = \epsilon_{ijk}\left(x_j \frac{\partial}{\partial x_k} + p_j \frac{\partial}{\partial p_k}\right)$$

and the translation group generated by $T_i = \frac{\partial}{\partial x_i}$, should act canonically.

Let us start with a bivector field of the general form

$$\Lambda = F_{ij}\frac{\partial}{\partial x_i} \wedge \frac{\partial}{\partial x_j} + G_{ij}\frac{\partial}{\partial p_i} \wedge \frac{\partial}{\partial p_j} + A_{ij}\frac{\partial}{\partial x_i} \wedge \frac{\partial}{\partial p_j},$$

satisfying that the Schouten bracket of Λ with itself vanishes:

$$[\Lambda, \Lambda] = 0\,,$$

and such that:

$$\mathcal{L}_\Gamma \Lambda = 0; \quad \mathcal{L}_{T_i} \Lambda = 0; \quad \mathcal{L}_{R_i} \Lambda = 0; \quad i = 1, 2, 3\,. \tag{4.132}$$

Invariance under translations, Eq. (4.132), implies that F_{ij}, G_{ij}, A_{ij} are independent of x. The invariance under dynamical evolution implies that G_{ij} is a constant of motion, moreover, $\dot{F}_{ij} = A_{ij} - A_{ji}$, and $G_{ij} = \dot{A}_{ij}$.

Because A is a function only of p, it follows that $\dot{A} = 0$, $G = 0$. By the same argument $\dot{F} = 0$ an we get $A_{ij} = A_{ji}$.

Because of invariance under rotation $A_{ij} = a(p^2)\delta_{ij} + b(p^2)p_i p_j$, but also $F_{ij} = f(p^2)\epsilon_{ijk} p_k$. Putting all this together we obtain that Λ must be of the form:

$$\Lambda = f(p^2)\epsilon_{ijk} p_k \frac{\partial}{\partial x_i} \wedge \frac{\partial}{\partial x_j} + a(p^2)\delta_{ij} \frac{\partial}{\partial x_i} \wedge \frac{\partial}{\partial p_j} + b(p^2) p_i \frac{\partial}{\partial x_i} \wedge p_j \frac{\partial}{\partial p_j}\,. \tag{4.133}$$

Let us make a few more considerations. The last term in the right-hand side of Eq. (4.133) defines a Poisson structure on its own. The compatibility condition $[\Lambda, \Lambda] = 0$ is equivalent to the differential equation:

$$3af + p^2(bf + 2\frac{da}{dp^2} f + 2bp^2 \frac{df'}{dp^2}) = 0\,.$$

Hence, for $a = 0$ we get:

$$f = f_0 \exp -\frac{1}{2} \int (\frac{1}{b} + 1) dp^2\,.$$

Finally, for $a \neq 0$ we find:

$$b = \frac{a'}{\frac{1}{2} - p^2 (\log|a|)'}; \qquad f = \frac{a^2 n}{p^2},$$

where $a' = da/dp^2$ and n is a constant of integration.

Then the Poisson brackets defined by Λ, become:

$$\{x_i, x_j\} = \frac{na(p^2)}{p^3} \epsilon_{ijk} p_k\,, \quad \{p_i, p_j\} = 0\,, \quad \{x_i, p_j\}$$
$$= a(p^2) \left[\delta_{ij} + \frac{(\log|a(p^2)|)' p_i p_j}{\frac{1}{2} - p^2 (\log|a(p^2)|)'} \right].$$

Upon replacing p_i by $P_i = p_i/a(p^2)$ we find the new brackets

$$\{x_i, x_j\} = \frac{n}{P^3}\epsilon_{ijk}P_k\,, \qquad \{P_i, P_j\} = 0\,, \qquad \{x_i, P_j\} = \delta_{ij}\,.$$

For $n = 0$, x_i, P_i are the standard canonical variables, while for $n \neq 0$, we must exclude the origin of the momentum $P = 0$ from the phase space. The latter case corresponds to the phase space description of a single helicity particle.

The significance of P_i, as opposed to p_i, is that it corresponds to the generator of spatial translation. The generators of the rotation subgroup are given by ($i = 1, 2, 3$):

$$J_i = \frac{1}{a}\epsilon_{ijk}x_j p_k - n\frac{p_i}{|p|} = \epsilon_{ijk}x_j P_k - n\frac{P_i}{|P|}\,.$$

The generator of time translation can be taken to be $|P|$. Of course as we are dealing with massless particles, a manifest relativistic description of our systems seems desirable. We shall proceed therefore to provide a relativistic description.

References

[Gi93] Giordano, M., Marmo, G., Rubano, C.: The inverse problem in the Hamiltonian formalism: integrability of linear Hamiltonian fields. Inv. Probl. **9**, 443–467 (1993)

[Ar76] Arnol'd, V.I.: Méthodes mathématiques de la mécanique classique. Ed. Mir, 1976. Mathematical methods of classical mechanics. Springer (1989)

[Sc53] Schouten, J.A.: On the differential operators of first-order in tensor calculus. Conv. Int. Geom. Diff., Italia (1953) (Ed. Cremonese, Roma 1954, pp. 1–7)

[Li77] Lichnerowicz, A.: Les variétés de Poisson et leurs algèbres de Lie associées. J. Diff. Geom. **17**, 253–300 (1977)

[Ki62] Kirillov, A.A.: Unitary representations of nilpotent Lie groups. Russ. Math. Surv. **17**, 57–110 (1962)

[Ib90] Ibort, A., López-Lacasta, C.: On the existence of local and globals for ordinary differential equations. J. Phys. A: Math. Gen. **23**, 4779–4792 (1990)

[Wi50] Wigner, E.P.: Do the equations of motion determine commutation relations and ambiguity in the Lagrangian formalism? Phys. Rev. **77**, 711–712 (1950)

[We83] Weinstein, A.: On the local structure of Poisson manifolds. J. Diff. Geom. **18**, 523–557 (1983)

[We87] Weinstein, A.: Symplectic groupoids. Bull. Am. Math. Soc. **16**, 101–04 (1987)

[Mo91] Morandi, G., Ferrario, C., Lo, G., Vecchio, G.M., Rubano, C.: The inverse problem in the calculus of variations and the geometry of the tangent bundle. Phys. Rep. **188**, 147–284 (1991)

[Dy90] Dyson, F.J.: Feynman's proof of the Maxwell equations. Am. J. Phys. **58**, 209–211 (1990)

[Ba82] Balachandran, A.P., Marmo, G., Stern, A.: A Lagrangian approach to the non-interaction theorem. Nuovo Cim. **69A**, 175–186 (1982)

[So92] Soni, S.K.: Classical spin and quantum algebras. J. Phys. A: Math. Gen. **25**, L837–L842 (1992)

[St93] Stern, A., Yakushin, I.: Deformed Wong particles. Phys. Rev. D **48**, 4974–4979 (1993)

[Ta89] Takhtajan, L.A.: Lectures on quantum groups. Introduction to quantum group and integrable massive models of quantum field theory. In: Ge, M.-L., Zhao, B.-H. (eds.). World Scientific (1990)

[Ma90] Majid, S.: On q-regularization. Int. J. Mod. Phys. **5**, 4689–4696 (1990)
[Tj92] Tjin, T.: Introduction to quantized Lie groups and Lie algebras. Int. J. Mod. Phys. A **7**, 6175–6214 (1992)
[Wo70] Wong, S.K.: Field and particle equations for the classical Yang-Mills field and particles with internal spin. Nuovo Cim. **65A**, 689–694 (1970)
[Ta92] Tanimura, S.: Relativistic generalization and extension to non-Abelian gauge theory of Feynman's proof of the Maxwell equations. Ann. Phys. **220**, 229–247 (1992)

Chapter 5
The Classical Formulations of Dynamics of Hamilton and Lagrange

On ne trouvera point de Figures dans set Ouvrage. Les méthodes que j'y expose ne demandent ni constructions, ni raisonnements géométriqus ou méchaniques, mais seulement des opérations algébriques, assujetties à une march réguliere et uniforme.
Joseph-Louis Lagrange, Mécanique Analytique, Avertissement de la premiére édition, 1788.

5.1 Introduction

The present chapter is perhaps the place where our discourse meets more neatly the classic textbooks on the subject. Most classical books concentrate on the description of the formalisms developed by Lagrange and Euler on one side, and Hamilton and Jacobi on the other and commonly called today the Lagrangian and the Hamiltonian formalism respectively. The approach taken by many authors is that of postulating that the equations of dynamics are derived from variational principles (a route whose historical episodes are plenty of lights and shadows [Ma84]).

Such a procedure is almost unquestioned because of its spectacular success in building the foundations of many modern physical theories. However, the current state of affairs is not satisfactory, as variational principles are rather poor when trying to explore the foundations of the theories of dynamics; our goal is to look even further forward to incorporate eventually quantum dynamics.

Thus we will approach this chapter as a continuation of our previous effort, that is trying to understand the properties of dynamical systems when they possess invariant structures, that are (completely or not) determined by it. Hence our first task will be to

---

'The reader will find no figures in this work. The methods which I set forth do not require either constructions or geometrical or mechanical reasonings, but only algebraic operations, subject to a regular and uniform rule of procedure'.

J.F. Cariñena et al., *Geometry from Dynamics, Classical and Quantum*,
DOI 10.1007/978-94-017-9220-2_5

explore non-degenerate Poisson structures, a question that is left from the previous chapter, and that we will conclude here arriving to the 'standard' presentation of Hamiltonian system. We will take a further step by exploring the consequences of a dynamics possessing not just a cotangent bundle description but a tangent bundle one. In this way we will arrive to the standard Lagrangian description of dynamical system and to a solution of the inverse problem for Lagrangian systems. It is most remarkable that we will do that without relinking in any way to variational principles or other 'metaphysical principles'. We will only look for more and more exigent physical properties to our systems until ttheir class is seen as being "simple".

5.2 Linear Hamiltonian Systems

In the previous chapter we introduced the Hamiltonian formalism by looking for Poisson tensors Λ compatible with a given dynamics, in the sense that Γ ($= X_A$ if the dynamics is linear), is a Hamiltonian vector field with respect to such dynamics, that is there is a function H such that $\Gamma = \Lambda(dH)$.

We will analyze now the particular case in which the Poisson structure considered is non-degenerate, that is, the only Casimir functions are constant functions or, if you wish, there is a one-to-one correspondence between exact 1-forms and Hamiltonian vector fields. Such particular instances of Poisson structures have been playing a most relevant role in the understanding of the structure of physical theories so they deserve a more detailed analysis. We will devote the first part of this chapter to them.

If the Poisson structure Λ is defined on a $2n$-dimensional linear space E and is constant, then it defines a non-degenerate bilinear form ω on E. Notice that since Λ is non-degenerate, the linear isomorphism associated to it, $\widehat{\Lambda}\colon E^* \to E$, given by $\langle \beta, \widehat{\Lambda}(\alpha)\rangle = \Lambda(\alpha, \beta)$, is invertible and $\widehat{\Lambda}^{-1}\colon E \to E^*$ defines a bilinear form ω on E by means of:

$$\omega(u, v) := \Lambda(\Lambda^{-1}(u), \Lambda^{-1}(v)) = \langle \widehat{\Lambda}^{-1}(v), u\rangle, \qquad u, v \in E,$$

where $\langle \cdot, \cdot \rangle$ denotes as usual the natural pairing between E and E^*. By selecting a linear basis $\{e_i \mid i = 1, \ldots 2n\}$ on E, if Λ^{ij} denotes the corresponding components of Λ, then the components ω_{ij} of the bilinear form ω above satisfy:

$$\Lambda^{ij}\omega_{jk} = -\delta^i_k. \tag{5.1}$$

It is also common to use the notation Λ^{-1} to denote $-\omega$ (or ω^{-1} to denote $-\Lambda$). Because of the geometrization principle, the bilinear form ω defines a non-degenerate closed differential form on E, denoted with the same symbol ω, and very often this setting constitutes the entrance door to the study of Hamiltonian systems.

We will start our analysis by a quick review of the structure and properties of this framework and will continue afterwards by providing dynamical systems possessing these invariant structures, the same treatment as in former chapters.

5.2.1 Symplectic Linear Spaces

Definition 5.1 Let E be a real linear space and σ a non-degenerate skew symmetric bilinear form on it, that is, $\sigma(u, v) = 0$ for all $v \in E$ implies that $u = 0$. The pair (E, σ) is then said to be a (real) symplectic linear space.

A slightly more general concept is that of a presymplectic linear space, that is a pair (E, σ) such that σ is a skew-symmetric, but maybe degenerate, 2-form in E.

Because of the way we were introducing the notion of symplectic linear spaces, it is clear that the canonical leaves of linear Poisson structures are automatically sympletic linear spaces. Thus all we have to do is go back to the catalogue of linear Poisson structures discussed in Sect. 4.3.3 and extract their leaves. We will not do that though. We will proceed by describing a class of examples that exhibits the main structure of symplectic linear spaces.

Let V be a linear space of dimension n, and V^* its dual space. We can define a skew-symmetric bilinear form σ in $E = V \oplus V^*$ by

$$\sigma_V((u, \alpha), (v, \beta)) = \langle \alpha, v \rangle - \langle \beta, u \rangle\,, \qquad u, v \in V, \alpha, \beta \in V^*, \tag{5.2}$$

that can easily be shown to be non-degenerate: if $\sigma_V((u, \alpha), (v, \beta)) = 0,\ \forall (v, \beta) \in V \oplus V^*$, it suffices to consider the subspace of vectors of the form either $(v, 0)$ or $(0, \beta)$, respectively, to convince ourselves that u and α should be the zero vector and the zero covector, respectively.

As a particular instance, if $V = \mathbb{R}^n$, then $E = \mathbb{R}^n \oplus \mathbb{R}^{n*} \cong \mathbb{R}^{2n}$, where the last identification is provided by choosing a linear basis in $\mathbb{R}^n$. Let us consider just the canonical basis $\{e_i \mid i = 1, \ldots, n\}$ of $\mathbb{R}^n$ and denote the corresponding coordinates as q^i, namely, if $v \in \mathbb{R}^n$, $v = q^i(v)e_i$. The corresponding dual basis $\{e^i \mid i = 1, \ldots, n\}$ will also be denoted $\{e^i = dq^i \mid i = 1, \ldots, n\}$, because $dq^i(v)$ is the i-th component of the vector v. Then, the coordinates of covectors in $\mathbb{R}^{n*}$ with respect to the dual basis will be denoted p_i, and this will allow us to rewrite the differential 2-forms defined by σ on $\mathbb{R}^{2n}$ as:

$$\sigma = \sum_{i=1}^{n} dq^i \wedge dp_i\ .$$

According to the tradition we will call it the canonical 2-form on $\mathbb{R}^{2n}$.

Another interesting instance of symplectic linear space is provided by the following construction. If (V_1, ω_1) and (V_2, ω_2) are symplectic linear spaces, then $V_1 \oplus V_2$ can be endowed with a skew-symmetric bilinear 2-form $\omega_1 \ominus \omega_2$, by means of

$$(\omega_1 \ominus \omega_2)((v_1, v_2), (u_1, u_2)) = \omega_1(v_1, u_1) - \omega_2(v_2, u_2)\,, \quad v_a, u_a \in V_a, \quad a = 1, 2\,,$$

or, in other words, if $\pi_i : V_1 \oplus V_2 \to V_i$, $i = 1, 2$, are the canonical projections on each factor, $\omega_1 \ominus \omega_2 = \pi_1^*\omega_1 - \pi_2^*\omega_2$.

The form $\omega_1 \ominus \omega_2$ is non-degenerate, because if we assume that $(\omega_1 \ominus \omega_2)((v_1, v_2), (u, 0)) = 0$ for all $u \in V_1$, as ω_1 is non-degenerate, v_1 must be zero, and the same reasoning works for the other argument.

Notice that the same arguments work for the standard direct sum bilinear form $\omega_1 \oplus \omega_2$, defined as:

$$(\omega_1 \oplus \omega_2)((v_1, v_2), (u_1, u_2)) = \omega_1(v_1, u_1) + \omega_2(v_2, u_2)\,, \quad v_a, u_a \in V_a, \quad a = 1, 2\,.$$

For reasons that will be obvious later on, it is the 'direct difference' $\omega_1 \ominus \omega_2$, of two symplectic forms that is going to be relevant for us now.

5.2.1.1 Symplectic Linear Spaces and Symplectic Structures

Because of the geometrization principle, we may consider the geometrical structure corresponding to a symplectic linear space.

Thus given a symplectic linear space (E, σ), we begin by thinking that σ defines a smooth differential 2-form on E. Actually because it is constant, σ is closed. Is this last property an integral part of the geometrization principle for symplectic linear spaces?

To elucidate this question, we recall that the Poisson structure $\Lambda = -\sigma^{-1}$ determined by σ defines a structure of Poisson algebra in the space of functions on E, in particular the Poisson bracket $\{f, g\}$ thus defined satisfies the Jacobi identity. Notice that the Poisson tensor Λ is defined as:

$$\Lambda(\alpha, \beta) = \sigma\left(\widehat{\sigma}^{-1}(\alpha), \widehat{\sigma}^{-1}(\beta)\right)\,,$$

where α and β are 1-forms on E and $\widehat{\sigma}$ is the natural isomorphism $\widehat{\sigma}_x \colon T_xE \to T_x^*E$, $\langle\widehat{\sigma}_x(u), v\rangle = \sigma(u, v)$. In particular, when $\alpha = df$ and $\beta = dg$ we find the expression for the Poisson bracket:

$$\{f, g\} = \Lambda(df, dg) = \sigma\left(\hat{\sigma}^{-1}(df),\ \hat{\sigma}^{-1}(dg)\right).$$

But now, if we denote by X_f the vector field on E defined by $\widehat{\sigma}^{-1}(df)$, it satisfies that $\Lambda(X_f) = -df$, hence it is just the Hamiltonian vector field defined by f (with respect to Λ). For this reason we will call also in this context X_f the Hamiltonian vector field defined by f. With this notation we conclude:

$$\{f, g\} = \sigma(X_f, X_g)\,, \quad f, g \in \mathcal{F}(E), \tag{5.3}$$

and then $X_f = \{\cdot, f\}$.

It is clear that if we try to define a Poisson bracket in the algebra of smooth functions on E by using Eq. (5.3), in general it would not define a Poisson algebra. Actually the only property that we have to check is the Jacobi identity because the Leibnitz rule (or derivation property will be guaranteed as X_f is a vector field, hence a derivation). Thus the geometrization principle for symplectic linear spaces should be completed by asking for that the bracket defined by (5.3) should be a Poisson bracket.

Now a simple computation shows that:

$$\{f, \{g, h\}\} + \{h, \{f, g\}\} + \{g, \{h, f\}\} = 2\, d\sigma(X_f, X_g, X_h)\,. \tag{5.4}$$

Of course, if σ is constant then $d\sigma = 0$ and the bracket above defines a Poisson bracket as we already know. However if we allow σ to be non constant, then the only requirement that σ must satisfy to define a Poisson bracket in the space of functions $\mathcal{F}(E)$ is that σ be closed. In this case Jacobi identity for the bracket follows and furthermore such identity implies that

$$[X_f, X_g] = X_{g,f}\,.$$

Thus the implementation of the extended geometrization principle for symplectic linear spaces leads us to the following definition:

Definition 5.2 We will say that a linear space E is endowed with a symplectic structure ω when ω is a non-degenerate closed 2-form on E.

Moreover, we will say that the symplectic structure is exact when the closed 2-form ω is exact, that is, there exists a 1-form θ such that $\omega = -d\theta$.

When ω is closed and of a constant rank, but maybe degenerate, we will say that E is endowed with a presymplectic structure.

Recall that the non-degeneracy of ω is equivalent, by definition, to the property that the maps $\hat{\omega}_x : T_xE \to T_x^*E$, given by:

$$\langle \hat{\omega}_x(v), v' \rangle = \omega_x(v, v'), \qquad v, v' \in T_xE\,,$$

have maximal rank. They induce a mapping between sections which, with a slight abuse of notation, we will also write $\hat{\omega}$ or again ω. Then, a non-degenerate Poisson structure Λ in E is equivalent to the symplectic structure ω so defined [Jo64].

Given a symplectic structure ω, we can also do conversely what we did in the introduction of this section: the regularity of the map $\hat{\omega}$ may be used to define a skew-symmetric contravariant tensor Λ by means of

$$\Lambda(\alpha, \beta) = \omega\left(\hat{\omega}^{-1}(\alpha), \hat{\omega}^{-1}(\beta)\right)\,,$$

where α and β are 1-forms on E. In particular, when $\alpha = df$ and $\beta = dg$ we again find the expression:

$$\{f, g\} = \Lambda(df, dg) = \omega\left(\hat{\omega}^{-1}(df), \hat{\omega}^{-1}(dg)\right)$$

for the Poisson bracket defined by the symplectic structure, and likewise the expression for the Hamiltonian vector field X_g.

The point emphasized here is that a symplectic structure is nothing but a, maximal, constant rank Poisson structure. That is, if $\{f, g\} = 0$ for any function g, then the function f must be locally constant. On the other hand, for the case of a degenerate Poisson structure on an n-dimensional linear space E, there will exist (at least locally defined) non-constant Casimir functions $\mathcal{C}$, such that $\{\mathcal{C}, g\} = 0$ for any $g \in \mathcal{F}(E)$. Actually, the image of Λ defines an involutive distribution, and therefore integrable. If the rank of Λ is r in a neighborhood of the point $x \in E$, then it is possible to choose adapted coordinates in a neighborhood of the point $x \in E$, $\{y^1, \ldots, y^n\}$, in such a way that $\{\frac{\partial}{\partial y^1}, \ldots, \frac{\partial}{\partial y^r}\}$ generate the distribution, while $\{y^{r+1}, \ldots, y^n\}$ are constants along the maximal integral submanifolds of the distribution and, therefore, they give locally defined Casimir functions.

Finally, we also recall that in case of a symplectic structure, ω, the condition $\mathcal{L}_X \Lambda = 0$ is equivalent to $\mathcal{L}_X \omega = 0$. This is also equivalent to the property of $i_X \omega$ being a closed 1-form. Hamiltonian vector fields correspond to exact 1-forms, $i_{X_H}\omega = dH$, and those X's for which $i_X \omega$ is closed, but may be non-exact, are said to be locally Hamiltonian.

5.2.2 The Geometry of Symplectic Linear Spaces

In this section we will study different properties of symplectic and presymplectic linear spaces that will be used in the following sections.

Let (E, σ) be a presymplectic vector space. Given any subspace $V \subset E$, it is possible to define an associated vector space $V^\perp$, called the reciprocal, symplectic polar or symplectic orthogonal of V, by

$$V^\perp = \{u \in E \mid \sigma(u, v) = 0, \quad \forall v \in V\}\,. \tag{5.5}$$

In other words, $V^\perp$ is the linear subspace of E whose image under $\widehat{\sigma}$ is in the annihilator $V^0 = \{f \in E^* \mid f(u) = 0, \forall u \in V\}$ of V.

Let i_V denote the canonical injection $i_V : V \to E$, and $i_V^* : E^* \to V^*$ the dual map of i_V. Then the kernel of the linear map i_V^* is just the annihilator V^0 of V and then $E^*/V^0 \cong V^*$.

If E is finite-dimensional, then from the previous identifications we get

$$\dim V^0 = \dim E - \dim V.$$

Notice that with these notations, the characteristic space of σ, that is $\ker\widehat{\sigma}$ is just $E^\perp$, thus if σ is non-degenerate $E^\perp = \{0\}$.

Let σ_V be the restriction of σ to V, that is $\sigma_V = i_V^*\sigma$. Then, $\ker \sigma_V = V \cap V^\perp$, because

$$\ker \sigma_V = \{u \in V \mid \sigma(u, v) = 0,\ \forall v \in V\} = V \cap V^\perp .$$

Another remarkable property is that if σ is degenerate the quotient space $E/\ker \sigma$ can be endowed with a symplectic form ω such $\pi^*\omega = \sigma$, where π is the natural projection $\pi : E \to E/\ker \sigma$. The 2-form ω so constructed is said to be the reduction of σ. In order to show this property, it suffices to observe that $\omega(v + \ker \sigma, u + \ker \sigma) = \sigma(v, u)$ is well defined and satisfies the desired conditions.

There is an important relation among the dimension of a subspace V and those of $V^\perp$ and $V \cap E^\perp$:

$$\dim E + \dim(V \cap E^\perp) = \dim V + \dim V^\perp , \tag{5.6}$$

because taking into account that $V^\perp = \{v \in E \mid \widehat{\sigma}(v)u = 0,\ \forall u \in V\}$, and using the identification of E with its bidual, $V^\perp = [\widehat{\sigma}(V)]^0$, we find that

$$\dim V^\perp = \dim E - \dim \widehat{\sigma}(V).$$

On the other hand, when considering the restriction of $\widehat{\sigma}$ onto the subspace V we obtain a homomorphism $\widehat{\sigma}_V : V \to E^*$, with kernel $V \cap E^\perp$, and then, from the isomorphism

$$\widehat{\sigma}_V(V) \cong V/(V \cap E^\perp) ,$$

we get the following relation

$$\dim \widehat{\sigma}_V(V) = \dim V - \dim(V \cap E^\perp) ,$$

that together with the first one gives us the mentioned relation (5.6). Notice that when σ is non-degenerate,

$$\dim V^\perp = \dim E - \dim V . \tag{5.7}$$

It is easy to check that in a symplectic inear space the association $V \mapsto V^\perp$ is nilpotent of order two, i.e., $(V^\perp)^\perp = V$, and that the following properties are satisfied by symplectic polars:

1. $V \subset W \Rightarrow W^\perp \subset V^\perp$,
2. $V^{\perp\perp} = V$,
3. $(V + W)^\perp = V^\perp \cap W^\perp$,
4. $(V \cap W)^\perp = V^\perp + W^\perp$,

Regarding (1) if $V \subset W$ and $z \in W^\perp$, then $\sigma(z, w) = 0$, for all $w \in W$, and in particular, $\sigma(z, v) = 0$, for all $v \in V$, which proves it.

To prove (2), notice that $V \subset V^{\perp\perp}$. Moreover, as we are in the finite-dimensional symplectic case, $E^\perp = \{0\}$, and therefore $\dim V + \dim V^\perp = \dim E$. Similarly,

$\dim V^{\perp} + \dim V^{\perp\perp} = \dim E$, from which we see that $\dim V = \dim V^{\perp\perp}$, what implies that $V = V^{\perp\perp}$.

Property (3) follows because if $\omega(v, u_1) = 0$ and $\omega(v, u_2) = 0$ for any pair of vectors u_1 of V and u_2 of W, then also $\omega(v, \lambda_1 u_1 + \lambda_2 u_2) = 0$ holds for any pair of real numbers λ_1 and λ_2, and then $V^{\perp} \cap W^{\perp} \subset (V + W)^{\perp}$. Conversely, if $v \in (V + W)^{\perp}$, then v must be in $V^{\perp} \cap W^{\perp}$.

Finally, for proving $(V \cap W)^{\perp} = V^{\perp} + W^{\perp}$ it suffices to take into account

$$(V \cap W)^{\perp} = (V^{\perp\perp} \cap W^{\perp\perp})^{\perp} = V^{\perp} + W^{\perp} ,$$

where use has been made of $V^{\perp\perp} = V$as well as the preceding relation $(V + W)^{\perp} = V^{\perp} \cap W^{\perp}$. □

So far the symplectic polar looks very much like considering the orthogonal to a given subspace in an inner product space, however since ω is alternating we might have $V \cap V^{\perp} \neq 0$. Indeed, for $\dim V = 1$, we have always $V \subset V^{\perp}$.

According with the intersection properties of a subspace V and its symplectic polar $V^{\perp}$, there are four natural classes of subspaces in a symplectic linear space. We say that:

Definition 5.3 Let (E, ω) be a symplectic linear space and V a linear subspace[1]:

1. V is isotropic if $V \subset V^{\perp}$, then $\dim V \leq \frac{1}{2} \dim E$.
2. V is coisotropic if $V^{\perp} \subset V$, then $\dim V \geq \frac{1}{2} \dim E$.
3. V is Lagrangian if $V = V^{\perp}$, $\dim V = \frac{1}{2} \dim E$.
4. V is symplectic if $V \cap V^{\perp} = 0$.

We remark that V is called symplectic when $\omega|_V$ is non-degenerate, but because $\ker \sigma_V = V \cap V^{\perp}$, that justifies to call V a symplectic subspace. Notice that for any subspace, V, $E = V \oplus V^{\perp}$.

If (E, σ) is a presymplectic linear space, i.e., $E^{\perp} \neq \{\mathbf{0}\}$, then because of Eq. (5.6) we obtain the following proposition:

Proposition 5.4 *Let (E, σ) be a presymplectic linear space. Then*:

1. *If W is an isotropic subspace*, $\dim E + \dim E^{\perp} \geq 2 \dim W$
2. *If W is a coisotropic subspace, then* $\dim E + \dim E^{\perp} \leq 2 \dim W$.
3. *Finally, if W is a Lagrangian subspace*, $\dim E + \dim E^{\perp} = 2 \dim W$.

Proof In fact, if W is isotropic,$W \subset W^{\perp}$, then $\dim W \leq \dim W^{\perp}$, and then $2 \dim \leq \dim W + \dim W^{\perp} = \dim E + \dim W \cap E^{\perp} \leq \dim E + \dim E^{\perp}$.

When W is coisotropic, $W^{\perp} \subset W$, and $E^{\perp} \subset W^{\perp}$, from which we see that $W \cap E^{\perp} = E^{\perp}$. Therefore,

$$2 \dim W \geq \dim W + \dim W^{\perp} = \dim E + \dim W \cap E^{\perp} = \dim E + \dim E^{\perp}.$$

[1] The same definitions hold for an infinite-dimensional symplectic linear space without the dimension relations.

Finally, if $W = W^\perp$, then $W \cap E^\perp = E^\perp$. Moreover,

$$2 \dim W = \dim W + \dim W^\perp = \dim E + \dim W \cap V^\perp = \dim E + \dim E^\perp. \quad \square$$

As a corollary of he previous proposition we obtain:

Corollary 5.5 *If* (E, ω) *is symplectic linear space, an* $W \subset E$ *is linear subspace then: if* W *is isotropic, then* $\dim E \geq 2 \dim W$; *if* W *it is coisotropic, then* $\dim E \leq 2 \dim W$ *and, if* W *is Lagrangian,* $\dim E = 2 \dim W$.

If ω is a linear symplectic form in E, then each isotropic space W is contained in a Lagrangian one, because if we assume that W is contained in $W^\perp$ and $W \neq W^\perp$, then there would exist a vector $v \in W^\perp$ such that $v \notin W$. In this case the subspace $W + \langle v \rangle$ will also be isotropic and contains W. In the finite-dimensional case, by iterating the procedure we will arrive to a Lagrangian subspace containing W.[2] Notice that every one-dimensional subspace is isotropic and therefore there always exist Lagrangian subspaces. The proof of our assertion also shows that a Lagrangian subspace is nothing but a maximal isotropic subspace. Therefore an isotropic subspace is Lagrangian if and only if it is coisotropic.

Our first example of a symplectic linear space was $E = W \oplus W^*$ with the canonical simplectic form σ_W, Eq. (5.2). It is easy to check that in this case both W and W^* are Lagrangian subspaces of E. We will show in the next few lines that this is actually the general structure of a symplectic space.

A linear map $\phi \colon E \to F$ between the symplectic spaces (E, ω_E), (F, ω_F) is called symplectic if $\omega_F(\phi(u), \phi(v)) = \omega_E(u, v)$, for all $u, v \in E$. It is simple to see that a symplectic linear map must be injective because $\ker \phi \subset E^\perp$. If a linear symplectic map is an isomorphism it will be called a symplectic isomorphism, or symplectomorphism for short.

The importance of the previous construction relies on the following proposition:

Theorem 5.6 (Structure theorem for symplectic spaces) *Let* (E, ω) *be a linear symplectic space and* W *be a Lagrangian subspace of* E. *Then, there exists another Lagrangian subspace* W' *supplementary of* W. *Moreover there is a symplectic isomorphism between* (E, ω) *an* $(W \oplus W^*, \sigma_W)$.

Proof Let us remark that $W = W^\perp$ because W is Lagrangian. There will be a vector v_1 such that $v_1 \notin W$ and as $W \cap \langle v_1 \rangle = \mathbf{0}$, we find that $(W \cap \langle v_1 \rangle)^\perp = E$, i.e., $W + \langle v_1 \langle^\perp = E$. Let choose $v_2 \in \langle v_1 \rangle^\perp$ such that $v_2 \notin W + \langle v_1 \rangle$; once again we can say that as $W \cap \langle v_1, v_2 \rangle = \mathbf{0}$, the space E is a sum $E = W + \langle v_1, v_2 \rangle^\perp$, and furthermore the subspace $\langle v_1, v_2 \rangle$ is isotropic, by construction. We iterate the process by choosing $v_3 \in \langle v_1, v_2 \rangle^\perp$ with $v_3 \notin W + \langle v_1, v_2 \rangle$. Finally we will arrive at a linear isotropic subspace W' such that $E = W + W'^\perp$ and $W \cap W' = \mathbf{0}$, and there is no vector $v \in W'^\perp$ which is not contained in $W + W'$. Now, as $W + W'^\perp = E$, we can conclude that $W + W' = E$, and therefore W' is Lagrangian.

---

[2] The argument will work unchanged in the infinite-dimensional instance applying Zorn's Lemma.

Lagrangian subspace with a supplementary Lagrangian subspace W'. Let now ϕ be the map $\phi : W' \to W^*$ defined by $\phi(v) = \hat{\omega}(v) \mid_W$. This map is an isomorphism because if $\phi(v) = 0$ then $w(v, u) = 0$, $\forall u \in W$, and consequently $v \in W' \cap W^\perp = W' \cap W = \mathbf{0}$. Thus, the map $\mathrm{id}_W \oplus \phi : E = W \oplus W' \to W \oplus W^*$, is an isomorphism and a symplectic map, because

$$\begin{aligned}\omega_0((v_1, \phi(u_1)), (v_2, \phi(u_2)) &= \phi(u_1)v_2 - \phi(u_2)v_1 = \omega(u_1, v_2) - \omega(u_2, v_1)\\ &= \omega(u_1 + v_1, u_2 + v_2)\ .\end{aligned}$$

□

It is clear from the construction that for a given Lagrangian subspace V we have many supplementary Lagrangian subspaces and therefore many symplectic isomorphisms $E \cong V \oplus V^*$.

Given V and W we can find all other supplementary spaces by considering $(v_i, w_j + A_{jk} v^k)$ with $A_{jk} = A_{kj}$.

Another remarkable property of Lagrangian subspaces is the following:

Theorem 5.7 *An isomorphism $\phi : V_1 \to V_2$ between symplectic linear spaces is symplectic if and only if the graph of the map is a Lagrangian subspace of $(V_1 \oplus V_2, \omega_1 \ominus \omega_2)$.*

Proof It suffices to observe that the graph of ϕ, $\mathrm{Graph}(\phi) = \{(v, \phi(v)) \mid v \in V_1\}$, is a linear subspace of dimension $\dim V_1$ and that

$$(\omega_1 \ominus \omega_2)((v, \phi(v), (u, \phi(u))) = \omega_1(v, u) - \omega_2(\phi(v), \phi(u)) = \omega_1(v, u) - (\phi^* \omega_2)(v, u).$$

Then $\omega_1 \ominus \omega_2 \mid_{\mathrm{Graph}(\phi)} = 0$ iff $\omega_1 - \phi^* \omega_2$. □

Another simple consequence of the previous results is the following statement:

Corollary 5.8 *Given a presymplectic linear space (E, ω), there is a basis $\mathcal{B} = \{e_i \mid i = 1, \ldots, n\}$ of E such that the expression of σ is*

$$\sigma = \sum_{j=1}^{r} \epsilon^j \wedge \epsilon^{j+r}\ ,$$

where $\mathcal{B}^* = \{\epsilon^i \mid i = 1, \ldots, n\}$ is the dual basis of $\mathcal{B}$ and $2r$ is the rank of σ. In particular, for a symplectic structure ω in a linear space of dimension $2\,n$, it should be $n = 2r$, and then

$$\omega = \sum_{j=1}^{n} \epsilon^j \wedge \epsilon^{j+n}\ .$$

Such linear basis will be called canonical (also sometimes standard).

5.2.3 Generic Subspaces of Symplectic Linear Spaces

A generic subspace V of a symplectic linear space (E, ω) will be neither isotropic, Lagrangian, coisotropic nor symplectic. We will analyze in this section, the structure induced from the symplectic structure ω on an arbitrary linear subspace $V \subset E$.

The restriction of ω to V defines ω_V. If V is symplectic, i.e., $V \cap V^{\perp} = 0$, then also $V^{\perp}$ is symplectic because $V^{\perp} \cap (V^{\perp})^{\perp} = V^{\perp} \cap V = \mathbf{0}$, and $E = V \oplus V^{\perp}$. The symplectic structure ω decomposes into

$$\omega = \omega_V \oplus \omega_{V^{\perp}} . \tag{5.8}$$

In terms of the associated Poisson brackets we have that the previous decomposition defines the subspace of linear functions $\mathcal{F}_V^{(1)}$; that is, any $\alpha \in V^*$ defines the linear function f_α given by $f_\alpha(u) = \alpha(v)$ where $u = v + v^{\perp}$, $v \in V$ and $v^{\perp} \in V^{\perp}$. Similarly $(V^{\perp})^*$ defines $\mathcal{F}_{V^{\perp(1)}}$, and because of Eq. (5.8), it follows that

$$\{\mathcal{F}_V^{(1)}, \mathcal{F}_{V^{\perp}}^{(1)}\} = 0 \tag{5.9}$$

Notice that we can also consider:

$$\mathcal{F}_V = \{f \in \mathcal{F}(E) \mid df\,(V^{\perp}) = 0\} \tag{5.10}$$

$$\mathcal{F}_{V^{\perp}} = \{f \in \mathcal{F}(E) \mid df\,(V) = 0\}, \tag{5.11}$$

of course these subalgebras are generated by $\mathcal{F}_{V^{(1)}}$ and $\mathcal{F}_{V^{\perp(1)}}$ respectively.

When V is not a symplectic subspace we get a degenerate 2-form ω_V. We can consider now the quotient space $V/\ker\omega_V = \tilde{V}$. By construction, this quotient space inherits a symplectic structure that we denote as $\tilde{\omega}_V$. If we consider $V^{\perp}$ then the restriction $\omega_{V^{\perp}}$ will be degenerate again with characteristic subspace $\ker\omega_{V^{\perp}}$. On the other hand, because $\ker\omega_V = V \cap V^{\perp}$, we get:

$$\dim\ker\omega_V = \dim\ker\omega_{V^{\perp}} . \tag{5.12}$$

Let us denote by V_σ the subspace image of any section $\sigma : \tilde{V} \to V_\sigma \subset E$. We can consider $(V_\sigma)^{\perp} \subset E$. Clearly $(V_\sigma)^{\perp}$ is symplectic and contains $V^{\perp}$, therefore $V^{\perp}$ is a coisotropic subspace of the symplectic linear space $(V_\sigma)^{\perp}$. Considering now the analogue construction, starting with $V^{\perp}$, i.e.,

$$\widetilde{(V^{\perp})} = V^{\perp}/\ker\omega_{V^{\perp}}, \tag{5.13}$$

and now, if $\mu \colon \widetilde{(V^{\perp})} \to V_\mu^{\perp} \subset E$ is any section, the image $V_\mu^{\perp}$ is a symplectic subspace of E. If we consider the associated polar subspace $(V_\mu^{\perp})^{\perp}$ we find a symplectic subspace of E which contains V as a coisotropic subspace. From this construction it is clear that symplectic subspaces of E which contain V as a coisotropic subspace will depend on the section μ.

By putting together V_σ and $V_\mu^\perp$, we get a symplectic decomposition of E

$$E = V_\sigma \oplus V_\mu^\perp \oplus (\ker \omega_V \oplus \ker \omega_{V^\perp}). \tag{5.14}$$

It is clear that $(V_\sigma)^\perp \cong V_\mu^\perp \oplus \ker \omega_V \oplus \ker \omega_{V^\perp}$ and $(V_\mu^\perp)^\perp \cong V_\sigma \oplus \ker \omega_V \oplus \ker \omega_{V^\perp}$. It is also possible to rewrite this decomposition in terms of Poisson subalgebras.

5.2.4 Transformations on a Symplectic Linear Space

Consider an isomorphism $\phi \in GL(E)$ of a symplectic linear space (E, ω). In general the pull-back $\phi^*\omega$ of ω under ϕ is different from ω. When $\phi^*\omega = \omega$, ϕ will be called a symplectic isomorphism (or symplectomorphism for short). We will denote by $\mathrm{Sp}(\omega)$ the group of symplectic isomorphisms of the linear symplectic structure ω.

Because of Corollary 5.8, in a canonical basis the symplectic form is represented by the matrix

$$J = \begin{pmatrix} 0 & I \\ -I & 0 \end{pmatrix} \tag{5.15}$$

and the matrix A representing a symplectic isomorphism ϕ in a canonical basis should satisfy

$$A^T J A = J. \tag{5.16}$$

Henceforth matrices A satisfying Eq. (5.16) will be called symplectic matrices and the collection of all of them form a group called the symplectic group of order $2n$, $\mathrm{Sp}(2n, \mathbb{R})$. It is clear that $\mathrm{Sp}(2n, \mathbb{R})$ is a closed subgroup of $GL(2n, \mathbb{R})$, hence it is a Lie group according with the analysis of matrix groups done in Sect. 2.6.1.

Thus, out of a given symplectic structure ω we get a new one $\phi^*\omega$ unless ϕ is a symplectic isomorphism. Given any two different symplectic structures, say ω_1 and ω_2, it is always possible to find a transformation ϕ that takes ω_1 into ω_2. Indeed, if we choose a canonical basis for ω_1 and another canonical basis for ω_2 then the linear map determined by these two linear basis is the required map (which is not unique). Therefore $GL(E)$ acts transitively on the space of non-degenerate 2-forms. Because $\mathrm{Sp}(\omega)$ is a subgroup of $GL(E)$, we have that the family of symplectic linear structures on E are in a one-to-one correspondence with the quotient space $GL(E)/\mathrm{Sp}(\omega)$.

Notice that in the simplest case, that is when $\dim E = 2$, the group of linear isomorphisms of E is isomorphic to $GL(2, \mathbb{R})$ which has dimension four. On the other hand $\mathrm{Sp}(2n, \mathbb{R})$ is isomorphic to $SL(2, \mathbb{R})$, then it has dimension three, and the homogeneous space of non-equivalent symplectic forms in the two-dimensional space is one dimensional; in fact, they are all the nonzero multiples of the 2-form

$\epsilon^1 \wedge \epsilon^2$. Note that if $A = \begin{pmatrix} a & b \\ c & d \end{pmatrix}$ then the simplecticity condition Eq. (5.16),

$$A = \begin{pmatrix} a & c \\ b & d \end{pmatrix} \begin{pmatrix} 0 & I \\ -I & 0 \end{pmatrix} \begin{pmatrix} a & b \\ c & d \end{pmatrix} = (ad - bc) \begin{pmatrix} 0 & 1 \\ -1 & 0 \end{pmatrix}, \tag{5.17}$$

shows that $ad - bc = 1$, which is just the condition $\det A = 1$.

To classify various subgroups in $\mathrm{Sp}(\omega)$, it is convenient to exploit the structure theorem, Theorem 5.6, and use the isomorphism $E \cong V \oplus V^*$ for some chosen Lagrangian subspace V. By using linear coordinate functions for V and V^*, say (x^i, p_i) and inducing them on E via the given isomorphism, we can write ω as

$$\omega = dx^i \wedge dp_i. \tag{5.18}$$

This shows that the symplectic form is not only closed but exact too, i.e., we can write $\omega = -d\theta$ with $\theta = p_i\, dx^i$.

We notice at this point that if ϕ is a symplectic isomorphism, it will take Lagrangian subspaces into Lagrangian subspaces, therefore we go from the isomorphism with $V \oplus V^*$ to some other isomorphism with $V' \oplus V'^*$. In terms of coordinate functions ω will be written $\omega = d\acute{x}^i \wedge d\acute{p}_i$. If we denote by θ' the pull-back of θ with respect to ϕ, then because ϕ is symplectic $d(\theta - \theta') = 0$. But in a linear space any closed form is exact, then there must be a function S such that

$$dS_\phi = \theta - \theta' = p_i\, dx^i - \acute{p}_i\, d\acute{x}^i\,,$$

that will be called a generating function for the symplectic map ϕ. It is more convenient to think of the symplectic map $\phi\colon V \oplus V^* \to V' \oplus V'^*$ as a Lagrangian subspace in the direct sum of $V \oplus V^*$ and $V' \oplus V'^*$. Then if we think of the difference $\theta - \theta'$ as defining a 1-form on $(V \oplus V^*) \times (V' \oplus V'^*)$ such 1-form (actually $\pi_1^*\theta - \pi_2^*\theta'$) will be closed (hence exact) when restricted to the graph of ϕ. The graph of ϕ is a $2n$-dimensional space and can be parametrized by an appropriate choice of coordinates. This is the idea behind the construction of generating functions of different types, Types I, II, III and IV, as one usually finds in classical textbooks (see for instance [Go81, Sa71]).

Thus, e.g., we can choose coordinates $\{x^i, \acute{p}_i\}$ and express $\theta - \theta' \mid_{\mathrm{graph}(\phi)}$ in terms of the generating function of Type II, S_ϕ^{II} as follows:

$$p_i\, dx^i + \acute{x}^i\, d\acute{p}_i = dS_\phi^{II}(x^i, \acute{p}_i)\,.$$

The transformation is then given by

$$p_i = \frac{\partial S_\phi}{\partial x^i}\,, \quad \acute{x}^i = \frac{\partial S_\phi}{\partial \acute{p}_i}\,. \tag{5.19}$$

In order for the transformation to be well defined it should be:

$$\det\left(\frac{\partial^2 S}{\partial x^i\,\partial \acute{p}_j}\right) \neq 0\,. \tag{5.20}$$

We should remark that the generating function depends on both ϕ and the variables which appear as differentials on the left-hand side, i.e., the Type. For instance, another choice would lead to

$$p_i\,dx^i - \acute{p}_i\,d\acute{x}^i = dS^I_\phi(x^i, \acute{x}^i)\,. \tag{5.21}$$

We will only develop the Type II case but other approaches follow similar lines.

As our transformation ϕ is linear by assumption, then for

$$S_\phi = \frac{1}{2}A_{ij}x^i x^j + \frac{1}{2}B^{ij}\acute{p}_i\acute{p}_j + C_i{}^j\acute{p}_j x^i$$

we find

$$p_i = \frac{\partial S_\phi}{\partial x^i} = A_{ij}x^j + C_i{}^j\acute{p}_j\,, \qquad \acute{x}^i = B^{ij}\acute{p}_j + C_j{}^i x^j\,.$$

By assuming that the matrix $C = (C_i{}^j)$ is invertible we find

$$\acute{p}_j = (C^{-1})_j{}^i p_i - (C^{-1})_j{}^k A_{ki}x^i\,,$$

$$\acute{x}^j = C_k{}^j x^k + B^{jm}(C^{-1})_m{}^i p_i - B^{jm}(C^{-1})_m{}^k A_{ki}x^i\,.$$

If the Lagrangian subspace described by x^i is required to be invariant under ϕ then $B = 0$, and our transformation is represented by

$$\begin{pmatrix}\acute{x}\\ \acute{p}\end{pmatrix} = \begin{pmatrix}C^T & 0\\ -C^{-1}A & C^{-1}\end{pmatrix}\begin{pmatrix}x\\ p\end{pmatrix}. \tag{5.22}$$

If the same space is to be preserved pointwise we get $C = I$ and the representative matrix becomes

$$\begin{pmatrix}I & 0\\ -A & I\end{pmatrix} \tag{5.23}$$

with A a symmetric matrix.

If C is not invertible, the condition (5.20) is not satisfied and the generating function must be expressed in different coordinates. We consider, for instance, one of the Type I, Eq. (5.21), and now the transformation is defined by

$$p_i\,\frac{\partial S}{\partial x^i}\,; \qquad \acute{p}_j = -\frac{\partial S}{\partial \acute{x}^j}\,,$$

and it is well defined if

$$\det\left(\frac{\partial^2 S}{\partial x^i\,\partial \acute{x}^j}\right) \neq 0\,.$$

We consider

$$S_\phi = C_{ij}x^i\acute{x}^j + \frac{1}{2}A_{ij}x^ix^j + \frac{1}{2}B_{ij}\acute{x}^i\acute{x}^j$$

to find

$$p_i = C_{ij}\acute{x}^j + A_{ij}x^j; \qquad \acute{p}_j = -C_{ij}x^i - B_{ji}\acute{x}^i .$$

Therefore,

$$\acute{x}^j = (C^{-1})^{ji}p_i - (C^{-1})^{ji}A_{ik}x^k ,$$

$$\acute{p}_j = -C_{ji}x^i - B_{ji}(C^{-1})^{ik}p_k + B_{jm}(C^{-1})^{mi}A_{ik}x^k .$$

It is possible to consider subalgebras of $\mathfrak{sp}(\omega)$. Here it is convenient again to use the vector field X_S of a matrix S associated to an element in the Lie algebra of $\mathrm{Sp}(\omega)$. We recall that if

$$\begin{pmatrix} \acute{x} \\ \acute{p} \end{pmatrix} = S \begin{pmatrix} x \\ p \end{pmatrix} = \begin{pmatrix} A & B \\ M & N \end{pmatrix} \begin{pmatrix} x \\ p \end{pmatrix} ,$$

we set

$$X_S = (x^T, p^T) \begin{pmatrix} A & B \\ M & N \end{pmatrix} \begin{pmatrix} \partial_x \\ \partial_p \end{pmatrix} ,$$

that is

$$X_S = (x^T A + p^T M)\frac{\partial}{\partial_x} + (x^T B + p^T N)\frac{\partial}{\partial_p} . \tag{5.24}$$

By using $-d(p^T\, dx) = \omega$ and imposing on the vector field the condition that $\mathcal{L}_{X_S}\omega = 0$ we will find

$$\mathcal{L}_{X_S}(p^T\, dx) = df_S(p, x) ,$$

where

$$\frac{\partial f_S}{\partial p} = p^T M^T , \qquad \frac{\partial f_S}{\partial x} = x^T B + p^T(A^T + N) ,$$

because

$$\mathcal{L}_{X_S}(p_k\, dx^k) = dx^j(x_iB^{ij} + p_iN^i{}_j + p_kA_{jk} + dp_j\, p_kM^{kj} .$$

Therefore, very much like finite transformations, infinitesimal transformations can be characterized by quadratic functions

$$f_S(p, x) = A^{ij}p_ip_j + B_{ij}x^ix^j + C_i{}^jx^ip_j .$$

Therefore the group of symplectic isomorphisms has dimension $n(2n + 1)$.

If we consider now a one-parameter group of transformations, say

$$\phi_t = e^{tZ} ,$$

then

$$\frac{dS_\phi(t)}{dt} = f_Z$$

gives the relation between generating functions of finite symplectic isomorphism ϕ and those of the infinitesimal ones. This gives us the opportunity to comment on the correspondence between transformations and generating functions. If Z is the vector field $Z = x^T A\partial_x - p^T A^T \partial_p$ we have

$$\mathcal{L}_Z(p^T dx) = p^T d(A^T x) - p^T A^T dx = 0\,,$$

i.e., for any choice of the matrix A the transformations of this form correspond to the constant generating function. We can consider the 1-form $-x^T dp$, now $\mathcal{L}_Z(x^T dp) = x^T Adp - x^T d(Ap) = 0$. We can take now any quadratic function F and consider a 1-form like $p^T dx + dF$ in this case we will find

$$\mathcal{L}_Z(p^T dx + dF) = d\mathcal{L}_Z F\,.$$

Therefore the infinitesimal generating function can be written as the Lie derivative of another function. This phenomenon is rooted in the cohomology of Lie algebras. In particular, as the 1-form θ such that $\omega = -d\theta$ is only defined up to addition of a closed 1-form, one may ask if for a given vector field Z it is always possible to find a 1-form θ such that $d\theta = dx^T \wedge dp$ and $\mathcal{L}_Z\theta = 0$. For vector fields with representative matrix having degenerate purely imaginary eigenvalues the answer is negative.

Let us recall now (see Sect. 3.2.5 and Appendix F for more details) that an $\mathcal{F}$-valued 1-cocycle on the group G is a map $c_1 : G \to \mathcal{F}$ such that

$$\delta c_1(g, g') = gc_1(g') - c_1(gg') + c_1(g)\,.$$

We can show in general that $S\colon \phi \mapsto S_\phi$ is a 1-cocycle from $\mathrm{Sp}(\omega) \to \mathcal{F}(E)$ while $f : \mathfrak{sp}(\omega) \to \mathcal{F}(E)$, $Z \mapsto f_Z$ is a 1-cocycle for the corresponding Lie algebra.

The question we are asking concerns the possibility that such cocycle becomes a coboundary when restricted to subgroups or subalgebras of the symplectic algebra.

5.2.5 On the Structure of the Group $Sp(\omega)$

We can use this result to show that $Sp(\omega)$ is a closed subgroup of $GL(E)$. Indeed, symplectic forms are a subset of skew-symmetric matrices, $\mathbb{R}^{n(2n-1)}$, which are a vector space, therefore a manifold. The map $\det\colon \mathbb{R}^{n(2n-1)} \to \mathbb{R}$ is continuous, therefore $\det^{-1}(\mathbb{R} - \{\,0\,\})$ is an open submanifold because non zero values of det are regular.

Proposition 5.9 *Every isotropic subspace is contained in a Lagrangian subspace. Every two Lagrangian subspaces can be mapped into each other via a symplectic transformation.*

Necessary and sufficient conditions for the existence of a symplectic transformation connecting two subspaces is that they have the same rank.

Proposition 5.10 *Necessary and sufficient conditions for a symplectic transformation to preserve the decomposition $V = W \oplus W^{\perp}$, into Lagrangian subspaces is that it has the form*

$$\begin{pmatrix} A & 0 \\ 0 & (A^T)^{-1} \end{pmatrix} \tag{5.25}$$

Symplectic transformations which preserve all elements of a given Lagrangian subspace have the form

$$\begin{pmatrix} I & S \\ 0 & I \end{pmatrix} \tag{5.26}$$

with $S = S^T$.

These transformations form an Abelian group, isomorphic to the additive group of symmetric matrices. It is possible to enquire about the equivalence of these transformations under symplectic conjugation. One finds that a necessary and sufficient condition is that $x^i (S_1)_{ij} x^j$ is equivalent to $x^i (S_2)_{ij} x^j$.

Symplectic Transvections

Any hyperplane H (of codimension 1) can be represented as $\ker \omega(a)$ for some vector a. The symplectic transformation $V \to V$ which preserves every element of H has the form $x \mapsto x - \lambda\omega(x, a)a$.

Symplectic transvections are symplectically conjugate iff λ_1/λ_2 is a square in $\mathbb{K}$. Thus for $\mathbb{K} = \mathbb{R}$, $\lambda_1\lambda_2 < 0$ will not provide conjugated transformations.

Proposition 5.11 *Every symplectic transformation is the product of transvections. (Indeed it is at most the product of 2n symplectic transvections if* $\dim E = n$*).*

We have $x \mapsto x + \sum_{i=1}^{q} \lambda_i \omega(x, a_i) a_i$. It would be interesting to find the minimum for q.

Proposition 5.12 *The centre of $Sp(n, \mathbb{K})$ contains only two elements, $x \mapsto x$ and $x \mapsto -x$. There are no additional normal subgroups.*

5.2.6 Invariant Symplectic Structures

Existence of Invariant Symplectic Structures

A generic linear dynamical system has a commuting algebra of symmetries but no constants of the motion. On the other hand, we will see that the existence of invariant structures on the carrier space of our dynamical system allows us to associate symmetries with constants of the motion and other properties that are not shared by generic ones (like Liouville's theorem on phase space [Ab78], energy-period theorem [Go69], etc.).

We would like to consider the restriction to symplectic structures of the problem of determining the existence of invariant Poisson structures for a given dynamical problem. More specifically, given a linear dynamical system, determine all, if any, linear symplectic structures $\omega = \Omega_{ij}\, dx^i \wedge dx^j$ which are preserved by our system.

The study is similar to that we did for Poisson structure (see Sect. 4.2.2).

As usual we will consider first a very simple example so that the ideas involved can be easily grasped. Let us consider the vector field

$$X_A = (mx + ny)\frac{\partial}{\partial x} + (px + qy)\frac{\partial}{\partial y} ,$$

corresponding to the matrix

$$A = \begin{pmatrix} m & n \\ p & q \end{pmatrix} .$$

The most general constant symplectic structure in $\mathbb{R}^2$ is

$$\omega = \alpha\, dx \wedge dy , \qquad \alpha \in \mathbb{R} ,$$

with associated matrix $\Omega = \alpha J$ (see Eq. (5.15)).

We have that $\mathcal{L}_{X_A}\omega = 0$ is equivalent to $m = -q$ because

$$\mathcal{L}_{X_A}\omega = -\alpha(q\, dy \wedge dx + m\, dy \wedge dx) = \alpha(m + q)dx \wedge dy .$$

In other words, X_A is Hamiltonian if and only if the trace of the matrix A vanishes. Setting $q = -m$ we compute the Hamiltonian function. Then

$$i_X\omega = dH = -\alpha(px\, dx - my\, dx - mx\, dy - ny\, dy) = -d\left(\frac{1}{2}px^2 - \frac{1}{2}ny^2 - mxy\right)\alpha ,$$

and the Hamiltonian H of the vector field X_A is given by a quadratic function defined by the matrix $G = -\Omega A$, which is symmetric because $m = -q$. Conversely, given a quadratic function defined by a symmetric matrix G, the corresponding Hamiltonian vector field is a vector field X_A defined by the matrix $A = -\Omega^{-1}G$. We can consider our quadratic form

$$2mxy + px^2 - ny^2$$

and reduce it to canonical form. We will get, depending on the value of the discriminant $m^2 + np$:

1. $\frac{1}{2}(p^2 + q^2)$: with $G = \frac{1}{2}\begin{pmatrix} 1 & 0 \\ 0 & 1 \end{pmatrix}, A = \begin{pmatrix} 0 & 1 \\ -1 & 0 \end{pmatrix}$.
2. $\frac{1}{2}(p^2 - q^2)$: with $G = \frac{1}{2}\begin{pmatrix} 1 & 0 \\ 0 & -1 \end{pmatrix}, A = \begin{pmatrix} 0 & 1 \\ 1 & 0 \end{pmatrix}$.
3. $\frac{1}{2}p^2$ or $\frac{1}{2}q^2$: with $G = \frac{1}{2}\begin{pmatrix} 0 & 0 \\ 0 & 1 \end{pmatrix}, A = \begin{pmatrix} 0 & 1 \\ 0 & 0 \end{pmatrix}$, or $G = \frac{1}{2}\begin{pmatrix} 1 & 0 \\ 0 & 0 \end{pmatrix}, A = \begin{pmatrix} 0 & 0 \\ 1 & 0 \end{pmatrix}$ respectively .

The associated vector fields will be in each case:

1. $X_1 = p\frac{\partial}{\partial q} - q\frac{\partial}{\partial p}$.
2. $X_2 = q\frac{\partial}{\partial p} + p\frac{\partial}{\partial q}$.
3. $X_3 = p\frac{\partial}{\partial q}$ or $q\frac{\partial}{\partial p}$.

Let us try to read in algebraic terms what we have done. First a simple computation shows that $\mathcal{L}_{X_A}\omega = 0$ if and only if

$$A^T\Omega + \Omega A = 0, \tag{5.27}$$

where $\Omega^T = -\Omega$ is the skew-symmetric matrix associated with ω as it was shown in Sect. 4.2.3, Eq. (4.7). Then we have that $(\Omega A)^T = \Omega A$, that is ΩA is symmetric which is equivalent to

$$A^T = -\Omega A \Omega^{-1} . \tag{5.28}$$

This equation says that the matrix associated with the Hamiltonian function is symmetric (our previous quadratic form). The corresponding Hamiltonian function H will be given by the quadratic function

$$H = -\frac{1}{2}(\Omega A)_{kj}\, x^k x^j .$$

By the same token, Eq. (5.28) implies that $\mathrm{Tr}\, A = 0$. we also find

$$(A^{2k+1})^T = -\Omega(A^{2k+1})\Omega^{-1} ,$$

therefore (see again Eq. (4.8))

$$\mathrm{Tr}\, A^{2k+1} = 0 , \tag{5.29}$$

and because of the preceding equation the vector field associated with any odd power of A is also Hamiltonian with respect to the same symplectic structure. On the other hand,

$$A^{2k}\Omega = \Omega A^{2k} .$$

This shows that the quadratic functions $f_k = (\Omega A^{2k+1})_{ij}\, x^i x^j$ are constants of motion for X_A, and moreover, for $X_{A^{2j+1}}$. In fact, if $N_k = \Omega A^{2k+1}$, the invariance condition

$$(A^{2j+1})^T N_k + N_k A^{2j+1} = 0$$

is a consequence of $A^{2k}\Omega = \Omega A^{2k}$, because

$$\begin{aligned}(A^{2j+1})^T N_k + N_k A^{2j+1} &= (-1)^{2j+1}\Omega A^{2j+1}\Omega^{-1}\Omega A^{2k+1} + \Omega A^{2k+1} A^{2j+1} \\ &= -\Omega A^{2(j+k)+1} + A^{2(j+k)+1}\Omega = 0\,.\end{aligned}$$

This means that the quadratic functions $f_k = \frac{1}{2}(\Omega A^{2k+1})_{ij}\, x^i x^j$ satisfy

$$\{f_k,\, f_j\} = 0. \tag{5.30}$$

with respect to the Poisson bracket associated with Ω.

Going back to our original problem it looks like that everything is contained in the starting relation Eq. (5.27) with $\det \Omega \neq 0$. As a matter of fact it can be shown that:

Theorem 5.13 *A necessary and sufficient condition for the matrix A without zero eigenvalues or even degenerate eigenvalues to define a Hamiltonian vector field with respect to some symplectic structure is that A is similar to the opposite of its transpose* $-A^T$*, i.e.,*

$$A = -\Omega A^T \Omega^{-1},$$

with Ω *being a skew-symmetric matrix.*

In one direction the statement is obvious, in the other direction can be proved using reduction to normal form.

5.2.6.1 On the Uniqueness of Invariant Symplectic Structures

Having considered the existence problem of admissible symplectic structures, we consider in this section how many different Hamiltonian descriptions are possible for the same linear vector field. In other terms we are looking for the set of all solutions for Ω in the Eq. (5.27).

The problem of finding alternative symplectic structures for the vector field defined by the matrix A is equivalent to looking for all different decompositions of A into the product of a skew-symmetric invertible matrix Λ, the opposite to the inverse of Ω, times a symmetric one H: $A = \Lambda \cdot H$, that we discussed at length in Sect. 4.2.3, Theorem 4.2. Given such a decomposition, if A undergoes the transformation $A \mapsto P^{-1}AP$, we have

$$P^{-1}AP = P^{-1}\Lambda HP = P^{-1}\Lambda (P^T)^{-1} P^T HP\,,$$

where $P^T HP$ is symmetric and $P^{-1}\Lambda(P^T)^{-1}$ is skew-symmetric. It is now clear that if $P^{-1}AP = A$ we will get new decompositions of A provided that either $P\Lambda P^T \neq \Lambda$ (i.e., $P^{-1}\Lambda P^{-1T}$, or $P^T HP \neq H$. Therefore, symmetries for our system will take us from one symplectic description to another.

Turning back to the symmetries of a vector field X_A, it is now clear that even powers of A will be associated with noncanonical transformations. The exponential $\exp sA^{2k}$ will take us from one symplectic description to another. For a generic Hamiltonian system, i.e., all eigenvalues are simple, we generate in this way all constant symplectic structures admissible for Γ; they will be parametrized by the parameters $s_0, s_1, \ldots, s_{2(n-1)}$ appearing in the exponentials.

Let us give, as it is customary by now, an example. Consider the linear system:

$$\Gamma = \omega_1 \left(x\frac{\partial}{\partial y} - y\frac{\partial}{\partial x} \right) + \omega_2 \left(z\frac{\partial}{\partial w} - w\frac{\partial}{\partial z} \right) \tag{5.31}$$

with $\omega_1 \neq \omega_2$. The associated matrix A is given by:

$$A = \begin{pmatrix} \omega_1 J & 0 \\ 0 & \omega_2 J \end{pmatrix} \tag{5.32}$$

with J the standard 2×2 symplectic matrix.

An admissible symplectic structure for Γ is provided by $\omega_0 = dx \wedge dy + dz \wedge dw$, which is obtained from the factorization

$$\begin{pmatrix} \omega_1 J & 0 \\ 0 & \omega_2 J \end{pmatrix} = \begin{pmatrix} \omega_1 I & 0 \\ 0 & \omega_2 I \end{pmatrix} \begin{pmatrix} J & 0 \\ 0 & J \end{pmatrix}, \tag{5.33}$$

where I denotes the identity matrix in two dimensions.

Even powers of the representative matrix are given by

$$A^{2k} = (-1)^k \begin{pmatrix} \omega_1^{2k} & 0 & 0 & 0 \\ 0 & \omega_1^{2k} & 0 & 0 \\ 0 & 0 & \omega_2^{2k} & 0 \\ 0 & 0 & 0 & \omega_2^{2k} \end{pmatrix}. \tag{5.34}$$

Therefore the most general constant invariant symplectic structure is $a\, dx \wedge dy + b\, dz \wedge dw$, with the condition $ab \neq 0$.

It should be noticed that it is possible to obtain more general symplectic structures by making a to be a function of $x^2 + y^2$ and b a function of $z^2 + w^2$, as long as $ab \neq 0$. This is coherent with our earlier statement that our linear symmetries can be made non linear by allowing the entries to depend on constants of the motion. Let us pursue the computations for the present example. A general symmetry for our vector field is provided by

$$B = \begin{pmatrix} a & b & 0 & 0 \\ -b & a & 0 & 0 \\ 0 & 0 & r & s \\ 0 & 0 & -s & r \end{pmatrix} \tag{5.35}$$

with $\det B \neq 0$, i.e., $(a^2 + b^2)(r^2 + s^2) \neq 0$.

If we substitute a, b, r, s by constants of motion while preserving the condition $\det B \neq 0$, we have

$$\begin{pmatrix} x \\ y \\ z \\ w \end{pmatrix} = \begin{pmatrix} a & b & 0 & 0 \\ -b & a & 0 & 0 \\ 0 & 0 & r & s \\ 0 & 0 & -s & r \end{pmatrix} \begin{pmatrix} q_1 \\ p_1 \\ q_2 \\ p_2 \end{pmatrix} = \begin{pmatrix} aq_1 + bp_1 \\ -bq_1 + ap_1 \\ rq_2 + sp_2 \\ -sq_2 + rp_2 \end{pmatrix}, \tag{5.36}$$

therefore,

$$\tilde{\omega} = d(aq_1 + bp_1) \wedge d(ap_1 - bq_1) + d(rq_2 + sp_2) \wedge d(rp_2 - sq_2) \tag{5.37}$$

with a, b, r, s arbitrary constants of motion.

For the n-dimensional isotropic harmonic oscillator with phase space $\mathbb{R}^{2n}$, symmetries are given by the real form of $Gl(n, \mathbb{C})$. Entries of each matrix can be made to be any function of $z_a z_b^*$, i.e., any element of $\mathcal{F}(\mathbb{C}P^n \times \mathbb{R}_+)$. It should be noticed that some of these transformations may be canonical and therefore they do not change the symplectic structure to which we apply it.

5.2.7 Normal Forms for Hamiltonian Linear Systems

From the general reduction to normal forms we have that a linear vector field Γ with a representative matrix that does not have imaginary or null eigenvalues can be reduced to the standard form

$$\Gamma = A^i{}_j x^j \frac{\partial}{\partial x^i} - (A^T)^j{}_i y_j \frac{\partial}{\partial y_i} \tag{5.38}$$

with Hamiltonian function $H = x^j A^i{}_j y_i$ with respect to the natural symplectic structure.

For the part which corresponds to imaginary eigenvalues or to an even number of zero eigenvalues, the situation is different. For imaginary eigenvalues, a typical elementary Jordan Block of Hamiltonian type will be

$$\begin{pmatrix} \begin{matrix} 0 & \lambda \\ -\lambda & 0 \end{matrix} & \begin{matrix} 1 & 0 \\ 0 & 1 \end{matrix} & 0 \\ 0 & \begin{matrix} 0 & \lambda \\ -\lambda & 0 \end{matrix} & \begin{matrix} 1 & 0 \\ 0 & 1 \end{matrix} \\ 0 & 0 & \begin{matrix} 0 & \lambda \\ -\lambda & 0 \end{matrix} \end{pmatrix} \tag{5.39}$$

with associated linear vector field

$$\begin{aligned} \Gamma = \lambda \left(y_1 \frac{\partial}{\partial x_1} - x_1 \frac{\partial}{\partial y_1} + y_2 \frac{\partial}{\partial x_2} - x_2 \frac{\partial}{\partial y_2} + y_3 \frac{\partial}{\partial x_3} - x_3 \frac{\partial}{\partial y_3} \right) \\ + x_2 \frac{\partial}{\partial x_1} + y_2 \frac{\partial}{\partial y_1} + x_3 \frac{\partial}{\partial x_2} + y_3 \frac{\partial}{\partial y_2} \,. \end{aligned} \tag{5.40}$$

This vector field is Hamiltonian with respect to the symplectic form

$$\Omega = dx_1 \wedge dy_3 + dy_2 \wedge dx_2 + dx_3 \wedge dy_1 \tag{5.41}$$

with associated matrix

$$[\Omega] = \begin{pmatrix} 0 & 0 & \begin{matrix} 0 & 1 \\ -1 & 0 \end{matrix} \\ 0 & \begin{matrix} 0 & 1 \\ -1 & 0 \end{matrix} & 0 \\ \begin{matrix} 0 & 1 \\ -1 & 0 \end{matrix} & 0 & 0 \end{pmatrix} \tag{5.42}$$

and Hamiltonian function

$$H = -\frac{\lambda}{2}(x_2^2 + y_2^2 - 2x_1x_3 - 2y_1y_3) + (x_2y_3 - x_3y_2) \,. \tag{5.43}$$

For even degeneracy, say $n = 2$, we have:

$$\Gamma = \lambda \left(y_1 \frac{\partial}{\partial x_1} - x_1 \frac{\partial}{\partial y_1} + y_2 \frac{\partial}{\partial x_2} - x_2 \frac{\partial}{\partial y_2} \right) + x_2 \frac{\partial}{\partial x_1} + y_2 \frac{\partial}{\partial y_1} \,, \tag{5.44}$$

with symplectic form:

$$\Omega = dx_1 \wedge dx_2 + dy_1 \wedge dy_2 \,, \tag{5.45}$$

and Hamiltonian:

$$H = \lambda(y_1x_2 - x_1y_2) + \frac{1}{2}(x_2^2 + y_2^2) \,. \tag{5.46}$$

For null eigenvalues n must be even and we simply put $\lambda = 0$ in the vector field and Hamiltonian corresponding to the imaginary case with even degeneracy.

According to the present classification, linear Hamiltonian systems fall in three main families.

1. The representative matrix does not have null or imaginary eigenvalues. In this case the carrier space can be given a structure of cotangent bundle and our vector field is a cotangent lift.
2. The representative matrix has only imaginary eigenvalues with odd degeneracy (minimal polynomial equal to the characteristic polynomial).
3. The representative matrix has only imaginary eigenvalues or null with even degeneracy (minimal polynomial equal to the characteristic polynomial).

According to our general formula for the flow of a linear vector field

$$e^{tA}x = \sum_k e^{t\lambda_k} P_k(t) x_k ,$$

in the last two cases we have an oscillatory part due to purely imaginary eigenvalues and a polynomial part of degree equal to the index of nilpotency minus one.

As an example we will consider the case of null eigenvalue which has even degeneracy.

Consider the kth order system of differential equations:

$$x^{(k)} = 0, \qquad y^{(k)} = 0 . \tag{5.47}$$

We can associate with them a first-order equation by setting

$$\begin{aligned} x &= x_1, \quad x' = x_2, \ldots, \quad x^{(k-1)} = x_k , \\ y &= y_1, \quad y' = y_2, \ldots, \quad y^{(k-1)} = y_k , \end{aligned} \tag{5.48}$$

and we get the first-order system of differential equations:

$$\begin{aligned} \frac{dx_1}{dt} &= x_2, \ \ldots, \ \frac{dx_{k-1}}{dt} = x_k, \ \frac{dx_k}{dt} = 0 \\ \frac{dy_1}{dt} &= y_2, \ \ldots, \ \frac{dy_{k-1}}{dt} = y_k, \ \frac{dy_k}{dt} = 0 . \end{aligned}$$

The associated Hamiltonian system will be

$$\Gamma = x_2 \frac{\partial}{\partial x_1} + x_3 \frac{\partial}{\partial x_2} + \cdots + x_k \frac{\partial}{\partial x_{k-1}} + y_2 \frac{\partial}{\partial y_1} + \cdots + y_k \frac{\partial}{\partial y_{k-1}} \tag{5.49}$$

with symplectic structure

$$\Omega = \sum_j (-1)^{j+1} dx_j \wedge dx_{k+1-j} + \sum_j (-1)^{j+1} dy_j \wedge dy_{k+1-j} , \tag{5.50}$$

and Hamiltonian

$$H = \sum_{j=1}^{1+k/2} \left((-1)^{j+1} x_{j-1} x_{k+1-j}\right) + \frac{1}{2}(-1)^{k/2} x_{k/2}^2$$
$$+ \sum_{j=1}^{1+k/2} \left((-1)^{j+1} y_{j-1} y_{k+1-j}\right) + \frac{1}{2}(-1)^{k/2} y_{k/2}^2 . \tag{5.51}$$

5.3 Symplectic Manifolds and Hamiltonian Systems

As it was discussed in Sect. 4.2.3, integral leaves of the canonical distribution defined by a Poisson structure inherit a non-degenerate Poisson structure, that is, a symplectic structure. If Λ is a constant Poisson structure on a linear space, its integral leaves are affine spaces and the induced symplectic structures are constant, however, even in the case of linear Poisson structures, i.e., homogeneous Poisson tensors of degree -1, the symplectic leaves are coadjoint orbits of Lie groups (see Sect. 4.3.3) and the induced symplectic structure is called the canonical Kirillov-Kostant-Souriau symplectic form. These symplectic manifolds are the fundamental model for the class of Hamilltonian systems with symmetry as will be discussed later on.

Thus we observe that in discussing the structure of Hamiltonian systems (with respect to the Poisson tensor), i.e., factorizable dynamics, we have to cope with symplectic structures on manifolds and not just linear spaces. We will devote this section to exploring the aspects of Hamiltonian dynamical systems which are specific to symplectic structures with respect to what was already discussed on the properties of Hamiltonian systems in the previous chapter. In particular we will discuss under what conditions a Hamiltonian system has the structure of a mechanical system, that is, the symplectic manifold is the cotangent bundle of a given configuration space.

5.3.1 Symplectic Manifolds

As discussed earlier, a symplectic manifold is a smooth manifold M together with a symplectic form, that is a non-degenerate closed 2-form ω on it. Any symplectic manifold is in particular a Poisson manifold with Poisson tensor $\Lambda = -\omega^{-1}$. A vector field Γ in M will be called Hamiltonian if it is Hamiltonian with respect to the Poisson tensor Λ or, in other words, if there exists a function H such that $i_\Gamma \omega = dH$. In such case, H will be called the Hamiltonian of the dynamical system Γ.

Notice that in a symplectic manifold (M, ω) the Poisson bracket of two functions f, g, defined by the Poisson tensor $\Lambda = -\omega^{-1}$ becomes:

$$\{f, g\} = \omega(\Lambda(df), \Lambda(dg)) . \tag{5.52}$$

As indicated above, the Jacobi identity satisfied by the Poisson brackets is equivalent to $d\omega = 0$. In fact, taking into account that for any triplet of functions $f, g, h \in \mathcal{F}(E)$,

$$X_f(\omega(X_g, X_h)) = X_f\{g, h\} = \{\{g, h\}, f\} ,$$

and using the above mentioned relation $[X_f, X_g] = X_{\{g,f\}}$, we find that $\omega([X_f, X_g], X_h)$ can be rewritten as

$$\omega([X_f, X_g], X_h) = \omega(X_{\{g,f\}}, X_h) = \{\{g,f\}, h\},$$

(along with the corresponding expressions obtained by cyclically permuting the indices), we find that

$$d\omega(X_f, X_g, X_h) = 2\left(\{\{g, h\}, f\} + \{\{h, f\}, g\} + \{\{f, g\}, h\}\right).$$

Therefore, as a local basis for the space of vector fields on M can be built from Hamiltonian vector fields, the Jacobi identity for the Poisson bracket is equivalent to ω being closed.

Notice that because of Gromov's theorem, Theorem 3.20, any open even-dimensional manifold possesses a symplectic structure. Thus symplectic structures are abundant. Not only that, symplectic structures satisfy the holonomic geometrization principle, which in this situation, is an involved way of saying that they are locally constant. More precisely:

Theorem 5.14 (Darboux Theorem) *Let (M, ω) be a symplectic manifold. For any $x_0 \in M$ there exists a local chart (φ, U) such that $x_0 \in U$ and*

$$\omega\mid_U = dq^i \wedge dp_i\,, \tag{5.53}$$

where $\varphi\colon U \subset \mathbb{R}^{2n}$ and $\varphi(x) = (q^1, \ldots, q^n, p_1, \ldots, p_n)$.

Proof Because ω is closed, it is locally exact. Let $x_0 \in M$ and let U be a contractible open neighborhood of x_0 small enough such that $\omega\mid_U = d\theta$ and U is contained in the domain of a chart centered at x_0 ($\varphi(x_0) = \mathbf{0}$). Let ω_0 be the constant form on U defined by the value of ω at x_0, i.e., $\omega_0(x) = \omega(x_0)$ (in other words, ω_0 is the pull-back of the canonical symplectic structure on $\mathbb{R}^{2n}$ with respect to φ). Then we show that there exists a family of diffeomorphisms ψ_t, $t \in [0, 1]$, on U, and an isotopy of symplectic 2-forms ω_t defined on U such that

$$\psi_t^*\omega_t = \omega_0\,.$$

The family ψ_t is such that $\psi_0 = \mathrm{id}_U$ and $\omega_1 = \omega$. Thus $\psi_1^*\omega = \omega_0$ and ω has the desired structure on the chart $\varphi \circ \psi_1^{-1}$.

Then, to show that there exists such a family of diffeomorphisms, let us consider $\theta_t = \theta_0 + t(\theta - \theta_0)$ where $d\theta_0 = \omega_0$, and $\omega_t = d\theta_t$. Let us compute the derivative of the previous equation. Then,

$$\psi_1^*\omega_1 - \omega_0 = \int_0^1 \frac{d}{dt}\psi_t^*\omega_t \, dt$$
$$= \int_0^1 \psi_{t*}(\mathcal{L}_{X_t}\omega_t + \omega - \omega_0)\, dt = \int_0^1 \psi_{t*}\left(d\left(i_{X_t}\omega_t - (\theta - \theta_0)\right)\right) dt.$$

Again, we may choose U small enough such that the equation

$$i_{X_t}\omega_t = \theta - \theta_0 \,,$$

has a unique solution X_t for each $t \in [0, 1]$ (it suffices to consider U small enough so that $\omega_t = \omega_0 + t(\omega - \omega_0)$ is invertible, i.e., such that the difference $\omega - \omega_0$ is small enough in U). Then the family of diffeomorphisms ψ_t can be obtained by integrating the differential equation:

$$\frac{d\psi_t}{dt} = X_t \circ \psi_t \,,$$

in U (again, if necessary we can 'squeeze' U to guarantee that there is a flow box around x_0 for the time-dependent vector field X_t, for $t \in [0, 1]$). □

Local coordinates (q^i, p_i) on a symplectic manifold (M, ω) such that ω takes the canonical form Eq. (5.53) will be called canonical (or Darboux) coordinates. The existence of local coordinates around any point on a symplectic manifold allows us to choose coordinates such that the Hamiltonian vector field X_H corresponding to a function H takes the canonical form:

$$\dot{q}^i = \frac{\partial H}{\partial p_i}, \qquad \dot{p}_i = -\frac{\partial H}{\partial q^i}, \quad i = 1, \ldots, n. \tag{5.54}$$

The previous equations are called Hamilton's equations.

Thus, locally all symplectic manifolds are 'equal', that is, symplectomorphic. However this is not so globally. Notice that because they are locally constant and holonomic there are no local invariants characteristic of symplectic manifolds in acute difference with Riemannian or pseudo-Riemannian manifolds.

On the other hand, it is easy to provide examples of symplectic manifolds of the same dimension which are not symplectomorphic. For instance any compact orientable Riemann surface carries a natural symplectic structure, its volume form, but any two Riemann surfaces with different genus cannot be diffeomorphic, hence they cannot be symplectomorphic either. Moreover, it is also possible to show that even in a given manifold it is possible that there exist symplectic structures which are not symplectically equivalent. Again, Gromov's theorem guarantees that there exists a symplectic form for each cohomology class on an open manifold. Thus symplectic forms corresponding to different cohomology classes cannot be symplectomorphic.

Another simple instance of this phenomenon is provided by $M = \mathbb{C}P^2$ the complex projective 2-dimensional space. Then there is a family of symplectic structures labelled by an integer k which are not symplectically equivalent (see Sect. 6.4.5 for a detailed study of the complex projective space in relation with the geometry of quantum evolution).

We do not pretend to discuss here the state of the art concerning global invariants of symplectic manifolds, a subject that has driven an intense research lately. Instead we will characterize a class of symplectic manifolds, cotangent bundles, that play a significant role in the description of dynamical systems and that will lead us eventually to characterize systems that can be described using a variational principle.

Before solving the previous question, let us discuss with more care the symplectic structure of a cotangent bundle. Thus if $\pi_Q \colon T^*Q \to Q$ is the cotangent bundle of Q, there is a canonical 1-form θ_Q defined on it and that is defined by the formula:

$$\theta_Q(q,p)(U) = \langle p, T\pi_Q(U)\rangle, \qquad U \in T_{(q,p)}(T^*Q)\,. \tag{5.55}$$

Notice that this definition is equivalent to the tautological property of θ_Q:

$$\alpha^*\theta_Q = \alpha,$$

where $\alpha \in \Omega^1(Q)$ is an arbitrary 1-form on Q, i.e., a section of T^*Q and the left-hand side of the formula above indicates the pull-back of θ_Q along $\alpha \colon Q \to T^*Q$. The canonical symplectic structure on T^*Q is defined as

$$\omega_Q = -d\theta_Q\,,$$

and in local bundle coordinates (q^i, p_i) we have the expressions:

$$\theta_Q = p_i\, dq^i, \qquad \omega_Q = dq^i \wedge dp_i\,,$$

hence local bundle coordinates are canonical coordinates for ω_Q and provide local Darboux trivializations of ω_Q.

Exercise 5.1 Prove that the graph of any closed 1-form is a Lagrangian submanifold in T^*Q with respect to the canonical symplectic form ω_Q. Find local canonical coordinates for (T^*Q, ω_Q) which are not bundle coordinates.

Now given a diffeomorphism $\phi \colon Q \to Q$, we may lift it to T^*Q as follows $\varphi^c \colon T^*Q \to T^*Q$, $\varphi^c(q,p) = (\varphi(q), (T\varphi^{-1})^*p)$. Then clearly

$$(\varphi^c)^*\theta_Q = \theta_Q\,, \tag{5.56}$$

as the simple computation below shows:

$$\begin{aligned}\langle(\varphi^c)^*\theta_Q(q,p), U\rangle &= \langle\theta_Q(\varphi^c(q,p), (\varphi^c)_*U)\rangle = \langle(\varphi^{-1})^*p, (\pi_Q)_*(\varphi^c)_*U\rangle\\ &= \langle(\varphi^{-1})^*p, \varphi_*(\pi_Q)_*U\rangle = \langle p, (\pi_Q)_*U\rangle = \langle\theta_Q(q,p), U\rangle\,.\end{aligned}$$

Moreover φ^c is a symplectic diffeomorphism because

$$(\varphi^c)\omega_Q = \omega_Q \,. \tag{5.57}$$

Given a vector field X on Q we can lift it to T^*Q as follows. Consider the local flow φ_t of X, then we may define the family of local diffeomorphisms φ_t^c of T^*Q that define a vector field X^c called the complete cotangent lift of X defined as:

$$\frac{d}{dt}\varphi_t^c = X^c \circ \varphi_t^c \,.$$

Notice that, by definition, the vector field X^c is π_Q-projectable on $(\pi_Q)_* X^c = X$, thus X^c and X are π_Q-related. It is a simple exercise to check that in local bundle coordinates we have the following expression for X^c:

$$X^c = X^i \frac{\partial}{\partial q^i} - p_i \frac{\partial X^i}{\partial q^j} \frac{\partial}{\partial p_j} \,,$$

provided that $X = X^i(q)\partial/\partial q^i$.

Another characterization of the complete lift X^c is that it is the only vector field in T^*Q projectable onto X and preserving the canonical 1-form θ_Q, i.e., such that

$$\mathcal{L}_{X^c}\theta_Q = 0 \,. \tag{5.58}$$

In fact, the local diffeomorphisms φ_t^c preserve the canonical 1-form θ_Q and therefore $\mathcal{L}_{X^c}\theta_Q = 0$. Conversely, if the local coordinate expression of X is $X = X^i(q)\frac{\partial}{\partial q^i}$, then a vector field projecting on X must be of the form $\widetilde{X} = X^i(q)\frac{\partial}{\partial q^i} + f_j(q,p)\frac{\partial}{\partial p_j}$, and taking into account that

$$\mathcal{L}_{\widetilde{X}}(p_i\, dq^i) = (\widetilde{X}p_i)\, dq^i + p_i\, d(\widetilde{X}q^i) = f_i(q,p)\, dq^i + p_i\, dX^i \,,$$

and therefore the condition $\mathcal{L}_{X^c} = 0$ fixes the values of f_i to be given by the coordinate expressions indicated above for the complete lift X^c:

$$f_i(q,p) = -p_j \frac{\partial X^j}{\partial q^i} \,.$$

The complete lifts of vector fields satisfy:

$$[X^c, Y^c] = [X, Y]^c, \qquad \forall X, Y \in \mathfrak{X}(Q) \,. \tag{5.59}$$

Finally, we must remark that because the flow φ_t^c is symplectic (Eq. (5.57)) the complete lifts X^c are (locally) Hamiltonian vector fields, even more because $\mathcal{L}_X^c$ $\theta_Q = 0$ (Eq. (5.56)) we conclude that

$$i_{X^c}\omega_Q = d(i_{X^c}\theta_Q)\,, \tag{5.60}$$

and consequently X^c is Hamiltonian with Hamiltonian the 'momentum' function P_X associated to X defined by:

$$P_X(q,p) = \langle \theta_Q(q,p), X^c(q,p)\rangle = \langle p, X(q)\rangle\,.$$

Then a simple computation shows that:

Proposition 5.15 *The family of momentum functions P_X, $X \in \mathfrak{X}(Q)$, close a Lie algebra with respect to the Poisson bracket isomorphic to $\mathfrak{X}(Q)$. That is:*

$$\{P_X, P_Y\}_{T^*Q} = P_{[X,Y]}, \qquad \forall X, Y \in \mathfrak{X}(Q)\,. \tag{5.61}$$

and the map $P\colon\ \mathfrak{X}(Q) \to \mathcal{F}^1(T^*Q)$ is a Lie algebra homomorphism.

Proof We first realize that $P_{[X,Y]}$ is the Hamiltonian of $[X^c, Y^c]$, because of Eq. (5.59) hence

$$i_{[X^c,Y^c]}\theta_Q = \mathcal{L}_{X^c} i_{Y_c}\theta_Q = X^c(P_Y) = \{P_X, P_Y\}_{T^*Q}\,.$$ □

5.3.2 *Symplectic Potentials and Vector Bundles*

5.3.2.1 Tensorial Characterization of Symplectic Linear Spaces

We want to characterize tensorially a symplectic linear space (E, ω), in a similar way as we did with the linear structure in terms of the Liouville vector field Δ in Sect. 3.1.3. Thus the linear structure on E is characterized by the vector field Δ and we get $\mathcal{L}_\Delta\omega = 2\omega$, or better:

$$d(i_\Delta\omega) = 2\omega\,. \tag{5.62}$$

Moreover ω is always exact in a linear space E, so there exists a symplectic potential 1-form θ such that $d\theta = \omega$. Hence because Eq. (5.62) $i_\Delta\omega = 2(\theta + df)$. Redefining the symplectic potential θ as $\theta + df$, we get that the Liouville vector field defining the linear structure satisfies:

$$i_\Delta\omega = 2\theta\,,$$

for some symplectic potential θ. Again because of the structure theorem for a symplectic linear space, there exists a Lagrangian subspace L such that $\theta\mid_L = 0$.

To show this we can argue as follows. Notice that $W = \ker\theta$ is a coisotropic subspace because codim $W = 1$, then $W/W^\perp$ is a symplectic linear space. Choose any symplectic subspace $\tilde{L}$ in $W/W^\perp$ and define $L = \pi^{-1}(\tilde{L})$ with $\pi :\ W \to W/W^\perp$ the canonical projection. The subspace L is a Lagrangian subspace with the required properties.

If we choose now linear canonical coordinates (q^k, p_k) on E adapted to L, then we will get that in these coordinates $\theta = p_k\, dq^k - q^k dp_k$ (notice that $i_\Delta \theta = 0$). We are now ready to prove the following theorem that characterizes tensorially symplectic linear structures.

Theorem 5.16 *Let (M, ω) be a symplectic manifold. Then there exists a linear structure on M such that ω is a linear symplectic structure iff there exists a Liouville vector Δ field on M such that $\theta = \frac{1}{2} i_\Delta \omega$ is a symplectic potential.*

Proof We have to prove the converse implication. Then suppose that there exists a Liouville vector field Δ, then M is a linear space. Now if $2\theta = i_\Delta \omega$ is a symplectic potential, then $\mathcal{L}_\Delta \omega = 2\omega$. Notice that we also have that $\mathcal{L}_\Delta \theta = 2\theta$ and $i_\Delta \theta = 0$. Hence θ is linear on a set of linear coordinates on M, i.e., $\theta = \Omega_{ij} x^i dx^j$, but then $\Omega_{ij} = -\Omega_{ij}$ and $\omega = d\theta = \Omega_{ij} dx^i \wedge dx^j$. □

5.3.2.2 Compatible Partial Linear Structures and Lagrangian Distributions

Cotangent bundles provide one of the most important examples of symplectic manifolds. Can we characterize them tensorially in a similar way as we did for linear spaces? The answer to this question is easily obtained after the following considerations. Suppose that $M = T^*Q$ is a cotangent bundle symplectic manifold with the canonical symplectic structure ω_Q. The cotangent bundle T^*Q is a vector bundle over Q, hence it is characterized tensorially be a partial linear structure, or a partial Liouville vector field Δ_{T^*Q}. Let us recall that such a vector field has as zero set the zero section of T^*Q and the fibers are the unstable manifolds of any zero point. Moreover the fibers are Lagrangian submanifolds (actually Lagrangian linear subspaces) and $\mathcal{L}_{\Delta_{T^*Q}} \omega_Q = \omega_Q$. Hence $d i_{\Delta_{TQ}} \omega_Q = \omega_Q$ and $\theta_Q = i_{\Delta_{T^*Q}} \omega_Q$ is a symplectic potential. Notice that in addition $\mathcal{L}_{\Delta_{TQ}} \theta_Q = \theta_Q$ and $\theta_Q \mid_{T_q^*Q} = 0$, that is, the vertical Lagrangian distribution of T^*Q is in $\ker \theta_Q$.

The vertical Lagrangian distribution of the cotangent bundle T^*Q is unperturbed if we add to the symplectic form ω_Q terms of the form $\pi_Q^* F$ with F a closed form on Q. That is, define the 'charged' symplectic form ω_F in T^*Q to be:

$$\omega_F = \omega_Q + \pi_Q^* F\,,$$

or, in local bundle coordinates (q^i, p_k), we have $\omega_F = dq^i \wedge dp_i + F_{ij}\, dq^i \wedge dq^j$. This is the situation when we describe a charged particle in a magnetic field A. Then the Hamiltonian system corresponds to consider the symplectic form ω_F with $F = dA$. See more about this in the examples below.

Suppose that we have a symplectic manifold (M, ω). In general our manifold M will not carry a linear structure. However it can carry a partial linear structure, i.e., a partial Liouville vector field Δ (recall Sect. 3.3.2). Then we have:

Theorem 5.17 (Characterization of cotangent bundles) *Let (M, ω) be a symplectic manifold, then (M, ω) is symplectomorphic to a natural alternative cotangent bundle*

T^*Q iff there exists a partial Liouville vector field Δ such that the zero manifold of Δ is diffeomorphic to Q (hence $\dim Q = \frac{1}{2}\dim M$), and $\mathcal{L}_\Delta \omega = \omega$.

Proof It remains to prove the converse. Thus if M carries a partial Liouville vector field Δ, then it has the structure of a vector bundle $\pi : M \to Q$, with Q the zero set of Δ. Moreover, the fibres of the map π carry the structure of a vector space whose linear functions are just the solutions to the equation $\Delta f = f$. The rank of the bundle $\pi : M \to Q$, i.e., the dimension of the standard fibre, is n, because $\dim Q = n$ and $\dim M = 2n$. Finally, from the homogeneity condition for ω, $\mathcal{L}_\Delta \omega = \omega$, we get that ω has a symplectic potential $\theta = i_\Delta \omega$.

Then, locally, if we choose a system of bundle coordinates, (q^i, w_i) for the bundle $\pi : M \to Q$, it is easy to show that ω has to have the local form:

$$\omega = w_i F^i_{jk} dq^j \wedge dq^k + \Sigma^k_j dq^j \wedge dw_k \,, \tag{5.63}$$

because of the homogeneity condition. Actually if we compute $\mathcal{L}_\Delta\left(\omega(\partial/\partial w_j, \partial/\partial w_k)\right)$ we get that it must be, if different from zero, a function of degree -1 on the coordinates w_k, but it must be a smooth function at 0 (the zero point of Δ) which is impossible. Moreover, repeating the argument with $\left(\omega(\partial/\partial q^j, \partial/\partial w_k)\right)$, we get:

$$\mathcal{L}_\Delta \left(\omega(\partial/\partial q^j, \partial/\partial w_k)\right) = 0$$

then, it is a function Σ^k_j depending just on the base coordinates q^i. Finally, the same computation for $\left(\omega(\partial/\partial q^j, \partial/\partial q^k)\right)$, leads to:

$$\mathcal{L}_\Delta \left((\omega(\partial/\partial q^j, \partial/\partial q^k)\right) = \omega(\partial/\partial q^j, \partial/\partial q^k)\,,$$

that shows that it must be a homogeneous function of degree 1, i.e., linear on w_k's. Moreover because of Eq. 5.63 the form ω will be non-degenerate iff the matrix Σ^k_j is invertible.

Now, we have that $d\omega = 0$, thus computing $d\omega$ with the expression given in Eq. (5.63), we get immediately that $F^i_{jk} = \partial \Sigma^k_j / \partial q^l$.

Then, defining for each $\alpha \in \Omega^1(Q)$, the vector fields α^V as

$$i_{\alpha^V}\omega = \pi^*\alpha \,,$$

we can check immediately (because Σ is non-degenerate) that α^V must be vertical, i.e., $\pi_*\alpha^V = 0$. Then the vertical distribution of the bundle $M \to Q$ is Lagrangian, i.e., the fibres of π are Lagrangian submanifolds of M. Notice that the function $\omega(\alpha^V, \beta^V)$ is homogeneous of degree -1 with respect to Δ, hence 0, and that the vertical vectors α^V generate the tangent space to the fibres.

Then we define a map $\Phi\colon T^*Q \to M$ given by $\Phi(q, \alpha_q) = \alpha^V_q(0)$ where we identify the vertical subspace of the vector bundle at $\mathbf{0}_q$ with the fibre $\pi^{-1}(q)$. This map clearly is an isomorphism of vector bundles. However, this map is not a symplectomorphism with respect to the canonical symplectic structure on T^*Q. Notice that the symplectic potential $\theta = i_\Delta \omega$ is just θ_Σ, i.e., the pull-back under Φ of the canonical 1-form θ_Q under the bundle morphism $\Sigma\colon T^*Q \to T^*Q$. Thus (M, ω) is symplectomorphic to (T^*Q, ω_Σ).

Moreover, if $\theta - \Phi^*\theta_Q$ is closed, this means that $\theta_\Sigma - \theta_Q$ is closed, then $\omega_\Sigma = \omega_Q$. □

Thus we want to point out that the existence of a cotangent bundle structure for a given Hamiltonian dynamics X_H on a symplectic manifold (M, ω) is determined just by the existence of a partial Liouville vector field Δ satisfying the properties of the previous theorem. We will complete this discussion incorporating in the argument the tangent bundle formulation of dynamical systems in Sect. 5.6.

Notice that we are not imposing that the dynamics have to leave invariant such a partial Liouville vector field. If this were the case, this would mean that X_H must be linear in the momenta, or in other words, that H must be quadratic in the momenta. This is exactly the situation we are going to discuss now.

5.3.3 Hamiltonian Systems of Mechanical Type

As we were commenting, for a vector field Γ to possess an invariant cotangent bundle structure, it is necessary and sufficient, first that it has a Hamiltonian description, that is, it admits an invariant symplectic structure and in addition it must commute with a partial Liouville vector field Δ as described in Theorem 5.17. If this is the case, the Hamiltonian must be a quadratic function on the momenta, thus homogeneous of degree 2 with respect to Δ. Hence the most general expression for such Hamiltonian $H\colon T^*Q \to \mathbb{R}$, is:

$$H(q, p) = \frac{1}{2} g^{ij} p_i p_j + V(q)\,.$$

Or using some pedantic notation, $H(q, p) = \frac{1}{2} g(p, p) + \pi_Q^* V$, with g a contravariant symmetric $(2, 0)$ tensor on Q and V a function on Q. Such Hamiltonian systems will be said to be of mechanical type. The term $K(q, p) = \frac{1}{2} g(p, p)$ is called the kinetic energy of the system. Notice that the kinetic energy defines a quadratic form along the fibres of T^*Q (that could be degenerate). The term $V(q)$ is called the potential energy of the system and even if we now assume it not to be regular, this is not the case in general (recall the Kepler problem Sect. 7.4.3).

We will not enter here a discussion of the specific properties of systems of mechanical type, as this subject has been masterfully treated in many classical textbooks.

There are some relevant considerations regarding the integrability and separability properties of such systems and its relation with the existence of alternative cotangent bundle structures for it. See a discussion of this phenomena at the end of Chap. 8.

Before entering a discussion of the existence and properties of symmetries for Hamiltonian systems in general, let us conclude this section by briefly commenting on the simplest of all Hamiltonian system: free motion on Riemannian manifolds.

5.3.3.1 Geodesic Motion and Sprays

As it was said before, free motion on a Riemannian manifold is the simplest class of homogeneous Hamiltonian systems. The dynamical system Γ is described on a cotangent bundle T^*Q, where Q is a Riemannian manifold with metric η, or in local coordinates $\eta = \eta_{ij}dx^i \otimes dx^j$. The Hamiltonian of the system is the 'kinetic energy' K_η defined above, i.e.

$$K_\eta(q, p) = \frac{1}{2}\eta^{ij}p_i p_j\,,$$

where $\eta^{ij}\,\eta_{jk} = \delta^i_k$.

It is now clear that the Hamiltonian vector Γ_η with Hamiltonian K_η is homogeneous of degree 1 and in coordinates the Hamilton equation take the form:

$$\dot{q}^i = \frac{\partial K_\eta}{\partial p_i} = \eta^{ij}p_j, \qquad \dot{p}_i = -\frac{\partial K_\eta}{\partial q^i} = -\frac{1}{2}\frac{\partial \eta^{jk}}{\partial q^i}p_j p_k\,.$$

The vector field Γ_η is called the geodesic spray of the metric η. Because of its homogeneity property, the geodesic spray satisfies:

$$\Gamma_\eta(q, \lambda p) = \lambda \Gamma_\eta(q, p)\,.$$

The projection $\gamma(t)$ to Q of integral curves of Γ_η are called geodesics of the metric η. They represent the projection on the 'space of configurations' Q of the actual motion of a system moving freely (no external forces acting upon it). Notice that a geodesic curve $\gamma(t)$ is characterized by a point on Q, for instance $q_0 = \gamma(t_0)$, and $p_0 \in T^*_{q_0}Q$, but because of the homogeneity condition, if we substitute $p_0 \mapsto \lambda p_0$, then the projection of the corresponding integral curve of Γ_η is $\gamma(\lambda t)$, the original geodesic parametrized now by λt instead of t.

Geodesics play an important role both in Geometry and Physics. It is worth to point out here that the space of geodesics carries a canonical symplectic structure. The description of such structure is left to this chapter where some related notions will be discussed.

5.4 Symmetries and Constants of Motion for Hamiltonian Systems

5.4.1 Symmetries and Constants of Motion: The Linear Case

As we said earlier, a generic linear dynamical system has a commuting algebra of symmetries but no constants of the motion. On the other hand, if our vector field preserves a symplectic structure we have seen that some of the previous symmetries (those associated with odd powers of the representing matrix A) are associated with constants of the motion. It is reasonable therefore, once a symplectic structure has been chosen, to consider those linear vector fields which preserve the given structure, i.e., $\mathcal{L}_{X_A}\omega = 0$. Because of this relation, it is easy to see now that $\omega(X_{A^p}, X_{A^q}) = f_{pq}$ is a constant of motion for any $X_{A^{2k+1}}$. It should be expected that some of these f_{mn} are actually vanishing, indeed $\omega(X_{A^{2k+1}}, X_{A^{2r+1}}) = 0$. We can examine closer this phenomenon.

From

$$\mathcal{L}_{X_{A^{2k+1}}}\omega = 0,$$

we get

$$d(i_{X_{A^{2k+1}}}\omega) = 0,$$

i.e., if we set $\alpha_{2k+1} = i_{X_{A^{2k+1}}}\omega$, we have a closed 1-form. Because $\mathbb{R}^n$ is contractible there will be a function f_k such that $df_k = \alpha_{2k+1}$, thus

$$i_{X_{A^{2k+1}}}\omega = df_k.$$

Now we take the Lie derivative of this relation with respect to $X_{A^{2r+1}}$ and find that $d(\mathcal{L}_{X_{A^{2r+1}}}f_k) = 0$, i.e.,

$$\mathcal{L}_{X_{A^{2r+1}}}f_k = c_k \in \mathbb{R}.$$

As f_k is quadratic and $X_{A^{2r+1}}$ is linear, the constant c_k must vanish. Thus we have shown that for vector fields X_A which preserve ω we can find a constant of motion directly. We have also shown that half of the symmetries of X_A (those associated with odd powers) are associated with constants of motion and moreover $\omega(X_{A^{2k+1}}, X_{A^{2r+1}}) = 0$.

If we consider submanifolds obtained by considering the exponentiated action of X_A, X_{A^3},... etc., we get in general a cylinder of dimension s, $T^r \times \mathbb{R}^{s-r}$ where $r \leq s \leq n$. If the minimal polynomial of A is of maximum degree, we have $s = n$. If all eigenvalues of A are purely imaginary our cylinder becomes a torus. As our dynamical vector field is a member of this action, we find that each trajectory with initial conditions on a cylinder will stay on the same cylinder.

As we have shown in the previous section, the constants of motion f_k that we obtained before for a linear Hamiltonian vector field commute among themselves, this is:

$$\{f_k, f_l\} = 0, \qquad \forall k, l,$$

because $\{f_k, f_l\} = X_l(f_k) = 0$. However if we consider a Lie algebra $\mathfrak{g}$ of infinitesimal linear symmetries of a given dynamics, then we will have that if they preserve the symplectic structure, that is:

$$\mathcal{L}_{X_C}\omega = 0,$$

then arguing as before we get that:

$$i_{X_C}\omega = df_C,$$

where f_C is a quadratic function. Notice that now, what we can show is that:

$$\{f_{C_k}, f_{C_l}\} = f_{[C_k, C_l]} + \sigma(k, l),$$

where the C_k's generate the Lie algebra $\mathfrak{g}$ and $\sigma(k, l)$ is a constant. Of course the Jacobi identities imply a 2-cocycle condition for σ.

5.4.2 Symplectic Realizations of Poisson Structures

Before continuing the discussion on symmetries it is worth addressing a different, though intimate related, problem, that will be of interest for us in the context of the general integration problem we have already addressed for Lie algebras. Once we have realized the relevance of the existence of Poisson structures for dynamics and the fact that symplectic manifolds provide a simpler description because of the existence of normal forms as discussed above, we may ask about the existence of 'symplectic realizations' for Poisson dynamics, that is, given a dynamics Γ on a manifold P possessing an invariant Poisson structure described by the Poisson tensor Λ, does there exist a symplectic manifold (M, ω) and a submersion $\pi : M \to P$ that is a Poisson map? If there exists a Hamiltonian vector field X_H on M such that Γ and X_H are π-related, $T\pi_\circ X_H = \Gamma \circ \pi$, we will say that X_H provides a symplectic realization of the dynamics Γ and that the projection map $\pi : M \to P$ is a symplectic realization of the Poisson structure (P, Λ). Notice that in such case integral curves of X_H project down onto integral curves of Γ, and then a way to describe integral curves of our dynamics will be achieved by selecting for each initial condition $x \in P$ a point $m \in M$ such that $\pi(m) = x$, then integrate the dynamics X_H finding the integral curve $\gamma(t)$ passing through m and then $\pi(\gamma(t))$ is the desired integral curve of X_H. Thus we may summarize the previous observations in the following definitions.

Definition 5.18 Let (P, Λ) be a Poisson manifold. A symplectic manifold (M, ω) and a submersion $\pi : M \to P$ such that

$$\{f \circ \pi, g \circ \pi\}_\omega = \{f, g\}_\Lambda \circ \pi, \qquad f, g \in C^\infty(P), \tag{5.64}$$

is called a symplectic realization of (P, Λ).

If Γ is a Hamiltonian vector field on P, i.e., $-\Lambda(df) = \Gamma$, then X_F with $F = f \circ \pi$ is a symplectic realization of the dynamics Γ.

Given a symplectic realization (M, ω, π) of a Poisson manifold (P, Λ), there is a natural map sending Hamiltonian vector fields $\Lambda_f = -\Lambda(df)$ on P, into Hamiltonian vector fields $X_f = \omega^{-1} d\pi \circ f$. Let us denote this map by $s\colon \mathrm{Ham}(P, \Lambda) \to \mathrm{Ham}(M, \omega)$. Then it is clear that because of Eq. (5.64),

$$s([\Lambda_f, \Lambda_g]) = [s(\Lambda_f), s(\Lambda_g)], \qquad \forall f, g \in C^\infty(P),$$

or, in other words, the symplectic lifting map s is a Lie algebra homomorphism.

We may ask ourselves if the map s will lift to a homomorphism of the groups underlying the Lie algebras of Hamiltonian vector fields on P and M respectively, $\mathrm{Ham}(P, \Lambda)$ and $\mathrm{Ham}(M, \omega)$, as it happens in the case of finite-dimensional Lie algebras and Lie groups that we discussed in Sect. 2.6.2. The answer is that in general this is not possible because of the existence of topological obstructions due to the structure of the group of symplectomorphisms, however we can solve this question rather easily if we restrict our map s to a finite-dimensional subalgebra $\mathfrak{g}$ of $\mathrm{Ham}(P, \Lambda)$. In such a case, $\mathfrak{h} := s(\mathfrak{g}) \subset \mathrm{Ham}(M, \omega)$ will be again another finite-dimensional subalgebra of the Lie algebra of Hamiltonian vector fields on M, hence because of Lie's theorem, Theorem 2.26, there will exist a group homomorphism $S\colon G \to H$ where G and H are the unique connected and simply connected Lie groups whose Lie algebras are $\mathfrak{g}$ and $\mathfrak{h}$ respectively. The groups G and H are coverings of groups of diffeomorphisms on P and M respectively and they can be interpreted as group actions on the manifolds P and M. We will not pursue these questions here but we concentrate on the map $s_{\mathfrak{g}}\colon \mathfrak{g} \to \mathrm{Ham}(M, \omega)$ obtained by restricting the map s to a non-trivial finite-dimensional Lie algebra of Hamiltonian vector fields (wherever it exists).

Notice that if such a finite-dimensional subalgebra $\mathfrak{g}$ exists, then we can define the dual map of s:

$$J\colon M \to \mathfrak{g}^*, \qquad \langle J(m), \xi\rangle = s(\xi)(m), \quad \forall m \in M, \xi \in \mathfrak{g}. \tag{5.65}$$

This map J will be called a momentum map and is called on to play a central role in the study of symplectic manifolds, hence, dynamical systems, with symmetry. Before, continuing the discussion of it, let us introduce a natural equivalence notion.

Given two symplectic realizations (M_a, ω_a, π_a), $a = 1, 2$ for (P, Λ) a morphism of symplectic realizations is a smooth map $\phi\colon M_1 \to M_2$ such that $\pi_1 \circ \phi = \pi_2$ and ϕ is symplectic. Two symplectic realizations (M_a, ω_a, π_a), $a = 1, 2$, will be equivalent if there is a morphism ϕ between them which is a symplectomorphisms. Notice that because the map ϕ defining a morphism between symplectic realizations is symplectic, then it must be locally injective, thus it makes sense to talk about a minimal symplectic realization (if it exists) for a given Poisson structure.

If the Poisson structure is constant, i.e., homogeneous of degree -2, the task of finding a symplectic realization is rather simple.

Exercise 5.2 Describe the natural symplectic realizations of constant Poisson structures.

More interesting is the case of homogeneity of degree -1, or if you wish, Lie-Poisson structures. Let us recall that a Lie–Poisson structure is the canonical Poisson structure on the dual $\mathfrak{g}^*$ of the Lie algebra $\mathfrak{g}$ of a Lie group G. The answer to this is very illustrative and we will devote the following section to its study.

5.4.3 Dual Pairs and the Cotangent Group

Consider as indicated before a Lie group G with Lie algebra $\mathfrak{g}$, which is just the tangent space at the identity of G, $\mathfrak{g} = T_eG$. Then we may consider the cotangent bundle T^*G equipped with its canonical symplectic structure ω_G and Liouville 1-form θ_G. Notice that the group G acts on G by left or right translations. Let us consider left translations in what follows. Thus to any element $g \in G$ we associate the diffeomorphism $L_g : G \to G$ defined by $L_g(h) = gh$, $h \in G$. This action lifts to T^*G by $\Lambda_g : T^*G \to T^*G$, with $\Lambda_g(h, \alpha_h) = (gh, (TL_g^*)^{-1}(\alpha_h))$. Notice that $TL_g^* : T^*_{gh}G \to T^*_hG$ by $TL_g^*\alpha_{gh}(v_h) = \alpha_{gh}(L_g v_h)$, with $v_h \in T_hG$, $\alpha_h \in T^*_hG$ and $\alpha_{gh} \in T^*_{gh}G$, and the definition of Λ_g acting on covectors by using the inverse of TL_g^* is required to get an action, that is, a representation of G on the group of diffeomorphisms of T^*G, or in other words to satisfy $\Lambda_g \circ \Lambda_{g'} = \Lambda_{gg'}$ and not the order of factors exchanged.

On the other hand any Lie group is a parallelizable manifold, that is $TG \cong G \times \mathfrak{g}$, hence $T^*G \cong G \times \mathfrak{g}^*$. There are two natural ways of achieving such trivializations of the (co-) tangent bundle: using left or right translations. We will use a similar notation as before to indicate the trivialization of TG, that is:

$$L : TG \to G \times \mathfrak{g}; \qquad L(v_g) = (g, TL_{g^{-1}}(v_g)), \quad v_g \in T_gG\,,$$

and similarly with using right translations. The induced trivializations for T^*G are given by:

$$R^* : T^*G \to G \times \mathfrak{g}; \qquad R^*(\alpha_g) = (g, TR_g^*(\alpha_g)), \quad \alpha_g \in T_g^*G\,.$$

It is worth noticing that TR_g^* maps T_g^*G into $T_e^*G = \mathfrak{g}^*$. Now, we may compose R^* with the projection into the second factor to get the map:

$$J : T^*G \to \mathfrak{g}^*; \qquad J(\alpha_g) = TR_g^*(\alpha_g), \quad \forall \alpha_G \in T_g^*G\,.$$

Proposition 5.19 *The map J is a momentum map and $J : T^*G \to \mathfrak{g}^*$ is a symplectic realization of the Lie–Poisson structure on $\mathfrak{g}^*$.*

Proof To check that J is a momentum map, we have to prove that the map $s\colon \mathfrak{g} \to \mathrm{Ham}(T^*G)$, $s(\xi) = X_\xi$, is a Lie algebra homomorphism. The vector field X_ξ is the Hamiltonian vector field whose Hamiltonian is $J_\xi = \langle J, \xi\rangle$, for any $\xi \in \mathfrak{g}$. Notice that this is equivalent to the map J being a Poisson morphism, Eq. (5.64):

$$\{F \circ J, H \circ J\}_{T^*G} = \{F, H\}_{\mathfrak{g}^*} \circ J, \qquad F, H \in \mathcal{F}(\mathfrak{g}^*)\,. \tag{5.66}$$

Notice that to prove the previous equation all is needed is to prove it for linear functions, that is, if $\xi \in \mathfrak{g}$, we define the linear function on $\mathfrak{g}^*$ by $F_\xi(\mu) = \langle \mu, \xi\rangle$, with any $\mu \in \mathfrak{g}^*$.

Then, by definition (see Eq. 4.28) we have that

$$\{F_\xi, F_\zeta\}_{\mathfrak{g}^*} = F_{[\xi,\zeta]}. \tag{5.67}$$

On the other hand we see that if ξ_G denotes the vector field defined on G by the left action of G on itself or, in other words, the right-invariant vector field defined by $\xi \in \mathfrak{g}$.[3]

Because of Eq. (5.60) we know that the complete lifting X^c to a the cotangent bundle T^*Q of a vector field X on Q is Hamiltonian with Hamiltonian function P_X. Hence the complete lifting ξ_G^c of the vector field ξ_G to T^*G is Hamiltonian with Hamiltonian $P_{\xi_Q}(\alpha_g) = \langle \alpha_g, \xi_G(g)\rangle$, but because ξ_G is right-invariant, we have that $\xi_G(g) = TR_g\xi$, then $\langle \alpha_g, \xi_G(g)\rangle = \langle TR_g^*\alpha_g, \xi\rangle = \langle J(\alpha_g), \xi\rangle = J_\xi(\alpha_g)$, then:

$$i_{\xi_G^c}\omega_G = dJ_\xi\,.$$

Finally, notice that $F_\xi \circ J(\alpha_g) = \langle J(\alpha_g), \xi\rangle = J_\xi(\alpha)$, thus because of Eq. (5.67), checking the result we are looking after, Eq. (5.66), amounts to showing that:

$$\{J_\xi, J_\zeta\}_{T^*G} = J_{[\xi,\zeta]}\,.$$

However again because of Eq. (5.59) we get $[X^c, Y^c] = [X, Y]^c$, hence $[\xi_G^c, \zeta_G^c] = [\xi_G, \zeta_G]^c = [\xi, \zeta]_G^c$. Then:

$$i_{[\xi_G^c,\zeta_G^c]}\omega_G = i_{[\xi,\zeta]_G^c}\omega_G = dJ_{[\xi,\zeta]}\,,$$

and on the other hand:

$$i_{[\xi_G^c,\zeta_G^c]}\omega_G = \mathcal{L}_{\xi_G^c} i_{\zeta_G^c}\omega_G = d\xi_G^c(J_{\zeta_G^c}) = d\{J_\xi, J_\zeta\}_{T^*G}\,,$$

because $\mathcal{L}_{\xi_G^c}\omega_G = 0$ (recall that ξ_G^c is Hamiltonian) and $\{J_\xi, J_\zeta\}_{T^*G} = \omega_G(\xi_G^c, \zeta_G^c)$. Hence we get that $c(\xi, \zeta) = \{J_\xi, J_\zeta\}_{T^*G} - J_{[\xi,\zeta]}$ is a 2-cocycle in $\mathfrak{g}$ with values in $\mathbb{R}$, but remember that this cocycle is obtained from the variation of the 1-form

[3] Notice that switch from 'left action' to 'right invariance' in the previous statement. It happens because of $TR_g\xi_G(h) = \frac{d}{ds}R_g((\exp s\xi)h)\mid_{s=0} = \frac{d}{ds}\exp s\xi(hg) = \xi_G(hg)$.

θ_G under the action of G. But remark that θ_G is invariant because this action is a complete lifting, and therefore $c = 0$. Alternatively, the direct computation offered in Eq. (5.61) gives the desired result. □

Remark 5.1 In the proof of Proposition 5.19 we have never used the explicit form of ω_G because it actually works the same for complete actions of groups on cotangent bundles. It is illustrative also as an exercise in the geometry of groups, to work the proof by developing an explicit formula for ω_G and the corresponding Poisson brackets (see for instance [Ab78], Proposition 4.4.1. and [Cu77]).

Exercise 5.3 Work out a group-theoretical expression for ω_G, i.e., in terms of left (or right) invariant vectors and forms. In other words, write down the formula for ω_G using the global trivialization $T^*G \cong G \times \mathfrak{g}^*$.

Hence we conclude these comments by observing that the cotangent bundle of the group G provides a symplectic realization of the Lie–Poisson structure. This realization has another relevant aspect: the symplectic manifold is a group itself, but even more important there is a canonical groupoid structure on T^*G.

A few more remarks are in order concerning dynamical systems and this example of a symplectic realization. The analysis of dynamics on groups or cotangent bundles of groups is extremely rich. The main example is the dynamics of rigid bodies in finite dimensions and of incompressible fluids in infinite dimensions. A lot of effort has been poured into clarifying the geometry of such systems. We can only cite here the works by V. Arnold, J.E. Marsden, etc. From our point of view, such dynamical systems are systems belonging to a (finite-) dimensional Lie algebra of vector fields. Then integrating the Lie algebra we obtain a group. We will devote a full chapter to exploring the dynamics of systems defined on groups, but instead of following the path of reduction, we will approach them from the point of view of their integrability properties, that in this case ‘integrability’ means the possibility of obtaining their solutions by using (nonlinear in general) superposition rules. Such systems will be called Lie-Scheffer systems (see Chap. 9).

5.4.3.1 Momentum Maps and Group Actions

As we indicated before, Proposition 5.19 is not specific of the cotangent bundle of a group. Actually let G be a Lie group acting on a smooth manifold Q, and consider the lifted action of G, or cotangent lifting of the action of G, to T^*Q. Then if ξ_Q denotes the fundamental vector field associated to $\xi \in \mathfrak{g}$ in Q, then the fundamental vector field associated to ξ in T^*Q is ξ_Q^c. The action of G on T^*Q preserves the symplectic structure because (Eq. (5.58)):

$$\mathcal{L}_{\xi_Q^c} \theta_Q = 0\,.$$

Hence because ξ_Q^c leaves θ_Q invariant it will leave ω_Q invariant too. Moreover, we have that the Hamiltonian of ξ_Q^c will be $P_\xi(q, p) = \langle p, \xi_Q \rangle$. But this is nothing but the momentum map corresponding to the Lie algebra homomorphism $s\colon \mathfrak{g} \to \mathrm{Ham}(T^*Q)$, $s(\xi) = \xi_Q^c = X_{J_\xi}$ with

$$J_\xi(q,p) = \langle J(q,p), \xi\rangle = \langle p, \xi_Q(q)\rangle = P_\xi(q,p)\,. \tag{5.68}$$

The same formula shows that

$$\{J_\xi, J_\zeta\}_{T^*Q} = J_{[\xi,\zeta]}, \qquad \forall \xi, \zeta \in \mathfrak{g}\,,$$

and the momentum map:

$$J\colon\; T^*Q \to \mathfrak{g}^*$$

provides another symplectic realization of $\mathfrak{g}^*$.

In this case, what we discover is that if we have a dynamics Γ such that it admits a cotangent bundle structure, and in particular an invariant cotangent bundle structure, hence it will be Hamiltonian with a quadratic Hamiltonian function in the momenta, then the symmetry group G of Γ consisting of transformations that commute with the partial Liouville vector field Δ_{T^*Q} will define a momentum $J\colon T^*Q \to \mathfrak{g}^*$ and the vector field Γ will be tangent to the level sets of this momentum map.

5.4.4 An Illustrative Example: The Harmonic Oscillator

We will start, as in many instances of this book, by discussing an illustrative example. In Sect. 1.2.7 we introduced complex coordinates to describe the harmonic oscillator and the canonical action of the $U(n)$ on it (in the isotropic case, other subgroups in the anisotropic situation). There was also a call for a further analysis of the topology of its space of orbits. We will proceed now to perform such analysis in the light provided by the previous discussions in this chapter and, at the same time, paving the way for further discussions in Chap. 6 when the harmonic oscillator will be discussed in the context of Hermitean systems and the geometry of quantum systems.

An important notion in discussing the Hamiltonian structure of the harmonic oscillator, actually for any dynamical system, is the bifurcation set of the system, that is the set where the topology of the level sets of the Hamiltonian function of the system changes. The bifurcation set of the system encodes most of the properties of the reduced system, i.e., the system that we will get once we have 'removed' the symmetries of the system or, in a slightly 'old fashioned' language, the cyclic variables [Sm70].

We will start with the isotropic one-dimensional harmonic oscillator whose configuration space is $\mathbb{R}$ and its phase space is $T^*\mathbb{R} \cong \mathbb{R}^2$ whose points will be denoted as usual by (q,p). The Hamiltonian function is just $H = \frac{1}{2}(p^2 + q^2)$. In complex coordinates $z = q + ip$ and $z^* = q - ip$ we have $H = \frac{1}{2}z^*z$ and the equations of motion turn out to be $\dot{z} = -iz$ and the phase space is just $\mathbb{C}$. The invariance group is now $U(1)$ which is topologically the unit circle S^1. The orbits of the action of $U(1)$ in $\mathbb{C}$ by complex multiplication are just the orbits of the dynamics of the harmonic oscillator and then the space of orbits, that is the quotient space $\mathbb{C}/U(1)$ is diffeomorphic to the half-line $\mathbb{R}^+$, the point 0 corresponding to the critical point (0, 0) (which happens to be the bifurcation set of the system), and any other point $r = \sqrt{q^2 + p^2}$ corresponds to the circle of radius r in $\mathbb{R}^2$.

5.4.5 The 2-Dimensional Harmonic Oscillator

Much more interesting is the isotropic harmonic oscillator in dimension 2. Now the configuration space is $\mathbb{R}^2$ with points (q_1, q_2) and the phase space is 4-dimensional $T^*\mathbb{R}^2 \cong \mathbb{R}^4$ with points $(q_1, 1_2, p_1, p_2)$. The Hamiltonian of the isotropic oscillator is $H = \frac{1}{2}(p_1^2 + p_2^2 + q_1^2 + q_2^2)$ which is the sum $H_1 + H_2$, with $H_a = \frac{1}{2}(p_a^2 + q_a^2)$, $a = 1, 2$, the Hamiltonian of a 1-dimensional oscillator. Notice that H_a are constants of motion for the dynamics defined by H.

If we denote by $T(r, \rho) = S^1_r \times S^1_\rho$ a torus corresponding to the points of $T^*\mathbb{R}^2$ such that $H_1 = h_1 = \frac{1}{2}r_1^2$ and $H_2 = h_2 = \frac{1}{2}\rho^2$, then we have that the level set $H = h = \frac{1}{2}R^2 =$ constant, is the union of all such sets when we vary r, ρ such that $0 \leq r^2 + \rho^2 \leq R^2$. Notice that the level set $H = \frac{1}{2}R^2$ is the 3-dimensional sphere S^3_R of radius R in $\mathbb{R}^4$, thus the union of all tori $T(r, \rho)$, $0 \leq r^2 + \rho^2 \leq R^2$ foliates the 3-dimensional sphere of radius R (the foliation is singular because there are two leaves, $T(r, 0)$ and $T(0, \rho)$ with different topologies). Notice that $T(0, \rho) = S^1_\rho$ is a circle as well as $T(r, 0) = S^1_r$. The family of tori $T(r, \rho)$ filling the sphere is depicted in Fig. 5.1.

The equations of motion in polar coordinates (φ, r) and (θ, ρ) are just:

$$\dot{r} = 0, \quad \dot{\varphi} = -1, \qquad \dot{\rho} = 0, \quad \dot{\theta} = -1\,,$$

which are integrated trivially. The previous observations allows us to describe the level surface $H = h$ as a double cone (see Fig. 5.2) obtained by mapping $S^3_R = \{H = h\}$ into $\mathbb{R}^3$ by $(q_1, q_2, p_1, p_2) \mapsto (x, y, z)$ with $x = q_1$, $y = p_1$ and $z = \theta\rho$. Notice that if we consider the circle $2r^2 = q_1^2 + p_1^2$, then $\rho^2 = R^2 - r^2$ and the torus $T(r, \rho)$ corresponds to the cylinder over the circle of radius $\sqrt{2r^2}$ in the xy plane of height z varying form $-\pi\rho$ to $\pi\rho$. Notice that the upper rim $(z = \pi\rho)$ and the lower rim $z = -\pi\rho$ of the cylinder $T(r, \rho)$ are identified.

The degenerate tori $T(r, 0)$ corresponds to the horizontal circle of radius $sqrt2r^2 = \sqrt{2}$, and the degenerate tori $T(0, \rho)$ correspond to the vertical segment along the z axis $-\pi R \leq z \leq \pi R$. Trajectories of the harmonic oscillator correspond to helicoids along the cylinders with pass ρ. Each trajectory cuts the equatorial disk $z = 0$ at exactly one point except the unit circle which is a trajectory by itself. Thus the space of trajectories can be described as a disk D^2 with its boundary identified to a point. This is the topological way of describing the sphere S^2. Thus we have obtained a

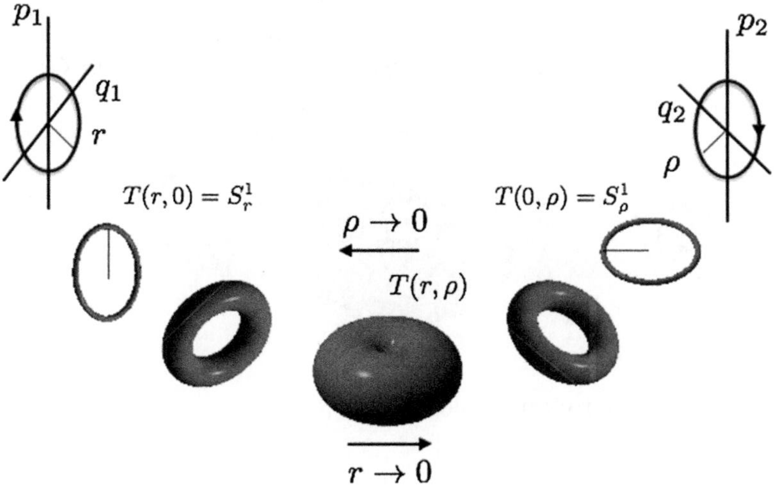

Fig. 5.1 The family of tori foliating the level surfaces of a 2D harmonic oscillator

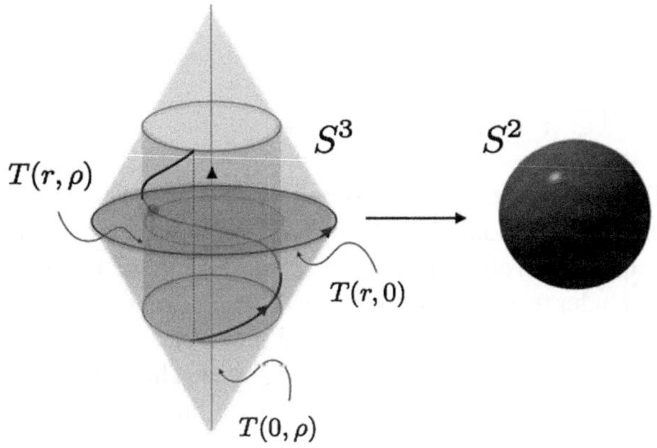

Fig. 5.2 The space of orbits of the 2D harmonic oscillator is a sphere S^2

map $\pi_H : S^3 \to S^2$ from the sphere S^3 describing an energy level set for the isotropic harmonic oscillator, and the sphere S^2 describing the space of trajectories of the given oscillator with fixed energy. This map is called the Hopf fibration and will appear in different occasions throughout the book. We will proceed in the next paragraphs to provide an analytical, rather than topological, description of it.

To finish this part we may also describe the trajectories and the level sets of the 2-dimensional isotropic oscillator, but this time using the stereographic projection of the 3-dimensional sphere. The (North pole) stereographic map $S\colon S^3 \to \mathbb{R}^3$ is defined as (see Fig. 5.3): $S(u_1, v_1, u_2, v_2) = (\zeta_1, \zeta_2, \zeta_3) = \zeta$, $(u_1, v_1, u_2, v_2) \in \mathbb{R}^4$ such that $u_1^2 + v_1^2 + u_2^2 + v_2^2 = 1$ and $\zeta = (\zeta_1, \zeta_2, \zeta_3) \in \mathbb{R}^3$, with:

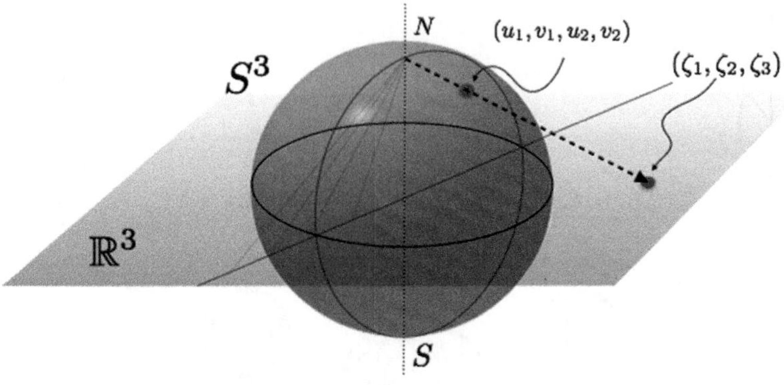

Fig. 5.3 The stereographic projection from S^3 into $\mathbb{R}^3$

$$\zeta_1 = \frac{u_1}{1-v_2}, \quad \zeta_1 = \frac{v_1}{1-v_2}, \quad \zeta_3 = \frac{u_2}{1-v_2},$$

or, conversely,

$$u_1 = \frac{2\zeta_1}{1+||\zeta||^2}, \quad v_1 = \frac{2\zeta_2}{1+||\zeta||^2}, \quad u_2 = \frac{2\zeta_3}{1+||\zeta||^2}, \quad u_1 = \frac{-1+||\zeta||^2}{1+||\zeta||^2}.$$

The South pole of S^3 is mapped into the origin $(0, 0, 0)$ and the North pole is mapped into ∞ that should be added to $\mathbb{R}^3$ to complete the picture. Notice that meridians are mapped into straight lines in $\mathbb{R}^3$ starting at the origin and all of them end up at ∞.

Under this projection the tori $T(r, \rho)$ are mapped into a family of tori foliating $\mathbb{R}^3$ where the tori $T(r, 0)$ is mapped into a circle of radius $\sqrt{2}R$ situated in the xy-plane and the tori $T(0, \rho)$ becomes the z-axis. The trajectories wind up around the tori and are depicted in Fig. 5.4 for various initial data.

5.4.5.1 The Complex Structure of the 2-Dimensional Harmonic Oscillator

As we did in Sect. 1.2.7, we may introduce complex coordinates to analyze the harmonic oscillator. In doing so we will gain a different perspective that will be very insightful when dealing with Hermitean systems. Thus we will define complex variables $z_1 = q_1 + ip_1$ and $z_2 = q_2 + ip_2$. The inner product in $\mathbb{C}^2$ is defined as $\langle z, w\rangle = \bar{z}_1 w_1 + \bar{z}_2 w_2$, with $z = (z_1, z_2)$ and $w = (w_1, w_2)$ vectors in $\mathbb{C}^2$. Then the Hamiltonian function of the 2-dimensional oscillator is written as: $H = \frac{1}{2}\langle z, z\rangle = \frac{1}{2}||z||^2$.

The complex coordinate vector fields are given by:

$$\frac{\partial}{\partial z} = \frac{1}{2}\left(\frac{\partial}{\partial q} - i\frac{\partial}{\partial p}\right), \quad \frac{\partial}{\partial \bar{z}} = \frac{1}{2}\left(\frac{\partial}{\partial q} + i\frac{\partial}{\partial p}\right),$$

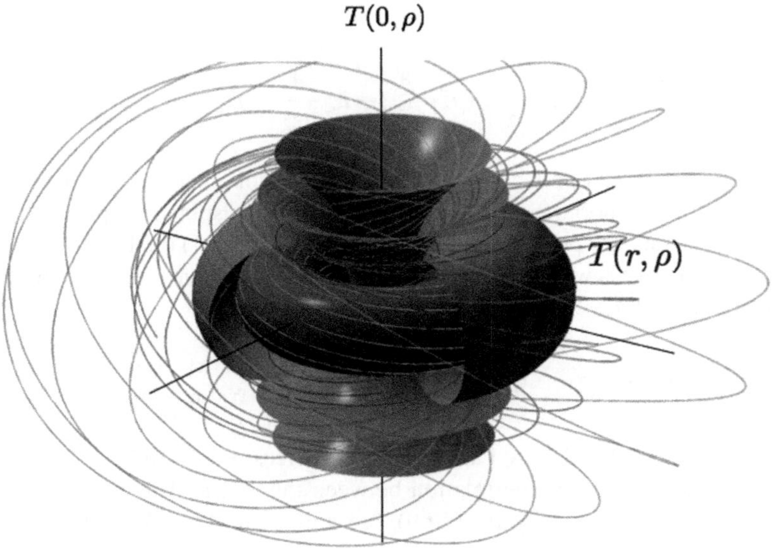

Fig. 5.4 The foliation of $\mathbb{R}^3$ by tori $T(r, \phi)$ under the stereographic projection and some trajectories of the 2D harmonic oscillator

and, conversely:

$$\frac{\partial}{\partial q} = \frac{\partial}{\partial z} + \frac{\partial}{\partial \bar{z}}, \qquad \frac{\partial}{\partial p} = i\left(\frac{\partial}{\partial z} - \frac{\partial}{\partial \bar{z}}\right).$$

It is convenient to write ∂_z instead of $\partial/\partial z$ and $\partial_{\bar{z}}$ for $\partial/\partial \bar{z}$ (or even ∂ and $\bar{\partial}$ respectively if there is no risk of confusion).

The canonical symplectic structure $\omega_0 = dq_1 \wedge dp_1 + dq_2 \wedge dp_2$ on $T^*\mathbb{R}^2$ becomes:

$$\omega_0 = \frac{i}{2}\left(dz_1 \wedge d\bar{z}_1 + dz_2 \wedge d\bar{z}_2\right),$$

hence the Poisson bracket of two functions $f(z, \bar{z})$, $g(z, \bar{z})$ becomes:

$$\{f, g\} = 2i \sum_{a=1,2} \frac{\partial f}{\partial z_a}\frac{\partial g}{\partial \bar{z}_a} - \frac{\partial f}{\partial \bar{z}_a}\frac{\partial g}{\partial z_a},$$

and the corresponding Hamiltonian vector fields X_f, become:

$$X_f = 2i \sum_{a=1,2} \frac{\partial f}{\partial z_a}\frac{\partial}{\partial \bar{z}_a} - \frac{\partial f}{\partial \bar{z}_a}\frac{\partial}{\partial z_a}.$$

Thus, for instance:

$$X_H = iz_1 \frac{\partial}{\partial z_1} - i\bar{z}_1 \frac{\partial}{\partial \bar{z}_1} + iz_2 \frac{\partial}{\partial z_2} - i\bar{z}_2 \frac{\partial}{\partial \bar{z}_2} .$$

The group $U(2)$ acts on $\mathbb{C}^2$ by matrix multiplication. This action of the group $U(2)$ is called the spinorial representation of $U(2)$ or the spin 1/2 representation of $U(2)$. With the standard presentation of $U(2)$ (see Sect. 2.5.4) the Lie algebra $\mathfrak{u}(2)^*$ of $U(2)$ consists of skew-Hermitean matrices iA with $A = A^\dagger$. We may write any Hermitean matrix A as:

$$A = \frac{1}{2}(x_0\sigma_0 + x_1\sigma_1 + x_2\sigma_2 + x_3\sigma_3) = \frac{1}{2}\begin{pmatrix} x_0 + x_3 & x_1 + ix_2 \\ x_1 - ix_2 & x_0 - x_3 \end{pmatrix} , \tag{5.69}$$

where σ_μ, $\mu = 0, 1, 2$ and 3 are Pauli's sigma matrices, Eq. (2.138).

If we want to emphasize the relation between the matrix A and the vector $x = (x_0, x_1, x_2, x_3) \in \mathbb{R}^4$ given by Eq. (5.69), we will write $A = A_x$ and, vice versa, $x = x_A$.

Again as we know form Sect. 1.2.7, Eq. (1.68), the real quadratic constants of motion have the form $f_A(z, \bar{z}) = \frac{1}{2}\langle z, Az\rangle$ with A Hermitean. Notice that the Hamiltonian of the system H is f_{σ_0}. Denoting by $J_i = f_{\sigma_i}$, $i = 1, 2, 3$, we get that any constant of motion f_A can be written as:

$$f_A = x_0 f_{\sigma_0} + x_1 f_{\sigma_1} + x_2 f_{\sigma_2} + x_2 f_{\sigma_3} = x^\mu J_\mu ,$$

which we denote by $J_\mu = f_{\sigma_\mu}$. If we collect the components J_1, J_2, J_3 as a 3-dimensional vector $\mathbf{J} = (J_1, J_2, J_3)$, then $\mathbf{J}$ defines a map:

$$\mathbf{J} \colon \mathbb{C}^2 \to \mathbb{R}^3, \qquad \mathbf{J}(z) = \langle z, \boldsymbol{\sigma} z\rangle ,$$

where $\boldsymbol{\sigma} = (\sigma_1, \sigma_2, \sigma_3)$ is the 3-dimensional vector whose components are Pauli matrices (Eq. 10.68).

We can think of the previous maps in a group-theoretical way as momentum maps. Recall that the group $U(2)$ can be identified with $U(1) \times SU(2)$ mod $\mathbb{Z}_2$, via the map $(e^{i\theta}, U) \mapsto e^{i\theta}U$, $U \in SU(2)$. Elements U of the special unitary group $SU(2) = \{U \in M_2(\mathbb{C} \mid U^\dagger U = \mathbb{I}_2\}$ are written as:

$$U = \begin{pmatrix} z_1 & -\bar{z}_2 \\ z_2 & \bar{z}_1 \end{pmatrix} , \qquad \text{with } |z_1|^2 + |z_2|^2 = 1 . \tag{5.70}$$

The Lie algebra of $SU(2)$ is the Lie subalgebra of $\mathfrak{u}_2^*$ consisting of traceless matrices, that is matrices of the form

$$A = \frac{1}{2}(x_1\sigma_1 + x_2\sigma_2 + x_3\sigma_3) = \frac{1}{2}\mathbf{x} \cdot \boldsymbol{\sigma} .$$

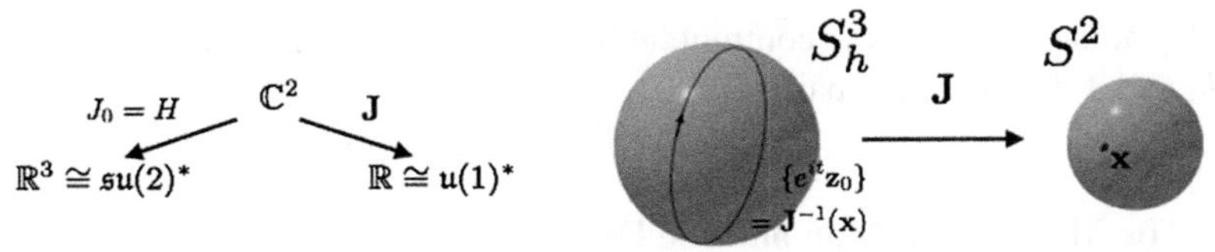

Fig. 5.5 *Left* The dual pair defined by the action of $U(1)$ and $SU(2)$ on $\mathbb{C}^2$. *Right* The Hopf fibration as a restriction of the momentum map $\mathbf{J}$

Now the action of $U2)$ on $\mathbb{C}^2$ restricts to actions of the subgroups $U(1)$ and $SU(2)$ and both groups act symplectically. Not only that, they will preserve the canonical symplectic structure ω_0. They define momentum maps which are given by the Hamiltonian itself and the map $\mathbf{J}$.

Theorem 5.20 *The spinorial action of* $U(2)$ *and its subgroups* $U(1)$ *and* $SU(2)$ *described above are Hamiltonian with momentum maps*:

$$J_0 \colon \mathbb{C}^2 \to \mathfrak{u}_1^* \cong \mathbb{R}, \quad J_0 = H$$

and

$$\mathbf{J} \colon \mathbb{C}^2 \to \mathfrak{su}_2^* \cong \mathbb{R}^3, \quad \mathbf{J}(z) = \langle z, \boldsymbol{\sigma} z\rangle, \quad z \in \mathbb{C}^2 .$$

Moreover, pair $(J_0, \mathbf{J})$ *define a dual pair* $\mathbf{J} = J_0^*$ *and it is called the energy-momentum map of the system (see Fig. 5.5).*

Proof It is a direct computation. In the case of $J_0 = H$, we already know that it is the momentum map of the action of the center $U(1)$ of $U(2)$ on $\mathbb{C}^2$. Its orbits being the trajectories of the harmonic oscillator.

The components of the momentum map $\mathbf{J}$ are given by:

$$J_1 = \frac{1}{2}\langle z, \sigma_1 z\rangle = \bar{z}_1 z_2 + \bar{z}_2 z_1 = q_1 q_2 - p_1 p_2 ,$$
$$J_2 = \frac{1}{2}\langle z, \sigma_2 z\rangle = -i\bar{z}_1 z_2 + i\bar{z}_2 z_1 = q_1 p_2 - q_2 p_1 ,$$
$$J_3 = \frac{1}{2}\langle z, \sigma_3 z\rangle = \bar{z}_1 z_1 - \bar{z}_2 z_2 = \frac{1}{2}(q_1^2 + p_1^2 - q_2^2 - p_2^2) .$$

We compute the Hamiltonian vector fields corresponding to J_1 to get:

$$X_{J_1}(z) = \{z, J_1\} = -i\sigma_1 z ,$$

which is just the linear vector field corresponding to the matrix $i\sigma_1$, $X_{J_1} = X_{i\sigma_1}$, and it is the fundamental vector field associated to the action of the generator $i\sigma_1$ of the Lie algebra of $SU(2)$. After similar computations we get the same results for J_2 and J_3.

Finally notice that J_0 is the commutant of $\mathbf{J}$ in the Poisson algebra of $\mathbb{C}^2$, thus the maps J_0 and $\mathbf{J}$ define a dual pair. □

5.4.5.2 The Momentum Map and the Hopf Fibration

The momentum map $\mathbf{J}$ corresponding to the canonical action of $SU(2)$ on $\mathbb{C}^2$ maps $\mathbb{C}^2$ into $\mathbb{R}^3$, but when restricted to the 3-dimensional sphere given by a level set of the Hamiltonian, we are going to show that its image is the 2-dimensional sphere S^2 describing the space of trajectories of the 2-dimensional harmonic oscillator, i.e.,. is the Hopf fibration (or just Hopf map) described in Sect. 5.4.5. To prove that we need a few results on products of σ matrices whose proofs are simple exercises.

Lemma 5.21 *For any* $z, u, w \in \mathbb{C}^2$ *we get*:

1. $\langle u, \boldsymbol{\sigma} z\rangle \cdot \boldsymbol{\sigma} w = 2\langle u, w\rangle z - \langle u, z\rangle w,$
2. $\mathrm{Re}\langle z, \boldsymbol{\sigma} u\rangle \cdot \mathrm{Re}\langle z, \boldsymbol{\sigma} w\rangle \cdot = \langle z, z\rangle \mathrm{Re}\langle u, \boldsymbol{\sigma} w\rangle - \mathrm{Im}\mathrm{Re}\langle z, \boldsymbol{\sigma} u\rangle \mathrm{Re}\langle z, \boldsymbol{\sigma} w\rangle,$
3. $||\mathbf{J}(z)|| = \frac{1}{2}\langle z, z\rangle = \frac{1}{2}||z||^2 = H(z, \bar{z}),$
4. $\mathbf{J}(e^{it}z) = \mathbf{J}(z).$

The next theorem follows immediately.

Theorem 5.22 *The restriction of the momentum map* $\mathbf{J}$ *to* $S^3_h = \{z \in \mathbb{C}^2 \mid H = \frac{1}{2}||z||^2 = h\}$ *is the Hopf fibration* π_H *mapping each point* z *into the trajectory of the harmonic oscillator passing through it. The preimage* $\mathbf{J}^{-1}(\mathbf{x})$ *is the set* $\{e^{it}z_0\}$, *for any* z_0 *such that* $\mathbf{J}(z_0) = \mathbf{x}$ *(see Fig. 5.5).*

Finally, we get:

Proposition 5.23 *The Hopf map* $\mathbf{J}$ *is a Poisson map, i.e.,* $\{J_1, J_2\} = J_3$ *and cyclic permutations.*

There is an alternative, though equivalent, description of the Hopf fibration that will be of interest in relation with the geometry of Hermitian systems and Kähler manifolds which is based on the geometry of the group $SU(2)$ itself. In this presentation we will obtain the formulas for the invariants defined by the Hopf map as a principal fibre bundle over S^2 with fibre $U(1)$. We will postpone this discussion until Sects. 6.4.2 and 6.4.4. However we may mention here that the basic idea consists in identifying the group $SU(2)$ itself with the sphere $S^3 \subset \mathbb{C}^2$ as we did before, that is the unitary matrix U given by Eq. (5.70) is identified with the point $z = (z_1, z_2)$, then we may define the map $\pi : SU(2) \to S^2$ given by:

$$\pi(U) = U\sigma_3 U^\dagger = \mathbf{x}\,,$$

where the vector $\mathbf{x} = (x_1, x_2, x_3)$ is defined as:

$$\mathbf{x} \cdot \boldsymbol{\sigma} = U\sigma_3 U^\dagger\,.$$

A direct computation shows:

$$x_1 = \langle z, \sigma_1 z\rangle, \quad x_2 = \langle z, \sigma_2 z\rangle, \quad x_3 = \langle z, \sigma_3 z\rangle\,,$$

and $||\mathbf{x}|| = 1$, i.e., $\mathbf{x} \in S^2 \subset \mathbb{R}^3$, in accordance with our previous results.

5.4.5.3 The n-dimensional Harmonic Oscillator

We can consider now the n-dimensional isotropic harmonic oscillator under the light of the previous discussions. Thus we will denote by $(q, p) \in T^*\mathbb{R}^n \cong \mathbb{R}^{2n}$ a point in phase space, The Hamiltonian of the system will be $H(q, p) = \frac{1}{2}\sum_{a=1}^n (p_a^2 + q_a^2)$.

We may as in the discussion of the 2-dimensional harmonic oscillator, introduce complex coordinates: $z_a = q_a + ip_a$, $a = 1, \ldots, n$, and then $H(z, \bar{z}) = \frac{1}{2}\langle z, z\rangle$, with $\langle z, w\rangle = \sum_{a=1}^n \bar{z}_a w_a$. The level set h of the Hamiltonian is the $2n-1$-dimensional sphere $S_h^{2n-1} \subset \mathbb{C}^n$ of radius $R = \sqrt{2h}$.

The equations of motion defined by H are given in complex coordinates by: $\dot{z} = -iz$, $\dot{\bar{z}} = i\bar{z}$, and the dynamical vector field is: $\Gamma = X_H = -iz\partial_z + i\bar{z}\partial_{\bar{z}}$.

The group $U(1)$ acts on $\mathbb{C}^2$ by complex multiplication, i.e., $(e^{it}, z) \mapsto e^{it}z$ and its infinitesimal generator defines the fundamental vector field $\Gamma = X_H$. Its orbits being the trajectories of the harmonic oscillator. The space of orbits of fixed energy h is then the quotient space $S_h^{2n-1}/U(1)$. Such a space is the $n-1$-dimensional complex projective space $\mathbb{C}Pn - 1$.

The description of the complex projective space $\mathbb{C}Pn - 1$ is done by introducing complex homogeneous coordinates $[z] = [z_1, \ldots, z_n]$ corresponding to a family of complex coordinate charts:

$$[z] \mapsto \begin{cases} \left(\frac{z_2}{z_1}, \ldots, \frac{z_n}{z_1}\right) \in \mathbb{C}^{n-1}, & \text{if } z_1 \neq 0\,, \\ \cdots & \\ \left(\frac{z_1}{z_n}, \ldots, \frac{z_{n-1}}{z_n}\right) \in \mathbb{C}^{n-1}, & \text{if } z_n \neq 0\,. \end{cases}$$

The projection map $\pi_H \colon S^{2n-1} \to \mathbb{C}Pn - 1$, $\pi_H(z) = [z]$ will be called the Hopf map, or Hopf fibration, in dimension $n-1$. Notice that there is a canonical identification between $\mathbb{C}P^1$ and the sphere S^2 that equips the sphere with a complex structure (in such case we call S^2 the Riemann sphere). The properties and geometry of this projection as well as of complex projective spaces will be discussed at length in Chap. 6, Sect. 6.4.3 because it provides the natural setting to describe pure quantum states.

The symmetry group of the system Γ is the group $U(n)$ which is defined exactly as the group of $n \times n$ complex matrices, such that $\langle Uz, Uz\rangle = \langle z, z\rangle$ for all $z \in \mathbb{C}^n$. Then because of the factorization $U(n) = (U(1) \times SU(n))/\mathbb{Z}_n$, where the subgroup $U(1)$ is the center of the group, we may restrict the action of $U(n)$ to the two subgroups $U(1)$ and $SU(n)$. The first has momentum map $J_0 = H$ and the second has a momentum map $\mathbf{J}\colon \mathbb{C}^n \to \mathfrak{su}(n)^*$. The trace bilinear form $(A, B) \mapsto -\mathrm{Tr}\, A^\dagger B$ restricted to the space of skew-Hermitean matrices allows us to identify $\mathfrak{su}(n)^*$ with $n \times n$ skew-Hermitean matrices iA. Thus we may consider the basis $N^{(ij)}$ introduced in Sect. 1.2.7, Eq. (1.71), and the corresponding component J_{ij} of the momentum map $\mathbf{J}$ given by:

$$J_{ij}(z, \bar{z}) = \frac{1}{2}(\bar{z}_i z_j - z_i \bar{z}_j)\,.$$

As it was proved in the case of $n = 2$, we also have that $\mathbf{J}$ maps S_h^{2n-1} into the complex projective space $\mathbb{C}P^{n-1}$ and the fibers of this map are the orbits of $U(1)$. It is interesting to understand the natural embedding of $\mathbb{C}P^{n-1}$ inside $\mathfrak{su}(n)^*$ as a coadjoint orbit. Recall that the coadjoint orbits of a group G are the symplectic leaves of the canonical linear Poisson structure on the dual $\mathfrak{g}^*$ of its Lie algebra (Sect. 4.3.3).

The complex projective space $\mathbb{C}P^{n-1}$ is the coadjoint orbit of $SU(n)$ corresponding to a diagonal matrix in $\mathfrak{su}(n)^*$ with diagonal $(\lambda, \ldots, \lambda, \mu)$, $\lambda \neq \mu$. The isotropy group of the action (which is now matrix conjugation $A \mapsto UAU^\dagger$ is the subgroup $S(U(n-1) \times U(1))$ and $\mathbb{C}^{n-1} = SU(n)/S(U(n-1) \times U(1))$.

Finally we conclude this section by observing that the pair $J_0 = H$ and $\mathbf{J}$ define a dual pair as in the 2-dimensional case. The map $(H, \mathbf{J})$ is called the energy-momentum map and the study of its topological properties play an important role in the analysis of mechanical systems with symmetry [Sm70]. We will come back to this subject in the study of Kepler's problem, Sect. 7.4.3.

5.5 Lagrangian Systems

In our quest for dynamically determined structures we arrive to the discussion of the problem that in a sense gave birth to all the previous one. This problem is called the inverse problem of the calculus of variations and can be stated in our context by asking when a given dynamical vector field Γ can be obtained as the Euler–Lagrange equations for some variational principle. In such generality the problem is hopeless and we have to qualify it by describing with precision which ones are the geometrical structures that arise naturally in the calculus of variations. Once this is done the inverse problem of calculus of variations for our vector field Γ amounts to a search for the appropriate geometrical structures.

It is difficult to emphasize the relevance and extraordinary importance that variational principles have played, and still are playing, in the derivation of dynamical laws for specific physical systems. In many occasions, and more often as we search for deeper structures, such variational principles become almost our only guide. This is particularly true of field theories such as Yang–Mills and other interaction theories. Thus, apart from historical reasons or the completeness of our exposition, we feel that devoting the next few sections to discuss this problem could be of relevance for future research.

5.5.1 Second-Order Vector Fields

The previous chapters dealt with (mostly linear) vector fields defining a dynamics on a Poisson manifold or, as a special case, on a symplectic manifold, actually a Poisson and/or a symplectic linear space in the linear case. In the whole of this section we will specialize to a particular class of vector fields, namely second-order vector fields (also called, for reasons that will become clear shortly, Second-Order Differential Equations, (SODE's for short) that we will try now to characterize.

Let us start then with a pair (V, Γ), where V is a vector space, with $\dim V = k$, and Γ a vector field in V. We shall assume here that V is endowed with a partial linear structure. Recalling what has been discussed in a previous section, we mean by this that there exists a vector field $\Delta \in \mathfrak{X}(V)$ (a (partial) dilation (or Liouville) vector field) such that the equations

$$\mathcal{L}_\Delta g = 0, \qquad g \in \mathcal{F}(V)\,, \tag{5.71}$$

and

$$\mathcal{L}_\Delta f = f, \qquad f \in \mathcal{F}(V)\,, \tag{5.72}$$

have respectively, say, n and $k-n$ functionally independent solutions $g^1, \ldots, g^n$ and $f^1, \ldots, f^{k-n}$, i.e., such that

$$dg^1 \wedge \cdots \wedge dg^n \wedge df^1 \wedge \cdots \wedge df^{k-n} \neq 0. \tag{5.73}$$

Notice that the g's span a linear space, denoted as $\mathcal{F}_0$ ($\dim(\mathcal{F}_0) = n$), and are an algebra under multiplication as well, while the f's span also a linear space that we will denote as $\mathcal{F}_1$ (with $\dim \mathcal{F}_1 = k-n$) and are also an $\mathcal{F}_0$-module (i.e., $f \in \mathcal{F}_1, g \in \mathcal{F}_0$ will entail $gf \in \mathcal{F}_1$ by the Leibnitz rule), but obviously not an algebra.

As it has been discussed elsewhere, the dilation field Δ can be immediately exponentiated in $\mathcal{F}_1$ as

$$\exp\{t\Delta\}f = \exp(t)f, \qquad \forall f \in \mathcal{F}_1\,, \tag{5.74}$$

whence the name of 'dilation' field. It can also be exponentiated in $\mathcal{F}_0$ as well, of course, but there the action is trivial. In this way V acquires the structure of a vector bundle with fibres of dimension $k-n$ spanned by the f^i's. The fibres will be vector spaces, and that is what is defined as a 'partial linear structure'. Taking then one set $(f^1, \ldots, f^{k-n})$ of independent solutions of Eq. (5.72) as basis on the fibres, the Liouville vector field can be written as

$$\Delta = f^i \frac{\partial}{\partial f^i}\,, \tag{5.75}$$

and it is clear that the zeroes of $\mathbf{f} =: (f^1, \ldots, f^{k-n})$ coincide with those of Δ. As the f's are functionally independent, the zero is a regular value of the map $\mathbf{f} : V \to \mathbb{R}^{k-n}$, and therefore $\mathbf{f}^{-1}(0)$ will be a regular submanifold M of V, the 'zero section' of the vector bundle.

Then we can state the following:

Definition 5.24 A vector field $\Gamma \in \mathfrak{X}(V)$ will be a second-order vector field with respect to the partial linear structure defined by the Liouville field Δ if

$$\mathcal{L}_\Gamma \mathcal{F}_0 = \mathcal{F}_1 \,. \tag{5.76}$$

Notice that this implies at once that $\dim \mathcal{F}_0 = \dim \mathcal{F}_1 = n$, and hence, $k = 2n$, i.e., V must be an even-dimensional space.

Let now $m \in M$ and choose local coordinates $g^1, , \ldots, g^n$, in a neighbourhood $U \subset M$, when viewed as functions, $g^1, \ldots, g^n \in \mathcal{F}_0$, of course (the choice will be global if M is also a vector space). Notice that whatever we say here will be only 'local' (and at most) as far as the 'base manifold' M is concerned. Indeed, Eq. (5.74) tells us that we can generate the whole fibre over any point of M via exponentiation. Therefore, our statements here will be valid (at least), in a 'tubular neighbourhood' in V of any neighbourhood $U \subset M$. By this we mean an open subset of TM that includes the whole of the fibres above a neighbourhood U in the base manifold. Then, it is not hard to see that the condition given by Eq. (5.76) implies for Γ the local expression

$$\Gamma = f^i \frac{\partial}{\partial g^i} + F^i(g, f) \frac{\partial}{\partial f^i} \,, \tag{5.77}$$

with the F^i's arbitrary (smooth) functions of the g^i's and f^i's. In other words, being second-order will fix unambiguously (only) the 'first components' of a vector field.

The equations of motion associated with Γ are then of the form:

$$\begin{cases} \dfrac{dg^i}{dt} = f^i \,, \\ \dfrac{df^i}{dt} = F^i(g, f) \,, \end{cases} \tag{5.78}$$

i.e., altogether

$$\frac{d^2 g^i}{dt^2} = F^i(g, \frac{dg}{dt}) \,, \tag{5.79}$$

i.e., they are second-order Newtonian equations, written in normal form, on the 'base manifold' M. This justifies the name SODE for second-order fields.

Proceeding further, we can define the $(1, 1)$-type tensor field

$$S = \frac{\partial}{\partial f^i} \otimes dg^i \tag{5.80}$$

It is not hard to prove that S and Δ enjoy the following properties:

(i) $\Delta \in \ker(S)$,
(ii) $\ker(S) = \mathrm{Im}(S)$ (then $S^2 = 0$),
(iii) $\mathcal{L}_\Delta S = -S$,
(iv) $\mathbb{N}_S = 0$.

where $\mathbb{N}_S$ denotes the Nijenhuis tensor associated with S (see Appendix E).

It was proved in [DeF89] that S and Δ define a 'unique' tangent bundle structure on V. Therefore, $V = TM$.

Remark 5.2

(i) It is clear that different choices of the g's and/or f's will (or can) lead in principle to different tangent bundle structures. Under this respect, it is perhaps more desirable (and more clarifying) to view the pair (V, Γ) (with no further qualifications) as the 'standard' reference structure, and to view any choice of the g's and f's as providing us with a (at least local) diffeomorphism between V and a tangent bundle ($V \approx TM$ then), and to say that Γ is 'related' to a second-order vector field on TM.
(ii) We can also use a slightly different construction, one that has the advantage of emphasizing the rôle that the dynamical vector field Γ plays in the construction of the tangent bundle structure(s).

Suppose we select, if $\dim V = 2n$, n functionally independent functions $g^1, \ldots, g^n$, i.e., such that $dg^1 \wedge \cdots \wedge dg^n \neq 0$, and assume (this is the really crucial assumption) that the n functions

$$f^i =: \mathcal{L}_\Gamma g^i$$

are also functionally independent among themselves and also with the g's, i.e.

$$dg^1 \wedge \cdots \wedge dg^n \wedge df^1 \wedge \cdots \wedge df^n \neq 0\,.$$

Then the g's and f's can be taken (at least locally) as a new set of coordinates for V. In this new coordinate system the dynamics will be given by

$$\Gamma = f^i \frac{\partial}{\partial g^i} + F^i \frac{\partial}{\partial f^i}$$

where, by construction, $F^i = \mathcal{L}_\Gamma f^i = \mathcal{L}_\Gamma \mathcal{L}_\Gamma g^i$. The g's and f's define then (again, at least locally) a tangent bundle structure, with the g's providing coordinates for the base manifold and the f's for the fibres, with respect to which Γ is a second-order vector field. The Liouville dilation along the fibres field will be given now by

$$\Delta = f^i \frac{\partial}{\partial f^i}\,.$$

Example 5.4 Let us take $V = \mathbb{R}^2$ with Cartesian coordinates (x_1, x_2) and consider the vector field

$$\Gamma = x_2 \frac{\partial}{\partial x_1} - x_1 \frac{\partial}{\partial x_2} . \tag{5.81}$$

It is obvious that, with no need of introducing any additional structures in V, Γ is, in appropriate units, just the standard 1D-harmonic oscillator.

Using the notation

$$q =: x_1 \quad \text{and} \quad \mathcal{L}_\Gamma q = x_2 =: v , \tag{5.82}$$

Γ acquires the form

$$\Gamma = v \frac{\partial}{\partial q} - q \frac{\partial}{\partial v} , \tag{5.83}$$

and V becomes the tangent bundle $V = T\mathbb{R}$, with the base space being $\mathbb{R}$ with coordinate q.

We might as well, however, have taken, e.g.

$$\bar{q} = \bar{q}(x_1, x_2) = \alpha x_1 - \beta x_2 , \qquad \alpha^2 + \beta^2 = 1 , \tag{5.84}$$

as the new 'coordinate', together with

$$\mathcal{L}_\Gamma \bar{q} =: \bar{v} = \alpha x_2 + \beta x_1 \tag{5.85}$$

as the new 'velocity'. Notice that $d\bar{q} \wedge d\bar{v} = dx_1 \wedge dx_2 \neq 0$. But then,

$$\Gamma = \bar{v} \frac{\partial}{\partial \bar{q}} - \bar{q} \frac{\partial}{\partial \bar{v}} , \tag{5.86}$$

which is again second-order in a new tangent bundle structure $V = T\mathbb{R}$, with $\mathbb{R}$ now the $\bar{q}$-axis, i.e., an axis rotated counterclockwise by an angle θ ($\tan\theta = \beta/\alpha$) w.r.t. the previous one.

As a less simple example, let us consider, e.g., new coordinates given by Q, U with

$$Q =: \sinh x_1 , \qquad U =: \mathcal{L}_\Gamma Q = x_2 \cosh x_1 , \tag{5.87}$$

and $dQ \wedge dU = (\cosh^2 x_1) dx_1 \wedge dx_2 \neq 0$. The transformation inverts to $x_1 = \operatorname{arcsinh}(Q)$, $x_2 = U/\sqrt{Q^2+1}$. As the branch-cuts of the square root are along the imaginary axis and away from the real axis, the latter is a perfectly smooth function on the real axis.

Then, a simple algebraic calculus yields

$$\Gamma = \left(x_2 \frac{\partial Q}{\partial x_1} - x_1 \frac{\partial Q}{\partial x_2} \right) \frac{\partial}{\partial Q} + \left(x_2 \frac{\partial U}{\partial x_1} - x_1 \frac{\partial U}{\partial x_2} \right) \frac{\partial}{\partial U} , \tag{5.88}$$

where, of course, $x_{1,2} = x_{1,2}(Q, U)$.

But $\partial Q/\partial x_2 = 0$ and $x_2 \partial Q/\partial x_1 = x_2 \cosh x_1 \equiv U$, and hence,

$$\Gamma = U\frac{\partial}{\partial Q} + f(Q, U)\frac{\partial}{\partial U}, \tag{5.89}$$

(where the expression for the 'force' $f(Q, U)$ can be read off from the previous equation and has the rather unappealing form $f(Q, U) = U^2 Q/(Q^2 + 1) - \sqrt{Q^2 + 1}\,\mathrm{arcsinh}(Q))$ will again be a second-order field, though not a simple harmonic oscillator anymore, in the new coordinates and, again $V = TM$, where the base M in the $\mathbb{R}^2$ plane will be no more a (1D) vector space but rather the graph of $\sinh x_1$, and the linear structure will survive only on the fibres spanned by the coordinate U.

Despite the nonlinearity of the transformation, it will be seen shortly that the result could have been anticipated in a single stroke, the map $(q, u) \to (Q, U)$ being what will be defined as a 'point transformation' (see below for that), i.e., one that maps diffeomorphically fibres into fibres, i.e. it is what is usually called a 'fibre-preserving' diffeomorphism. It has the only advantage of showing that a second-order field can in general be related only to a 'partial linear structure'.

To conclude the discussion of this example, let us consider a different coordinate transformation defined by

$$Q = x^1\left(1 + (x^1)^2 + (x^2)^2\right), \qquad U = x^2\left(1 + (x^1)^2 + (x^2)^2\right).$$

This is a perfectly invertible transformation, actually a diffeomorphism. Indeed,

$$dQ \wedge dU = \left(1 + 4E + 3E^2\right) dx^1 \wedge dx^2,$$

where $E = (x^1)^2 + (x^2)^2$. Explicitly, the inversion is given by

$$x^1 = 2Q/\left(\sqrt{1 + Q^2 + U^2} + 1\right), \qquad x^2 = 2U/\left(\sqrt{1 + Q^2 + U^2} + 1\right).$$

The fact that $(x^1)^2 + (x^2)^2$ is a constant of motion implies that, here too,

$$U = \mathcal{L}_\Gamma Q.$$

The harmonic oscillator is then a second-order field in the new coordinate system as well. However the transformation $(x^1, x^2) \leftrightarrow (Q, U)$ does not maps fibres into fibres, i.e., it is not a fibre-preserving diffeomorhism.

5.5.2 The Geometry of the Tangent Bundle

From now on we shall assume that a given (and fixed) tangent bundle structure has been defined on the 'evolution space' $V \approx \mathbb{R}^{2n}$, and we will denote henceforth the (local or global) coordinates on the base manifold M as $(q^1, \ldots, q^n)$ and the (global) coordinates on the fibres as $(v^1, \ldots, v^n)$. In this section we shall discuss various geometrical objects that can be associated with a tangent bundle structure and that characterize the geometry of a tangent bundle.

5.5.2.1 Vertical and Dilation Fields: Homogeneous Tensors

A tangent bundle TM, with M the base manifold (usually, but not necessarily –see the example in the previous section– a vector space itself) is endowed with a natural projection $\pi : TM \to M, \pi : (q, v) \mapsto v$. We will denote by $\mathfrak{X}(TM)$ the set of vector fields on TM, and by $\Omega^1(TM)$ the set of 1-forms on TM. The set of 0-forms (smooth functions) will be denoted as $\mathcal{F}(TM)$.

Vertical vector fields are those projectable on the zero vector field in M, and they can be viewed then as vector fields that are 'along the fibres', i.e., fields acting only on the fibre coordinates. In other words, a vector field X is vertical if $\mathcal{L}_X(\pi^* f) = 0$, $\forall f \in \mathcal{F}(M)$.

Note that homogenous functions of degree zero, i.e., functions that are pull-backs of functions on the base manifold, are constant along the fibres, and vice-versa.

The local coordinate expression of a vertical field is

$$X = X^i \frac{\partial}{\partial v^i}, \qquad X^i \in \mathcal{F}(TM) \,. \tag{5.90}$$

and the set $\mathfrak{X}^v(TM)$ of vertical vector fields is a $C^\infty(TM)$-module.

An immediate consequence of the definition is that

$$\mathcal{L}_{[X,Y]}(\pi^* f) = 0\,, \qquad \forall X, Y \in \mathfrak{X}^v(TM), \qquad \forall f \in \mathcal{F}(M)\,, \tag{5.91}$$

and $\mathfrak{X}^v(TM)$ is therefore an infinite-dimensional Lie algebra generating an involutive distribution and hence, by Frobenius' theorem, integrable distribution (Theorem 2.16, Sect. 2.5.2; see also [Mm85]).

The dilation along the fibres vector field, also-called the 'Liouville' or the 'star field', introduced in the previous section and that characterizes the partial linear structure has the local expression

$$\Delta = v^i \frac{\partial}{\partial v^i}\,, \tag{5.92}$$

and it can be used to characterize tensor fields in the following manner.

Definition 5.25 A tensor field τ on TM will be said to be homogeneous of degree p ($p \in \mathbb{R}$) in the fibre coordinates if

$$\mathcal{L}_\Delta \tau = p\tau \, . \tag{5.93}$$

Example 5.5

1. Homogeneous functions of degree zero are constant along the fibres. This follows from the fact that, as far as the dependence on the variables on which it operates goes (the fibre coordinates in the present case), a dilation field has no non trivial constants of motion. Therefore, homogeneous functions of degree zero are pull-backs of functions on the base manifold.
2. Homogeneous functions of negative degree cannot be smooth. Indeed, $\mathcal{L}_\Delta f = pf$, $p < 0$, implies that f must be singular on the base manifold (the zero section).
3. Let $\theta = a_i \, dq^i + b_i \, dv^i$ be a 1-form, with $a_i, b_i \in C^\infty(TM)$. Then, $\mathcal{L}_\Delta \theta = p\theta$ is easily seen to imply $\mathcal{L}_\Delta a_i = 0$ and $\mathcal{L}_\Delta b_i = (p-1)b_i$. In particular, a 1-form will be homogeneous of degree zero iff (see ii)) the b_i's are all zero and the a_i's are pull-backs of functions on the base manifold, $a_i = \pi^* \alpha_i$, $\alpha_i \in \mathcal{F}(M)$, i.e., iff $\theta = \pi^* \alpha$, $\alpha = \alpha_i \, dq^i \in \Omega^1(M)$. The 1-form θ is forced then to be the pull-back of a 1-form on the base manifold.
4. A second-order vector field has the local expression

$$\Gamma = v^i \frac{\partial}{\partial q^i} + F^i \frac{\partial}{\partial v^i} \, , \qquad F^i \in \mathcal{F}(TM) \, . \tag{5.94}$$

Observing that the 'first components' of Γ are homogeneous of degree one in the fibre coordinates, while the $\partial/\partial v^i$'s are homogeneous of degree -1, and one infers immediately that, in general,

$$\mathcal{L}_\Delta \Gamma = \Gamma + [(\mathcal{L}_\Delta - 2)F^i] \frac{\partial}{\partial v^i} \, . \tag{5.95}$$

Second-order homogeneous fields are also-called 'sprays'. Necessarily they are of degree one and correspondingly, the 'forces' F^i will have to be homogeneous of degree two in the fibre coordinates.

Second-order fields allow for a simple characterization of zero vector fields on TM: $X \in \mathfrak{X}(TM) = 0$ iff

$$\mathcal{L}_X(\pi^* f) = 0 \, , \quad \text{and} \quad \mathcal{L}_X(\mathcal{L}_\Gamma \pi^* f) = 0 \, , \qquad \forall f \in \mathcal{F}(M), \tag{5.96}$$

and with Γ any second-order field in the set $\mathfrak{X}(TM)$.

The proof is an easy exercise in coordinates, so we will omit details.

Remark 5.3 Of course a vector field is entirely characterized by its action on functions. What the preceding property tells us is that it is enough to consider functions that are constant along the fibres and functions that are homogeneous of degree one in the fibre coordinates.

5.5.2.2 Vertical and Tangent Lifts

Let now $X = X^i(q)\,\partial/\partial q^i \in \mathfrak{X}(M)$ be a vector field on the base manifold. For any $q \in M$, $X(q) \in T_qM$ (the fibre through q), using the identification of T_qQ with $T_{(q,v)}(T_qQ)$ for any vector $(q, v) \in TQ$ we can define the curve in TM

$$\mathbb{R} \ni t \to (q, v + tX(q)) \tag{5.97}$$

through (q, v). This defines also a one-parameter of diffeomorphisms, that are just translations along the fibres. This one-parameter group defines a new vector field, the vertical lift X^v of X, that is easily seen to be given, in local coordinates, by

$$X^v =: (\pi^* X^i)\frac{\partial}{\partial v^i} \in \mathfrak{X}^v(TM)\,. \tag{5.98}$$

Then one can see that a vertical vector field $X \in \mathfrak{X}(TM)$ is a vertical lift of a vector field in the base if and only if $\forall f \in \mathcal{F}(M)$, there exists a $g \in \mathcal{F}(M)$ such that

$$\mathcal{L}_X(\mathcal{L}_\Gamma \pi^* f) = \pi^* g\,, \tag{5.99}$$

where Γ is any second-order field.

We shall denote $v : X \in \mathfrak{X}(M) \to \mathfrak{X}^v(TM)$ the map associating with any vector field $X \in \mathfrak{X}(M)$ the corresponding vertical lift i.e.,

$$v\colon \mathfrak{X}(M) \to \mathfrak{X}(TM)\,, \tag{5.100}$$

and correspondingly, we also denote $v : TM \to T(TM))$ the corresponding map.

Given now a vector field $X \in \mathfrak{X}(M)$, this will generate a (local) one-parameter group of transformations on M that we will denote as $\{\phi_t\} : \phi_t : M \mapsto M, \phi_{t=0} = \mathrm{Id}_M$ and $\phi_t \circ \phi_\tau = \phi_{t+\tau}$ whenever both ϕ_t, ϕ_τ and $\phi_{t+\tau}$ are defined. This will be granted if X is a 'complete' field (see e.g. [Ar76]). If this is not the case, we will refer to the flow generated by X as to a 'pseudo-one-parameter group'.

We can now 'lift' ϕ_t to the tangent map $\phi_t^c =: T\phi_t \in \mathrm{Diff}\,(TM)$. We recall that, given any (smooth) map $\phi : M \to M$, the tangent map $T\phi$ is defined, in local coordinates, as $T\phi : (q^i, v^i) \mapsto (\phi^i(q), v^j \partial\phi^i/\partial q^j)$, and acts linearly on the fibres. Correspondingly:

Definition 5.26 The infinitesimal generator $X^c \in \mathfrak{X}(TM)$ of the one-parameter group (or pseudogroup) of diffeomorphisms ϕ_t^c will be called the tangent, or complete, lift of the vector field $X \in \mathfrak{X}(M)$ to TM.

It is a simple matter to show that, in local coordinates, the tangent lift of $X = X^i(q)\,\partial/\partial q^i \in \mathfrak{X}(M)$ is given by

$$X^c = (\pi^* X^i)\frac{\partial}{\partial q^i} + (\mathcal{L}_\Gamma \pi^* X^i)\frac{\partial}{\partial v^i}\,, \tag{5.101}$$

with Γ any second-order vector field. Note that this does not depend on the choice of the SODE Γ

For reasons that will become clear later, diffeomorphisms of TM that are tangent lifts of diffeomorphisms on the base manifold are called also 'Newtonian' diffeomorphisms.

Again, one can prove easily in coordinates that:

Proposition 5.27 *A vector field $X \in \mathfrak{X}(TM)$ is a tangent lift of a vector field $Y \in \mathfrak{X}(M)$ iff, $\forall f \in \mathcal{F}(M)$, there exists $g \in \mathcal{F}(M)$ such that:*

$$\begin{aligned} \mathcal{L}_X(\pi^* f) &= \pi^* \mathcal{L}_Y f\,, \\ \mathcal{L}_X(\mathcal{L}_\Gamma \pi^* f) &= \mathcal{L}_\Gamma(\pi^* g\,. \end{aligned} \tag{5.102}$$

Proof The first condition is a necessary and sufficient condition for X to be projectable onto Y, with $g = \mathcal{L}_Y f)$. Therefore X must be in local coordinates of the form

$$X = Y^i(q)\frac{\partial}{\partial q^i} + Z^i(q,v)\frac{\partial}{\partial v^i}\,.$$

Taking into account that

$$\begin{aligned} \mathcal{L}_X(\mathcal{L}_\Gamma \pi^* f) &= \left(X = Y^i(q)\frac{\partial}{\partial q^i} + Z^i(q,v)\frac{\partial}{\partial v^i}\right)\left(v^k\frac{\partial f}{\partial q^k}\right) \\ &= v^k Y^i(q)\frac{\partial^2 f}{\partial q^i \partial q^k} + Z^k\frac{\partial f}{\partial q^k}, \end{aligned}$$

while

$$v^j\frac{\partial}{\partial q^j}\left(Y^l\frac{\partial f}{\partial q^l}\right) = v^j Y^l\frac{\partial^2 f}{\partial q^j \partial q^l} + v^j\frac{\partial Y^l}{\partial q^j}\frac{\partial f}{\partial q^l},$$

and then the second condition says that

$$Z^i = v^j\frac{\partial Y^i}{\partial q^j}$$

which means that X is the complete lift of Y. □

Remark 5.4 'Newtonian' diffeomorphisms.

In local language, a second-order vector field will be of the form $\Gamma = v^i\partial/\partial q^i + F^i\partial/\partial v^i$, $F^i \in \mathcal{F}(TM)$, and the 'first half' of the associated equations of motion will be $dq^i/dt = v^i$ (and vice-versa, of course), i.e., as already said before, they are of the (standard) Newtonian form. If now ϕ_t^c is the tangent lift of ϕ_t, then, $\phi_t^c : (q^i, v^i) \mapsto (Q^i, U^i)$, with $Q^i = \phi_t^i(q)$, $U^i = v^j \partial\phi_t^i/\partial q^j$. But then, $dQ^i/dt = (\partial\phi_t^i/\partial q^j)(dq^j/dt) \equiv v^i$, and the Newtonian form of the equations of motion is preserved. This justifies the name 'Newtonian' given to diffeomorphisms that are tangent lifts of diffeomorphisms on the base manifold. These are also-called 'point transformations'. One can also show that a diffeomorphism preserving the second-order character of any SODE is forced to be a Newtonian diffeomorphism (see e.g. [Mm85])

It is useful to compute the Lie brackets of both vertical and tangent lifts. They are given by the following:

Proposition 5.28 *The Lie brackets of vertical and tangent lifts satisfy*

$$
\begin{aligned}
[X^v, Y^v] &= 0\,,\\
[X^c, Y^c] &= [X, Y]^c\,,\\
[X^v, Y^c] &= [X, Y]^v\,.
\end{aligned}
$$

Moreover,

$$[X^c, \Gamma] \in \mathfrak{X}^v(TM)\,, \tag{5.103}$$

as well as

$$[X^v, \Gamma] - X \in \mathfrak{X}^v(TM) \tag{5.104}$$

for any SODE Γ where, with some abuse of notation, we have used the symbol X for the vector field $(\pi^* X^i)\partial/\partial q^i$.

Remark 5.5 (i) The second of the preceding equations tells us that the map $X \mapsto X^c$ is a Lie algebra homomorphism. This means that tangent lifts are a Lie subalgebra of the Lie algebra of vector fields on TM and that vertical lifts are an Abelian ideal thereof (see e.g., [SW86]).

(ii) Equation (5.103) expresses in a compact way, and at the infinitesimal level, precisely the contents of the previous Remark, namely that tangent lifts send second-order vector fields into second-order ones.

Proof The proof of the first property is immediate using the local coordinate expressions of X^v and Y^v. Remark also that X^c and Y^c are π-related with X and Y, respectively, and then $[X^c, Y^c]$ is π-projectable onto $[X, Y]$. Moreover, using the local coordinate expressions of X^c and Y^c we see that

$$
\begin{aligned}
[X^c, Y^c] &= \left[X^i \frac{\partial}{\partial q^i} + \frac{\partial X^i}{\partial q^j} v^j \frac{\partial}{\partial v^i},\, Y^k \frac{\partial}{\partial q^k} + \frac{\partial Y^k}{\partial q^l} v^l \frac{\partial}{\partial v^k}\right]\\
&= \left(X^k \frac{\partial Y^i}{\partial q^k} - Y^k \frac{\partial X^i}{\partial q^k}\right)\frac{\partial}{\partial q^i} + \left[X^c\left(\frac{\partial Y^k}{\partial q^l} v^l\right) - Y^c\left(\frac{\partial X^k}{\partial q^l} v^l\right)\right]\frac{\partial}{\partial v^k}
\end{aligned}
$$

which can be rewritten as

$$[X^c, Y^c] = [X, Y]^i \frac{\partial}{\partial q^i} + \left(X^c \mathcal{L}_\Gamma Y^k - Y^c \mathcal{L}_\Gamma X^k\right) \frac{\partial}{\partial v^k}$$

for any second-order differential equation vector field Γ. Now, using the above mentioned property, this can be written as

$$[X^c, Y^c] = [X, Y]^i \frac{\partial}{\partial q^i} + \mathcal{L}_\Gamma (XY^k - YX^k) \frac{\partial}{\partial v^k} = [X, Y]^i \frac{\partial}{\partial q^i} + \mathcal{L}_\Gamma([X, Y]^k) \frac{\partial}{\partial v^k}$$

i.e., $[X^c, Y^c] = [X, Y]^c$.

Finally the proof of the last property is trivial, because

$$\left[X^i \frac{\partial}{\partial v^i}, Y^k \frac{\partial}{\partial q^k} + \frac{\partial Y^k}{\partial q^l} v^l \frac{\partial}{\partial v^k}\right] = X^i \frac{\partial Y^k}{\partial q^i} \frac{\partial}{\partial v^k} - Y^j \frac{\partial X^i}{\partial q^j} \frac{\partial}{\partial v^i} = [X(Y^k) - Y(X^k)] \frac{\partial}{\partial v^k},$$

which shows that $[X^v, Y^c] = [X, Y]^v$. □

It is clear that vertical and tangent lifts of a basis of vector fields in $\mathfrak{X}(M)$ are clearly a local basis for $\mathfrak{X}(TM)$.

5.5.2.3 The Vertical Endomorphism Again

We shall discuss here in more detail the vertical endomorphism S, that we have already introduced in Sect. 5.5.1. On general grounds, a $(1, 1)$ tensor field is a fibrewise linear map $T\colon TM \to TM$, and it induces a dual bundle map $T^*\colon T^*M \to T^*M$ via

$$\langle TX, \alpha\rangle = \langle X, T^*\alpha\rangle =: T(X, \alpha)\,, \tag{5.105}$$

where $\langle\cdot,\cdot\rangle$ denotes the natural (point-wise) pairing between vector fields and 1-forms. If there is no risk of confusion we will denote both bundle maps with the same symbol T.

The vertical endomorphism S is the $(1, 1)$-tensor field in TM defined in intrinsic terms (see e.g., [Mo91]) from the action on vector fields as $S = v \circ T\pi$, where v is the map associating with any vector field on the base manifold its vertical lift and $T\pi$ is the tangent map of the canonical projection $\pi : TM \to M$.

In local tangent bundle coordinates,

$$S = \frac{\partial}{\partial v^i} \otimes dq^i\,. \tag{5.106}$$

Note that by definition it vanishes on vertical vector field and also

$$S\left(\frac{\partial}{\partial q^i}\right) = \frac{\partial}{\partial v^i}$$

from where the local expression of S is obtained.

Another coordinate-free characterization of the vertical endomorphism is that the vertical endomorphism S is the unique $(1, 1)$-tensor field in TM satisfying

$$S(\mathcal{L}_\Gamma \pi^* df) = \pi^* df, \tag{5.107}$$

$\forall f \in \mathcal{F}(M)$ and for any SODE Γ, because in particular for $f = q^i$ this condition means

$$S(\mathcal{L}_\Gamma \pi^* dq^i) = S(dv^i) = dq^i,$$

and from

$$\mathcal{L}_\Gamma \pi^* df = v^j \left[\pi^* \left(\frac{\partial^2 f}{\partial q^i \partial q^j}\right)\right] dq^i + \left[\pi^* \left(\frac{\partial f}{\partial q^i}\right)\right] di^i$$

the assumed condition implies $S(dq^i) = 0$, from the local-coordinate form follows.

We explicitly mention here some of the most important properties of the vertical endomorphism, namely:

$$\begin{aligned} &S^2 = 0\,, \\ &\operatorname{Im} S = \ker S = \mathfrak{X}^v(TM)\,, \\ &\Gamma \in \mathfrak{X}(TM) \text{ is a SODE iff } S(\Gamma) = \Delta, \end{aligned} \tag{5.108}$$

where Δ is the Liouville field.

As another remarkable property, both vertical and tangent lifts have the properties of leaving the vertical endomorphism unchanged, i.e., for any vector field $Z \in \mathfrak{X}(TM)$, such that there exists $X \in \mathfrak{X}(M)$ and $Z = X^v$, we have that $\mathcal{L}_Z S = 0$ and similarly, if $Z = X^c$ also $\mathcal{L}_Z S = 0$.

Of course, if $Z = X^v$, then $\mathcal{L}_Z(\partial/\partial v^i) = 0$ and $\mathcal{L}_Z(dq^i) = d(Zq^i) = 0$. Similarly, if $X = f^i(q)\partial/\partial q^i$ we have that

$$\mathcal{L}_Z\left(\frac{\partial}{\partial v^i}\right) = \left[X^c, \frac{\partial}{\partial v^i}\right] = -\frac{\partial f^j}{\partial q^i}\frac{\partial}{\partial v^j}, \qquad \mathcal{L}_Z(dq^i) = df^i,$$

from where

$$\mathcal{L}_Z S = \mathcal{L}_Z\left(\frac{\partial}{\partial v^i}\right) \otimes dq^i + \frac{\partial}{\partial v^i} \otimes df^i = 0.$$

5.5.2.4 Exterior Derivatives Associated with (1, 1) Tensors

Tensors of (1, 1)-type have been defined through their action on vector fields and 1-forms. Here we want to discuss the extension of their action to higher-order forms.

As discussed in the literature (see e.g., [Mo91]), there are various possible extensions of the action of a tensor $T \in \mathcal{T}_1^1(TM)$ mapping forms of a given order onto forms of the same order. They all share the fact that the resulting action 'goes through' the coefficients of any form. It is therefore natural, to begin with, to define the action of T on any function $f \in \mathcal{F}(TM)$ (a 0-form) as $Tf = 0$.

Now we can give the following:

Definition 5.29 With any tensor field $T \in \mathcal{T}_1^1(TM)$ we can associate a derivation δ_T of degree zero on the graded algebra of forms $\Omega(TM)$ whose action on 0- and 1-forms is defined as

$$\begin{aligned} \delta_T f &= 0\,, & \forall f \in \mathcal{F}(TM)\,, \\ \delta_T \theta &= T\theta, & \forall \theta \in \Omega^1(TM)\,. \end{aligned} \tag{5.109}$$

We recall (see e.g., [ChB82]) that graded derivations on forms are entirely determined by their action on 0- and 1-forms. Given that, one can prove at once that:

Proposition 5.30 *If $\omega \in \Omega^k(TM)$, $1 \leq k \leq 2n$, then:*

$$\begin{aligned} \delta_T\omega(X_1, X_2, \ldots, X_k) =& \,\omega(TX_1, X_2, \ldots, X_k) + \omega(X_1, TX_2, \ldots, X_k) \\ &+ + \cdots + \omega(X_1, X_2, \ldots, TX_k)\,. \end{aligned}$$

This is the extension of the action of T on higher order forms we will concentrate on here. Notice that, if $T = \mathbb{I}$ (the identity tensor), then $\delta_{\mathbb{I}}$ will act simply by multiplying a form by its degree.

The exterior derivative d is an derivation of degree one in $\Omega(TM)$. A known result states that the graded commutator of a derivation of degree p and a derivation of degree q is a derivation of degree $p+q$ [ChB82]. We can then state the following:

Definition 5.31 The operation d_T on the algebra of forms defined as

$$d_T =: \delta_T \circ d - d \circ \delta_T \tag{5.110}$$

is a derivation of degree one that will be called the exterior derivative associated with the (1, 1)-type tensor T.

The action of d_T on 0- and 1-forms is easily seen to be given explicitly by

$$\begin{aligned} d_T f &= Tdf \\ d_T\theta(X, Y) &= (\mathcal{L}_{TX}\theta)(Y) - (\mathcal{L}_{TY}\theta)(X) + \theta(T[X, Y]) \end{aligned} \tag{5.111}$$

The first one of these equations is an immediate consequence of the definition of d_T. The second one can be proved using either one of the identities

$$
\begin{aligned}
d\theta(X, Y) &= \mathcal{L}_X(\theta(Y)) - \mathcal{L}_Y(\theta(X)) - \theta([X, Y]) \,, \\
d\theta(X, Y) &= (\mathcal{L}_X\theta)(Y)) - (\mathcal{L}_Y\theta)(X)) + \theta([X, Y]),
\end{aligned} \tag{5.112}
$$

so we will omit details. In particular, it is not hard to check that, if $T = \mathbb{I}$, then $d_{\mathbb{I}} \equiv d$.

Example 5.6 It can be useful to have the explicit expression of d_T in local coordinates. To this effect, let us introduce the collective coordinates ξ^i, $i = 1, \ldots, 2n$, with

$$
\xi^i = q^i, \qquad \xi^{i+n} = v^i, \quad i = 1, \ldots, n \,. \tag{5.113}
$$

Then the tensor T will be represented as

$$
T = T^i{}_j \frac{\partial}{\partial \xi^i} \otimes d\xi^j \,, \quad T^i{}_j \in \mathcal{F}(TM) \,, \tag{5.114}
$$

and

$$
d_T f = \frac{\partial f}{\partial \xi^k} T^k{}_j d\xi^j \,, \tag{5.115}
$$

while, with $\theta = \theta_i \, d\xi^i$,

$$
d_T\theta = \frac{1}{2} \left(\frac{\partial \theta_j}{\partial \xi^k} T^k{}_i + \theta_k \frac{\partial T^k{}_i}{\partial \xi^j} - (i \leftrightarrow j) \right) d\xi^i \wedge d\xi^j \,. \tag{5.116}
$$

In particular, for $T = S = (\partial/\partial v^i) \otimes dq^i$, $S^i{}_j = \delta_j^{i-n}$ for $n < i \le 2n$, and zero otherwise, and we find, going back from the ξ's to the q's and v's,

$$
\begin{aligned}
d_S f &= \frac{\partial f}{\partial v^i} dq^i \,, \\
d_S \theta &= \frac{\partial \alpha_j}{\partial v^i} dq^i \wedge dq^j + \frac{\partial \beta_j}{\partial v^i} dq^i \wedge dv^j \,,
\end{aligned} \tag{5.117}
$$

if $\theta = \alpha_i \, dq^i + \beta_i \, dv^i$.

A relevant property of the exterior derivative associated with T, with an easy proof, is contained in the following:

Proposition 5.32 *The degree one derivation d_T commutes with the exterior derivative d and the graded commutator of i_X and d_T satisfies the generalized Cartan identity,*

$$
i_X \circ d_T + d_T \circ i_X = \mathcal{L}_{TX} + [\delta_T, \mathcal{L}_X] \,, \tag{5.118}
$$

which reduces to the standard Cartan identity when $T = \mathbb{I}$. Moreover,

$$\mathcal{L}_X \circ d_T - d_T \circ \mathcal{L}_X = d_{\mathcal{L}_X T} . \tag{5.119}$$

5.5.2.5 Cohomology of d_T and Nijenhuis Tensors

Given the exterior derivative d_T, we may enquire under which conditions we can associate a DeRham-type cohomology with d_T for a general $(1, 1)$-type tensor T. In an essentially equivalent way, we may enquire: (i) under which conditions the (coboundary) property $d_T^2 = 0$ holds, and (ii) whether or not a Poincaré-type lemma holds.

First of all, let us remark that it follows at once from the definition (and from $d^2 = 0$) that

$$d \circ d_T + d_T \circ d = 0 , \tag{5.120}$$

and hence that

$$d_T^2 \circ d = d \circ d_T^2 \tag{5.121}$$

i.e., that d_T^2 commutes with the external derivative. As d_T is a degree-one derivation, d_T^2 is a (degree-two) derivation commuting with d. Therefore its action on forms is entirely determined by that on functions [ChB82]. A long but otherwise straightforward algebra leads then to

$$(d_T^2 f)(X, Y) = df(N_T(X, Y)) , \tag{5.122}$$

where the Nijenhuis tensor associated with the tensor field $T \in \mathcal{T}_1^1(TM)$ is the $(1, 2)$-type tensor N_T given by (see Appendix E):

$$N_T(X, Y) =: [TX, TY] + \widehat{T^2}[X, Y] - T[TX, Y] + \widehat{T}[TY, X] . \tag{5.123}$$

It can be useful to have the explicit expression of the Nijenhuis tensor in local coordinates. Introducing again collective coordinates it is a long but straightforward exercise to show that

$$N_T = \frac{1}{2}(N_T)^i_{km} \frac{\partial}{\partial \xi^i} \otimes d\xi^k \wedge d\xi^m , \tag{5.124}$$

where

$$(N_T)^i_{km} = \frac{\partial T^i{}_m}{\partial \xi^j} T^j{}_k + \frac{\partial T^j{}_k}{\partial \xi^m} T^i{}_j - (k \leftrightarrow m) . \tag{5.125}$$

So, $N_T = 0$ whenever the components of T are constant. This will be the case in particular for the vertical endomorphism and, therefore, $N_S = 0$, for the vertical endomorphism. One could also use Eq. (5.117) to prove at once that $d_S^2 f = (\partial^2 f/\partial v^i \partial v^j) dq^i \wedge dq^j = 0$.

Coming back to the problem posed at the beginning of this section, we can state:

Proposition 5.33 *The exterior derivative d_T associated with a $(1, 1)$-type tensor T satisfies the coboundary property $d_T \circ d_T = 0$ iff T satisfies the Nijenhuis condition:*

$$N_T = 0 \tag{5.126}$$

Definition 5.34 Tensors satisfying the Nijenhuis condition will be said to be of the Nijenhuis type.

It follows then that if T is of the Nijenhuis type, then 'T-exact' forms, i.e., forms of the type $\alpha = d_T \beta$ for some β, are also 'T-closed', i.e. $d_T \alpha = 0$. The converse property, i.e. that $d_T \alpha = 0$ implies that there exists a β such that $\alpha = d_T \beta$, at least locally), i.e., the analog of the Poincaŕe Lemma, requires however the Nijenhuis type tensor T to be invertible.

Remark 5.6 The invertibility condition is not simply a technical requirement for the construction of the proof, as the following example shows. The vertical endomorphism S is of the Nijenhuis type but, as $S^2 = 0$, it is not invertible. Consider then, e.g., the 2-form

$$\alpha = \frac{1}{2} \alpha_{ij} dv^i \wedge dv^j, \tag{5.127}$$

with $\alpha_{ij} + \alpha_{ji} = 0$, $\alpha_{ij} \in \mathbb{R}$. As $d_S dv^i = 0$, then $d_S \alpha = 0$, but $\alpha = d_S \theta$ with θ a 1-form fails to be true even locally. For instance, $d_S \theta(\partial/\partial v^i, \partial/\partial v^j) = 0$ for any 1-form θ, while $\alpha(\partial/\partial v^i, \partial/\partial v^j) = \alpha_{ij}$.

5.5.2.6 Horizontal Lifts Associated with Second-Order Vector Fields: Nonlinear Linear Connections on *TM*

Given a local basis $X_1, \ldots, X_n$ of vector fields on M, their vertical lifts $X_1^v, \ldots, X_n^v$, yield a holonomic local basis for $\mathfrak{X}^v(TM)$. We recall that 'holonomic' means that the vector fields pairwise commute. Notice that the correspondence $X \in \mathfrak{X}(M) \mapsto X^v \in \mathfrak{X}^v(TM)$ is $\mathcal{F}(M)$-linear, i.e., if $X \mapsto X^v$, then $f X \mapsto (f X)^v = f X^v$, $\forall f \in \mathcal{F}(M)$.

In the absence of additional structures, there are many ways of finding, at every point of TM, an n-dimensional subspace, transversal to the vertical one, to supplement the latter. Any such subspace will have the right to be called a 'horizontal space', and the choice of such a subspace in each point of TM is called a 'horizontal distribution'.

We have already a possible candidate for a supplementary subspace at every point, namely that one provided by the tangent lifts $X_1^c, \ldots, X_n^c$. However, as it is easy to check that $(fX)^c = fX^c + (\mathcal{L}_\Gamma f) X^v$, for any $X \in \mathfrak{X}(M)$ and $f \in \mathcal{F}(M)$, and then this association is not $\mathcal{F}(M)$-linear.

A different (and intrinsic) way to find an $\mathcal{F}(M)$ -linear distribution, is found by singling out a given (but otherwise arbitrary) second-order field Γ. A preliminary result to this effect is contained in the following:

Proposition 5.35 *If* $X \in \mathfrak{X}(M)$, Γ *is a SODE and S is the vertical endomorphism, then,*

$$S[X^v, \Gamma] = X^v\,. \tag{5.128}$$

Indeed, from $S(\Gamma) = \Delta$ and $\mathcal{L}_{X^v}S = 0$, we obtain $S[X^v, \Gamma] = S(\mathcal{L}_{X^v}\Gamma) = \mathcal{L}_{X^v}(S\Gamma) = [X^v, \Delta] = X^v$, the last passage following from vertical lifts being homogeneous of degree -1 in the fibre coordinates. Notice that this result could have been inferred also directly from Eq. (5.104).

Now we state (and will justify in a moment) the following:

Definition 5.36 The horizontal lift of a vector field $X \in \mathfrak{X}(M)$ associated with a given second-order vector field Γ is the vector field

$$X^h =: \frac{1}{2}\{[X^v, \Gamma] + X^c\} \in \mathfrak{X}(TM)\,. \tag{5.129}$$

Proposition 5.37 *The horizontal lift* X^h *satisfies (and is entirely characterized by)*

$$\mathcal{L}_{X^h}(\pi^* f) = \pi^* \mathcal{L}_X f\,, \qquad f \in \mathcal{F}(M)\,, \tag{5.130}$$

and

$$\mathcal{L}_{X^h}(\mathcal{L}_\Gamma \pi^* f) = \frac{1}{2}\mathcal{L}_{X^v}(\mathcal{L}_\Gamma \mathcal{L}_\Gamma \pi^* f)\,. \tag{5.131}$$

It follows from this result and the fact that the association $X \in \mathfrak{X}(M) \mapsto X^v \in \mathfrak{X}^v(TM)$ $X \mapsto X^v$ is $\mathcal{F}(M)$-linear, i.e., that $(fX)^h = fX^h$, that the correspondence $X \in \mathfrak{X}(M) \mapsto X^h \in \mathfrak{X}(TM)$ is $\mathcal{F}(M)$-linear as well. Moreover, using the definition of X^h and (5.128) we see that the horizontal lift satisfies

$$S(X^h) = X^v\,, \tag{5.132}$$

which follows from $S(X^c) = X^v$.

Remark that if the local expression of the given second-order vector field Γ in natural coordinates (q^i, v^i) on the tangent bundle, is

$$\Gamma = v^i \frac{\partial}{\partial q^i} + F^i(q, v)\frac{\partial}{\partial v^i}\,,$$

then, having in mind that

$$\mathcal{L}_\Gamma \frac{\partial}{\partial v^i} = -\frac{\partial}{\partial q^i} - \frac{\partial F^j}{\partial v^i}\frac{\partial}{\partial v^j}\,,$$

and $\mathcal{L}_\Gamma dq^i = dv^i$, we find that

$$\mathcal{L}_\Gamma S = -\frac{\partial}{\partial q^i} \otimes dq^i - \frac{\partial F^j}{\partial v^i} \frac{\partial}{\partial v^j} \otimes dq^i + \frac{\partial}{\partial v^i} \otimes dv^i ,$$

i.e., the matrix representing $\mathcal{L}_\Gamma S$ in the basis $\{\partial/\partial q^i, \partial/\partial v^i \mid i = 1, \ldots, n\}$ is

$$\begin{pmatrix} -\delta^i_j & 0 \\ -\frac{\partial F^i}{\partial v^j} & \delta^i_j \end{pmatrix} ,$$

and then $(\mathcal{L}_\Gamma S)^2 = I$. Therefore it allows us to construct two projection operators, the first one $P_V = \frac{1}{2}(I + \mathcal{L}_\Gamma S)$, with range in the vertical subbundle, and the second one $P_H = I - P_V = \frac{1}{2}(I - \mathcal{L}_\Gamma S)$. They give us a splitting of the tangent space at each point, because $P_H + P_V = I$ and $P_H P_V = P_V P_H = 0$, providing the abovementioned generalized nonlinear connection.

The explicit expression of the projectors are

$$P_V = \begin{pmatrix} 0 & 0 \\ -\frac{1}{2}\frac{\partial F^i}{\partial v^j} & \delta^i_j \end{pmatrix} , \quad P_H = \begin{pmatrix} \delta^i_j & 0 \\ \frac{1}{2}\frac{\partial F^i}{\partial v^j} & 0 \end{pmatrix} ,$$

and then the coefficients of the associated nonlinear connection are given by

$$\Gamma^i_j = -\frac{1}{2}\frac{\partial F^i}{\partial v^j} .$$

In local coordinates, if $X = X^i(q)\, \partial/\partial q^i$ and $\Gamma = v^i\, \partial/\partial q^i + F^i\, \partial/\partial v^i$, then,

$$X^h = (\pi^* X^i) H_i , \tag{5.133}$$

where the 'horizontal basis' H_i is given by

$$H_i = \left(\frac{\partial}{\partial q^i}\right)^h = \frac{\partial}{\partial q^i} - \Gamma^j_i \frac{\partial}{\partial v^j} = \frac{\partial}{\partial q^i} + \frac{1}{2}\frac{\partial F^k}{\partial v^i}\frac{\partial}{\partial v^k} . \tag{5.134}$$

The local basis for the linear space of vector fields on TQ adapted to this connection is given by the previous vector fields and

$$V_i = \frac{\partial}{\partial v^i} ,$$

the corresponding dual basis for $\Omega^1(TQ)$ being

$$H^i = dq^i \qquad V^i = dv^i + \Gamma^i_j\, dq^j = dv^i - \frac{1}{2}\frac{\partial F^i}{\partial v^j}\, dq^j .$$

We will throughout use this basis. Of course, if the second-order vector field Γ is a spray, the connection so defined is linear.

Let us take now, in natural coordinates, the $\partial/\partial v^i$'s and the H_i's as a basis. Recall that $\partial/\partial v^i = (\partial/\partial q^i)^v$ and that $H_i = (\partial/\partial q^i)^h$. A corresponding dual basis of 1-forms will be given by forms $H^i = dq^i$, $V^i =$, $i = 1, \ldots, n$ such that

$$\left\langle V^i \middle| \frac{\partial}{\partial v^j} \right\rangle = \left\langle V^i \middle| H_j \right\rangle = \delta^i_j, \quad i, j = 1, \ldots, n, \tag{5.135}$$

and

$$\left\langle V^i | H_j \right\rangle = \left\langle H^i | \frac{\partial}{\partial v^j} \right\rangle = 0, \qquad i, j = 1, \ldots, n. \tag{5.136}$$

It will be also useful to have at our disposal the explicit form of the Lie brackets among the vectors of the basis. A long but straightforward algebra yields then the following:

Proposition 5.38 *The vertical and horizontal bases satisfy the commutation rules*:

$$\left[\frac{\partial}{\partial v^i}, \frac{\partial}{\partial v^j} \right] = 0,$$

$$\left[\frac{\partial}{\partial v^i}, H_j \right] = \frac{1}{2} \frac{\partial^2 F^k}{\partial v^i \partial v^j} \frac{\partial}{\partial v^k}, \tag{5.137}$$

$$\left[H_i, H_j \right] = -R^m_{ij} \frac{\partial}{\partial v^m},$$

where

$$R^m_{ij} = -\frac{1}{2} \left(\frac{\partial^2 F^m}{\partial q^i \partial v^j} - \frac{\partial^2 F^m}{\partial v^i \partial q^j} + \frac{1}{2} \left[\frac{\partial^2 F^m}{\partial v^j \partial v^l} \frac{\partial F^l}{\partial v^i} - \frac{\partial^2 F^m}{\partial v^i \partial v^{ql}} \frac{\partial F^l}{\partial uj} \right] \right). \tag{5.138}$$

Remark 5.7 (i) It follows from the previous result that, if $X = X^i(q)\partial/\partial q^i$ and $Y = Y^i(q)\partial/\partial q^i$, then,

$$[X^h, Y^h] - [X, Y]^h = -X^i Y^j R^m_{ij} \frac{\partial}{\partial v^m}, \tag{5.139}$$

i.e., that the horizontal distribution fails to be involutive (and a Lie algebra homomorphism) to the extent that the $(1, 2)$-type tensor

$$R = \frac{1}{2} R^m_{ij} \frac{\partial}{\partial v^m} \otimes dq^i \wedge dq^j \tag{5.140}$$

is nonzero.

(ii) It follows also from (5.139) and (5.132) that

$$S[X^h, Y^h] = [X, Y]^v \tag{5.141}$$

(iii) It is also easy to show that

$$\Gamma = v^i H_i + \frac{1}{2}(\Gamma - [\Delta, \Gamma]), \tag{5.142}$$

i.e., Γ will be a horizontal field iff it is a spray ($\Gamma = [\Delta, \Gamma]$).

(iv) Given a vector field $X \in \mathfrak{X}(M)$, we can define the 'horizontal lifts' of its integral curves as the integral curves of the horizontal lift X^h. Explicitly, if $X = X^i \partial/\partial q^i$, the horizontal lifts will be solutions of the equations:

$$\begin{cases} \dfrac{dq^i}{dt} = X^i \\ \dfrac{dv^i}{dt} = \dfrac{1}{2} X^k \dfrac{\partial F^i}{\partial v^k}, \end{cases} \tag{5.143}$$

with the appropriate initial conditions.

5.5.2.7 Nonlinear Connections and SODE's

In order to make contact with a somewhat more familiar context, let us assume that a Riemannian (or pseudo-Riemannian) metric

$$g = g_{ij}\, dq^i \otimes dq^j \tag{5.144}$$

has been given on M, and consider the second-order vector field:

$$\begin{aligned} \Gamma &= v^i \frac{\partial}{\partial q^i} + F^i \frac{\partial}{\partial v^i}, \\ F^i &= -\Gamma^i_{km} v^k v^m, \\ \Gamma^i_{km} &= \frac{1}{2} g^{ij} \left(\frac{\partial g_{jk}}{\partial q^m} + \frac{\partial g_{jm}}{\partial q^k} - \frac{\partial g_{km}}{\partial q^j} \right), \end{aligned} \tag{5.145}$$

associated with g. The Γ^i_{km}'s are of course the familiar Christoffel symbols defining the Levi-Civita connection, and the second-order (in M) equations of motion associated with Γ are

$$\frac{d^2 q^i}{dt^2} + \Gamma^i_{km} v^k v^m = 0 \tag{5.146}$$

i.e., Γ describes geodesic motion on M. Notice that: (i) Γ is a spray ($[\Delta, \Gamma] = \Gamma$) and (ii) if ∇ (see the Appendix E on connections and covariant derivatives) denotes the covariant derivative associated with the Levi-Civita connection, then,

$$\frac{d^2q^i}{dt^2} + \Gamma^i_{km} v^k v^m = 0 \Longleftrightarrow \nabla_{\dot{\gamma}} \dot{\gamma} = 0 , \tag{5.147}$$

i.e., the equations for geodesic motion γ express just the condition of vanishing of such covariant derivative. It will be seen in the next section that the equations of motion associated with Γ are precisely the Euler-Lagrange equations that can be derived from the Lagrangian $L = \frac{1}{2} g_{ij} v^i v^j \equiv \frac{1}{2} (\pi^* g)(\Delta, \Delta) \in \mathcal{F}(TM)$.

The horizontal distribution associated with Γ is easily found to be

$$H_i = \frac{\partial}{\partial q^i} - \Gamma^k_{im} v^m \frac{\partial}{\partial v^k} . \tag{5.148}$$

Also,

$$\theta^i = dv^i + \Gamma^i_{jm} v^m dq^j , \tag{5.149}$$

and

$$R^m_{ij} = \widetilde{R}^m_{kij} v^k , \tag{5.150}$$

where $\widetilde{R}$ is the Riemann curvature tensor of the Levi-Civita connection.

So far for linear connections associated with a metric tensor. Recognizing however that the basic concept involved in the definition of a connection is that of a lifting procedure for curves, with the ensuing prescription for parallel transport, then we can conclude that the association of a horizontal distribution with a given second-order field Γ (not a spray in general) will define for us a 'nonlinear' connection on M with its associated procedure of parallel transport. We refer to the literature [Cr83, Cr87] for further developments concerning this specific point.

5.5.3 Lagrangian Dynamics

Up to now we have basically studied the geometry of the tangent bundle structure that can be attributed to a given vector space V ($V \cong TM$, then). Now we begin a systematic description of Lagragian dynamics on TM.

5.5.3.1 Lagrangian one- and Two-Forms: Regular and Singular Lagrangians

Let $L \in \mathcal{F}(TM)$ be a smooth function on TM. L will be called a 'Lagrangian' from now on. Then we define the Lagrangian, or Cartan, 1-form θ_L associated with L as the (semi-basic) 1-form

$$\theta_L =: d_S L . \tag{5.151}$$

We recall that a 1-form α is 'semibasic' iff $S^*\alpha = 0$, i.e., iff it is of the local form $\alpha = \alpha_i\, dq^i$, with the α_i's functions on TM, in general. Such a 1-form α will be, in particular, a 'basic' form iff $\alpha = \pi^*\beta$, with $\beta \in \Omega^1(M)$, i.e., iff the α_i's are functions of the q's alone. Semibasic forms are also called 'horizontal'. The 2-form

$$\omega_L = -d\theta_L = -d_S dL \tag{5.152}$$

will be called the Lagrangian 2-form associated with L. The function L will be said to be a 'regular Lagrangian' iff ω_L is nondegenerate, hence a symplectic form on TM, and 'singular' otherwise.

In local coordinates (q^i, v^i), we have the explicit expressions

$$\theta_L = \frac{\partial L}{\partial v^i}\, dq^i\,, \tag{5.153}$$

and

$$\omega_L = \frac{\partial^2 L}{\partial v^i \partial v^j} dq^i \wedge dv^j + \frac{1}{2}\left(\frac{\partial^2 L}{\partial v^i \partial q^j} - \frac{\partial^2 L}{\partial q^i \partial v^j}\right) dq^i \wedge dq^j\,. \tag{5.154}$$

If X is a vector field on TM having the local expression $X = A^i\partial/\partial q^i + B^i\partial/\partial v^i$, then,

$$i_X\omega_L = \frac{\partial^2 L}{\partial v^i \partial v^j} A^i dv^j + \left[\left(\frac{\partial^2 L}{\partial v^i \partial q^j} - \frac{\partial^2 L}{\partial q^i \partial v^j}\right) A^i - \frac{\partial^2 L}{\partial v^i \partial v^j} B^i\right] dq^j\,. \tag{5.155}$$

Therefore, ω_L will be non-degenerate (i.e., $i_X\omega_L = 0$ implies that $X = 0$) iff the 'Hessian matrix'

$$H_{ij} =: \frac{\partial^2 L}{\partial v^i \partial v^j} \tag{5.156}$$

is of maximum rank (i.e., equals to dim M).

Remark 5.8 Although formulated in local language, the non-degeneracy condition is clearly an intrinsic one. Note that $\omega_L^{\wedge n} \neq 0$ iff the Hessian matrix is regular.

It is interesting to inquire if the map associating θ_L with L is injective and same for the map associating ω_L to θ_L. As to the first map, since the fact that θ_L vanishes on vertical fields, its kernel is made up clearly of functions that are constant along the fibres, i.e., $\theta_L = 0$ if and only if L is of the form $L = \pi^*g$ for some $g \in \mathcal{F}(M)$. As to the second, ω_L will be zero if the semi-basic 1-forms θ_L is closed. This can be only satisfied if θ_L is a basic form. Hence, $\omega_L = 0$ iff $\theta_L = \pi^*\alpha$, with $\alpha \in \Omega^1(M)$ and $d\alpha = 0$. But this implies that L be of the form $L = i_\Gamma\pi^*\alpha$ with Γ any second-order field. Locally (and also globally if M is a vector space) this means that L has to be a 'total time derivative', $L = df/dt$ for some $f \in \mathcal{F}(M)$.

All in all, we have proved that the most general Lagrangian that makes the associated Lagrangian 2-form to vanish identically has the general expression

$$L = \pi^* g + i_\Gamma \pi^* \alpha \,, \qquad g \in \mathcal{F}(M), \quad \alpha \in \Omega^1(M), \; d\alpha = 0 \,, \tag{5.157}$$

i.e., it is the sum of a 'pure potential term' ($\pi^* g$) and of what is often called a 'gauge term', the 'total time derivative' term $i_\Gamma \pi^* \alpha$. This can be rewritten in a different way, because a 1-form $\alpha \in \Omega^1(M)$ defines a function $\widehat{\alpha} : TM \to \mathbb{R}$ as follows $\widehat{\alpha}(v) = \langle \alpha_{\pi(v)}, v \rangle$, and then $i_\Gamma \pi^* \alpha = \widehat{\alpha}$.

Let $Y = C^i \partial/\partial q^i + D^i \partial/\partial v^i$ be a vector field in $\mathfrak{X}(TM)$. Then, using Eq. (5.155) and the definition of the vertical endomorphism, one proves immediately, in coordinates, that $\omega_L(X, SY) = -\omega_L(SX, Y) = (\partial^2 L/\partial v^i \partial v^j) A^i C^j$. Hence,

$$\omega_L(X, SY) + \omega_L(SX, Y) = 0 \,, \; \forall X, Y,$$

and, remembering the definition (see subsection 2.4) of the zeroth-order derivation associated with a (1, 1)-type tensor, we have found that any Lagrangian 2-form ω_L satisfies

$$\delta_S \omega_L = 0 \,, \tag{5.158}$$

and, as a consequence of $S^2 = 0$,

$$\omega_L(SX, SY) = 0 \,, \qquad \forall X, Y \in \mathfrak{X}(TM) \,, \tag{5.159}$$

i.e., ω_L will vanish on any pair of vertical fields.

Recall that δ_S is a derivation δ_S of degree zero such that

$$(\delta_S \omega_L)(X_1, X_2) = \omega(S(X_1), X_2) + \omega_L(X_1, S(X_2), \ldots, X_k)$$

Noticing that $\omega_L(SX, Y) = i_Y i_{SX} \omega_L$ and $\omega_L(X, SY) = i_{SY} i_X \omega_L \equiv i_Y(S i_X \omega_L)$, this last result can be rewritten as

$$i_{SX} \omega_L + S i_X \omega_L = 0 \,, \qquad \forall X \in \mathfrak{X}(TM) \,. \tag{5.160}$$

5.5.3.2 Conditions Under Which a Closed 2-Form Is a Lagrangian 2-Form

We have seen in the previous Subsection that to every function (let it be regular or not) $L \in \mathcal{F}(TM)$ we can associate (a Cartan 1-form and) a Lagrangian 2-form ω_L which is closed (actually exact) and vanishes on pairs of vertical fields. A relevant question is then under which conditions such a 2-form on TM is actually a 2-form for some $L \in \mathcal{F}(M)$. This is settled by the following [Ca95]:

Theorem 5.39 *Let ω be a closed 2-form, $\omega \in \Omega^2(TM)$. Then, ω will be a Lagrangian 2-form, $\omega = \omega_L$, for some $L \in \mathcal{F}(TM)$ iff*

$$\delta_S \omega = 0\,, \tag{5.161}$$

where S is the vertical endomorphism.

The 'only if' part of the theorem has already been proved. Let us turn then to the 'if' part. Let ω and $X, Y \in \mathfrak{X}(M)$ have the local expressions

$$\begin{aligned} \omega &= \frac{1}{2} F_{ij}\, dq^i \wedge dq^j + G_{ij}\, dq^i \wedge dv^j + \frac{1}{2} H_{ij}\, dv^i \wedge dv^j\,, \\ X &= A^i \frac{\partial}{\partial q^i} + B^i \frac{\partial}{\partial v^i}\,, \\ Y &= C^i \frac{\partial}{\partial q^i} + D^i \frac{\partial}{\partial v^i}\,. \end{aligned} \tag{5.162}$$

Then, $\delta_S \omega(X, Y) = (G_{ij} - G_{ji}) A^i C^j + H_{ij}(B^i C^j + A^i D^j)$, and $\delta_S \omega = 0$ implies then $H_{ij} = 0$ and $G_{ij} = G_{ji}$, and hence,

$$\omega = \frac{1}{2} F_{ij}\, dq^i \wedge dq^j + G_{ij}\, dq^i \wedge dv^j\,. \tag{5.163}$$

Closure of ω implies, by Poincaré's Lemma, that there will exist, locally at least, a 1-form α such that $\omega = -d\alpha$. Such an α will be defined, of course, 'modulo' a closed 1-form. If it has the local expression $\alpha = M_i\, dq^i + N_i\, dv^i$, then $H_{ij} = 0$ implies $\partial N_i/\partial v^j = \partial N_j/\partial v^i$, which is an integrability conditions along the fibres. There will exist therefore a function $f \in \mathcal{F}(TM)$ such that $N_i = \partial f/\partial v^i$. Taking then $\theta = \alpha - df$, θ will be semi-basic, $\theta = A_i\, dq^i$, where $A_i = M_i - \partial f/\partial q^i$. Then, $\omega = -d\theta$ and, more explicitly,

$$\omega = \frac{\partial A_i}{\partial v^j} dq^i \wedge dv^j + \frac{1}{2}\left(\frac{\partial A_i}{\partial q^j} - \frac{\partial A_j}{\partial q^i}\right) dq^i \wedge dq^j\,. \tag{5.164}$$

Symmetry of G_{ij} will imply then $\partial A_i/\partial v^j = \partial A_j/\partial v^i$, which is again an integrability condition along the fibres, implying the existence of a function $L \in \mathcal{F}(TM)$ such that $A_i = -\partial L/\partial v^i$. All in all this leads to

$$\omega = -d\left(\frac{\partial L}{\partial v^i}\, dq^i\right) = \omega_L, \tag{5.165}$$

and this achieves the proof of the theorem.

Remark 5.9 The way the proof of the theorem has been constructed, the use of the Poincaré Lemma and of the integrability conditions indicate that, in general, the proof is only a local one. This is perhaps a purely academic remark if the tangent bundle structure is constructed over a ($2n$-dimensional) vector space V, where there are no topological obstructions and the proof becomes global. However, we would like to stress that even in more general cases the fibres remain vector spaces with no topological obstructions. Therefore the functions B and L will exist (at least) in a 'tubular neighbourhood' of a neighbourhood in the base manifold, and the proof will be 'local' only to this extent.

5.5.3.3 Euler-Lagrange Equations in Hamiltonian and in Standard Form

The Liouville field Δ is vertical and the vertical endomorphism S is homogeneous of degree minus one in in the fibre coordinates, i.e., $\mathcal{L}_\Delta S = -S$. Therefore, as $\theta_L = S(dL)$, we see that

$$\mathcal{L}_\Delta \theta_L = (\mathcal{L}_\Delta S)^\vee (dL) + S(d\mathcal{L}_\Delta L) = \theta_{\Delta(L)} - \theta_L = \theta_{\Delta(L)-L}.$$

The function

$$E_L =: (\mathcal{L}_\Delta - 1)L\,, \tag{5.166}$$

is called, for reasons that will become clear shortly, the 'energy function' associated with the Lagrangian L. Note that, if L is a 'pure potential', and hence it is homogeneous of degree zero in the fibre coordinates, then, $E_L = -L$, while if L is a 'pure gauge', homogeneous of degree one, $E_L \equiv 0$.

On the other side, $i_\Delta \theta_L = 0$ because θ_L is semibasic and the vector field Δ is vertical. Using the homotopy identity, $\mathcal{L}_X = i_X \circ d + d \circ i_X$, valid for any vector field X, for the vector field Δ we see that

$$i_\Delta \omega_L = -i_\Delta d\theta_L = -\mathcal{L}_\Delta \theta_L = -d_S E_L\,. \tag{5.167}$$

Let now Γ be a vector field such that

$$i_\Gamma \omega_L = dE_L\,. \tag{5.168}$$

Then, using Eq. (5.160), we find at once

$$S i_\Gamma \omega_L = -i_{S\Gamma} \omega_L \equiv d_S E_L\,, \tag{5.169}$$

which implies that, if Γ exists at all, $S\Gamma - \Delta$ will be in the kernel of ω_L. If L is regular (and hence ω_L is non-degenerate), Γ will exist, will be unique, and $S\Gamma = \Delta$ will force Γ to be a 'second-order vector field. If L is not regular, there is no guarantee that Eq. (5.168) has any solutions for Γ. This happens, e.g., if L is a 'pure potential' ($\mathcal{L} = \pi^* g$, $g \in \mathcal{F}(\mathcal{M})$), which implies $\omega_L = 0$ but $E_L = -L$. At the other extreme, if L is a 'pure gauge' (loosely speaking, $\mathcal{L} = df/dt$ for some $f \in \mathcal{F}(M)$), then $\omega_L = E_L = 0$ and the search becomes pointless, as any vector field would trivially fit the job. Finally, even if there are solutions, it may well be that among the possible solutions no second-order one can be found.

Let us remark also that any Γ solving Eq. (5.168) will leave ω_L invariant, i.e.

$$\mathcal{L}_\Gamma \omega_L = 0 . \tag{5.170}$$

Then we can summarize the discussion:

(i) Any regular Lagrangian L will define a (its own) symplectic structure on the tangent bundle TM.
(ii) The (unique) second-order field Γ of Eq. (5.168) will be ω_L-Hamiltonian w.r.t. the symplectic structure, with the energy function E_L playing the rôle of the Hamiltonian.

Remark 5.10 (i) If L is of the standard form, also called a Lagrangian 'of mechanical type', $L = T - V$, with T a kinetic term quadratic in the velocities and $V = V(q)$ a potential energy term, then, $E_L = T + V$, and this justifies the name 'energy function' given to E_L.

(ii) At variance with the canonical (and/or Poisson) formalism on cotangent bundles studied in previous chapters, where the symplectic structure is given geometrically once and for all independently from the dynamics, here the Lagrangian defines both the dynamics and the symplectic structure.

(iii) If Γ has the local form $\Gamma = v^i\partial/\partial q^i + F^i\partial/\partial v^i$, it is left as a simple exercise in local coordinates to show that Eq. (5.168) yields, explicitly, the equations

$$\frac{\partial^2 L}{\partial v^i \partial q^j} v^j + F^j \frac{\partial^2 L}{\partial v^i \partial v^j} - \frac{\partial L}{\partial q^i} = 0, \qquad i = 1, \ldots, n \tag{5.171}$$

which, upon using $v^i = dq^i/dt$ and $F^i = dv^i/dt$, are just the familiar Euler-Lagrange equations:

$$\frac{d}{dt}\left(\frac{\partial L}{\partial v^i}\right) - \frac{\partial L}{\partial q^i} = 0 , \qquad i = 1, \ldots, n . \tag{5.172}$$

Let us elaborate now a bit on these equations. With the familiar identification of the Lie derivative w.r.t. a vector field with a total time derivative, they can be rewritten in the equivalent form $\mathcal{L}_\Gamma(\partial L/\partial v^i) - \partial L/\partial q^i = 0$. Multiplying then by dq^i (summing over i) and observing that

$$\mathcal{L}_\Gamma \theta_L = \mathcal{L}_\Gamma \left(\frac{\partial L}{\partial v^i}\right) dq^i + \left(\frac{\partial L}{\partial v^i}\right) dv^i ,$$

we obtain the equivalent set of equations

$$\mathcal{L}_\Gamma \theta_L - dL = 0 . \tag{5.173}$$

This equation as well is written in intrinsic terms, and is what is usually called the 'standard form' of the Euler-Lagrange equations. It is obviously fully equivalent to the Hamiltonian Eq. (5.168). Indeed, $\mathcal{L}_\Gamma \theta_L = -i_\Gamma \omega_L + d i_\Gamma \theta_L$, and $i_\Gamma \theta_L = i_\Gamma d_S L = i_\Gamma S dL = i_{S\Gamma} dL = i_\Delta dL = \mathcal{L}_\Delta L$. Putting things together, we recover Eq. (5.168). With a completely similar set of manipulations one can easily show that Eq. (5.168) leads indeed to Eq. (5.173).

In this way we have obtained the intrinsic (both Hamiltonian and 'standard' or 'Lagrangian') versions of the Euler-Lagrange equations. These are the forms that we will use from now on and until the end of this chapter.

In this as well as in the previous subsection the starting point has been the datum of a Lagrangian function. From that we have, so-to-speak, 'deduced' the associated dynamical vector field Γ which, if the Lagrangian is regular, turns out to be unique and second-order.

To keep however with the spirit of the present book, where the 'primary' object is always assumed to be a dynamics (i.e., a vector field), and to anticipate to some extent the discussion of the Inverse Problem that will be carried on (hopefully) in full detail in Sect. 5.5, we have to invert to some extent the line of reasoning that has led us to Eqs. (5.168) and (5.173). Let then Γ be a dynamical vector field (a dynamical system, for short). Then,

Definition 5.40 The dynamical system Γ on a tangent bundle TM will be said to admit a Lagrangian description with the admissible Lagrangian $L \in \mathcal{F}(TM)$ iff the equation:

$$\mathcal{L}_\Gamma \theta_L - dL = 0\,, \tag{5.174}$$

holds.

We know already that if the Lagrangian is regular, then Γ has to be second-order, so we will concentrate on second-order fields, and hence on dynamical systems of the Newtonian type, from now on.

Not all second-order fields will admit of a Lagrangian description (some specific examples will be discussed in Sect. 5.5) and the latter, if and when it exists, need not be unique. As a simple example, let Γ be the dynamics of an n-dimensional isotropic harmonic oscillator. Setting the frequency and the mass of the oscillator equal to one, Γ is then given by

$$\Gamma = v^i \frac{\partial}{\partial q^i} - q^i \frac{\partial}{\partial v^i}\,, \tag{5.175}$$

and it is a simple exercise to show that any Lagrangian of the form

$$L_B = B_{ij}\{v^i v^j - q^i q^j\}\,, \tag{5.176}$$

with $B = ||B_{ij}||$ being a symmetric $n \times n$ matrix (an invertible one, if we insist that L_B be a regular Lagrangian), will be a perfectly admissible Lagrangian for the isotropic harmonic oscillator.

Remark 5.11 If no further restrictions are imposed on the admissible Lagrangians, then the problem posed in the above definition has always (at least) one trivial solution. Indeed, let L be a 'pure gauge', i.e., $L = i_{\Gamma_0}\pi^*\alpha,\ \alpha \in \Omega^1(M),\ d\alpha = 0$, where Γ_0 is a second-order field. As α is basic, Γ_0 can be any second-order field, and therefore we are free to replace it with Γ itself. Then, $\theta_L = \pi^*\alpha$ and $dL = d(i_\Gamma \pi^*\alpha)$. This leads to $\mathcal{L}_\Gamma \theta_L - dL = i_\Gamma \pi^* d\alpha$, and the closure of α will ensure that L will be actually an admissible Lagrangian for any second-order field Γ. The converse is also true, i.e., a Lagrangian that is admissible for any second-order field is necessarily a pure gauge, but we will not insist on this point (For details, see [Mo91]).

To close this section, we may inquire under which conditions a 2-form ω is a Lagrangian 2-form for a given second-order field $\Gamma = v^i\partial/\partial q^i + F^i\partial/\partial v^i$, i.e., what is the form that Theorem 15 acquires when the dynamics (which is, after all, the 'primary' object for us) is assigned. If it is so, besides being closed, ω has to be invariant under the dynamics, $\mathcal{L}_\Gamma\omega = 0$. Let us show briefly that this condition and the additional condition of vanishing on pairs of vertical fields, $\omega(SX, SY) = 0$, $\forall X, Y$, become then equivalent to the condition $\delta_S\omega = 0$ of Theorem 15. It will be useful to use here the horizontal and vertical basis $(H_i, \partial/\partial v^i)$ and cobasis (dq^i, θ^i) 'adapted' to Γ. Then, in local coordinates, a vector field $X = a^i\,\partial/\partial q^i + b^i\,\partial/\partial v^i$ can be rewritten as $X = a^i H_i + \widetilde{b^i}\,\partial/\partial v^i$, where

$$\widetilde{b^i} = b^i - \frac{1}{2}\,a^j\,\frac{\partial F^i}{\partial v^j}\,.$$

By taking the Lie derivative w.r.t. Γ of the condition $\omega(SX, SY) = 0$ one finds at once that this implies that

$$\omega((\mathcal{L}_\Gamma S)X, SY) + \omega(SX, (\mathcal{L}_\Gamma S)Y) = 0\,. \tag{5.177}$$

In the adapted bases,

$$\mathcal{L}_\Gamma S = \theta^i \otimes \frac{\partial}{\partial v^i} - dq^i \otimes H_i \tag{5.178}$$

It is then a long but straightforward exercise in coordinates to show that, if $Y = c^i\partial/\partial q^i + f^i\partial/\partial v^i$, then,

$$\omega((\mathcal{L}_\Gamma S)X, SY) + \omega(SX, (\mathcal{L}_\Gamma S)Y) = (a^if^j - a^jf^i)\omega\left(\frac{\partial}{\partial v^i}, H_j\right)\,,$$

and therefore,

$$\omega((\mathcal{L}_\Gamma S)X, SY) + \omega(SX, (\mathcal{L}_\Gamma S)Y) \equiv \omega(SX, Y) + \omega(X, SY) \tag{5.179}$$

and this shows that an equivalent set of assumptions is:

(i) ω is closed and $\mathcal{L}_\Gamma\omega = 0$.

(ii) ω vanishes on pairs of vertical fields, $\omega(SX, SY) = 0, \forall X, Y \in \mathfrak{X}(TM)$.

5.5.3.4 Point Transformations

Point-transformations of *TM*, i.e., transformations that are tangent lifts of transformations on the base manifold, and that are represented, at the infinitesimal level, by tangent lifts of vector fields in $\mathfrak{X}(M)$, play a privileged rôle in the framework of Lagrangian dynamics. This is due to the fact that, as stated in almost all standard textbooks on Classical Dynamics, the Euler-Lagrange equations are 'form-covariant' under point-transformations. By this we mean the following:

Consider the Euler-Lagrange equations in the usual form,

$$\frac{d}{dt}\left(\frac{\partial L}{\partial v^i}\right) - \frac{\partial L}{\partial q^i} = 0\,. \tag{5.180}$$

If we perform a point-transformation $(q^i, v^i) \to (\bar{q}^i, \bar{v}^i)$, given by

$$\bar{q}^i = \bar{q}^i(q)\,, \tag{5.181}$$
$$\bar{v}^i = \bar{v}^i(q, u) =: v^j \frac{\partial \bar{q}^i}{\partial q^j}$$

(the last one expressing simply $\bar{v}^i = d\bar{q}^i/dt$, in more conventional language), then, defining

$$\widetilde{L}(\bar{q}, \bar{v}) =: L(q(\bar{q}), v(\bar{q}, \bar{v})) \tag{5.182}$$

one finds, with some simple algebra

$$\frac{d}{dt}\left(\frac{\partial \widetilde{L}}{\partial \bar{v}^i}\right) - \frac{\partial \widetilde{L}}{\partial \bar{q}^i} = \frac{\partial q^j}{\partial \bar{q}^i}\left[\frac{d}{dt}\left(\frac{\partial L}{\partial v^j}\right) - \frac{\partial L}{\partial q^j}\right], \tag{5.183}$$

and this expresses the well-known 'form-covariance' property of the Euler-Lagrange equations with respect to point-transformations.

In a more intrinsic way, denoting by, say, ϕ the map $\phi : (\bar{q}, \bar{v}) \mapsto (q, v)$, then this equation can be rewritten simply as $\widetilde{L} = \phi^* L$.

In a more intrinsic language, let $X^c \in \mathfrak{X}(TM)$ be the tangent lift of a vector field $X \in \mathfrak{X}(M)$. As discussed in a previous Section, X^c will be the infinitesimal generator of a one-parameter group $\{\phi_t\}_{t\in\mathbb{R}}$, $\phi_t \in \text{Diff}\,(TM)$ of point transformations of *TM*. Recall that

$$\mathcal{L}_{X^c} S = 0\,, \tag{5.184}$$

and

$$S([X^c, \Gamma]) = 0\,, \tag{5.185}$$

when X^c is a tangent lift and Γ any second-order field, which states that $[\Gamma, X^c]$ is vertical. This expresses at the infinitesimal level the fact that tangent lifts of diffeomorphisms of the base manifold map second-order fields into second-order ones. Given a Lagrangian function, we have that

$$\mathcal{L}_{X^c}\theta_L = \mathcal{L}_{X^c}(SdL) = S\mathcal{L}_{X^c}dL\ ,$$

and hence,

$$\mathcal{L}_{X^c}\theta_L = \theta_{\mathcal{L}_{X^c}L}\ , \tag{5.186}$$

and

$$\mathcal{L}_{X^c}\omega_L = \omega_{\mathcal{L}_{X^c}L}\ , \tag{5.187}$$

as well. At the finite level this implies that

$$\phi_t^*\theta_L = \theta_{\phi_t^*L}\ , \tag{5.188}$$

as well as

$$\phi_t^*\omega_L = \omega_{\phi_t^*L}\ , \tag{5.189}$$

and this is specific of point transformations. Therefore, taking the pull-back of Eq. (5.173), we obtain

$$\mathcal{L}_{\phi_t^*\Gamma}\theta_{\phi_t^*L} - d(\phi_t^*L) = 0 \tag{5.190}$$

where the transformed field $\phi_t^*\Gamma =: (\phi_t^{-1})_*\Gamma \equiv (\phi_{-t})_*\Gamma$ will be again, as we already know, a second-order field, and will admit of a Lagrangian description in terms of the transformed Lagrangian L. This is the intrinsic way of expressing the form-covariance property of the Euler-Lagrange equations w.r.t. point transformations.

At the infinitesimal level we find, as $\mathcal{L}_{X^c}\mathcal{L}_\Gamma\theta_L = (\mathcal{L}_{[X^c,\Gamma]} + \mathcal{L}_\Gamma\mathcal{L}_{X^c})\theta_L$, the infinitesimal version of the preceding equation

$$\mathcal{L}_{X^c}(\mathcal{L}_\Gamma\theta_L - dL) = \mathcal{L}_\Gamma\theta_{\mathcal{L}_{X^c}L} - d\mathcal{L}_{X^c}L + \mathcal{L}_{[X^c,\Gamma]}\theta_L = 0\ , \tag{5.191}$$

i.e., with $L' =: \mathcal{L}_{X^c}L$,

$$\mathcal{L}_\Gamma\theta_{L'} - dL' = \mathcal{L}_{[\Gamma,X^c]}\theta_L\ . \tag{5.192}$$

Now, using Eq. (5.119) and the Cartan identity, one can prove easily that

$$i_{[X^c,\Gamma]}\theta_L = i_{[X^c,\Gamma]}(SdL) = i_{S[X^c,\Gamma]}dL = 0\ , \tag{5.193}$$

and hence the previous equation can be rewritten as

$$\mathcal{L}_\Gamma\theta_{L'} - dL' = i_{[\Gamma,X^c]}\omega_L\ . \tag{5.194}$$

Again using the Cartan identity and the fact that, by virtue of $\widehat{S}(\Gamma) = \Delta$, $i_\Gamma\theta_{L'} = \mathcal{L}_\Delta L'$, and this last equation can be rewritten as

$$i_\Gamma\omega_{L'} - dE_{L'} = i_{[X^c,\Gamma]}\omega_L \ , \tag{5.195}$$

which is the Hamiltonian version of Eq. (5.168), where, as before, $E_{L'} = (\mathcal{L}_\Delta - 1)L'$.

As a consequence of this, we see immediately that the following holds:

Proposition 5.41 *Let be $\Gamma \in \mathfrak{X}(TM)$ a second-order field and L a regular, admissible Lagrangian for Γ, and let X^c be the infinitesimal generator of a one-parameter group of point transformations (hence the tangent lift of a vector field $X \in \mathfrak{X}(M)$). Then,*

$$L' = \mathcal{L}_{X^c}L \tag{5.196}$$

will be an admissible Lagrangian for Γ iff Γ and X^c commute, i.e., iff

$$[X^c, \Gamma] = 0 \ . \tag{5.197}$$

Remark 5.12 (i) Of course, nothing guarantees that L' will be regular even if L is.

(ii) At the finite level, $[X^c, \Gamma] = 0$ implies that Γ will be left unaltered by the one-parameter group $\phi_t^*\Gamma = \Gamma$. The vector field X^c will be then the infinitesimal generator of a symmetry for the dynamics represented by Γ. More on this in the next Subsection.

5.5.4 Symmetries, Constants of Motion and the Noether Theorem

The concept of symmetries (and of infinitesimal symmetries) for a second-order field has been introduced in the previous Subsection in the context, which is the most relevant for Lagrangian dynamics, of point-transformations, that are represented, at the infinitesimal level, by tangent lifts of vector fields on the base manifold. Here we will enlarge somewhat the discussion by not requiring transformations to be necessarily point-transformations.

We begin with some general and perhaps well-known definitions that we recall here mainly for the sake of completeness. A 'constant of motion' for a vector field (a second-order one for us, but the definition is of course more general) $\Gamma \in \mathfrak{X}(TM)$ is any function $F \in \mathcal{F}(TM)$ such that

$$\mathcal{L}_\Gamma F = 0 \ . \tag{5.198}$$

This is of course nothing but the familiar notion of vanishing of the 'time' derivative of F along the trajectories of Γ, $dF/dt = 0$.

An 'infinitesimal symmetry' for Γ will be instead any vector field $X \in \mathfrak{X}(TM)$ such that

$$[X, \Gamma] = 0 , \tag{5.199}$$

and this implies, as already stated, that Γ will be left invariant by the one-parameter group $\phi_t \in \text{Diff}\,(TM)$ generated by X. Otherwise stated, ϕ_t will send trajectories into trajectories (it 'will permute the trajectories' among themselves).

Recall the very useful relation among graded derivations valid for any pair of vector fields X and Y:

$$i_{[X,Y]} = \mathcal{L}_X \circ i_Y - i_Y \circ \mathcal{L}_X . \tag{5.200}$$

If Γ admits a Lagrangian description with a regular Lagrangian L, then there will be a unique ω_L-Hamiltonian vector field X_F associated with F via $i_{X_F}\omega_L =: dF$. Using then the previous identity and $\mathcal{L}_\Gamma \omega_L = 0$ we obtain $i_{[X_F,\Gamma]}\omega_L = -\mathcal{L}_\Gamma i_{X_F}\omega_L = -\mathcal{L}_\Gamma dF = -d(\mathcal{L}_\Gamma F)$, and hence,

$$\mathcal{L}_\Gamma F = 0 \Rightarrow [X_F, \Gamma] = 0 , \tag{5.201}$$

i.e., if F is a constant of motion, then the Hamiltonian vector field X_F will be an infinitesimal symmetry for Γ.

Vice-versa, let X be an infinitesimal symmetry for Γ, $[X, \Gamma] = 0$ (not necessarily an ω_L-Hamiltonian vector field). Then, the function

$$F_X =: i_X i_\Gamma \omega_L = \omega_L(\Gamma, X) \tag{5.202}$$

is a constant of motion. Indeed, using $i_\Gamma \omega_L = dE_L$, one can also write $F_X = \mathcal{L}_X E_L$, and then using $[X, \Gamma] = 0$,

$$\mathcal{L}_\Gamma F_X = \mathcal{L}_\Gamma(\mathcal{L}_X E_L) = \mathcal{L}_X(\mathcal{L}_X E_L) = 0$$

In the particular case of X being a Hamiltonian vector field $X = X_G$, then $F_{X_G} = \omega_L(\Gamma, X_G) = \{E_L, G\} = X_G E_L$.

Remark 5.13 This seems to be a general procedure for associating symmetries with constants of motion, and vice-versa. However, there is no control on how effective the procedure may be, and in many relevant cases it may well turn to be empty, as the following example shows. Let us consider the isotropic harmonic oscillator with the standard Lagrangian, in the appropriate units,

$$L = \frac{1}{2}\delta_{ij}(v^i v^j - q^i q^j) , \qquad i, j = 1, \ldots , n > 1 .$$

Then, the vector fields

$$X^i{}_j = v^i \frac{\partial}{\partial q^j} + v^j \frac{\partial}{\partial q^i} - q^i \frac{\partial}{\partial v^j} - q^j \frac{\partial}{\partial v^i} , \qquad i \neq j \tag{5.203}$$

are all symmetries for the dynamics of the isotropic oscillator (see [Mm85]), and are associated, in the sense specified above, with the components

$$Q^{ij} = v^i v^j + q^i q^j \tag{5.204}$$

of the 'quadrupole tensor' of the harmonic oscillator, that are all constants of motion, but $\mathcal{L}_{X^i{}_j} E_L = 0$, and the procedure we have just outlined for associating 'backwards' constants of the motion with symmetries is completely empty.

Let us turn now to Noether's theorem. We begin with the following:

Definition 5.42 Let Γ be a second-order vector field admitting a (or at least of one) Lagrangian description with a regular Lagrangian L. A Noether symmetry for Γ is, at the infinitesimal level, a tangent lift $X^c \in \mathfrak{X}(TM)$ such that

$$\mathcal{L}_{X^c} L = i_\Gamma \pi^* \alpha \ , \tag{5.205}$$

for a closed 1-form $\alpha \in \mathfrak{X}^*(M)$.
Otherwise stated, $\mathcal{L}_{X^c} L$ is a 'total time derivative', i.e., locally at least, $\mathcal{L}_{X^c} L = \mathcal{L}_\Gamma h$, $h = \pi^* g$, $g \in \mathcal{F}(M)$ (loosely speaking, $\mathcal{L}_{X^c} L = dg/dt$).

Observe that then $\mathcal{L}_{X^c} \theta_L = \theta_{\mathcal{L}_{X^c} L} = \theta_{\widehat{\alpha}}$ and similarly $\mathcal{L}_{X^c E} = E_{\mathcal{L}_{X^c} L} = E_{\widehat{\alpha}} = 0$ and consequently

$$i([X^c, \Gamma])\omega_L = \mathcal{L}_{X^c}(i_\gamma \omega_L) - i_\Gamma(i\gamma\omega_L) = d(\mathcal{L}_{X^c} E_L)$$

that shows that X^c is a symmetry for Γ, i.e.

$$[X^c, \Gamma] = 0. \tag{5.206}$$

Proposition 5.43 (*E. Noether*) *If $X^c \in \mathfrak{X}(TM)$ is a Noether symmetry for Γ such that $\mathcal{L}_{X^c} L = \mathcal{L}_\Gamma h$, then,*

$$F_{X^c} = i_{X^c} \theta_L - h \tag{5.207}$$

is a constant of motion.

Indeed,

$$\mathcal{L}_\Gamma F_{X^c} = \mathcal{L}_\Gamma i_{X^c} \theta_L - \mathcal{L}_\Gamma h = i_{X^c} \mathcal{L}_\Gamma \theta_L + i_{[\Gamma, X^c]} \theta_L - \mathcal{L}_{X^c} L = i_{X^c} dL - \mathcal{L}_{X^c} L = 0,$$

and this achieves the proof.

Remark 5.14

(i) By simply differentiating the equation defining F_{X^c} we find that

$$dF_{X^c} = i_{X^c} \omega_L + \mathcal{L}_{X^c} \theta_L - dh.$$

But $\mathcal{L}_{X^c}\theta_L = \theta_{\mathcal{L}_{X^c}L} = \theta_{\mathcal{L}_\Gamma h} = dh$, and hence,

$$i_{X^c}\omega_L = dF_{X^c} \, , \tag{5.208}$$

i.e., X^c is ω_L-Hamiltonian with F_{X^c} as Hamiltonian function. This implies, of course,

$$\mathcal{L}_{X^c}\omega_L = 0 \, , \tag{5.209}$$

as well.

(ii) Spelling out explicitly the equation $i_{[X^c,\Gamma]}\omega_L = 0$, we obtain also easily

$$\mathcal{L}_{X^c}E_L = 0 \, . \tag{5.210}$$

Noether symmetries are therefore a conspicuous example of the case in which Eq. (5.202) yields only a trivial constant of motion.

As an example, let us consider again the n-dimensional harmonic oscillator. The associated vector field is, in appropriate units,

$$\Gamma = v^i \frac{\partial}{\partial q^i} - q^i \frac{\partial}{\partial v^i} \, , \tag{5.211}$$

which, introducing once again the collective coordinates ξ^i, $i = 1, \dots, 2n$, with $\xi^i = q^i$, $i = 1, \dots, n$, and $\xi^i = v^{i-n}$, $i = n+1, \dots, 2n$, can also be written as

$$\Gamma = \widetilde{\Gamma}^i{}_j \, \xi^j \frac{\partial}{\partial \xi^i} \, , \tag{5.212}$$

where the matrix $\widetilde{\Gamma} = (\Gamma^i{}_j)$ is given by

$$\widetilde{\Gamma} = \begin{pmatrix} 0 & \mathbb{I}_n \\ -\mathbb{I}_n & 0 \end{pmatrix} , \tag{5.213}$$

with each entry an $n \times n$ matrix, 0 the null matrix and $\mathbb{I}_n$ the $n \times n$ identity matrix.

Consider then the linear vector field

$$X_A = A^i{}_j \, q^j \frac{\partial}{\partial q^i} \in \mathfrak{X}(M) \, , \qquad M = \mathbb{R}^n \, , \tag{5.214}$$

with $A = (A^i{}_j)$ any $n \times n$ numerical matrix with real entries. As A varies in the set of $n \times n$ real matrices, the X_A's will generate the action (actually the representation) of the general linear group $GL(n, \mathbb{R})$ on $M = \mathbb{R}^n$. The associated tangent lifts will be given by

$$X_A^c = \widetilde{A}^i{}_j \, \xi^j \frac{\partial}{\partial \xi^i} \, , \tag{5.215}$$

where

$$\widetilde{A} = (\widetilde{A}^i{}_j) = \begin{pmatrix} A & 0 \\ 0 & A \end{pmatrix}, \tag{5.216}$$

and they will generate the tangent lift of the action of $GL(n, \mathbb{R})$ to $TM = \mathbb{R}^{2n}$.

As, given any two linear vector fields defined by matrices A and B, $X_A = A^i{}_j\, \xi^j \frac{\partial}{\partial \xi^i}$, $X_B = B^i{}_j\, \xi^j \frac{\partial}{\partial \xi^i}$, we have $[X_A, X_B] = X_{[B,A]}$, and in view of the fact that, with the structure of the above matrices, $[\widetilde{A}, \widetilde{\Gamma}] = 0\ \forall A$, we find at once that the full general linear group $GL(n, \mathbb{R})$ is a symmetry group for the dynamics of the linear harmonic oscillator.

Let us inquire now what is the rôle of this group in the Lagrangian framework when the harmonic oscillator is described by a Lagrangian of the type

$$L_B(q, v) = \frac{1}{2} B_{ij} \left(v^i v^j - q^i q^j \right) ,$$

as discussed in a previous Subsection. In terms of collective coordinates,

$$L_B(q, v) = \frac{1}{2} \widetilde{B}_{ij}\, \xi^i\, \xi^j$$

with

$$\widetilde{B} = \begin{pmatrix} -B & 0 \\ 0 & B \end{pmatrix}. \tag{5.217}$$

Recalling that, by construction, B has to be a symmetric matrix, some long but straightforward algebra leads to

$$\mathcal{L}_{X_A^c} L_B = \frac{1}{2}\, \widetilde{C}_{ij}\, \xi^i\, \xi^j , \tag{5.218}$$

where

$$\widetilde{C} = \begin{pmatrix} -C & 0 \\ 0 & C \end{pmatrix} \tag{5.219}$$

and

$$C = BA + A^t B. \tag{5.220}$$

It is pretty obvious that, unless $C = 0$, $\mathcal{L}_{X_A^c} L_B$ will be a 'quadratic' form in the velocities, and hence it cannot be a total time derivative. Therefore:

Proposition 5.44 *The vector field X_A^c will be a (an infinitesimal) Noether symmetry for the isotropic harmonic oscillator with the Lagrangian L_B iff*

$$C = BA + A^t B = 0 \tag{5.221}$$

in which case, $\mathcal{L}_{X_A^c} L_B = 0$*, and the associated Noether constant of motion will be given by*

$$F_A = i_{X_A^c} \theta_{L_B} = v^h B_{hk} A^k{}_j q^j \tag{5.222}$$

Remark 5.15

(i) If we consider the orthogonal subgroup $O(n)$ of the general linear group, which is generated, at the infinitesimal level, by vector fields X_A associated with skew-symmetric matrices $(A + A^t = 0)$, then the orthogonal transformations that are also Noether symmetries for L_B will be those (and only those) that are generated by matrices A that are in the 'commutant' of B,

$$\mathcal{C}(B) = \{A \mid [A, B] = 0\} .$$

If $B = I$ (the 'standard' Lagrangian for the isotropic harmonic oscillator), then the whole of the orthogonal group will be also a group of Noether symmetries, and the associated constants of motion will be given by $F_A = v^h\, \delta_{hk}\, A^k{}_j\, q^j$. This will not be the case when $B \neq I$.

(ii) Of course, B being symmetric it can always be diagonalized via an orthogonal transformation followed, if necessary, by a scale transformation. So, without loss of generality, we can always consider B to be of the form

$$B = \mathrm{diag}\{\underbrace{1, 1, \ldots, 1}_{n-m \text{ times}}, \underbrace{-1, -1, \ldots, -1}_{m \text{ times}}\} \tag{5.223}$$

in which case the invariance group of the corresponding Lagrangian, and hence of Noether symmetries, is easily seen to be $O(n-m, m)$. The consequences on the association of symmetries with constants of motion will be investigated in the following Subsection.

(iii) Notice that Eq. (5.221) expresses in general the (necessary and sufficient) condition under which the linear transformations (a one-parameter group obtained by exponentiation) generated by A leave invariant, say, the quadratic form:

$$B = \frac{1}{2} B_{ij} q^i q^j . \tag{5.224}$$

As the standard orthogonal matrices are those leaving invariant the quadratic form corresponding to $B = I$, it is natural to generalize the notion of a linear transformation being orthogonal to that of being 'B-orthogonal' whenever it satisfies (5.221) for a general symmetric matrix B. We can then rephrase the previous theorem by saying that the vector field X_A^c will be an infinitesimal Noether symmetry for the fharmonic oscillator with the Lagrangian L_B iff A is B-orthogonal.

If B is positive, the B-orthogonal matrices will provide us with a realization of the orthogonal group $O(n)$ different from the standard one, a realization of $O(n-m, m)$ if B has the signature as in Eq. (5.223). These realizations will be of course isomorphic to the standard one, actually they will provide us with a different 'placement' of $O(n)$ (or $O(n-m, m)$) within the full general linear group, the two realizations being related by the conjugation defined by the orthogonal transformation that diagonalizes B.

5.5.4.1 Consequences on Noether's Theorem of the Existence of Alternative Lagrangians

Noether's theorem is commonly assumed to give a unique prescription for associating constants of motion with symmetries. This is not so if the same dynamical system admits of different and genuinely alternative Lagrangian description, as the following example shows.

Let us consider a three-dimensional isotropic harmonic oscillator with the standard Lagrangian

$$L = \frac{1}{2}[(v^1)^2 + (v^2)^2 + (v^3)^2 - (q^1)^2 - (q^2)^2 - (q^3)^2] , \tag{5.225}$$

which is $O(3)$-invariant. The Noether theorem associates then the three conserved components of the angular momentum with the infinitesimal generators of the rotation group $O(3)$, and this is all standard material.

Consider instead the alternative, $O(2, 1)$-invariant, Lagrangian function:

$$L' = \frac{1}{2}[(v^1)^2 + (v^2)^2 - (v^3)^2 - (q^1)^2 - (q^2)^2 + (q^3)^2] . \tag{5.226}$$

The $(2 + 1)$ Lorentz group $O(2, 1)$ has the infinitesimal generators:

$$K_1 = q^3\frac{\partial}{\partial q^1} + q^1\frac{\partial}{\partial q^3}, \qquad K_2 = q^3\frac{\partial}{\partial q^2} + q^2\frac{\partial}{\partial q^3} , \tag{5.227}$$

and

$$J = q^1\frac{\partial}{\partial q^2} - q^2\frac{\partial}{\partial q^2} , \tag{5.228}$$

corresponding respectively to the two 'boosts' in the q^1 and q^2 directions and to the rotations in the (q^1, q^2) plane. They close on the Lie algebra $\mathfrak{o}(2, 1)$ of $O(2, 1)$, namely,

$$[K_1, J] = K_2, \qquad [K_2, J] = -K_1 , \tag{5.229}$$

and

$$[K_1, K_2] = J . \tag{5.230}$$

The tangent lifts K_1^c, K_2^c and J^c will close on the same Lie algebra, of course.

As already stated, the Lagrangian L' is invariant under the action of K_1^c, K_2^c and J^c. The associated Noether constants of motion turn then out to be

$$\begin{aligned} F_1 &= i_{K_1^c}\theta_{L'} = q^3 v^1 - q^1 v^3 \\ F_2 &= i_{K_2^c}\theta_{L'} = q^3 v^2 - q^2 v^3 \end{aligned} \tag{5.231}$$

and, as in the previous case, $i_{J^c}\theta_{L'} = q^1 v^2 - q^2 v^1$. We see therefore that the (three) components of the angular momentum become the Noether constants of motion associated with the generators of the Lorentz group if the Lagrangian of the harmonic oscillator is taken to be the $O(2, 1)$-invariant Lagrangian L'.

Remark 5.16 If the standard Lagrangian L is assumed in place of L', two out of the three generators of the $(2 + 1)$ Lorentz group (the generators of the Lorentz boosts) are of course still symmetries for the dynamics, but no more Noether symmetries. We can use then Eq. (5.202) to evaluate the constants of motion associated with the Lorentz boosts. They turn out to be

$$G_1 =: \mathcal{L}_{K_1^c} E_L = Q^{13}, \qquad G_2 =: \mathcal{L}_{K_2^c} E_L = Q^{23}, \tag{5.232}$$

where

$$Q^{13} = v^1 v^3 + q^1 q^3, \qquad Q^{23} = v^2 v^3 + q^2 q^3, \tag{5.233}$$

i.e., the constants of motion turn out to be now two of the components of the quadrupole tensor of the harmonic oscillator.

What this admittedly simple example shows is that an essential ingredient in the association of constants of motion with symmetries via Noether's theorem is the choice of a Lagrangian and that the association can become ambiguous whenever the latter is not unique.

5.5.5 A Relativistic Description for Massless Particles

Following Wigner's program [Wi39], we take advantage of the fact that upon quantization our system should provide us with a unitary irreducible representation of the Poincaré group $\mathcal{P}$. This suggest that this time we look for a Lagrangian description written on the group manifold of $\mathcal{P}$.

Now the carrier space will be the connected component of the Poincaré group $\mathcal{P}_+^\uparrow \cong \mathbb{R}^4 \rtimes \mathcal{L}_+^\uparrow$ where $\mathcal{L}_+^\uparrow$ is the proper orthochronous Lorentz group (see Sect. 4.6), Eq. (4.108). We make the following identification of variables:

1. $x = (t, \mathbf{x}) \in \mathbb{R}^4$ represents the physical coordinate in space-time.
2. p is the 4-momentum of the massless particle. It is defined in such a way that we take into account that the 3-momentum cannot be transformed to zero by any Lorentz transformation.

We introduce a basis (e_0, e_1, e_2, e_3) on $\mathbb{R}^4$ such that the Minkowski metric is defined as:

$$e_0^2 = -1; \qquad e_i \cdot e_j = \delta_{ij}; \qquad e_0 \cdot e_i = 0.$$

We set:

$$p = \omega(\Lambda e_0 + \Lambda e_3),$$

where $\Lambda \in \mathcal{L}_+^\uparrow$ and ω is positive. With the help of the above coordinates the Lagrangian is given by:

$$L = p_\mu \dot{x}^\mu + \lambda \mathrm{Tr}\,(T_{21}\Lambda^{-1}\dot{\Lambda}),$$

where T_{12}, T_{23}, T_{31}, are the generators of the subgroup of the space rotations. The trace operation is actually a scalar product in $\mathcal{L}_+^\uparrow$. By writing the Lagrangian in more detail:

$$L = \dot{x}^\mu(\Lambda(e_0 + e_3))_\mu + \lambda \mathrm{Tr}\,(T_{21}\Lambda^{-1}\dot{\Lambda})$$

we see that L is invariant under the left action of $\mathcal{P}$, once we notice that x transforms like the dual of $\Lambda(e_0+e_3) = p$. To find the equations of motion we proceed as usual, that is consider the Euler-Lagrange equations:

$$\frac{d}{dt} i_{X^t}\theta_L = \mathcal{L}_{X^t} L.$$

The 1–form θ_L associated with the Lagrangian is

$$\theta_L = p_\mu dx^\mu + \lambda \mathrm{Tr}\,(T_{21}\Lambda^{-1} d\Lambda).$$

We use $X = a^\mu \partial/\partial x^\mu$, $a^\mu \in \mathbb{R}$, and we find for any $a \in \mathbb{R}^4$,

$$\frac{d}{dt}(p_\mu a^\mu) = 0,$$

which implies that $dp_\mu/dt = 0$. By using vector fields Y's, infinitesimal generators of the Lorentz group, and defining matrices T_Y, in the Lie algebra of the Lorentz group, as

$$T_{\mu\nu} = i_{Y_{\mu\nu}} d\Lambda \cdot \Lambda^{-1},$$

we find

$$\mathcal{L}_Y \Lambda = i_Y d\Lambda = (i_Y d\Lambda\Lambda^{-1})\Lambda = T_Y \Lambda,$$

and also

$$i_Y \Lambda^{-1} d\Lambda = \Lambda^{-1}(i_Y d\Lambda\Lambda^{-1})\Lambda = \Lambda^{-1} T_Y \Lambda.$$

These right invariant generators define the left action. Therefore $\mathcal{L}_{Y'}L = 0$ and we get

$$\frac{d}{dt} i_Y \theta_L = 0.$$

We notice that $\mathcal{L}_Y(p_\mu x^\mu) = 0$ because it preserves L. Therefore $(\mathcal{L}_Y p_\mu)x^\mu = -p_\mu \mathcal{L}_Y x^\mu = -p_\mu i_Y dx^\mu$ and we can compute now $i_Y \theta_L$ to be

$$\begin{aligned} i_Y \theta_L &= i_Y \lambda \mathrm{Tr}\,(T_{21}\Lambda^{-1} d\Lambda) - (\mathcal{L}_Y p) \cdot x \\ &= \lambda \mathrm{Tr}\,(T_{21}\Lambda^{-1} i_Y d\Lambda \Lambda^{-1}\Lambda) - x \cdot T_Y \Lambda(e_0 + e_3) \\ &= \lambda \mathrm{Tr}\,(T_{21}\Lambda^{-1} T_Y \Lambda) - x \cdot T_Y(e_0 + e_3)\,, \end{aligned}$$

and we find also

$$i_{Y_{\mu\nu}} \theta_L = \lambda \mathrm{Tr}\,(T_{21}\Lambda^{-1} T_{\mu\nu}\Lambda) - x \cdot T_{\mu\nu}\Lambda(e_0 + e_3),$$

i.e.,

$$\frac{d}{dt}(\lambda(\Lambda_{\mu 1}\Lambda_{\nu 2} - \Lambda_{\mu 2}\Lambda_{\nu 1})) + z_\mu p_\nu - z_\nu p_\mu) = 0.$$

As a^μ and $T_{\mu\nu}$ are a basis for the algebra of the Poincaré group, the equations of motion are provided by the conservation of p_μ, and the conservation of $J_{\mu\nu} = \lambda(\Lambda_{\mu 1}\Lambda_{\nu 2} - \Lambda_{\mu 2}\Lambda_{\nu 1})) + z_\mu p_\nu - z_\nu p_\mu$. By looking at θ_L and $d\theta_L$ it is immediately clear that the dynamics actually takes place on the coadjoint orbit of the Poincaré group passing through $(e_0 + e_3, T_{21})$. The stability group of this element under the coadjoint representation is provided by the translation group generated by the following elements

$$\pi_1 = T_{10} + T_{13}; \qquad \pi_2 = T_{20} + T_{23}$$

and the rotation associated with T_{21}. We then get the Euclidean group $E(2)$. Thus we obtain

$$\mathcal{L}_+^\uparrow / E(2) = \mathbb{R} \times S^2.$$

5.6 Feynman's Problem and the Inverse Problem for Lagrangian Systems

5.6.1 Feynman's Problem Revisited

We examine now Feynman's problem in the framework of tangent bundles. As was discussed in the previous section, this is the natural setting for Lagrangian mechanics. Thus it makes sense to try to describe all dynamical vector fields which are

Hamiltonian with respect to some Poisson structure subject to certain fundamental conditions. Then, we can ask whether all of them are associated with an Euler–Lagrange SODE defined by a Lagrangian. Finally, we consider the local structure of such Lagrangian.

5.6.1.1 Localizability

We will be discussing systems whose classical phase space is the tangent bundle of a configuration space Q. The first condition that we will impose on Poisson tensors Λ defined on TQ is that they will describe localizable systems, i.e.,

$$\{ q^i , q^j \} = 0 \,. \tag{5.234}$$

This condition was used in the Feynman procedure as presented in Sect. 4.5.2, and later on was a crucial ingredient for the introduction of interactions Eq. (4.73). From a physical point of view, this condition reflects that the classical limit of the quantum theory defines a configuration space whose points can be used to label the quantum states of the system. This notion is thus borrowed from quantum theory and another way of stating it is simply that position operators commute. Geometrically, this means that the algebra of observables on Q is an Abelian subalgebra of the full algebra $\mathcal{F}(TQ)$.

Definition 5.45 We will say that a Poisson tensor Λ on TQ is localizable if $\tau^*(\mathcal{F}(Q))$ is an Abelian Poisson subalgebra of $\mathcal{F}(TQ)$.

Notice that if Λ is localizable, then the Hamiltonian vector fields corresponding to coordinates on the configuration space Q commute. On the other hand it is clear that if a Poisson dynamical system Γ has a Lagrangian realization, the Poisson tensor Λ is localizable. Because of this, this condition is called 'variational admissibility' in [No82] and was used to show the existence of a (local) Lagrangian description for certain Poisson dynamical systems.

In other words, if the Poisson tensor Λ is given by

$$\Lambda = a^{ij}\frac{\partial}{\partial q^i}\wedge\frac{\partial}{\partial q^j} + b^{ij}\frac{\partial}{\partial q^i}\wedge\frac{\partial}{\partial v^j} + c^{ij}\frac{\partial}{\partial v^i}\wedge\frac{\partial}{\partial v^j}\,, \tag{5.235}$$

localizability amounts to $a^{ij} = 0$. But these conditions are equivalent to imposing the requirement that the Hamiltonian vector fields X_{q^i} corresponding to the coordinates on Q be vertical. In fact,

$$X_{q^i} = b^{ij}\frac{\partial}{\partial v^j} + 2a^{ij}\frac{\partial}{\partial q^k}\,, \tag{5.236}$$

which means that X_{q^i} are vertical if and only if $a^{ij} = 0$. Moreover, the vector fields X_{q^i} commute among themselves, then

$$[X_{q^i}, X_{q^i}] = 0 \iff b^{ij}\frac{\partial b^{kl}}{\partial v^j} - b^{kj}\frac{\partial b^{il}}{\partial v^j} = 0 \ .$$

5.6.1.2 Hamiltonian Second-Order Equations

Another assumption which is often implicitly assumed is that there are dynamical systems which are SODE's on TQ and which are Hamiltonian with respect to a given Poisson tensor Λ. For instance, in the derivation of Feynman's results in Sect. 4.5.2 and later on Sect. 4.5.3, the existence of a SODE which was Hamiltonian with respect to Λ was assumed. In general, this need not to be the case. In this regard, consider the example of $T\mathbb{R}^2$, with coordinates $(q^1, q^2; v^1, v^2)$ and the Poisson tensor $\Lambda = \partial/\partial q^1 \wedge \partial/\partial v^1$. For this case, there is no SODE Γ such that, $\Gamma = -\Lambda(dH)$.

Definition 5.46 We will say that a Poisson tensor Λ in TQ is Newtonian if there exists a SODE Γ such that $\Gamma = X_H$ for some Hamiltonian function H on TQ, in other words,

$$\Gamma = -\widehat{\Lambda}(dH) \ . \tag{5.237}$$

Notice that there is a weaker assumption on the Newtonian character of Λ, which is as follows: Suppose that there is a SODE Γ such that $\mathcal{L}_\Gamma \Lambda = 0$. Then we will say that Γ is canonical. If Λ is Newtonian then the latter condition automatically holds, but the converse is not necessarily true. In the previous example on $\mathbb{R}^2$ it is easy to check that the SODE $\Gamma = v^1\partial/\partial q^1 + v^2\partial/\partial q^2$ is canonical.

From Eq. (4.33), Γ being canonical implies that

$$\Gamma\{F, G\} = \{\Gamma(F), G\} + \{F, \Gamma(G)\} \ , \tag{5.238}$$

i.e., the Poisson bracket is compatible with the dynamics defined by Γ, but this does not guarantee the existence of a Hamiltonian function for Γ. However if Λ is nondegenerate, it is sufficient that Γ be canonical to insure the existence of a (local) Hamiltonian function. Notice that if Γ has a regular Lagrangian description, then automatically is a SODE and Λ is Newtonian.

Then we can state the result as follows:

Lemma 5.47 *If Γ is a canonical SODE in TQ for a localizable Poisson tensor Λ, then the bracket $\{q^i, v^j\} = b^{ij}$ is symmetric. In particular if Λ is Newtonian and localizable the same result holds.*

Proof If Λ is localizable we have that $\{q^i, q^j\} = 0$. However because of Eq. (5.238), we get

$$0 = \Gamma\{q^i, q^j\} = \{\Gamma(q^i), q^j\} + \{q^i, \Gamma(q^j)\} = -b^{ji} + b^{ij} \ . \qquad \square$$

Another important observation concerning Hamiltonian SODE's is the following:

Lemma 5.48 *If Γ is a Hamiltonian SODE and α is a non-null basic 1-form, then $\widehat{\Lambda}(\alpha) \neq 0$. In particular, there are no nonconstant basic Casimir functions.*

Proof Let $\alpha = \alpha_i \, dq^i$ a basic form. Then,

$$i_{\widehat{\Lambda}(\alpha)} dH = \Lambda(\alpha, dH) = i_\Gamma \alpha = \alpha_i v^i = \widehat{\alpha}(q, v),$$

and consequently, if $\widehat{\Lambda}(\alpha) = 0$ then $\alpha = 0$.

In particular, if there is a basic Casimir function ϕ, then $\widehat{\Lambda}(d\phi) = 0$ and therefore $d\phi = 0$. □

Conversely, it can be easily seen that if Γ is a canonical SODE for Λ and it has basic Casimir functions, then Λ must be nonregular. Otherwise, a canonical SODE Γ would be Hamiltonian and no basic Casimir functions would exist.

5.6.1.3 Regular Poisson Tensors

Let us recall the important rôle played by the condition $i_S \omega = 0$ in the geometry of closed 2-forms on TQ. As it was indicated in Theorem 5.39, this condition is necessary and sufficient to insure the existence of a (local) Lagrangian L such that $\omega = \omega_L$. Has this condition similar implications for Poisson tensors?

To answer this question we first make the following observation:

Lemma 5.49 *If Λ is localizable and there is a SODE Γ which is canonical (in particular if it is Newtonian), then $i_S \Lambda = 0$.*

Proof Consider Λ as in Eq. (5.235). Then, a simple computation shows that,

$$i_S \Lambda = 2a^{il} \frac{\partial}{\partial v^i} \wedge \frac{\partial}{\partial q^l} + b^{il} \frac{\partial}{\partial v^i} \wedge \frac{\partial}{\partial v^l} \, .$$

Thus, if Λ is localizable, $a^{il} = 0$, and if Λ is canonical, (Lemma 5.47) b^{il} is symmetric. □

Notice that the converse is not true. (To prove this result one can use the above example.)

Thus, the condition $i_S \Lambda = 0$ is too weak to insure the existence of a local Lagrangian since it does not first insure the regularity of Λ. On the other hand, we have seen that the condition that it is localizable plus Newtonian implies $i_S \Lambda = 0$. In fact, we can show that these conditions are sufficient to insure that Λ is regular.

Theorem 5.50 *If Λ is a localizable and Newtonian Poisson tensor on TQ, then Λ is regular, and hence it defines a symplectic structure on TQ.*

Proof Notice that $\Lambda^{\wedge n} = \Lambda \wedge \cdots \wedge \Lambda$ is proportional to

$$\det(b^{ij}) \frac{\partial}{\partial q^1} \wedge \cdots \wedge \frac{\partial}{\partial q^n} \wedge \frac{\partial}{\partial v^1} \wedge \cdots \wedge \frac{\partial}{\partial v^n} .$$

Thus, Λ will be regular iff $\det(b^{ij}) \neq 0$.

We shall show it by proving that Λ defines a monomorphism when restricted to closed basic 1-forms. In fact, let $F \in \mathcal{F}(\mathcal{U})$, where $\mathcal{U}$ is an open set in Q, and $f = \tau^* F \in \mathcal{F}(\tau^{-1}(\mathcal{U}))$. Computing $\widehat{\Lambda}(df)$ we find

$$\widehat{\Lambda}(df) = b^{ij} \frac{\partial f}{\partial q^i} \frac{\partial}{\partial v^j} , \tag{5.239}$$

where we have used that Λ is localizable, and hence $a^{ij} = 0$. On the other hand, if Λ were not a monomorphim on closed basic 1-forms, this would imply the existence of a function ϕ on an open set $\mathcal{U}$ in Q such that $\widehat{\Lambda}(d\phi) = 0$, i.e., that ϕ is a (locally defined) Casimir basic function. However, because of Lemma 5.48, this cannot occur if Λ is Newtonian. Thus $\det b^{ij} \neq 0$, because of Eq. (5.239). □

It is interesting to point out that if Λ is nondegenerate, it does not necessarily follow that Λ is localizable and/or Newtonian. If fact, the Poisson tensor $\Lambda = \partial/\partial q^1 \wedge \partial/\partial q^2 + \partial/\partial v^1 \wedge \partial/\partial v^2$ is neither one nor the other.

Therefore, if there exists such a SODE vector field Γ the Poisson tensor is invertible. When writing it in block matrix form

$$(\widehat{\Lambda}) = \left(\begin{array}{c|c} 0 & b \\ \hline -b & c \end{array} \right) , \tag{5.240}$$

we find the inverse matrix given by

$$(\widehat{\omega}) = \left(\begin{array}{c|c} b^{-1} c b^{-1} & -b^{-1} \\ \hline b^{-1} & 0 \end{array} \right) , \tag{5.241}$$

and we have a closed non-degenerate 2-form ω by inverting the Poisson structure Λ as described in Sect. 5.2.1, Eq. (5.1). We can write ω as

$$\omega = G_{ij} dq^i \wedge dv^j + F_{ij} dq^i \wedge dq^j , \tag{5.242}$$

where

$$G_{ij} = b_{ik} c^{kl} b_{lj}; \quad F_{ij} = -b_{ij} ,$$

and $b_{ik} b^{kj} = \delta^j_i$.

On the other hand, if $i_S \Lambda = 0$ and ω is the inverse of Λ, then $i_S \omega = 0$ and conversely. Consequently, by Them 1, if there is a second order differential equation vector field Γ which is Hamiltonian with respect to a localizable Poisson structure

Λ, then there will be a (locally defined) Lagrangian function L in TQ such that Γ is the dynamical vector field defined by L. In this sense, as indicated by Dyson, Feynman's approach does not go beyond the Lagrangian formalism and internal degrees of freedom should be considered for such a purpose. Notice that we have made no use of further conditions, such as $\{ q^i, v^j \} = \delta^{ij}$ [Hu92]. These results also provide the general setting for the remarks in [No82]. Thus the impossibility of going beyond the Lagrangian formalism using Feynman's procedure on tangent bundles is a fundamental restriction. Nevertheless, the Lagrangian functions arising in this manner can be quite different from the standard 'mechanical' type (in contrast with the result in [Mo93]). We will see however that this is not the case, if we impose an additional physical condition on Λ.

5.6.1.4 Dynamical Independence of the Poisson Brackets from the State of Motion of Test Particles

From a physical point of view, it is reasonable to assume that the fields acting on a particle are external. This means that if we are using a test particle to probe the field, then the field itself will not depend on the state of motion of the test particle, i.e., it will not depend on the velocity of the particle. In this sense, we can assume that the Poisson tensor Λ will not depend on the velocities. In other words, translation along the fibres of TQ will be Poisson maps for Λ. Yet another way of stating this is: The coefficients Λ^{ij} of the Poisson tensor Λ in natural coordinates q^i, v^i are defined on the configuration space.

Definition 5.51 We will say that a Poisson tensor is velocity independent if $\mathcal{L}_{\partial/\partial v^i} \Lambda = 0$, for all $i = 1, \ldots, n$, i.e., if translations along the velocities are Poisson maps.

We have seen that the condition of localizability and that Λ is Newtonian are not equivalent to that of the regularity of Λ. In spite of this, we can show the following:

Theorem 5.52 *If Λ is a non-degenerate Poisson tensor such that $i_S \Lambda = 0$ and it is velocity independent, then Λ is Newtonian.*

Proof The matrix b^{ij} is non-degenerate because Λ is regular. We will denote the inverse of b^{ij} by b_{ij}. Because $i_S \Lambda = 0$, then b_{ij} is symmetric, and because Λ is velocity independent $b_{ij} = b_{ij}(q)$ is a basic function. Then, $H = \frac{1}{2} b_{ij}(q) v^i v^j$ is such that

$$X_H = -\widehat{\Lambda}(dH) = v^i \frac{\partial}{\partial q^i} + \left(\frac{1}{2} v^i v^j b^{lm} \frac{\partial b_{ij}}{\partial q^l} + 2c^{lm} b_{il} v^i \right) \frac{\partial}{\partial v^m} ,$$

which is a Hamiltonian SODE. □

Notice that if the Poisson tensor Λ is velocity independent, then the transformation from one reference frame to another, having uniform velocity with respect to the original frame, is canonical. If the Poisson tensor Λ is velocity independent, then the functions G_{ij} and F_{ij} depend only on the coordinates q^i. Using the results in the previous section (Theorem 5.39), we know that there exists a semibasic 1-form $\theta = A_i(q, v)\, dq^i$ such that $d\theta = \omega$. But then

$$G_{ij} = \frac{\partial A_i}{\partial v^j} ,$$

and,

$$F_{ij} = \frac{\partial A^i}{\partial q^j} - \frac{\partial A^j}{\partial q^i} .$$

Because G_{ij} depends on just the q's, then $A_i(q, v) = G_{ij}(q)v^j + a_i(q)$. On the other hand (again from Theorem 5.39), we get that $A_i = \partial L/\partial v^i$. Then integrating once more, we get

$$L = \frac{1}{2} G_{ij}(q)v^i v^j + a_i(q)v^i - V(q) ,$$

which is a mechanical type Lagrangian.

We thus conclude:

Theorem 5.53 *If Λ is a localizable, Newtonian and velocity independent Poisson tensor on TQ, then Λ defines a family of Hamiltonian SODE's whose Lagrangians are of mechanical type differing on the potential term, and whose Cartan 2-form is the inverse of Λ.*

5.6.2 Poisson Dynamics on Bundles and the Inclusion of Internal Variables

So far we have proved that the generalization of Feynman's attempt at constructing localizable second-order Hamiltonian dynamics on tangent bundles always leads (locally) to Lagrangian systems with a well defined structure. Nevertheless, in Sects. 4.5.3 and 4.5.4 we have constructed examples of dynamics which are not of the Lagrangian type (at least at first sight). This was done by introducing internal degrees of freedom. We will discuss such a situation in what follows. For that purpose we will first need to understand some geometrical aspects of Poisson structures on bundles, which now follow.

5.6.2.1 Poisson Structures on Bundles: Localizability

We shall consider a bundle $E \xrightarrow{\pi} B$ over the base manifold B. We can assume for simplicity that E is a vector bundle, but this is not strictly necessary for what follows. If we are given a Poisson tensor Λ_B on the base space B, we can ask for the existence of a Poisson tensor Λ_E on the total space E such that $\pi_*(\zeta)\Lambda_E(\zeta) = \Lambda_B(\pi(\zeta))$, $\zeta \in E$. We will then say that Λ_E is π-related to Λ_B. This is also equivalent to saying that the map π is a Poisson map from E to B. In fact, upon introducing local coordinates b^i for B and ζ_a along the fibres of E, we can then write

$$\Lambda_E = \Lambda_B^{ij}(b)\frac{\partial}{\partial b^i} \wedge \frac{\partial}{\partial b^j} + A_a^i(b,\zeta)\frac{\partial}{\partial b^i} \wedge \frac{\partial}{\partial \zeta_a} + C_{ab}(b,\zeta)\frac{\partial}{\partial \zeta_a} \wedge \frac{\partial}{\partial \zeta_b}\,, \tag{5.243}$$

where the first term on the r.h.s. of Eq. (5.243) is simply the local expression for Λ_B.

Equivalently, the fact that Λ_E is π-related to Λ_B implies that the map $\pi^*\colon \mathcal{F}(B) \to \mathcal{F}(E)$ induced by π is a Lie algebra homomorphism.

A particular case arises when $\Lambda_B = 0$, i.e., the base manifold carries a trivial Poisson bracket. In such a case, the possible Poisson tensors on E which are π-related to it have the general form

$$\Lambda_E = A_a^i(b,\zeta)\frac{\partial}{\partial b^i} \wedge \frac{\partial}{\partial \zeta_a} + C_{ab}(b,\zeta)\frac{\partial}{\partial \zeta_a} \wedge \frac{\partial}{\partial \zeta_b}\,. \tag{5.244}$$

For instance, if B is the configuration space Q for a physical system, and E is the corresponding tangent bundle, $E = TQ$, then Poisson tensors Λ_{TQ} π-related to $\Lambda_Q = 0$ are those such that Λ_{TQ} is localizable.

Another example that will concern us here is the introduction of internal variables for a system defined on a configuration space Q. Internal variables are modeled by defining a bundle $\rho\colon F \to Q$. Attached to each base point $m \in Q$ there is a fibre $\rho^{-1}(m)$ that can be a representation space M for a group of gauge symmetries. Again the Poisson bracket on Q is trivial, and a Poisson tensor Λ_I which is ρ-related to it will have the form

$$\Lambda_I = A_a^i(q,I)\frac{\partial}{\partial q^i} \wedge \frac{\partial}{\partial I_a} + C_{ab}(q,I)\frac{\partial}{\partial I_a} \wedge \frac{\partial}{\partial I_b}\,, \tag{5.245}$$

where I_a denote local coordinates along the fibres of F.

If the I_a's are to be considered as internal variables, they should be localized at points in configuration space. This means that they must be 'measurable' simultaneously with position operators, i.e.,

$$A_a^i = \{\, q^i, I_a \,\}_I = 0\,. \tag{5.246}$$

We will call such a localizability property for Poisson tensors on F 'localizability on the bundle', or just 'localizability', which we trust will not cause any confusion. In this case, Eq. (5.245) reduces to

$$\Lambda_I = C_{ab}(q,I)\frac{\partial}{\partial I_a} \wedge \frac{\partial}{\partial I_b}\,. \tag{5.247}$$

Notice that Λ_I is a Poisson tensor iff $[\Lambda_I, \Lambda_I] = 0$. This condition is true for all points $q \in Q$. This then implies that on each fibre $\rho^{-1}(q)$, the Poisson tensor Λ_I

induces a Poisson tensor structure which is obtained from Λ_I by fixing q. We can then conclude that a localizable Poisson structure on a bundle $F \to Q$ is equivalent to a bundle of Poisson manifolds, but the Poisson structure can change from fibre to fibre.

Definition 5.54 A localizable Poisson tensor on a fibre bundle $\rho : F \to Q$ is a bundle Poisson tensor, i.e., a smooth assignment to each fibre of a Poisson tensor Λ_I.

Notice that Hamiltonian vector fields on such a bundle are vertical vector fields, with their restriction to each fibre being Hamiltonian in the corresponding Poisson structure. A special case of such localizable Poisson tensors on F occurs when the transition functions of the bundle F are Poisson maps. Then we could say that the bundle $F \to Q$ is Poisson and the standard fibre is a Poisson manifold modeling all the others.

Examples of this situation occur when we consider a principal fibre bundle P over Q with structure group G. If Q is simply a point, then P is the group itself, and the Poisson structure Λ_P is nothing but a Poisson structure on G. This case will be discussed in detail in Sect. 5.6.4.

Let us instead consider the coadjoint representation of G on the dual $\mathfrak{g}^*$ of the Lie algebra $\mathfrak{g}$ of G. Then we can construct the associated bundle $\mathrm{ad}\,^*P \to Q$, with standard fibre $\mathfrak{g}^*$. Recall that this bundle is obtained as the quotient of $P \times \mathfrak{g}^*$ by the natural action of G on it. On $\mathfrak{g}^*$ we consider the linear Poisson tensor,

$$\Lambda_{\mathfrak{g}} = c_{ij}{}^k x_k \frac{\partial}{\partial x_i} \wedge \frac{\partial}{\partial x_j}, \tag{5.248}$$

where x_i denote linear coordinates on $\mathfrak{g}^*$ and $c_{ij}{}^k$ are the structure constants of $\mathfrak{g}$ on a basis E_i which is dual to the basis defined by x_i on $\mathfrak{g}^*$. We then define a Poisson tensor Λ on $P \times \mathfrak{g}^*$ as the direct sum of the zero tensor in P and $\Lambda_{\mathfrak{g}}$. This Poisson tensor is G-invariant. Then it will induce a localizable Poisson tensor on $\mathrm{ad}\,^*P$. If the bundle $P \to Q$ were already carrying a G-invariant localizable Poisson tensor Λ_P, we could form $\Lambda_P + \Lambda_{\mathfrak{g}}$ to obtain a new localizable Poisson tensor on $\mathrm{ad}\,^*P$.

Alternatively, we can proceed as follows: Let Λ_P be a G-invariant localizable Poisson tensor on P and $T\colon G \to V$ a linear representation of G on V. Then in the associated vector bundle $E_T = P \times V/G$, there is a localizable Poisson tensor induced from Λ_P. Some specific examples of this situation will be discussed in Sect. 5.6.4.

5.6.2.2 Extension of Poisson Structures and Internal Variables

We note that if $E \to B$ is a bundle, then it could happen that there are no smooth sections. For instance, if $E \to B$ is a principal bundle, then there are no smooth, nor continuous, sections unless the bundle is trivial. In physical terms, this means that it is not possible to globally fix the state of the inner variables in a continuous way. In this sense, there is no way to 'switch off' the inner variables and to restrict our attention to a system without them. On the other hand, if $E \to B$ is a vector bundle, i.e., if the inner variables are defined in a vector space, then there are always smooth sections. In fact, there is the zero section which assigns to each inner variable the zero value, thereby switching off the internal variables.

Choosing a section $\sigma : B \to E$ of the projection map π provides a different link between a Poisson tensor Λ_E on E and a Poisson tensor Λ_B on B. We will say that Λ_E is an extension of Λ_B if the map σ is Poisson, or equivalently if the induced map $\sigma^* \colon \mathcal{F}(E) \to \mathcal{F}(B)$ is a Lie algebra homomorphism. This will mean in particular that if E is a vector bundle and σ is the zero section, then in bundle coordinates ξ^i and ζ^a, the Poisson brackets will have the form

$$\{\xi^i, \xi^j\}(\xi, 0) = \Lambda_B(\xi), \quad \{\xi^i, \zeta^a\}(\xi, 0) = 0, \quad \{\zeta^a, \zeta^b\}(\xi, 0) = 0.$$

Whenever this accretive condition on inner variables is required from physical principles, we will impose in addition that the Poisson tensor Λ_E will be an extension of Λ_B with respect to the zero section.

5.6.2.3 Feynman's Problem in Tangent Bundles with Inner Degrees of Freedom

We are now prepared to try to determine all Poisson structures on a tangent bundle with internal degrees of freedom and satisfying some fundamental physical restrictions. We will follow a similar chain of reasoning as that in Sect. 5.6.1.

Our setting will be that of a system with configuration space Q and possessing internal degrees of freedom modeled on a bundle $\rho \colon F \to Q$. We can pull back the bundle $\tau_Q \colon TQ \to Q$ along the map ρ, to obtain a new bundle $E = \rho^*(TQ)$ over F. The pull-back bundle $\rho^*(TQ)$ is defined as the set of pairs (I, v) where $I \in F$, $v \in TQ$, and their base points agree, i.e., $\rho(I) = q = \tau_Q(v)$. There is a natural map $p_1 \colon \rho^*(TQ) \to F$, sending each pair (I, v) into its first component $p_1(v, I) = v$. Then the fibre of $\rho^*(TQ)$ at the point $I \in F$ is simply the vector space T_qQ with $q = \rho(I)$. Thus $p_1 \colon \rho^*(TQ) \to F$ is a vector bundle with the fibres of $TQ \to Q$, $p_1^{-1}(I) = T_{\rho(I)}Q$ (see Fig 5.6.). Notice that if we take the pull-back of the bundle $F \to Q$ along the map $\tau_Q \colon TQ \to Q$, then we obtain the bundle $\tau_Q^*(F)$ made up of pairs (v, I) with the same property as before, and projection map $p_2 \colon \tau_Q^*(F) \to TQ$ given by $p_2(v, I) = v$. The total spaces $\rho^*(TQ)$ and $\tau_Q^*(F)$ are the same (this will be denoted by $TQ \bowtie F$), but the fibre structures are different. The fibre of $TQ \bowtie F$ as a bundle over TQ with projection map p_2 is given by $F_q = \rho^{-1}(q)$.

We will be searching for Poisson tensors Λ on $TQ \bowtie F = \rho^*(TQ)$ that will be p_1-related to a localizable Poisson tensor Λ_I on $F \to Q$ given by Eq. (5.247). Such Poisson tensors will be called localizable on $TQ \bowtie F$. This requirement is equivalent to Λ to having the form

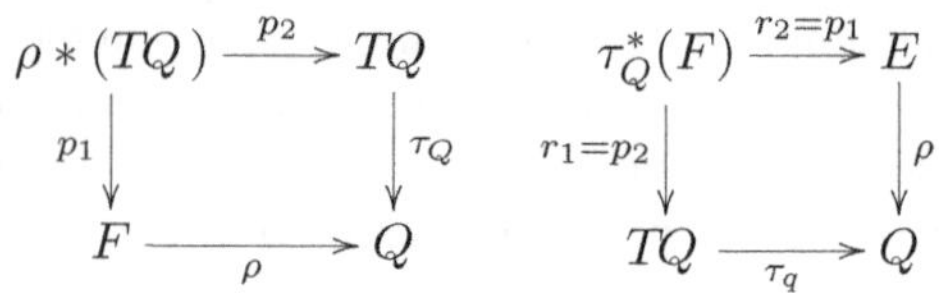

Fig. 5.6 Internal variables and configuration bundles

$$\Lambda = b^{ij}(q, v, I)\frac{\partial}{\partial q^i} \wedge \frac{\partial}{\partial v^j} + c^{ij}(q, v, I)\frac{\partial}{\partial v^i} \wedge \frac{\partial}{\partial v^j}$$
$$+ A_a^i(q, v, I)\frac{\partial}{\partial v^i} \wedge \frac{\partial}{\partial I_a} + C_{ab}(q, I)\frac{\partial}{\partial I_a} \wedge \frac{\partial}{\partial I_b} . \tag{5.249}$$

Notice that the fundamental Poisson brackets of the variables q^i, v^i and I_a can all depend on q, v and I, except for C_{ab} which depends only on the q's and I's.

We have discovered in the previous section that the existence of a Hamiltonian SODE for a Poisson tensor has important consequences. Something similar happens now. Even though the bundle $p_1 \colon TQ \bowtie F \to F$ is not a tangent bundle, there is a natural tensor S on it that plays the same rôle as the tangent structure on TQ discussed in Sect. 5.5.3. Locally, this tensor S has the same form as the tangent structure on TQ, i.e.,

$$S = \frac{\partial}{\partial v^i} \otimes dq^i . \tag{5.250}$$

It is easy to show that S is well defined, either by direct computation or by noticing that there is a natural projection Π from TF to $E = \rho^*(TQ)$, given by

$$\Pi \colon (q^i, I^a; v^i, \dot{I}^a) \mapsto (q^i, I^a; v^i), \tag{5.251}$$

and that the natural tangent structure S_F on TF is projectable along this map (see Fig. 5.7). The projected tensor results from S given by Eq. (5.250).

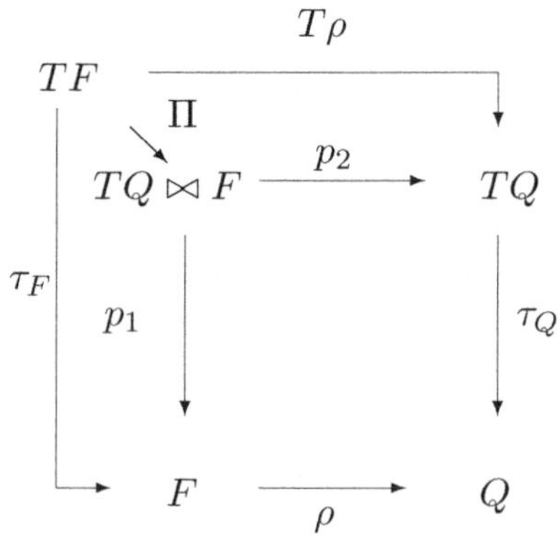

Fig. 5.7 The almost tangent structure on $TQ \bowtie F$

We will be considering dynamical systems which are second-order in the variables q^i and v^i and first-order on the variables I_a. The corresponding vector fields will again be called second-order differential equations (SODE's) on E, or also driven SODE's [Ib91]. These vector fields Γ are characterized by $S(\Gamma) = \Delta$, where Δ is the dilation field along the fibres of $\rho^*(TQ)$, i.e., Δ has the form $v^i\partial/\partial v^i$. The most general form of a driven SODE is given by

$$\Gamma = v^i \frac{\partial}{\partial q^i} + F^i(q, v, I)\frac{\partial}{\partial v^i} + f_a(q, v, I)\frac{\partial}{\partial I_a} , \tag{5.252}$$

and its integral curves satisfy the coupled system of differential equations:

$$\ddot{q}^i = F^i(q, v; I) , \tag{5.253}$$

$$\dot{I}_a = f_a(q, v; I) . \tag{5.254}$$

Compare these equations with Eq. (4.72).

We will denote by $\mathcal{F}_{TQ\bowtie F}(Q)$ the functions on $E = TQ \bowtie F$ obtained by pull-back of functions on Q, $\mathcal{F}_{TQ\bowtie F}(Q) = (\rho \circ p_1)^*\mathcal{F}(Q)$, i.e., functions depending only on q's. We will denote by $\mathcal{F}_{TQ\bowtie F}(F)$ the functions on E obtained by pull-back of functions on F by the map p_1, i.e., functions on E depending on both q's and I's.

Proposition 5.55 *Let Λ be a localizable Poisson tensor on $TQ \bowtie F$ (Eq. 5.249). If Γ is a Hamiltonian SODE, then there are no nonconstant locally defined Casimir functions on $\mathcal{F}_{TQ\bowtie F}(Q)$. Moreover, if the Poisson tensor Λ_I is regular along the fibres of ρ, then there are no nonconstant locally defined Casimir function s on $\mathcal{F}_{TQ\bowtie F}(F)$ either.*

Proof Let ϕ be a locally defined Casimir function on $\mathcal{F}_{TQ\bowtie F}(Q)$. Then $\Gamma(\phi) = 0$ if Γ is a Hamiltonian SODE on E. But clearly this implies that $\partial\phi/\partial q^i = 0$, as in Lemma 5.48.

On the other hand, if ϕ is a locally defined Casimir function on $\mathcal{F}_{TQ\bowtie F}(F)$, then $\Gamma(\phi) = 0$ again because Γ is Hamiltonian. Because of Eq. (5.252), this implies that

$$0 = v^i \frac{\partial\phi}{\partial q^i} + f_a(q, v, I)\frac{\partial\phi}{\partial I_a} .$$

Since ϕ does not depend on v, $f_a(q, v, I) = f_{ia}(q, I)v^i$ and

$$\frac{\partial\phi}{\partial q^i} + f_{ia}(q, I)\frac{\partial\phi}{\partial I_a} = 0 . \tag{5.255}$$

On the other hand, if ϕ is a locally defined Casimir function on $\mathcal{F}_{TQ\bowtie F}(F)$, then $\widehat{\Lambda}(d\phi) = 0$, and this implies that

$$\left(b^{ij}\frac{\partial\phi}{\partial q^j} - A^i_a\frac{\partial\phi}{\partial I_a}\right)\frac{\partial}{\partial v^i} + C_{ab}\frac{\partial\phi}{\partial I_a}\frac{\partial}{\partial I_b} = 0 .$$

Then $C_{ab}(q, I)\partial\phi/\partial I_a = 0$. But if Λ_I is non-degenerate along the fibres of ρ, there cannot be nonconstant locally defined Casimir functions along the fibres of ρ, i.e., $\partial\phi/\partial I_a = 0$. Hence because of Eq. (5.255), $\partial\phi/\partial q^i = 0$. □

As in the previous section we will say that a Poisson tensor on $TQ \bowtie F$ is Newtonian if it possesses a Hamiltonian SODE.

What about the regularity of Λ on $TQ \bowtie F$? To answer this question, we can repeat the discussion leading to Theorem 5.50, but first we notice that:

Lemma 5.56 *A localizable Poisson tensor Λ on $TQ \bowtie F$ of the form given by Eq. (5.249) will be regular iff b^{ij} and C_{ab} are invertible.*

Proof The matrix defined by Λ on the basis constructed from $\partial/\partial q^i$, $\partial/\partial I_a$ and $\partial/\partial v^i$ is

$$\left(\begin{array}{c|c|c} 0 & 0 & b \\ \hline 0 & C & A \\ \hline -b^t & -A^t & c \end{array}\right),$$

which is invertible iff b and C are. □

Theorem 5.57 *If Λ is a localizable Poisson tensor on $TQ \bowtie F$ such that the Poisson tensor Λ_I on the configuration bundle $F \to Q$ is regular along the fibres, and possessing a Hamiltonian SODE, then Λ is regular.*

Proof It is easy to show that b^{ij} is invertible using locally defined Casimirs on $\mathcal{F}_{TQ\bowtie F}(Q)$. In fact, Λ must be a monomorphism acting on $\mathcal{F}_{TQ\bowtie F}(Q)$. (If not there would be locally defined Casimir functions on $\mathcal{F}_{TQ\bowtie}(Q)$. This is impossible because of Proposition 5.55.) On the other hand, it is easy to check that $\widehat{\Lambda}(d\phi)$ is again $b^{ij}\partial\phi/\partial q^i \partial/\partial v^i$, implying the invertibility of b^{ij}.

Finally, Λ is also a monomorphism on $\mathcal{F}_E(F)$, because if Λ_I is regular along the fibres there are no locally defined Casimir functions on $\mathcal{F}(F)$ (Proposition 5.55 again). Then, computing $\Lambda(d\phi)$ we find

$$\widehat{\Lambda}(d\phi) = C_{ab}\frac{\partial\phi}{\partial I_a}\frac{\partial}{\partial I_b} + \left(b^{ij}\frac{\partial\phi}{\partial q^j} - A_a^i\frac{\partial\phi}{\partial I_a}\right)\frac{\partial}{\partial v^i},$$

then, C^{ab} must be invertible. □

From this result we cannot conclude the existence of a Lagrangian function on $TQ \bowtie F$, because it is not a tangent bundle. However we could try to construct a singular Lagrangian L on TF, such that after some reduction the Cartan form ω_L induces a form ω on E that is the inverse of Λ. The necessary and sufficient conditions for this to happen were established in [Ib91] in a situation slightly different from the one we are dealing with. We will then state it, adapting it to the present context, as follows:

Theorem 5.58 *Let $p_1 : TQ \bowtie F \to TQ$ be the vector bundle obtained by pull-back from a bundle F over Q, and let $p_2 : TQ \bowtie F \to F$ be the natural projection on the second factor. Let S be the natural $(1,1)$-tensor field on $TQ \bowtie F$ given by Eq. (5.250) that projects onto the natural tangent structure S_Q on TQ along the map Π defined by Eq. (5.251). If ω is a symplectic form on $TQ \bowtie F$, then there will exist a (locally defined) function L on TF such that the Cartan 2-form ω_L will project onto ω along Π, if and only if the following two conditions are satisfied:*

i. $i_S\omega = 0$.
ii. $\omega(V_1, V_2) = 0$ *for all vector fields vertical with respect to p_2, i.e., vectors tangent to the tangent spaces $T_qQ \subset TQ \bowtie F$.*

Proof The necessity of conditions (i) and (ii) is almost obvious. Let L be a Lagrangian function on TF such that ω_L projects onto ω along Π. Then, $\ker \omega_L = \ker T\Pi$. From the fact that ω_L vanishes on vertical bivectors on TF, we get condition (ii). In general, because of Eq. (5.158), ω_L is such that $i_{S_F}\omega_L = 0$, where S_F is the tangent structure on TF. We get S by projecting S_F along the map Π, and consequently condition (i) is satisfied.

Conversely, let us assume that there is a symplectic form ω on $TQ \bowtie F$ satisfying (i) and (ii). Let $\Omega = \Pi^*\omega$ be the pull-back of ω to TF. Then we will show that $i_{S_F}\Omega = 0$. In fact,

$$i_{S_F}\Omega(X, Y) = \Omega(S_F(X), Y) + \Omega(X, S_F(Y)) ,$$

for any two vector fields X and Y on TF. Taking $X = \partial/\partial q^i$ and $Y = \partial/\partial q^j$, we get

$$\begin{aligned} i_{S_F}\Omega\left(\frac{\partial}{\partial q^i}, \frac{\partial}{\partial q^j}\right) &= \Omega\left(\frac{\partial}{\partial v^i}, \frac{\partial}{\partial q^j}\right) + \Omega\left(\frac{\partial}{\partial q^i}, \frac{\partial}{\partial v^j}\right) \\ &= \omega\left(\frac{\partial}{\partial v^i}, \frac{\partial}{\partial q^j}\right) + \omega\left(\frac{\partial}{\partial q^i}, \frac{\partial}{\partial v^j}\right) = i_S\omega\left(\frac{\partial}{\partial q^i}, \frac{\partial}{\partial q^j}\right) = 0 . \end{aligned}$$

If we take now $X = \partial/\partial q^i$ and $Y = \partial/\partial I_a$, we get

$$\begin{aligned} i_{S_F}\Omega\left(\frac{\partial}{\partial q^i}, \frac{\partial}{\partial I_a}\right) &= \Omega\left(\frac{\partial}{\partial v^i}, \frac{\partial}{\partial I_a}\right) + \Omega\left(\frac{\partial}{\partial q^i}, \frac{\partial}{\partial \dot{I}_a}\right) \\ &= \omega\left(\frac{\partial}{\partial v^i}, \frac{\partial}{\partial I_a}\right) = i_S\omega\left(\frac{\partial}{\partial q^i}, \frac{\partial}{\partial I_a}\right) = 0 . \end{aligned}$$

Finally consider $X = \partial/\partial I_a$ and $Y = \partial/\partial I_b$. Then,

$$i_{S_F}\Omega\left(\frac{\partial}{\partial I_a}, \frac{\partial}{\partial I_b}\right) = \Omega\left(\frac{\partial}{\partial I_a}, \frac{\partial}{\partial \dot{I}_b}\right) + \Omega\left(\frac{\partial}{\partial \dot{I}_a}, \frac{\partial}{\partial I_b}\right) = 0 .$$

The evaluation of $i_{S_F}\Omega$ on all the remaining pairs of basis vectors vanishes because they always contain a vertical vector field. Then because of Theorem 5.39, there will exist a (locally) defined Lagrangian function L on TF such that $\Omega = \omega_L$, completing the proof. □

Thus, we can finally state:

Theorem 5.59 *If Λ is a Newtonian localizable Poisson tensor on $TQ \bowtie F$ such that the Poisson tensor Λ_I on the configuration bundle $F \to Q$ is regular along the fibres, then there is a Lagrangian realization of it on TF, in the sense that there exists a locally defined Lagrangian function L on $F \to Q$, such that Λ is obtained by direct reduction from ω_L.*

In this way we have obtained another no-go theorem analogous to that of Feynman. Thus even with the inclusion of inner degrees of freedom we do not escape the Lagrangian formalism whenever the Poisson brackets on the inner degrees of freedom are regular. The latter need not be the case as was seen in Sects. 2 and 3. In fact, for particles in a Yang-Mills field or for spinning particles, the inner degrees of freedom (the isospin and spin, respectively) are defined in a space with degenerate Poisson brackets. However, in both situations, we can restrict the classical inner degrees of freedom to a subbundle of F made up of symplectic leaves on each fibre; for example, spheres for the spinning particle and coadjoint orbits for the Yang-Mills particle, with the restriction that Λ_I is regular. Then we will apply the previous theorem to this bundle obtaining a Lagrangian representation for the system. Notice that the Lagrangian will be singular and only locally defined if the topology of the fibres of the symplectic bundle is complicated enough.

5.6.3 The Inverse Problem for Lagrangian Dynamics

5.6.3.1 The Inverse Problem of the Calculus of Variations

In previous Sections we have given examples of second-order vector fields that admit more than one Lagrangian description with genuinely nonequivalent, i.e., not differing merely by a 'total time derivative' (a gauge term), Lagrangians, and we have also mentioned that there are second-order fields for which no non trivial Lagrangian description is available.

The problem of whether or not a given second-order field Γ admits of Lagrangian descriptions and, in the affirmative case, whether or not the description is unique constitutes what is commonly known as the 'Inverse Problem'. The full name that one finds in the literature is actually 'Inverse Problem in the Calculus of Variations', and that for the historical reason that the problem was stated for the first time by Helmoltz back in 1887 within the setting of Hamilton's principle.

It has been established long ago [CS55] that the Inverse Problem for one-dimensional systems has actually infinitely many solutions. So, we will consider here second-order dynamical systems in two or higher space dimensions.

We will discuss briefly here only two relatively modern approaches to the Inverse Problem. The first one has to do with the uniqueness of the Lagrangian description once a (regular) Lagrangian is available. The second one deals instead also with the very existence of Lagrangian descriptions (if any).

To illustrate the first approach, let us assume that a given second-order field admits of a Lagrangian description with a regular Lagrangian L, and let us denote by ω_L the associated Lagrangian 2-form and by $H = (H_{ij})$ the Hessian matrix of L (hence: $\det H \neq 0$ by assumption).

If a second Lagrangian exists, say L' with associated Lagrangian 2-form $\omega_{L'}$, we can define a $(1, 1)$ tensor R, called also the recursion operator associated with the two Lagrangians via:

$$\omega_L(RX, Y) =: \omega_{L'}(X, Y) \tag{5.256}$$

for any pair X and Y of vector fields on the tangent bundle.

The main properties of the recursion operator are reviewed in Appendix F, and we will only summarize them here, referring to the Appendix (and to the literature cited therein) for more details:

(i) If ω_L is nondegenerate, R is uniquely defined, and will be invertible iff $\omega_{L'}$ is non-degenerate as well. In other words: $\ker(R) \equiv \ker(\omega_{L'})$.

(ii) In local coordinates, R can be written as

$$R = M_j{}^i \left(\frac{\partial}{\partial q^i} \otimes dq^j + \frac{\partial}{\partial v^i} \otimes dv^j \right) + (\mathcal{L}_{\Gamma j} M)^i \frac{\partial}{\partial v^i} \otimes dq^j \,, \tag{5.257}$$

where

$$M = H' \cdot H^{-1} \,, \tag{5.258}$$

H' being the Hessian matrix of L'. This implies that $M \cdot H$ be a symmetric matrix.

(iii) The matrix M satisfies

$$\mathcal{L}_\Gamma M = [A, M] \,, \tag{5.259}$$

where, if $\Gamma = v^i \partial/\partial q^i + F^i \partial/\partial v^i$, the matrix $A = (A_i{}^j)$ is given by

$$A_i{}^j =: -\frac{1}{2} \frac{\partial F^j}{\partial v^i} \,. \tag{5.260}$$

(iv) Finally, M satisfies also the constraint

$$[M, \mathcal{L}_\Gamma A + B] = A[A, M] + [A, M]A \,, \tag{5.261}$$

where

$$B = (B_i{}^j) \,, \qquad B_i{}^j =: \frac{\partial F^j}{\partial q^i} \tag{5.262}$$

With the usual identification of Γ with a total time derivative with respect to an evolution parameter, we can view Eq. 5.4 as a 'Heisenberg-type' evolution equation for M (which is now assumed to be unknown), supplemented by the constraints given by Eq. 5.6 and by the request that M be a symmetric matrices, i.e. by the additional constraint:

$$M = M^t. \tag{5.263}$$

Of course both the evolution equation and the constraints are trivially solved by $M \propto I$ (the identity matrix), and the Lagrangian description will be unique (up to equivalence) if this is the case. Both constraints have to be always satisfied in order that the 'dynamical' evolution generates admissible M's, and this is somehow reminiscent of the situation one encounters in Dirac's theory of constraints. Taking successive Lie derivatives with respect to Γ of the constraints will (or may) generate therefore new constraints that have to be satisfied, thereby further restricting the admissible form of M. If it happens that the full set of constraints generated in this way forces M to be the trivial solution of the problem, i.e., a multiple of the identity, then, up to equivalence, the Lagrangian description will be unique.

For example, for velocity-independent forces, $A_i{}^j \equiv 0$, and the relevant set of equations reduces to:

$$\begin{gathered} \mathcal{L}_\Gamma M \\ [M, B] = 0 \\ M = M^t. \end{gathered} \tag{5.264}$$

This will be the case, e.g., in two space dimensions, for simple Lagrangians 'of the mechanical type', i.e., of the form:

$$L = \frac{1}{2}(u_x^2 + u_y^2) - V(x, y) \tag{5.265}$$

Then, it has been shown that the Lagrangian description for two-dimensional systems of the mechanical type is indeed unique unless V is a separable potential, and this is the case only for the free particle and the harmonic oscillator.

Remark 5.17 This conclusion seems to be at variance with a result of Henneaux and Shepley [HS84], according to which a family of admissible (and alternative) Lagrangians for the Kepler problem is given by:

$$L_\gamma = \frac{1}{2}\delta_{ij} v^i v^j + \frac{c}{q} + \gamma \frac{J}{q^2} \tag{5.266}$$

where: $q =: \sqrt{\delta_{ij} q^i q^j}$, c is a fixed real constant,

$$J =: \sqrt{q^2 \delta_{ij} v^i v^j - (\delta_{ij} q^i v^j)^2} \tag{5.267}$$

is the absolute value of the angular momentum and the family is parametrized by the real parameter γ. There is however no contradiction, as, in two space dimensions (this is nothing but Kepler's second law, of course)

$$\frac{J}{q^2} = i_\Gamma d\theta \tag{5.268}$$

with θ the polar angle, and the additional term reduces to a 'gauge term', which is not the case, as Henneaux and Shepley have shown, in higher space dimensions.

We turn now to the second method, that is due to M. Henneaux. First of all, let us introduce some preliminary notations and conventions. Denoting then by $X \wedge Y$ the bivector (i.e., the skewsymmetrized tensor product) associated with a pair of vector fields X and Y,

$$X \wedge Y =: X \otimes Y - Y \otimes X, \tag{5.269}$$

it is not hard to check that, if Ω is a 2-form, then:

$$\Omega(X, Y) = \frac{1}{2} \langle \Omega | X \wedge Y \rangle \tag{5.270}$$

where $\langle \cdot | \cdot \rangle$ stands for the fully-contracted multiplication of Ω and the bivector.

We can now introduce an equivalence relation (denoted by the symbol $\simeq$) between pairs of bivectors by declaring two of them to be equivalent iff a $\mathcal{F}(TM)$-linear combination of them (their difference in particular) vanishes upon contraction with a given, and fixed, 2-form.

If now a given second-order field Γ admits of a Lagrangian description with an admissible Lagrangian L, then the associated Lagrangian 2-form ω_L, besides being invariant under the dynamics and closed, will have to vanish on pairs of vertical fields. This amounts to the constraint

$$\omega_L(SX \wedge SY) = 0\,, \quad \forall X, Y \in \mathfrak{X}(TM)\,, \tag{5.271}$$

i.e., $SX \wedge SY \simeq 0$ in the notation established above. This constraint too has to be preserved by the dynamical evolution. Taking then successive derivatives with respect to. Γ will generate the set of constraints

$$\omega_L((\mathcal{L}_\Gamma)^k (SX \wedge SY)) = 0, \qquad k = 0, 1, 2, \ldots \tag{5.272}$$

that have to be obeyed by the (a priori unknown) Lagrangian 2-form ω_L. If $n = \dim(M)$, the dimension of the space of bivectors is the same as that of 2-forms, i.e., $n(2n-1)$, and this will be also the maximum number of independent constraints that can be generated in the above way.

An important observation is that if for a certain value of k the $(k+1)$-th bivector depends on the previous ones, then the same will be true for all the subsequent ones. The generation of independent bivectors will stop then, and k will be the maximum number of independent constrains. If $k = n(2n - 1)$, the bivectors generated in this way will form a basis, implying $\omega_L = 0$, and the Inverse Problem will have no solution at all.

Let us prove then the statement we have just made. Let us denote by simplicity as $v^0, v^1, \ldots, v^k, \ldots$ the set of bivectors generated by taking successive Lie derivatives, i.e.

$$v^{i+1} = \mathcal{L}_\Gamma v^i , \quad i \geq 0 . \tag{5.273}$$

Independence of the first k bivectors and dependence of the $(k+1)$-th from the previous ones are expressed by:

$$\begin{aligned} &v^0 \wedge v^1 \wedge \cdots \wedge v^k \neq 0 \\ &v^0 \wedge v^1 \wedge \cdots \wedge v^k \wedge v^{k+1} = 0 . \end{aligned} \tag{5.274}$$

Then,

$$0 = \mathcal{L}_\Gamma (v^0 \wedge v^1 \wedge \cdots \wedge v^k \wedge v^{k+1}) = \sum_{i=0}^{k+1} v^0 \wedge \cdots \wedge \mathcal{L}_\Gamma v^i \wedge \cdots \wedge v^{k+1} . \tag{5.275}$$

But all the terms but the last vanish on account of the definition of the v^i's. We are thus left with:

$$0 = v^0 \wedge v^1 \wedge \cdots \wedge v^k \wedge \mathcal{L}_\Gamma v^{k+1} \equiv v^0 \wedge v^1 \wedge \cdots \wedge v^k \wedge v^{k+2} , \tag{5.276}$$

and this completes the proof.

Returning then to the set of constraints that have been generated with this procedure, we see that the number of independent 2-forms that satisfy all the constraints is given by the 'codimension' of the set of bivectors that has been generated. If the codimension is zero, there will be no possible Lagrangian description. If it is greater than zero those forms in the set that are also closed and that are invariant under the dynamics (and this will have to be checked separately) will be admissible Lagrangian 2-forms for Γ and will correspond to genuinely alternative Lagrangians. The codimension of the set of bivectors will give then the 'maximum' number of possible alternative Lagrangians.

To illustrate the method, let us consider $T\mathbb{R}^2$, $(n = 2)$, with coordinates (x, y, u_x, u_y) and the second-order field:

$$\Gamma = u_x \frac{\partial}{\partial x} + u_y \frac{\partial}{\partial y} + (x + y) \frac{\partial}{\partial u_x} + xy \frac{\partial}{\partial u_y} \tag{5.277}$$

Let us denote by X^i, $i = 0, \ldots, 5$, the standard basis of bi vector fields on $T\mathbb{R}^2$,

$$X^{(0)} = \frac{\partial}{\partial u_x} \wedge \frac{\partial}{\partial u_y}, \qquad X^{(1)} = \frac{\partial}{\partial x} \wedge \frac{\partial}{\partial y}, \qquad X^{(2)} = \frac{\partial}{\partial x} \wedge \frac{\partial}{\partial u_x}$$
$$X^{(3)} = \frac{\partial}{\partial x} \wedge \frac{\partial}{\partial u_y}, \qquad X^{(4)} = \frac{\partial}{\partial y} \wedge \frac{\partial}{\partial u_x}, \qquad X^{(5)} = \frac{\partial}{\partial y} \wedge \frac{\partial}{\partial u_y}.$$

Taking then Lie derivatives with respect to Γ starting from $v^0 =: X^{(0)}$ will generate the set of $n(2n-1) = 6$ bivectors:

$$v^i = a^i{}_j X^{(j)}, \qquad i, j = 0, \ldots, 5, \tag{5.278}$$

where

$$a = (a^i{}_j) = \begin{pmatrix} 1 & 0 & 0 & 0 & 0 & 0 \\ 0 & 0 & 1 & -1 & 0 & 0 \\ 0 & 1 & 0 & 0 & 0 & 0 \\ 0 & 0 & -1 & -x & 1 & y \\ 0 & 0 & 0 & -u_x & 0 & u_y \\ 0 & 0 & 0 & -(x+y) & 0 & xy \end{pmatrix} \tag{5.279}$$

Now: $v^0 \wedge \ldots \wedge v^5 = DX^{(0)} \wedge \ldots \wedge X^{(5)}$, where : $D = \det a$, and will vanish iff D does. But $D = xyu_x - (x+t)u_y \neq 0$. Therefore the v^i's are a basis, codim span $\{(v^0, \ldots, v^5)\} = 0$, and the present criterion shows that Γ does not admit non trivialLagrangian descriptions.

At the opposite extreme, let us consider the n-dimensional isotropic harmonic oscillator. Then,

$$\Gamma = v^i \frac{\partial}{\partial q^i} - q^i \frac{\partial}{\partial v^i}. \tag{5.280}$$

Starting from the bivectors

$$\Lambda^{(0)}_{ij} = \frac{\partial}{\partial v^i} \wedge \frac{\partial}{\partial v^j}, \qquad i, j = 1, \ldots, n, \tag{5.281}$$

and taking Lie derivatives, by making use of $\mathcal{L}_\Gamma \partial/\partial q^i = \partial/\partial v^i$ and $\mathcal{L}_\Gamma \partial/\partial v^i = -\partial/\partial q^i$, one obtains

$$\Lambda^{(1)}_{ij} =: \mathcal{L}_\Gamma \Lambda^{(0)}_{ij} = \frac{\partial}{\partial q^j} \wedge \frac{\partial}{\partial v^i} - \frac{\partial}{\partial q^i} \wedge \frac{\partial}{\partial v^j}. \tag{5.282}$$

But then,

$$\Lambda^{(2)}_{ij} =: \mathcal{L}_\Gamma \Lambda^{(1)}_{ij} \equiv 0 \tag{5.283}$$

and the procedure terminates. As there are $n(n-1)/2$ independent bivectors in the set of the $\Lambda^{(0)}_{ij}$'s and as many in that of the $\Lambda^{(1)}_{ij}$'s (that are by construction skew-symmetric), the codimension of the space of bivectors that has been generated turns out to be n^2.

Consider then the family of 2-forms with constant coefficients:

$$\omega_{B,C} = B_{ij}\, dq^i \wedge dv^j + \frac{1}{2}\, C_{ij}\, dq^i \wedge dq^j . \tag{5.284}$$

They vanish on vertical bivectors and are closed (actually exact). If they have to be Lagrangian 2-forms, then $B = (B_{ij})$ must be the Hessian matrix of the Lagrangian, hence a symmetric matrix, while $C = (C_{ij})$ is skew-symmetric by construction. The total number of independent $\omega_{B,C}$'s is then precisely n^2 (recall that there are $n(n+1)/2$ independent symmetric $n \times n$ matrices and $n(n-1)/2$ skew-symmetric ones) . However, imposing invarince under the dynamics we obtain $\mathcal{L}_\Gamma \omega_{B,C} = C_{ij} dv^i \wedge dq^j$ (all the other terms vanishing by symmetry) and, in the next order, $\mathcal{L}_\Gamma(\mathcal{L}_\Gamma \omega_{B,C}) = C_{ij}\left(dv^i \wedge dv^j - dq^i \wedge dq^j\right)$, which forces the matrix C to be zero. We are then left with 2-forms that are parametrized only by symmetric matrices and that are generated, of course, by the family $\{L_B\}$ of Lagrangians that were introduced and discussed in a previous Section.

These simple examples suggest that generically, i.e., unless the dynamical vector field admits of a fairly large symmetry group (as is the case for the harmonic oscillator), the set of constraints that is generated in both methods will be enough either to prove the the nonexistence of a Lagrangian description or it will force the latter to be unique. We conclude that generic second-order dynamical systems either will not admit of a Lagrangian description at all, or the Lagrangian description will be generically unique.

These conclusions seem to suggest that, in the case of non-uniqueness of the Lagrangian description, one might try to add a small perturbation to the dynamics, find out the (hopefully unique) Lagrangian description for the perturbed dynamics and then switch off the perturbation to recover the 'right' Lagrangian for the original dynamics. However, as the space of second-order vector fields is not endowed with any natural topology, what 'small' means in this context is a highly ambiguous concept. Moreover, a generic perturbation may well lead to a new vector field that simply does not admit of any Lagrangian description, and we are not aware of any sensible procedure to select perturbations of the appropriate form (what we might call 'Lagrangian perturbations'). Last but not least, in physical problems one is almost always confronted not with 'generic' but with very 'specific' dynamical systems, and if there are ambiguities in their Lagrangian description they will remain there.

Remark 5.18 As a final remark, let us stress what may seem an obvious fact, but need not be such, namely that all the methods for tackling the Inverse Problem that have been described here assume implicitly that the tangent bundle structure has been given once and for all. So, all the conclusions are true relative to a given tangent bundle structure, but might well not hold true under alternative tangent bundle structures like, e.g., the one that was used in the last example. In other words, and although we do not have at the moment any specific examples, we cannot rule out the possibility for a vector field Γ on a vector space V to be second-order with respect to two different tangent bundle structures on V and to admit, say, of Lagrangian descriptions with one or more alternative Lagrangians with respect to one of the structures and of fewer or none at all with respect to the second one.

Before addressing the central discussion on this paper, we will review its classical counterpart from a variety of different perspectives.

5.6.3.2 The Hamiltonian Inverse Problem for Classical Systems

Given a vector field Γ defined on a state space modelled by the manifold with local coordinates x^k, the Hamiltonian inverse problem for Γ consists in finding a Poisson bracket $\{\cdot,\cdot\}$ and a smooth function H such that

$$\{H, x^k\} = \Gamma^k(x), \tag{5.285}$$

and $\Gamma = \Gamma^k(x)\partial/\partial x^k$. In other words, the Hamiltonian inverse problem for Γ is equivalent to determine whether or not Γ is a Hamiltonian vector field with respect to some Poisson brackets. It is remarkable that in this formulation, neither the Poisson brackets nor the Hamiltonian are determined a priori, this is, if they carry physical information it has to be determined by further information not contained in the vector field Γ alone.

The inverse problem can be considered locally or globally, i.e., we can discuss the existence of a Poisson structure in the neighborhood of an ordinary point for the vector field Γ such that $\mathcal{L}_\Gamma\{\cdot,\cdot\} = 0$, and in a second step the existence of a globally defined Poisson structure. We will not pursue this aspect of the problem here (see [Ib90] for a discussion of the related issue of global aspects of the inverse problem in the Lagrangian formalism).

5.6.3.3 The Symplectic Inverse Problem for Classical Systems

The inverse Hamiltonian problem can be restricted requiring that the Poisson tensor Λ defined by the fundamental commutation relations

$$\{x^i, x^j\} = \Lambda^{ij}(x), \tag{5.286}$$

will be nondegenerate, i.e., $\det \Lambda^{ij} \neq 0$. In such case, inverting Λ we will get a 2-form ω, defined by

$$\omega = \omega_{ij}\, dx^i \wedge dx^j,$$

whith $\Lambda^{ik}\omega_{kj} = \delta^i_j$, such that ω is closed and,

$$i_\Gamma \omega = dH, \tag{5.287}$$

that is, Γ is the Hamiltonian vector field defined by the function H and the symplectic form ω. Notice that in particular solving the inverse problem for Γ, i.e., finding a symplectic form ω and a function H such that Eq. (5.287) holds, requires that the phase space of the system will be even dimensional.

5.6.3.4 The Lagrangian Inverse Problem

If the state space manifold has the form (locally at least) of the tangent bundle of a configuration space Q with local coordinates x^k, v^k, then the symplectic inverse problem for Γ would be solved if we were able to find a regular Lagrangian for it, that is, a Lagrangian L such that its Euler–Lagrange equations will be equivalent to the vector field Γ and $\det \partial^2 L/\partial v^k \partial v^j \neq 0$. It is well-known that the Euler–Lagrange vector field obtained from a regular Lagrangian defines a second-order differential equation (SODE). Thus, if Γ is a SODE,

$$\Gamma = v^k \frac{\partial}{\partial x^k} + f^j(q, v) \frac{\partial}{\partial v^j},$$

i.e., the equations deining its integral curves are

$$\frac{dx^k}{dt} = v^k, \qquad \frac{dv^j}{dt} = f^j(q, v),$$

then the Lagrangian inverse problem for Γ will be to find a function L such that it satisfies the linear partial differential equation,

$$\frac{\partial^2 L}{\partial v^k v^j} f^j - \frac{\partial^2 L}{\partial v^k x^j} v^j - \frac{\partial L}{\partial x^k} = 0,$$

and the regularity condition on L,

$$\det \frac{\partial^2 L}{\partial v^k v^j} \neq 0,$$

which spoils the linearity of the previous equation.

There is also a Poisson version of the previous discussion, as reflected by the following equations,

$$\{H, x^i\} = v^i, \{H, \{H, x^i\}\} = f^i(x, \{H, x\}).$$

Adding the localizability condition,

$$\{x^i, x^j\} = 0,$$

we recover Feynman's problem as reported by Dyson [Dy90]. It is possible to go beyond the Lagrangian description while preserving the commutator description? The answer to this question is contained in the following theorem (see [Ca95] for an updated revision of Feynman's problem and references therein).

Theorem 5.60 *Let* $\Gamma = v^i\partial/\partial x^i + f^i\partial/\partial v^i$ *be a SODE, then it has a Poisson description in terms of a function* $H(x, v)$ *and a localizable Poisson bracket iff it can be given a (local) Lagrangian description.*

A simple proof of the previous theorem is obtained if we make the additional assumption that,

$$\frac{\partial}{\partial v^k}\{\cdot,\cdot\} = 0.$$

Then, if we start with $v^k = \{H, x^k\}$, taking the derivative with respect to v^j we find,

$$\delta_j^k = \{\frac{\partial H}{\partial v^j}, x^k\} = \frac{\partial^2 H}{\partial v^j \partial v^l}\{v^l, x^k\}.$$

Thus we can define the Legendre transform and there will be a Lagrangian description.

5.6.4 *Feynman's Problem and Lie Groups*

5.6.4.1 Standard Lagrangian Realizations for External Variables

We know that Feynman's procedure gives a negative answer in many instances; negative in the sense that there is always a Lagrangian description for dynamics compatible with the requirements described in Theorem 5.57. In this section we will discuss Feynman's procedure from a group theoretical setting, showing its relation to the construction of Lagrangian realizations of systems defined on group manifolds or on associated bundles to principal fibre bundles. This construction again includes the cases of Yang-Mills particles and spinning particles.

We will start by considering Feynman's procedure for the case where the configuration space is a Lie group G.

5.6.4.2 Feynman's Procedure on Group Manifolds

The classical velocity phase space of the system we will examine here is TG with the Lie group G as the configuration space of the system. We will be looking for all dynamics and Poisson brackets on TG which are localizable and Newtonian. We know already because of Theorem 5.50 that the relevant Poisson tensors must be regular and that there will exist Lagrangian descriptions for such systems. If the

Poisson tensor is velocity independent, we know that the only possible non trivial dynamics are due to interactions with a background gravitational field arising from some non constant metric tensor η on G, and with an electromagnetic field, defined on the group by a vector and scalar potentials A and ϕ. If we impose G-invariance on the dynamical system and on the Poisson tensor, then all fields must be G-invariant. We will show how Feynman's procedure can be applied to this situation.

We first need to introduce some notation and facts about the geometry of Lie groups, their tangent bundles and SODE's on them.

Let L (R) denote the action of G on itself by left (right) translations, i.e., $L\colon G \times G \to G$, $L(g,h) = L_g(h) = gh$, $(R(g,h) = R_g(h) = hg)$ for all $g, h \in G$. A vector field X on G is called left (right) invariant if $TL_gX(h) = X(gh)$ $(TR_gX(h) = X(hg))$. If X is left (right) invariant then $X(g) = TL_gX(e)$ $(X(g) = TR_gX(e))$. The Lie algebra $\mathfrak{g}$ of G is the Lie subalgebra of left (right) invariant vector fields on G. These invariant vector fields can be identified with their values at the identity element of the group. Then fixing a basis E_i of T_eG we define the corresponding basis of invariant vector fields $X_i^L(g) = TL_gE_i$, $X_i^R(g) = TR_gE_i$. From now on we will identify $\mathfrak{g}$ with T_eG and we fix a basis E_i of $\mathfrak{g}$.

We next define left (right) invariant forms on G. We shall denote by θ_L^i (θ_R^i) a basis of left (right) invariant 1-forms on G dual to X_i^L (X_i^R). They obviously satisfy $\langle X_i^L, \theta_L^j\rangle = \delta_i^j$, $(\langle X_i^R, \theta_R^j\rangle = \delta_i^j)$. The canonical Maurer-Cartan $\mathfrak{g}$-valued 1-forms Θ_L and Θ_R on G can be written as

$$\Theta_L = E_i \otimes \theta_L^i; \quad \Theta_R = E_i \otimes \theta_i^R . \tag{5.288}$$

The structure constants $c_{ij}{}^k$ of the Lie algebra $\mathfrak{g}$ are defined from

$$[X_i^L, X_j^L] = c_{ij}{}^k X_k^L ,$$

or from the analogous relations using right invariant vector fields. The vector bundle TG is trivial and can be identified with $G\times\mathfrak{g}$, using left or right translations as follows: $l\colon TG \to G \times \mathfrak{g}$, $l(g,\dot g) = (g, TL_g^{-1}\dot g)$, $(r\colon TG \to G \times \mathfrak{g}$, $r(g,\dot g) = (g, TR_g^{-1}\dot g))$, for any $\dot g \in T_gG$. We will denote by $v_L = TL_g^{-1}(\dot g)$ $(v_R = TR_g^{-1}(\dot g))$ and $v_L = v_L^iE_i$. Coordinates on TG obtained using the map l will be called body or convective coordinates, and coordinates defined by means of r will be called spatial coordinates. It is clear that body and spatial coordinates are related by $(g, v_L) = (g, \mathrm{Ad}_g v_R)$. From now on, we will be using only left-invariant or body coordinates, as it is equivalent to the description in terms of right-invariant ones. We will then suppress the indexes 'L'.

It is clear that the canonical 1-tensor S_G on TG has the following form when written in body coordinates:

$$S_G = \frac{\partial}{\partial v^i} \otimes \theta^i ,$$

and a SODE D will be written as

$$D = v^i X_i + f^i(g, v)\frac{\partial}{\partial v^i} \,.$$

If L is a Lagrangian function on TG which is left-invariant, i.e., $L = l^* L_{\mathfrak{g}}$, or $L(g, \dot g) = L_{\mathfrak{g}}(TL_g^{-1}\dot g)$, for a function $L_{\mathfrak{g}}$ defined on $\mathfrak{g}$, then the Cartan 1-form is

$$\theta_L = \frac{\partial L}{\partial v^i}\theta^i \,,$$

and the Cartan 2-form,

$$\omega_L = \frac{\partial^2 L}{\partial v^i v^j}\theta^i \wedge dv^j - \frac{1}{2}c_{jk}{}^i \frac{\partial L}{\partial v^i}\theta^j \wedge \theta^k \,.$$

If it is regular, the Lagrangian SODE defined by L is given by

$$\Gamma = v^i X_i + W^{lk} v^j c_{jk}{}^i \frac{\partial L}{\partial v^i}\frac{\partial}{\partial v^l} \,,$$

where as usual W^{ij} are the elements of the inverse matrix of W with $W_{ij} = \frac{\partial^2 L}{\partial v^i \partial v^j}$.

5.6.4.3 Free Motion

By free motion we mean geodetic motion with respect to a metric η on G. Then $\eta = \eta^L_{ij}\theta^i_L \otimes \theta^j_L = \eta^R_{ij}\theta^i_R \otimes \theta^j_R$, where η^L_{ij} and η^R_{ij} are functions on G. If η is left (right) invariant, then the functions η^L_{ij} (η^R_{ij}) are constant. If the metric η is biinvariant then it satisfies:

$$\eta_{ir} c_{jk}{}^r + \eta_{kr} c_{ji}{}^r = 0 \,.$$

Let L_η be the kinetic energy defined by the left-invariant metric η,

$$L_\eta = \frac{1}{2}\eta_{ij}(g)\dot g^i \dot g^j = \frac{1}{2}\eta_{ij} v^i v^j \,. \tag{5.289}$$

The Cartan 1-form will be

$$\theta_\eta = \eta_{ij} v^i \theta^j \,,$$

while the Cartan 2-form is

$$\omega_\eta = \eta_{ij}\theta^i \wedge dv^j - \frac{1}{2}\eta_{ij} v^i c_{kl}{}^j \theta^k \wedge \theta^l \,.$$

The energy of the system is the Lagrangian itself, $E_{L_\eta} = L_\eta$, and the Euler–Lagrange SODE Γ_η is obtained from

$$\Gamma_\eta = v^i X_i + v^k v^j c_{kj}{}^l \frac{\partial}{\partial v^l} \,. \tag{5.290}$$

Thus the equations of motion look like:

$$\frac{dg}{dt} = TL_g v; \quad \frac{dv^i}{dt} = c_{jk}{}^i v^j v^k \,, \tag{5.291}$$

which defines the geodesic flow on G associated with η. If the metric is biinvariant, then the equations of motion reduce to

$$\frac{dv}{dt} = 0 \,. \tag{5.292}$$

The Poisson tensor defined by ω_{L_η} is given by

$$\Lambda_\eta = \eta^{ij} \frac{\partial}{\partial v^i} \wedge X_j - c_i^{jk} v^i \frac{\partial}{\partial v^j} \wedge \frac{\partial}{\partial v^k} \,,$$

and $c_i^{jk} = \eta_{il}\eta^{jm}\eta^{kn} c_{mn}{}^l$. Because the Hamiltonian vector field with Hamiltonian v^i is given by $\eta^{ij} X_j$, the fundamental Poisson brackets defined by Λ_η are

$$\{ g^i, g^j \} = 0, \quad \{ g^i, v^j \} = \eta^{ik} X_k(g^i), \quad \{ v^i, v^j \} = c^{ij}{}_k v^k \,, \tag{5.293}$$

where g^i represent a parametrization of the elements $g \in G$.

We will now discuss Feynman's procedure in this setting. For this we examine the particular case of $G = SU(2)$.

5.6.4.4 The Isotropic Rotator and ^3He Superfluid

It is well known that $SU(2)$ is diffeomorphic to S^3. Thus we will be considering Feynman's problem on the sphere S^3. We will first examine free motion on S^3, i.e., geodetic motion with respect to the biinvariant metric on $SU(2)$. Physically, this corresponds to a noninteracting isotropic rotator. After discussing the free system, we introduce interactions by modifying the Poisson structure and equations of motion in a similar manner to what was done in Sect. 4.5.3.

Let g denote an $SU(2)$ matrix, $g^{-1} = g^\dagger$, where $\dagger$ denotes hermitean conjugation. Matrices g in $SU(2)$ will be parametrized as

$$g = \begin{pmatrix} x_0 + ix_3 & -x_2 + ix_1 \\ x_2 + ix_1 & x_0 - ix_2 \end{pmatrix} = x_0 I + ix_k\sigma_k, \tag{5.294}$$

with $x_0^2 + x_1^2 + x_2^2 + x_3^2 = 1$ and σ_k are the Pauli matrices. In spatial coordinates, free motion means that, Eq. (5.292),

$$\frac{d}{dt}(\dot{g}g^\dagger) = 0 \,, \tag{5.295}$$

This is the equation of motion for a noninteracting isotropic rotator. These equations are invariant under left and right multiplication by $SU(2)$

$$g \to ug, \quad g \to gv \,, \quad u, v \in SU(2) \,. \tag{5.296}$$

The associated velocity phase space of the system is $TSU(2)$ which we identify with $SU(2) \times \mathfrak{su}(2)$ by means of right translations. We will denote the spatial (angular) velocities (formerly v_R) by ω_i, $i = 1, 2, 3$, this is $\dot{g}g^\dagger = w = \frac{i}{2}w_k\sigma_k$. From Eq. (5.293), the Poisson brackets between $SU(2)$ matrix elements are zero, with the remaining brackets given by

$$\{\, g, \omega_i \,\} = -\frac{i}{2}\sigma_i g \,, \tag{5.297}$$

$$\{\, \omega_i, \omega_j \,\} = \epsilon_{ijk}\, \omega_k \,, \tag{5.298}$$

For this choice ω_i generates right translations on $SU(2)$. The equation of motion (5.295) is recovered using the Hamiltonian function

$$H_0 = -\frac{1}{2}\sum_{i=1}^{3} \omega_i^2 = -\frac{1}{2}\omega^2 \,. \tag{5.299}$$

which is the standard free Hamiltonian for an isotropic rotator. (We have set the moment of inertia equal to one.)

The chiral transformations (5.296) can be canonically implemented for the above Poisson structure. For this we must specify the following action of u and v on the angular velocities:

$$\omega_i\sigma_i \to u\omega_i\sigma_i v \,. \tag{5.300}$$

Thus ω_i is invariant under the left action of $SU(2)$ parametrized by u, while right $SU(2)$ parametrized by v induces rotations. The Hamiltonian (5.299) is invariant with respect to both left and right $SU(2)$, and the transformations are associated with canonical symmetries.

5.6.4.5 Interactions on S^3

We next apply the analogue of Feynman's assumptions in order to introduce interactions on S^3. As on $\mathbb{R}^3$, we only obtain the coupling to electromagnetism.

We consider equations of motion of the form

$$\dot{g}g^\dagger = \frac{i}{2}\sigma_i\omega_i \; , \tag{5.301}$$

$$\dot{\omega}_i = f_i(g, \omega) \; . \tag{5.302}$$

They are the analogues of Eqs. (4.39) and (4.40). As always we demand that the equations of motion are Hamiltonian. For the Poisson brackets we again assume (5.297) and that the Poisson brackets between matrix elements of g are zero, implying localizability on S^3. As with the case on $\mathbb{R}^3$, we generalize the Poisson brackets of the velocities (or here, angular velocities) with themselves. We take

$$\{\,\omega_i, \omega_j\,\} = \epsilon_{ijk}(\omega_k + B_k) \; . \tag{5.303}$$

Particle interactions are due to the functions f_i and B_k which were zero for the free system. We next determine f_i and B_k, assuming that they depend on the coordinates g and the velocities ω_i. (For simplicity we shall assume that they have no explicit dependence on the time.) The procedure is identical to that carried out on $\mathbb{R}^3$.

It is not hard to see that B_k is a function of only the spatial coordinates, as was the case for $\mathbb{R}^3$. For this we can expand the Jacobi identity

$$\{\,g, \{\,\omega_i, \omega_j\,\}\,\} + \{\,\omega_i, \{\,\omega_j, g\,\}\,\} + \{\,\omega_j, \{\,g, \omega_i\,\}\,\} = 0,$$

to find that $\{\,g, B_k\,\} = 0$, and hence B_k is independent of ω_i. As on $\mathbb{R}^3$, B_k is divergenceless. This follows from the Jacobi identity involving three angular velocities and can be expressed by

$$\{\,\omega_i, B_i\,\} = 0 \; .$$

Next we take the time derivatives of Poisson brackets (5.297) and (5.303). From (5.297) we get

$$\{\,\dot{g}, \omega_i\,\} + \{\,g, \dot{\omega}_i\,\} = -\frac{i}{2}\sigma_i\dot{g} \; .$$

Upon substituting the equations of motion (5.301) and (5.302), we find that

$$\{\,g, f_i\,\} = \frac{i}{2}\epsilon_{ijk}\sigma_j B_k(g) g \; ,$$

and as a result, the force can be at most linear in the angular velocity ω_i,

$$f_i = \epsilon_{ijk} B_j(g)\omega_k + E_i(g) \; . \tag{5.304}$$

Equation (5.304) is the Lorentz force equation on S^3 and it defines the electric field $E_i(g)$.

Conditions on $E_i(g)$ result from taking the time derivative of the Poisson bracket (5.303). We obtain

$$\epsilon_{ijk}\{\,\dot{\omega}_i, \omega_j\,\} = \dot{\omega}_k + \dot{B}_k(g) \; . \tag{5.305}$$

Next we substitute (5.304) and use the condition that the magnetic fields have no explicit time dependence. Implicit time dependence through g means that $\dot{B}_i$ does not vanish, but instead

$$\dot{B}_i + \{ B_i, \omega_j \} \omega_j = 0 \ .$$

From (5.305) we are then left with

$$\epsilon_{ijk} \{ E_i, \omega_j \} - E_k = 0 \ , \tag{5.306}$$

which is the statement that the electric field has zero curl in the presence of static magnetic fields.

The Hamilton function for this system is

$$H = -\frac{1}{2}\omega^2 + \phi(g) \ . \tag{5.307}$$

In order to recover the equations of motion (5.301), (5.302) and (5.304) using this Hamiltonian we need to define the electric field E_i in terms of the scalar potential $\phi(g)$ as follows:

$$E_i(g) = \{ \omega_i, \phi(g) \} \ . \tag{5.308}$$

Furthermore, using this definition we can obtain the condition (5.306) after applying the Jacobi identity

$$\{ \omega_i, \{ \omega_j, \phi \} \} + \{ \omega_j, \{ \phi, \omega_i \} \} + \{ \phi, \{ \omega_i, \omega_j \} \} = 0 \ .$$

It is easy to generalize the above treatment of a particle moving on $SU(2)$ to the case of a particle on an arbitrary Lie group manifold. For this we need only replace the Pauli matrices by the appropriate group generators and ϵ_{ijk} by the structure constants. For instance we can apply the previous procedure to a closely related group $SO(3)$. Then, we will obtain as Poisson brackets the following:

$$\{ R_{ij}, R_{kl} \} = 0, \qquad \{ w_i, R_{jk} \} = \epsilon ijl R_{lk}, \qquad \{ w_i, w_j \} = \epsilon_{ijk}(w_k + B_k), \tag{5.309}$$

where we denote by $R = (R_{ij})$ the elements of $SO(3)$ with.

$$(\dot{R}R^t)_{ij} = \epsilon_{ijk} w_k,$$

and $R_{ij} = \cos\theta\delta_{ij} + (1 - \cos\theta)n_i n_j + \sin\theta\epsilon_{ijk}n_k$, $n^2 = 1$. The dynamics has the form of Eq. (5.133), with

$$H(R, w) = -\frac{1}{2}w^2 + V(R),$$

and $V(R) = V(\cos\theta) = C(\frac{1}{2} + 2\cos\theta)^2$. We obtain in this way the Legget equation for the B-phase of the ^3He superfluid [Br78]. Thus we can construct a Lagrangian for such system on $TSO(3)$.

5.6.4.6 Standard Formulation for Internal Variables

If the external configuration space is trivial, i.e., the system has only internal degrees of freedom, we can call it a pure isospin system. Such systems are generally associated with first-order equations of motion as was the case in Sect. 4.5.3. The isospin variables I_a parametrizing the system will be defined on the Lie algebra $\mathfrak{g}$ of a compact semisimple Lie group G, typically $SU(n)$. We can identify $\mathfrak{g}$ with its dual space $\mathfrak{g}^*$ using a biinvariant metric on G, the Killing-Cartan metric for instance. In any case, we can assume that the inner space is the dual $\mathfrak{g}^*$ of the Lie algebra $\mathfrak{g}$ of the Lie group G. The standard Poisson structure for a pure isospin system will be the linear one defined by Eq. (5.248). In order to apply to it the results on Sect. 5.6.2 we can think of it as a trivial bundle F with fibre $\mathfrak{g}^*$ over a single point $Q = \{\text{point}\}$. The Poisson tensor Λ_I on it will again be chosen as the natural linear Poisson tensor $\Lambda_{\mathfrak{g}}$, Eq. (5.248), induced by the Lie group G. Notice that because TQ is a single point, the bundle $TQ \bowtie F$ is simply $\mathfrak{g}^*$, and the Poisson tensor Λ will be $\Lambda_{\mathfrak{g}}$. But this tensor satisfies all conditions from Theorem 5.57 except that is in general nonregular. If we select a coadjoint orbit $\mathcal{O} \subset \mathfrak{g}^*$, then $\Lambda_{\mathcal{O}} = \Lambda_{\mathfrak{g}}|_{\mathcal{O}}$ becomes regular and the associated symplectic form is the well-known Kirillov–Kostant–Souriau form $\omega_{\mathcal{O}}$ on $\mathcal{O}$. Then, Theorem 5.57 applied to $\Lambda_{\mathcal{O}}$ in $\mathcal{O}$, shows that there must be a Lagrangian (possibly local) realization on $T\mathcal{O}$. This is evident because locally $\omega_{\mathcal{O}}$ must be exact. Then we can choose any locally defined symplectic potential A such that $dA = \omega_{\mathcal{O}}$ and define the Lagrangian $L(\xi, \dot{\xi}) = A_i(\xi)\dot{\xi}^i$, where ξ^i denote local coordinates on $\mathcal{O}$. This Lagrangian is locally defined and singular. For instance, for the standard formalism for spin variables S_i discussed in §3 and §4 with the standard bracket defined by Eq. (67), the coadjoint orbits spanned by S_i are spheres S_r^2 of radius r. The Kirillov–Kostant–Souriau 2-form is a multiple of the area form on S^2 which is locally exact. This allows us to define local Lagrangians for the spin dynamics Eq. (24). In this particular example, a global (singular) Lagrangian realization can be constructed using the Hopf fibration $S^3 \to S^2$ (see for instance [Ba82]). In general then, the Lagrangians constructed in this way will be singular. It would be desirable to have a Lagrangian realization by a regular Lagrangian. Such Lagrangian regularization exists and will be described in the following paragraph.

5.6.4.7 A Lagrangian Realization of the Natural Poisson Bracket on $\mathfrak{g}^*$

The space $\mathfrak{g}^*$ can be considered as the Poisson reduction of T^*G, the cotangent bundle of the group G. We will identify T^*G with $G \times \mathfrak{g}^*$ using left translations on the group, i.e., $l^* \colon T^*G \to G \times \mathfrak{g}^*$, $l^*(\alpha_g) = (g, TL_g^*\alpha_g)$, for all $\alpha_g \in T_g^*G$, and we will denote by $x = TL_g^*\alpha_g$. With respect to a basis θ^i of left-invariant 1-forms we will have $x = x_i\theta^i$. In these coordinates the Poisson bracket on $\mathfrak{g}^*$ is again, $\{x_i, x_j\} = c_{ij}{}^k x_k$.

On $T^*G \cong G \times \mathfrak{g}^*$ we consider the canonical 1-form $\theta_0 = x_i\theta^i$. The canonical symplectic 2-form $\omega_0 = -d\theta_0$, given by

$$\omega_0 = \theta^i \wedge dx_i - \frac{1}{2}x_i c_{jk}{}^i \theta^j \wedge \theta^k \,,$$

on the cotangent bundle T^*G is biinvariant with respect to the lifting of the action of G by left or right invariant translations. For instance considering a basis of left invariant vector fields $X_1, \ldots, X_n$ on G, we can consider their cotangent lifts $X_1^*, \ldots, X_n^*$ to T^*G by setting

$$\mathcal{L}_{X_j^*}(\pi^* f) = \pi^*(\mathcal{L}_{X_j} f), \qquad \forall f \in \mathcal{F}(G),$$

with $\pi : T^*G \to G$ the cotangent bundle projection, and

$$\mathcal{L}_{X_j^*}\theta_0 = 0.$$

We notice that $\mathcal{L}_{X_j^*}\theta_0 = 0$ implies that $i_{X_j^*}d\theta_0 = -d\langle X_j^*, \theta_0\rangle$ and $\langle X_j^*, \theta_0\rangle$ depends only on X_j. Then the nondegeneracy of ω_0 uniquely defines X_j^*. The vector fields X_i^* are thus Hamiltonian with Hamiltonian $J_i = \langle X_j^*, \theta_0\rangle$. Similar results occur for right-invariant vector fields. Because the infinitesimal generators of the left action of G on itself are right-invariant vector fields, and vice versa, we have that the map $J_R = J_i\,\theta_L^i : T^*G \to \mathfrak{g}^*$, represents the momentum map of the Hamiltonian action defined by right-translations on the group. Then it is easy to see that J_R, the momentum map corresponding to right translations, is defined by $J_R(\alpha_g) = TL_g^*\alpha_g$, where $\alpha_g \in T_g^*G$. Notice that J_R corresponds to the projection π_L of T^*G onto $\mathfrak{g}^*$ by the left action of G on T^*G.

Elements of $\mathfrak{g}$ define linear functions on $\mathfrak{g}^*$ and the elements X_i define the linear functions x_i. Thus the projection $J_R : T^*G \to \mathfrak{g}^*$ gives $J_R^*(x_i) = J_i$. The natural symplectic structure ω_0 on T^*G defines a Poisson structure Λ_0 which can be written as

$$\Lambda_0 = X_i^L \wedge \frac{\partial}{\partial x_i} + c_{jk}{}^i x_i \frac{\partial}{\partial x_j} \wedge \frac{\partial}{\partial x_k} \,. \tag{5.310}$$

The most important fact about J_R is that it is a Poisson map with respect to the Poisson structure Λ_0 defined on T^*G by ω_0 and the Poisson structure $\Lambda_{\mathfrak{g}}$. This is easily seen from the particular expression of the Poisson Brackets on Eqs. (5.310) and (5.248).

Thus, T^*G provides a symplectic realization for a pure isospin system, however, because T^*G is not a tangent bundle, it will not provide a Lagrangian realization. For that, we just need to choose a right invariant Riemannian metric η on G and define the kinetic energy Lagrangian Equation (5.289) on TG. The Cartan 2-form ω_{L_η} will then be the pull-back of ω_0 by the identification between TG and T^*G provided by the metric η. Then, the action of G on (TG, ω_{L_η}) will define a Lagrangian momentum map $J : TG \to \mathfrak{g}^*$, given by $J(g, v) = \eta_{ij} v^i \theta^j$, that will be Poisson. More generally, we can consider any Lagrangian function $L \in \mathcal{F}(TG)$ which is nonsingular, i.e., ω_L is symplectic. We consider the left or right action on G and define a lift to TG which a priori does not coincide with the tangent lift. We define

$$\mathcal{L}_{\tilde{X}_i^L}(\tau_G^* f) = \tau_G^*(\mathcal{L}_{X_i^L} f); \qquad \mathcal{L}_{\tilde{X}_i^L} \theta_L = 0,$$

where $\tau_G : TG \to G$ is the tangent bundle projection, and similarly, for the right action.

It is clear that $P_i = \langle \tilde{X}_i^L, \theta_L \rangle$ allows us to define a projection $P : TG \to \mathfrak{g}^*$ which is a Lagrangian realization of the Poisson manifold $(\mathfrak{g}^*, \Lambda_{\mathfrak{g}})$.

At this point we may wonder if the Lagrangian vector field associated with L will be P-projectable onto $\mathfrak{g}^*$. Of course a necessary and sufficient condition for this is that $\{ E_L, P_i \}_L$ must be P-projectable onto $\mathfrak{g}^*$ for any P_i. This will be the case if the Lagrangian function is invariant under the flow lift of right invariant translations. When this is the case the flow lift coincides with the Liouville lift. (We call Liouville lift the one defined by using θ_L).

What is the relation between the regular Lagrangian realization of $\mathfrak{g}^*$ on TG described above and the (local) singular Lagrangian realization on $T\mathcal{O}$? The answer is easily obtained if we extend the coadjoint orbit to the full group. That is, we consider $\mathcal{O}$ as an homogeneous space of G, $\mathcal{O} = G/G_{\mathcal{O}}$ where $G_{\mathcal{O}}$ is the isotropy group of an arbitrary element $\mu \in \mathcal{O}$. Then μ will define a left-invariant 1-form α_μ on G by the formula

$$\alpha_\mu(g) = TL_{g^{-1}}^* \mu, \qquad \forall g \in G\,;$$

Namely, $\alpha_\mu = \mu_i \theta^i$. A simple computation shows that $d\alpha_\mu = \mu_i c_{jk}{}^i \theta^j \wedge \theta^k$ is a closed presymplectic 2-form on G which is G_μ-invariant, i.e., the 2-form $d\alpha_\mu$ is projectable to $\mathcal{O}$. Moreover, the characteristic distribution of $d\alpha_\mu$ is spanned by the elements on $\mathfrak{g}_\mu$, the Lie algebra of G_μ. Then, the projected 2-form on $\mathcal{O}$ will be symplectic and it coincides with the Kirillov–Kostant–Souriau symplectic form $\omega_{\mathcal{O}}$.

Consider now the linear Lagrangian on TG defined by the 1-form α_μ, i.e.,

$$L_\mu(g, v) = \mu_i v^i. \tag{5.311}$$

It is clear that $\theta_{L_\mu} = \tau_G^* \alpha_\mu$ and the Cartan 2-form is given by,

$$\omega_{L_\mu} = \tau_G^* d\alpha_\mu.$$

The Lagrangian L_μ is of course degenerate and the kernel K of its Cartan 2-form is given by $K = \mathfrak{g}_\mu \times \mathfrak{g}$, the factor $\mathfrak{g}$ corresponding to all vertical vectors on $TG \cong G \times \mathfrak{g}$ and $\mathfrak{g}_\mu$ corresponding to the kernel of $d\alpha_\mu$ on G. It is then clear that $TG/K = \mathcal{O}$, and in this sense, the singular Lagrangian L_μ, Eq. (5.311), provides a singular Lagrangian realization for the coadjoint orbit $\mathcal{O}$.

5.6.4.8 Lagrangian Realization of Hydrodynamical Systems of Kirchhoff Type

We can apply the previous ideas to a system quite different to the spin dynamics considered above. The equations of motion for a rigid body in an ideal incompressible fluid at rest at infinity are given by Kirchhoff equations [Mi60] assuming the motion of the fluid to be potential,

$$\dot{\mathbf{p}} = \mathbf{p} \times \mathbf{w}; \quad \dot{\mathbf{M}} = \mathbf{M} \times \mathbf{w} + \mathbf{p} \times \mathbf{u}, \tag{5.312}$$

$\mathbf{M}$ and $\mathbf{p}$ are the total angular and linear momentum and $u_i = \partial H/\partial p_i, w_i = \partial H/\partial M_i$, where H is the energy of the system and is assumed to be positive and quadratic in both variables $\mathbf{M}$ and $\mathbf{p}$. This equations can be put in Hamiltonian form considering the variables $\mathbf{M}, \mathbf{p}$ lying in the dual of the Lie algebra of the Euclidean group $E(3)$. The Poisson tensor has the form

$$\Lambda_{E(3)} = \epsilon_{ijk} p_k \frac{\partial}{\partial M_i} \wedge \frac{\partial}{\partial p_j} + \epsilon_{ijk} M_k \frac{\partial}{\partial M_i} \wedge \frac{\partial}{\partial M_j}. \tag{5.313}$$

Notice that if we think of the dual of the Lie algebra of $E(3)$ as a trivial bundle $\mathbb{R}^3 \times \mathbb{R}^3 \to \mathbb{R}^3$, the tensor $\Lambda_{E(3)}$ is not localizable on the bundle because of the term corresponding to the bracket $\{p_i, M_j\} = \epsilon_{ijk} p_k$ (see Eqs. 5.104 and 5.107). Another system that can be reduced to a Poisson dynamics using the tensor (5.313) is the Legget equation of spin dynamics in the A-phase of the superfluid ^3He (compare with the B-phase discussed in section before). The construction discussed above permits to lift this dynamics to a Lagrangian setting on $TE(3)$. For this we shall use a metric η on $E(3)$ and the momentum map $J_R \colon TE(3) \to \mathbb{R}^6$ and the Lagrangian will be simply the kinetic energy L_η.

The ideas developed in the previous paragraphs can be extended to the following situation: Let $P(G, Q)$ be a principal fibre bundle over the base manifold Q with fibre the group G. We shall consider the Poisson bundle $\mathrm{ad}^* P \to Q$ described in Sect. 5.6.2. This bundle can be obtained by symplectic reduction of the universal phase space for particles in a Yang-Mills field [We78, St77]. The details of the computations have been exhaustively described elsewhere (see for instance [Gu84]), which we summarize below. Consider the cotangent bundle T^*P. In addition, consider the cotangent lift of the action of G on P to T^*P. The action is Hamiltonian with momentum map $J \colon T^*P \to \mathfrak{g}^*$. The quotient of T^*P by the action of G is diffeomorphic to the bundle $p \colon E = \pi^*(\mathrm{ad}^* P) \to T^*Q$. The bundle E carries a natural Poisson structure which is localizable, Newtonian and velocity independent. A Lagrangian realization can be easily obtained using a regular Lagrangian function L on TP and lifting the action of G to TP using θ_L as indicated in the previous subsection. (See [Al94] for details on the tangent bundle setting.)

5.6.4.9 Nonstandard Formulations

In this section we will analyze some solutions to the inverse problem of Poisson dynamics on group manifolds. We will rediscover Lie-Poisson structures on group manifolds that will be used to provide an alternative Hamiltonian description of the isotropic rotator. The equations of motion (5.295) are SODE's and consequently are associated with external variables. In general, solutions to the inverse problem will provide different models for variables, beside the linear Poisson structure used in the two previous section.

5.6.4.10 The Inverse Problem of Poisson Dynamics on Lower Dimensional Spaces

To deal with Poisson structures we consider a general approach which is very useful for lower dimensional problems. The strategy has been developed in general terms in [Gr93, Ko85] and [Ca94], and consists of using the exterior differential calculus on forms, instead of the graded Lie bracket on multivector fields. This can be done by associating $(n-k)$-forms with k-vector fields V on $\mathbb{R}^n$ (or on an arbitrary manifold) by setting

$$\Psi_V = i_V \Omega\,, \tag{5.314}$$

Ω being a volume form on $\mathbb{R}^n$. The map Ψ provides an isomorphism between k-vector fields and $(n-k)$-forms. On monomial multivector fields we have,

$$(i_V\Omega)(Y_1,\ldots,Y_{n-k}) = \Omega(X_1,\ldots,X_k,Y_1,\ldots,Y_{n-k})\,,$$

for $V = X_1 \wedge \cdots \wedge X_k$. For a bivector field $\Lambda = \Lambda^{ij}\partial/\partial x^i \wedge \partial/\partial x^j$ and $\Omega = dx^1 \wedge \cdots \wedge dx^n$, we have

$$\Psi_\Lambda = 2\sum_{i<j}(-1)^{i+j}\Lambda^{ij}dx^1 \wedge \cdots \wedge \widehat{dx^i} \wedge \cdots \wedge \widehat{dx^j} \wedge \cdots \wedge dx^n\,,$$

where the symbol $\widehat{\ }$ stands for omission. We should notice that this isomorphism is not natural and depends on Ω.

By using the Schouten bracket (see Sect. 4.3.1) on multivector fields V_1 and V_2, we find

$$i_{[V_1,V_2]} = -i_{V_1}i_{V_2}d - di_{V_2\wedge V1} + i_{V_1}di_{V_2} + i_{V_2}di_{V_1}\,,$$

which generalizes the analogous relation,

$$i_{[X,Y]} = i_X i_Y d - di_{X\wedge Y} + i_X di_Y - i_Y di_X\,,$$

for vector fields X and Y.

A relevant proposition for our purposes is the following:

Proposition 5.61 [Gr93] *A bivector field* Λ *is a Poisson structure if and only if* $2i_\Lambda d\Psi_\Lambda = d\Psi_{\Lambda\wedge\Lambda}$. *Two Poisson structures* Λ_1 *and* Λ_2 *are compatible if and only if* $d\Psi_{\Lambda_1\wedge\Lambda_2} = i_{\Lambda_1}d\Psi_{\Lambda_2} + i_{\Lambda_2}d\Psi_{\Lambda_1}$.

In low dimensions we have some immediate results:

On $\mathbb{R}^2$, with coordinates (x, y), any bivector field $\Lambda = f(x, y)\partial/\partial x \wedge \partial/\partial y$ defines a Poisson structure because there $[\Lambda, \Lambda]$ is a skew-symmetric three tensor on a two-dimensional manifold, which therefore vanishes identically.

On $\mathbb{R}^3$, with coordinates (x^1, x^2, x^3), we have $\Lambda = \Lambda^{ij}\partial/\partial x^i \wedge \partial/\partial x^j$ and with $\Omega = dx^1 \wedge dx^2 \wedge dx^3$, we get $\Psi_\Lambda = \epsilon_{ijk}\,\Lambda^{ij}\,dx^k$. The Jacobi identity gives $\Psi_\Lambda \wedge d\Psi_\Lambda = 0$, i.e., the 1-form Ψ_Λ must admit an integrating factor. Locally this implies that any Poisson structure is parametrized by two functions, f_0, f_1, such that $\Psi_\Lambda = f_0\, df_1$ and the associated Poisson bracket has the form

$$\{x^i, x^j\} = f_0\,\epsilon^{ijk}\,\frac{\partial f_1}{\partial x^k}\ .$$

This result was used in Sect. 4.5.3 to write Eq. (4.49). If we have a dynamical vector field Γ and we look for a Hamiltonian description, we need to solve the equation $\Lambda(dH) = -\Gamma$. For this problem it is better to start with a volume in contravariant form, say $\widetilde{\Omega} = g\,\partial/\partial x_1 \wedge \ldots \wedge \partial/\partial x_n$. Then instead of $i_\Lambda\Omega = \Psi$, we have $i_\Psi\widetilde{\Omega} = \Lambda$. If it is possible to select Ω such that $\mathcal{L}_\Gamma\Omega = 0$, we have $\Lambda(dH) = i_{\Psi\wedge dH}\widetilde{\Omega} = \Gamma$ or $i_\Gamma\Omega = \Psi \wedge dH$. Therefore if Γ admits an invariant volume, a necessary and sufficient condition for Γ to admit a Hamiltonian description is that $i_\Gamma\Omega$ is the product of two 1-forms $\alpha \wedge \beta$ with $d\beta = 0$ and $\alpha \wedge d\alpha = 0$. Thus, if $\alpha = f_0\,df_1$, the function f_1 must be a constant of motion for Γ.

Because of $d\Psi_{\Lambda\wedge\Lambda} = 0$, we find that

$$i_{\Lambda_1}d(f_2\,df_3) + i_{\Lambda_2}d(f_0\,df_2) = 0\ ,$$

reduces to

$$d(f_0\,df_1 + f_2\,df_3) \wedge (f_2\,df_3 + f_0\,df_1) = 0\ ,$$

i.e.,

$$(f_0\,df_2 - f_2\,df_0) \wedge df_1 \wedge df_3 = 0\ .$$

where we have used our formula on compatible Poisson structures.

If the bracket is an admissible bracket for Γ, then f_1 and f_3 are two constants of motion and we get a family of admissible brackets by taking f_0 and f_2 to also be constants of motion.

In general,

$$(f_0\,df_2 - f_2\,df_0)(\Gamma) = 0$$

gives $d(\log|f_2| - \log|f_0|)(\Gamma) = 0$, i.e., the ratio of f_0 and f_2 is a constant of motion for Γ. Thus, if f_1 and f_3 are two independent constants of motion, we get

$$\Psi = f_0 \, df_1 + f_0 \, F(f_1, f_3) \, df_3 \, ,$$

i.e.,

$$\{ x^i, x^j \} = f_0 \, \epsilon^{ijk} \left(\frac{\partial f_1}{\partial x^k} + F(f_1, f_3) \, \frac{\partial f_3}{\partial x^k} \right)$$

is the most general expression for Poisson brackets compatible with Γ.

5.6.4.11 Poisson Structures on Three-Dimensional Lie Groups

We now consider a matrix Lie group G. In this way if $g = (g_{ij})$, we have $dg = (dg_{ij})$ and the left and right Maurer–Cartan 1-forms Eq. (5.288) valued 1-form are given by,

$$\Theta_L = g^{-1} \cdot dg; \qquad \Theta_R = dg \cdot g^{-1},$$

respectively. If we consider a basis E_i for the Lie algebra $\mathfrak{g}$ of G, the associated left invariant vector fields $X_1^L, \ldots, X_n^L$, and right invariant ones $X_1^R, \ldots, X_n^R$, satisfy,

$$i_{X_j^L}(g^{-1} \cdot dg) = E_j, \qquad \text{or} \quad i_{X_j^R}(dg \cdot g^{-1}) = E_j.$$

Any bivector field on G can be written as $\Lambda = \Lambda_L^{ij} X_i^L \wedge X_j^L = \Lambda_R^{ij} X_i^R \wedge X_j^R$, where Λ_L^{ij}, $\Lambda_R^{ij} \in \mathcal{F}(G)$. By using a basis of 1-forms we can also define a volume $\Omega_L = \theta_L^1 \wedge \theta_L^2 \wedge \cdots \wedge \theta_L^n$ or $\Omega_R = \theta_R^1 \wedge \theta_R^2 \cdots \wedge \theta_R^n$ which coincide for unimodular groups. Then we can associate a $(n-2)$-form Ψ_Λ with any bivector field Λ by using Eq. (5.314).

We shall now restrict our considerations to 3-dimensional Lie groups. We consider $\Omega_L = \theta_L^1 \wedge \theta_L^2 \wedge \theta_L^3$ and $\Lambda = \Lambda_L^{ij} X_i^L \wedge X_j^L$ to find the 1-form,

$$\Psi_\Lambda = \epsilon_{ijk} \, \Lambda_L^{ij} \, \theta_L^k = f_k \theta_L^k = \alpha \, ,$$

with $f_k = \epsilon_{ijk} \Lambda_L^{ij}$. The Jacobi identity once more gives $\alpha \wedge d\alpha = 0$, i.e.,

$$f_j \, df_k \wedge \theta_L^j \wedge \theta_L^k = \frac{1}{2} f_k f_m \, c_{ij}{}^k \, \theta_L^m \wedge \theta_L^i \wedge \theta_L^j \, .$$

On the other hand, it is possible to write the differential of any function f in the form $df = (\mathcal{L}_{X_i^L} f) \theta_L^i$. Then by using this expression we get

$$f_j \, \mathcal{L}_{X_m^L} f_k \, \theta_L^m \wedge \theta_L^j \wedge \theta_L^k + \frac{1}{2} f_k f_m \, c_{ij}{}^k \, \theta_L^m \wedge \theta_L^i \wedge \theta_L^j = 0 \, .$$

Therefore the equation for (f_1, f_2, f_3) reads as:

$$f_i\,\mathcal{L}_{X_m^L} f_j - f_j\,\mathcal{L}_{X_m^L} f_i = c_{ij}{}^k f_k f_m\,, \tag{5.315}$$

and gives all Poisson structures on the 3-dimensional Lie group G.

To be more specific, we give some examples:

Example 5.7 The group $SU(2)$. The group $SU(2)$ is unimodular, and thus $\Omega_L = \Omega_R$. If we look for a left-invariant 1-form $\alpha = f_j\,\theta_L^j, f_j \in \mathbb{R}$, then Eq. (5.315) gives, $f_j = 0$ for all $j = 1, 2, 3$. Thus there are no Poisson brackets on $SU(2)$ which are left-invariant. Similarly, one shows that there are no right-invariant Poisson brackets.

The simplest non trivial case would be to consider

$$\Lambda = \Lambda_L^{ij} X_i^L \wedge X_j^L + \Lambda_R^{ij} X_i^R \wedge X_j^R\,.$$

Because of the unimodularity of $SU(2)$ and carrying on the computation of $\alpha \wedge d\alpha$, we find a solution for Λ_L^{ij} and Λ_R^{ij} as

$$\Lambda_L^{ij} = -\Lambda_R^{ij} \in \mathbb{R}\,.$$

Therefore,

$$\Lambda = X_i^L \wedge X_j^L - X_i^R \wedge X_j^R \tag{5.316}$$

is a Poisson structure on $SU(2)$. This bracket defines a Lie-Poisson bracket on $SU(2)$ [Sk82, Ya92, Ahl93] (see [Ta89, Ma90, Tj92] for a review).

It may be useful to find the precise Poisson structure associated with Λ when we realize $SU(2)$ as a group of 2×2 complex matrices, Eq. (5.294)

$$g = \begin{pmatrix} x_0 + ix_3 & -x_2 + ix_1 \\ x_2 + ix_1 & x_0 - ix_3 \end{pmatrix}, \tag{5.317}$$

with the condition $x_0^2 + x_1^2 + x_2^2 + x_3^2 = 1$. If we consider the left and right invariant vector fields $X_1^L, X_1^R, X_2^L, X_2^R$, associated with $E_1 = \begin{pmatrix} 0 & -1 \\ 1 & 0 \end{pmatrix}$ and $E_2 = \begin{pmatrix} 0 & i \\ i & 0 \end{pmatrix}$, we find

$$\begin{aligned} \Lambda_{SU(2)} &= (x_2^2 + x_1^2)\frac{\partial}{\partial x_3} \wedge \frac{\partial}{\partial x_0} + x_3 x_1 \frac{\partial}{\partial x_0} \wedge \frac{\partial}{\partial x_1} + x_0 x_2 \frac{\partial}{\partial x_0} \wedge \frac{\partial}{\partial x_2} \\ &\quad + x_0 x_1 \frac{\partial}{\partial x_1} \wedge \frac{\partial}{\partial x_3} + x_0 x_2 \frac{\partial}{\partial x_2} \wedge \frac{\partial}{\partial x_3} \end{aligned} \tag{5.318}$$

$$= \left(x_1 \frac{\partial}{\partial x_1} + x_2 \frac{\partial}{\partial x_2}\right) \wedge \left(x_3 \frac{\partial}{\partial x_0} - x_0 \frac{\partial}{\partial x_3}\right) + (x_2^2 + x_1^2)\frac{\partial}{\partial x^3} \wedge \frac{\partial}{\partial x_0}, \tag{5.319}$$

along with

$$\Psi_{\Lambda_{SU(2)}} = (x_2^2 + x_1^2)\frac{x_2\,dx_1 - x_1\,dx_2}{x_2^2 + x_1^2} \wedge d\left(\frac{1}{2}(x_0^2 + x_1^2 + x_2^2 + x_3^2)\right).$$

Example 5.8 The group $SL(2,\mathbb{R})$. It is possible repeat the above procedure for $SL(2,\mathbb{R})$. In this case we let our group matrices g have the form,

$$g = \begin{pmatrix} x_1 & x_2 \\ x_3 & x_4 \end{pmatrix}, \tag{5.320}$$

with the condition $x_1x_4 - x_2x_3 = 1$. We find,

$$\begin{aligned}\Lambda_{SL(2,\mathbb{R})} =& x_2x_1\frac{\partial}{\partial x_2} \wedge \frac{\partial}{\partial x_1} + 2x_2x_3\frac{\partial}{\partial x_4} \wedge \frac{\partial}{\partial x_1} + x_1x_3\frac{\partial}{\partial x_3} \wedge \frac{\partial}{\partial x_1} \\ &+ x_2x_4\frac{\partial}{\partial x_4} \wedge \frac{\partial}{\partial x_2} + x_3x_4\frac{\partial}{\partial x_4} \wedge \frac{\partial}{\partial x_3},\end{aligned}$$

with

$$\Psi_{\Lambda_{SL(2,\mathbb{R})}} = (x_2^2 + x_3^2)\frac{x_3\,dx_2 - x_2\,dx_3}{x_2^2 + x_3^2} \wedge d(x_1x_4 - x_2x_3).$$

Let us now consider the inverse problem for Γ on a three dimensional group. We consider $\Omega_L = \theta_L^1 \wedge \theta_L^2 \wedge \theta_L^3$, $i_\Lambda\Omega_L = \alpha$ and look for solutions of $\mathcal{L}_\Gamma\Lambda = 0$. By setting

$$\alpha = f_i\,\theta_L^i, \quad \Gamma = g^i\,X_i^L,$$

we get:

i. $f_ig^i = 0$, i.e., $i_\Gamma\alpha = 0$,
ii. $f_k \operatorname{div}\Gamma = -g^i(\mathcal{L}_{X_k^L}f_i - \mathcal{L}_{X_i^L}f_k + f_jc_{ik}{}^j)$ along with the conditions on functions f_j's given by Eq. (5.315).

When $\operatorname{div}\Gamma = 0$ we obtain some particular solutions by using constants of motion for Γ, say F with $\mathcal{L}_\Gamma F = 0$. Then we set $\alpha = dF$ and find $i_{\Lambda_I}\Omega_L = dF$. From $\operatorname{div}\Gamma = 0$, we get $\mathcal{L}_\Gamma\Lambda_I = 0$, with Γ in the image of Λ_I because Γ is tangent to the foliation induced by F.

5.6.4.12 Alternative Hamiltonian Description of the Isotropic Rotator

We have seen that the solution of the inverse problem for Poisson dynamics on Lie groups leads to families of Poisson brackets, the simplest ones being Lie-Poisson brackets on the group. Thus we can relax the localizability condition in the classical treatment of Feynman's problem to assume that the Poisson tensor in the total space is π-related to a Lie-Poisson tensor on the configuration group manifold. We shall explore such assumption in the simple situation of the group $SU(2)$, but most of the ideas can be extended to arbitrary Lie–Poisson structures. The main result we will obtain is that the Hamiltonian description of a particle moving on $S^3 \cong SU(2)$ is not unique [Mr93, St94]. We end up with a Hamiltonian formulation which is canonically inequivalent to the free system given in the Sect. 5.6.4.1, yet it yields the same equations of motion, i.e., (5.295).

We begin by examining the spatial coordinates. As our spatial manifold is S^3 we can once again utilize $SU(2)$ matrices g as coordinates Eq. (5.317). Then we introduce the weak assumption of nonlocalizability which means that the Poisson brackets between matrix elements take the quadratic relations Eq. (5.319). It will be convenient in what follows to change the notation and to use tensor product notation for the Poisson brackets on $SU(2)$. Then the Poisson brackets defined by the Poisson tensor Eq. (5.319), reads as [Sk82, Ya92, Ahl93]

$$\{ g_1, g_2 \} = [r_{12}, g_1 g_2] \,, \tag{5.321}$$

The indices 1 and 2 refer to two separate vector spaces on which the $SU(2)$ matrices act. The matrix r_{12}, known as the classical r-matrix, acts non trivial ly on both vector spaces 1 and 2, while g_1 and g_2 are defined by the tensor products:

$$g_1 = g \otimes 1 \quad g_2 = 1 \otimes g \,.$$

As usual the Poisson bracket should be antisymmetric and satisfy the Jacobi identity. Antisymmetry of (5.321) requires that

$$[r_{12} + r_{21}, g_1 g_2] = 0 \,, \tag{5.322}$$

or $r_{12}+r_{21}$ is an adjoint invariant. The Jacobi identity involving three $SU(2)$ elements g_1, g_2 and g_3 leads to certain quadratic relations for the classical r matrix known as the modified classical Yang-Baxter relations. They are:

$$[r_{23}, r_{31}] + [r_{31}, r_{12}] + [r_{12}, r_{23}] = \quad \text{adjoint invariant} \,. \tag{5.323}$$

Solutions to (5.322) and (5.323) are well known. If we express g in terms of its defining matrix representation then r can be written in terms of 4×4 matrices. A one parameter family of such matrices satisfying (5.322) and (5.323) is

$$r = \frac{i\lambda}{4} \begin{pmatrix} 1 & & & \\ & & -1 & \\ & 4 & & -1 \\ & & & 1 \end{pmatrix}, \tag{5.324}$$

where we assume the parameter λ to be real. Only in the limit $\lambda \to 0$ do we recover the ordinary condition of localizability.

As the dimension of the phase space should be six, we must introduce three more variables in addition to the spatial coordinates g. We denote the new variables by S_3 and S_-, where S_3 is real and S_- is complex. From them we shall later define the analogue of the angular velocity ω_i of section Sect. 5.6.4.1. For the Poisson brackets of S_i with themselves we take

$$\{ S_\pm, S_3 \} = \pm\frac{i\lambda}{2} S_\pm S_3 \;, \quad \{ S_+, S_- \} = -i\lambda\Big((S_3)^2 - (S_3)^{-2} \Big) \;, \tag{5.325}$$

where S_+ is the complex conjugate of S_-. These brackets were utilized previously in an alternative description of spin, Eq. (4.64). They were seen to satisfy the Jacobi identity, and to have the classical Casimir function

$$f_1 = S_+ S_- + (S_3)^2 + (S_3)^{-2} \;. \tag{5.326}$$

Thus $\{ S_i, f_1 \} = 0$. Moreover the Poisson brackets (5.325) were shown to correspond to the classical limit of the $U_q(\mathfrak{sl}(2))$ Hopf algebra.

The Poisson brackets (5.325) can be re-expressed so that they resemble the brackets (5.321) for g. For this define the 2×2 lower triangular matrices $\ell^{(-)}$,

$$\ell^{(-)} = \begin{pmatrix} S_3 & 0 \\ S_- & S_3^{-1} \end{pmatrix} . \tag{5.327}$$

The set of such matrices forms the $SB(2, \mathbb{C})$ group. Furthermore, if we denote their conjugate inverses by $\ell^{(+)} = \ell^{(-)\dagger -1}$, then (5.325) can be written according to

$$\{\ell_1^{(\pm)}, \ell_2^{(\pm)}\} = [\; r_{21} \;, \; \ell_1^{(\pm)} \ell_2^{(\pm)} \;] \;, \tag{5.328}$$

$$\{\ell_1^{(+)}, \ell_2^{(-)}\} = [\; r_{21} \;, \; \ell_1^{(+)} \ell_2^{(-)} \;] \;. \tag{5.329}$$

Eqs. (5.328) and (5.329) satisfy the requirement of antisymmetry and the Jacobi identity for the same reasons as does (5.321). In terms of the matrix $\ell^{(-)}$ the classical Casimir function can be written $f_1 = \mathrm{Tr}\; \ell^{(-)} \ell^{(-)\dagger}$.

It remains to specify the Poisson bracket relations for S_i with the spatial coordinates g. This choice must of course be consistent with the Jacobi identity and the Poisson brackets (5.321) and (5.325). One such choice is:

$$\begin{aligned} \{ S_3, g \} &= -\frac{i\lambda}{4} S_3 \sigma_3 g \;, \\ \{ S_\pm, g \} &= -i\lambda \left(\frac{1}{4} S_\pm \sigma_3 + S_3^{-1} \sigma_\mp \right) g \;, \end{aligned} \tag{5.330}$$

where $\sigma_\pm = \frac{1}{2}(\sigma_1 \pm i\sigma_2)$. These relations can be re-expressed in terms of the matrices $\ell^{(\pm)}$ as follows:

$$\begin{aligned} \{\ell_1^{(-)}, g_2\} &= -\ell_1^{(-)} \; r_{12} \; g_2 \;, \\ \{\ell_1^{(+)}, g_2\} &= \ell_1^{(+)} \; r_{21} \; g_2 \;. \end{aligned} \tag{5.331}$$

Next we introduce the Hamilton function which we denote by $H^{(\lambda)}$. The equations of motion for the free system (5.295) state that the angular velocities are constants of motion, so they should have zero Poisson brackets with $H^{(\lambda)}$. We will construct the angular velocities from S_i alone. Then $H^{(\lambda)}$ should have zero Poisson bracket with S_i, and the Hamiltonian can therefore depend only on the Casimir function (5.326). We take

$$H^{(\lambda)} = -\frac{1}{2\lambda^2}(f_1 - 2) . \tag{5.332}$$

We will justify this choice later.

It then remains to compute the equation of motion for g. From (5.330) and (5.332) we get

$$-2i\dot{g}g^\dagger = -2i\{g, H^{(\lambda)}\}\,g^\dagger = \Omega^3\sigma_3 + \Omega_-\sigma_+ + \Omega_+\sigma_- . \tag{5.333}$$

Ω_i are angular velocities, analogous to ω_i of section Sect. 5.6.4.1. Here they are defined according to

$$\begin{aligned} \Omega_3 &= -\frac{1}{2\lambda}\left((S_3)^2 - (S_3)^{-2} + S_-S_+\right), \\ \Omega_\pm &= \Omega_1 \pm i\Omega_2 = -\frac{1}{\lambda}S_\mp S_3^{-1} , \end{aligned} \tag{5.334}$$

or

$$\Omega_i = -\frac{1}{2\lambda}\mathrm{Tr}\,\ell^{(-)}\sigma_i\ell^{(-)\dagger} .$$

Since we also have that

$$\dot{\Omega}_i = \{\Omega_i, H^{(\lambda)}\} = 0 ,$$

we have recovered the equation of motion (5.295) of the isotropic rigid rotator.

The above formulation of the isotropic rotator is a one parameter deformation of the standard formulation given in Sect. 5.6.4.1. Thus, it is possible to recover the standard formulation as a limiting case. The relevant parameter here is λ and the limiting value for the parameter is $\lambda = 0$. In this regard, we have already seen that the Poisson bracket (5.321) of matrix elements of g with themselves vanishes when $\lambda \to 0$. The remaining Poisson brackets reduce to the canonical ones (i.e., (5.297) and (5.298)) when $\lambda \to 0$ provided that we make the variables S_i functions of the angular velocity ω_i of section Sect. 5.6.4.1, as well as the parameter λ,

$$S_i = S_i^{(\lambda)}(\omega) .$$

For this, we need to assume the following behavior for small λ

$$S_3 = 1 - \frac{\lambda}{2}\omega_3 + \frac{\lambda^2}{8}(\omega_3)^2 + O(\lambda^3) ,$$
$$S_\pm = -\lambda\omega_\mp + O(\lambda^3) . \tag{5.335}$$

Then (5.297) and (5.298) result from (5.325) and (5.330), respectively, upon taking the limit $\lambda \to 0$. Because the variables S_i are functions of angular velocities ω_i and the parameter λ, then so are angular velocities Ω_i functions of ω_i and λ:

$$\Omega_i = \Omega_i^{(\lambda)}(\omega) .$$

From (5.334) and (5.335) it follows that the two angular velocities coincide in the limit $\lambda \to 0$,

$$\Omega_i^{(\lambda\to 0)}(\omega) = \omega_i .$$

Finally we need to show that the Hamiltonians (5.299) and (5.332) of the two different formulations of the rotator agree in the $\lambda \to 0$ limit. For this we see that the classical Casimir function f_1 has the following small λ behaviour

$$f_1 \to 2 + \lambda^2\omega_i\omega_i + O(\lambda^3) ,$$

from which it follows that

$$H^{(\lambda\to 0)} = H_0 ,$$

H_0 given by (5.299). This of course was the reason for choosing $H^{(\lambda)}$ of the form given in (5.332).

For $\lambda \neq 0$, the treatment given here is canonically inequivalent to that given in section Sect. 5.6.4.1. Here, for instance, we see that the Hamiltonian (5.332) is not proportional to the square of the angular velocity (unlike (5.299)), nor does the Poisson bracket algebra of the angular velocity Ω_i correspond to an $SU(2)$ algebra (unlike (5.298)). Perhaps the most distinguishing feature of the formulation presented here concerns the symmetries. In the standard formulation we saw that chiral transformations were canonical symmetries. This is not the case in the alternative formulation. In fact, chiral transformations can not be canonically implemented for the Poisson structure given by (5.321), (5.325) and (5.330). For example, under the right action of $SU(2)$,

$$g \to gv , \quad \ell^{(\pm)} \to \ell^{(\pm)} , \quad v \in SU(2) , \tag{5.336}$$

the brackets (5.321) undergo the following transformation:

$$\{g_1, g_2\} \to \{g_1v_1, g_2v_2\} \neq [r_{12}, g_1v_1g_2v_2] . \tag{5.337}$$

Instead of being canonical the right action (5.336) is Lie–Poisson. Poisson Lie group transformations are classical analogues of quantum group transformations. The former are defined so that group multiplication is a Poisson map. Thus with regard to transformation (5.336) we want that $\{ v_1, v_2 \} \neq 0$, $v_1 = v \otimes 1$, $v_2 = 1 \otimes v$. If one chooses the following Poisson bracket for matrix elements of v with themselves

$$\{v_1, v_2\} = [\, r_{12}\,,\ v_1 v_2\,]\,, \tag{5.338}$$

(and zero with g and $\ell^{(\pm)}$) then the inequality in (5.337) can be replaced by an equality. In fact, from (5.338) it follows that all Poisson brackets (5.321), (5.325) and (5.330) are unchanged by this transformation. Then right multiplication is a Poisson map and (5.336) is a Poisson Lie group transformation. Further, since $\ell^{(\pm)}$ are unchanged by this transformation, and the Hamiltonian (5.332) depends only on these variables, then this transformation leaves the dynamics invariant. We then say that (5.336) is a Poisson Lie group symmetry.

There exists another Poisson Lie group symmetry and it is the analogue of the left $SU(2)$ canonical symmetry of the standard formalism. Unlike right $SU(2)$ given by (5.336) it has a non trivial action on $\ell^{(\pm)}$ (as well as g). In order to specify this action it is convenient to introduce another set of variables $d^{(\pm)}$, known as classical double variables [Dr83, Dr86]. These variables are defined by

$$d^{(\pm)} = \ell^{(\pm)} g\,. \tag{5.339}$$

It follows that $d^{(\pm)}$ is an element of $SL(2, \mathbb{C})$, (5.339) being an Iwasawa decomposition. In terms of these variables the right action of $SU(2)$ is given by $d^{(\pm)} \to d^{(\pm)} v$. We now define the left action of $SU(2)$ by

$$d^{(\pm)} \to u d^{(\pm)}\,, \quad u \in SU(2)\,. \tag{5.340}$$

The Poisson brackets for $d^{(\pm)}$ can be constructed from those of g and $\ell^{(\pm)}$. One finds

$$\begin{aligned} \{d_1^{(\pm)}, d_2^{(\pm)}\} &= -d_1^{(\pm)} d_2^{(\pm)}\, r_{12} + r_{21}\, d_1^{(\pm)} d_2^{(\pm)}\,, \\ \{d_1^{(-)}, d_2^{(+)}\} &= -d_1^{(-)} d_2^{(+)}\, r_{12} - r_{12}\, d_1^{(-)} d_2^{(+)}\,. \end{aligned} \tag{5.341}$$

These relations are invariant under (5.340) upon insisting that $u \in SU(2)$ has the following Poisson bracket with itself

$$\{u_1, u_2\} = [\, r_{12},\ u_1 u_2\,]\,,$$

and zero Poisson bracket with $d^{(\pm)}$. Then $SU(2)$ left multiplication is a Poisson map and (5.340) is a Poisson Lie group transformation. Further, since the classical Hamiltonian can be written

$$H^{(\lambda)} = -\frac{1}{2\lambda^2}\left(\mathrm{Tr}\ d^{(-)} {d^{(+)}}^{-1} - 2\right),$$

we can use the cyclic property of trace to show that it is unchanged under (5.340), and hence $SU(2)$ left multiplication is a Lie Poisson symmetry.

The Hamiltonian formalism presented here of free motion on $SU(2)$ can be quantized and the result is free motion on a noncommuting space, namely the quantum group $SU_q(2)$ [St94]. Furthermore, upon quantization, the two Lie Poisson symmetries turn into quantum group symmetries.

An intriguing question concerns whether or not there exists an analogous alternative Hamiltonian description for the interacting system given in Sect. 5.6.4.1, Eqs. (5.301–5.308). It would involve introducing electromagnetic fields, which when quantized would be defined on a non-commutative space, namely $SU_q(2)$.

References

[Ma84] Marmo, G., Mukunda, N., Sudarshan, E.C.G.: Relativistic particle dynamics–Lagrangian proof of the no-interaction theorem. Phys. Rev. D **30**, 2110–2116 (1984)

[Jo64] Jost, R.: Poisson brackets (an unpedagogical lecture). Rev. Mod. Phys. **36**, 572–579 (1964)

[Go81] Goldstein, H.: Classical Mechanics, 2nd edn. Addison-Wesley Publication, Reading (1981)

[Sa71] Saletan, E.J., Cromer, A.H.: Theoretical Mechanics. J. Wiley, New York (1971)

[Ab78] Abraham, R., Marsden, J.E.: Foundations of Mechanics, 2nd ed.n Benjamin, Reading (1978)

[Go69] Gordon, W.B.: On the relation between period and energy in periodic dynamical systems. J. Math. Mech. **19**, 111–114 (1969)

[Cu77] Cushman, R.H.: Notes on Topology and Mechanics. Mathametical Institute, Utrecht (1977)

[Sm70] Smale, S.: Topology and mechanics. Invent. Math. **10**, 305–331 (1970)

[DeF89] de Filippo, S., Landi, G., Marmo, G., Vilasi, G.: Tensor fields defining a tangent bundle structure. Ann. Inst. H. Poincaré **50**, 205–218 (1989)

[Mm85] Marmo, G., Saletan, E.J., Simoni, A., Vitale, B.: Dynamical Systems: A Differential Geometric Approach to Symmetry and Reduction. Wiley, Chichester (1985)

[Ar76] Arnol'd, V.I.: Méthodes mathématiques de la mécanique classique. Ed. Mir, 1976. Mathematical Methods of Classical Mechanics. Springer, Berlin (1989)

[SW86] Sattinger, D.H., Weaver, O.L.: Lie Groups and Algebras with Applications to Physics, Geometry and Mechanics. Springer, New York (1986)

[Mo91] Morandi, G., Ferrario, C., Lo, G.: Vecchio, G. Marmo and C. Rubano. The inverse problem in the calculus of variations and the geometry of the tangent bundle. Phys. Rep. **188**, 147–284 (1991)

[ChB82] Choquet-Bruhat, Y., DeWitt-Morette, C.: Analysis, Manifolds and Physics, 2nd edn. North-Holland, Amsterdam (1982)

[Cr83] Crampin, M.: Tangent bundle geometry for Lagrangian dynamics. J. Phys. A Math. Gen. **16**, 3772–3775 (1983)

[Cr87] Crampin, M., Ibort, L.A.: Graded Lie algebras of derivations and Ehresmann connections. J. Math. Pures et Appl. **66**, 113–125 (1987)

[Ca95] Cariñena, J.F., Ibort, A., Marmo, G., Stern, A.: The Feynman problem and the inverse problem for Poisson dynamics. Phys. Rep. **263**, 153–212 (1995)

[Wi39] Wigner, E.P.: On unitary representations of the inhomogeneous Lorentz group. Ann. Math. **40**, 149–204 (1939)

[No82] Novikov, S.P.: The Hamiltonian formalism and a many valued analogue of Morse theory. Russ. Math. Surveys. **37**, 1–56 (1982)

[Hu92] Hughes, R.J.: On Feynman's proof of the Maxwell equations. Amer. J. Phys. **60**, 301–306 (1992)

[Mo93] Moreira, I.C.: Comment on 'Feynman proof of the Maxwell equations' by R. J. Hughes. Am. J. Phys. **61**, 853 (1993)

[Ib91] Ibort, A., Marín-Solano, J.: On the inverse problem of the calculus of variations for a class of coupled dynamical systems. Inverse Prob. **7**, 713–725 (1991)

[CS55] Currie, D.I., Saletan, E.J.: q-equivalent particle Hamiltonians. I. The classical one-dimensional case. J. Math. Phys. **7**, 967–974 (1966)

[HS84] Henneaux, M., Shepley, L.C.: Lagrangians for spherically symmetric potentials. J. Math. Phys. **23**, 2101–2107 (1984)

[Ib90] Ibort, A., López-Lacasta, C.: On the existence of local and Global's for ordinary differential equations. J. Phys. A: Math. Gen. **23**, 4779–4792 (1990)

[Dy90] Dyson, F.J.: Feynman's proof of the Maxwell equations. Amer. J. Phys. **58**, 209–211 (1990)

[Br78] Brinkman, W.F., Cross, M.C.: Spin and orbital dynamics of superfluid ^3He. In: Brewer, D.F. (ed.) Progress in low temperature physics. North-Holland, Amsterdam (1978)

[Ba82] Balachandran, A.P., Marmo, G., Stern, A. : A Lagrangian approach to the non-interaction theorem. Nuovo Cim. **69 A**, 175–186 (1982)

[Mi60] Milne-Thompson, L.M.: Theoretical Hydrodynamics. MacMillan, London (1960)

[We78] Weinstein, A.: A universal phase space for particles in Yang-Mills fields. Lett. Math. Phys. **2**, 417–20 (1978)

[St77] Sternberg, S.: On minimal coupling and the symplectic mechanics of a classical particle in the presence of a Yang-Mills field. Proc. Nat. Acad. Sci. **74**, 5253–5254 (1977)

[Gu84] Guillemin, V., Sternberg, S.: Symplectic Techniques in Physics. Cambridge University Press, Cambridge (1984)

[Al94] Alekseevsky, D., Grabowski, J., Marmo, G., Michor, P.W.: Poisson structures on the cotangent bundle of a Lie group or a principal bundle and their reductions. J. Math. Phys. **35**, 4909–4927 (1994)

[Gr93] Grabowski, J., Marmo, G., Perelomov, A.: Poisson structures: towards a classification. Modern Phys. Lett. A **8**, 1719–1733 (1993)

[Ko85] J.L. Koszul. Crochets de Schouten-Nijenhuis et cohomologie. Astérisque, Soc. Math. de France, hors of série, 257–71 (1985)

[Ca94] Cariñena, J.F., Ibort, L.A., Marmo, G., Perelomov, A.M.: The geometry of Poisson manifolds and Lie algebras. J. Phys. A: Math. Gen. **27**, 7425–7449 (1994)

[Sk82] Sklyanin, E.K.: Some algebraic structures connected with the YangBaxter equation. Funct. Anal. Appl. **16**, 263–270 (1982)

[Ya92] Ya. I. Granovski, I. M. Lutzenko, A. S. Zhedanov. Mutual integrability, quadratic algebras, and dynamical symmetry. Ann. of Phys. 217, 1–20 (1992)

[Ahl93] Ahluwalia, K.S.: Fundamentals of Poisson Lie Groups with Application to the Classical Double. Cambridge preprint DAMTP 93–53 (1993). hep-th/9310068

[Ta89] Takhtajan, L.: Introduction to quantum group and integrable massive models of quantum field theory. In: Ge, M.-L., Zhao, B.-H. (eds.) World scientific, Singapore (1990)

[Ma90] Majid, S.: On q-regularization. Int. J. Mod. Phys. **5**, 4689–4696 (1990)

[Tj92] Tjin, T: Introduction to quantized Lie groups and Lie algebras. Int. J. Mod. Phys. A **7**, 6175–6214 (1992)

[Mr93] Marmo, G., Simoni, A., Stern, A.: Poisson Lie group symmetries for the isotropic rotator. Int. J. Mod. Phys. A **10**, 99–114 (1995)

[St94] Stern, A., Yakushin, I.: Deformation quantization of the isotropic rotator. Mod. Phys. Lett. A **10**, 399–408 (1995)

[Dr83] Drinfel'd, V.G.: Hamiltonian structures on Lie groups, Lie Bialgebras and the geometric meaning of the classical Yang-Baxter equations. Sov. Math. Doklady **27**, 68–71 (1983)

[Dr86] Drinfel'd, V.G.: Hamiltonian structures on Lie groups, Lie Bialgebras and the geometric meaning of the classical Yang-Baxter equations. In: Proceedings of the International Congress on Mathematicians, Berkeley, vol. 1. Academic Press, New York (1986)

Chapter 6
The Geometry of Hermitean Spaces: Quantum Evolution

A measurement on one property can produce unavoidable changes in the value previously assigned to another property, and it is without meaning to speak of a microscopic system possessing precise values for all its attributes. This contradicts the classical representation of all physical quantities by numbers. The laws of atomic physics must be expressed, therefore, in a nonclassical mathematical language that constitutes a symbolic expression of the properties of microscopic measurement.

Julian Schwinger, *Quantum Kinematics and Dynamics* 1970.

6.1 Summary

The basic geometrical structures arising in Quantum Mechanics are analyzed as in previous chapters, that is, we ask when a given dynamics possesses simultaneously invariant symplectic and metric stuctures. After solving this inverse problem in the linear case, we will discuss some of its implications. All the fundamental mathematical structures needed to discuss quantum mechanical systems will unfold in front of our eyes. However we should warn the reader that the systems obtained so far are treated as classical ones, i.e., the physical epistemology of them is not inferred from the mathematical structures. Some questions will be raised in this sense and a discussion of the physical meaning of the emerging mathematical structures will be considered at the end of the chapter.

6.2 Introduction

In Chap. 5 we concentrated our attention on studying contravariant geometrical structures compatible with a given dynamical system: vector fields and contravariant 2-tensors (Poisson tensors) were our primary concern, even though at the same time zeroth order tensors (constants of motion) and covariant 2-tensors

J.F. Cariñena et al., *Geometry from Dynamics, Classical and Quantum*,
DOI 10.1007/978-94-017-9220-2_6

(symplectic forms) appear naturally into the picture as well as the geometrical structures needed to develop an intrinsic approach to the calculus of variations.

We have pointed out at various times that even if skew-symmetric structures played a significant role, symmetric ones will also play their part in the description of physical world. This chapter is devoted to them. In it we will discover that the existence of compatible symmetric structures for a given dynamical system imposes extremely tight conditions on it, to the point of almost determining its structure uniquely. Not only that, the geometry of systems admitting a compatible Hermitean structure is exactly the geometry needed to describe quantum evolution in a very specific form that will be explained later on.

We must point out here that the so-called 'geometry of quantum systems' can be shared by many systems, quantum or not; that is, no attempt will be made here to describe the epistemological and phenomenological intricacies of quantum systems, least of all to try to present the 'theory of measurements' founding the physical theory of quantum systems. However it is remarkable that the extremely elegant and comprehensive view offered by the geometrical picture of quantum systems offers a crisp picture of many aspects of quantum physics that on many occasions are hindered by obscure reasonings and that, from this point of view will simply constitute straightforward consequences of the geometry of the system and do not depend on its classical or quantum physical character.

A one-parameter group of transformations has many features which do not depend on the specific space on which it acts. When the space is the space of states or the space of observables, the evolution should be compatible with additional structures dictated by the physical setting we are considering, i.e., classical or quantum.

To be more precise, in the quantum case we are going to study the properties of dynamical systems on Hilbert spaces exploring the consequences of the existence of invariant structures that this time is going to be an Hermitean structure. We shall stress here that we do not pretend to describe the geometrical foundations of Quantum Mechanics, which is something very different. Neither we are going to approach the problem of the axiomatic foundation of quantum mechanics, i.e., we are not going to describe the phenomenological and physical arguments leading to construction of the Hilbert space of states of a quantum system and their unitary evolution, together with some extremely subtle aspects of the physical interpretation of observables and the measuring process. In this sense this chapter deals more with what can be called quantum-like systems, i.e., systems whose dynamics takes place in Hilbert spaces and preserves certain Hermitean structures. But we do not imply that the system is quantum, that is, such a system could very well be a classical system, hence the states of the system will simply be phase space like classical states and the space of functions will be the space of classical observables (this idea has been used often, see for instance [St66, Du90, Ma96]).

However we must point out that a big deal of the results we are going to describe can be applied directly to actual quantum systems, hence this chapter can also be read as an exposition on the geometrical properties of the dynamics of quantum systems. The structures we are considering do have a physical interpretation and are coherent with the experimental findings and therefore may be falsified.

To avoid technical difficulties we will deal mostly with finite dimensional carrier spaces V, i.e., only 'finite level quantum systems' will be considered, an assumption which is not far from experimental constraints.

6.3 Invariant Hermitean Structures

6.3.1 Positive-Factorizable Dynamics

We will assume as customary that Γ is a linear vector field on a finite-dimensional linear space E. While studying conditions for a linear vector field $\Gamma = X_A$ to be Hamiltonian with constant symplectic form Ω, see Sect. 5.2.6, we found, Eq. (5.27):

$$A^T\Omega + \Omega A = 0; \qquad \Omega^T = -\Omega, \tag{6.1}$$

with $H = \Omega A$ a symmetric matrix, whereas, the condition for the same vector field to be compatible with a constant symmetric (2, 0)-tensor, see Sect. 4.2.2, it was found to be, Eq. (4.2):

$$A^T G + GA = 0; \qquad G^T = G, \tag{6.2}$$

with the matrix $B = GA$ being skew-symmetric. If we had $A = \Lambda H$, with Λ skew-symmetric, it is clear that $\det A \neq 0$ would imply $\det \Omega \neq 0$ and $\det H \neq 0$.

Recall that the previous conditions are distilled from the factorization problem for a linear dynamical system, i.e., the existence of a skew symmetric contravariant, i.e., (2, 0), tensor Λ on E and a covariant symmetric tensor G on E such that $A = \Lambda \circ G$. Let us also recall that Λ can be thought as a linear map $\Lambda\colon E^* \to E$ such that $\langle \beta, \Lambda(\alpha)\rangle = -\langle \Lambda(\beta), \alpha\rangle$ for all $\alpha, \beta \in E^*$, and the symmetric covariant tensor G can be thought as a linear map $G\colon E \to E^*$, such that $\langle G(u), v\rangle = \langle u, G(v)\rangle$ for all $u, v \in E$. Then the composition $\Lambda \circ G$ is an endomorphism $A\colon E \to E$. Clearly if A is invertible both maps Λ and G must be invertible too, thus the linear map $\Omega = \Lambda^{-1}$ will define a non-degenerate 2-form on E, i.e., a linear symplectic structure on E and the dimension of E must be even. Moreover

$$\Omega \circ A = G. \tag{6.3}$$

Replacing the linear maps A, Ω and G by the corresponding matrices once a linear basis has been chosen (and keeping the same notation for the corresponding matrices) we obtain that equation (6.3) becomes the matrix equation $\Omega A = G$ and then $(\Omega A)^T = G^T = G = \Omega A$, hence Eq. (6.1). Moreover, from the fundamental factorization condition $A = \Lambda G$, we obtain that $A^T = -G\Lambda$ and then $A^T G = -G\Lambda G = -GA$, that is Eq. (6.2).

Therefore we conclude that any such a linear factorizable system Γ with invertible representative matrix A will preserve both, a symplectic structure Ω and a non-degenerate symmetric tensor G.

If we assume now that the tensor G is definite positive we can obtain more information about the structure of Ω and G satisfying Eq. (6.3). Certainly, consider a Lagrangian subspace $L \subset E$ for Ω. The symmetric tensor G defines a scalar product on E, hence on L. Then consider now the subspace $L^{\perp_G}$ orthogonal to L with respect to the metric G. Clearly

$$AL^{\perp_G} = L \tag{6.4}$$

because if $u \in L^{\perp_G}$ and $v \in L$, we have $\Omega(Au, v) = G(u, v) = 0$. Thus $Au \in L^{\perp_\Omega} = L$ because L is Lagrangian, then $AL^{\perp_G} \subset L$. Moreover because A is invertible we have that both subspaces must be the same. Before addressing the general structure theorem for such situation we can recapitulate some of the previous discussion as:

Proposition 6.1 *Suppose that on a real vector space E of dimension m we have $A = \Lambda G$ with Λ skew-symmetric,* $\ker A = 0$ *and G positive definite symmetric. Then E is even dimensional, $m = 2n$ and:*

1. *If we use G as a scalar product on E, then the adjoint $A^\dagger$ of the operator A with respect to G satisfies $A^\dagger = -A$, in other words, A is G-skew-symmetric.*
2. *When considering the polar decomposition of the matrix A in the form $A = |A|\, J$ with respect to the scalar product G, then the matrix J is a complex structure commuting with A and $|A|$.*

Proof We have shown before that E must be even dimensional. Item 1 is just another way of expressing the compatibility condition Eq. (6.2). Notice that if we define a scalar product $\langle u, v\rangle_G = G(u, v)$ on E and A is a linear operator $A \colon E \to E$, then (E finite-dimensional) $\langle A^\dagger u, v\rangle_G = \langle u, Av\rangle_G$. Then by choosing a basis and denoting the corresponding matrices with the same symbols, we get $A^\dagger = G^{-1}A^T G$. Then $A^\dagger = -A$ is equivalent to $A^T G = -GA$.

Finally to prove 2 let us consider the polar decomposition of the matrix A as $A = |A|\, J$ where J is an orthogonal matrix with respect to the scalar product G, i.e., $J^T G J = G$, and $|A|$ is a positive G-symmetric operator (uniquely defined because $\ker A = 0$).

Again, because $\ker A = 0$ such a factorization is unique with $|A| = \sqrt{A^\dagger A} > 0$. As $A^\dagger = -A$, $A^\dagger A = -A^2 > 0$. Therefore $J = A/|A|$ is a function of A and then J commutes with A. Finally, a left multiplication by J in both members of the relation corresponding to $A^\dagger = -A$, leads to $|A| = -J^2\, |A|$, and as $|A|$ is invertible we get $J^2 = -1$. □

In the proof of 2 in the previous theorem, we derived two additional facts that are convenient to highlight.

Corollary 6.2 *Under the conditions of the previous theorem, Theorem 6.1, we get that the complex structure J obtained from the dynamics is orthogonal with respect to the scalar product G, i.e.*

$$G(Ju, Jv) = G(u, v), \qquad \forall u, v \in E, \tag{6.5}$$

and

$$\Omega(Ju, u) > 0, \qquad \forall u \neq 0. \tag{6.6}$$

Proof The G-orthogonality of J, Eq. (6.5), follows from its definition $J = A\,|A|^{-1}$; thus:

$$G(Ju, Jv) = G(A|A|^{-1}u, A|A|^{-1}v) = G(u, A^\dagger A|A|^{-2}v) = G(u, v),$$

because $|A|^2 = A^\dagger A = -A^2$ and A commutes with $|A|$.

The positivity condition, Eq. (6.6), follows from the positivity of $|A|$, that is $A^\dagger A$ is G-positive because:

$$G(A^\dagger Au, u) = \langle A^\dagger Au, u\rangle_G = \langle Au, Au\rangle_G > 0, \quad u \neq \mathbf{0}.$$

The operator $A^\dagger A$ is self-adjoint. We can extend $A^\dagger A$ to the complexification $E^{\mathbb{C}} = E \otimes \mathbb{C}$ of the real linear space E, then because of the spectral theorem $A^\dagger A = \sum_{k=1}^n \lambda_k P_k$ where $\lambda_k > 0$ are the eigenvalues of $A^\dagger A$ and P_k the orthogonal projectors on the corresponding eigenspaces. Then $|A| = \sqrt{A^\dagger A} = \sum_{k=1} \sqrt{\lambda_k} P_k$ and

$$G(|A|u, u) = \langle |A|u, u\rangle_G = \sum_{k=1} \sqrt{\lambda_k}||P_k u||^2 > 0,$$

for all $u \neq 0$. □

Definition 6.3 Let Ω be a linear symplectic structure on E. We will say that a complex structure J on E is tame with respect to Ω, or Ω-tame, if:

$$\Omega(Ju, u) > 0, \qquad \forall u \neq 0.$$

Thus because of the previous corollary, the complex structures obtained from factorizable invertible dynamics are tame with respect to the corresponding symplectic structure.

We notice that it is also possible to consider a degenerate situation with $\ker A \neq \mathbf{0}$.

Proposition 6.4 *Suppose that on a real vector space E of dimension m we have $A = \Lambda H$, with $\ker A \neq \mathbf{0}$, Λ skew-symmetric and H positive definite symmetric; then H defines a scalar product on E and:*

1. $E = \ker A \bigoplus (\ker A)^\perp$ *is a decomposition of E that is invariant under the action of A.*

2. *On* $(\ker A)^{\perp}$ *there exists an orthogonal operator* J_1 *such that* $J_1^2 = -I$ *and commutes with A (when restricted to* $(\ker A)^{\perp}$*); as a by-product, this implies that the dimension of* $(\ker A)^{\perp}$ *is even.*
3. *On* $\ker A$, *A vanishes and we can define* J_0 *with* $J_0^2 = -1$ *only when* $\dim(\ker A)$ *is even.*

Proof To prove 1, notice that as in Proposition 6.1, we have $A^{\dagger} = -A$, with the adjoint operator $A^{\dagger}$ defined with respect to the scalar product determined by H (this property is just another way of writing the compatibility conditions, Eq. (6.2), derived from the factorization property). Now if $v \in (\ker A)^{\perp}$ we have that $\langle Av, u\rangle = \langle v, A^{\dagger}u\rangle = -\langle v, Au\rangle = 0$ for all $u \in \ker A$, thus $Av \in (\ker A)^{\perp}$.

To prove 2 and 3, notice that because of the invariance of $(\ker A)^{\perp}$ with respect to A we can apply Proposition 6.1 to the restriction of A to $(\ker A)^{\perp}$ and obtain an orthogonal operator J_1 such that $J_1^2 = -I$ which commutes with A (hence the dimension of $(\ker A)^{\perp}$ must be even). □

Collecting these results, when $\dim E = 2n$, we can define $J = J_1 \bigoplus J_0$ which commutes with A, is orthogonal, and $J^2 = -1$. Moreover, from $\Lambda = AH^{-1}$ it follows that Λ also decomposes according to the orthogonal decomposition $E = \ker A \bigoplus (\ker A)^{\perp}$.

Before starting to unfold the consequences of these observations, we would like to end this section by stating a structure theorem for linear dynamics on linear spaces satisfying the previous conditions.

Notice that selecting an orthonormal basis $\{u_i \mid i = 1, \ldots n\}$ on an arbitrary Lagrangian subspace $L \subset E$ with $\dim E = 2n$, then $\{Ju_i\}$ is an orthonormal basis of JL, because J is G-orthogonal. If L were a real subspace, i.e., $JL \cap L = \mathbf{0}$ and JL were Lagrangian, then the collection $\{u_i, Ju_k\}$ would be an orthonormal basis, and rescaling the vectors suitably, a symplectic basis. In general, it is not true that any Lagrangian subspace is real; neither is JL a Lagrangian subspace (in Sect. 6.3.2 we will establish under what conditions that is true). However we can prove that such a Lagrangian subspace always exists.

Theorem 6.5 *Let E be a even-dimensional real linear space,* Ω *a linear symplectic structure and G a scalar product on it. Then there exists a G-orthogonal symplectic basis.*

Proof Consider the invertible linear operator A defined by Ω and G, i.e., $\Omega A = G$. We are in the situation of Proposition 6.1 and Corollary 6.2. If we consider as in the proof of Corollary 6.2 the complexfied space $E^{\mathbb{C}} = E \otimes \mathbb{C}$, then the extended liner operator iA is Hermitian because $A^{\dagger} = -A$. Because of the spectral theorem, there will exists an orthonormal basis of eigenvectors W_k, $k = 1, \ldots, n$, with eigenvalues $\pm\sqrt{\lambda_k}$ where $\lambda_k > 0$ are the egivenvalues of $A^{\dagger}A = -A^2$ as in the proof of Corollary 6.2. The eigenvectors $W_k \in E^{\mathbb{C}}$ will have the form:

$$W_k = u_k + iv_k, \qquad k = 1, \ldots, n,$$

and $u_k, v_k \in E$. Then it is easy to show that

$$G(u_i, u_j) = G(v_i, v_j) = \frac{1}{2}\delta_{ij}; \qquad G(u_i, v_j) = 0, \quad \forall i, j = 1, \dots, n. \tag{6.7}$$

Notice that because

$$\begin{aligned}\delta_{jk} &= \langle W_j, W_k\rangle_G = \langle u_j + iv_j, u_k + iv_k\rangle_G \\ &= \langle u_j, u_k\rangle_G + \langle v_j, v_k\rangle_G + i\left(\langle u_j, v_k\rangle_G - \langle v_j, u_k\rangle_G\right),\end{aligned}$$

then:

$$\langle u_j, u_k\rangle_G + \langle v_j, v_k\rangle_G = \delta_{ij} \tag{6.8}$$

and

$$\langle u_j, v_k\rangle_G = \langle v_j, u_k\rangle_G \tag{6.9}$$

for all u_j, v_k. We also have that $(AW_k)^* = A^*W_k^*$ but $A^* = A$ because A is a real matrix and * denotes complex conjugation. Then $AW_k^* = -i\sqrt{\lambda_k}W_k^*$, if the eigenvalue of A with eigenvector W_k were $i\sqrt{\lambda_k}$. The simple computation

$$\langle AW_j^*, W_k\rangle_G = \langle W_j, A^\dagger W_k\rangle_G = -\langle W_j, AW_k\rangle_G,$$

shows:

$$\langle -i\sqrt{\lambda_j}(u_j - iv_j), u_k + iv_k\rangle_G = -\langle u_j + iv_j, i\sqrt{\lambda_k}(u_k + iv_k)\rangle_G.$$

that is,

$$\sqrt{\lambda_j}\left(\langle u_j, u_k\rangle_G - \langle v_j, v_k\rangle_G\right) = -\sqrt{\lambda_k}\left(\langle u_j, u_k\rangle_G - \langle v_j, v_k\rangle_G\right),$$

and,

$$\sqrt{\lambda_j}\left(\langle u_j, v_k\rangle_G + \langle v_j, u_k\rangle_G\right) = -\sqrt{\lambda_k}\left(\langle u_j, v_k\rangle_G + \langle v_j, u_k\rangle_G\right).$$

Then because $\sqrt{\lambda_k} > 0$ for all k, we get:

$$\langle u_j, u_k\rangle_G = \langle v_j, v_k\rangle_G \tag{6.10}$$

and

$$\langle u_j, v_k\rangle_G = -\langle v_j, u_k\rangle_G \tag{6.11}$$

for all j, k. Equation (6.10) combined with Eq. (6.8) gives the first expression in Eq. (6.7). finally Eq. (6.11) together with Eq. (6.9) implies that $\langle u_j, v_k\rangle_G = 0$ for all i, j, concluding the proof of relations Eq. (6.7). Then the basis $\{u_j, v_k\}$ is G-orthogonal.

Now recall that $AW_k = i\sqrt{\lambda_k}W_k$ and $|A|W_k = \sqrt{\lambda_k}W_k$. Then $JW_k = iW_k$, and consequently, $Ju_k = -v_k$ and $Jv_k = u_k$, $k = 1, \ldots, n$. Then:

$$\Omega(u_j, u_k) = \Omega(-\sqrt{\lambda_j}^{-1} Av_j, u_k) = -\sqrt{\lambda_j}^{-1} G(v_j, u_k) = 0 \tag{6.12}$$

and

$$\Omega(v_j, v_k) = \Omega(\sqrt{\lambda_j}^{-1} Au_j, v_k) = \sqrt{\lambda_j}^{-1} G(u_j, v_k) = 0 \tag{6.13}$$

and finally,

$$\Omega(u_j, v_k) = \Omega(-\sqrt{\lambda_j}^{-1} Av_j, v_k) = -\sqrt{\lambda_j}^{-1} G(v_j, v_k) = -\frac{2}{\sqrt{\lambda_j}}$$

thus scaling the vectors v_k (for instance) with the factor $-\frac{\sqrt{\lambda_j}}{2}$ we get the desired symplectic basis. □

Because of the previous theorem, linear dynamics Γ with operators determined by linear operators A possessing the previous factorization properties, i.e., $A = \Lambda \circ G$, with Λ a non-degenerate constant Poisson structure and G a definite positive symmetric $(2, 0)$ tensor, i.e., a scalar product, have a very definite structure. We will obtain a more detailed description of the dynamics defined by such operators in the forthcoming sections and because of the central role that they play in the analysis of Hermitean dynamics we will call them positive factorizable dynamics.

Definition 6.6 A linear dynamical system Γ on a linear space E with associated linear operator A, will be called positive-factorizable if it is factorizable, that is $A = \Lambda G$ with Λ a skew symmetric $(0, 2)$ tensor and G a symmetric $(2, 0)$ tensor, invertible, i.e., both Λ and G are nondegenerate, and G defines a scalar product in E, i.e., it defines a positive definite, symmetric bilinear form on E.

We conclude this analysis by stating the following theorem about the structure of A, Ω and G, that recollects most of the previous results.

Theorem 6.7 *Let A be positive factorizable linear operator on the vector space E, $A = \Lambda G$, then there exists a complex structure J on E which is orthogonal with respect to the scalar product defined by G and Ω-tame. Moreover, there exists a real Lagrangian subspace L, that is $JL \cap L = \mathbf{0}$, with respect to the symplectic structure $\Omega = \Lambda^{-1}$ such that $L^{\perp_G}$ is Lagrangian too and $AL = JL = L^{\perp_G}$. Hence the linear space E decomposes as an orthogonal sum as $E = L \oplus JL$ and the operator A is anti-diagonal with respect to this decomposition.*

Proof Because of Proposition 6.1 there exists a complex structure J on E which is orthogonal with respect to the scalar product defined by G. We also showed, Eq. (6.4) that $AL^{\perp_G} = L$ for any Lagrangian subspace L. Because of Theorem 6.5 there exists a G-orthogonal symplectic basis $\{u_j, v_k\}$ on E adapted to J that satisfies Eqs. (6.7). Then consider the subspace $L = \text{span}\{u_1, \ldots, u_n\}$. Notice that the vectors u_k are non zero and orthogonal, then $\dim L = n$. Because J is G-orthogonal then the space JL has dimension n and the vectors v_k provide an orthogonal basis for it. Again the orthogonality conditions (6.7) imply that $JL = L^{\perp_G}$ because the vectors u_j and v_k are orthogonal to each other. Thus $JL \cap L = \mathbf{0}$ and L is a real subspace.

Moreover $Au_k = J|A|u_k = \sqrt{\lambda_k} Ju_k = \sqrt{\lambda_k} v_k$. Thus $AL = JL$ and A is anti-diagonal with respect to the orthogonal decomposition $E = L \oplus L^{\perp_G}$. Notice that we always have that $AL^{\perp_G} = L$; actually the previous computation shows that $Av_k = \sqrt{\lambda_k}^{-1} A^2 u_k = -\sqrt{\lambda_k} u_k$ for all k.

Finally because of Eqs. (6.12), (6.13) both subspaces L and JL are Lagrangian. □

When the symmetric bilinear form associated with Γ is not positive definite, G becomes a pseudo-scalar product and the inner product will be pseudo-Hermitean. It is now clear that a given linear dynamical system admits more than one Hamiltonian description and we may associate with it Hermitean or pseudo-Hermitean structures. The flow associated with the dynamical evolution will preserve all of them.

The complex structure J associated to a positive factorizable dynamics is also a symplectic map for the symplectic structure Ω as the following computation shows.

$$\Omega(Ju, Jv) = \Omega(AA^{-1}Ju, Jv) = G(JA^{-1}u, Jv) = G(A^{-1}u, v) = \Omega(u, v); \quad \forall u, v \in E,$$

where we have used that J and A commute. Complex structures satisfying the previous condition are called compatible with the symplectic structure.

Definition 6.8 Given a symplectic linear space (E, Ω), a complex structure J such that:

$$\Omega(Ju, Jv) = \Omega(u, v), \qquad \forall u, v \in E, \tag{6.14}$$

and

$$\Omega(Ju, u) > 0, \qquad \forall u \neq \mathbf{0}, \tag{6.15}$$

is said to be a complex structure compatible with the symplectic structure Ω, or Ω-compatible for short. In other words J is Ω-compatible if it is a symplectic map and is Ω-tame.

If J is Ω-compatible we will denote by g_J (or simply by g if there is no risk of confusion) the associated scalar product, i.e., $g(u, v) = \Omega(Ju, v)$.

If we concentrate our attention now on the structures Ω and J alone, we will get the following result.

Proposition 6.9 *Let (E, Ω) be a symplectic linear space and J a complex structure compatible with Ω, then a subspace $L \subset E$ is Lagrangian iff JL is Lagrangian. Moreover JL is the orthogonal complement of L with respect to the scalar product g associated to J.*

Proof It is easy to prove that JL is the orthogonal complement of L. Let $u, v \in L$, then

$$g(u, Jv) = \Omega(Ju, Jv) = \Omega(u, v) = 0.$$

Moreover because J is symplectic JL is Lagrangian iff L is Lagrangian. □

If J is Ω-compatible, the combination of Ω, J and g defines an Hermitean structure on E.

Definition 6.10 Let E be a complex linear space, then a Hermitean structure (also an inner product) is a positive definite sesquilinear map on E, that is, a map $\mathbf{h}\colon E \times E \to \mathbb{C}$ such that it is complex linear in the second argument, satisfies:

$$\mathbf{h}(u, v) = \overline{\mathbf{h}(v, u)}, \qquad \forall u, v \in E\,,$$

and $\mathbf{u}, \mathbf{u} > 0$ for all $u \neq \mathbf{0}$.

Notice that if $\mathbf{h}$ is an Hermitean structure $\mathbf{h}(iu, v) = -i\mathbf{h}(u, v)$. It is a simple computation to check that if the complex structure J is Ω-compatible, then

$$\mathbf{h}(u, v) = g(u, v) + i\Omega(u, v) = \Omega(Ju, v) + i\Omega(u, v)\,,$$

defines a Hermitean structure on E. Conversely, if $\mathbf{h}$ defines a Hermitean structure then the structures:

$$g(u, v) = \mathrm{Re}\,\mathbf{h}(u, v), \qquad \Omega(u, v) = \mathrm{Im}\,\mathbf{h}(u, v) \tag{6.16}$$

define a compatible pair.

Then we may summarize the previous discussion by stating that when the conditions in the preceding propositions hold true, we have a scalar product g defined by H, a constant Poisson structure defined by Λ along with a complex structure J, orthogonal with respect to g, which commutes with A. This allows us to complexify the real vector space E and we may also define an Hermitean scalar product h on it, and $i\,A$ will be self-adjoint on the resulting n-dimensional Hilbert space. Thus we have again all the ingredients to associate with A a unitary flow on the complex vector space we have constructed out of E. Notice that any positive factorizable linear dynamics gives raise to such structure. We will analyze their properties in the following section.

Before embarking on a detailed analysis of such dynamics we would like to stress that most of the previous results are also valid in infinite dimensions, in the following sense. Let E be a real linear space, not necessarily finite-dimensional. Let A be an invertible linear operator on E and we will assume as before that A is factorizable, i.e., $A = \Lambda \circ G$. If A is invertible Λ must be invertible too, then $\Omega = \Lambda^{-1}$ is called a strong symplectic form on E. We will say now that A is positive factorizable if G defines a scalar product on E. If we assume now that E is complete with respect to the norm defined by G, or in other words that E is a (real) Hilbert space with the scalar product G and that A is a bounded linear operator such that A^{-1} is bounded too, then the results stated in Proposition 6.1, Corollary 6.2, Theorem 6.5, and Theorem 6.7 are true with some obvious changes in the statements. Proposition 6.9 is also true if Ω is a strong symplectic structure and J is a bounded symplectic map. We will not pay too much attention right now to the possibility of extending to infinite dimensions the previous results, but we will come back to them at the end of this chapter when dealing with quantum mechanical systems.

6.3.2 Invariant Hermitean Metrics

We will summarize from a slightly different perspective the results obtained at the end of the Sect. 6.3.1. Let E be a real linear space. We may use the tensorialization principle to promote the linear tensorial objects, Ω, J, G studied in the previous section, to smooth tensors on E. Thus we introduce the geometrical tensors ω, g and J corresponding to positive factorizable linear dynamical system $\Gamma = X_A$:

$$\omega = \omega_{ij} dx^i \wedge dx^j, \quad g = g_{ij} dx^i \otimes dx^j, \quad J = J_i^j dx^i \otimes \frac{\partial}{\partial x^j}. \tag{6.17}$$

At this stage we will not be worried about further integrability conditions imposed with the tensorialization principle as our structures are constant and only their algebraic properties are relevant, that is: $J : E \to E$ is a complex structure and ω a constant symplectic structure.

Definition 6.11 We will say that ω is J-invariant if

$$\omega(JX, JY) = \omega(X, Y),$$

for all vector fields X, Y in E.

In this case

$$\omega(X, JY) = \omega(JX, J^2 Y) = -\omega(JX, Y),$$

i.e.

$$\omega(X, JY) + \omega(JX, Y) = 0; \tag{6.18}$$

for all vector fields X, Y in E.

Notice that conversely if ω and J satisfy the preceding relation, Eq. (6.18), then,

$$\omega(JX, JY) = -\omega(X, J^2 Y) = \omega(X, Y),$$

i.e., ω is J-invariant. Moreover, using J-invariance we see that the bilinear form g defined as:

$$g(X, Y) = \omega(JX, Y) \tag{6.19}$$

defines a bilinear symmetric form, because

$$g(Y, X) = \omega(JY, X) = -\omega(Y, JX) = \omega(JX, Y) = g(X, Y).$$

Note that

$$g(JX, JY) = \omega(J^2 X, JY) = -\omega(X, JY) = \omega(JY, X) = g(Y, X) = g(X, Y),$$

i.e., J is g-orthogonal:

$$g(JX, Y) = g(JX, Y). \tag{6.20}$$

We say that (ω, J) is a compatible pair (as after Definition 6.10, Eq. (6.16), in the linear case) if ω is J-invariant and $g(X, Y) = \omega(JX, Y)$ defines a symmetric positive definite form. In such case J is ω-compatible using the terminology in the Sect. 6.3.1. Then we can define a Hermitean product in E, considered now as a complex space because of the complex structure J, by means of

$$h(X, Y) = g(X, Y) + i\,\omega(X, Y) = \omega(JX, Y) + i\,\omega(X, Y).$$

Recall also that the real linear space E can be complexified and the complex extension of J to $E^{\mathbb{C}}$, $J^{\mathbb{C}}$, admits two eigenvalues, $\pm i$. Moreover,

$$E^{\mathbb{C}} = E^+ \oplus E^-$$

where

$$E^{\pm} = \{X \in E^{\mathbb{C}} \mid J^{\mathbb{C}} X = \pm i\, X\}.$$

A tensorial triple (g, J, ω), with g a metric, J a complex structure and ω a symplectic structure satisfying the conditions (6.18), (6.19) and (6.20) will be called an admissible triple. Because admissible triples are in one-to-one correspondence with Hermitean products we will use both notions interchangeably. For reasons that will be clear later on, we may also refer to an admissible triple as a linear Kähler structure.

Definition 6.12 A linear vector field Γ will be called Hermitean if it preserves a compatible pair (ω, J), i.e.

$$\mathcal{L}_\Gamma \omega = 0, \qquad \mathcal{L}_\Gamma J = 0, \tag{6.21}$$

or equivalently, if $\mathcal{L}_\Gamma h = 0$

Then, the inverse problem for Hermitean dynamics can be stated as follows: Given a linear dynamics Γ find a Hermitean product h such that $\mathcal{L}_\Gamma h = 0$.

Notice that in using the Kähler geometry on E defined by the Hermitean product h (see later on Sect. 6.4.3 and [Er10], [Ib96a] for more details on the Kähler geometry of Hermitean systems), we are using a classical picture to describe the dynamics of a system that later on we may be able to identify with a quantum system Γ. This was precisely the point of view adopted in [St66].

Thus, if we represent Γ in a real basis for E we get, $\Gamma = A^l{}_k \xi^k \, \partial/\partial \xi^l$, and the first invariance condition in Eq. (6.16), $\mathcal{L}_\Gamma \omega = 0$, is equivalent to the factorization property for A. This forces us to look first for the inverse symplectic problem for linear vector fields [Gi93]. Because of the discussion on Sect. 5.2.6, we obtain Eq. (5.29), i.e., all odd powers of A are traceless matrices, $\mathrm{Tr}\, A^{2k+1} = 0$ for all $k = 0, 1, \ldots$. The second invariance condition in Eq. (6.21) becomes

$$[A, J] = 0, \tag{6.22}$$

or, in other words, A is a complex linear map with respect to the complex structure defined in E by J.

How can we determine all compatible pairs ω, J, that exist for a given linear field Γ? Let us consider any invertible linear map $T : E \to E$ and denote by A_T the transformed linear operator, that is, $A_T = T^{-1} A T$. If we apply T to the previous relations, Eq. (6.21) or in matrix form, to the factorization property of A, and Eq. (6.22). Then we get,

$$A_T = \Lambda_T \cdot g_T, \qquad [A_T, J_T] = 0 \tag{6.23}$$

with

$$\Lambda_T = T^{-1} \Lambda (T^{-1})^t; \qquad g_T = T^t g T\,, \tag{6.24}$$

and

$$J_T = T^{-1} J T\,. \tag{6.25}$$

Therefore, if T commutes with A, that is $A_T = A$, then we obtain a new Hamiltonian description provided that $\Lambda_T \neq \Lambda$ or $g_T \neq g$. Notice that because of the general factorization theorem, Theorem 4.2, even powers of A provide such transformations in a natural way: $T_t = \exp(tA^{2k})$, $k = 1, 2, \ldots$. In general we have the following result:

Proposition 6.13 *Let Γ a Hermitean linear dynamics with associated linear operator A. Any invertible nonunitary transformation T which is a symmetry for A takes a Hermitean description of Γ onto an alternative one and any alternative Hermitean structure compatible with Γ is obtained by such a transformation.*

Proof Let (ω, J, g) or $h = g + i\omega$, be a Hermitean structure invariant under Γ. The group of unitary transformations for the Hermitean structure h denoted by $U(h)$ is the intersection of the group of symplectic transformations $Sp(\omega)$ and the group of isometries of g, $O(g)$, then T is non unitary, it is either non-symplectic, non-isometric or both. In any case $\Lambda_T \neq \Lambda$ or $g_T \neq g$.

Conversely, by using a Gram-Schmidt procedure, each Hermitean structure is reduced to the standard one in $\mathbb{C}^n$. Therefore composing one map with the inverse of the other, we get the required T. In the symplectic case this amounts to the so-called Darboux theorem and, in the Riemannian case it amounts to the diagonalization procedure. □

This shows that the space of alternative compatible pairs is a homogeneous space $C_A = G_A/(U(h) \cap G_A)$, where G_A is the isotropy group at A with respect to the adjoint action of $GL(E)$ on the space of linear operators on E, i.e., the group of linear symmetries of A, and $U(h)$ is the unitary group defined by the Hermitean structure h.

How can we describe the space C_A? or, more especifically, which are the symmetries of A which are not unitary?

Proposition 6.14 *For a generic matrix A, the symmetry group G_A of A is Abelian and is generated by the flows corresponding to all powers of A, A^j, $j = 0, 1, \ldots,$* dim V *while $U(h) \cap G_A$ is generated by the flows corresponding to the odd powers of A. The homogeneous space is the subgroup of G_A generated by the flows corresponding to the even powers of A.*

Thus even powers of A generate symmetries which are not unitary. Then, any finite level quantum system admits alternative quantum descriptions parametrized by the Abelian subalgebra generated by A^2.

6.3.3 Hermitean Dynamics and Its Stability Properties

Among the linear Hamiltonian vector fields preserving an invertible Poisson structure, we concentrate, in this section, on the subset of dynamical systems which are stable linear.

In particular, we show that:

Proposition 6.15 *A stable linear Hamiltonian vector field* Γ *has at least one Hamiltonian description with positive definite Hamiltonian function.*

Proof Let, as usual, A be the representative matrix of Γ, with the property $A = \Lambda\, H$. We decompose (the complex extension of) A as a sum, $A = S + N$, of a semisimple matrix S plus a nilpotent one N, with $N\, S = S\, N$. Then, for some integer k,

$$e^{tA} = e^{tS}\, e^{tN} = e^{tS} \left(\sum_{m=0}^{k} \frac{t^m}{m!} N^m \right),$$

since $N^{k+1} = 0$. The evolution of the initial condition x_0 is given by $x(t) = e^{tA} x_0$. To have a finite norm for $x(t)$ when $t \to \infty$ the nilpotent part must be absent, i.e., $k = 0$. Moreover, the boundness of the orbit implies that the semisimple part S must have only imaginary eigenvalues.

With the help of the associated eigenvectors we can construct an invariant Hermitean structure. The imaginary part is a symplectic structure and the contraction with the dynamical vector field Γ provides the positive definite Hermitean function. □

Thus in the Hamiltonian decomposition we have:

Corollary 6.16 *A linear Hamiltonian vector field which is stable, must have a representative matrix with purely imaginary eigenvalues and, therefore, will admit a positive definite Hamiltonian function.*

Summarizing: stable linear Hamiltonian systems may be associated with unitary flows on complex Hilbert spaces of finite dimension. We may call them 'quantum-like' systems. It is possible to formulate an inverse problem for these systems directly on complex vector spaces.

6.3.4 Bihamiltonian Descriptions

In Sect. 6.3.1 (Definition 6.10) we have seen that on a complex linear space E the Hermitean structure is associated with a triple (g, ω, J) on the real vector space corresponding to E. There we have seen that $g(u, v) = \omega(Ju, v)$. At the group level, this association is translated into the statement that transformations preserving the Hermitean structure (unitary transformations) coincide with the intersection of transformations that preserve the metric tensor g (orthogonal transformations) with the transformations which preserve ω (symplectic transformations) or, also, with those preserving the complex structure of J. Thus the intersection of the groups of transformations preserving any two of the involved structures, gives the unitary group of transformations.

In this section, starting with two triples (g_1, ω_1, J_1) and (g_2, ω_2, J_2) on the real vector space E, $\dim_{\mathbb{R}} E = 2n$, we would like to define when two triples are compatible. This definition arises from our desire to establish a situation similar to the one we have for compatible Poisson brackets in the canonical setting when we consider dynamical systems which are biHamiltonian.

As the space of physical states is not the vector space itself but rather the complex projective space, or the ray space, the analogy with the classical situation should be established not on the vector space but on the ray space. We notice that on the vector space E we have a linear structure represented by Δ. The reduction to the ray space requires also the generator of the shift in the phase, i.e., $J(\Delta)$, as we have two triples we are going to have a quotienting procedure associated with Δ and $J_1(\Delta)$ for the first triple, and Δ and $J_2(\Delta)$ for the second triple. In conclusion, if we denote by $P_1(E)$ and $P_2(E)$ the two quotient ray spaces, we could like to find conditions to be able to project tensor fields associated with (g_2, ω_2, J_2), expressed into contravariant form, onto $P_1(VE$ (see more details in Sects. 6.5.4, 6.5.5).

To be able to follow the subsequent arguments we consider first a simple example in two (real) dimensions. Here we are not considering the associated ray space, it would be just a point, we examine instead the meaning of the definition which we now give.

Definition 6.17 Two Hermitean structures h_1 and h_2, or their associated triples, will be said to be compatible if

$$\mathcal{L}_{J_1(\Delta)} h_2 = \mathcal{L}_{J_2(\Delta)} h_1 = 0 . \tag{6.26}$$

6.3.4.1 A Simple Example in Two Dimensions

Starting from the observation that two quadratic forms, one of which is assumed to be positive, can always be diagonalized simultaneously (at the price of using a non-orthogonal transformation, if necessary) we can assume from start g_1 and g_2 to be of the form

$$g_1 = \begin{pmatrix} \varrho_1 & 0 \\ 0 & \varrho_2 \end{pmatrix}, \qquad g_2 = \begin{pmatrix} \sigma_1 & 0 \\ 0 & \sigma_2 \end{pmatrix} . \tag{6.27}$$

The more general J such that $J^2 = -1$ will be of the form

$$J = \begin{pmatrix} a & b \\ -(1+a^2)/b & -a \end{pmatrix} . \tag{6.28}$$

Compatibility with g_1 requires that J be anti-Hermitean (with respect to g_1), and this leads to

$$J = J_{1\pm} = \begin{pmatrix} 0 & \pm\sqrt{\frac{\varrho_2}{\varrho_1}} \\ \mp\sqrt{\frac{\varrho_1}{\varrho_2}} & 0 \end{pmatrix} \tag{6.29}$$

and similarly

$$J = J_{2\pm} = \begin{pmatrix} 0 & \pm\sqrt{\frac{\sigma_2}{\sigma_1}} \\ \mp\sqrt{\frac{\sigma_1}{\sigma_2}} & 0 \end{pmatrix} \tag{6.30}$$

from the requirement of admissibility with g_2.

As a consequence:

$$\omega = \omega_{1\pm} = \begin{pmatrix} 0 & \pm\sqrt{\varrho_2\varrho_1} \\ \mp\sqrt{\varrho_2\varrho_1} & 0 \end{pmatrix}, \qquad \omega = \omega_{2\pm} = \begin{pmatrix} 0 & \pm\sqrt{\sigma_2\sigma_1} \\ \mp\sqrt{\sigma_2\sigma_1} & 0 \end{pmatrix}. \tag{6.31}$$

Now we have all the admissible structures, i.e., $(g_1, \omega_{1\pm}, J_{1\pm})$ and $(g_2, \omega_{2\pm}, J_{2\pm})$.

Let us compute the invariance group for the first triple, having made a definite choice for the possible signs (say: $J = J_+$). The group is easily seen to be

$$O_1(t) = \cos(t)\mathbb{I} + \sin(t)J_1 = \begin{pmatrix} \cos(t) & \sqrt{\frac{\varrho_2}{\varrho_1}}\sin(t) \\ -\sqrt{\frac{\varrho_1}{\varrho_2}}\sin(t) & \cos(t) \end{pmatrix}, \tag{6.32}$$

while for the second triple we obtain

$$O_2(t) = \cos(t)\mathbb{I} + \sin(t)J_2 = \begin{pmatrix} \cos(t) & \sqrt{\frac{\sigma_2}{\sigma_1}}\sin(t) \\ -\sqrt{\frac{\sigma_1}{\sigma_2}}\sin(t) & \cos(t) \end{pmatrix}, \tag{6.33}$$

and in general we obtain two different realizations of $SO(2)$.

The two realizations will commute when $\rho_2/\rho_1 = \sigma_2/\sigma_1$. The latter condition is easily seen (by imposing e.g., $[J_{1,2}, T] = 0$) to be precisely the condition of compatibility of the two triples.

To conclude the discussion of the example, let us see what happens in the complexified version of the previous discussion.

To begin with we have to define multiplication by complex numbers on $\mathbb{R}^2$, thus making it a complex vector space, and this can be done in two ways, namely as:

$$(x + iy)\begin{pmatrix} a \\ b \end{pmatrix} = (x\mathbb{I} + J_1 y)\begin{pmatrix} a \\ b \end{pmatrix}, \tag{6.34}$$

or as

$$(x + iy)\begin{pmatrix} a \\ b \end{pmatrix} = (x\mathbb{I} + J_2 y)\begin{pmatrix} a \\ b \end{pmatrix}. \tag{6.35}$$

Correspondingly, we can introduce two different Hermitean structures on $\mathbb{R}^2$ as

$$h_1 = g_1 + i\omega_1, \quad \text{or as} \quad h_2 = g_2 + i\omega_2 . \tag{6.36}$$

They are antilinear in the first factor and in each case the corresponding multiplication by complex numbers must be used. The $O_1(t)$ and $O_2(t)$ actions both coincide with the multiplication of points of $\mathbb{R}^2$ by the complex numbers e^{it} (i.e., with different realizations of $U(1)$), but the definition of multiplication by complex numbers is different in the two cases.

6.3.5 The Structure of Compatible Hermitean Forms

Having clarified the setting by means of a previous example, let us proceed now to the general situation and the various consequences emerging from the compatibility condition.

Consider two different Hermitean structures on E, $h_1 = g_1 + i\omega_1$ and $h_2 = g_2 + i\omega_2$. We define the associated quadratic functions $\mathbf{g}_1 = g_1(\Delta, \Delta)$ and $\mathbf{g}_2 = g_2(\Delta, \Delta)$. Their associated Hamiltonian vector fields will be exactly $\Gamma_1 = J_1(\Delta)$ and $\Gamma_2 = J_2(\Delta)$, if we use $\frac{1}{2}\mathbf{g}_1$ and $\frac{1}{2}\mathbf{g}_2$ as Hamiltonians. This follows easily from $\omega(\Gamma, \Delta) = g(\Delta, \Delta)$ and $i_\Gamma \Omega = dH$, and,

$$\mathcal{L}_\Delta H = i_\Delta i_\Gamma \Omega = dH(\Delta) = 2H = g(\Delta, \Delta) .$$

From the definition of compatible Hermitean structures we get:

$$\mathcal{L}_{\Gamma_{1,2}} \Omega_{2,1} = \mathcal{L}_{\Gamma_{1,2}} g_{2,1} = 0 .$$

Remark 6.1 Notice that, if $\omega = \omega_{ij} dx^i \wedge dx^j$ is a constant symplectic structure and $X = A^i{}_j x^j \partial/\partial x^i$ is a linear vector field, then the condition: $\mathcal{L}_X \omega = 0$ can be written in terms of the representative matrices as the requirement that the matrix ωA be symmetric, i.e., (recall Eq. (6.1)):

$$\omega A - (\omega A)^T = \omega A + A^T \omega = 0 , \tag{6.37}$$

while the condition $\mathcal{L}_X g = 0$ is equivalent to the matrix gA being skew-symmetric, i.e., (recall now Eq. (6.2)):

$$gA + (gA)^T = gA + A^T g = 0 . \tag{6.38}$$

Now because $\mathcal{L}_{\Gamma_1} g_2 = 0$ we get that the matrix $g_2 A_1$ will be skew-symmetric (Eq. (6.38)), but then

$$\mathcal{L}_{\Gamma_1} \mathbf{g}_2 = x^T (g_2 A_1 + A_1^T g_2) x = 0 , \tag{6.39}$$

and we obtain:

$$0 = \mathcal{L}_{\Gamma_1}(i_{\Gamma_2}\omega_2) = i_{\Gamma_2}\mathcal{L}_{\Gamma_1}\omega_2 + i_{[\Gamma_1,\Gamma_2]}\omega_2\,,$$

and as $\mathcal{L}_{\Gamma_1}\omega_2 = 0$, we conclude:

$$i_{[\Gamma_1,\Gamma_2]}\omega_2 = 0,$$

and similarly for ω_1. As neither ω_1 nor ω_2 are degenerate, this implies that Γ_1 and Γ_2 commute, i.e. that

$$[\Gamma_1, \Gamma_2] = 0.$$

Moreover, remembering that the Poisson bracket of any two functions f and g, $\{f, g\}$ is defined as $\{f, g\} = \omega(X_g, X_f)$, where X_f and X_g are the Hamiltonian vector fields associated with f and g respectively, we find $0 = \mathcal{L}_{\Gamma_1}\mathbf{g}_2 = d\mathbf{g}_2(\Gamma_1) = 2\omega_2(\Gamma_2, \Gamma_1)$. Hence we find:

$$\{\mathbf{g}_1, \mathbf{g}_2\}_2 = 0\,,$$

where $\{\cdot, \cdot\}_2$ is the Poisson bracket associated with ω_2 and similarly for ω_1.

Proposition 6.18 *Two Hermitean structures h_1 and h_2 with associated quadratic forms $\mathbf{g}_1 = g_1(\Delta, \Delta)$, $\mathbf{g}_2 = g_2(\Delta, \Delta)$, and Hamiltonian vector fields Γ_1 and Γ_2 are compatible iff $\{\mathbf{g}_1, \mathbf{g}_2\}_{1,2} = 0$ and $\mathcal{L}_{\Gamma_1}\omega_2 = \mathcal{L}_{\Gamma_2}\omega_1 = 0$.*

Proof It remains to be proved that $\mathcal{L}_{\Gamma_1}g_2 = 0$. Notice that this is equivalent to the matrix g_2A_1 being skew-symmetric. Then the condition $\{\mathbf{g}_1, \mathbf{g}_2\}_2 = 0$ implies exactly that because of Eq. (6.39). □

Remembering that the weak form of the tensorialization principle states that statements concerning linear vector fields translate into equivalent statements for the $(1, 1)$-type tensor fields having the same representative matrices, and recalling that the defining matrices of Γ_1 and Γ_2 are precisely those of the corresponding complex structures, we see at once that:

$$[\Gamma_1, \Gamma_2] = 0 \Longleftrightarrow [J_1, J_2] = 0\,,$$

i.e., that the two complex structures will commute as well.

In general, given any two $(0, 2)$ (or $(2, 0)$) tensor fields g_1 and g_2 one (at least) of which, say g_1, is invertible, the composite tensor $g_1^{-1} \circ g_2$ will be a $(1, 1)$-type tensor. Then, out of the two compatible structures we can build up the two $(1, 1)$-type tensor fields

$$G = g_1^{-1} \circ g_2\,,$$

and

$$T = \omega_1^{-1} \circ \omega_2.$$

Actually one can prove at once that the two are related, and a direct calculation proves that:

$$G = J_1 \circ T \circ J_2^{-1} = -J_1 \circ T \circ J_2$$

or, in other words,

$$T = -J_1 \circ G \circ J_2 \,. \tag{6.40}$$

It turns out that T (and hence G) commutes with both complex structures i.e.,

$$[G, J_a] = [T, J_a] = 0, \qquad a = 1, 2 \,. \tag{6.41}$$

This follows from the fact that both G and T are Γ-invariant, i.e.,

$$\mathcal{L}_{\Gamma_{1,2}} G = \mathcal{L}_{\Gamma_{1,2}} T = 0 \,,$$

and from Eq. (6.40).

It follows also from Eqs. (6.40) and (6.41) that G and T commute, i.e.:

$$[G, T] = 0 \,.$$

Moreover, G enjoys the property that

$$g_a(Gu, v) = g_a(u, Gv), \qquad a = 1, 2 \,. \tag{6.42}$$

Indeed one can prove by direct calculation that

$$g_1(Gu, v) = g_1(u, Gv) = g_2(u, v) \,,$$

while

$$g_2(Gu, v) = g_2(u, Gv) = g_1^{-1}(g_2(u, \cdot), g_2(v, \cdot)) \,,$$

and this completes the proof. Notice that Eq. (6.42) can be read as G being self-adjoint with respect to both metrics.

Notice that the derivation of this result does not require the compatibility condition to hold. If the latter is assumed, however, one can prove also that T is self-adjoint with respect to both metrics, and that both $J_1$and J_2 are instead skew-adjoint with respect to both structures, i.e., that

$$g_a(Tu, v) = g_a(u, Tv), \qquad a = 1, 2,$$

and that

$$g_1(u, J_2 v) + g_1(J_2 u, v) = 0, \qquad \forall u, v \in E,$$

with a similar equation with the indices interchanged.

Indeed, from, e.g.: $\mathcal{L}_{\Gamma_1}\omega_2 = 0$ we obtain, in terms of the representative matrices and using: $J_1 = g_1^{-1}\omega_1$:

$$\omega_2 g_1^{-1}\omega_1 = \omega_1 g_1^{-1}\omega_2 \Leftrightarrow \omega_2\omega_1^{-1} g_1 = g_1\omega_1^{-1}\omega_2.$$

Remembering the definition of T, that is equivalent to: $g_1 T = (g_1 T)^T$, and this leads to

$$T = g_1^{-1} T^T g_1 = (T^\dagger)_1$$

where $(T^\dagger)_1$ is the adjoint of T with respect to g_1. Interchanging indices, one can prove that: $(T^\dagger)_2 = T$ as well.

Concerning the J's (that have already been proved to be skew-adjoint with respect to the respective metric tensors), consider, e.g.

$$(J_1^\dagger)_2 =: g_2^{-1} J_1^T g_2 = -g_2^{-1} g_1 J_1 ,^T g_1^{-1} g_2 = -G^{-1} J_1 G = -J_1$$

as G and the J's commute. A similar result holds of course for J_2.

Summarizing what has been proved up to now, we have found that G, T, J_1 and J_2 are a set of mutually commuting linear operators. G and T are self-adjoint, while J_1 and J_2 are skew-adjoint, with respect to both metric tensors.

If we now diagonalize G, the $2n$-dimensional vector space $E \cong \mathbb{R}^{2n}$ will split into a direct sum of eigenspaces: $E = \oplus_k V_{\lambda_k}$, where the λ_k's $(k = 1, \ldots, r \leq 2n)$ are the distinct eigenvalues of G. According to what has just been proved, the sum will be an orthogonal sum with respect to both metrics, and, in V_{λ_k}, $G = \lambda_k \mathbb{I}_k$, with $\mathbb{I}_k$ the identity matrix in V_{λ_k}. Assuming compatibility, T will commute with G and will be self-adjoint. Therefore we will get a further orthogonal decomposition of each V_{λ_k} of the form

$$V_{\lambda_k} = \bigoplus_r W_{\lambda_k, \mu_{k,r}}$$

where the $\mu_{k,r}$'s are the (distinct) eigenvalues of T in V_{λ_k}.The complex structures commute in turn with both G and T. Therefore they will leave each one of the $W_{\lambda_k, \mu_{k,r}}$'s invariant.

Now we can reconstruct, using the g's and the J's, the two symplectic structures. They will be block-diagonal in the decomposition of E, and on each one of the $W_{\lambda_k, \mu_{k,r}}$'s they will be of the form

$$g_1 = \lambda_k g_2, \qquad \omega_1 = \mu_{k,r}\omega_2,$$

where, with some abuse of notation, we have written $g_1, g_2, \omega_1, \omega_2$ instead of their restrictions to the subspaces $W_{\lambda_k, \mu_{k,r}}$. Therefore, in the same subspaces:

$$J_1 = g_1^{-1}\omega_1 = \frac{\mu_{k,r}}{\lambda_k} J_2 .$$

It follows from: $J_1^2 = J_2^2 = -1$ that: $(\frac{\mu_{k,r}}{\lambda_k})^2 = 1$, whence: $\mu_{k,r} = \pm\lambda_k$ (and $\lambda_k > 0$). The index r can then assume only two values, corresponding to $\pm\lambda_k$ and at most V_{λ_k} will have the decomposition of V_{λ_k} into the orthogonal sum: $V_{\lambda_k} = W_{\lambda_k,\lambda_k} \oplus W_{\lambda_k,-\lambda_k}$. All in all, what we have proved is the following:

Lemma 6.19 *If the two Hermitean structures $h_1 = (g_1, \omega_1)$ and $h_2 = (g_2, \omega_2)$, coming from admissible triples (g_1, ω_1, J_1) and (g_2, ω_2, J_2), are compatible, then the vector space $E \cong \mathbb{R}^{2n}$ will decompose into the (double) orthogonal sum:*

$$E \cong \bigoplus_{k=1,\ldots,r\ ,\alpha=\pm 1} W_{\lambda_k,\alpha\lambda_k} , \tag{6.43}$$

where the index $k = 1, \ldots, r \leq 2n$ labels the eigenspaces of the $(1, 1)$-type tensor $G = g_1^{-1} \circ g_2$ corresponding to its distinct eigenvalues $\lambda_k > 0$, while $T = \omega_1^{-1} \circ \omega_2$ will be diagonal (with eigenvalues $\pm\lambda_k$) on the $W_{\lambda_k,\pm\lambda_k}$'s, on each one of which:

$$g_1 = \lambda_k g_2, \qquad \omega_1 = \pm\lambda_k\omega_2, \qquad J_1 = \pm J_2 . \tag{6.44}$$

As neither symplectic form is degenerate, the dimension of each one of the $W_{\lambda_k,\pm\lambda_k}$'s will be necessarily even.

Now we can further qualify and strengthen the compatibility condition by stating the following:

Definition 6.20 Two (compatible) Hermitean structures will be said to be in a generic position iff the eigenvalues of G and T have minimum (i.e., double) degeneracy.

In general, two appropriate geometrical objects like two (0, 2) or (2, 0)-type tensor fields are said to be in a generic position if they can be 'composed' to yield a 1-1 tensor whose eigenvalues have minimum degeneracy. For instance g_1 and g_2 are in a generic position if the eigenvalues of $G = g_1^{-1} \circ g_2$ have minimum degeneracy, which possibly depends on further conditions: when compatibility is required, this degeneracy is double. The results that we have just proved will imply that each one of the W_{λ_k,λ_k}, $W_{\lambda_k,-\lambda_k}$ will have the minimum possible dimension, that is two.

Denoting them by E_k (these now two-dimensional subspaces $k = 1, \ldots, n$), all that has been said up to now can be summarized in the following:

Theorem 6.21 (Structure theorem for generic compatible Hermitean structures) *If the Hermitean structures $\mathbf{h}_1$ and $\mathbf{h}_2$ are compatible and in a generic position, then the*

2n-dimensional carrier linear space E splits into a sum of n mutually bi-orthogonal (i.e., orthogonal with respect to both metrics g_1 and g_2) two-dimensional vector subspaces $E = E_1 \oplus E_2 \oplus \ldots \oplus E_n$. All the structures g_a, J_a, ω_a decompose accordingly into a direct sum of structures on these two-dimensional subspaces, and on each one of the E_k's they can be written as

$$
\begin{aligned}
g_1|_{E_k} &= \lambda_k(e_1^* \otimes e_1^* + e_2^* \otimes e_2^*);\ \lambda_k > 0 \quad & g_2|_{E_k} &= \varrho_k\, g_1|_{E_k};\ \varrho_k > 0 \\
J_1|_{E_k} &= (e_2 e_1^* - e_1 e_2^*) & J_2|_{E_k} &= \pm J_1|_{E_k} \\
\omega_1|_{E_k} &= \lambda_k(e_1^* \wedge e_2^*) & \omega_2|_{E_k} &= \pm\varrho_k\, \omega_1|_{E_k}
\end{aligned}
$$

where $e_2 = J_1 e_1$, e_1 is any given vector in E_k and the e^*'s are the dual basis of the e''s. In other words on each subspace g_1 and g_2 are proportional, while $J_1 = \pm J_2$ and accordingly $\omega_2 = \pm\varrho\omega_1$.

Every linear vector field preserving both $h_1 = (g_1, \omega_1)$ and $h_2 = (g_2, \omega_2)$ will have a representative matrix commuting with those of T and G , and it will be block-diagonal in the same eigenspaces E_k. Therefore, in the generic case, the analysis can be restricted to each 2-dimensional subspace E_k in which the vector field will preserve both a symplectic structure and a positive-definite metric. Therefore it will be in $Sp(2) \cap SO(2) = U(1)$ and, on each E_k, it will represent a harmonic oscillator with frequencies depending in general on the V_k's .

Going back to the general case, we can make contact with the theory of complete integrability of a biHamiltonian system by observing that T plays here the role of a recursion operator. Indeed, we show now that it generates a basis of vector fields preserving both the Hermitean structures h_a given by:

$$
\Gamma_1, T\Gamma_1, \ldots, T^{n-1}\Gamma_1 \,. \tag{6.45}
$$

To begin with, these fields preserve all the geometrical structures, commute pair-wise and are linearly independent. In fact these properties follow from the observation that T, being a constant 1-1 tensor, satisfies the Nijenhuis condition $[T, T] = 0$. Therefore, for any vector field X:

$$
\mathcal{L}_{TX} T = T \mathcal{L}_X T \tag{6.46}
$$

that, T being invertible, amounts to

$$
\mathcal{L}_{TX} = T \mathcal{L}_X \tag{6.47}
$$

So, $\forall k \in \mathbb{N}$:

$$
\mathcal{L}_{T^k\Gamma_1} = T \mathcal{L}_{T^{k-1}\Gamma_1} = \ldots = T^k \mathcal{L}_{\Gamma_1} \tag{6.48}
$$

and

$$
T^k \mathcal{L}_{\Gamma_1} \omega_a = 0 = T^k \mathcal{L}_{\Gamma_1} g_a \quad ; \tag{6.49}
$$

Moreover, $\forall s \in \mathbb{N}$:

$$[T^{k+s}\Gamma_1, T^k\Gamma_1] = \mathcal{L}_{T^{k+s}\Gamma_1} T^k\Gamma_1 = T^s \mathcal{L}_{T^k\Gamma_1} T^k\Gamma_1 = T^s[T^k\Gamma_1, T^k\Gamma_1] = 0. \tag{6.50}$$

Besides, the assumption of minimal degeneracy of T implies that the minimal polynomial of T be of degree n. Indeed, we have shown that the diagonal form of T is

$$T = \bigoplus_{k=1,\ldots,n} \{\pm\rho_k \mathbb{I}_k\} \tag{6.51}$$

where $\mathbb{I}_k$ is the identity on V_k . Any linear combination

$$\sum_{r=0}^{m} \alpha_r T^r = 0 \quad , \quad m \leq n-1 \quad , \tag{6.52}$$

yields a linear system for the α_r's of n equations in $m+1$ unknowns whose matrix of coefficients is of maximal rank and that, for $m = n-1$, coincides with the full Vandermonde matrix of the ρ_k's .

Then, we can conclude that the n vector fields $T^r\Gamma_1, r = 0, 1, \ldots, n-1$ form a basis.

6.4 Complex Structures and Complex Exterior Calculus

In Sect. 6.3.5 we discussed the inverse problem for Hermitean dynamics. As we discussed there, Hermitean structures combine metric and symplectic structures and define in a natural way a complex linear structure. Before continuing our analysis and the discussion of an extremely interesting family of systems fitting into this geometrical description, quantum systems, we will digress on the geometry of complex linear spaces. In the same way as in Chap. 3 we were provided with a purely tensorial characterization of real linear structures, we would like to proceed for complex structures, that is to a geometrization of complex calculus. This task will be accomplished in the next few sections.

6.4.1 The Ring of Functions of a Complex Space

Throughout this book the dual approach to dynamical systems, that is evolution considered as derivations on an algebra of observables, has been stressed. When we encounter complex spaces new interesting phenomena occur.

Thus suppose that E is a complex linear space, i.e., a real space equipped with a linear complex structure J, that is a $(1, 1)$-tensor such that $J^2 = -I$. The ring $\mathcal{F}(E)$ of smooth real functions defined on E, as was discussed in Chap. 3, is generated by its linear part, the eigenvectors with eigenvalue 1 of the Liouville vector field Δ_E. The linear part of $\mathcal{F}(E)$ is the real dual space E^* of E. Now, it happens that E is equipped with a complex structure, so it is a complex space, thus we shall consider $\mathbb{C}$-valued functions instead. However remember that in the real case we selected smooth functions on E as the natural space capturing the differential calculus on E. If we have a complex linear space, what shall we consider instead of smooth functions?

Let us first study the linear part of what this algebra would be. The space of real linear maps E^* can be complexified to obtain $E^* \otimes \mathbb{C} = (E^*)^{\mathbb{C}}$, that is the space of real linear maps from E to $\mathbb{C}_{\mathbb{R}}$ (i.e., $\mathbb{C}$ is equipped with its natural real space structure). The natural notion of transformation between complex linear spaces is a set of complex linear maps. We can show immediately that any real linear map $\alpha\colon E \to \mathbb{C}$ decomposes as the sum of a complex linear and a complex antilinear map:

$$\alpha = \alpha^{(1,0)} + \alpha^{(0,1)}. \tag{6.53}$$

In fact, let

$$\alpha^{(1,0)}(v) = \frac{1}{2}(\alpha(v) - i\alpha(iv)), \qquad \alpha^{(0,1)} = \frac{1}{2}(\alpha(v) + i\alpha(iv)).$$

It is clear that

$$\alpha^{(1,0)}(iv) = i\alpha^{(1,0)}(v), \quad \text{and} \quad \alpha^{(0,1)}(iv) = -i\alpha^{(0,1)}(v).$$

We denote the complex linear maps $\alpha^{(1,0)}$ as $\Lambda^{(1,0)}(E^*)$ and by $\Lambda^{(0,1)}(E^*)$ the complex antilinear maps $\alpha^{(0,1)}$. Hence,

$$(E^*)^{\mathbb{C}} = \Lambda^{(1,0)}(E^*) \oplus \Lambda^{(0,1)}(E^*). \tag{6.54}$$

If we are given a smooth map $f\colon E \to \mathbb{C}$, its differential at each point $x \in E$ defines a linear map $df(x)\colon T_xE = E \to \mathbb{C}$. But, according to Eq. (6.53), $df(x) = df(x)^{(1,0)} + df(x)^{(0,1)}$. Notice that the complex structure J, being a $(1,1)$ tensor, acts on df in a natural way, $(J^*df)(X) = df(J(X))$ for any vector field X. Thus the complex linear and antilinear parts of df are eigenvectors of J with eigenvalues $\pm i$ respectively.

We will say that a function f is holomorphic at x if $df(x)^{(0,1)} = 0$ (respectively, we will say that f is anti-holomorphic at x if $df^{(1,0)}(x) = 0$). We will say that f is holomorphic if it is holomorphic at x for all $x \in E$ (respect. antiholomorphic). Notice that f being holomorphic is equivalent to the expression $df(x) \circ J(x) = J_0 \circ df(x)$, for all $x \in E$, or simply

$$df \circ J = J_0 \circ df. \tag{6.55}$$

where J_0 denotes the complex structure on $\mathbb{C}$, i.e., multiplication by i.

If we introduce complex coordinates z_k in E, with real structure $z^k = x^k + iy^k$, $k = 1, \ldots, n$, $\dim_{\mathbb{R}} E = 2n$, and we write $f = u + iv$, then the previous expression $df^{(0,1)} = 0$ takes the familiar form of Cauchy-Riemann equations for f:

$$\frac{\partial u}{\partial x^k} = \frac{\partial v}{\partial y^k}; \qquad \frac{\partial u}{\partial y^k} = -\frac{\partial v}{\partial x^k}, \qquad k = 1, \ldots, n\,. \tag{6.56}$$

Now notice that the space of complex linear maps $\alpha_x : T_x E \rightarrow \mathbb{C}$ is just the complexification of the cotangent space $T_x^* E$, that is $(T_x^* E)^{\mathbb{C}} = T_x^* E \otimes \mathbb{C}$. But $(T_x^* E)^{\mathbb{C}}$ can be decomposed into its holomorphic and antiholomorphic parts, i.e., $(T_x^* E)^{\mathbb{C}} = \Lambda^{(1,0)}(T_x^* E) \oplus \Lambda^{(0,1)}(T_x^* E)$ following Eq. (6.54). Often we will simply write $(T_x^* E)^{\mathbb{C}} = T_x^* E^{(1,0)} \oplus T_x^* E^{(0,1)}$ for every $x \in E$. Globally we get:

$$(T^* E)^{\mathbb{C}} = T^* E^{(1,0)} \oplus T^* E^{(0,1)}\,.$$

The previous decomposition introduces a similar decomposition on the space of complex-valued 1-forms on E, i.e., if we denote by $\Omega^1(E, \mathbb{C})$ the space of complex-valued 1-forms on E, that is, $\Omega^1(E, \mathbb{C}) = \Omega^1(E) \otimes \mathbb{C}$, they may be identified with sections of the bundle $(T^* E)^{\mathbb{C}}$. Then we may decompose a 1-form α into its holomorphic and antiholomorphic part: $\alpha = \alpha^{(1,0)} + \alpha^{(0,1)}$ (pointwise as in Eq. (6.53)), then

$$\Omega^1(E, \mathbb{C}) = \Omega^{(1,0)}(E) \oplus \Omega^{(0,1)}(E).$$

The exterior differential d maps a smooth function f into a 1-form df, then because of the decomposition of df into its holomorphic and antiholomorphic parts, we can write:

$$df = df^{(1,0)} + df^{(0,1)}\,,$$

Thus we can define the Cauchy–Riemann first-order differential operators ∂ and $\bar{\partial}$ as the holomorphic and antiholomorphic parts of the exterior differential respectively, that is:

$$\partial f = df^{(1,0)}, \qquad \bar{\partial} f = df^{(0,1)}\,. \tag{6.57}$$

Hence a function f is holomorphic iff $df = \partial f$ or equivalently, $\bar{\partial} f = 0$ (respectively, anti-holomorphic if $\partial f = 0$).

Using complex coordinates z^k, the Cauchy–Riemann operators take the simple form:

$$\partial f = \frac{\partial f}{\partial z^k} dz^k, \qquad \bar{\partial} f = \frac{\partial f}{\partial \bar{z}^k} d\bar{z}^k$$

with $\bar{z}^k$ denoting the complex conjugate of z_k.

We may consider now arbitrary complex differential forms on the complex linear space E. The differentials $dz^1, \ldots, dz^n, d\bar{z}^1, \ldots, d\bar{z}^n$ constitute a basis for the module of (complex) differential forms of degree 1. If follows that a differential form of degree k can be expressed as a linear combination with complex-valued coefficients of exterior products of order k of the 1-forms dz^i and $d\bar{z}^j$, $i, j = 1, \ldots, n$. An homogeneous term containing p factors dz^i and $q = k - p$ factors $d\bar{z}^j$ is said to be of bidegree (p, q). A differential form of bidigree (p, q) is a sum of homogeneous terms of bidegree (p, q).

The notion of bi-degree is well defined in the realm of complex linear spaces and holomorphic changes of coordinates, i.e., a smooth change of coordinates $\phi\colon E \to E$ is holomorphic if the differential map $d\phi(x)\colon T_x E = E \to T_\phi(x)E = E$ is linear complex for all $x \in E$. In other words the map $\phi\colon E \to F$ is a holomorphic map if:

$$d\phi \circ J = J \circ d\phi .$$

It is clear now that if we introduce complex coordinates z^i as before and $w^i = z^i \circ \phi$ are the transformed coordinates, then w^i are holomorphic functions iff ϕ is holomorphic and $dw^i = dz^i \circ \phi$, $d\bar{w}^i = d\bar{z}^i \circ \phi$. We will denote the module of differential forms of bidegree (p, q) as $\Omega^{(p,q)}(E)$. Clearly:

$$\Omega^k(E, \mathbb{C}) = \bigoplus_{p+q=k,\ p,q\geq 0} \Omega^{(p,q)}(E) .$$

Notice that, in particular, $\Omega^2(E, \mathbb{C}) = \Omega^{(2,0)}(E) \oplus \Omega^{(1,1)}(E) \oplus \Omega^{(0,2)}(E)$. A similar bidegree structure can be introduced in the tensor algebra over E and in the corresponding algebra of differentiable tensors over E.

6.4.2 Complex Linear Systems

From the discussion before we conclude that a linear complex structure is defined by the constant (1,1)-tensor

$$T_J = J_i{}^k \frac{\partial}{\partial x^k} \otimes dx^i ,$$

that we will keep denoting by J. Notice that in complex coordinates

$$J = dz^i \frac{\partial}{\partial \bar{z}^i} - d\bar{z}^i \frac{\partial}{\partial z^i} .$$

Thus if a given dynamical system Γ on E is possess an invariant linear complex structure J, then

$$\mathcal{L}_\Gamma J = 0,$$

which, is equivalent to:

$$[A, J] = 0, \tag{6.58}$$

if Γ is linear with $\Gamma = X_A$. But Eq. (6.58) means that the linear map A is linear complex, i.e., the matrix A can be identified with an $n \times n$ complex matrix Z and the vector field Γ defines a complex linear vector field on $\mathbb{C}^n$,

$$\Gamma = Z^j{}_i z^i \frac{\partial}{\partial z^j}$$

whose flow is given by the one-parameter family of complex isomorphisms,

$$\phi_t = e^{tZ} .$$

In the real space E, if we consider a basis $u_1, \ldots, u_n, v_1, \ldots, v_n$, adapted to J, i.e., such that $J(u_k) = v_k$ and $J(v_k) = -u_k$, and denoting the corresponding set of linear coordinates by x^k, y^k, we will have:

$$\Gamma = X^j{}_i \left(x^i \frac{\partial}{\partial x^j} + y^i \frac{\partial}{\partial y^j} \right) + Y^j{}_i \left(x^i \frac{\partial}{\partial y^j} - y^i \frac{\partial}{\partial x^j} \right) ,$$

with $Z = X + iY$.

In this sense the simplest dynamical system possessing an invariant complex structure is the vector field defined by the complex structure itself, i.e., if J is a given complex structure $\Gamma = X_J$ has $T_J = J$ as a invariant complex structure. The vector field X_J is just the isotropic harmonic oscillator.

Notice that the Liouville vector field Δ_E characterizing the linear space E is complex with respect to any complex structure on E because $\Delta_E = X_I$ and $[I, J] = 0$, hence a linear complex structure is characterized by the two objects, a vector field Δ_E and the $(1, 1)$-tensor J. However it can be shown (see later) that the complex linear vector space structure E can be described by two commuting vector fields Δ, Γ with the properties of $\Delta_E = X_I$ and $\Gamma = X_J = J(\Delta_E)$.

We denote by $\mathcal{J}_l(E)$ the space of linear complex structures on E. It is immediate to see that the group $GL(E)$ of real linear automorphisms of E acts transitively on this space. In fact, if J_1, J_2 are two complex linear structures, choose adapted bases to both, thus, if $\{u_k, v_k\}$ and $\{u'_k, v'_k\}$ denote such basis for J_1, J_2 respectively, define the linear isomorphism $P \colon E \to E$, by $P(u_k) = u'_k$, $P(v_k) = v'_k$. Then, $J_2 \circ P = P \circ J_1$ and the action of $GL(E)$ on $\mathcal{J}_l(E)$ by conjugation is transitive. In terms of the model $\mathbb{R}^{2n}$, the group $GL(2n, \mathbb{R})$ acts on $\mathcal{J}_l(\mathbb{R}^{2n})$ by $J_0 \mapsto P^{-1} J_0 P$, and the stabilizer group at the linear complex structure J_0 is the set of matrices such that $[P, J_0] = 0$, i.e., the set of complex matrices $GL(n, \mathbb{C})$.

Proposition 6.22 *The space $\mathcal{J}_l(E)$ of linear complex structures on E is diffeomorphic to the homogeneous space $GL(2n,\mathbb{R})/GL(n,\mathbb{C})$ and has dimension $2n^2$. This space has two connected components corresponding to matrices with positive or negative determinant. The component $\mathcal{J}_l^+(E)$ is diffeomorphic to the homogeneous space $GL^+(2n,\mathbb{R})/GL(n,\mathbb{C})$ and contains the complex structure J_0.*

Proof Notice that $GL(2n,\mathbb{R})$ has two components depending on the sign of the determinant. We shall denote by $GL^+(2n,\mathbb{R})$ those matrices with positive determinant. Notice that $\det J_0 = 1$ and thus J_0 belongs to the component $GL^+(2n,\mathbb{R})/GL(n,\mathbb{C})$. □

Exercise 6.1 Show that if $\dim E = 2$, then $\mathcal{J}_l(E) \cong S^2$.

Consider the space of compatible linear complex structures with a given linear dynamics. We shall denote such space by $\mathcal{J}_l(E,\Gamma)$. Thus, if $\mathcal{J}_l(E,\Gamma) \neq \emptyset$, then $\Gamma = X_A$ where A is a complex linear map with respect to a given complex structure. Using the model space $(\mathbb{R}^{2n}, J_0)$ for this complex structure, then $[A, J_0] = 0$, and the space $\mathcal{J}_l(E,\Gamma)$ is diffeomorphic to the space of matrices such that $J^2 = -I$ and $[A, J] = 0$.

For a complex generic A, $\mathcal{J}_l(\mathbb{R}^{2n}, A) = \pm J_0$. On the opposite side, we find however $\mathcal{J}_l(E, \Delta_E) = \mathcal{J}_l(E)$.

6.4.3 Complex Differential Calculus and Kähler Manifolds

We can now consider the geometrical notion of complex structure on E given by a non-necessarily constant, (1,1)-tensor $J = J_i{}^k(x)\partial/\partial x^k \otimes dx^i$ that at each point defines a complex structure on the tangent space $T_xE \cong E$ of E.

Definition 6.23 A (1,1)-tensor J on E such that for each $x \in E$, $J(x)^2 = -I$ will be called an almost complex structure.

An *almost complex* manifold is an even-dimensional real manifold M endowed with a (1, 1)-type tensor field J, called an *almost complex structure,* satisfying

$$J^2 = -I\,. \tag{6.59}$$

We will discuss later the strong holonomy principle for linear complex structures, i.e., under what conditions an almost complex structure has a local normal form given by a constant linear complex structure. We will start introducing the notion of complex and almost complex manifold.

A complex manifold [Ch67, Sc86] is a smooth manifold M that can be locally modeled on $\mathbb{C}^n$ for some n, and for which the chart-compatibility conditions are required to be holomorphic maps.

A complex manifold as a real smooth manifold M has dimension $2n$, hence it can be embedded in $\mathbb{R}^{2N}$ with N large enough. The local charts define holomorphic transition functions. It is not necessarily true that M can be embedded as a holomorphic submanifold of $\mathbb{C}^N$.

Alternatively, the algebra of holomorphic functions on $\mathbb{C}^N$ is not a differentiable algebra, hence we cannot extend the theory developed in Chap. 3 to complex manifolds.

Complex manifolds that can be embedded holomorphically in $\mathbb{C}^N$ are called Stein manifolds. A compact complex manifold cannot be embedded holomorphically in $\mathbb{C}^N$ (notice that compact complex manifolds have only constant holomorphic functions).

Nirenberg's theorem [Ni57] shows that an almost complex manifold is a complex manifold iff the almost complex structure J satisfies the integrability condition $N_J = 0$, where N_J is the Nijenhuis torsion associated with J (see Appendix E).

Finally, let M be a real, even-dimensional, smooth manifold with a complex structure J and closed 2-form ω which is J-invariant, that is:

$$\omega(X, JY) + \omega(JX, Y) = 0; \qquad X, Y \in TM\,. \tag{6.60}$$

Notice that as we discussed in the case of linear Hermitean structures, this implies that the bilinear form:

$$g(X, Y) = \omega(JX, Y)\,;$$

is symmetric and nondegenerate. When g is positive, that is $g(X, X) > 0$ for all $X \neq 0$, then (M, g, J, ω) is a said to be Kähler manifold [Ch67, Sc86, We58].[1] Also, $J^2 = -I$ implies

$$\omega(JX, JY) = \omega(X, Y)\,; \qquad g(JX, JY) = g(X, Y)\,, \tag{6.61}$$

for all $X, Y \in TM$. Notice that Eq. (6.61) implies the analog of Eq. (6.60) for g, namely

$$g(X, JY) + g(JX, Y) = 0\,. \tag{6.62}$$

A tensorial triple (g, J, ω), with g a metric, J a complex structure and ω a symplectic structure satisfying conditions (6.60)-(6.61) will be called an *admissible triple.*

Moreover a Kähler manifold satisfies $\nabla J = 0$ with ∇ the Levi-Civita connection of the metric g.

[1] If g is not positive definite, then M is also-called a pseudo-Kähler manifold.

6.4.4 Algebras Associated with Hermitean Structures

We will conclude the analysis of the geometry of invariant Hermitean structures by discussing various algebraic structures which are naturally associated with such structures and that are going to play a relevant role in the structure of quantum systems.

We start again with a given Hermitean tensor $\mathbf{h} = g + i\omega$. Associated with it we have a Poisson bracket $\{\cdot, \cdot\}$ defined by the symplectic structure $\omega = \operatorname{Im} \mathbf{h}$. Moreover the symmetric part of the tensor $\mathbf{h}$ allows us to define also a Riemannian bracket

$$(f, g) = G(df, dg) = G^{kj} \frac{\partial f}{\partial \xi^k} \frac{\partial g}{\partial \xi^j}, \tag{6.63}$$

with G the contravariant symmetric $(2, 0)$ tensor obtained inverting the real part g of $\mathbf{h}$. The symmetric product $(\cdot, \cdot)$ is obviously commutative but not associative.

We can define a Kähler bracket combining both products as:

$$[\![f, g]\!] = (f, g) + i\{ f, g \}. \tag{6.64}$$

However in the spirit of deformations is natural to consider the products,

$$f \circ_\nu g = \frac{\nu}{2}(f, g) + fg$$

and

$$f \star_\nu g = fg + \frac{\nu}{2}[\![f, g]\!]. \tag{6.65}$$

The product $\circ_\nu$ defines a nonassociative commutative real deformation of the commutative associative algebra $\mathcal{F}(E)^{\mathbb{C}}$, whereas $\star_\nu$ determines a complex non-commutative deformed algebra that reduces to fg when $\nu = 0$. By using the natural complex conjugation they become involutive algebras. In addition we have the relations,

$$\{ f, g \} = \frac{1}{i\nu}(f \star_\nu g - g \star_\nu f),$$
$$f \circ_\nu g = \frac{1}{2}(f \star_\nu g + g \star_\nu f).$$

The previous definitions can be extended without pain to Kähler manifolds. A distinguished class of functions on a Kähler manifold are those real functions f such that

$$\mathcal{L}_{X_f} G = 0,$$

where X_f is the Hamiltonian vector field defined by f. This definition implies that the natural derivation corresponding to the Hamiltonian vector field X_f defined by f is also a Killing vector and its flow are isometries of the metric structure.

Definition 6.24 Let $(M, \mathbf{h})$ be a Kähler manifold. If f is a complex-valued function on M, we say that f is Kählerian if the real and the imaginary parts u, v of f satisfy that $\mathcal{L}_{X_u} G = 0$, $\mathcal{L}_{X_u} G = 0$.

Kählerian functions capture in a geometrical way the class of Hermitean (or self-adjoint) operators on a space with inner product. Thus if $\mathcal{H}$ is a finite-dimensional complex space with inner product[2] $\langle \cdot, \cdot \rangle$, that is $\mathcal{H}$ carries a Hermitean structure $\mathbf{h}$ given by $\mathbf{h}(u, v) = \langle u, v \rangle$, $u, v \in \mathcal{H}$, a Hermitean (or self-adjoint) operator is a complex linear map $A \colon \mathcal{H} \to \mathcal{H}$ such that $A = A^\dagger$ or, equivalently, $\langle Au, v \rangle = \langle u, Av \rangle$ for all $u, v \in \mathcal{H}$. Clearly, if A is a Hermitean operator then $f_A(u) = \langle u | Au \rangle$ is a Kählerian function. Notice that the function f_A is real, i.e., $\bar{f}_A = f_A$. Moreover the Hamiltonian vector field X_{f_A} is the linear vector field iAu, then because A is Hermitean, its flow $\Phi_t = \exp itA$ is a one-parameter group of unitary operators, that is $\langle \Phi_t u, \Phi_t v \rangle = \langle u, v \rangle$, for all $u, v \in \mathcal{H}$. As a consequence, Φ_t preserves both the real and the imaginary part of $\mathbf{h}$, that is they are orthogonal transformations for the metric g and isometries of the tensor G, then $\mathcal{L}_{X_{f_A}} G = 0$.

Notice that if T is any complex linear map on $\mathcal{H}$, then $T = A + iB$ with $A = (T + T^\dagger)/2$, $B = (T - T^\dagger)/2i$ are Hermitean operators. Then the complex function $f_T(u) = \langle u, Tu \rangle = f_A(u) + if_B(u)$ is Kählerian, in other words, the quadratic functions f_T with T a complex linear operator on any finite-dimensional Hilbert space $\mathcal{H}$ are Kählerian.

However on a generic Kähler manifold, Kählerian functions are not abundant. A set of real Kähler functions is said to be complete if they separate derivations. We have:

Proposition 6.25 *Let* $(M, \mathbf{h})$ *be a Kähler manifold. The following statements are equivalent:*

1. *The function* f *is Kählerian.*
2. *The Hamiltonian vector field* X_f *is a derivation of the Riemannian product* $(\cdot, \cdot)$.
3. *The Hamiltonian vector field* X_f *is a derivation of the Kähler product* $[\![\cdot, \cdot]\!]$.
4. *The Hamiltonian vector field* X_f *is a derivation of the* $\circ_\nu$*-product.*
5. *The Hamiltonian vector field* X_f *is a derivation of the* $\star_\nu$*-product.*

The structure of the space of Kählerian functions is described in the following theorem [Ci90, Ci91, Ci94].

Theorem 6.26 *The set of Kähler functions is complete and closed under the* $\star_\nu$ *product if the holomorphic sectional curvature,*[3] *is constant and equal to* $2/\nu$ *and, moreover for any triple of Kähler functions, then:*

---

[2] In other words, $\mathcal{H}$ is a finite-dimensional instance of a complex Hilbert space.

[3] See for instance [Go82], pp. 200–201 Theorems 6.1.2 and 6.1.3.

$$(f \star_\nu g) \star_\nu h = f \star_\nu (g \star_\nu h)\,, \tag{6.66}$$

The statement is also true if $\nu \to \infty$.

6.5 The Geometry of Quantum Dynamical Evolution

So far we have discussed dynamical systems admitting a compatible Hermitean structure and explored some of their properties, but we have made no effort to connect them to any particular physical model. We have already indicated though, that such systems are related to Quantum Physics. In this section we will make the connection explicit and we will show how a geometrical picture of Quantum Mechanics nicely emerges from it.

Of course it will not be possible to offer in such a limited space a self-contained description of the foundations of Quantum Mechanics, so that references to the appropriate literature will be offered when needed. The emphasis here will be, as in the rest of this book, to offer the reader a new perspective where, starting always from the dynamics, a geometrical picture emerges.

This geometrical picture in the case of quantum systems has a number of aspects that have both a mathematical and also a physical interest. We will try to exhibit some of their most important features in what follows.

6.5.1 On the Meaning of Quantum Dynamical Evolution

We will refine first the notion of dynamics for a physical system with respect to the 'naive' approach taken in the beginning of Chap. 2.

From a minimalist point of view, a description of any physical system, be it 'classical' or 'quantum', requires the introduction of a family of observable quantities, say $\mathcal{A}$, a family of 'states' representing the maximal information we can gather on our system, say $\mathcal{S}$, and a pairing between them:

$$\mu\colon \mathcal{A} \times \mathcal{S} \to \mathcal{P}\,,$$

with $\mathcal{P}$ the space of (Borelian) probability measures on the real line $\mathbb{R}$.

The interpretation of the pairing μ being that given a state $\rho \in \mathcal{S}$, an observable $A \in \mathcal{A}$, that is, the number $\mu(A, \rho)(\Delta)$ for any Borel set $\Delta \subset \mathbb{R}$, is the probability that a measurement of A while the system is in the state ρ gives a result in Δ.

The evolution of the physical system characterized by $(\mathcal{A}, \mathcal{S}, \mu)$ will be described either by a one-parameter group of automorphisms Φ_t of the state space $\mathcal{S}$, or of the space of observables $\mathcal{A}$, or of the space probability measures. That is, either we may consider that the state of the system actually changes with respect to a certain time parameter t while the observables are detached from the dynamics, the so-called

'Schrödinger picture', or alternatively, we may consider that the observables attached to the system are actually evolving in time, the 'Heisenberg picture'. We may even think that the probability measures $\mu(A, \rho)$ change in time instead.

We will accept first as a fundamental fact that the dynamical description of a quantum system will be given by a 'vector field' Γ on a carrier space that could be the space of states, observables or probability measures. The realization of the vector field Γ and of the carrier space will depend on the picture we will choose to describe the system under investigation.

At the dawn of Quantum Mechanics, the founding fathers elaborated two main pictures: one, which may be associated with the names of Schrödinger and Dirac, starts by associating a Hilbert space with any physical system and such Hilbert space is related to the space of states $\mathcal{S}$. The second one, associated with the names of Heisenberg, Born, von Neumann and Jordan, starts by associating with any physical system a $*$-algebra related to the space of observables $\mathcal{A}$. Later on, also probabilistic approaches have been developed. We may quote the names of Weyl, Wigner and Ehrenfest in this sense.

In what follows we shall concentrate on the Hilbert space approach but before that we would like to outline how the various pictures are related.

Hilbert spaces were introduced and used in a systematic way in the description of quantum systems first by Dirac [Di45] as a consequence of the fact that one needs a superposition rule (and hence a linear structure) in order to accommodate a consistent description of the interference phenomena that are fundamental for Quantum Mechanics, i.e., the double-slit experiment. Parenthetically, we should note that a complex Hilbert space carries with it in a natural way a complex structure (multiplication of vectors by the imaginary unit). The rôle of the latter was discussed in the early 1940s by Reichenbach [Re44]. Later on Stückelberg [St60] emphasized the rôle of the complex structure in deducing in a consistent way the uncertainty relations of Quantum Mechanics (see also the discussion in [Es04] and [Ma05b]).

In a naive way however the Hilbert space approach started with Schrödinger who introduced the space of square integrable functions on $\mathbb{R}^3$ depending on a time parameter, say $\Psi(x, t)$. These functions were called wave functions and were connected to the evolution of the system by solving Schrödinger's wave equation:

$$i\hbar \frac{d}{dt}\Psi = H\Psi. \tag{6.67}$$

Here the Hamiltonian operator H is required to be Hermitean with respect to the L^2 inner scalar product on $\mathbb{R}^3$:

$$\langle \Psi, \Phi \rangle = \int_{\mathbb{R}^3} \overline{\Psi(x)}\Phi(x) d^3 x.$$

Born suggested that $|\Psi(x)|^2 = \overline{\Psi(x)}\Psi(x)$ should be interpreted as a probability density and the wave function Ψ as an amplitude of probability. The linearity of the equation of motion allowed one to easily accommodate the description of interference phenomena which were observed in diffraction experiments with electrons. Nowadays interference phenomena are interpreted as paradigmatic examples of the essence of Quantum Mechanics.

Remark 6.2 Physical interpretation of the wave function.

Let us consider briefly the physical meaning that can be attributed to $\Psi(x, t)$. Born's proposal was to regard $\rho(\mathbf{x}, t) = \overline{\Psi(\mathbf{x}, t)}\Psi(\mathbf{x}, t)$ as a probability measure. In other words, the quantity $\int_\Delta \rho(x, t) d^3x$ should represent the probability of observing a particle at time t within the volume element Δ.

This interpretation is possible if Ψ is square-integrable and normalized. Born was led to his postulate by comparing how the scattering of a particle by a potential is described in classical and Quantum Mechanics. For instance, if an electron is allowed to interact with a short-range potential V, that is V has support on a compact set, and if a screen is placed at a distance large enough from the interaction region, the electron will be detected at a fixed point on the screen. On repeating the experiment, the electron will be detected at a different point. After several experiments, the fraction of the number of times the electron is detected at $\mathbf{x}$ at time t is proportional to $|\Psi(\mathbf{x}, t)|^2$, which is therefore the probability density of such an event. Remarkably, the electron exhibits both a corpuscular and a wavelike nature by virtue of the probabilistic interpretation of its wave function. This means that the 'quantum electron' does not coincide with the notion of 'classical electron' we may have developed in a classical framework.

Ultimately, it is necessary to give up the attempt to describe the particle motion as if we could use what we are familiar with from everyday's experience. We have instead to limit ourselves to an abstract mathematical description, which makes it possible to extract information *from the experiments that we are able to perform*. This crucial point (which is still receiving careful consideration in the current literature) is well emphasized by the interference from a double slit.

6.5.1.1 Dirac's Insight and Heisenberg's Equation

Out of the description in terms of wave functions Dirac suggested using an abstract Hilbert space to be associated with any physical system, that is, a possible infinite-dimensional, complex separable linear space $\mathcal{H}$ carrying an Hermitean structure (or inner product) $\langle \cdot, \cdot \rangle$ which is complete with respect to the topology defined by the norm $|| \cdot ||$ associated to the inner product, i.e., $||\Psi||^2 = \langle \Psi, \Psi \rangle$.

Dirac introduced the braket notation for vectors ('ket') $| \cdot \rangle$ in $\mathcal{H}$ and forms ('bra') $\langle \cdot |$ in $\mathcal{H}'$. Because of Riesz theorem there is a natural isomorphism between a Hilbert space and its dual, thus any vector $|\Psi\rangle$ in $\mathcal{H}$ defines a (bounded) complex linear map on $\mathcal{H}$, denoted by $\langle \Psi |$ such that $\langle \Psi | (|\Phi\rangle) = \langle \Psi, \Phi \rangle$ for any $|\Phi\rangle \in \mathcal{H}$, and conversely, hence the notation for the inner product: $\langle \Psi | \Phi \rangle := \langle \Psi, \Phi \rangle$.

The wave function was replaced by a vector $|\Psi(t)\rangle \in \mathcal{H}$ in the Hilbert space $\mathcal{H}$ and we will call them (pure) quantum states. Observables were introduced as a derived concept, namely as Hermitean operators on $\mathcal{H}$. Mean values or expectation values of an observable A in a given state $|\Psi\rangle$ are given by:

$$e_A(\Psi) = \frac{\langle\Psi|A|\Psi\rangle}{\langle\Psi|\Psi\rangle}. \tag{6.68}$$

Given a state $|\Psi\rangle$ a probability distribution is defined with the help of a resolution of the identity operator in $\mathcal{H}$, say:

$$I = \sum_k P_k = \sum_k \frac{|\Psi_k\rangle\langle\Psi_k|}{\langle\Psi_k|\Psi_k\rangle\langle\Psi|\Psi\rangle},$$

for a countable family of rank-one orthogonal projectos P_k. Each operator P_k has the form $P_k = \frac{|\Psi_k\rangle\langle\Psi_k|}{\langle\Psi_k|\Psi_k\rangle}$ for some vector $|\Psi_k\rangle \in \mathcal{H}$. Then the probability distribution for the vector $|\Psi\rangle$ is given by the non-negative numbers:

$$P_k(\Psi) = \frac{\langle\Psi|\Psi_k\rangle\langle\Psi_k|\Psi\rangle}{\langle\Psi_k|\Psi_k\rangle}.$$

For a continuous index a labeling the projectors P_a, say:

$$I = \int da \, \frac{|a\rangle\langle a|}{\langle a|a\rangle},$$

with da an appropriate measure in the index space, we obtain a probability density:

$$P_a(\Psi) = \frac{\langle\Psi|a\rangle\langle a|\Psi\rangle}{\langle a|a\rangle\langle\Psi|\Psi\rangle}. \tag{6.69}$$

When the rank-one projectors $|a\rangle\langle a|/\langle a|a\rangle$ are associated with the spectral decomposition of the operator A, say:

$$A = \int a \frac{|a\rangle\langle a|}{\langle a|a\rangle} da,$$

we have the relation between the expectation value function and the probability distribution Eq. (6.69):

$$e_A == \int a P_a(\Psi) da = \int a \frac{\langle\Psi|a\rangle\langle a|\Psi\rangle}{\langle a|a\rangle\langle\Psi|\Psi\rangle} da.$$

Thus a description in terms of expectation value functions may be considered to be a probabilistic description. In a search for an implementation of the so-called 'correspondence principle' or the 'quantum-to-classical' transition, Ehrenfest elaborated a description of the evolution equations in terms of expectation value functions known today as the 'Ehrenfest picture'.

From what we have said, it is clear that the probabilistic description depends not on the Hilbert space vector Ψ but in the 'ray' defined by it, i.e., the one-dimensional complex subspace

$$[\Psi] = \{\lambda|\Psi\rangle \mid \lambda \in \mathbb{C}\}$$

in $\mathcal{H}$. Even more, it is well known that a "complete" measurement in Quantum Mechanics (a simultaneous measurement of a complete set of commuting observables[4] [Di58, Es04, Me58]) does not provide us with a uniquely defined vector in some Hilbert space, but rather with a "ray", i.e., an equivalence class of vectors differing by multiplication through a non zero complex number. Even fixing the normalization, an overall phase will still remain unobservable.

In explicit terms, states are not elements of the Hilbert space $\mathcal{H}$, but rather points of the complex projective space $P(\mathcal{H}) = \mathcal{H}_0/\mathbb{C}_0$, with $\mathcal{H}_0 = \mathcal{H}\backslash\{\mathbf{0}\}$. The space of rays $P(\mathcal{H})$ will be also referred to as the ray space of $\mathcal{H}$.

Of course this transition from a linear Hilbert space to a Hilbert manifold raises the question of the description of superposition and interference phenomena (see later the discussion in Sect. 6.5.2). Moreover the linear equations of motion on a linear space like Schrödinger's equation (6.67), should be replaced by equations of motion on a manifold.

It is not difficult to show that out of Schrödinger's evolution equation Eq. (6.67) for an Hermitean operator H we obtain the evolution equation:

$$i\hbar\frac{d}{dt}\rho_\Psi = [H, \rho_\Psi]\,,$$

for the rank-one operator $\rho_\Psi = |\Psi\rangle\langle\Psi|/\langle\Psi|\Psi\rangle$ associated to the ray $[\Psi]$ defined by $|\Psi\rangle$. This immersion of the space of rays into the space of Hermitean operators acting on $\mathcal{H}$, i.e.,

$$[\Psi] \mapsto \rho_\Psi = \frac{|\Psi\rangle\langle\Psi|}{\langle\Psi|\Psi\rangle}\,, \tag{6.70}$$

allows us to extend the previous equation to any convex combination of rank-one projectors $\rho_k = P_k = |\Psi_k\rangle\langle\Psi_k|/\langle\Psi_k|\Psi_k\rangle$ (not necessarily orthogonal) by means of probability vectors $p = (p_1, p_2, \ldots, p_n, \ldots)$, $p_k \geq 0$, $\sum_k p_k = 1$. Then the evolution equation for the Hermitean operator $\rho = \sum_k p_k \rho_k$ is given by:

$$i\hbar\frac{d}{dt}\rho = [H, \rho]. \tag{6.71}$$

This equation is usually called von Neumann's equation and the convex combinations ρ are usually called 'density states' or density operators. They were introduced in

[4] We will not worry at this stage about the technical complications that can arise, in the infinite-dimensional case, when the spectrum of an observable has a continuum part.

the description of quantum physical systems by L. Landau and J. von Neumann independently [La27, Ne27].

Clearly, if we replace the convex combinations of rank-one orthogonal projectors with probability vectors $\{p_k\}$, by any sequence of complex numbers $\{c_1, c_2, \ldots, c_n, \ldots\}$ and arbitrary rank-one Hermitean operators ρ_k, we get the class of Hermitean (generically unbounded) operators $C = \sum_k c_k \rho_k$ with equation of motion given by:

$$i\hbar \frac{d}{dt} C = [H, C]. \tag{6.72}$$

This equation of motion on the space of operators is quite often called the Heisenberg form of the equations of motion.

Then, in the case of quantum systems, it is the previous equation, Eq. (6.72), the one that we will consider as defining the dynamics of the system, even though as it has been discussed already, there are other 'pictures' equivalent to it.

In the subsequent parts of this chapter we are going to spell out some details of what was already said. To avoid the complications, and why not, embarrassments, due to the subtle and in many cases difficult mathematical problems connected with the topology of functional spaces, we should restrict our considerations for the most part to finite-dimensional quantum systems, even if in many instances the derived results extend to the infinite-dimensional situation.

6.5.2 The Basic Geometry of the Space of Quantum States

Following Dirac we will consider that associated to a quantum system there is a complex Hilbert space $\mathcal{H}$ and that a class of states of the system are given by rays in $\mathcal{H}$. Such states will be called in what follows pure states [Ha70] and they are points of the projective Hilbert space $P(\mathcal{H})$.

If $\mathcal{H}$ has finite complex dimension, say $n+1$, the projective Hilbert space $P(\mathcal{H})$ is just the n-dimensional complex projective space $\mathbb{C}P^n = \{[z] \mid z \in \mathbb{C}^{n+1},\ z' \in [z] \Leftrightarrow z' = \lambda z, 0 \neq \lambda \in \mathbb{C}\}$.

Quotienting with respect to multiplication by non zero complex numbers $\mathbb{C}_0 \cong \mathbb{R}^+ \times U(1)$, gives raise to the following double fibration:

$$\begin{array}{ccc} \mathbb{R}^+ & \longrightarrow & \mathcal{H}_0 \\ & & \downarrow \\ U(1) & \longrightarrow & \mathbb{S} \\ & & \downarrow \\ & & P(\mathcal{H}) \end{array} \tag{6.73}$$

where the first row indicates the action of the radial part $\mathbb{R}^+$ of $\mathbb{C}$, while the second row is the action of the angular part $U(1)$ of $\mathbb{C}$ on the sphere $\mathbb{S}$ of unitary vectors (which in the finite-dimensional case is the sphere $S^{2n+1} \subset \mathbb{C}^{n+1}$), and whose final result is the projective Hilbert space $P(\mathcal{H})$.

Remark 6.3 Notice that the Hilbert space $\mathcal{H}$ acquires the structure of a principal fiber bundle [Hus75, Ch09, St51], with base $P(\mathcal{H})$ and typical fiber $\mathbb{C}_0$, both in finite and infinite dimensions.

Even if we are dealing with finite-dimensional Hilbert spaces, in this section we will keep the abstract notation $|\Psi\rangle$ for vectors instead of z, just showing the most of the computations work in infinite dimensions too. The natural identification of $\mathcal{H}$ and its dual, allows for the (unique) association of every equivalence class $[\Psi]$ with the rank-one orthogonal projector ρ_Ψ defined by Eq. (6.70) with the known properties:

$$\rho_\Psi^\dagger = \rho_\Psi , \qquad \rho_\Psi^2 = \rho_\Psi , \qquad \mathrm{Tr}\, \rho_\Psi = 1 .$$

It is clear by construction that the association depends on the Hermitean structure we consider (see Sect. 6.5.3 for further comments on this point).

The space of rank-one projectors is usually denoted as $\mathcal{D}_1^1(\mathcal{H})$ (see Sect. 6.6.1 [Gr05] for additional information). It is then clear that in this way we can identify $\mathcal{D}_1^1(\mathcal{H})$ with the projective Hilbert space $P(\mathcal{H})$. Hence, what the best of measurements will yield will be always, no more and no less, than a rank-one projector.

Also, transition probabilities that, together with the expectation values of self-adjoint linear operators representing dynamical variables, are among the only observable quantities one can think of, will be insensitive to overall phases, i.e., they will depend only on the (rank-one) projectors associated with the states. If $A = A^\dagger$ is any such observable, then its expectation value $\langle A \rangle_\Psi$ in the state $|\Psi\rangle$ will be given by

$$\langle A \rangle_\Psi = \frac{\langle \Psi | A | \Psi \rangle}{\langle \Psi | \Psi \rangle} \equiv \mathrm{Tr}\,(\rho_\Psi A) . \tag{6.74}$$

Transition probabilities are in turn expressed via a binary product that can be defined on pure states. Again, if $|\Psi\rangle$ and $|\Phi\rangle$ are any two states, then the (normalized) transition probability from $|\Psi\rangle$ to $|\Phi\rangle$ will be given by:

$$\frac{|\langle \Phi | \Psi \rangle|^2}{\langle \Psi | \Psi \rangle \langle \Phi | \Phi \rangle} = \mathrm{Tr}\,(\rho_\Psi \rho_\Phi) \tag{6.75}$$

and the trace on the right-hand side of Eq. (6.75) will define the binary product among pure states (but more on this shortly below). Therefore the most natural setting for Quantum Mechanics is not primarily the Hilbert space itself $\mathcal{H}$ but rather the projective Hilbert space $P(\mathcal{H})$ or, equivalently, the space of rank-one projectors $\mathcal{D}_1^1(\mathcal{H})$, whose convex hull will provide us with the set of all density states [Ne27, Neu2, Fa57].

On the other hand, the superposition rule, which leads to interference phenomena, remains one of the fundamental building blocks of Quantum Mechanics, one that, among other things, lies at the very heart of the modern formulation of Quantum

Mechanics in terms of path integrals [Br05], [Gl81], an approach that goes actually back to earlier suggestions by Dirac [Di58, Di33].

To begin with, if we consider, for simplicity, two orthonormal states

$$|\Psi_1\rangle, |\Psi_2\rangle \in \mathcal{H}, \qquad \langle\Psi_i|\Psi_j\rangle = \delta_{ij}, \qquad i, j = 1, 2 \tag{6.76}$$

with the associated projection operators

$$\rho_1 = |\Psi_1\rangle\langle\Psi_1|, \qquad \rho_2 = |\Psi_2\rangle\langle\Psi_2|\,. \tag{6.77}$$

A linear superposition with (complex) coefficients c_1 and c_2 with: $|c_1|^2 + |c_2|^2 = 1$ will yield the normalized vector

$$|\Psi\rangle = c_1|\Psi_1\rangle + c_2|\Psi_2\rangle \tag{6.78}$$

and the associated projector

$$\rho_\Psi = |\Psi\rangle\langle\Psi| = |c_1|^2\,\rho_1 + |c_2|^2\,\rho_2 + \left(c_1c_2^*\rho_{12} + h.c.\right) \tag{6.79}$$

where: $\rho_{12} = |\Psi_1\rangle\langle\Psi_2|$, which cannot however be expressed directly in terms of the initial projectors.

A procedure to overcome this difficulty by retaining at the same time the information concerning the relative phase of the coefficients can be summarized as follows [Cl07, Ma05, Ma05b].

Consider a third, fiducial vector $|\Psi_0\rangle$ with the only requirement that it is not orthogonal[5] neither to $|\Psi_1\rangle$ nor to $|\Psi_2\rangle$. It is possible to associate normalized vectors $|\Phi_i\rangle$ with the projectors ρ_i $(i = 1, 2)$ by setting

$$|\Phi_i\rangle = \frac{\rho_i|\Psi_0\rangle}{\sqrt{\mathrm{Tr}\,(\rho_i\rho_0)}}, \qquad i = 1, 2\,. \tag{6.80}$$

Remark 6.4 Note that, as all the ρ's involved are rank-one projectors[6]:

$$\mathrm{Tr}\,(\rho_i\rho_0)\,\mathrm{Tr}\left(\rho_j\rho_0\right) = \mathrm{Tr}\left(\rho_i\rho_0\rho_j\rho_0\right); \qquad \forall i, j = 1, 2\,, \tag{6.81}$$

and:

$$|\Phi_i\rangle\langle\Phi_i| = \frac{\rho_i\rho_0\rho_i}{\sqrt{\mathrm{Tr}\,(\rho_i\rho_0\rho_i\rho_0)}} \equiv \rho_i, \qquad i = 1, 2\,. \tag{6.82}$$

Forming now the linear superposition: $|\Phi\rangle = c_1|\Phi_1\rangle + c_2|\Phi_2\rangle$ and the associated projector: $\rho = |\Phi\rangle\langle\Phi|$, one finds easily, using also Eqs. (6.81) and (6.82), that:

---

[5] In terms of the associated rank-one projections: $\mathrm{Tr}\,(\rho_i\rho_0) \neq 0$, $i = 1, 2$, with: $\rho_0 = |\Psi_0\rangle\langle\Psi_0|$.

[6] The proof of Eqs. (6.81) and (6.82) is elementary and will not be given here.

$$\rho = |c_1|^2 \rho_1 + |c_2|^2 \rho_2 + \frac{c_1 c_2^* \rho_1 \rho_0 \rho_2 + h.c.}{\sqrt{\mathrm{Tr}\,(\rho_1 \rho_0 \rho_2 \rho_0)}}, \tag{6.83}$$

which can be written in a compact form as

$$\rho = \sum_{i,j=1}^{2} c_i c_j^* \frac{\rho_i \rho_0 \rho_j}{\sqrt{\mathrm{Tr}\left(\rho_i \rho_0 \rho_j \rho_0\right)}}. \tag{6.84}$$

The results (6.83) and (6.84) are now written entirely in terms of rank-one projectors. Thus, a superposition of rank-one projectors which yields another rank-one projector is possible, but requires the arbitrary choice of the fiducial projector ρ_0.

Remark 6.5 This procedure is equivalent to the introduction of a connection on the bundle, usually called the Pancharatnam connection [Ma95].

Remark 6.6 If the (normalized) probabilities $|c_1|^2$ and $|c_2|^2$ are given, Eq. (6.78) describes a one-parameter family of linear superposition of states, and the same will be true in the case of Eq. (6.83). Both families will be parametrized by the relative phase of the coefficients.

Remark 6.7 Comparison of Eqs. (6.79) and (6.83) shows that, while the first two terms on the right-hand side of both are identical, the last terms of the two differ by an extra (fixed) phase, namely:

$$\frac{\rho_1 \rho_0 \rho_2}{\sqrt{\mathrm{Tr}\,(\rho_1 \rho_0 \rho_2 \rho_0)}} = \rho_{12} \exp\left\{i\left[\arg\left(\langle\psi_1|\psi_0\rangle - \arg\left(\langle\psi_2|\psi_0\rangle\right)\right)\right]\right\}. \tag{6.85}$$

Remark 6.8 The result of Eq. (6.84) can be generalized in an obvious way to the case of an arbitrary number, say n, of orthonormal states none of which is orthogonal to the fiducial state. The corresponding family of rank-one projectors will be parametrized in this case by the $(n-1)$ relative phases.

If now we are given two (rank-one) projectors, or more with an obvious generalization, and only the relative probabilities are given, we are led to conclude that the system is described by the convex combination (a rank-two density matrix): $\rho = |c_1|^2 \rho_1 + |c_2|^2 \rho_2$, which is again Hermitean and of trace one, but now: $\rho - \rho^2 > 0$ (strictly). The procedure leading from this "impure"state to one of the pure states given by, say, Eq. (6.84), i.e., the procedure that associates a pure state with a pair of pure states, is a composition law for pure states that has been termed in the literature as a "purification" of "impure"states [Man00].

In the Hilbert space formulation of Quantum Mechanics one needs also to find the spectral family associated with any observable, represented by a self-adjoint operator on the Hilbert space. Limiting ourselves for simplicity to observables with pure point-spectrum, these notions can be made easily to "descend" to the projective Hilbert space $P\mathcal{H}$ by noticing that, if $A = A^\dagger$ is an observable, and considering from now on only normalized vectors, the expectation value (6.74) associates with the observable A a (real) functional on $P\mathcal{H}$. We will develop this approach fully in the coming sections.

Unitary (and, as a matter of fact, also anti-unitary) operators play also a relevant rôle in Quantum Mechanics [Me58]. In particular, self-adjoint operators can be thought as infinitesimal generators of one-parameter groups of unitaries. Both unitary and anti-unitary operators share the property of leaving all transition probabilities invariant. At the level of the projective Hilbert space they represent isometries of the binary product (6.75).

The converse is also true. Indeed, it was proved long ago by Wigner [Wi32, Wi59] that bijective maps on $P\mathcal{H}$ that preserve transition probabilities (i.e., isometries of the projective Hilbert space) are associated with unitary or anti-unitary transformations on the original Hilbert space.[7]

6.5.3 The Hermitean Structure on the Space of Rays

We have mentioned already that the complex projective Hilbert space $P(\mathcal{H})$ we have defined as the space of pure states does not depend on the particular Hermitean structure we have in the Hilbert space $\mathcal{H}$, that is, it just depends on the vector space structure of $\mathcal{H}$ and the action of $\mathbb{C}_0 = \mathbb{C}\backslash\{0\}$. This action defines $\mathcal{H}_0 = \mathcal{H}\backslash\{\mathbf{0}\}$, the Hilbert space without the zero vector as a principal bundle with base space the complex projective space $P(\mathcal{H})$.

The identification, or immersion, of $P(\mathcal{H})$ with the space of rank-one orthogonal projectors depends, on the contrary, on the specific Hermitean product we use to define the scalar product in the Hilbert space.

As an instance of this let us consider the simplest possible situation where $\mathcal{H} = \mathbb{C}^2$. We could define a family of inner products by setting:

$$\langle z \mid w\rangle_\lambda = (\bar{z}_1, \bar{z}_2)\begin{pmatrix} \lambda_1^2 & 0 \\ 0 & \lambda_2^2 \end{pmatrix}\begin{pmatrix} w_1 \\ w_2 \end{pmatrix} = \lambda_1^2\bar{z}_1 w_1 + \lambda_2^2\bar{z}_2 w_2 \,. \tag{6.86}$$

where λ_1, λ_2 are now real numbers. With this 2-parameter family of scalar products, we would have a 2-parameter family of realizations of the abstract unitary group $U(2)$. Its elements would be given by 2×2 matrices

$$U = \begin{pmatrix} u_{11} & u_{12} \\ u_{21} & u_{22} \end{pmatrix}$$

with complex entries satisfying:

[7] The association being up to a phase, this may lead to the appearance of “ray"(or “projective") representations [Ba54, Ha92, Mac68, Mac04, Me58, Sa97] of unitary groups on the Hilbert space instead of ordinary ones, a problem that we will not discuss here though.

$$(\bar{u}_{11}, \bar{u}_{21}) \begin{pmatrix} \lambda_1^2 & 0 \\ 0 & \lambda_2^2 \end{pmatrix} \begin{pmatrix} u_{11} \\ u_{21} \end{pmatrix} = \lambda_1^2 |u_{11}|^2 + \lambda_2^2 |u_{21}|^2 = 1\,.$$

This means that the bra-vector associated with the ket-vector $\begin{pmatrix} 1 \\ 0 \end{pmatrix}$ should be $(0,\ 1/\lambda_2^2)$. We would write now:

$$[\Psi] \mapsto \rho_{\Psi,\lambda} = \frac{|\Psi\rangle_\lambda\, {}_\lambda\langle\Psi|}{\langle\Psi|\Psi\rangle_\lambda}\,,$$

to make sure this rank-one projector is orthogonal with respect to the λ-scalar product $\langle\cdot|\cdot\rangle_\lambda$ given by Eq. (6.86).

As it will be discussed at length later on, this will become relevant in the framework of the GNS construction. Actually the family of products $\langle\cdot|\cdot\rangle_\lambda$ are nothing but the inner products constructed in a family of Hilbert spaces associated with a 2-parameter family of states using the GNS construction (see Sect. 6.6.2).

To summarize the content of this section, we have argued that all the relevant building blocks of Quantum Mechanics have to be formulated in terms of objects that "live"in the projective Hilbert space $P(\mathcal{H})$. The latter, however, is no longer a linear vector space. As will be discussed in the following sections, it carries instead a rich manifold structure. In this context, the very notion of linear transformations looses meaning, and we are led in a natural way to consider it as a nonlinear manifold.

This given, only objects that have a tensorial character will be allowed. We will do that by following the steps laid in Sect. 6.4 in the first part of this chapter. In the last part of the chapter, having achieved this goal, we will turn back to discuss the problem of alternative Hermitean structures in the context of Quantum Mechanics.

6.5.4 Canonical Tensors on a Hilbert Space

Given a linear space $\mathcal{H}$ carrying a Hermitean structure $\langle\cdot,\cdot\rangle$, we have seen how we can construct the tensor fields g, J and ω on $\mathcal{H}_\mathbb{R}$, the realification of $\mathcal{H}$ (see Sect. 6.3.2).

The $(0,2)$-tensors g and ω define maps from $T\mathcal{H}_\mathbb{R}$ to $T^*\mathcal{H}_\mathbb{R}$. The two of them being non-degenerate, we can also consider their inverses, i.e., the $(2,0)$ contravariant tensors $S = g^{-1}$ (a metric tensor) and $\Lambda = \omega^{-1}$ (a Poisson tensor) mapping $T^*\mathcal{H}_\mathbb{R}$ to $T\mathcal{H}_\mathbb{R}$. The contravariant $(2,0)$ tensors S and Λ can be combined together to define an Hermitean product for any two forms α, β in the dual $\mathcal{H}_\mathbb{R}^*$:

$$\langle\alpha,\beta\rangle_{\mathcal{H}_\mathbb{R}^*} = S(\alpha,\beta) + i\,\Lambda(\alpha,\beta). \tag{6.87}$$

This expression induces the two (non-associative) real brackets on the space of smooth, real-valued functions on $\mathcal{H}_\mathbb{R}$ that were already introduced in Sect. 6.4.3 (here $\nu = 1$), that is:

1. The (symmetric) Jordan bracket $(f, h)_g = S(df, dh)$ (recall Eq. (6.63)) and,
2. The (antisymmetric) Poisson bracket $\{f, h\}_\omega = \Lambda(df, dh)$.

By extending both these brackets to complex functions by linearity we obtain eventually a complex bracket $[\![\cdot, \cdot]\!]$ defined as (recall Eq. (6.64)):

$$[\![f, h]\!] = \langle df, dh\rangle_{\mathcal{H}^*_{\mathbb{R}}} = (f, h)_g + i\{f, h\}_\omega. \tag{6.88}$$

To obtain explicit expressions for these structures in finite dimensions, we may introduce complex coordinates z^k and the corresponding orthogonal-symplectic coordinates (q^k, p_k), that is we select an orthonormal basis $\{e_k\}_{k=1,\dots,n}$ in $\mathcal{H}$ and the global linear coordinates (q^k, p_k) for $k = 1, \dots, n$ on $\mathcal{H}_{\mathbb{R}}$ defined as

$$\langle e_k, u\rangle = z^k = q^k + ip_k, \qquad \forall u \in \mathcal{H}.$$

Now we will change our notation and to make it closer to the notation used when dealing with calculus on manifolds, we denote points in the Hilbert space $\mathcal{H}$ by x, u, v, etc. Then after simple computations[8] we get the coordinate expressions for the tensors J, g and ω:

$$J = dp_k \otimes \frac{\partial}{\partial q^k} - dq^k \otimes \frac{\partial}{\partial p_k}, \quad g = dq^k \otimes dq^k + dp_k \otimes dp_k, \quad \omega = dq^k \wedge dp_k,$$

as well as the corresponding contravariant expressions for G and Λ:

$$G = \frac{\partial}{\partial q^k} \otimes \frac{\partial}{\partial q^k} + \frac{\partial}{\partial p_k} \otimes \frac{\partial}{\partial p_k}, \qquad \Lambda = \frac{\partial}{\partial p_k} \wedge \frac{\partial}{\partial q^k}. \tag{6.89}$$

Hence,

$$(f, h)_g = \frac{\partial f}{\partial q^k}\frac{\partial h}{\partial q^k} + \frac{\partial f}{\partial p_k}\frac{\partial h}{\partial p_k}, \qquad \{f, h\}_\omega = \frac{\partial f}{\partial p_k}\frac{\partial h}{\partial q^k} - \frac{\partial f}{\partial q^k}\frac{\partial h}{\partial p_k}.$$

Using complex coordinates: $z^k = q^k + ip_k$, $\bar{z}^k = q^k - ip_k$, we can also write:

$$G + i\Lambda = 4\frac{\partial}{\partial z^k} \otimes \frac{\partial}{\partial \bar{z}^k}, \tag{6.90}$$

where

$$\frac{\partial}{\partial z^k} = \frac{1}{2}\left(\frac{\partial}{\partial q^k} - i\frac{\partial}{\partial p_k}\right), \qquad \frac{\partial}{\partial \bar{z}^k} = \frac{1}{2}\left(\frac{\partial}{\partial q^k} + i\frac{\partial}{\partial p_k}\right). \tag{6.91}$$

Then we have :

[8] Summation over repeated indices being understood here and in the rest of the Section.

$$[\![f, h]\!] = 4\frac{\partial f}{\partial z^k}\frac{\partial h}{\partial \bar{z}^k},$$

or, in more detail:

$$(f, h)_g = 2\left(\frac{\partial f}{\partial z^k}\frac{\partial h}{\partial \bar{z}^k} + \frac{\partial h}{\partial z^k}\frac{\partial f}{\partial \bar{z}^k}\right), \qquad \{f, h\}_\omega = -2i\left(\frac{\partial f}{\partial z^k}\frac{\partial h}{\partial \bar{z}^k} - \frac{\partial h}{\partial z^k}\frac{\partial f}{\partial \bar{z}^k}\right). \tag{6.92}$$

Notice also that:

$$J = -i\left(dz^k \otimes \frac{\partial}{\partial z^k} - d\bar{z}^k \otimes \frac{\partial}{\partial \bar{z}^k}\right). \tag{6.93}$$

In particular, for any linear operator A in $\mathcal{H}$ we can define the quadratic function:

$$f_A(u) = \frac{1}{2}\langle u, Au\rangle = \frac{1}{2}z^\dagger Az \tag{6.94}$$

where z is the column vector $(z_1, \ldots, z_n)$ with the complex coordinates of u. It follows immediately from Eq. (6.92) that, for any linear operators A, B we have:

$$(f_A, f_B)_g = f_{AB+BA}, \qquad \{f_A, f_B\}_\omega = -if_{AB-BA}. \tag{6.95}$$

The Jordan (or symmetric) bracket of any two quadratic functions f_A and f_B is related to the (commutative) Jordan product of A and B, $[A, B]_+$, defined[9] as:

$$[A, B]_+ = AB + BA, \tag{6.96}$$

while their Poisson bracket is related to the commutator product $[A, B]_-$ defined as:

$$[A, B]_- = -i\ (AB - BA)\ . \tag{6.97}$$

In particular, if A and B are Hermitean, their Jordan product (6.96) and their Lie bracket will be Hermitean as well. Hence, the set of Hermitean operators on $\mathcal{H}_\mathbb{R}$, equipped with the binary operations (6.96) and (6.97), becomes a Lie-Jordan algebra [Jo34, Jo34b, Fa12], and the binary product:

$$A \cdot B = \frac{1}{2}\left([A, B]_+ + i\,[A, B]_-\right) \tag{6.98}$$

is just the associative composition of linear operators. We remark parenthetically that all this extends without modifications to the infinite-dimensional case when we consider bounded operators.

Returning our attention to quadratic functions, it is not hard to check that:

---

9 That is actually twice the Jordan Bracket $A \circ B$ as it is usually defined in the literature [Em84], but we find here more convenient to use here this slightly different definition.

$$[\![f_A, f_B]\!] = 2 f_{AB}\,, \tag{6.99}$$

which proves the associativity of the Kähler bracket (6.88) on quadratic functions, i.e.,

$$[\![[\![f_A, f_B]\!], f_C]\!] = [\![f_A, [\![f_B, f_C]\!]]\!] = 4 f_{ABC}, \qquad \forall A, B, C \in \mathrm{End}(\mathcal{H}). \tag{6.100}$$

We will study now smooth real-valued functions on $\mathcal{H}_{\mathbb{R}}$. First of all, it is clear that f_A will be a real function iff A is Hermitean. The Jordan and Poisson brackets will define then a Lie-Jordan algebra structure on the set of real, quadratic functions, and, according to Eq. (6.100), the bracket $[\![\cdot, \cdot]\!]$ will be an associative bracket.

For any such $f \in \mathcal{F}(\mathcal{H}_{\mathbb{R}})$ we may define two vector fields, the gradient ∇f of f and the Hamiltonian vector field X_f associated with f, given respectively by:

$$\begin{matrix} g(\nabla f, \cdot) = df \\ \omega(X_f, \cdot) = df \end{matrix} \quad \text{or} \quad \begin{matrix} G(df, \cdot) = \nabla f, \\ \Lambda(df, \cdot) = X_f \end{matrix}\,,$$

which allow to write down the Jordan and the Poisson brackets as:

$$(f, h)_g = g(\nabla f, \nabla h), \qquad \{f, h\}_\omega = \omega(X_f, X_h)\,. \tag{6.101}$$

Explicitly, in the coordinates (q^k, p_k) before:

$$\nabla f = \frac{\partial f}{\partial q^k}\frac{\partial}{\partial q^k} + \frac{\partial f}{\partial p_k}\frac{\partial}{\partial p_k} = 2\left(\frac{\partial f}{\partial z^k}\frac{\partial}{\partial \bar{z}^k} + \frac{\partial f}{\partial \bar{z}^k}\frac{\partial}{\partial z^k}\right) \tag{6.102}$$

and,

$$X_f = \frac{\partial f}{\partial p_k}\frac{\partial}{\partial q^k} - \frac{\partial f}{\partial q^k}\frac{\partial}{\partial p_k} = 2i\left(\frac{\partial f}{\partial z^k}\frac{\partial}{\partial \bar{z}^k} - \frac{\partial f}{\partial \bar{z}^k}\frac{\partial}{\partial z^k}\right), \tag{6.103}$$

which shows that

$$J(\nabla f) = X_f\,, \qquad J(X_f) = -\nabla f\,.$$

Let us recall that to any (bounded) linear operator $A \colon \mathcal{H} \to \mathcal{H}$ we can associate:

1. A quadratic function f_A as in Eq. (6.94),
2. A vector field $X_A \colon \mathcal{H} \to T\mathcal{H},\ x \longmapsto (x, Ax)$ as in Eq. (2.112),
3. A (1, 1) tensor field: $T_A \colon T_x\mathcal{H} \ni (x, y) \longmapsto (x, Ay) \in T_x\mathcal{H}$ as in (2.107).

Clearly, as remarked already, f_A is real if and only if A is Hermitean. In this case the vector field X_A becomes the gradient vector field associated to f_A:

$$\nabla f_A = X_A\,, \tag{6.104}$$

and

$$X_{f_A} = J(X_A)\,. \tag{6.105}$$

Indeed, using also the notation $\langle \cdot, \cdot \rangle$ for the standard pairing between vectors and covectors, Eq. (6.104) holds because

$$g\,(y, X_A\,(x)) = g\,(y, Ax) = \frac{1}{2}\,(\langle y, Ax\rangle_{\mathcal{H}} + \langle Ax, y\rangle_{\mathcal{H}}) = \langle df_A\,(x)\,, y\rangle\,,$$

while Eq. (6.105) follows from the second expression in Eq. (6.15), i.e., from: $g\,(y, Ax) = \omega\,(y, (JX_A)(x)) = \omega\,(y, iAx)$. Thus, we will also write, with a slight abuse of notation,

$$\nabla f_A = A, \quad \text{and} \quad X_{f_A} = iA \tag{6.106}$$

In particular, for the identity operator $\mathbb{I}$, we obtain the Liouville vector field $\Delta = X_{\mathbb{I}}$:

$$\nabla f_{\mathbb{I}} = \Delta, \qquad X_{f_{\mathbb{I}}} = J\Delta\,,$$

with the last vector in the previous equation called the phase vector field and denoted:

$$\Gamma = J(\Delta) = p_k \frac{\partial}{\partial q^k} - q^k \frac{\partial}{\partial p_k}\,. \tag{6.107}$$

6.5.5 The Kähler Geometry of the Space of Pure Quantum States

In Sect. 6.5.4 we have described the canonical tensors associated to an Hermitean structure, however as it was discussed before, the space of pure states of a quantum system are the elements on a projective Hilbert space. In this section we would like to discuss in some detail the geometrical structure of the complex projective Hilbert space $P\mathcal{H}$. We will restrict ourselves to the finite-dimensional case even if all of our results extend naturally to the infinite-dimensional situation.

Notice first that we may describe the quotient $\pi : \mathcal{H}_0 \to P(\mathcal{H})$ by considering the distribution generated by the Liouville vector field Δ and the phase vector field Γ. Clearly both vector fields commute $[\Delta, \Gamma] = 0$ (notice that Γ is homogeneous of degree 0), then we can integrate trivially the distribution generated by Δ and Γ by using their respective flows: $\varphi_t^{\Delta}(u) = e^t u$ and $\varphi_s^{\Gamma}(u) = e^{is}u$. Thus if we consider the action of $\mathbb{C}_0$ on $\mathcal{H}$ as $\lambda u = re^{is}u$ with $r = |\lambda|$ and $s = \arg\lambda$, it corresponds to the element $\varphi_t^{\Delta} \circ \varphi_s^{\Gamma}$ with $\log r = t$. In other words, the fields Δ and Γ are the vector fields defined by the natural action of the multiplicative group $\mathbb{C}_0$ on $\mathcal{H}_0$.

The orbits of such action correspond to the orbits of the 2-parameter group of transformations $\varphi_t^{\Delta} \circ \varphi_s^{\Gamma}$ and they are just the fibres of the projection map π onto the projective Hilbert space $P(\mathcal{H})$. They are also the leaves of the foliation described by the integrable distribution defined by Δ and Γ.

Contravariant tensorial objects on $\mathcal{H}$ will pass to the quotient (i.e., will be projectable with respect to the projection map π) if and only if they are left invariant by both Δ and Γ, namely if they are homogeneous of degree zero (which means invariance with respect to Δ) and invariant under multiplication of vectors by a phase (invariance with respect to Γ).

Typical quadratic functions that pass to the quotient will be normalized expectation values of the form (recall Eq. (6.68)):

$$e_A(u) = \mathrm{Tr}(\rho_u A) = \frac{\langle u|A|u\rangle}{\langle u|u\rangle}\,, \tag{6.108}$$

with A any linear operator and any Hermitean structure on $\mathcal{H}$.

Concerning projectability of tensors, the complex structure J, being (cfr., Eq. (6.93)) homogeneous of degree zero and phase-invariant, will be a projectable tensor, while it is clear that the Jordan and Poisson tensors S and Λ defined respectively in Eq. (6.89) or, for that matter, the complex-valued tensor of Eq. (6.90) will *not* be projectable (as they are phase-invariant but homogeneous of degree -2). To turn them into projectable objects we will have to multiply them by the a phase invariant quadratic factor, e.g., $\theta\,(z) = z^\dagger z$, thus defining new tensors

$$\widetilde{\Lambda}\,(z) = \theta\,(z)\,\Lambda\,(z)\,, \qquad \widetilde{S}\,(z) = \theta\,(z)\,S\,(z)\,. \tag{6.109}$$

The induced tensors in the quotient space $P(\mathcal{H})$ will be denoted with the same symbols, $\widetilde{\Lambda}$, $\widetilde{S}$ and $\widetilde{J}$ respectively, and they will define a Hermitian structure on $P(\mathcal{H})$. Such Hermitean structure is called the Fubini-Study Hermitean structure and will be denoted by $\mathbf{h}_{FS}$.

Accordingly the inverses of the contravariant projected tensors will be denoted by ω_{FS} and g_{FS} respectively, that is $\omega_{FS} = \widetilde{\Lambda}^{-1}$, $g_{FS} = \widetilde{S}^{-1}$. Notice that the tensor $\widetilde{\Lambda}$ on $\mathcal{H}$ is not a Poisson tensor anymore because of the multiplicative factor θ, however the projected tensor is Poisson. We will prove this, along with the fact that $P(\mathcal{H})$ inherits a Kähler structure, by a direct computation.

6.5.5.1 Local Expressions in the Projective Space

Thus, let us examine the structures induced on $P(\mathcal{H})$ providing explicit expressions for them in local coordinates. Recall that, in the finite-dimensional case, $P(\mathcal{H}) = \mathbb{CP}^n$ and it is therefore made up of equivalence classes of vectors $z = (z^0, z^1, \cdots, z^n) \in \mathbb{C}^{n+1}$ with respect to the equivalence relation $z' \cong \lambda z$, $\lambda \in \mathbb{C}_0$.

The complex chart defined by the local coordinates z^k will be denoted by $\mathbf{z}$. Coordinates on $\mathbb{CP}^n$ induced from the complex coordinates z^k are called homogeneous coordinates and are denoted by $[\mathbf{z}] = [z^0 : z^1 : \ldots : z^n]$. The pull-back under the map π of the Fubini-Study metric g_{FS} to $\mathbb{C}^{n+1}$ is given by [Be06]:

$$\pi^* g_{\mathrm{FS}} = \frac{(\mathbf{z}\cdot\bar{\mathbf{z}})d\mathbf{z}\otimes_S d\bar{\mathbf{z}} - (d\mathbf{z}\cdot\bar{\mathbf{z}})\otimes_S(\mathbf{z}\cdot d\bar{\mathbf{z}})}{(\mathbf{z}\cdot\bar{\mathbf{z}})^2},$$

where $\mathbf{z}\cdot\bar{\mathbf{z}} = \sum_k z^k \bar{z}^k$, $d\mathbf{z}\cdot\bar{\mathbf{z}} = \sum_k dz^k\bar{z}^k$, $d\mathbf{z}\otimes_S d\bar{\mathbf{z}} = \sum_k dz^k d\bar{z}^k + d\bar{z}^k dz^k$, and so on. Similarly the pull-back of the symplectic form gives:

$$\pi^*\omega_{\mathrm{FS}} = i\,\frac{(\mathbf{z}\cdot\bar{\mathbf{z}})\,d\mathbf{z}\wedge d\bar{\mathbf{z}} - (d\mathbf{z}\cdot\bar{\mathbf{z}})\wedge(\mathbf{z}\cdot d\bar{\mathbf{z}})}{(\mathbf{z}\cdot\bar{\mathbf{z}})^2} = -d\theta_{\mathrm{FS}}\,,$$

where:

$$\theta_{\mathrm{FS}} = \frac{1}{2i}\frac{\bar{\mathbf{z}}\cdot d\mathbf{z} - \mathbf{z}\cdot d\bar{\mathbf{z}}}{\mathbf{z}\cdot\bar{\mathbf{z}}}\,.$$

The isometries of the Fubini-Study structure are just the usual unitary transformations which, in infinitesimal form, are written as:

$$\dot{z}^k = i A^k{}_j z^j$$

where $A = (A^k_j)$ is a Hermitean matrix. These are the equations for the flow of a generic Killing vector field, which therefore has the form [10]:

$$X_A = \dot{z}^k\partial_{z^k} - \dot{\bar{z}}^k\partial_{\bar{z}^k} = i A^k{}_j(z^j\partial_{z^k} - \bar{z}^j\partial_{\bar{z}^k})\,.$$

A straightforward calculation shows that

$$\omega_{\mathrm{FS}}(X_A,\cdot) = d\left(i_{X_A}\theta_{\mathrm{FS}}\right)\,,$$

i.e., X_A is the Hamiltonian vector field associated with the expectation value function (6.68):

$$e_A(\mathbf{z},\bar{\mathbf{z}}) = \frac{\bar{\mathbf{z}}\cdot A\mathbf{z}}{\mathbf{z}\cdot\bar{\mathbf{z}}} = i_{X_A}\theta_{\mathrm{FS}} \tag{6.110}$$

for the Hermitean matrix A, that is:

$$\omega_{\mathrm{FS}}(\cdot, X_{e_A}) = de_A\,.$$

Again a simple algebraic computation shows that, given any two expectation functions e_A, e_B (A, B being Hermitean matrices), their corresponding Hamiltonian

---

[10] Notice that these are exactly the Killing vector fields of the unit sphere S^{2n+1}. In particular, for $A = \mathbb{I}$ we obtain $X_k = \Gamma$ which is a vertical vector field with respect to the Hopf projection [St51] $\pi_H : S^{2n+1} \to \mathbb{CP}^n$.

vector fields satisfy:

$$\omega_{\mathrm{FS}}(X_{e_A}, X_{e_B}) = X_{e_A}(de_B) = -ie_{AB-BA}\,.$$

Therefore, the Poisson brackets associated with the symplectic form ω_{FS}:

$$\{f, g\}_{\omega_{\mathrm{FS}}} = \omega(X_f, X_g)\,,$$

are such that:

$$\{e_A, e_B\}_{\omega_{\mathrm{FS}}} = -ie_{AB-BA}\,.$$

In a similar way, one can prove that the gradient vector field ∇_{e_A} has the form:

$$\nabla_{e_A} = A^k{}_j(z^j\partial_{z^k} + \bar{z}^j\partial_{\bar{z}^k})$$

so that

$$g_{\mathrm{FS}}(\nabla_{e_A}, \nabla_{e_B}) = \nabla_{e_A}(de_B) = e_{AB+BA} - e_A \cdot e_B\,.$$

Remember that a real function f on $P\mathcal{H}$ is Kählerian iff its Hamiltonian vector field X_f is also Killing (Definition 6.24). Such functions represent quantum observables. The above calculations show that the space $\{e_A \mid A = A^\dagger\}$ of expectation values of Hermitean operators on $P(\mathcal{H})$ consists exactly of all real Kählerian functions. To extend this concept to the complex case, one says that a complex-valued function on $P(\mathcal{H})$ is Kählerian iff its real and imaginary parts are Kählerian. Clearly, any such f is a function of the form (6.110) with A a (bounded) linear operator on $\mathcal{H}$. Also, on the space, $\mathcal{K}^{\mathbb{C}}(P(\mathcal{H}))$, of complex Kählerian functions one can define an associative bilinear product (also-called a star-product) via (recall Eq. (6.65)):

$$f \star g = f \cdot g + \frac{1}{2}\mathbf{h}_{\mathrm{FS}}(df, dg) = f \cdot g + \frac{1}{2}\left[(f, g)_{g_{\mathrm{FS}}} + i\{f, g\}_{\omega_{\mathrm{FS}}}\right]\,, \qquad (6.111)$$

under which the space $\mathcal{K}^{\mathbb{C}}(P\mathcal{H})$ is closed since $e_A \star e_B = e_{AB}$. Thus we have obtained another realization of the $\mathbb{C}^*$-algebra of bounded linear operators $\mathcal{B}(\mathcal{H})$.[11]

6.5.6 The Momentum Map and the Jordan–Scwhinger Map

We will study now the action of the unitary group $U(\mathcal{H})$ on $\mathcal{H}$. The unitary group $U(\mathcal{H})$ is the group of linear isomorphisms that preserve the admissible tensorial triple (g, ω, J). In the following, we will denote as $\mathfrak{u}(\mathcal{H})$ the Lie algebra of $U(\mathcal{H})$ which

[11] Actually both algebras are isomorphic as $*$-algebras and as C^*-algebras where the norm in $\mathcal{K}^{\mathbb{C}}(P\mathcal{H})$ is the supremum norm.

is just the space of anti-Hermitean operators on $\mathcal{H}$.[12] Thus, in finite dimensions, we will identify the real linear space of all Hermitean operators with the dual $\mathfrak{u}(\mathcal{H})^*$ of $\mathfrak{u}(\mathcal{H})$ via the non-degenerate pairing:

$$\langle A, T\rangle = \frac{i}{2}\mathrm{Tr}(AT), \tag{6.112}$$

with A a Hermitean operator and $T \in \mathfrak{u}(\mathcal{H})$. Having identified $\mathfrak{u}(\mathcal{H})^*$ with the real linear space of Hermitean operators, we can define on it a Lie bracket (cfr. Sect. 6.5.4):

$$[A, B]_- = \frac{1}{i}(AB - BA), \tag{6.113}$$

with respect to which it becomes a Lie algebra, and also a Jordan bracket:

$$[A, B]_+ = AB + BA\,, \tag{6.114}$$

and, as we have discussed before, both structures together equip $\mathfrak{u}(\mathcal{H})^*$ with the structure of a Lie-Jordan algebra.

In addition, $\mathfrak{u}(\mathcal{H})^*$ is equipped with the scalar product

$$\langle A, B\rangle_{\mathfrak{u}^*} = \frac{1}{2}\mathrm{Tr}(AB) \tag{6.115}$$

which satisfies

$$\begin{aligned}\langle [A,\xi]_-, B\rangle_{\mathfrak{u}^*} &= \frac{1}{2}\mathrm{Tr}([A,\xi]_- B) = \frac{1}{2}\mathrm{Tr}(A[\xi, B]_-)\\ &= \frac{1}{2}\mathrm{Tr}(\xi[B, A]_-) = \langle A, [\xi, B]_-\rangle_{\mathfrak{u}^*}\,,\end{aligned} \tag{6.116}$$

and,

$$\begin{aligned}\langle [A,\xi]_+, B\rangle_{\mathfrak{u}^*} &= \frac{1}{2}\mathrm{Tr}([A,\xi]_+ B) = \frac{1}{2}\mathrm{Tr}(A, [\xi, B]_+)\\ &= \frac{1}{2}\mathrm{Tr}(\xi[A, B]_+) = \langle A, [\xi, B]_+\rangle_{\mathfrak{u}^*}\,.\end{aligned} \tag{6.117}$$

With any $A \in \mathfrak{u}(\mathcal{H})^*$, we can associate the fundamental vector field X_A on the Hilbert space corresponding to the element $\frac{1}{i}A \in \mathfrak{u}(\mathcal{H})$ defined by the formula[13]:

---

[12] At this point the finite and infinite-dimensional situations separate as in the infinite-dimensional case, because of Stone's theorem, the Lie algebra of $U(\mathcal{H})$ will consists of, in general unbounded, anti-Hermitean operators.

[13] In what follows we will denote by x the vectors in the real Hilbert space $\mathcal{H}_{\mathbb{R}}$.

$$\frac{d}{dt}e^{itA}(x)|_{t=0} = iAx, \qquad \forall x \in \mathcal{H}_{\mathbb{R}}.$$

In other words, $X_A = iA$. We already know from Sect. 6.5.4 that iA is a Hamiltonian vector field with Hamiltonian function $f_A\colon \omega(\cdot, X_A) = df_A$. Thus, for any $x \in \mathcal{H}_{\mathbb{R}}$ we obtain an element $\mu(x) \in \mathfrak{u}(\mathcal{H})^*$ such that:

$$\langle \mu(x), iA \rangle = f_A(x) = \frac{1}{2}\langle x, Ax \rangle_{\mathcal{H}} \tag{6.118}$$

We obtain a mapping:

$$\mu\colon \mathcal{H}_{\mathbb{R}} \to \mathfrak{u}(\mathcal{H})^*, \tag{6.119}$$

which is just the momentum map of the action of the group $U(\mathcal{H})$ on $\mathcal{H}$ [Ch09].

More explicitly, it follows from Eq.(6.112) that:

$$\mu(x) = |x\rangle\langle x|. \tag{6.120}$$

We may therefore conclude that the unit sphere in $\mathcal{H}$ can be mapped into $\mathfrak{u}(\mathcal{H})^*$ in an equivariant way with respect to the coadjoint action of $U(\mathcal{H})$. Actually we have:

$$\mu(Ux) = U^\dagger|x\rangle\langle x|U = \mathrm{Ad}^*_{U^\dagger}\mu(x).$$

In finite dimensions, the unit sphere is odd dimensional and the orbit in $\mathfrak{u}(\mathcal{H})^*$ is symplectic, hence even dimensional.

With every $A \in \mathfrak{u}(\mathcal{H})^*$ we can associate, with the familiar identification of the tangent space at every point of $\mathfrak{u}(\mathcal{H})^*$ with $\mathfrak{u}(\mathcal{H})^*$ itself, the linear function (hence a one-form) $\hat{A}\colon \mathfrak{u}(\mathcal{H})^* \to \mathbb{R}$ defined as: $\hat{A} = \langle A, \cdot \rangle_{u^*}$. Then, we can define two contravariant tensors, a symmetric (Jordan) tensor:

$$R(\hat{A}, \hat{B})(\xi) = \langle \xi, [A, B]_+ \rangle_{\mathfrak{u}^*} \tag{6.121}$$

and a Poisson tensor, which is just the Kostant-Kirillov-Souriau [Ki76, Ko70]) tensor $\Lambda_{\mathfrak{u}^*}$ (see Sect. 4.3.4):

$$\Lambda_{\mathfrak{u}^*}(\hat{A}, \hat{B})(\xi) = \langle \xi, [A, B]_- \rangle_{\mathfrak{u}^*} \tag{6.122}$$

$A, B, \xi \in \mathfrak{u}(\mathcal{H})^*$. We notice that the exterior derivative of the quadratic function f_A is the pull-back of $\hat{A}$ via the momentum map since, for all $x \in \mathcal{H}$:

$$\mu^*(\hat{A})(x) = \hat{A} \circ \mu(x) = \langle A, \mu(x) \rangle_{\mathfrak{u}^*} = \frac{1}{2}\langle x, Ax \rangle_{\mathcal{H}} = f_A(x)$$

i.e.,

$$\mu^* \widehat{A} = df_A . \tag{6.123}$$

This means that if $\xi = \mu(x)$:

$$(\mu_* G)(\hat{A}, \hat{B})(\xi) = G(df_A, df_B)(x) = \{f_A, f_B\}_g(x) = f_{[A,B]_+}(x) = R(\hat{A}, \hat{B})(\xi)$$

i.e.,

$$\mu_* G = R .$$

Similarly, using now Eq. (6.95), we find:

$$(\mu_* \Lambda)(\hat{A}, \hat{B})(\xi) = \Lambda(df_A, df_B)(x) = \{f_A, f_B\}_\omega(x) = f_{[A,B]_-}(x) = \Lambda_{\mathfrak{u}^*}(\hat{A}, \hat{B})(\xi)$$

i.e.,

$$\mu_* \Lambda = \Lambda_{\mathfrak{u}^*} .$$

Thus, the momentum map μ relates the contravariant metric tensor G and the Poisson tensor Λ with the corresponding contravariant tensors R and $\Lambda_{\mathfrak{u}^*}$. Together they form the complex tensor:

$$(R + i\Lambda_{\mathfrak{u}^*})(\hat{A}, \hat{B})(\xi) = 2\langle \xi, AB \rangle_{\mathfrak{u}^*} \tag{6.124}$$

which provides the contravariant form of the Hermitean product on $\mathfrak{u}(\mathcal{H})^*$.

6.5.7 A Simple Example: The Geometry of a Two-Level System

Let $\mathcal{H} = \mathbb{C}^2$ be the Hilbert space appropriate to describe a two-level system (also referred sometimes in the quantum information literature as a q-bit). We can write any $A \in \mathfrak{u}^*(\mathcal{H}) = U(2)$ as

$$A = y_A^0 \mathbb{I} + \mathbf{y}_A \cdot \boldsymbol{\sigma} \tag{6.125}$$

where $\mathbb{I}$ is the 2×2 identity, $\mathbf{y}_A \cdot \boldsymbol{\sigma} = y_A^1 \sigma_1 + y_A^2 \sigma_2 + y_A^3 \sigma_3$ and $\boldsymbol{\sigma} = (\sigma_1, \sigma_2, \sigma_3)$ are the well-known Pauli matrices (see Sect. 10, Eq. (10.68)), that satisfy the well-known identities [Me58]:

$$\sigma_h \sigma_k = \delta_{hk} \mathbb{I} + i \varepsilon_{hkl} \sigma_l \tag{6.126}$$

$(h, k, l = 1, 2, 3)$ and

$$\sigma_j\sigma_k\sigma_l = i\varepsilon_{jkl}\mathbb{I} + \sigma_j\delta_{kl} - \sigma_k\delta_{jl} + \sigma_l\delta_{jk}\,. \tag{6.127}$$

Hence, explicitly

$$A = \begin{pmatrix} y_A^0 + y_A^3 & y_A^1 - iy_A^2 \\ y_A^1 + iy_A^2 & y_A^0 - y_A^3 \end{pmatrix} \tag{6.128}$$

Every $A \in \mathfrak{u}(2)^*$ is then represented by the (real) 'four-vector' $(y_A^\mu) = \left(y_A^0, \mathbf{y}_A\right)$:

$$y_A^\mu = \langle A|\sigma_\mu\rangle, \qquad \mu = 0, 1, 2, 3\,.$$

If we represent vectors in $\mathbb{C}^2$ as two-components complex spinors, i.e.,

$$|x\rangle = \begin{pmatrix} z_1 \\ z_2 \end{pmatrix}, \qquad z_1, z_2 \in \mathbb{C} \tag{6.129}$$

we find, using Eq. (6.128):

$$f_A(z_1, z_2) = \frac{1}{2}\left\{\left(y_A^0 + y_A^3\right)|z_1|^2 + \left(y_A^0 - y_A^3\right)|z_2|^2 + \left(y_A^1 - iy_A^2\right)\bar{z}_1 z_2 + \left(y_A^1 + iy_A^2\right)\bar{z}_2 z_1\right\}\,.$$

Exercise 6.2 Show that rank-one projectors $\rho = |x\rangle\langle x|$, $\left(\rho^\dagger = \rho, \operatorname{Tr}\rho = 1, \rho^2 = \rho\right)$, can be parametrized as [Mo01]:

$$\rho = \rho\,(\theta, \phi) = \begin{pmatrix} \sin^2\frac{\theta}{2} & \frac{1}{2}e^{i\phi}\sin\theta \\ \frac{1}{2}e^{-i\phi}\sin\theta & \cos^2\frac{\theta}{2} \end{pmatrix}; \qquad 0 \leq \theta < \pi, \quad 0 \leq \phi < 2\pi\,. \tag{6.130}$$

Then, because of Exercise 6.2, rank-one projectors correspond to vectors y_ρ^μ of the form:

$$y^0 = \frac{1}{2},\ y^1 = \frac{1}{2}\sin\theta\cos\phi,\ y^2 = -\frac{1}{2}\sin\theta\sin\phi,\ y^3 = -\frac{1}{2}\cos\theta\,, \tag{6.131}$$

thus $\mathbf{y}^2 = (y^0)^2 = 1/4$ for all rank-one projectors.

Using Eqs. (6.128) and (6.129), the momentum map $\mu = |x\rangle\langle x|$ can be written explicitly in the form (6.125).

Exercise 6.3 Show that the components of the vector y_ρ^μ of the momentum map μ are given by:

$$\begin{aligned} y^0 &= \frac{|z_1|^2 + |z_2|^2}{2}, & y^1 &= \frac{z_1\bar{z}_2 + z_2\bar{z}_1}{2} \\ y^2 &= i\,\frac{z_1\bar{z}_2 - z_2\bar{z}_1}{2}, & y^3 &= \frac{|z_1|^2 - |z_2|^2}{2} \end{aligned} \tag{6.132}$$

We can associate with every vector $(y_A^\mu) = \left(y_A^0, \mathbf{y}_A\right)$ the vector field

$$X_A = y_A^0 \partial_0 + y_A^1 \partial_1 + y_A^2 \partial_2 + y_A^3 \partial_3 ,$$

with $\partial_\mu = \partial/\partial y^u$, $\mu = 0, 1, 2, 3$. Also (see the discussion immediately before Eq. (6.121)), $\hat{A} = \langle A, \cdot \rangle_{u^*}$ will be represented by the one-form:

$$\hat{A} = y_A^0 dy^0 + y_A^1 dy^1 + y_A^2 dy^2 + y_A^3 dy^3 . \tag{6.133}$$

Exercise 6.4 Show, using Eq. (6.125)m that:

$$AB = \left(y_A^0 y_B^0 + \mathbf{y}_A \cdot \mathbf{y}_B\right) \mathbb{I} + \left(y_A^0 \mathbf{y}_B + y_B^0 \mathbf{y}_A + i \mathbf{y}_A \times \mathbf{y}_B\right) \cdot \boldsymbol{\sigma}$$

where "×" denotes the standard cross-product of three-vectors.

Hence

$$\langle AB \rangle_{u^*} = \frac{1}{2} \mathrm{Tr}\ (AB) = y_A^0 y_B^0 + \mathbf{y}_A \cdot \mathbf{y}_B .$$

In particular: $\langle \rho\,(\theta, \phi)\, \rho\,(\theta', \phi') \rangle_{u^*} = \left\{1 + \sin\theta \sin\theta' \cos\left(\phi - \phi'\right) + \cos\theta \cos\theta'\right\}$ /4 for rank-one projectors. Moreover

$$[A, B]_+ = 2\left(y_A^0 y_B^0 + \mathbf{y}_A \cdot \mathbf{y}_B\right) \mathbb{I} + \left(y_A^0 \mathbf{y}_B + y_B^0 \mathbf{y}_A\right) \cdot \boldsymbol{\sigma} ,$$

while

$$[A, B]_- = 2(\mathbf{y}_A \times \mathbf{y}_B) \cdot \boldsymbol{\sigma} .$$

Also, we check, using Eqs. (6.132) and (6.133) that $\mu^* \hat{A} = df_A$, in agreement with Eq. (6.123).

Proceeding further, we find the explicit expressions for the tensors R and Λ_{u^*}:

$$\begin{aligned} R(\hat{A}, \hat{B})\,(\xi) &= \langle \xi, [A, B]_+ \rangle_{u^*} = \langle [\xi, A]_+ , B \rangle = \\ &= 2\xi^0 \left(y_A^0 y_B^0 + \mathbf{y}_A \cdot \mathbf{y}_B\right) + 2\left(y_A^0 \mathbf{y}_B + y_B^0 \mathbf{y}_A\right) \cdot \xi = \\ &= 2\left(y_A^0 \xi^0 + \mathbf{y}_A \cdot \xi\right) y_B^0 + 2\left(y_A^0 \xi + \xi^0 \mathbf{y}_A\right) \cdot \mathbf{y}_B \end{aligned} \tag{6.134}$$

and hence, explicitly [Gr05]:

$$\begin{aligned} R\,(\xi) = {} & 2\partial_0 \otimes \left(\xi^1 \partial_1 + \xi^2 \partial_2 + \xi^3 \partial_3\right) + 2\left(\xi^1 \partial_1 + \xi^2 \partial_2 + \xi^3 \partial_3\right) \otimes \partial_0 \\ & + 2\xi^0 \left(\partial_0 \otimes \partial_0 + \partial_1 \otimes \partial_1 + \partial_2 \otimes \partial_2 + \partial_3 \otimes \partial_3\right) \end{aligned} \tag{6.135}$$

Quite similarly, one finds:

$$\Lambda_{\mathfrak{u}^*}(\hat{A}, \hat{B})(\xi) = 2(\xi \times \mathbf{y}_A) \cdot \mathbf{y}_B = 2(\mathbf{y}_A \times \mathbf{y}_B) \cdot \xi \tag{6.136}$$

and:

$$\Lambda_{\mathfrak{u}^*}(\xi) = 2\left(\xi^1 \partial_2 \wedge \partial_3 + \xi^2 \partial_3 \wedge \partial_1 + \xi^3 \partial_1 \wedge \partial_2\right). \tag{6.137}$$

We thus find the tensor:

$$R + i\Lambda_{\mathfrak{u}^*} = 2\left[\partial_0 \otimes y^k \partial_k + y^k \partial_k \otimes \partial_0 + y^0(\partial_0 \otimes \partial_0 + \partial_k \otimes \partial_k) + i\epsilon_{hkl} y^h \partial_k \otimes \partial_l\right].$$

Remark 6.9 To conclude this section, we notice that we can define two additional (1, 1) tensors, $\widetilde{R}$ and $\widetilde{J}$, that will appear below in Sect. 6.6.1, via

$$\widetilde{R}_\xi(A) = [\xi, A]_+ = R(\widehat{A}, \cdot)(\xi),$$

and

$$\widetilde{J}_\xi(A) = [\xi, A]_- = \Lambda_{\mathfrak{u}^*}(\widehat{A}, \cdot)(\xi),$$

for any $A \in T_\xi \mathfrak{u}^*(\mathcal{H}) \cong \mathfrak{u}^*(\mathcal{H})$, the last passage in both equations following from Eqs. (6.116) and (6.117).

Exercise 6.5 Show that in coordinates y^μ, we have:

$$\widetilde{R}_\xi = 2\left(\xi^0 dy^0 + \boldsymbol{\xi} \cdot d\mathbf{y}\right) \otimes \partial_0 + 2\left(\xi^i dy^0 + \xi^0 dy^i\right) \otimes \partial_i$$

and

$$\widetilde{J}_\xi = 2\varepsilon_{ijk}\, \xi^i dy^j \otimes \partial_k .$$

6.6 The Geometry of Quantum Mechanics and the GNS Construction

In the previous sections of this chapter, we have worked out the geometrical structures that naturally arise in the standard approach to Quantum Mechanics, which starts from a complex separable Hilbert space and identifies the space of physical states with its associated complex projective space. In this framework, algebraic notions, such as that of the $\mathbb{C}^*$-algebra whose real elements are the observables of the system, arises only as a derived concept.

In this section, we would like to show how geometrical structures emerge also in a more algebraic setting, where one starts from the very beginning with an abstract $\mathbb{C}^*$-algebra to obtain the Hilbert space of pure states as a derived concept via the so-called Gelfand-Naimark-Segal (GNS) construction [Br87]. A more detailed discussion can be found in [Ch09].

6.6.1 The Space of Density States

We have seen in Sect. 6.5.5 that it is possible to obtain $\mathcal{P}(\mathcal{H})$ as a quotient of $\mathcal{H}_0$ with respect to the involutive distribution associated with Δ and Γ. Equation (6.120) shows that the image of $\mathcal{H}_0$ under the momentum map associated to the natural action of $U(\mathcal{H})$ consists of the set of all non-negative Hermitean operators of rank one, that will be denoted as $\mathcal{P}^1(\mathcal{H})$, i.e.,[14]

$$\mathcal{P}^1(\mathcal{H}) = \{|x\rangle\langle x| \mid x \in \mathcal{H}, \quad x \neq 0\} .$$

On the other hand, the coadjoint action of $U(\mathcal{H})$ on $\mathcal{P}^1(\mathcal{H})$:

$$(U, \rho) \mapsto U\rho U^\dagger , \qquad \rho \in \mathcal{P}^1(\mathcal{H}), U \in \mathcal{U}(\mathcal{H}) ,$$

foliates $\mathcal{P}^1(\mathcal{H})$ into the spaces $\mathcal{D}^1_r(\mathcal{H}) = \{|x\rangle\langle x| \mid \langle x, x\rangle_{\mathcal{H}} = r^2\}$. In particular we will denote by $\mathcal{D}^1_1(\mathcal{H})$ the space of rank-one orthogonal projection operators, which is the image via the momentum map of the sphere $\mathbb{S} = \{x \in \mathcal{H} \mid \langle x, x\rangle_{\mathcal{H}} = 1\}$ and can be identified with the complex projective space $P(\mathcal{H})$ via:

$$[x] \subset P(\mathcal{H}) \leftrightarrow \frac{|x\rangle\langle x|}{\langle x, x\rangle} \in \mathcal{D}^1_1(\mathcal{H}) .$$

We have also discussed the geometry of $P(\mathcal{H})$ as a Kähler manifold. In the following we will examine this fact in more detail, by showing explicitly that $\mathcal{D}^1_1(\mathcal{H})$ carries a natural Kähler manifold structure.

Let $\xi \in \mathfrak{u}(\mathcal{H})^*$ be the image through the momentum map of a unit vector $x \in S_{\mathcal{H}}$, i.e., $\xi = |x\rangle\langle x|$ with $\langle x|x\rangle = 1$, so that $\xi^2 = \xi$. The tangent space of the coadjoint $U(\mathcal{H})$-orbit at ξ is generated by vectors of the form $[A, \xi]_-$, for any Hermitean A. From Eq. (6.116), it follows that the Poisson tensor Λ defined in (6.122) satisfies:

$$\Lambda(\hat{A}, \hat{B})\,(\xi) = \langle \xi, [A, B]_-\rangle_{u^*} = \langle [\xi, A]_-, B\rangle_{u^*} \qquad (6.138)$$

This defines an invertible map $\tilde{\Lambda}$ that associates to any 1-form $\hat{A}$ the tangent vector at ξ: $\tilde{\Lambda}(\hat{A}) = \Lambda(\hat{A}, \cdot) = [\xi, A]_-$. We will denote with $\tilde{\eta}_\xi$ its inverse: $\tilde{\eta}_\xi([\xi, A]_-) = \hat{A}$. This allows us to define, on each $U(\mathcal{H})$ orbit in $\mathfrak{u}(\mathcal{H})^*$, a canonical 2-form η which is given by:

[14] Note that here the vectors are not necessarily normalized.

$$\eta_\xi([A,\xi]_-,[B,\xi]_-) =: (\tilde{\eta}_\xi([\xi,A]_-),[B,\xi]_-) = (\hat{A},[B,\xi]_-) \tag{6.139}$$

for all $[A,\xi]_-, [B,\xi]_- \in T_\xi \mathfrak{u}(\mathcal{H})^*$.

It is also easy to check that η satisfies the equalities:

$$\begin{aligned}\eta_\xi([A,\xi]_-,[B,\xi]_-) &= -(\hat{A},[B,\xi]_-) = -\langle A,[B,\xi]_-\rangle_{u^*} = -\langle \xi,[A,B]_-\rangle_{u^*} \\ &= \langle [A,\xi]_-, B\rangle_{u^*},\end{aligned}$$

for any $A, B \in \mathfrak{u}(\mathcal{H})^*$.

Exercise 6.6 Compute $\ker \eta_\xi$.

We can summarize these results in the following:

Theorem 6.27 *The restriction of the 2-form (6.139) to the $U(\mathcal{H})$-orbit $\mathcal{D}_1^1(\mathcal{H})$ defines a canonical symplectic form η characterized by the property*

$$\eta_\xi([A,\xi]_-,[B,\xi]_-) = \langle [A,\xi]_-, B\rangle_{u^*} = -\langle \xi,[A,B]_-\rangle_{u^*} \tag{6.140}$$

In a very similar way, starting from the symmetric Jordan tensor R given in (6.121), one can construct a (1, 1) tensor $\tilde{R}(\hat{A}) = R(\hat{A},\cdot) = [\xi, A]_+$ and its inverse: $\tilde{\sigma}([\xi,A]_+) = \hat{A}$. Thus we obtain a covariant tensor σ such that:

$$\sigma_\xi([A,\xi]_+,[B,\xi]_+) = \langle [A,\xi]_+, B\rangle_{u^*} = \langle \xi,[A,B]_+\rangle_{u^*}. \tag{6.141}$$

Notice that, at this stage, σ_ξ is only a partial tensor, being defined on vectors of the form $[A,\xi]_+$, which belong to the image of the map $\tilde{R}$. However, on $T_\xi\mathcal{D}_1^1(\mathcal{H})$, we have $[A,\xi]_- = [A,\xi^2]_- = [[A,\xi],\xi]_+$, so that, after some algebra, one can also prove that:

$$\begin{aligned}\sigma_\xi([A,\xi]_-,[B,\xi]_-) &= \sigma_\xi([[A,\xi]_-,\xi]_+,[[B,\xi]_-,\xi]_+) = \langle \xi,[[A,\xi]_-,[B,\xi]_-]_+\rangle_{u^*} \\ &= \frac{1}{2}\mathrm{Tr}(\xi[[A,\xi]_-,[B,\xi]_-]_+) = \frac{1}{2}\mathrm{Tr}(\xi[A,\xi]_-[B,\xi]_-) \\ &= \langle [A,\xi]_-,[B,\xi]_-\rangle_{u^*}.\end{aligned}$$

Therefore we have also the following:

Corollary 6.28 *On the $U(\mathcal{H})$-orbit $\mathcal{D}_1^1(\mathcal{H})$ we can define a symmetric covariant tensor σ such that:*

$$\sigma_\xi([A,\xi]_-,[B,\xi]_-) = \langle [A,\xi]_-,[B,\xi]_-\rangle_{u^*}. \tag{6.142}$$

Moreover, turning back to the the (1, 1) tensor $\tilde{\Lambda}$ given above, one has the following result [Gr05]:

Theorem 6.29 *When restricted to $\mathcal{D}_1^1(\mathcal{H})$, the $(1,1)$ tensor $\tilde{\Lambda}$, which satisfies:*

$$\tilde{\Lambda}^3 = -\tilde{\Lambda} \tag{6.143}$$

will become invertible. Hence: $\tilde{\Lambda}^2 = -\mathbb{I}$ and therefore it will define a complex structure I such that

$$\eta_\xi([A,\xi]_-, I_\xi([B,\xi]_-)) = \sigma_\xi([A,\xi]_-, [B,\xi]_-)\,, \tag{6.144}$$

and

$$\eta_\xi(I_\xi([A,\xi]_-), I_\xi([B,\xi]_-)) = \eta_\xi([A,\xi]_-, [B,\xi]_-)\,.$$

Proof Eq. (6.143) follows from a direct calculation by taking into account that $\xi^2 = \xi$. The last two expressions follow by combining Eqs. (6.140) and (6.142).

To prove that I is a complex structure one has first to show that it defines an almost complex structure (which follows easily from the fact that $[[[A,\xi]_-,\xi]_-,\xi]_- = -[A,\xi]_-$) and then that its Nijenhuis torsion vanishes. Detailed calculations of this can be found in [Gr05]. □

Putting everything together, we can now conclude that, as expected:

Theorem 6.30 *$(\mathcal{D}_1^1(\mathcal{H}), I, \sigma, \eta)$ is a Kähler manifold.*

At last, we notice that there is an identification of the orthogonal complement of any unit vector $x \in \mathcal{H}$ with the tangent space of the $U(\mathcal{H})$-orbit in $\mathfrak{u}(\mathcal{H})^*$ at $\xi = |x\rangle\langle x|$. Indeed, for any y perpendicular to x ($\|x\|^2 = 1$) the operators:

$$P_y^x -: (\mu_*)_x(y) = |y\rangle\langle x| + |x\rangle\langle y| \tag{6.145}$$

can be written as $P_y^x = [A_y, \xi]$, where A_y is a Hermitean operator such that $A_y x = iy$, $A_y y = -i\|y\|^2 x$ and $A_y z = 0$ for any z perpendicular to both x and y, as it can be directly checked by applying both expressions to a generic vector in $\mathcal{H}$ which can be written as $ax + by + cz$ with $a, b, c \in \mathbb{C}$. Then, from Eqs. (6.140) and (6.142), it follows immediately that, for any y, y' orthogonal to x

$$\eta_\xi(P_y^x, P_{y'}^x) = -\frac{1}{2}\mathrm{Tr}\,(\xi[A_y, A_{y'}]_-) = -\frac{1}{2i}(\langle y, y'\rangle - \langle y', y\rangle) = -\omega(y, y') \tag{6.146}$$

$$\sigma_\xi(P_y^x, P_{y'}^x) = \frac{1}{2}\mathrm{Tr}\,(\xi[A_y, A_{y'}]_-) = -\frac{1}{2}(\langle y, y'\rangle + \langle y', y\rangle) = g(y, y') \tag{6.147}$$

In conclusion, we have the following:

Theorem 6.31 *For any $y, y' \in \mathcal{H}$, the vectors $(\mu_*)_x(y)$, $(\mu_*)_x(y)$ are tangent to the $U(\mathcal{H})$-orbit in $\mathfrak{u}(\mathcal{H})^*$ at $\xi = \mu(x)$ and:*

$$\sigma_\xi((\mu_*)_x(y), (\mu_*)_x(y)) = g(y, y') \tag{6.148}$$

$$\eta_\xi((\mu_*)_x(y), (\mu_*)_x(y)) = -\omega(y, y') \tag{6.149}$$

$$I_\xi(\mu_*)_x(y)) = (\mu_*)_x(Jy) \tag{6.150}$$

where the last formula follows from Eq. (6.144).

More generally, with minor changes, we can reconstruct similar structures for any $\mathcal{D}^1_r(\mathcal{H})$, obtaining Kähler manifolds $(\mathcal{D}^1_r(\mathcal{H}), I^r, \sigma^r, \eta^r)$. An analog of the above theorem shows then that the latter can be obtained from a sort of 'Kähler reduction' starting from the original linear Kähler manifold $(\mathcal{H}_\mathbb{R}, J, g, \omega)$.

Example 6.7 Let us go back to the previous example of rank-one projectors on $\mathcal{H} = \mathbb{C}^2$. According to (6.131), the latter are described by 3-dimensional vectors $\xi = (y^1, y^2, y^3)$ such that $\xi^2 = 1/4$ ($y_0 = 1/2$ always), which form a 2-dimensional sphere of radius $1/2$. A generic tangent vector X_A and a generic one form $\hat{A}$ at ξ are of the form $X_A = y^0_A\partial_0 + y^1_A\partial_1 + y^2_A\partial_2 + y^3_A\partial_3$ and $\hat{A} = y^0_A dy^0 + y^1_A dy^1 + y^2_A dy^2 + y^3_A dy^3$ with $y^0_A = 0$ and $\mathbf{y}_A \cdot \xi = 0$.

It is clear from (6.136) that the map $\tilde{\Lambda}$ that associates to any 1-form $\hat{A}$ the tangent vector at ξ: $\tilde{\Lambda}(\hat{A}) = \Lambda(\hat{A}, \cdot) = [A, \xi]_-$ is manifestly invariant and given by: $\tilde{\Lambda}(\hat{A}) = 2(\xi \times \mathbf{y}_A) \cdot \boldsymbol{\partial}$, where we have set $\boldsymbol{\partial} = (\partial_1, \partial_2, \partial_3)$. It follows that the 2-form η_ξ is such that:

$$\eta_\xi([A, \xi]_-, [B, \xi]_-) = 2\xi \cdot (\mathbf{y}_A \times \mathbf{y}_B) \tag{6.151}$$

so that

$$\eta_\xi = 2\epsilon^{ijk} y^i dy^j \wedge dy^k \tag{6.152}$$

which is proportional by a factor $\left(y_1^2 + y_2^2 + y_3^2\right)^{-\frac{3}{2}}$ to the symplectic 2-form on a 2-dimensional sphere,[15] when pulled back to the sphere.

In a similar way, from (6.134), one can prove that $\tilde{R}(\hat{A}) = R(\hat{A}, \cdot) = [\xi, A]_+ = 2(y^0_A y^0 + \mathbf{y}_A \cdot \xi)\partial_0 + 2(y^0_A \xi + y_0 \mathbf{y}_A) \cdot \boldsymbol{\partial}$. Thus, because of (6.142), we have:

$$\sigma_\xi([A, \xi]_-, [B, \xi]_-) = 4(\xi \times \mathbf{y}_A) \cdot (\xi \times \mathbf{y}_B) = \mathbf{y}_A \cdot \mathbf{y}_B \tag{6.153}$$

where the last equality follows from the fact that $\xi^2 = 1/4$ and ξ is orthogonal to both $\mathbf{y}_A$ and $\mathbf{y}_B$.

Finally, starting for example from Eq. (6.144), it is not difficult to check that

$$I_\xi([B, \xi]_-) = y'_B \cdot \boldsymbol{\partial} \;\; with \;\; \mathbf{y}'_B = \xi \times \mathbf{y}_B \,. \tag{6.154}$$

A direct calculation shows that $I^3_\xi = -I_\xi$.

---

15 that is also the volume element of a 2-dimensional sphere of radius $r = 1/2$, as it should be.

6.6.2 The GNS Construction

The algebraic approach to quantum physics started, among others, with the work of Haag and Kastler [Ha64] which is also at the basis of the mathematical approach to quantum field theory [Ha92], relies on the so-called GNS construction which goes back to the work of Gelfand and Naimark on one side and Segal on the other.

The starting point of this construction is an abstract C^*-algebra $\mathcal{A}$ [Br87, Em84] with unity, the latter being denoted as $\mathbb{I}$. The elements $a \in \mathcal{A}$ such that $a = a^*$ constitute the set $\mathcal{A}_{re}$ (a vector space over the reals) of the real elements[16] of the algebra. In particular: $\mathbb{I} \in \mathcal{A}_{re}$. The obvious decomposition: $a = a_1 + i a_2$, with

$$a_1 = \frac{1}{2}(a + a^*); \quad a_2 = \frac{1}{2i}(a - a^*), \tag{6.155}$$

means that, as a vector space, $\mathcal{A}$ is the direct sum of $\mathcal{A}_{re}$ and of the set $\mathcal{A}_{im}$ (also a vector space over the reals) of the imaginary elements, i.e., of the elements of the form ia, $a \in \mathcal{A}_{re}$. The subspace of real elements $\mathcal{A}_{re}$ can be given the structure of a Lie-Jordan algebra [Ch09, Em84, Fa12], where, using here the conventions of Sect. 6.5.4, the Lie product is defined as

$$[a, b] = \frac{1}{2i}(ab - ba), \tag{6.156}$$

while the Jordan product is given by

$$a \circ b = \frac{1}{2}(ab + ba), \tag{6.157}$$

for all $a, b \in \mathcal{A}_{re}$. The product in the algebra is then recovered as

$$ab = a \circ b + i[a, b]. \tag{6.158}$$

Remark 6.10 The main example of a C^*-algebra is the algebra $\mathcal{B}(\mathcal{H})$ of all bounded operators on a Hilbert space $\mathcal{H}$. In this case [Em84]: $\mathcal{A}_{re}$ is the set of all bounded self-adjoint operators on $\mathcal{H}$.

Definition 6.32 A state on a C^*-algebra $\mathcal{A}$ with unit is a linear functional $\omega\colon \mathcal{A} \to \mathbb{C}$ satisfying:

1. Real: $\omega(a^*) = \overline{\omega(a)}$, $\forall a \in \mathcal{A}$,
2. Positive: $\omega(a^* a) \geq 0$, $\forall a \in \mathcal{A}$ and
3. Normalized: $\omega(\mathbb{I}) = 1$.

The space of all states of the C^*-algebra $\mathcal{A}$ will be denoted by $\mathcal{D}(\mathcal{A})$.

---

[16] Also-called the observables.

The space of states $\mathcal{D}(\mathcal{A})$ is convex, that is $\mu_1\omega_1 + \mu_2\omega_2$ is a state for any $\omega_a \in \mathcal{D}(\mathcal{A})$, $\mu_a \geq 0$, $a = 1, 2$, and $\mu_1 + \mu_2 = 1$.

Each state ω defines a non-negative sesquilinear $\langle \cdot \mid \cdot \rangle_\omega$ between pairs of elements $a, b \in \mathcal{A}$ via:

$$\langle a|b\rangle_\omega := \omega(a^* b)\,. \tag{6.159}$$

Reality and positivity of the state guarantee that the pairing (6.159) satisfies the Schwartz inequality, i.e.,

$$|\langle a|b\rangle_\omega| \leq \sqrt{\langle a|a\rangle_\omega}\sqrt{\langle b|b\rangle_\omega} \tag{6.160}$$

Exercise 6.8 Prove Schwartz' inequality Eq. (6.160).

The form $\langle \cdot, \cdot \rangle_\omega$ however might be degenerate. We are thus led to consider the Gelfand ideal [Em84, Ha92] $\mathcal{J}_\omega$ consisting of all elements $j \in \mathcal{A}$ such that $\omega(j^* j) = 0$ and to define the set $\mathcal{A}/\mathcal{J}_\omega$ of equivalence classes:

$$\Psi_a = [a + \mathcal{J}_\omega] \tag{6.161}$$

Exercise 6.9 Prove that $\mathcal{J}_\omega$ is a bilateral ideal of the C^*-algebra $\mathcal{A}$.

Now it is immediate to see that $\mathcal{A}/\mathcal{J}_\omega$ is a pre-Hilbert space with respect to the scalar product:

$$\langle \Psi_a, \Psi_b \rangle = \omega(a^* b)\,, \tag{6.162}$$

because the Schwartz inequality (6.160) implies: $\langle j|a\rangle_\omega = \langle a|j\rangle_\omega = 0$ for all $a \in \mathcal{A}$, $j \in \mathcal{J}_\omega$, and hence the scalar product Eq. (6.162) does indeed depend only on the equivalence classes Ψ_a, Ψ_b, of a and b, and not on the specific representatives chosen.

Completing this space with respect to the topology defined by the scalar product, one obtains a Hilbert space $\mathcal{H}_\omega$ on which the original C^*-algebra $\mathcal{A}$ acts via the natural representation:

$$\pi_\omega(a)\Psi_b = \Psi_{ab}\,, \qquad \forall a, b \in \mathcal{A}\,. \tag{6.163}$$

Notice that if such a representation is faithful, i.e., the map $\pi_\omega \colon a \mapsto \pi_\omega(a)$ is an isomorphism, the operator norm of $\pi_\omega(a)$ equals the $\mathbb{C}^*$-norm of a [Br87].

Clearly the equivalence class of the unit element in $\mathcal{A}$, denoted by $\Omega = \Psi_{\mathbb{I}}$, satisfies: $\|\Psi_{\mathbb{I}}\| := \sqrt{\langle \Psi_{\mathbb{I}}|\Psi_{\mathbb{I}}\rangle} = 1$ and provides a cyclic vector for the representation π_ω. We recall [Ha92] that a vector $\Omega \in \mathcal{H}_\omega$ is called *cyclic* if $\pi_\omega(\mathcal{A})\,\Omega$ is dense in $\mathcal{H}_\omega$. Moreover:

$$\langle \Omega|\pi_\omega(a)|\Omega\rangle = \omega(a)\,. \tag{6.164}$$

This tells us that, if we consider that $\mathcal{A}$ acts by duality on $\mathcal{D}(\mathcal{A})$, the Hilbert space $\mathcal{H}_\omega$ corresponding to a given state ω is the orbit of $\mathcal{A}$ through ω itself. Notice that any other element $b \in \mathcal{A}$ such that the vector $\Psi = \pi_\omega(b)\Omega$ is of unit norm, defines a new state ω_Ψ by:

$$\omega_\Psi(a) = \langle\Psi|\pi_\omega(a)|\Psi\rangle = \omega(b^*ab)\,. \tag{6.165}$$

These states are called vector states of the representation π_ω, and are particular examples of more general states of the form:

$$\omega_\rho(a) = \mathrm{Tr}\;(\rho\,\pi_\omega(a))\;, \tag{6.166}$$

where $\rho \in \mathcal{B}(\mathcal{H}_\omega)$ is a density operator [Em84, Ha92]. States of the form (6.166) are called a "folium" of the representation π_ω. Also, one says that a state ω is pure if it cannot be written as a convex combination of other states in $\mathcal{D}(\mathcal{A})$, so that the set of pure states defines a set of extremal points in $\mathcal{D}(\mathcal{A})$.

The universality and uniqueness of the GNS construction is guaranteed [Br87] by the following:

Theorem 6.33

1. *If π_ω is a cyclic representation of A on a Hilbert space $\mathcal{H}$, any vector representation ω_Ψ for a normalized Ψ, see Eq. (6.166), is equivalent to π_ω.*
2. *A GNS represenation π_ω of $\mathcal{A}$ is irreducible iff ω is a pure state.*

Example 6.10 The GNS construction can be very simple for finite-dimensional C^*-algebras. Consider, e.g., the algebra $\mathcal{A} = \mathrm{End}(\mathbb{C}^n)$ of linear operators on $\mathbb{C}^n$, i.e., of the $n \times n$ matrices with complex entries. Any non-negative operator $\omega \in \mathrm{End}(\mathbb{C}^n)$ defines a state by:

$$\omega(A) = \mathrm{Tr}\,(\omega A)\;, \qquad \forall A \in \mathcal{A}\,, \tag{6.167}$$

while we can define the scalar product in $\mathcal{H}_\omega$ as:

$$\langle A|B\rangle = \omega(A^*B) = \mathrm{Tr}\,(B\omega A^*)\,. \tag{6.168}$$

If we write ω as $\omega = RR^*$, we find:

$$\langle B \mid B\rangle_\omega = \mathrm{Tr}\,(BR)(BR)^*\;,$$

showing that the Gelfand ideal consists of the elements that annihilate R.

If ω is a rank-one projector and $\{e_k\}$ is an orthonormal basis for which $\omega = |e_1\rangle\langle e_1|$, writing A_{km} for the matrix elements of A in such a basis, the scalar product assumes the form:

$$\langle A|B\rangle = \sum_{k=1}^{n} \bar{A}_{k1} B_{k1} \tag{6.169}$$

while the Gelfand ideal $\mathcal{J}_\omega$ is given by:

$$\mathcal{J}_\omega = \{X \in \mathcal{A} \mid X_{k1} = 0\,,\ k = 1, \cdots, n\} \tag{6.170}$$

Thus $\mathcal{H}_\omega = \mathcal{A}/\mathcal{J}_\omega$ is nothing but $\mathbb{C}^n$ itself and π_ω is the defining representation.

If ω is a rank-m density operator: $\omega = p_1|e_1\rangle\langle e_1| + \cdots + p_m|e_m\rangle\langle e_m|$ *with* $p_1, \cdots, p_m > 0$ and $p_1 + \cdots p_m = 1$, the scalar product is given by:

$$\langle A|B\rangle = \sum_{k=1}^{n}\sum_{j=1}^{m} p_m \bar{A}_{kj} B_{kj} \tag{6.171}$$

and the Gelfand ideal is given by:

$$\mathcal{J}_\omega = \{X \in \mathcal{A} \mid X_{kj} = 0\,,\ k = 1, \cdots, n;\ j = 1, \cdots, m\} \tag{6.172}$$

showing that $\mathcal{H}_\omega$ is the direct sum of m copies of $\mathbb{C}^n$. Now the representation π_ω is no longer irreducible, decomposing into the direct sum of m copies of the defining representation:

$$\pi_\omega(A) = \mathbb{I}_m \otimes A \tag{6.173}$$

where $\mathbb{I}_m$ is the $m \times m$ identity matrix.

Let us go back now to the GNS construction and consider first a pure state ω over $\mathcal{A}$, which gives raise to the irreducible representation π_ω in the Hilbert space $\mathcal{H}_\omega$. We have already seen that self-adjoint operators, that correspond to the real elements of $\mathcal{A}$, may be identified with the dual $\mathfrak{u}^*(\mathcal{H}_\omega)$ of the Lie algebra $\mathfrak{u}(\mathcal{H}_\omega)$ of the unitary group $U(\mathcal{H}_\omega)$ and how the momentum map

$$\mu_\omega : \mathcal{H}_\omega \to \mathfrak{u}^*(\mathcal{H}_\omega)\,, \qquad \mu_\omega(\psi) = |\psi\rangle\langle\psi| \tag{6.174}$$

relates the Poisson tensor on $\mathfrak{u}^*(\mathcal{H}_\omega)$ with that on $\mathcal{H}_\omega$, via the pull-back.

We have also seen that the unit sphere in $\mathcal{H}_\omega \backslash \{\mathbf{0}\}$ can be projected onto $\mathfrak{u}^*(\mathcal{H}_\omega)$ in an equivariant way, in such a way that the Poisson and the Riemann tensor in $\mathcal{P}(\mathcal{H}_\omega)$ are both related to the same tensors defined on $\mathfrak{u}^*(\mathcal{H}_\omega)$ by using the Lie and the Jordan product that are defined on it. Thus the momentum map provides a symplectic realization of $\mathfrak{u}^*(\mathcal{H}_\omega)$, which we call a Kählerian (or Hermitean) realization on the complex projective space. If G is a group represented unitarily on $U(\mathcal{H}_\omega)$, then there is a natural projection $\nu\colon \mathfrak{u}^*(\mathcal{H}_\omega) \to \mathfrak{g}^*$ and the composition $\nu \circ \mu\colon \mathcal{P}(\mathcal{H}_\omega) \to \mathfrak{g}^*$ is called a classical Jordan-Schwinger map.

6.7 Alternative Hermitean Structures for Quantum Systems

6.7.1 Equations of Motion on Density States and Hermitean Operators

As we showed before, the equations of motion on density states are given by:

$$i\hbar\frac{d}{dt}\rho = [H, \rho],$$

out of

$$i\hbar\frac{d}{dt}|\Psi\rangle = H|\Psi\rangle, \qquad \text{and} \quad -i\hbar\frac{d}{dt}\langle\Psi| = H\langle\Psi|,$$

i.e., assuming that H is Hermitean. When we consider, for some Hermitean operator A, the expectation values functional:

$$e_A(\Psi) = \frac{\langle\Psi|A|\Psi\rangle}{\langle\Psi|\Psi\rangle} = \mathrm{Tr}(A\rho_\Psi) = \mathrm{Tr}\left(A\frac{|\Psi\rangle\langle\Psi|}{\langle\Psi|\Psi\rangle}\right)$$

and consider

$$0 = \frac{d}{dt}\frac{\langle\Psi|A|\Psi\rangle}{\langle\Psi|\Psi\rangle} = \frac{d}{dt}\mathrm{Tr}(A\rho_\Psi) = \mathrm{Tr}(\frac{d}{dt}A\rho_\Psi) + \mathrm{Tr}(A\frac{d}{dt}\rho_\Psi)$$

We will consider for simplicity finite-dimensional carrier spaces V, i.e., 'finite level systems', an assumption which is not far from experimental constraints, where only systems on finite volumes and finite energy are considered, the infinite-dimensional systems being idealizations of 'actual' systems.

6.7.2 The Inverse Problem in Various Formalisms

6.7.2.1 The Schrödinger Picture

We will review briefly some of the relevant aspects of the geometry of the Schrödinger picture [Du90, St66]. Given a Schrödinger-type equation (recall the examples in Sect. 6.6.2):

$$i\,\frac{d\psi}{dt} = H\,\psi\,, \qquad \psi(x) \in \mathcal{H} = \mathbb{C}^n\,, \tag{6.175}$$

with H being a linear operator on the complex vector space $\mathcal{H}$, the inverse problem consists of finding invariant scalar products under the associated time evolution, if they exist.

Given the vector field $-i\,H$, previous results show that the following statements are equivalent:

1. There exists an (invariant) scalar product such that $H = H^\dagger$.
2. H is diagonalizable and has a real spectrum.
3. All the orbits $e^{iHt}\Psi$, for any initial condition ψ are bounded sets.

Remark 6.11 It is clear that if H fulfills any of the previous conditions, the same will be true for the operator $\widetilde{H} = T^{-1}\,H\,T$, where T is any invertible operator; the operator $\widetilde{H}$ will have the same spectrum of H and be self-adjoint with respect to the new scalar product $h_T(\phi, \psi) = h(\phi, T^*T\psi)$. In particular, if T commutes with H we get alternative 'quantum descriptions' for the same equations of motion. Indeed, we have:

If H is diagonalizable and has a real spectrum, the same properties are satisfied for all operators in the similarity class of H, i.e., $T^{-1}\,H\,T$) for $T \in GL(n, \mathbb{C})$.

The family of alternative Hamiltonian structures for H is parametrized by the commutant of H in $GL(n, \mathbb{C})$ quotiented by the commutant of H in $U(n, \mathbb{C})$.

6.7.2.2 The Ehrenfest Picture

The Ehrenfest picture of Quantum Mechanics describes the dynamical evolution of expectation values of linear operators A on V, that is in this picture the dynamics is defined as a derivation in the Lie algebra $^{\mathcal{Q}}$ of real quadratic functions on ^{H}, i.e., functions of the form $f_B(\psi) = \frac{1}{2}\langle\psi|B|\psi\rangle$ with B a symmetric operator. The Lie algebra structure is defined by the Poisson bracket $\{\cdot, \cdot\}$ associated to the symplectic structure ω. It is easily noticed that $^{\mathcal{Q}}$ is a Lie subalgebra with respect to $\{\cdot, \cdot\}$. It is easy to show that $\{f_A, f_B\} = f_{[A,B]}$.

Then the dynamical behaviour of the system with Hamiltonian H is defined by the inner derivation $D_H = \{f_H, \cdot\}$, i.e.,

$$i\frac{d}{dt}f_B = \{f_H, f_B\}. \tag{6.176}$$

Thus, the inverse problem in the Ehrenfest picture can be put up as the search for a Poisson bracket on quadratic functions such that there exists f_H such that,

$$\mathcal{L}_\Gamma f_A = \{f_H, f_A\}.$$

The combined use of the Poisson bracket structure on the space of quadratic real functionals together with the complex structure in ^{H} to describe the dynamics of quantum systems were already discussed in [St66].

The geometrical setting for the previous discussion is the category of Kähler manifolds, and the dynamical sytems we are describing preserve Kähler structures.

Consider now the Hermitean tensor h in contravariant form,

$$h = h^{kj} \frac{\partial}{\partial \xi^k} \otimes \frac{\partial}{\partial \xi^j} = \frac{1}{2} \left(s^{kj} \frac{\partial}{\partial \xi^k} \overset{s}{\otimes} \frac{\partial}{\partial \xi^j} \right) + \frac{i}{2} \left(\Lambda^{kj} \frac{\partial}{\partial \xi^k} \wedge \frac{\partial}{\partial \xi^j} \right) . \quad (6.177)$$

As indicated previously, the fundamental Poisson brackets defined by Λ, define a Lie bracket on quadratic functions.

Defining

$$(A_f)_{jk} = \frac{\partial^2 f}{\partial \xi^j \partial \xi^k},$$

we get,

$$[A_f, A_g]_{jk} = \frac{\partial^2 \{ f, g \}}{\partial \xi^j \partial \xi^k},$$

and

$$[A_f, A_g] = A_f \cdot \Lambda \cdot A_g - A_g \cdot \Lambda \cdot A_f .$$

From here we get immediately Heisenberg's picture. In fact, if Γ leaves invariant h, i.e., $\mathcal{L}_\Gamma h = 0$, then necessarily,

$$\frac{d}{dt} f_B = \{ f_H, f_B \}, \quad (6.178)$$

as in Eq. (6.176). Then acting on both sides of the equation above, Eq. (6.178), we get

$$\frac{d}{dt} B = [H, B],$$

because obviously, $A_{f_B} = B$.

6.7.2.3 The Heisenberg Picture

A different approach to Quantum Mechanics is given by what is known as the Heisenberg picture. Here dynamics is encoded in the algebra of observables, considered as the real part of an abstract $\mathbb{C}^*$-algebra.

First, we have to consider observables as associated with Hermitean operators (finite dimensional matrices if the system is finite-dimensional). These matrices do not define an associative algebra because the product of two Hermitean matrices is not Hermitean. However we may complexify this space by writing a generic matrix as the sum of a real part A and an imaginary part iB, A and B being Hermitean. In this way we find that:

Proposition 6.34 *The complexification of the algebra of observables allows us to write an associative product of operators $A = A_1 + iA_2$, where A_1 and A_2 are real Hermitean. We shall denote by $\mathcal{A}$ the corresponding associative algebra.*

Finally we can proceed to define the equations of motion on this complexified algebra of observables. It is introduced by means of the Heisenberg equation:

$$i\hbar\frac{d}{dt}A = [A, H], \quad A \in \mathcal{A}, \tag{6.179}$$

where H is called the Hamiltonian of the system we are describing. To take into account an explicit time-dependence of the observable we may also write the equation of motion in the form

$$\frac{d}{dt}A = -\frac{i}{h}[A, H] + \frac{\partial A}{\partial t}, \qquad A \in \mathcal{A}. \tag{6.180}$$

From a formal point of view, this expression is similar to the expression of Hamilton equation written on the Poisson algebra of classical observables (i.e., on the algebra of functions representing the classical quantities with the structure provided by the Poisson bracket we assume our classical manifold is endowed with). This similarity is not casual and turns out to be very useful in the study of the quantum-classical transition. We shall come back to this point later on.

Remark 6.12 The equations of motion written in this form are necessarily derivations of the associative product and can therefore be considered as 'intrinsically Hamiltonian'. In the Schrödinger picture, however, if the vector field is not anti-Hermitean, the equation still makes sense, but the dynamics need not be Kählerian. To recover a similar treatement, one has to give up the requirement that the evolution preserves the product structure on the space of observables.

Let us recall that the inverse problem as it was posed in the introduction intended to find a Lie product $[\cdot, \cdot]$ in the space of observables of a quantum system an a function H such that the equations of motion will have the Hamiltonian form $\dot{B} = [B, H]$. We should mention first that the search for Lie products $[\cdot, \cdot]$ ends up very fast if we consider the associative product on the space of operators on a Hilbert space defined by their composition because of the following theorem by Dirac [Fo76]:

Theorem 6.35 *Any Lie algebra bracket on the space of operators of a complex Hilbert space satisfying the derivation property,*

$$[A, B \cdot C] = [A, B] \cdot C + B \cdot [A, C],$$

is necessarily of the form $[A, B] = \mu(A \cdot B - B \cdot A)$, with μ a complex number.

Thus, if we use the associative product on the space of operators defined by the composition of linear operators acting on a Hilbert space, the Lie product must be the ordinary commutator up to a constant factor, leaving open only the problem of finding the Hamiltonian function H. However, it is not necessary to use the defining representation of the operators to equip them with an algebra structure. In fact, we can proceed in an alternative way, thinking of the space of observables as a primary notion and representing them afterwards. This point of view is very close to Schwinger's approach in [Sc70]. Thus we are led to consider as a fundamental entity the measurement algebra of a physical system, a linear space generated by abstract operators $X(\alpha)$ representing selective measurements defining the states of the system. An associative algebra structure $X(\alpha)X(\beta) = \sum_\gamma (\alpha\beta\gamma)X(\gamma)$ is introduced on them, that will depend in principle on the physical realization we choose for the selective measurements $X(\alpha)$. Then we can explore alternative associative products defined by the structure constants $(\alpha\beta\gamma)$ and such that the dynamics Γ will be Hamiltonian with respect to the Lie product defined by it.

A natural way of obtaining associative products in the space of operators is suggested by the following argument. If f is a quadratic function on V, we can define

$$(A_f)_{kl} = \frac{\partial^2 f}{\partial \xi^k \partial \xi^l} ,$$

then we get a Lie product by the formula,

$$[A_f, A_g]_{kl} = \frac{\partial^2}{\partial \xi^k \partial \xi^l} \{ f, g \} .$$

If the fundamental commutation relations on our quantum space V are given by $\{ \xi^j, \xi^k \} = \Lambda^{jk}$, we obtain, $[A, B]_{kl} = A_{ki} \Lambda^{ij} B_{jl} - B_{ki} \Lambda^{ij} A_{jl}$. Then, inspired by this product we define a new Lie product on the space of linear operators on V introducing the associative product,

$$A \circ_K B = A e^{\lambda K} B,$$

with λ a real number. Hence a new Lie product is defined as

$$[A, B]_K = A \circ_K B - B \circ_K A.$$

We find also, $[A, B \circ_K C]_K = [A, B]_K \circ_K C + B \circ_K [A, C]_K$. Now if we define,

$$\phi_K(A) = e^{\frac{1}{2}\lambda K} A e^{\frac{1}{2}\lambda K} = A_K,$$

we get

$$[\phi_K(A), \phi_K(B)] = \phi_K([A, B]_K). \tag{6.181}$$

More generally, we can define

$$F_{K_1,K_2}(A) = e^{\lambda K_1} A e^{\lambda K_2},$$

gives

$$[F_{K_1,K_2}(A), F_{K_1,K_2}(B)] = F_{K_1,K_2}([A, B]_K)$$

with $\exp \lambda K = \exp \lambda K_1 \exp \lambda K_2$.

Proposition 6.36 *Any linear vector field that is a derivation of the associative product $\circ_K$ for $\lambda = 0$, is also a derivation for $\lambda \neq 0$ if K is a constant of motion for the linear vector field.*

Then, also in the Heisenberg picture we have many alternative quantum descriptions. In fact, the Lie products $[\cdot, \cdot]_K$ define alternative structures for our dynamics. However this creates a problem because the operators are composed with respect to the new associative product $\circ_K$, which is not the natural product induced by the tautological representation of the operators on V. But we have already observed, Eq. (6.181), that the map ϕ_K intertwines the defining representation of the algebra of operators with a representation defining the Lie product $[\cdot, \cdot]_K$ (even though the map is not induced by a linear isomorphism of the underlying Hilbert space). This suggests that we should define a new Hilbert space structure by,

$$\langle \psi_1 | \psi_2 \rangle_K = \langle \psi_1 | e^{\frac{1}{2}\lambda K} e^{\frac{1}{2}\lambda K} | \psi_2 \rangle.$$

Notice that the unity for the deformed associative product is given by $I_K = e^{-\lambda K}$. Thus, $\langle \psi_1 | I_K | \psi_2 \rangle_K = \langle \psi_1 | \psi_2 \rangle$.

6.7.2.4 The Dirac Picture

What we call here the Dirac picture has its origin in what in Quantum Mechanics goes under the name of Dirac's interaction picture. It starts from the consideration that the propagator $U(t, t_0)$ maps by construction the initial wave function or state, $\psi(t_0)$, into $\psi(t)$:

$$\psi(t) = U(t, t_0)\psi(t_0),$$

and hence satisfies the equation

$$i\hbar \frac{d}{dt} U(t, t_0) = H(t)\, U(t, t_0),$$

with the initial condition $U(t_0, t_0) = 1$.

In Quantum Mechanics one starts from here to write $U = U_0\,W$, with W another unitary operator which encodes all the effects resulting from the interaction and it allows us to write U in terms of the comparison 'free evolution' described by U_0.

One proceeds to write an equation for W, which is usually solved by perturbation procedures. For us it is simply a way to complete our description of linear systems:

1. We have then on the vector space, Schrödinger picture,

$$i\hbar\frac{d}{dt}\psi = H\,\psi,$$

2. On the algebra of observables or operators, Heisenberg picture,

$$i\hbar\frac{d}{dt}\,B = [H, B],$$

3. On quadratic functions, Ehrenfest picture,

$$i\hbar\frac{d}{dt}\,f_B = \{f_H, f_B\}$$

4. On the group of unitary transformations, Dirac picture,

$$i\hbar\frac{d}{dt}\,U = H\,U\ .$$

All these pictures are equivalent and in finite dimensions it is only a matter of convenience to use one or the other, even though some aspects might be more transparent in a picture with respect to another.

In infinite dimensions, that we rarely consider in this book, their potentiality are radically different. When domains problems are relevant, Heisenberg picture and Ehrenfest picture may be quite different when one deals with unbounded operators.

The Dirac picture, which is the equation for the propagator, when evaluated in the position representation becomes an equation for the Green function and becomes highly relevant when the Schrödinger picture is dealing with operators H with continuous spectrum and 'eigenfunctions' which would be square integrable only locally.

For the aim of this book this picture becomes very much relevant when we consider Lie–Scheffers systems. Moreover, the equation on the group might be the 'abstract version' of a nonlinear equation as it happens for Riccati-type equations. In other terms, the equation on the group would be the same even if the group would act nonlinearly on a manifold M and therefore give raise to a nonlinear equation. This would be usually the case when the Hamiltonian matrix H is invariant under some closed subgroup K and we would consider the associated equation on the homogeneous space G/K.

From our point of view the most relevant aspect of these equations is that they only rely on the properties of the group G, algebraical or geometrical, and do not depend on the specific action or representation. For instance, on a given vector space, say three-dimensional, we would use different metric tensors, say

$$a_1^2\, dx_1 \otimes dx_1 + a_2^2\, dx_2 \otimes dx_2 + a_3^2\, dx_3 \otimes dx_3,$$

with $a_1\, a_2\, a_3 \neq 0$.

Each choice of these coefficients would give raise to different realizations of the rotation group, say,

$$R_3 = \frac{a_1}{a_2} x_1 \frac{\partial}{\partial x_2} - \frac{a_2}{a_1} x_2 \frac{\partial}{\partial x_1}, \quad R_2 = \frac{a_3}{a_1} x_3 \frac{\partial}{\partial x_1} - \frac{a_1}{a_3} x_1 \frac{\partial}{\partial x_3},$$
$$R_1 = \frac{a_2}{a_3} x_2 \frac{\partial}{\partial x_3} - \frac{a_3}{a_2} x_3 \frac{\partial}{\partial x_3}.$$

The corresponding linear equation associated to $H = b^1\, R_1 + b^2\, R_2 + b^3\, R_3$ would be

$$\dot{x}_1 = -\frac{b_1 a_2}{a_1} x_2 + \frac{b_2 a_3}{a_1} x_3\ ,$$

and similarly for other coordinates.

The Poisson bracket for the Ehrenfest picture would be

$$\{x_j, x_k\} = \epsilon_{jkl}\, a_l^2\, x_l\ .$$

At the group level we would have

$$\frac{dS}{dt} S^{-1} = b^1\, R_1 + b^2\, R_2 + b^3\, R_3\ ,$$

and then

$$S(t) = e^{t(b^1\, R_1 + b^2\, R_2 + b^3\, R_3)}\ ,$$

independently of the particular realization of the operators R_j.

Similarly at the abstract Lie algebra level, the Heisenberg picture, not on the operators but on the level of the $\mathbb{C}^*$-algebra, we would have

$$\frac{d}{dt}(Y^j\, R_j) = [Y^k\, R_k, Y^l\, R_l]\ ,$$

independently of the realization.

6.7.2.5 The Fundamental Example Again: The Harmonic Oscillator

We will revisit the inverse problem for the harmonic oscillator (see [Ib97] for a detailed description). Thus, following the ideas in the last paragraphs of the Sect. 6.7.2.4, we can obtain alternative quantum descriptions by looking for solutions of the functional equation,

$$[\tilde{H}, a]_K = -i\omega a,$$

with $K = K(a^\dagger a)$. Thus, any decomposition of $H = \omega(a^\dagger a + 1/2)$ as $H = \tilde{H} e^{\lambda K}$, gives us alternative commutation relations and alternative Hamiltonians because $[H, a] = [\tilde{H}, a]_K$.

By setting

$$\phi_K(a) = e^{\frac{1}{2}\lambda K(a^\dagger a)} a e^{\frac{1}{2}\lambda K(a^\dagger a)} = A, \quad \text{and} \quad \phi_K(a^\dagger) = e^{\frac{1}{2}\lambda K(a^\dagger a)} a^\dagger e^{\frac{1}{2}\lambda K(a^\dagger a)} = A^\dagger,$$

we can construct a new Fock space,

$$|N\rangle = \frac{1}{\sqrt{n!}}(A^\dagger)^n |0\rangle,$$

with $a|0\rangle = 0$ and with scalar product defined by $\langle M|N\rangle = \delta_{mn}$.

We shall consider as an example of the previous discussion the one dimensional quantum harmonic oscillator. The equations of motion are given by

$$i\frac{d}{dt}\begin{pmatrix} a \\ a^\dagger \end{pmatrix} = \omega \begin{pmatrix} 1 & 0 \\ 0 & -1 \end{pmatrix} \begin{pmatrix} a \\ a^\dagger \end{pmatrix},$$

where the operators $a, a^\dagger$ verify the canonical commutation relations

$$[a, a^\dagger] = 1\,,$$

and the Hamiltonian H of the system is given by

$$H = \omega\left(a^\dagger a + \frac{1}{2}\right),$$

and the equations of motion are written as

$$\frac{d}{dt}a = i[H, a]\,.$$

Let K be now an arbitrary function of $a^\dagger a$,

$$K = K(a^\dagger a) = K^\dagger,$$

which is a constant of motion $[H, K] = 0$. Then, we define the following family of alternative associative products,

$$a \circ_K a^\dagger = a e^{\lambda K} a^\dagger = a e^{\lambda K/2} e^{\lambda K^\dagger/2} a^\dagger = A_K A_K^\dagger \,,$$

where $A_K = a\, e^{\lambda K/2}$.

The new 'creation' and 'anihilation' operators A_K and $A_K^\dagger$ verify the commutation relations

$$[a, a^\dagger]_K = a e^{\lambda K} a^\dagger - a^\dagger e^{\lambda K} = A_K A_K^\dagger - A_K^\dagger A_K = 1 \,.$$

The search for alternative Hamiltonians H_K for the dynamics implies the equations

$$[H_K, a]_K = -a, \quad [H_K, a^\dagger]_K = a^\dagger \,,$$

namely,

$$H_K e^{\lambda K} a - a e^{\lambda K} H_K = -a \,,$$

that has a family of solutions,

$$H_K = \frac{\omega}{2}(a^\dagger a + 1) e^{-\lambda K} .$$

We can represent this family in the Hilbert space constructed from the vacuum state $|0\rangle$ defined by,

$$A|0\rangle = 0 = a|0\rangle \,,$$

and the N-particle states defined by

$$|N\rangle = \frac{1}{\sqrt{n!}} (A^\dagger)^n |0\rangle \,.$$

The inner product is defined as

$$\langle M|N\rangle = \delta_{mn} \,.$$

This new product induces a new Poisson structure on the Hilbert space of quadratic functions.

6.7.3 Alternative Hermitean Structures for Quantum Systems: The Infinite-Dimensional Case

We now analyze the same kind of problems in the framework of Quantum Mechanics, taking advantage of the experience and results we have obtained in the previous sections where we dealt with a real $2n$-dimensional vector space.

In Quantum Mechanics the Hilbert space $\mathcal{H}$ is given as a vector space over the field of complex numbers. Now we assume that two Hermitean structures are given on it, which we will denote as $\langle\cdot,\cdot\rangle_1$and $\langle\cdot,\cdot\rangle_2$ (both linear, for instance, in the second factor). As in the real case, we look for the group that leaves invariant both structure, that is the group of unitary transformations with respect to both Hermitean structures. We call them bi-unitary operators.

In order to assure that $\langle\cdot,\cdot\rangle_1$ and $\langle\cdot,\cdot\rangle_2$ do not define different topologies on $\mathcal{H}$ it is necessary that there exists $A, B \in \mathbb{R}$, $0 < A, B$ such that

$$A\,\|x\|_2 \leq \|x\|_1 \leq B\,\|x\|_2\,, \qquad \forall x \in \mathcal{H}\,.$$

The use of Riesz's theorem on bounded linear functionals immediately implies that there exists an operator F defined implicitly by the equation

$$\langle x, y\rangle_2 = \langle Fx, y\rangle_1\,, \qquad \forall x, y \in \mathcal{H}\,,$$

and F replaces the previous G and T tensors of the real vector space situation, i.e., now it contains both the real and imaginary parts of the Hermitean structure and, in fact

$$F = (g_1 + i\omega_1)^{-1} \circ (g_2 + i\omega_2) \tag{6.182}$$

It is trivial to show that F is bounded, positive, and self-adjoint with respect to both Hermitean structures and that

$$\frac{1}{B^2} \leq \|F\|_1 \leq \frac{1}{A^2}\,;\ \frac{1}{B^2} \leq \|F\|_2 \leq \frac{1}{A^2}\,. \tag{6.183}$$

If $\mathcal{H}$ is finite-dimensional, F can be diagonalized, the two Hermitean structures decompose in each eigenspace of F, where they are proportional and we get immediately that the group of bi-unitary transformations is indeed,

$$U(n_1) \times U(n_2) \times \ldots \times U(n_k)\,, \qquad n_1 + n_2 + \ldots + n_k = n = \dim \mathcal{H}$$

where n_i denotes the degeneracy of the ith eigenvalue of F.

In the infinite-dimensional case F may have a point part of the spectrum and a continuum part. From the point part of the spectrum one gets $U(n_1) \times U(n_2) \times \ldots$ where now n_i can be also infinite. The continuum part is more delicate to discuss. It will contain for sure the commutative group U_F of bi-unitary operators of the form $e^{if(F)}$ where f is any measurable real-valued function.

The concept of genericity in the infinite-dimensional case can not be given as easily as in the finite-dimensional case. One can say that the eigenvalues should be non-degenerate but what do we say for the continuous spectrum? We give here an alternative definition that works for the finite and infinite case as well.

Note first that any bi-unitary operator must commute with F. Indeed: $\langle x, U^\dagger FUy\rangle_2 = \langle Ux, FUy\rangle_2 = \langle FUx, Uy\rangle_2 = \langle Ux, Uy\rangle_1 = \langle x, y\rangle_1 = \langle Fx, y\rangle_2 = \langle x, Fy\rangle_2$, from this:

$$U^\dagger FU = F\,, \qquad [F, U] = 0\,.$$

The group of bi-unitary operators therefore belongs to the commutant F' of the operator F. The genericity condition can be restated in a purely algebraric form as follows:

Definition 6.37 Two Hermitean forms are in a generic position iff $F^{''} = F'$, i.e., the bicommutant of F coincides with the commutant of F.

In other words this means that F generates a complete set of observables.

This definition reduces, for the case of a pure point spectrum, to the condition of nondegeneracy of the eigenvalues of F and, in the real case, to the minimum possible degeneracy of the eigenvalues of T and G, which is two.

To grasp how the definition works, we will give some simple examples. Consider: $(F\psi)(x) = x^2\psi(x)$ on the space $L_2([-b, -a] \cup [a, b])$ with $0 < a < b$: then the operator x , its powers x^n and the parity operator P belong to F' while $F^{''}$ does not contain x (and any odd power of x) because they do not commute with P. So if $F = x^2$ the two Hermitean structure are not in a generic position because $F^{''} \subset F'$. On the contrary, on the space $L_2([a, b])$, $F^{''} = F'$ because a parity operator P does not exist in this case, so the two Hermitean structure are now in a generic position. In this case the group of bi-unitary operators is $\left\{e^{if(x^2)t}\right\}$ for the appropriate class of functions f. In some sense, when a continuous part of the spectrum is considered, there appears a continous family of $U(1)$'s as a counterpart of the discrete family of $U(1)$'s corresponding to the discrete part of the spectrum.

Remark 6.13 i) Suppose that complex Hilbert spaces with two Hermitean structures have been constructed from a given real vector space V using two compatible and admissible triples (g_1, ω_1, J_1) and (g_2, ω_2, J_2). Then, by complexification, we get two different Hilbert space, each one with its proper multiplication by complex numbers and with its proper Hermitean structure. The previous case we have just studied is obtained if we assume $J_1 = J_2$. It is easy to show that this is a sufficient condition for compatibility. That is the reason why in the quantum-mechanical case the group of bi-unitary transformations is never empty, and the compatibility condition is encoded already in the assumptions.

ii) If $J_1 \neq J_2$ but the compatibility condition still holds, we know that V splits into $V_+ \oplus V_-$, where $J_1 = \pm J_2$ on $V_\pm$ respectively. On V_+ we have the previous case, while on V_- we get two Hermitean structures, one $\mathbb{C}$-linear and one anti-$\mathbb{C}$-linear in the second factor (which one is which depending on the complexification we have decided to use). From the point of view of the group of unitary transformations, this circumstance is irrelevant, because the set of unitary transformations does not change from being defined with respect to an Hermitean structure or with respect to its complex conjugate. We conclude from this that our analysis goes through in general, provided the compatibility condition holds.

We will try now to summarize our main result, by restating it at the same time in a more concise group-theoretical language. What we have shown is, to begin with, that once two admissible triples: (g_1, ω_1, J_1) and (g_2, ω_2, J_2) are given on a real, even-dimensional vector space $E \cong \mathbb{R}^{2n}$, they define two $2n$-dimensional real representations, $U_r(2n; g_1, \omega_1)$ and $U_r(2n; g_2, \omega_2)$ of the unitary group $U(n)$, $U_r(2n; g_a, \omega_a)$ $(a = 1, 2)$ being the group of transformations that leave simultaneously g_a and ω_a (and hence J_a) invariant. Their intersection

$$W_r = \{U_r(2n; g_1, \omega_1) \cap U_r(2n; g_2, \omega_2)\} \tag{6.184}$$

will be their common subgroup that is an invariance group for both triples. The assumption of compatibility[17] implies that W_r should not reduce to the identity alone.

If the two triples are in a generic position, then

$$W_r = \underbrace{SO(2) \times SO(2) \times \ldots \times SO(2)}_{n \text{ factors}} \tag{6.185}$$

where, here: $SO(2) \cong U(1)$ or, more generally if the genericity assumption is dropped

$$W_r = U_r(2r_1; g, \omega) \times U_r(2r_2; g, \omega) \times \ldots \times U_r(2r_k; g, \omega) \tag{6.186}$$

where: $r_1 + r_2 + \ldots + r_k = n$ and (g, ω) is any one of the two pairs (g_1, ω_1) or (g_2, ω_2).

The real vector space $V \approx \mathbb{R}^{2n}$ will decompose then into a direct sum of even-dimensional subspaces that are mutually orthogonal with respect to both metrics, and on each subspace the corresponding (realization of the) special orthogonal group will act irreducibly.

Alternatively, we can complexify $E \cong \mathbb{R}^{2n}$ in two different ways, using the two complex structures that are at our disposal. The equivalent statement in the complex framework will be then:

Given two Hermitean structures h_a, $a = 1, 2$ on a complex n-dimensional vector space $\mathbb{C}^n$, they define two representations $U(n; h_a)$, $a = 1, 2$ of the group $U(n)$ on

[17] As the previous two-dimensional example shows explicitly, but it should be clear by now also in general.

the same $\mathbb{C}^n$. Then $U(h_1, n)$ (resp. $U(h_2, n)$) will be the group of transformations that are unitary with respect to h_1 (resp. h_2). The group W of simultaneous invariance for both Hermitean structures :

$$W \equiv \{U(h_1, n) \cap U(h_2, n)\} . \tag{6.187}$$

will be a subgroup of both $U(h_1, n)$ and $U(h_2, n)$, and our assumption of compatibility of the h_a 's implies that the component of W connected to the identity should not reduce to the identity alone.

The assumption of genericity implies that

$$W = \underbrace{U(1) \times U(1) \times \ldots \times U(1)}_{n \text{ factors}} \tag{6.188}$$

If the assumption of genericity is dropped, one can easily show, along the same lines as in the generic case, that W will be of the form:

$$W = U(r_1) \times U(r_2) \ldots \times U(r_k) , \tag{6.189}$$

with $r_1 + r_2 + \ldots + r_k = n$. The space $\mathbb{C}^n$ will decompose accordingly into a direct sum of subspaces that will be mutually orthogonal with respect to both h_a 's, and on each subspace the appropriate $U(r)$ will act irreducibly.

We have also shown that these results generalize to the infinite-dimensional case as well. Some extra assumptions must be added on the Hermitean structures in order that they define the same topology in $\mathbb{H}$ and an appropriate definition of genericity must also be given. Then, a decomposition like in Eqs.(6.188) and (6.189) is obtained, possibly with denumberable discrete terms and a continuum part as well. We note that, in the spirit of this work where two Hermitean structures are given from the very beginning, it is natural to supplement the compatibility condition, in the infinite-dimensional case, with a topological equivalence condition. However from the point of view of the study of bi-hamiltonian systems, where a fixed dynamics is given, it would be more natural to assume some weaker regularity condition, for instance that the given dynamics should be continuous with respect to both structures.

Bi-Hamiltonian systems 'generated' out of a pencil of compatible Poisson structures have been considered in connection with the separability problem [Ib00]. It should be noticed that our compatible structures would give raise to a pencil of compatible triples defined by:

$$g_\gamma = g_1 + \gamma g_2 \ , \quad \omega_\gamma = \omega_1 + \gamma \omega_2 \ , \quad J_\gamma \tag{6.190}$$

We will conclude this section by stressing again the fact that the auxiliary mathematical structures used in the description of physical systems, classical or quantum, usually are not uniquely determined by the physical evidence collected from the observation of the system. This situation is relevant because it reflects an intrinsic ambiguity, for instance in the correspondence between symmetries and constants of motion. At the classical level, some of these ambiguities have been already discussed in relation with the quantization problem but at the quantum level they have received little attention.

Even more fundamental than all that is the possibility of using alternative linear structures to describe quantum systems. A different path pointing to test the linearity of Quantum Mechanics was taken for instance by S. Weinberg [We89], where the linearity of the equations of motion was questioned. Alternative linear structures describe a more fundamental aspect, the nonexistence of an absolute linear structure on the space of quantum states.

Finally, we should point it out that many of the previous considerations can be extended to systems admitting higher order geometrical structures, such as Nambu Mechanical systems [Na73, Ma81, Ta94, Ma95], generalized Poisson structures [Az96, Az97, Iba98] etc.

References

[St66] Strocchi, F.: Complex coordinates and quantum mechanics. Rev. Mod. Phys. **38**, 36–40 (1966)

[Du90] Dubrovin, B.A., Marmo, G., Simoni, A.: Alternative Hamiltonian descriptions for quantum systems. Mod. Phys. Lett. A **5**, 1229–1234 (1990)

[Ma96] Marmo, G., Vilasi, G.: Symplectic structures and quantum mechanics. Mod. Phys. Lett. B **10**, 545–553 (1996)

[Er10] Ercolessi, E., Marmo, G., Morandi, G.: From the equations of motion to the canonical commutation relations. Riv. Nuovo Cim. **33**, 401–590 (2010)

[Ib96a] Ibort, A., Marmo, G.: From the equations of motion to quantum commutation relations. Proceedings of the V Workshop on Geometry and Physics, Jaca (1996)

[Gi93] Giordano, M., Marmo, G., Rubano, C.: The inverse problem in the Hamiltonian formalism: integrability of linear Hamiltonian fields. Inv. Probl. **9**, 443–467 (1993)

[Ch67] Chern, S.S.: Complex Manifolds without Potential Theory. Van Nostrand, Princeton (1967)

[Sc86] Schwartz, L.: Lectures on Complex Analytic Manifolds. Narosa, New Dehli (1986)

[Ni57] Newlander, A., Nirenberg, L.: Complex analytic coordinates in almost complex manifolds. Ann. Math. **65**, 391–404 (1957)

[We58] Weil, A.: Introduction a' l'Étude des Varietés Kähleriennes. Hermann, Paris (1958)

[Ci90] Cirelli, R., Maniá, A., Pizzocchero, L.: Quantum mechanics as an infinite-dimensional Hamiltonian system with uncertainty structure I. J. Math. **31**, 2891–2897 (1990) (ibid. Quantum mechanics as an infinite-dimensional hamiltonian system with uncertainty structure II. J. Math. Phys. 31, 2898–2903 (1990))

[Ci91] Cirelli, R., Maniá, A., Pizzocchero, L.: Quantum phase-space formulation of Schrödinger mechanics. Int. J. Mod. Phys. A **6**, 2133–2146 (1991)

[Ci94] Cirelli, R., Maniá, A., Pizzocchero, L.: A functional representation for non-commutative $\mathbb{C}^*$-algebras. Rev. Math. Phys. **6**, 675–697 (1994)

[Go82] Goldberg, S.I.: Curvature and Homology. Dover Publications, New York (1982)

[Di45] Dirac, P.A.M.: On the analogy between classical and quantum mechanics. Rev. Mod. Phys. **17**, 195–199 (1945)

[Re44] Reichenbach, H.: Philosophical Foundations of Quantum Mechanics. University of California Press, California (1944)

[St60] Stueckelberg, E.C.G.: Quantum Theory in Real Hilbert Space I, II. Helv. Phys. Acta **33**, 727–752 (1960) (ibid. 34, 621–28 (1961))

[Es04] Esposito, G., Marmo, G., Sudarshan, G.: From Classical to Quantum Mechanics: an introduction to the formalism. Cambridge Univ,Press(2004)

[Ma05b] Marmo, G., Scolarici, G., Simoni, A., Ventriglia, F.: The quantum-classical transition: the fate of the complex structure. Int. J. Geom. Meth. Mod. Phys. **2**, 127–145 (2005)

[Di58] Dirac, P.A.M.: The Principles of Quantum Mechanics. Clarendon Press, Oxford (1958)

[Me58] Messiah, A.: Mecanique Quantique, vol. *I*. Dunod, Paris (1958)

[La27] Landau, L.: Das Dampfungsproblem in Wellen-mechanik (The damping problem in wave mechanics). Z. Phys. **45**, 430–441 (1927)

[Ne27] von Neumann, J.: Wahrscheinlichkeitstheoretischer aufban der Quantenmechanik. Goettingenische Nachrichten **11**, 245–272 (1927)

[Ha70] Haefliger, A.: Homotopy and integrability. Lect. Note Math. **197**, 133–164, Springer (1970)

[Hus75] Husemoller, D.: Fibre Bundles, 2nd ed. Springer, New York (1975) (3d edn, Springer, 1994)

[Ch09] Chruściński, D., Marmo, G.: Remarks on the GNS representation and the geometry of quantum states. Open Syst. Inf. Dyn. **16**, 157–167 (2009)

[St51] Steenrod, N.: The Topology of Fibre Bundles. Princeton University Press, Princeton (1951)

[Gr05] Grabowski, J., Kuś, M., Marmo, G.: Geometry of Quantum Systems: Density States and Entanglement. J. Phys. A: Math. Gen. **38**, 10127–10244 (2005)

[Neu2] von Neumann, J.: Die Mathematische Grundlagen der Quantenmechanik. Springer, Berlin (1932) (English translation: Mathematical Foundations of Quantum Mechanics. Princeton University Press 1955)

[Fa57] Fano, U.: Description of states in quantum mechanics by density matrix and operator techniques. Rev. Mod. Phys. **29**, 74–93 (1957)

[Br05] Brown, L.M. (ed.) Feynman's Thesis. World Scientific, Hackensack (2005)

[Gl81] Glimm, J., Jaffe, A.: Quantum Physics. Springer, Berlin (1981). (A Functional Integral Point of View)

[Di33] Dirac, P.A.M.: The Lagrangian in Quantum Mehanics.

[Cl07] Clemente-Gallardo, J., Marmo, G.: The space of density states in geometrical quantum mechanics. In: Cantrijn, F., Crampin, M., Langerock, B. (eds.) Differential Geometric Methods in Mechanics and Field Theory, pp. 35–56. Ghent University Press, Ghent (ISBN 9789038211282) (2007)

[Ma05] Man'ko, V.I., Marmo, G., Sudarshan, E.C.G., Zaccaria, F.: Differential geometry of density states. Rep. Math. Phys. **55**, 405–422 (2005)

[Ma95] Marmo, G., Morandi, G.: Some geometry and topology. In: Lundqvist, S., Morandi, G., Lu, Y. (eds.) Low-Dimensional Quantum Field Theories for Condensed-Matter Physicists. World Scientific, Singapore (1995)

[Man00] Man'ko, V.I., Marmo, G., Sudarshan, E.C.G., Zaccaria, F.: Inner composition law for pure states as a purification of impure states. Phys. Lett. A **273**, 31–36 (2000)

[Wi32] Wigner, E.P.: Über die Operation der Zeitumkehr in der Quantenmechanik. Nachr. Akad. Ges. Wiss. Göttingen **31**, 546–559 (1932)

[Wi59] Wigner, E.P.: Group Theory and its Applications to the Quantum Mechanics of Atomic Spectra. Academic Press, New York (1959)

[Ba54] Bargmann, V.: On unitary ray representations of continuous groups. Ann. Math. **59**, 1–46 (1954)

[Ha92] Haag, R.: Local Quantum Physics. Springer, Heidelberg (1992)

[Mac68] Mackey, G.W.: Induced Representations of Groups and Quantum Mechanics. Benjamin, New York (1968)

[Mac04] Mackey, G.W.: Mathematical Foundations of Quantum Mechanics. Benjamin, New York (1963) (Dover 2004)

[Sa97] Samuel, J.: The geometric phase and ray space isometries. Pramana J. Phys. **48**, 959–967 (1997)

[Em84] Emch, G.G.: Mathematical and Conceptual Foundations of 20th Century Physics. North-Holland, Amsterdam (1984)

[Jo34] Jordan, P.: Uber die multiplikation quantenmechanischer grossen II. Zeitschrift f. Physik **87**, 505 (1934)

[Jo34b] Jordan, P., von Neumann, J., Wigner, E.P.: On an Algebraic Generalization of the Quantum Mechanical Formalism. Ann. Math. **35**, 29–64 (1934)

[Fa12] Falceto, F., Ferro, L., Ibort, A., Marmo, G.: Reduction of Lie-Jordan Banach algebras and quantum states. J. Phys. A: Math. Theor. **46**, 015201 (2013)

[Be06] Bengtsson, I., Zyczkowski, K.: Geometry of Quantum States: An Introduction to Quantum Entanglement. Cambridge University Press, Cambridge (2006)

[Ki76] Kirillov, A.A.: Elements of the Theory of Representations. Springer, Berlin (1976)

[Ko70] Konstant, B.: Quantization and unitary representations Part I. Prequantization. In: Taam, C. T. (ed.) Lecture Notes in Mathematics, vol. 170. Springer, Berlin (1970)

[Mo01] Morandi, G., Napoli, F., Ercolessi, E.: Statistical Mechanics. World Scientific, Singapore (2001). (An Intermediate Course)

[Br87] Bratteli, O., Robinson, D.W.: Operator Algebras and Quantum Statistical Mechanics. Springer, New york (1987)

[Ha64] Haag, R., Kastler, D.: An algebraic approach to quantum field theory. J. Math. Phys. **5**, 848–861 (1964)

[Fo76] Fock, V.A.: Fundamentals of Quantum Mechanics, 2nd edn. URSS Publications, Moscow (1976)

[Sc70] Schwinger, J.: Quantum Kinematics and Dynamics. Frontiers in Physics, Benjamin, New York (1970) (Proc. Nat. Acad. Sci. 46, 261–265 (1960))

[Ib97] Ibort, A., Man'ko, V., Marmo, G., Zaccaria, F.: The inverse problem for quantum and classical systems: from the classical equations of motion to the quantum commutation relations. Group 21. In: Doebner, H.-D., Nattermann, P., Scherer, W. (eds.) Physical Applications and Mathematical Aspects of Geometry, Groups and Algebras, vol. I, pp. 265–274. World Scientific, Singapore (1997)

[Ib00] Ibort, A., Magri, F., Marmo, G.: Bihamiltonian structures and Stäckel separability. J. Geom. Phys. **33**, 210–228 (2000)

[We89] Weinberg, S.: Testing quantum mechanics. Ann. Phys. **194**, 336–386 (1989)

[Na73] Nambu, Y.: Generalized Hamiltonian mechanics. Phys. Rev. D **7**, 499–510 (1973)

[Ma81] Marmo, G., Saletan, E.J., Simoni, A., Zaccaria, F.: Liouville dynamics and Poisson brackets. J. Math. Phys. **22**, 835–842 (1981)

[Ta94] Takhtajan, L.: On foundation of the generalized Nambu mechanics. Commun. Math. Phys. **160**, 295–315 (1994)

[Ma95] Marmo, G., Ibort, A.: A generalized reduction procedure for dynamical systems. In: M. Salgado and Elena Vázquez (eds) Proceedings of the IV Fall Workshop: Differential Geometry and its Applications, RSEF Monografías, vol. 3, pp. 55–70 (1995)

[Az96] Azcárraga, J.A., Perelomov, A.M., Pérez, J.C.: Bueno. New generalized Poisson structures. J. Phys. A: Math. Gen. **29**, 627–649 (1996)

[Az97] Azcárraga, J.A., Izquierdo, J.M., Pérez, J.C.: Bueno. On the higher order generalizations of Poisson structures. J. Phys. A: Math. Gen. **30**, L607–L616 (1997)

[Iba98] Ibáñez, R., de León, M., Marrero, J.C., Padrón, E.: Nambu-Jacobi and generalized Jacobi manifolds. J. Phys. A: Math. Gen. **31**, 1267–1286 (1998)

References

[illegible]

Chapter 7
Folding and Unfolding Classical and Quantum Systems

The unmanifest world is simple and linear, it is the manifest world which is 'folded' and nonlinear.
Giuseppe Marmo, *XXI International Workshop on Differential Geometric Methods in Theoretical Mechanics*, 2006

7.1 Introduction

Reduction procedures, the way we understand them today (i.e. in terms of Poisson reduction) can be traced back to Sophus Lie in terms of function groups, reciprocal function groups and indicial functions [Ei61, Fo59, Lie93, Mm85]. Function groups provide an algebraic description of the cotangent bundle of a Lie group but are slightly more general because they can arise from Lie group actions which do not admit a momentum map [Ma83]. Function groups are also know today as 'dual pairs' [Ho85] as discussed already in Sect. 4.3.4. The present chapter exhibits many instances of reduction procedures appearing in a variety of physical situations, both classical and quantum. This choice may give the impression of an episodic description, however it contains an illustration of the essential aspects of any reduction procedure, both in the algebraic and geometrical setting, pointing out analogies and differences between the classical and the quantum situation.

7.2 Relationships Between Linear and Nonlinear Dynamics

The aim of this chapter is to analyze the relationships between linear and nonlinear dynamics. We show that some nonlinear dynamics enjoy interesting properties, like some sort of integrability or separability, arising as reduction from linear dynamics. Conversely, starting from linear 'free' dynamics, we can obtain by appropriate

J.F. Cariñena et al., *Geometry from Dynamics, Classical and Quantum*,
DOI 10.1007/978-94-017-9220-2_7

reduction procedures well known dynamics that appear quite often in the physics literature.

7.2.1 Separable Dynamics

Physicists have used reduction procedures as an aid in trying to integrate the dynamics 'by quadratures'. Dealing, as usual, with a coordinate formulation, reduction and coordinate separability have overlapped a good deal. A first approach to the notion of separability was offered at the linear level in Sect. 3.6.2. From the point of view of the integration of the equations of motion, separability or the so called decomposition into independent motions, may be formalized as follows.

Definition 7.1 Consider a dynamical vector field Γ on a carrier manifold M and a decomposition

$$\Gamma = \sum_{i=1}^{r} \Gamma_i \,,$$

with the requirement that:

1. $[\Gamma_i, \Gamma_j] = 0$ for all $i, j = 1, \ldots, r$, and
2. $\mathrm{span}\,\{\Gamma_i(x) \mid i = 1, \ldots, r\} = T_x M\,, \quad \forall x \in N \subset M$, where N is an open and dense submanifold in M.

When such a decomposition exists, the evolution is given by the composition of the one-parameter groups associated with each Γ_j and the system Γ will be said to be separable.

Looking for superposition rules which would generalize the usual linear superposition rule of linear systems, Lie [Lie93] introduced dynamical systems admitting a decomposition of the form:

$$\Gamma = \sum_i a^j(t)\Gamma_j \,,$$

with $[\Gamma_i, \Gamma_j] = \sum_k c_{ij}{}^k \Gamma_k$ and $c_{ij}{}^k \in \mathbb{R}$ (i.e., the vector fields Γ_k span a finite-dimensional real Lie algebra) and still $\{\mathrm{span}\,\Gamma_i(x)\} = T_x M\,, \ \forall x \in N \subset M$, where N is open and dense in M. The integration of these systems may be achieved by finding a fundamental set of solutions: they admit a superposition rule even if the system is nonlinear. These systems have been called Lie–Scheffers systems and will be extensively studied in Chap. 9. An important representative of Lie-Scheffers systems is given by the Riccati equation. It is worth illustrating this example because it is an instance of a nonlinear equation obtained as a reduction from a linear one.

7.2.2 The Riccati Equation

In 1720, when he proposed his today famous equation, Riccati was interested in the following problem: Given a linear system of differential equations describing the dynamical evolution of a point (x_1, x_2) of the Euclidean plane,

$$\frac{d}{dt}\begin{pmatrix} x_1 \\ x_2 \end{pmatrix} = \begin{pmatrix} b_{11} & b_{12} \\ b_{21} & b_{22} \end{pmatrix}\begin{pmatrix} x_1 \\ x_2 \end{pmatrix} \tag{7.1}$$

where the b_{ij} are t-dependent functions, find the dynamical evolution of $x = x_2/x_1$. He found that such evolution is given by:

$$\dot{x} = c_0(t) + c_1(t)\,x + c_2(t)\,x^2\,, \tag{7.2}$$

where

$$c_0 = b_{21} \qquad c_1 = b_{22} - b_{11}, \qquad c_2 = -b_{12}\,.$$

Equation (7.2) is known as the Riccati equation and it is a first-order equation that only involves one variable, while the linear system Eq. (7.1) involves two unknown variables. In this sense we have reduced the number of degrees of freedom. Note also that the final equation does not depend on b_{11} and b_{22} but only on their difference. Therefore we can restrict ourselves to the case in which the matrix

$$A = \begin{pmatrix} b & c \\ a & -b \end{pmatrix} \tag{7.3}$$

defining the linear system is traceless and then belongs to the Lie algebra $\mathfrak{sl}(2, \mathbb{R})$. That is, rewriting the coordinates in $\mathbb{R}^2$ by (x, y) and considering the linear system defined by the matrix A:

$$\begin{cases} \dfrac{dx}{dt} = b\,x + c\,y \\ \dfrac{dy}{dt} = a\,x - b\,y \end{cases},$$

the vector field for such dynamics is given now by:

$$\Gamma_A = ax\frac{\partial}{\partial y} + b\left(x\frac{\partial}{\partial x} - y\frac{\partial}{\partial y}\right) + cy\frac{\partial}{\partial x} \tag{7.4}$$

with $a, b, c \in \mathbb{R}$ time-dependent functions.

We remark now that because the system is linear, Γ_A commutes with the Euler vector field $\Delta = x\partial/\partial x + y\partial/\partial y$, the dilation vector field on $\mathbb{R}^2$. Therefore, the algebra of functions f such that $\Delta(f) = 0$ on $\mathbb{R}^2\backslash\{0, 0\}$ will be invariant under Γ_A, i.e., $\Delta(f) = 0$ implies that $\Delta(\Gamma_A(f)) = 0$. Such functions f, being homogeneous of degree zero, depend only on the variable x/y or y/x. If we consider first the dynamics in terms of $\xi = x/y$, $y \neq 0$, we find

$$\frac{d\xi}{dt} = \frac{\dot{x}}{y} - \frac{x\dot{y}}{y^2} = \frac{cy + bx}{y} - \frac{x}{y}\left(\frac{ax}{y} - \frac{by}{y}\right) = c + 2b\xi^2 - a\xi^2 \tag{7.5}$$

i.e.,

$$\frac{d\xi}{dt} = c + 2b\xi - a\xi^2 \tag{7.6}$$

The transformation: $(x, y) \mapsto x/y$ is often called the Riccati transformation. Therefore we have obtained the Riccati equation, which is nonlinear, by reducing our linear dynamics with an invariant equivalence relation, namely

$$(x_1, y_1) \sim (x_2, y_2) \quad iff \quad x_1y_2 - y_1x_2 = 0\,,$$

which is the equivalence relation defined by the integral curves of the vector field Δ. Notice that the equivalence relation defined by such set of curves is invariant because, as indicated above, Γ_A is a linear vector field and then $[\Delta, \Gamma_A] = 0$.

We should notice here that the integral curves of Δ are open half-lines starting at $(0, 0)$, and the origin itself, therefore the set of such integral curves of Δ is singular at the origin. This implies that in the quotient space the integral curve $(0, 0)$ is not separated from all the others and the quotient space is not a Hausdorff space. Removing it, the quotient space is simply the circle S^1. The two choices for parametrizing its points $\xi = x/y$ and $\zeta = y/x$ define two charts covering it. In the language of differentiable spaces, we have obtained that the subalgebra of Δ-invariant functions on $\mathcal{F}(\mathbb{R}^2\backslash\{0, 0\})$ is isomorphic to the algebra of smooth functions on the circle S^1.

Equation (7.6) provides in fact the local expression of the vector field Γ on S^1 induced by Γ_A in the chart ξ. Because the range of this chart is $\mathbb{R}$ we can say that the induced vector field reduces to the Riccati equation on the line. Notice that in the second chart ζ the projected vector field has the expression

$$\frac{d\zeta}{dt} = a - 2b\zeta - c\zeta^2\,, \tag{7.7}$$

which again is a Riccati equation obtained from the previous one (7.6) by the change of variable $\zeta = 1/\xi$.

We might comment a little more on this last example. It is clear that the vector fields $x\partial/\partial y$, $y\partial/\partial x$ and $x\partial/\partial x - y\partial/\partial y$ close on the Lie algebra of the group $SL(2, \mathbb{R})$ (see Sect. 2.6.1). In fact, they are the linear vector fields X_e, X_f, X_h, associated with the traceless matrices

$$e = \begin{pmatrix} 0 & 0 \\ 1 & 0 \end{pmatrix}, \quad f = \begin{pmatrix} 0 & 1 \\ 0 & 0 \end{pmatrix}, \quad h = \begin{pmatrix} 1 & 0 \\ 0 & -1 \end{pmatrix}, \tag{7.8}$$

which are the usual generators of the Lie algebra $\mathfrak{sl}(2, \mathbb{R})$ with the commutation relations

$$[e, h] = 2e\,, \qquad [f, h] = -2f\,, \qquad [e, f] = h\,. \tag{7.9}$$

The vector field Γ is a linear combination of these three vector fields with functions of time as coefficients and constitutes the prototype of a Lie-Scheffers system. Still, using this representation we can say that Eq. (7.2) is associated with a 'free' motion on the group $SL(2, \mathbb{R})$:

$$\dot{g}\, g^{-1} = -\sum_j c_j(t) A_j\,,$$

where $A_0 = e$, $A_1 = f$, $A_2 = h$ is an appropriate basis of the Lie algebra of SL$(2, \mathbb{R})$. Associated to it we find a nonlinear superposition rule for the solutions: if $x_{(1)}, x_{(2)}, x_{(3)}$ are independent solutions, every other solution x is obtained from a constant value k of the following double ratio:

$$\frac{\left(x - x_{(1)}\right)\left(x_{(2)} - x_{(3)}\right)}{\left(x - x_{(2)}\right)\left(x_{(1)} - x_{(3)}\right)} = k\,. \tag{7.10}$$

Exercise 7.1 Derive the double ratio relation from invariants (constants of motion) of the linear system Γ_A.

Notice the geometric meaning of this construction. Starting from a linear equation in $\mathbb{R}^2$, we consider a map $\pi : \mathbb{R}^2 \to \mathbb{R}$, given by $\pi(x_1, x_2) = x_2/x_1$, and then the t-dependent vector field describing the given linear system of first-order differential equations projects on a corresponding one which describes a Riccati equation, a differential equation involving only one degree of freedom and simpler in this sense than the original system. The knowledge of the general solution of the differential equation allows us to find the general solution of the given system.

But we can also use this example in the opposite direction, as a suggestion that specific dynamics, for instance exhibiting superposition rules, can be deduced from linear ones under an appropriate reduction process.

Riccati type equations arise also in the reduction of the Schrödinger equation to the space of pure states (see also [Ch07]).

Another example, but for partial differential equations, is provided by the following variant of the Burgers equation.

7.2.3 Burgers Equation

To illustrate the procedure for partial differential equations in one space x and one time t variables, we consider the following variant of the Burgers equation

$$\frac{\partial w}{\partial t} + \frac{1}{2}\left(\frac{\partial w}{\partial x}\right)^2 - \frac{k}{2}\frac{\partial^2 w}{\partial x^2} = 0\,, \tag{7.11}$$

where k is an arbitrary but fixed real parameter.

This equation admits a (at least partial) superposition rule of the following kind: for any two solutions, $w_1(t,x)$ and $w_2(t,x)$, the function:

$$w = -k\log\left(\exp\left(-\frac{w_1+\ell_1}{k}\right) + \exp\left(-\frac{w_2+\ell_2}{k}\right)\right),$$

is again a solution of Eq. (7.11) with ℓ_1 and ℓ_2 arbitrary real constants.

The existence of a superposition rule might suggest that the equation may be related to a linear one. That is indeed the case and we find that the heat equation

$$\frac{\partial u}{\partial t} = \frac{k}{2}\frac{\partial^2 u}{\partial x^2}\,,$$

is indeed related to the nonlinear equation by the replacement of the dependent variable w by the new variable $u = \exp(-w/k)$.

Out of this experience one may consider the possibility of integrating more general evolution systems of differential equations by looking for a simpler system ("simpler" here meaning that it is an explicitly integrable system) whose reduction provides the system that we would like to integrate. In some sense, in a short sentence, we could say that the reduction procedure provides us with interacting systems out of free (or Harmonic) ones.

The great interest for new completely integrable systems boosted the research in this direction in the past 25 years and many interesting physical systems, both in finite and infinite dimensions were shown to arise in this way (see for instance [OlPe81] and references therein).

In the same ideology one may also put the attempts for the unification of all the fundamental interactions in Nature by means of Kaluza-Klein theories. In addition, the attempt to quantize theories described by degenerate Lagrangians called for a detailed analysis of reduction procedures connected with constraints. These techniques came up again when considering geometric quantization as a procedure to construct unitary irreducible representations for Lie groups by means of the orbit method [Ki76, Ki99].

We have already discussed a few instances in Sect. 1.2.10 of nonlinear systems arising from linear ones. Of course if our concern is primarily with the equations of motion we have to distinguish the various available descriptions: Newtonian, Lagrangian, Hamiltonian. Each description carries additional structures with respect to the equations of motion and one has to decide whether the reduction should be performed within the chosen category or if the reduced dynamics will be allowed to belong to another one. We will devote the rest of the chapter illustrating and discussing the reduction of linear systems compatible with certain geometrical structures. We will start by reviewing the simplest example that shows how 'nonlinearities' arise from reduction of a linear system: the three dimensional free particle.

7.2.4 Reducing the Free System Again

We consider again the reduction of the free system on $\mathbb{R}^3$ discussed in Sect. 1.2.10. Consider the equations of motion of a free particle of unit mass in Newtonian form:

$$\ddot{\mathbf{r}} = 0\,.$$

This system is associated to the second-order vector field in $T\mathbb{R}^3$, $\Gamma = \dot{\mathbf{r}} \cdot \nabla$ and has as constants of motion $\dot{\mathbf{r}}$ and $\boldsymbol{\ell} = \mathbf{r} \times \dot{\mathbf{r}}$:

$$\frac{d}{dt}(\mathbf{r} \times \dot{\mathbf{r}}) = 0, \qquad \frac{d}{dt}\dot{\mathbf{r}} = 0\,.$$

By making use of constants of motion, we showed how to choose invariant submanifolds for Γ such that taking the restrictions of the dynamics on such submanifolds, we can associate with it an equation of motion involving only $r = \|\mathbf{r}\|$, $\dot{r}$ and some 'coupling constants' related to the values of the constants of motion to get for instance Eq. (1.93):

$$\ddot{r} = \frac{\alpha\ell^2 + (1-\alpha)(2E - \dot{r}^2)r^2}{r^3}\,.$$

The geometrical interpretation of this is rather simple: we have selected an invariant submanifold $\Sigma \subset \mathbb{R}^3$ (the level set of a constant of motion), we have restricted the dynamics to it, and then we have used the rotation group to foliate Σ into orbits. The reduced dynamics is a vector field acting on the space of orbits $\widetilde{\Sigma} = \Sigma/SO(3)$. It should be remarked that even if Σ is selected in various ways, the choice we have made is compatible with the action of the rotation group. It should be clear now that our presentation goes beyond the standard reduction in terms of the momentum map, which involves additional structures. Indeed this reduction, when carried out with the canonical symplectic structure, would give us only the first solution in the example above.

There is another way to undertake the reduction of the free system above. On $T^*\mathbb{R}^3$ with coordinates $(\mathbf{r}, \mathbf{p})$, we can consider the functions:

$$\xi_1 = \frac{1}{2}\langle \mathbf{r}, \mathbf{r}\rangle\,, \qquad \xi_2 = \frac{1}{2}\langle \mathbf{p}, \mathbf{p}\rangle\,, \qquad \xi_3 = \langle \mathbf{r}, \mathbf{p}\rangle\,.$$

Here $\langle \mathbf{a}, \mathbf{b}\rangle$ denotes the scalar product $\mathbf{a} \cdot \mathbf{b}$, but it can be extended to a non definite positive scalar product.

The equations of motion of the free system (1.90) using ξ_1, ξ_2 and ξ_3 as coordinate functions become

$$\frac{d}{dt}\xi_1 = \xi_3\,, \qquad \frac{d}{dt}\xi_2 = 0\,, \qquad \frac{d}{dt}\xi_3 = 2\xi_2.$$

Note that any constant of motion of this system is then a function of ξ_2 and $(2\xi_1\xi_2 - \xi_3^2)$. Consider first the invariant submanifold $\xi_2 = k/2 \in \mathbb{R}$. Then we find,

$$\frac{d}{dt}\xi_1 = \xi_3\,, \qquad \frac{d}{dt}\xi_3 = k\,,$$

i.e., a uniformly accelerated motion in the variable ξ_1. It may be described by the Lagrangian

$$L(x, v) = \frac{1}{2}v^2 + kx,$$

where $x = \xi_1$, $v = \dot{\xi}_1 = \xi_3$.

Had we selected a different invariant submanifold, for instance,

$$2\xi_1\xi_2 - \xi_3^2 = \ell^2,$$

the restricted dynamics would have been:

$$\frac{d}{dt}\xi_1 = \xi_3\,, \qquad \frac{d}{dt}\xi_3 = \frac{\xi_3^2 + \ell^2}{\xi_1}\,.$$

A corresponding Lagrangian description is then provided by the function

$$L(x, v) = \frac{1}{2}\frac{v^2}{x} - \frac{2\ell^2}{x},$$

with $x = \xi_1$ and $\dot{x} = v = \xi_3$.

If we start with the dynamics of the isotropic harmonic oscillator, say $\dot{\mathbf{r}} = \mathbf{p}$ and $\dot{\mathbf{p}} = -\mathbf{r}$, we obtain in terms of the coordinates above:

$$\frac{d}{dt}\xi_1 = \xi_3\,, \qquad \frac{d}{dt}\xi_2 = -\xi_3\,, \qquad \frac{d}{dt}\xi_3 = 2(\xi_2 - \xi_1).$$

Therefore $\xi_1 + \xi_2$ is a constant of motion and if we introduce the functions $\eta_1 = \xi_1 - \xi_2$, ξ_3 and $\eta_2 = \xi_1 + \xi_2$, we would get

$$\dot{\eta}_1 = 2\xi_3, \qquad \dot{\xi}_3 = -2\eta_1\,,$$

together with $\dot{\eta}_2 = 0$, i.e., we get a one dimensional oscillator.

We would like to stress that the 'position' η_1 of this reduced system is not a function depending only on the initial position variables $\mathbf{r}$.

Let us point out a general aspect of the example we just considered. We first notice that the functions

$$\xi_1 = \frac{1}{2}\sum_a x_a x^a, \qquad \xi_2 = \frac{1}{2}\sum_a p_a p^a, \qquad \text{and} \qquad \xi_3 = \sum_a x_a p^a\,,$$

may be defined on any phase space $\mathbb{R}^{2n} = T^*\mathbb{R}^n$, with $\mathbb{R}^n$ an Euclidean space. If we consider the standard Poisson brackets, say

$$\{x^a, p_b\} = \delta^a_b, \quad \{p_a, p_b\} = 0 = \{x^a, x^b\}, \qquad a, b = 1, \ldots, n,$$

we find that for the new variables we get:

$$\{\xi_1, \xi_2\} = \xi_3, \quad \{\xi_1, \xi_3\} = 2\xi_1, \quad \{\xi_2, \xi_3\} = -2\xi_2. \tag{7.12}$$

Thus the functions ξ_i we are considering close on the real Lie algebra $\mathfrak{sl}(2, \mathbb{R})$ (see Eq. (7.9)). The infinitesimal generators $X_i = \{\xi_i, \cdot\}$ are complete vector fields and integrate to a symplectic action of $SL(2, \mathbb{R})$ on $\mathbb{R}^{2n}$ endowed with its natural symplectic structure given by the identification of $\mathbb{R}^{2n}$ with $T^*\mathbb{R}^n$.

Then, in the stated conditions there is always a symplectic action of $SL(2, \mathbb{R})$ on $T^*\mathbb{R}^n \simeq \mathbb{R}^{2n}$ with a corresponding momentum map $J\colon T\mathbb{R}^n \to \mathfrak{sl}(2, \mathbb{R})^* \cong \mathbb{R}^3$.

If we denote again the coordinate functions on this three dimensional vector space by $\{\xi_1, \xi_2, \xi_3\}$, with the Poisson brackets given by (7.12), then the momentum map J provides a symplectic realization of the Poisson manifold $\mathfrak{sl}(2, \mathbb{R})^*$. In the language used by S. Lie we would say that the coordinate functions $\{\xi_1, \xi_2, \xi_3\}$ along with all the smooth functions of them $\{f(\xi_1, \xi_2, \xi_3)\}$ define a function group. The Poisson subalgebra of functions of $\mathcal{F}(T^*\mathbb{R}^n)$ Poisson commuting with all the functions $f(\xi_1, \xi_2, \xi_3)$, constitute the reciprocal function group, and all functions in the intersection of both sets, say functions of the form $F(\xi_1\xi_2 - \frac{1}{4}\xi_3^2)$, constitute the indicial functions.

By setting $\xi_1 = \frac{1}{2}, \xi_3 = 0$ we identify a submanifold in $T\mathbb{R}^n$ diffeomorphic with TS^{n-1}, the tangent bundle of the $(n-1)$-dimensional sphere. It is clear that the reciprocal function group is generated by functions $J_{ab} = p_a x^b - p_b x^a$.

Thus, the reduced dynamics which we usually associate with the Hamiltonian

$$H = \frac{1}{2}p_r^2 + \frac{l^2}{2r^2}$$

is actually a dynamics on $\mathfrak{sl}(2, \mathbb{R})^*$ and therefore it has the same form independently of the dimension of the space $T\mathbb{R}^n$ we start with.

Symplectic leaves in $\mathfrak{sl}^*(2, \mathbb{R})$ are diffeomorphic to $\mathbb{R}^2$ and pairs of conjugated variables may be introduced, for instance as $\mu = \frac{1}{2}\xi_1$, $\nu = \xi_3/\xi_1$, or $\mu = \frac{1}{2}\xi_2$,

$\nu = -\xi_3/\xi_2$, and then:

$$\{\mu, \nu\} = 1\,.$$

We see in all these examples that the chosen invariant submanifold appears eventually as a 'coupling constant' in the reduced dynamics. Moreover, the final 'second-order description' may be completely unrelated to the original one, that is, what we will call now 'position' is actually a function of the old positions and velocities.

A few remarks are pertinent now.

Remark 7.1 We have not specified the signature of our scalar product on $\mathbb{R}^3$. It is important to notice that the final result does not depend on it. However, because in the reduced dynamics ξ_1 appears in the denominator, when the scalar product is not positive definite we have to remove the full algebraic variety $\langle \mathbf{r}, \mathbf{r}\rangle = 0$ to get a smooth vector field. If the signature were $(+, +, -)$, the relevant group would not be SO(3) anymore but it would be replaced by $SO(2, 1)$.

Remark 7.2 If we consider the Lagrangian description of the free particle, that is, we consider the Lagrangian:

$$L = \frac{1}{2}\langle \dot{\mathbf{r}}, \dot{\mathbf{r}}\rangle,$$

in polar coordinates it becomes

$$L = \frac{1}{2}\left(\dot{r}^2 + r^2(\dot{\mathbf{n}} \cdot \dot{\mathbf{n}})\right),$$

which restricted to the submanifold $\ell^2 = r^4(\dot{\mathbf{n}} \cdot \dot{\mathbf{n}})$ would give

$$L = \frac{1}{2}\left(\dot{r}^2 + \frac{\ell^2}{r^2}\right),$$

which is not the Lagrangian giving raise to the dynamics $\ddot{r} = \ell^2/r^3$. Therefore, we must conclude that the reduction, if done in the Lagrangian formalism, must be considered as a symplectic reduction in terms of the symplectic structure of Lagrangian systems, i.e., in terms of the symplectic Cartan 2-form ω_L and the energy function E_L (recall Sect. 5.5.3).

Remark 7.3 The free particle admits many alternative Lagrangians, therefore once an invariant submanifold Σ has been selected, we have many alternative symplectic structures to pull-back to Σ and define alternative involutive distributions to quotient Σ. The possibility of endowing the quotient manifold with a tangent bundle structure has to be investigated separately because the invariant submanifold Σ does not need to have a particular behaviour with respect to the tangent bundle structure. A recent generalization consists of considering that the quotient space may not have a tangent bundle structure but may have a Lie algebroid structure. Further examples and additional comments on the previous examples may be found in [La91, Ma92].

We shall close now these preliminaries by recalling that another example is the generalization of this procedure to free systems on higher dimensional spaces discussed in Sect. 1.2.12 where the Calogero-Moser potential was obtained starting from the free motion on the space of 2×2 symmetric matrices.

Other examples, as the Toda system and other well-known systems, obtained starting with free or harmonic motions on the space of $n \times n$ Hermitian matrices, free motions on $U(n)$, or free motions on the coset space $GL(n,)/U(n, \mathbb{C})$, can be found in the literature (see for instance [OlPe81]).

We can summarize the previous discussion by saying that the reduction of the various examples that we have considered are based on the selection of an invariant submanifold and the selection of an invariant subalgebra of functions.

7.2.5 Reduction and Solutions of the Hamilton-Jacobi Equation

To illustrate the relation between reduction of linear systems and finding solutons to the Hamilton-Jacobi equation we will choose the obtention of the Calogero-Moser system out of free motion on the space of symmetric 2×2 matrices performed in Sect. 1.2.12.

Let us consider the space of symmetric matrices X and the Lagrangian function

$$L = \frac{1}{2} \mathrm{Tr}\,(\dot{X}^2)\,.$$

This Lagrangian gives raise to the Euler-Lagrange equations of motion Eq. (1.130):

$$\ddot{X} = 0\,.$$

Moreover, the symplectic structure associated to it is defined by

$$\omega_L = \mathrm{Tr}\,(dX \wedge d\dot{X})\,,$$

and the energy is given by:

$$E_L = L\,.$$

The invariance of the Lagrangian under translations and rotations implies the conservation of the linear momentum $P = \dot{X}$ and the angular momentum $M = [X, \dot{X}]$. The corresponding explicit solutions of the dynamics are thus given by:

$$X(t) = X_0 + t P_0\,. \tag{7.13}$$

for given initial data X_0 and P_0.

It is possible to find easily a solution of the corresponding Hamilton-Jacobi equation. Indeed, by integrating the Lagrangian L along the solutions given by Eq. (7.13) or by solving

$$P_t\, dX_t - P_0\, dX_0 = dS(X_t, X_0; t)$$

we find that Hamilton's principal function S is written as

$$S = \frac{1}{2t}\mathrm{Tr}\,(X_t - X_0)^2 .$$

By fixing a value $\ell^2 = \frac{1}{2}\mathrm{Tr}\, M^2$ we select an invariant submanifold Σ. The corresponding reduced dynamics gives the Calogero-Moser equations (1.134). Therefore if we restrict S to those solutions which satisfy

$$\frac{1}{2}\mathrm{Tr}\,(X_t^2 X_0^2 - (X_t X_0)^2) = \ell^2 ,$$

we will find a solution for the Hamilton-Jacobi equation associated with the reduced dynamics [Ca07b, Ca09].

Remark 7.4 For any invertible constant symmetric matrix K, the Lagrangian function $L_K = \frac{1}{2}\mathrm{Tr}\,(\dot{X} K \dot{X})$ would describe again the free motion. More generally, for any monotonic function f, the composition $f(L_K)$ would be a possible alternative Lagrangian. The corresponding Lagrangian symplectic structure ω_L could be used to find alternative Hamilton-Jacobi equations. For those aspects we refer to [Ca06].

7.3 The Geometrical Description of Reduction

We will provide first a geometric description of the various examples of reduction discussed before. We denote by M the manifold which is the carrier space for the equations of motion of our system that will be represented by a vector field Γ. We suppose that its flow gives raise to a one-parameter group of transformations

$$\Phi\colon \mathbb{R} \times M \to M .$$

Occasionally, when we want our map to keep track of the infinitesimal generator Γ we will write Φ_Γ or $\Phi_\Gamma\colon \mathbb{R} \times M \to M$.

To apply the general reduction procedure we need:

1. A submanifold $\Sigma \subset M$, invariant under the evolution, i.e.,

$$\Phi(\mathbb{R} \times \Sigma) \subset \Sigma , \quad \text{or} \quad \Phi_t(m) = \Phi(t, m) \in \Sigma , \ \forall t \in \mathbb{R}, \ m \in \Sigma .$$

2. An invariant equivalence relation among points of Σ, that is we consider equivalence relation $\sim$ for which if $m \sim m'$ then $\Phi_t(m) \sim \Phi_t(m')$, for all $t \in \mathbb{R}$.

The reduced dynamics $\tilde{\Phi}$ or 'reduced evolution' is defined on the manifold $\tilde{\Sigma} = \Sigma/\sim$, of equivalence classes of points on Σ (assumed to be endowed with a differentiable structure) by means of the natural definition:

$$\widetilde{\Phi}_t(\tilde{m}) = \widetilde{\Phi_t(m)}\,.$$

The invariance of the equivalence relation $\sim$ guarantees that $\widetilde{\Phi}_t$ is well defined, that is will not depend on the choice of the representative m we choose in the class $\tilde{m}$.

One may also start the other way around: we could first consider an invariant equivalence relation $\sim$ on the whole manifold M and then select an invariant submanifold $\widetilde{\Sigma}$ for the reduced dynamics $\widetilde{\Phi}_t$ defined on the quotient space $\widetilde{M} = M/\sim$, to further reduce the dynamical evolution.

In real physical situations the invariant submanifolds arise as level set of functions. These level sets were called invariant relations by Levi-Civita [Am26] to distinguish them from level sets of constants of the motion. Usually, equivalence classes will be generated by orbits of Lie group actions or leaves of involutive distributions. 'Closed subgroup' theorems are often employed to guarantee the regularity of the quotient manifold [Pa57].

We can collect the previous arguments in the form of a theorem. The theorem cannot provide in general conditions guaranteeing that the quotient spaces arising on it are smooth manifolds, something that must be assumed as part of the conditions on each particular instance. The meaning of the theorem can be pictorially visualized in Fig. 7.1.

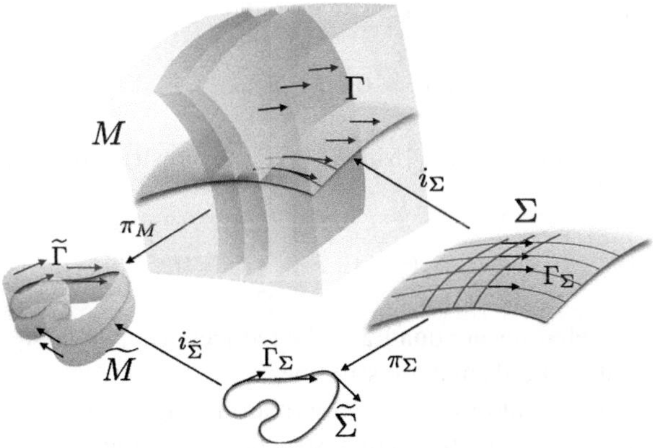

Fig. 7.1 A pictorial representation of geometrical reduction: the '*purple*' leaves in M represent the Γ-invariant equivalence relation while the '*green*' submanifold is the Γ-invariant submanifold Σ. We can reach the reduced space $\widetilde{\Sigma}$ either by restricting to Σ first and then quotienting by the induced equivalence relation, or we can quotient first and then restrict to the submanifold $\widetilde{\Sigma}$

Theorem 7.2 (Geometrical reduction theorem) *Let Γ be a dynamical system defined on the carrier manifold M. Let Σ be a Γ-invariant submanifold and $\sim$ a Γ-invariant equivalence relation. Let us assume that the quotient spaces $\widetilde{M} = M/\sim$ and $\widetilde{\Sigma} = \Sigma/\sim$ are smooth manifolds and the projection maps $\pi_M : M \to \widetilde{M}$, $\pi_\Sigma : \Sigma \to \widetilde{\Sigma}$ are smooth submersions. We denote by Γ_Σ the restriction of the dynamics Γ to Σ. The dynamical vector field Γ is π_M-projectable and its projection on $\widetilde{M}$ will be denoted by $\widetilde{\Gamma}$. Similarly, the vector field Γ_Σ is π_Σ-projectable and its projection to $\widetilde{\Sigma}$ will be denoted by $\widetilde{\Gamma}_\Sigma$. Then $\widetilde{\Sigma}$ is a $\widetilde{\Gamma}$-invariant submanifold in $\widetilde{M}$ and the restriction of $\widetilde{\Gamma}$ to it coincides with $\widetilde{\Gamma}_\Sigma$, that is:*

$$\widetilde{\Gamma}_\Sigma = \widetilde{\Gamma} \mid_{\widetilde{\Sigma}} .$$

Exercise 7.2 Prove the previous theorem (the proof is an exercise in 'diagram chasing').

When additional structures are present, like Poisson or symplectic structures, it is possible to get involutive distributions out of a family of invariant relations. The so-called 'symplectic reduction' is an example of this particular situation.

When the space is endowed with additional structures, say a tangent or a cotangent bundle, with the starting dynamics being, for instance, second-order (in the tangent case), we may also ask for the reduced one to be second order, once we ask the reduced space to be also endowed with a tangent space structure. This raises natural questions on how to find appropriate tangent or cotangent bundle structures on a given manifold obtained as a reduced carrier space. Similarly, we may start with a linear dynamics, perform a reduction procedure (perhaps by means of quadratic invariant relations) and enquire about possible linear structures on the reduced carrier space. A simple example of this situation is provided by the Maxwell equations. These equations may be written in terms of the Faraday 2-form F encoding the electric field E and the magnetic field B, as:

$$dF = 0 \quad d * F = 0 ,$$

when considered in the vacuum [Ma05]. We may restrict these equations to the invariant submanifold

$$F \wedge F = 0, \, F \wedge *F = 0 .$$

Even though these relations are quadratic the reduced Maxwell equations provide as solutions the radiation fields and are still linear.

In conclusion, when additional structures are brought into the picture, we may end up with extremely rich mathematical structures and quite difficult mathematical problems.

7.3.1 A Charged Non-relativistic Particle in a Magnetic Monopole Field

This system was considered by Dirac [Di31] and a variant of it, earlier by Poincaré [Po96]. To describe it in terms of a Lagrangian, Dirac introduced a 'Dirac string'. The presence of this unphysical singularity leads to technical difficulties in the quantization of this system. Several proposals have been made to deal with these problems.

Here we would like to show how our reduction procedure allows us to deal with this system and provides a clear way for its quantization. In doing this we shall follow mainly [Ba80, Ba83, Ba91].

The main idea is to replace $\mathbb{R}_0^3$ with $\mathbb{R}_0^4$ described as the product $\mathbb{R}_0^4 = S^3 \times \mathbb{R}_+$, and to get back our space of relative coordinates for the charge-monopole by means of a reduction procedure.

We set first $\mathbf{x} \cdot \boldsymbol{\sigma} = r s \sigma_3 s^{-1}$, where $r^2 = x_1^2 + x_2^2 + x_3^2$ and $s \in SU(2)$ (realized as 2×2 matrices of the defining representation; while $\{\sigma_1, \sigma_2, \sigma_3\}$ are the Pauli matrices Eq. (10.68)). We write the Lagrangian function on $\mathbb{R}_0^4$ as

$$L = \frac{1}{2} m \mathrm{Tr} \left(\frac{d}{dt} \left(r s \sigma_3 s^{-1} \right) \right)^2 - k \left(\mathrm{Tr}\, \sigma_3 s^{-1} \dot{s} \right)^2 .$$

This expression for the Lagrangian shows clearly the invariance under the left action of $SU(2)$ on itself and an additional invariance under the right $U(1)$-action $s \mapsto s e^{i\sigma_3 \theta}$ for $\theta \in [0, 2\pi)$. It is convenient to introduce left invariant 1-forms θ^a by means of $i\sigma_a \theta^a = s^{-1} ds$ and related left invariant vector fields X_a which are dual to them $\theta^a(X_b) = \delta^a_b$. If Γ denotes any second-order vector field on $\mathbb{R}_0^4$ we set $\dot{\theta}^a = \theta^a(\Gamma)$, where, with some abuse of notation, we are using the same symbol for θ^a on $\mathbb{R}_0^4$ and its pull-back to $T\mathbb{R}_0^4$. It is also convenient to use the unit vector $\mathbf{n}$ defined by $\mathbf{x} = \mathbf{n} r$, i.e., $\mathbf{n} \cdot \boldsymbol{\sigma} = s \sigma_3 s^{-1}$.

After some computations, the Lagrangian becomes

$$L = \frac{1}{2} m \dot{r}^2 + \frac{1}{4} m r^2 (\dot{\theta}_1^2 + \dot{\theta}_2^2) + k \dot{\theta}_3^2. \tag{7.14}$$

It is not difficult to find the canonical 1- and 2-forms for the Lagrangian symplectic structure. For instance $\theta_L = m\dot{r}\, dr + \frac{1}{2} m r^2 (\dot{\theta}_1\, d\theta_1 + \dot{\theta}_2\, d\theta_2) + 2k\dot{\theta}_3\, d\theta_3$; and of course $\omega_L = -d\theta_L$. The energy function E_L coincides with L.

If we fix the submanifold Σ_c by setting

$$\Sigma_c = \left\{ (\mathbf{r}, \mathbf{v}) \in T\mathbb{R}_0^4 \mid \dot{\theta}_3 = c \right\},$$

the submanifold turns out to be invariant because $\dot{\theta}_3$ is a constant of the motion.

On Σ_c, $\theta_L = m\dot{r}\, dr + \frac{1}{2} m r^2 (\dot{\theta}_1\, d\theta_1 + \dot{\theta}_2\, d\theta_2) + 2kc\, d\theta_3$. If we then use the foliation associated with X_3^T (the tangent lift of X_3 to $T\mathbb{R}_0^4$), we find that ω_L is the pull-back of a 2-form on the quotient manifold because $d\theta_3 = \theta_1 \wedge \theta_2$, and hence contains X_3^T in its kernel. The term $d\theta_3$ is exactly proportional to the magnetic field of the magnetic monopole sitting at the origin. Thus on the quotient space of Σ_c by the action of the left flow of X_3^T we recover the dynamics of the electron-monopole system on the (quotient) space $T(S^2 \times \mathbb{R}_+) = T\mathbb{R}_0^3$. It is not difficult to show that

$$\frac{d}{dt}\left(-\frac{i}{2}\left[\mathbf{n}\cdot\boldsymbol{\sigma}, mr^2\dot{\mathbf{n}}\cdot\boldsymbol{\sigma}\right] + k\mathbf{n}\cdot\boldsymbol{\sigma}\right) = 0; \quad k = \frac{eg}{4\pi}.$$

These constants of motion are associated with the rotational invariance and replace the usual angular momentum functions.

This example shows that the reduction of theLagrangian system of Kaluza-Klein type on $T\mathbb{R}^4$ does not reduce to a Lagrangian system on $T\mathbb{R}^3$ but just to a symplectic system.

7.4 The Algebraic Description

We can recapitulate the examples of reduction before and identify the two basic ingredients present on all of them.

1. A Γ invariant subalgebra (of functions) $\mathcal{R}$.
2. A Γ invariant submanifold of the carrier space $\Sigma \subset M$.

The evaluation map $\mathrm{ev}\colon M \times \mathcal{F} \to \mathbb{R}$ defined as $(m, f) \mapsto f(m)$, allows us to dualize the basic ingredients from the manifold to the associative and commutative algebra of functions on M, $\mathcal{F}$, the set of observables.

We first notice that to any submanifold $i_\Sigma : \Sigma \hookrightarrow M$ we can associate a short exact sequence of associative algebras

$$0 \longrightarrow \mathcal{I}_\Sigma \longrightarrow \mathcal{F} \xrightarrow{i_\Sigma^*} \mathcal{F}_\Sigma \longrightarrow 0\,,$$

where $\mathcal{F}_\Sigma$ is the associative and commutative algebra of functions in Σ, defined in terms of the identification map $i_\Sigma : \Sigma \hookrightarrow M, m \in \Sigma \mapsto m \in M$. Here $\mathcal{I}_\Sigma$ is given by:

$$\mathcal{I}_\Sigma = \left\{f \in \mathcal{F} \mid i_\Sigma^*(f) = 0\right\}$$

Since i_Σ^* is an algebra epimorphism, we find that $\mathcal{I}_\Sigma = \ker i_\Sigma^*$ is a bilateral ideal in $\mathcal{F}$. The algebra $\mathcal{F}_\Sigma$ is then the quotient algebra $\mathcal{F}/\mathcal{I}_\Sigma$.

Any derivation Γ acting on the set of functions of $\mathcal{F}$ will define a derivation on the set of functions $\mathcal{F}_\Sigma$ if and only if $\mathcal{L}_\Gamma\mathcal{I}_\Sigma \subset \mathcal{I}_\Sigma$, so that we can define an action of Γ on the set of equivalence classes by $\mathcal{L}_X(f + \mathcal{I}_\Sigma) = \mathcal{L}_X f + \mathcal{I}_\Sigma$, which defines a derivation on the reduced carrier space $\mathcal{F}_\Sigma$.

A simple example illustrates the procedure. On $T\mathbb{R}^3$ we can consider the submanifold Σ defined by the vanishing of the funtions

$$f_1 = \mathbf{r} \cdot \mathbf{r} - 1 \quad f_2 = \mathbf{r} \cdot \mathbf{v}. \tag{7.15}$$

Then $\mathcal{I}_\Sigma$ is the corresponding bilateral ideal.

We get in this way the submanifold Σ as TS^2. The algebra of functions $\mathcal{F}_\Sigma$ is obtained by pull-back from $\mathcal{F}(T\mathbb{R}^3)$, i.e., simply by using in the argument of $f(\mathbf{r}, \mathbf{v})$ the constraints (7.15), i.e., $f_1 = 0 = f_2$. A vector field X on $T\mathbb{R}^3$ will be tangent to $\Sigma = TS^2$ if and only if

$$\mathcal{L}_X(\mathbf{r} \cdot \mathbf{r} - 1) = \alpha(\mathbf{r} \cdot \mathbf{r} - 1) + \beta \mathbf{r} \cdot \mathbf{v},$$

for arbitrary functions α, β and also

$$\mathcal{L}_X(\mathbf{r} \cdot \mathbf{v}) = \alpha'(\mathbf{r} \cdot \mathbf{r} - 1) + \beta' \mathbf{r} \cdot \mathbf{v}.$$

for arbitrary functions α', β'.

It is not difficult to show that the module of such derivations is generated by

$$\mathcal{R}_l = \sum_{j,k} \epsilon_{ljk} \left(x_j \frac{\partial}{\partial x_k} + v_j \frac{\partial}{\partial v_k} \right); \quad \mathcal{V}_l = \sum_{i,j} \epsilon_{lij} x_i \frac{\partial}{\partial v_j}$$

A Γ-invariant under Γ subalgebra in $\mathcal{F}$, say $\widetilde{\mathcal{F}}$, for which $\mathcal{I}_\Sigma$ is an ideal, defines a Γ-invariant under Γ equivalence relation in M by setting

$$m' \sim m'' \quad \text{iff} \quad f(m') = f(m''), \quad \forall f \in \widetilde{\mathcal{F}} \tag{7.16}$$

It follows that $\widetilde{\mathcal{F}}$ defines a subalgebra in $\mathcal{F}_\Sigma$ and corresponds to a possible quotient manifold of Σ by the equivalence relation defined by $\widetilde{\mathcal{F}}$.

In general, a subalgebra of $\mathcal{F}$, say $\mathcal{F}_Q$, defines a short exact sequence of Lie algebras

$$0 \longrightarrow \mathfrak{X}^v \longrightarrow \mathfrak{X}^N \longrightarrow \mathfrak{X}_Q \longrightarrow 0 \tag{7.17}$$

where $\mathfrak{X}^v$ is the Lie algebra of vector fields annihilating $\mathcal{F}_Q$, $\mathfrak{X}^N$ is the normalizer of $\mathfrak{X}^v$ in $\mathfrak{X}(M)$, and $\mathfrak{X}_Q$ is the quotient Lie algebra. This sequence of Lie algebras may be considered a sequence of Lie modules with coefficients in $\mathcal{F}_Q$. In the previous case, $\mathcal{F}_Q$ would be the invariant subalgebra in $\mathcal{F}_\Sigma$ and the equivalence relation would be defined by the leaves of the involutive distribution $\mathfrak{X}^v$ (regularity requirements should be then imposed on $\mathcal{F}_Q$). See [La90] for details.

From the dual point of view it is now clear that reducible evolutions will be defined by one-parameter groups of transformations which are automorphisms of the corresponding short exact sequences. The corresponding infinitesimal versions will be defined in terms of derivations of the appropriate short exact sequence of algebras.

To illustrate this aspect, we consider the associative subalgebra of $\mathcal{F}(T\mathbb{R}^3)$ generated by $\{\mathbf{r}\cdot\mathbf{r}, \mathbf{v}\cdot\mathbf{v}, \mathbf{r}\cdot\mathbf{v}\}$. For this algebra it is not difficult to see that the vector fields

$$X_c = \sum_{a,b} \epsilon_{abc} \left(x^a \frac{\partial}{\partial x_b} + v^a \frac{\partial}{\partial v_b} \right)$$

generate $\mathfrak{X}^v$, while $\mathfrak{X}^N$ is generated by $\mathfrak{X}^v$ and

$$\mathbf{r}\cdot\nabla_{\mathbf{v}}, \quad \mathbf{v}\cdot\nabla, \quad \mathbf{r}\cdot\nabla, \quad \mathbf{v}\cdot\nabla_{\mathbf{v}}.$$

The quotient $\mathfrak{X}_Q$, with a slight abuse of notation, can also be considered to be generated by the vector fields

$$\mathbf{r}\cdot\nabla_{\mathbf{v}}, \quad \mathbf{v}\cdot\nabla, \quad \mathbf{r}\cdot\nabla, \quad \mathbf{v}\cdot\nabla_{\mathbf{v}},$$

which however are not all independent. Any combination of them with coefficients in the subalgebra may be considered a 'reduced dynamics'.

7.4.1 Additional Structures: Poisson Reduction

When a Poisson structure on the manifold M is available, we can further qualify the previous picture. We can consider associated short exact sequences of Hamiltonian derivations, i.e., derivations given by Hamiltonian vector fields. Hence, a Poisson reduction can be formulated in the following way: we start with $\mathcal{I}_\Sigma$, again an ideal in the commutative and associative algebra $\mathcal{F}$. We consider then the set of Hamiltonian derivations which map $\mathcal{I}_\Sigma$ into itself:

$$W(\mathcal{I}_\Sigma) = \{ f \in \mathcal{F} \mid \{f, \mathcal{I}_\Sigma\} \subset \mathcal{I}_\Sigma \}.$$

Then we consider $\mathcal{I}'_\Sigma = I_\Sigma \cap W(\mathcal{I}_\Sigma)$ and get the exact sequence of Poisson algebras

$$0 \longrightarrow \mathcal{I}'_\Sigma \longrightarrow W(\mathcal{I}_\Sigma) \longrightarrow \mathcal{Q}_\Sigma \longrightarrow 0.$$

When the ideal $\mathcal{I}_\Sigma$ is given by constraint functions as in the Dirac approach, $W(\mathcal{I}_\Sigma)$ is the set of first-class functions while $\mathcal{I}'_\Sigma$ is that of the first-class constraints.

We give here an example of an iterated reduction. We consider a parametrization of $T\mathbb{R}^4$ in terms of the identity matrix in dimension 2 σ_0, and the 2×2 Pauli matrices Eq. (10.68) as follows:

$$\pi = p_0\sigma_0 + \sum_{a=1}^{3} p_a\sigma_a = \begin{pmatrix} p_0 + p_3 & p_1 - i\,p_2 \\ p_1 + i\,p_2 & p_0 - p_3 \end{pmatrix},$$

$$g = y_0\sigma_0 + \sum_{a=1}^{3} y_a\sigma_a = \begin{pmatrix} y_0 + y_3 & y_1 - i\,y_2 \\ y_1 + i\,y_2 & y_0 - y_3 \end{pmatrix}.$$

A preliminary 'constraint' manifold is selected by requiring that

$$\mathrm{Tr}\,(g^\dagger g) = 1, \quad \mathrm{Tr}\,(\pi^\dagger g) = 0.$$

Recall that $(\mathbf{a}\cdot\boldsymbol{\sigma})(\mathbf{b}\cdot\boldsymbol{\sigma}) = \mathbf{b}\cdot\mathbf{a}\,\sigma_0 + (\mathbf{a}\times\mathbf{b})\cdot\boldsymbol{\sigma}$ from where we easily obtain that

$$\mathrm{Tr}\,g^\dagger g = y_0^2 + \mathbf{y}\cdot\mathbf{y}, \qquad \mathrm{Tr}\,\pi^\dagger g = y_0 p_0 + \mathbf{y}\cdot\mathbf{p}.$$

Therefore the sub manifold defined by the constraints is diffeomorphic to the tangent bundle of S^3, i.e., TS^3. The Hamiltonian

$$H = \frac{1}{2}\left(p_\mu p^\mu + y_\mu y^\mu\right)$$

defines a vector field tangent to the constraint manifold.

Similarly for the 'potential' function $V = \frac{1}{2}(y_0^2 + y_3^2 - y_1^2 - y_2^2)$. The Hamiltonian function $H + V$, when restricted to TS^3 with a slight abuse of notation acquires the suggestive form

$$H = \frac{1}{2}\left(p_0^2 + p_3^2 + y_0^2 + y_3^2\right) + \frac{1}{2}\left(p_1^2 + p_2^2 - y_1^2 - y_2^2\right).$$

By using the relation $y_0^2 + y_1^2 + y_2^2 + y_3^2 = 1$ we may also write it in the form

$$H = \frac{1}{2}\left(p_0^2 + p_3^2 + 2\left(y_0^2 + y_3^2\right)\right) + \frac{1}{2}\left(p_1^2 + p_2^2\right) - \frac{1}{2}.$$

Starting now with TS^3 we may consider a further reduction by fixing

$$\Sigma_K = \left\{(y_\mu, p^\mu) \in TS^3 \mid y_0 p_3 - p_0 y_3 + y_1 p_2 - y_2 p_1 = K\right\},$$

and quotienting by the vector field

$$X = y_0\frac{\partial}{\partial y_3} - y_3\frac{\partial}{\partial y_0} + y_1\frac{\partial}{\partial y_2} - y_2\frac{\partial}{\partial y_1} + p_0\frac{\partial}{\partial p_3} - p_3\frac{\partial}{\partial p_0} + p_1\frac{\partial}{\partial p_2} - p_2\frac{\partial}{\partial p_1},$$

which is tangent to the foliation because $\mathcal{L}_X(y_0 p_3 - p_0 y_3 + y_1 p_2 - y_2 p_1) = 0$.

The final reduced manifold will be $TS^2 \subset T\mathbb{R}^3$, with projection $TS^3 \to TS^2$ provided by the tangent of the Hopf fibration $\pi : S^3 \to S^2$, defined as

$$x_1 = 2(y_1y_3 - y_0y_2) \quad x_2 = 2(y_2y_3 - y_0y_1) \quad x_3 = y_0^2 + y_3^2 - y_1^2 - y_2^2.$$

The final reduced dynamics will be associated with the Hamiltonian function of the spherical pendulum.

The spherical pendulum is thus identified by

$$S^2 \subset \mathbb{R}^3 = \left\{\mathbf{x} \in \mathbb{R}^3 \mid \langle \mathbf{x}, \mathbf{x}\rangle = x_1^2 + x_2^2 + x_3^2 = 1\right\}$$

$$TS^2 \subset T\mathbb{R}^3 = \left\{(\mathbf{x}, \mathbf{v}) \in \mathbb{R}^3 \times \mathbb{R}^3 \mid \langle \mathbf{x}, \mathbf{x}\rangle = 1, \ \langle \mathbf{x}, \mathbf{v}\rangle = 0\right\}$$

The dynamics is given by means of $\omega = \sum_i dx_i \wedge dv_i$ when restricted to TS^2, in terms of $E = \frac{1}{2}\langle \mathbf{v}, \mathbf{v}\rangle + x_3$. The angular momentum is a constant of motion corresponding to the rotation around the Ox_3 axis. The energy momentum map

$$\mu : TS^2 \to \mathbb{R}^2 : (\mathbf{x}, \mathbf{v}) \mapsto (E(\mathbf{x}, \mathbf{v}), L(\mathbf{x}, \mathbf{v}))$$

has quite interesting properties as shown by [Cu97, Du80].

7.4.2 Reparametrization of Linear Systems

7.4.2.1 Reparametrization by Constants of Motion

Consider now, on a linear vector space E, dim $E = n$, with coordinates $\left(x^1, \ldots, x^n\right)$, a general, homogeneous linear dynamical system of the form

$$\frac{dx^j}{dt} = A^j{}_i x^i, \qquad j = 1, \ldots, n. \tag{7.18}$$

With the matrix $A \in \mathrm{End}\ (E)$ we can associate, as already discussed, the (linear) vector field

$$\Gamma_A = A^j{}_i x^i \frac{\partial}{\partial x^j}\,. \tag{7.19}$$

We can reparametrize the linear system (7.18) by using a constant of motion, and still we will obtain an explicitly integrable nonlinear system. Let h be a constant of motion for the dynamical system (7.18), $dh/dt = 0$, or equivalently

$$A^j{}_i\, x^i \frac{\partial h}{\partial x^j} = 0. \tag{7.20}$$

Then if we assume h to be non zero in an open dense set of E the dynamical system with the reparametrized evolution equations

$$\frac{dx^j}{dt} = h\,(x)\,A^j{}_i x^i = C(x)^j{}_i x^i \tag{7.21}$$

has the same constants of motion. In fact a simple computation shows that now

$$\frac{dF}{dt} = h(x)A^j{}_i\, x^i\, \frac{\partial F}{\partial x^j}, \tag{7.22}$$

and hence, if $h \neq 0$ in an open dense set of E, this implies that F is a constant of the reparametrized motion iff $A^j{}_i\, x^i\, \partial F/\partial x^j = 0$ in an open dense set of E and by continuity, it vanishes on all E and therefore F is a constant of motion determined by (6.91). Let us recall that the constants of motion are an associative and commutative real algebra, because if f and g are constants and $\lambda, \mu \in \mathbb{R}$, then

$$\frac{d}{dt}(\lambda f + \mu g) = \lambda \frac{df}{dt} + \mu \frac{dg}{dt} = 0 \tag{7.23}$$

and

$$\frac{d}{dt}(fg) = f\frac{dg}{dt} + g\frac{df}{dt} = 0. \tag{7.24}$$

Thus we conclude that if h is a constant of motion for the system (7.18) and $h \neq 0$ in an open dense set of E, then the system (7.22) has the same algebra of constants of motion as the original one.

This implies that if we denote by ϕ_t the flow of (7.22) and by $\gamma(t)$ the solution with initial data x_0, then

$$h(\gamma(t)) = h(\phi_t(x_0)) = h(x_0), \tag{7.25}$$

for all $t \in \mathbb{R}$. Hence, the integral curve $\gamma(t)$ of (7.22) starting at x_0 will be $\left(A = \left[A^j{}_i\right]\right)$

$$\gamma\,(t) = \exp\,(t h\,(x_0)\,A)\,x_0. \tag{7.26}$$

In fact, if $\gamma(t)$ is given by (7.26),

$$\frac{d}{dt}\gamma(t) = \frac{d}{dt}\exp(th(x_0)A)x_0 = h(x_0)A\gamma(t) = h(\gamma(t))A\gamma(t). \tag{7.27}$$

Denoting by Γ the dynamical system (7.18) and by $\Gamma_h = h\Gamma$ the system (7.22), if Γ has a maximal set of constants of motion the flow will be found by quadratures (see next chapter). If Γ is nonlinear, what will be interesting for us will be to find out whether there exists a constant of the motion h such that $h\Gamma$ is linear (notice that it can be linear with respect to the given linear structure or with respect to an alternative linear structure, see the discussion of this point in one of the following chapters).

The reparametrization procedure discussed above can be extended to a more general setting that allows us to construct nonlinear systems whose solutions can be found explicitly in terms of solutions of linear systems. Let Γ be the dynamical system defined by the nonlinear equations

$$\frac{dx^j}{dt} = C(x)^j{}_i\, x^i \tag{7.28}$$

and such that the entries of the matrix field $C(x) = [C^j{}_i(x)]$ are constants of motion for Γ, i.e.:$dC^j{}_i/dt = 0$, for all $i, j = 1, \ldots, n$. In other words, the functions $C^j{}_i$ verify the partial differential equations

$$\frac{\partial C^j{}_i}{\partial x^k} C^k{}_l\, x^l = 0, \tag{7.29}$$

for all i, j. In such case, if $\gamma(t)$ is the integral curve for Γ starting at x_0, it is clear that $\gamma(t)$ is given by $\left(C = \left[C^j{}_i\right]\right)$

$$\gamma(t) = \exp\left(tC\left(x_0\right)\right) x_0. \tag{7.30}$$

In fact, a simple computation shows that

$$\frac{d}{dt}\gamma(t) = C(x_0)\gamma(t) = C(\gamma(t))\gamma(t), \tag{7.31}$$

because C is a matrix of constants of motion for Γ.

Thus, for each set of Cauchy data x_0, $C(x_0)$ will be a numerical matrix and the flow of our system, when restricted to the surface:

$$\Sigma = \{x \in E \mid C(x) = C(x_0)\} \tag{7.32}$$

will be given by

$$x(t) = \exp\left(tC\left(x_0\right)\right) x_0 \tag{7.33}$$

Notice that the curve $\gamma(t)$ above is contained in Σ. It is clear that for given values of the constants of motion $C^j{}_i$ the surface Σ can be empty.

This situation is illustrated by the following example.

Example 7.3 On $\mathbb{R}^4$ with coordinates (x_1, x_2, v_1, v_2), consider the linear system Γ corresponding to two uncoupled harmonic oscillators :

$$\begin{cases} \dfrac{dx_1}{dt} = v_1 \\ \dfrac{dx_2}{dt} = v_2 \\ \dfrac{dv_1}{dt} = -x_1 \\ \dfrac{dv_2}{dt} = -x_2 \end{cases} \tag{7.34}$$

We notice that as we have two copies of the same second-order differential equation, $\ddot{x} + x = 0$, the Wronskian of two solutions is a constant of the motion,

$$\frac{d}{dt}(x_1 v_2 - v_1 x_2) = 0, \tag{7.35}$$

and hence $x_1 v_2 - v_1 x_2$ is a constant of motion. According to the previously described procedure, we can construct the nonlinear system

$$\begin{cases} \dfrac{dx_1}{dt} = v_1 \\ \dfrac{dx_2}{dt} = v_2 \\ \dfrac{dv_1}{dt} = -(x_1 v_2 - v_1 x_2)x_1 \\ \dfrac{dv_2}{dt} = -(x_1 v_2 - v_1 x_2)x_2 \end{cases} \tag{7.36}$$

and for any initial conditions $(x_1(0), x_2(0), v_1(0), v_2(0))$ our system will behave like a linear one, because it corresponds to a time reparametrization of the latter by a constant of motion. However there is a difference with the reparametrization discussed previously. In this case the vector field Γ is not modified by a product with a constant of motion but only its vertical part is. In other words, as pointed out before, we are multiplying not by a function but by a matrix.

7.4.2.2 Reparametrization by Arbitrary Functions

With reference to the previous section, it turns out that it is not strictly necessary that h be a constant of motion for the dynamical system Γ to be used as a reparametrization function. Consider again a linear equation of motion on the vector space E given by Γ, and an arbitrary function h, and the reparametrized equation $\Gamma_h = h\,\Gamma$. We know that the integral curve of Γ starting from the point x_0 is $\gamma(t) = \exp(t\,A)\;x_0$. We are interested in finding a new parametrization given by $t = \varphi(s)$ such that the curve

$$\gamma_h(s) = \gamma(\varphi(s)) \tag{7.37}$$

be an integral curve of the vector field Γ_h, i.e., that

$$\frac{d}{dt}\gamma(t) = \Gamma(\gamma(t)) \Longrightarrow \frac{d}{ds}\gamma_h(s) = \Gamma_h(\gamma_h(s))\,.$$

Taking into account that

$$\frac{d}{ds}\gamma_h(s) = \frac{d\varphi(s)}{ds}\left(\frac{d\gamma(t)}{dt}\right)_{t=\varphi(s)} = \frac{d\varphi(s)}{ds}\,A\,\gamma(\varphi(s))\,,$$

we see that in order for γ_h be an integral curve of Γ_h, the function φ should be such that

$$\frac{d\varphi(s)}{ds} = h(\gamma_h(s))\,,$$

or when written in a different way,

$$\frac{dt}{ds} = h(\gamma(t))\,,$$

whose solutions are obtained by integrating along γ the function h, namely,

$$s - s_0 = \int_{t_0}^{t} \frac{dt'}{h(\exp(t'A)x_0)}\,, \tag{7.38}$$

and by performing the integral we have

$$s - s_0 = s(t, x_0).$$

Notice that

$$\frac{ds}{dt} = \frac{1}{h(\exp((t-t_0)A)x_0)} \neq 0, \tag{7.39}$$

and thus we can solve this equation with respect to t, say $t = t(s, x_0)$. Then, we can write the integral curve of $\Gamma_h = h\Gamma$ through x_0 as

$$\gamma_h(s) = \gamma(t(s, x_0))\,. \tag{7.40}$$

In the particular case in which h is a constant of motion we have $h((\exp tA)x_0) = h(x_0)$ and therefore Eq. (7.39) yields

$$s - s_0 = \frac{t - t_0}{h(x_0)}. \tag{7.41}$$

It follows then that $t - t_0 = h(x_0)(s - s_0)$ and $\gamma(t) = (\exp(t - t_0)A)x_0$ becomes $x(s) = \exp(h(x_0)(s - s_0)A)x_0$, according to what we have found in Eq. (7.26).

7.4.2.3 A Cubic Dynamical System

We will discuss here as an application of the generalized reparametrization method described above the explicit integrability of a given cubic vector field.

Let us consider $\mathbb{R}^3$ with the diagonal metric or pseudo-metric η with signature $(\pm, +, +)$. We will denote the product of two vectors $\mathbf{q} = (q_0, q_1, q_2)$ and $\mathbf{q}' = (q_0', q_1', q_2')$ as

$$\langle \mathbf{q}, \mathbf{q}' \rangle = \eta_{\alpha\beta} q^\alpha \, q'^\beta = \pm q_0 q_0' + q_1 q_1' + q_2 q_2'. \tag{7.42}$$

Let us consider the Hamiltonian system (with phase space $\mathbb{R}^6$, then) associated with the quartic Hamiltonian

$$H = \frac{1}{2}\left(\langle \mathbf{p}, \mathbf{p} \rangle \langle \mathbf{q}, \mathbf{q} \rangle - \langle \mathbf{p}, \mathbf{q} \rangle^2 \right). \tag{7.43}$$

If the metric is Euclidean this can be written also as

$$H = \frac{1}{2} \left(\langle \mathbf{p} \times \mathbf{q}, \mathbf{p} \times \mathbf{q} \rangle \right) \tag{7.44}$$

with '$\times$' denoting the usual exterior product (vector product) of vectors in $\mathbb{R}^3$. The dynamical vector field is given by Hamilton's equations of motion

$$\dot{\mathbf{q}} = \frac{\partial H}{\partial \mathbf{p}}, \qquad \dot{\mathbf{p}} = -\frac{\partial H}{\partial \mathbf{q}} \tag{7.45}$$

We can compute the previous equations easily by using the standard Poisson brackets on phase space, and we get

$$\begin{aligned} \dot{\mathbf{p}} &= \{\mathbf{p}, H\} = \langle \mathbf{q}, \mathbf{p} \rangle \mathbf{p} - \langle \mathbf{p}, \mathbf{p} \rangle \, \mathbf{q} \\ \dot{\mathbf{q}} &= \{\mathbf{q}, H\} = \langle \mathbf{q}, \mathbf{q} \rangle \, \mathbf{p} - \langle \mathbf{q}, \mathbf{p} \rangle \mathbf{q} \end{aligned} \tag{7.46}$$

Then,

$$\ddot{\mathbf{q}} = \{\{\mathbf{q}, H\}, H\} = -2H \, \mathbf{q} \tag{7.47}$$

We also find the following constants of motion:

$$\{H, \langle \mathbf{q}, \mathbf{p} \rangle\} = 0, \qquad \{H, \langle \mathbf{p}, \mathbf{p} \rangle\} = 0, \qquad \{H, \langle \mathbf{q}, \mathbf{q} \rangle\} = 0\,.$$

Hence, the equations of motion can be written in the matrix form

$$\frac{d}{dt}\begin{bmatrix} \mathbf{q} \\ \mathbf{p} \end{bmatrix} = \begin{bmatrix} -\langle \mathbf{q}, \mathbf{p} \rangle & \langle \mathbf{q}, \mathbf{q} \rangle \\ -\langle \mathbf{p}, \mathbf{p} \rangle & \langle \mathbf{q}, \mathbf{q} \rangle \end{bmatrix} \begin{bmatrix} \mathbf{q} \\ \mathbf{p} \end{bmatrix} = C(\mathbf{q}, \mathbf{p}) \begin{bmatrix} \mathbf{q} \\ \mathbf{p} \end{bmatrix} \tag{7.48}$$

with C a matrix of constants of motion. Then this system can be integrated by exponentiation using Eq. (7.30).

7.4.3 Regularization and Linearization of the Kepler Problem

We shall discuss here the Kepler problem taking the point of view of the ideas developed in the previous sections, i.e., it will be shown that the 1-dimensional Kepler problem is related to a linear system and how this relation allows for an immediate and explicit integration of it. We discuss first the one dimensional Kepler problem and we leave the discussion of the 2 and 3-dimensional Kepler problems to one of the following sections.

Using natural coordinates (x, v) in $\mathbb{R}^+ \times \mathbb{R}$, where $\mathbb{R}^+$ is the set of positive real numbers, we consider the dynamical system described by the second-order differential equation:

$$\Gamma = v\frac{\partial}{\partial x} + f(x, v)\frac{\partial}{\partial v} \tag{7.49}$$

where in our case f is Kepler's force

$$f(x, v) = -\frac{k}{x^2}, \qquad k > 0\,,$$

derivable from the potential function

$$V = V(x) = -\frac{k}{x}\,,$$

i.e., our dynamical vector field is

$$\Gamma = v\frac{\partial}{\partial x} - \frac{k}{x^2}\frac{\partial}{\partial v}\,, \tag{7.50}$$

which is well defined because we have removed the point 0 in the domain of Γ. We do not discuss at this moment the choice of the name for this system. We only point out here that our system is not the 1-D reduction of the 3-D Kepler problem by using constants of motion and the natural equivalence relation defined by its symmetry group. The equations of motion for the 1-D Kepler problem are thus,

$$\begin{cases} \dot{x} = v \\ \dot{v} = -\dfrac{k}{x^2} \end{cases}, \tag{7.51}$$

which shows that they are singular at $x = 0$ (that is why we removed this point in the definition of the domain of Γ). These equations can be derived from the Lagrangian

$$L = \frac{1}{2}v^2 + \frac{k}{x} \,,$$

which gives raise to the energy

$$E(x, v) = \frac{1}{2}v^2 - \frac{k}{x} \,, \tag{7.52}$$

that is a constant of motion.

The usual Legendre transformation

$$p = \frac{\partial L}{\partial v} = v$$

leads to the Hamiltonian function:

$$H(x, p) = \frac{1}{2}p^2 - \frac{k}{x} \tag{7.53}$$

The symplectic structures to be used are either $\omega_L = dx \wedge dv$ on the velocity phase space or $\omega_0 = dx \wedge dp$ on the phase space, the configuration space being $\mathbb{R}^+$.

It is also obvious that the trajectories with energy $E \leq 0$ drop toward $x = 0$ because of the attractive character of the force f and then, they will explode (the velocity diverges) at $x = 0$ (see Fig. 7.2). In fact $v \cong x^{-1/2}$ as $x \to 0^+$.

Using the energy relation Eq. (7.52) it is immediate to integrate the equations of motion to get

$$t - t_0 = \int_{x_0}^{x} \frac{dx}{\sqrt{2E + (2k/x)}} \,.$$

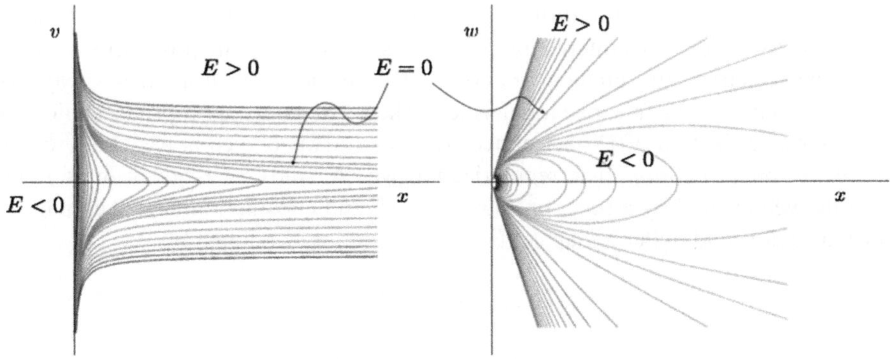

Fig. 7.2 *Left* Trajectories of the 1d Kepler problem. *Right* the trajectories of the regularized flow

We have displayed the energy levels of E on Fig. 7.2 showing the existence of three types of trajectories. Those contained in the region $E < 0$ exit and return to $x = 0$ with infinite velocity after reaching the maximum $x_{\max} = -k/E$. The trajectories contained in the region $E > 0$ exit from $x = 0$ with infinite velocity and they escape to infinity with a residual escape velocity $v_\infty = \sqrt{2E}$. Finally, the trajectories separating these two regions correspond to $E = 0$ and they describe the system escaping to infinity with $v_\infty = 0$.

All this suggests a natural way to obtain a better description of this system by multiplying Eq. (7.52) by x. Then we obtain for a fixed E,

$$k = \frac{1}{2} x\, v^2 - x\, E\,,$$

which shows that in the level set E of the energy, if we redefine the velocity as $w = \sqrt{x}\, v$, i.e., we define a new time function τ such that

$$\frac{dt}{d\tau} = \sqrt{x}\,,$$

we obtain a new system whose equations of motion are

$$\begin{cases} x' = w \\ w' = \frac{1}{2}\dfrac{w^2}{x} - \dfrac{k}{x} \end{cases}$$

where $x' = dx/d\tau$ and $w' = dw/d\tau$.

From the geometric viewpoint this reparametrization corresponds to first replacing the vector field Γ given by (7.50) by a new vector field $\sqrt{x}\,\Gamma$, i.e., multiplying the original vector field Γ by the function $h(x, v) = \sqrt{x}$, and second redefining the tangent structure in such a way that the new vector field is again a SODE with respect to the new tangent structure.

In any case we see that unfortunately we have not removed the singular character of the vector field Γ. An interesting problem is the study of the possibility of using as a reparametrization function $h(x, v) = x^\alpha$, i.e., $w = x^\alpha v$ in such a way that the new vector field free of singularities. There exists an appropriate choice for the reparametrization of Γ determined by $\alpha = 1$. This choice is frequently called *Sundman regularization*. The regularization we are studying is given multiplying first the vector field by a function $h(x, v) = x^\alpha$ and considering later the image of such a vector field under a map $\phi : \mathbb{R}^+ \times \mathbb{R} \to \mathbb{R}^+ \times \mathbb{R}$ defined by

$$\phi(x, v) = (x, x^\alpha\, v)\,,$$

necessary for the image vector field to be a SODE. We must remark that such ϕ is not the derivative of any map $\varphi : \mathbb{R}^+ \to \mathbb{R}^+$ that allows us to restore the SODE character of the field. We can compute the vector field $\widetilde{\Gamma} = \Phi_*(x^\alpha\, \Gamma)$ obtained from the reparametrized vector field $X = x^\alpha\,\Gamma$ via this transformation by a direct computation, but we better consider the problem in a perspective that can easily be generalized and will be shown to be useful in the future. We are looking for a vector field

$$\widetilde{\Gamma} = w\frac{\partial}{\partial x} + \tilde{f}(x, w)\frac{\partial}{\partial w}$$

that is the image of X under the transformation ϕ. Notice that even if X is not a SODE anymore, $\widetilde{\Gamma}$ is again a SODE, or in other words, X can be seen as a SODE but with respect to a different tangent space structure (see Sect. 5.6.3).

We recall that X and $\widetilde{\Gamma}$ are said to be *ϕ-related* if and only if

$$X(\phi^* g) = \phi^*(\widetilde{\Gamma} g)$$

for any function $g(x, w)$, and in particular, for $g(x, w) = x$ and $g(x, w) = w$:

$$X(\phi^*(x)) = \phi^*(\widetilde{\Gamma} x) = \phi^*(w)\,,$$
$$X(\phi^*(w)) = \phi^*(\widetilde{\Gamma} w) = \phi^*(\tilde{f})\,.$$

We find that

$$\phi^*(\tilde{f}) = \alpha\,\frac{w^2}{x} + x^{2\alpha} f(x, w/x^\alpha),$$

therefore,

$$\tilde{\Gamma} = w\frac{\partial}{\partial x} + \left(\alpha\,\frac{w^2}{x} + x^{2\alpha} f(x, w/x^\alpha)\right)\frac{\partial}{\partial w}$$

whose associated system of differential equations is

$$\begin{cases} \dfrac{dx}{ds} = w \\ \dfrac{dw}{ds} = \alpha\,\dfrac{w^2}{x} + x^{2\alpha} f(x, w/x^\alpha) \end{cases},$$

It is clear that $w = x^\alpha\, v$ behaves like $x^{\alpha-1/2}$ when $x \to 0^+$ and therefore if we want $v \to 0^+$ we should choose $\alpha > 1/2$. As we will see later on, the most appropriate choice is $\alpha = 1$. In such a case,

$$\tilde{f}(x, w) = \frac{w^2}{x} + x^2 f(x, w/x)\,,$$

and

$$\widetilde{\Gamma} = w\frac{\partial}{\partial x} + \left(\frac{w^2}{x} + x^2 f(x, w/x)\right)\frac{\partial}{\partial w}$$

and in our particular case, for which $f(x, v) = -k/x^2$, the associated system of differential equations is

$$\begin{cases} \dfrac{dx}{ds} = w \\ \dfrac{dw}{ds} = \dfrac{w^2}{x} - k \end{cases}$$

In spite of the appearance of the factor $1/x$ this system does not explode and it is easily seen that $w \to 0$ as $x \to 0^+$. See Fig. 7.2 for the trajectories of $\widetilde{\Gamma}$.

It seems that we have found a better way of handling the 1-D Kepler problem, but, as we will see in a moment, we can do still better. We will show that, actually, for $\alpha = 1$ this system is related to a linear one.

To show this we add an extra dimension, to be denoted h, i.e., we consider $\mathbb{R}^+ \times \mathbb{R} \times \mathbb{R}$ with coordinates (x, w, h) and the vector field obtained extending trivially $\widetilde{\Gamma}$ and that is denoted with the same symbol. The new equations of motion on M are,

$$\begin{cases} \dfrac{dx}{ds} = w \\ \dfrac{dw}{ds} = \alpha \dfrac{w^2}{x} + x^{2\alpha} f(x, w/x^\alpha) \\ \dfrac{dh}{ds} = 0 \end{cases}$$

which for $\alpha = 1$ become:

$$\begin{cases} \dfrac{dx}{ds} = w \\ \dfrac{dw}{ds} = \dfrac{w^2}{x} + x^2 f \,, \\ \dfrac{dh}{ds} = 0 \end{cases}$$

and we are interested in the case $f = -k/x^2$. Now we select the invariant level set of constant energy given by

$$\Sigma_\alpha = \left\{ (x, w, h) \in M \mid h = -V - \frac{w^2}{2x^{2\alpha}} \right\}, \qquad f = -dV/dx \,,$$

in the general case, or simply,

$$\Sigma_\alpha = \left\{ (x, w, h) \in M; \quad h = \frac{k}{x} - \frac{w^2}{2x^2} \right\},$$

in the Kepler case we are dealing with. On this surface our vector field $\widetilde{\Gamma}$ has the expression, using (x, w) as coordinates on M,

$$\widetilde{\Gamma} = w\frac{\partial}{\partial x} + \left(2\alpha x^{2\alpha-1}(-h - V) + x^{2\alpha} f(x, w/x^{\alpha})\right)\frac{\partial}{\partial w}$$

and in the Kepler case:

$$\widetilde{\Gamma} = w\frac{\partial}{\partial x} + 2\alpha x^{2\alpha-1}\left(k/x - h - kx^{2\alpha-2}\right)\frac{\partial}{\partial w}$$

which shows that we obtain a linear system when $\alpha = 1$, and in this case it becomes

$$\widetilde{\Gamma} = w\frac{\partial}{\partial x} + (k - 2hx)\frac{\partial}{\partial w}$$

Therefore we have been able to associate a linear non-homogeneous vector field with the nonlinear equation describing the Kepler system in 1D. The equations of motion

$$\begin{cases} \dfrac{dx}{ds} = w \\ \dfrac{dw}{ds} = k - 2hx \end{cases}$$

can be easily solved and we obtain for initial conditions $x(0) = 0$, $w(0) = 0$,

$$x(s) = \frac{k}{2h}\left(1 - \cos\sqrt{2h}s\right).$$

Using these solutions we can also find the parameter for the initial vector field

$$t = \int x(s)ds = \frac{k}{2h}\left(s - \frac{1}{\sqrt{2h}}\sin\sqrt{2h}s\right).$$

It is now possible to perform a further transformation suggested by the particular form of the solutions we have found. We consider now the transformation $\psi : \mathbb{R}^+ \to \mathbb{R}^+$ such that $x = \psi(y) = y^2$, and we will extend it to $\mathbb{R}^+ \times \mathbb{R} \to \mathbb{R}^+ \times \mathbb{R}$ as $\psi_*(y, u) = (y^2, 2yu)$, i.e., $x = y^2$ and $w = 2yu$. Hence we find the new vector field

$$Z = u\frac{\partial}{\partial y} - \frac{h}{2}y\frac{\partial}{\partial u}$$

Therefore, for positive values of h, this vector field represents the harmonic oscillator with frequency $w = \sqrt{h/2}$. Thus, after this long periple we have returned to the dynamical system used to motivate this chapter, the 1-D harmonic oscillator, Equation.

Summarizing the previous discussion we conclude that we have reduced the integration of the Kepler problem to the integration of an associated linear differential equation. We should notice that the energy relation $k/x - v^2/2 = h$ in (x, v) coordinates or $k/x - w^2/2x^2 = h$ in (x, w) coordinates, becomes

$$u^2 - \frac{1}{2}(k - hy^2) = 0\,,$$

in (y, u) coordinates. Therefore, we could have started with a linear vector field on $\mathbb{R}^3$ and after performing the various operations of reparametrization, nonlinear changes of coordinates and restriction to surfaces described above we would obtain the 1D Kepler problem.

7.5 Reduction in Quantum Mechanics

Having defined Poisson reduction we are now on good track to define a reduction procedure for quantum systems. After all, according to deformation quantization the Poisson bracket provides us with a first order approximation to Quantum Mechanics. We will use in the remaining the various formalisms to describe quantum dynamical evolution presented in Chap. 6, Sects. 6.6.2, 6.4. Simply recall here that the description of quantum systems is done basically by means of either the Hilbert space of states, where we define dynamics by means of the Schrödinger equation, or by means of the algebra of observables, where dynamics is defined by means of the Heisenberg equation. We may also consider other pictures like the Ehrenfest picture, the phase-space or Wigner picture, the tomographic picture, etc.

7.5.1 The Reduction of Free Motion in the Quantum Case

The description of the free quantum evolution is rather simple because the semi-classical treatment is actually exact [Es04]. In what follows we set $\hbar = 1$ for simplicity.

The Hamiltonian operator for free motion in two dimensions, given by:

$$H = -\frac{1}{2}\Delta, \qquad \Delta = \frac{\partial^2}{\partial x^2} + \frac{\partial^2}{\partial y^2}\,, \tag{7.54}$$

written in polar coordinates (Q, ϕ) $(x = Q\sin\phi,\ y = Q\cos\phi)$ it becomes:

$$H = -\frac{1}{2}\frac{1}{Q}\frac{\partial}{\partial Q}Q\frac{\partial}{\partial Q} - \frac{1}{Q^2}\frac{\partial^2}{\partial \phi^2}\,.$$

By a similarity transformation $H' = Q^{\frac{1}{2}} H Q^{-\frac{1}{2}}$ (notice that $Q > 0$) we get rid of the linear term and we obtain:

$$H' = -\frac{1}{2}\left[\frac{\partial^2}{\partial Q^2} + \frac{1}{Q^2}\left(\frac{1}{4} + \frac{\partial^2}{\partial \phi^2}\right)\right]\,.$$

Restricting H' to the subspace $\mathcal{S}_m \subset L^2(\mathbb{R}^2)$ of square integrable functions of the form

$$\mathcal{S}_m = \{\psi = \mathrm{e}^{im\phi} f(Q)\},$$

we find that on this particular subspace

$$H'\psi = -\frac{1}{2}\left[\frac{\partial^2}{\partial Q^2} - \frac{1}{Q^2}\left(m^2 - \frac{1}{4}\right)\right]\psi .$$

This determines a Hamiltonian operator along the radial coordinate and setting $g^2 = m^2 - \frac{1}{4}$ we have:

$$\widetilde{H} = -\frac{1}{2}\frac{\partial^2}{\partial Q^2} + \frac{1}{2}\frac{g^2}{Q^2} .$$

Suppose that we consider now the space of Hermitean matrices X (of real dimension n^2) and the quantum free motion on it. Solutions of the free problem (7.54) are given by wave-packets formed out of 'plane-waves':

$$\psi_P(X) = A\mathrm{e}^{i\mathrm{Tr}\, XP},$$

where A is a normalization constant, chosen in such a way as to give a delta function normalization.

By decomposing X into a 'radial' part Q and an 'angular' part G, say $X = G^{-1}QG$, like in the reduction of the classical free motion on matrices to get the Calogero-Moser system, Sect. 1.2.12, Eq. (1.131), we can write the wave function in the form:

$$\psi(Q, G) = A\mathrm{e}^{i\mathrm{Tr}\,(G^{-1}QGP)} = \psi_P(X).$$

In this particular case it is not difficult to show that $I_j(X, P) = \mathrm{Tr}\,(P^j)$ are constants of motion in involution for the classical system and give raise to the operators $(-i)^j \mathrm{Tr}\,\left(\frac{\partial}{\partial X}\right)^j$.

To perform specific computations let us go back to the two-dimensional situation. We consider $\psi_P(X) = A\mathrm{e}^{i\mathrm{Tr}\, PX}$ and project it along the eigenspace $\mathcal{S}_m$ of the angular momentum corresponding to the fixed value m.

We recall that (in connection with the unitary representations of the Euclidean group)

$$\int_0^{2\pi} d\phi\, \mathrm{e}^{im\phi}\, \mathrm{e}^{iPQ\cos\phi} = 2\pi J_m(PQ),$$

where J_m is the Bessel function of order m. Thus we conclude

$$\psi_P(Q) = 2\pi\sqrt{PQ}\,J_m(PQ).$$

In the particular case we are considering free motion is described by a quadratic Hamiltonian in $\mathbb{R}^2$. Therefore the Green function becomes

$$G(X_t - X_0, 0; t) = \frac{C}{2t}\mathrm{e}^{i\frac{\mathrm{Tr}\,(X_t - X_0)^2}{t}}.$$

The Green function can be written in terms of the action, that is, the solution of the Hamilton-Jacobi equation and the Van Vleck determinant ([Es04], Appendix 4.B).

By using polar coordinates the kernel of the propagator is

$$\begin{aligned} G(Q_t, Q; t) &= \sqrt{Q_t Q_0}\int_0^{2\pi} d\phi \mathrm{e}^{im\phi} K(X_t, X_0; t) \\ &= \frac{1}{2\pi i t}\sqrt{Q_t Q_0}\,\mathrm{e}^{i\frac{(Q_t^2+Q_0)^2}{2t}}\int_0^{2\pi} d\phi\;\mathrm{e}^{im\phi}\mathrm{e}^{-i\frac{Q_t Q_0\cos\phi}{t}} \\ &= \frac{1}{2\pi i t}\sqrt{Q_t Q_0}\mathrm{e}^{i\frac{(Q_t^2+Q_0)^2}{2t}} J_m\left(\frac{Q_t Q_0}{t}\right), \end{aligned}$$

where the angle is coming from the scalar product of X_t with X_0.

7.5.2 Reduction in Terms of Differential Operators

This simple example has shown that the reduction procedure in wave mechanics involves differential operators (see Chap. 10) and their eigenspaces. Let us therefore consider some general aspects of reduction procedures for differential operators.

In general, the Hamiltonian operator defining the Schrödinger equation on $L^2(D, d\mu)$ is a differential operator, which may exhibit a complicated dependence in the potential. It makes sense thus to study a general framework for the reduction of differential operators acting on some domain D, when we assume that the reduction procedure consists in the suitable choice of some 'quotient' domain D'.

Let us recall that if we consider $\mathcal{F} = C^\infty(\mathbb{R}^n)$, the algebra of smooth functions on $\mathbb{R}^n$. A differential operator of degree at most r is defined as a linear map $P: \mathcal{F} \to \mathcal{F}$ of the form

$$P = \sum_{|\sigma|\leq r} g_\sigma \frac{\partial^{|\sigma|}}{\partial x_\sigma}, \quad g_\sigma \in \mathcal{F} \tag{7.55}$$

where $\sigma = (i_1, \cdots i_n)$, $|\sigma| = \sum_k i_k$ and

$$\frac{\partial^{|\sigma|}}{\partial x_\sigma} = \frac{\partial^{|\sigma|}}{\partial x_1^{i_1} \cdots \partial x_n^{i_n}}$$

This particular way of expressing differential operators relies on the generator of 'translations', $\partial/\partial x_k$. Therefore, when the reduced space does not carry an action of the translation group this way of writing differential operators is not very convenient. There is an intrinsic way to define differential operators which does not depend on coordinates (see Chap. 10, Appendix G, for more details). One starts from the following observation

$$\left[\frac{\partial}{\partial x_j}, m_f\right] = m_{\frac{\partial f}{\partial x_j}},$$

where m_f is the multiplication by the function f operation, i.e., $m_f(g) = fg$, with $f, g \in \mathcal{F}$. It follows that

$$[P, m_f] = \sum_{|\sigma| \le r} g_\sigma \left[\frac{\partial^{|\sigma|}}{\partial x_\sigma}, m_f\right],$$

is of degree at most $r - 1$. Iterating for a set of $r + 1$ functions $f_0, \ldots, f_k \in \mathcal{F}$, one finds that

$$[\ldots, [[P, \hat{f}_0], \hat{f}_1], \ldots, \hat{f}_r] = 0;$$

This algebraic characterization allows for a definition of differential operators on any manifold.

The algebra of differential operators of degree 1 is a Lie subalgebra with respect to the commutator $[\cdot, \cdot]$ and splits into a direct sum

$$D^1 = \mathcal{F} \oplus D_c^1$$

where D_c^1 are derivations, i.e., differential operators of degree one which give zero on constants. We can endow the set with a Lie algebra structure by setting

$$[(f_1, X_1), (f_2, X_2)] = (X_1 f_2 - X_2 f_1, [X_1, X_2])$$

If we consider $\mathcal{F}$ as an Abelian Lie algebra, D_c^1 is the algebra of its derivations and then D^1 becomes what is known in the literature as the 'holomorph' of $\mathcal{F}$ [Ca94]. In this way the algebra of differential operators becomes the enveloping algebra of the holomorph of $\mathcal{F}$.

The set of differential operators on M, denoted as $\mathcal{D}(M)$, can be given the structure of a graded associative algebra and it is also a module over $\mathcal{F}$. Notice that this property would not make sense at the level of abstract operator algebra. To consider the problem of reduction of differential operators we consider the problem of reduction of first-order differential operators. Because the zeroth order ones are just functions, we restrict our attention to vector fields, i.e., the set D^1_c.

Given a projection $\pi : M \to N$ between smooth manifolds, we say that a vector field X_M projects onto a vector field X_N if

$$\mathcal{L}_{X_M}\pi^* f = \pi^*(\mathcal{L}_{X_N} f) \qquad \forall f \in \mathcal{F}(N).$$

We say thus that X_M and X_N are π-related.

Thus if we consider the subalgebra $\pi^*(\mathcal{F}(N)) \subset \mathcal{F}(M)$, a vector field is projectable if it defines a derivation of the subalgebra $\pi^*(\mathcal{F}(N))$. More generally, for a differential operator D^k, we shall say that it is projectable if

$$D^k\pi^*(\mathcal{F}(N)) \subset \pi^*(\mathcal{F}(N)).$$

It follows that projectable differential operators of degree zero are elements in $\pi^*(\mathcal{F}(N))$. Therefore projectable differential operators are given by the enveloping algebra of the holomorph of $\pi^*(\mathcal{F}(N))$, when the corresponding derivations are considered as belonging to $\mathfrak{X}(M)$.

Remark 7.5 Given a subalgebra of differential operators in $\mathcal{D}(M)$ it is not said that it is the enveloping algebra of the first-order differential operators it contains. When this happens, we cannot associate a corresponding quotient manifold with an identified subalgebra of differential operators. An example of this situation arises with angular momentum operators when we consider the 'eigenvalue problem' in terms of J_z and J^2. It is clear that this commuting subalgebra of differential operators cannot be generated by its 'first-order content'.

In the quantization procedure, this situation gives raise to anomalies [Al98].

7.5.3 The Kustaanheimo–Stiefel Fibration

In this section we would like to consider the reduction of differential operators associated with the KS projection $\pi_{KS} : \mathbb{R}^4_0 \to \mathbb{R}^3_0$, where $\mathbb{R}^j_0 = \mathbb{R}^j - \{0\}$, for $j = 3, 4$, and show that the Hydrogen atom operator may be obtained as a reduction of the operators associated with a family of harmonic oscillators.

Let us recall first how this map is defined. We first notice that $\mathbb{R}^4_0 = S^3 \times \mathbb{R}^+$ and identifying S^3 with $SU(2)$, $\mathbb{R}^4_0 \sim SU(2)\times\mathbb{R}^+$, while $\mathbb{R}^3_0 = S^2\times\mathbb{R}^+$. By introducing polar coordinates

$$g = Rs \quad s \in SU(2), \quad R \in \mathbb{R}^+,$$

if we take into account that $s\sigma_3 s^{-1}$ is a traceless hermitian matrix, we can define $\pi_{KS} : \mathbb{R}^4_0 \to \mathbb{R}^3_0$ as

$$\pi_{KS} : g \mapsto g\sigma_3 g^+ = R^2 s\sigma_3 s^{-1} = x^k \sigma_k,$$

where $\{\sigma_k \mid k = 1, 2, 3\}$ are Pauli matrices Eq. (10.68). In a Cartesian coordinate system one has

$$x_1 = 2(y_1 y_3 + y_2 y_0) \quad x_2 = 2(y_2 y_3 - y_1 y_0) \quad x_3 = y_1^2 + y_2^2 - y_3^3 - y_0^2,$$

where $g = \sum_{i=1}^{3} y_i \sigma^i$. Moreover, $\sqrt{x^j x_j} = r = R^2 = y^k y_k$.

The KS projection defines a principal fibration with structure group $U(1)$.

By the definition of π_{KS} it is easy to see that acting with $e^{i\lambda\sigma_3}$ on $SU(2)$ does not change the projected point on $\mathbb{R}_0^3$. The associated fundamental vector field X_3 is the left-invariant infinitesimal generator associated with σ_3, i.e., $i_{X_3} s^{-1} ds = i\sigma_3$. In coordinates it reads

$$X_3 = y^0 \frac{\partial}{\partial y^3} - y^3 \frac{\partial}{\partial y^0} + y^1 \frac{\partial}{\partial y^2} - y^2 \frac{\partial}{\partial y^1}$$

We consider the Lie algebra of differential operators generated by X_3 and $\pi_{KS}^*(\mathcal{F}(\mathbb{R}_0^3))$. Projectable differential operators with respect to π_{KS} are given by the normalizer of this algebra in the algebra of differential operators $\mathcal{D}(\mathbb{R}_0^4)$. As we already remarked this means that this subalgebra must map $\pi^*(\mathcal{F}(\mathbb{R}_0^3))$ into itself. If we denote this subalgebra by $\mathcal{D}^\pi$ we may also restrict our attention to the operators in $\mathcal{D}^\pi$ commuting with X_3.

In order to explicitly construct this algebra of differential operators we use the fact that $SU(2) \times \mathbb{R}_+$ is a Lie group and therefore it is parallelizable. Because the KS map has been constructed with the left-invariant vector field X_3, we consider the generators of the left-action of $SU(2)$, say right-invariant vector fields Y_1, Y_2, Y_3, and a central vector field along the radial coordinate, say $\mathcal{R}$. All these vector fields are projectable and therefore along with $\pi_{KS}^*(\mathcal{F}(\mathbb{R}_0^3)$ generate a projectable subalgebra of differential operators which covers the algebra of differential operators on $\mathbb{R}_0^3$.

This map is surjective and we can ask to find the 'inverse image' of the operator

$$\hat{H} = -\frac{\Delta_3}{2} - \frac{k}{r},$$

which is the operator associated with the Schrödinger equation of the hydrogen atom (Δ_3 denotes the Laplacian in the three dimensional space). As this operator is invariant under the action of $\mathfrak{so}(4) \sim \mathfrak{su}(2) \oplus \mathfrak{su}(2)$, associated with the angular momentum and the Runge-Lenz vector, we may look for a representative in the inverse image which shares the same symmetries. As the pull-back of the potential k/r creates no problems, we may concentrate our attention on the Laplacian. Because of the invariance requirements, our candidate for the inverse image will have the form

$$D = f(R)\frac{\partial^2}{\partial R^2} + g(R)\frac{\partial}{\partial R} + h(R)\Delta_3^s + c(R),$$

where R is the radial coordinate in $\mathbb{R}_0^4$, and f, g, h are functions to be determined. We recall that in polar coordinates the Laplacian Δ_3 has the expression

$$\Delta_3 = \frac{\partial^2}{\partial r^2} + \frac{2}{r}\frac{\partial}{\partial r} + \frac{1}{r^2}\Delta_2^s,$$

where we denote by Δ_n^s the Laplacian on the n-dimensional sphere.

By imposing $D\pi_{KS}^*\phi = \pi_{KS}^*(\hat{H}_3\phi)$ for any $\phi \in \mathcal{F}(\mathbb{R}_0^3)$ we find that the representative in the inverse image has the expression

$$H' = -\frac{1}{2}\frac{1}{4R^2}\Delta_4 - \frac{k}{R^2}.$$

This operator is usually referred to as the conformal Kepler Hamiltonian [Av05].

Now, with this operator we may try to solve the eigenvalue problem

$$\left(-\frac{1}{2}\frac{1}{4R^2}\Delta_4 - \frac{k}{R^2}\right)\psi - E\psi = 0$$

It defines a subspace in $\mathcal{F}(\mathbb{R}_0^4)$ which coincides with the one determined by the equation

$$\left(-\frac{1}{2}\Delta_4 - 4ER^2 - 4k\right)\psi = 0.$$

This implies that the subspace is given by the eigenfunctions of the Harmonic oscillator with frequency $\omega(E) = \sqrt{-8E}$. We notice then that a family of oscillators is required to solve the eigenvalue problem associated with the Hydrogen atom. To find the final wave functions on $\mathbb{R}^3$ we must require that $\mathcal{L}_{X_3}\psi = 0$ in order to find eigenfunctions for the three dimensional problem. Eventually we find the correct relations for the corresponding eigenvalues

$$E_, = -\frac{k^2}{2(m+1)^2}, \quad m \in \mathbb{N}.$$

Of course, dealing with Quantum Mechanics we should ensure that the operator

$$H' = -\frac{1}{2}\frac{1}{4R^2}\Delta_4 - \frac{k}{R^2}$$

is essentially self-adjoint to be able to associate with it a unitary dynamics. One finds that the Hilbert space should be constructed as a space of square integrable functions on $\mathbb{R}_0^4$ with respect to the measure $4R^2d^4y$ instead of the Euclidean measure on $\mathbb{R}^4$.

We shall not go into the details of this, but the problem of a different scalar product is strictly related to the reparametrization of the classical vector field, required to turn it into a complete one. This would be a good example for J. Klauder's saying: 'these are classical symptoms of a quantum illness' (see [Zu93]). Further details can be found in [Av05, Av05b]. As for the reduction of the Laplacian in Quantum Mechanics see also [Fe06, Fe07, Fe08].

7.5.4 Reduction in the Heisenberg Picture

The Heisenberg picture of Quantum Mechanics relies on the non-commutative algebra of observables (see Sect. 6.6.2), therefore it is instructive to consider a reduction procedure for non-commutative algebras.

The example of reduction procedure in a non-commutative setting that we are going to discuss reproduces the Poisson reduction in the 'quantum-classical' transition and goes back to the celebrated example of the quantum $SU(2)$ written by Woronowicz [Wo87] and is adapted from [Gr94].

We consider the space $S^3 \subset \mathbb{R}^4$, identified with the group $SU(2)$ represented in terms of matrices. The $\star$-algebra $\mathcal{A}$ generated by matrix elements is dense in the algebra of continuous functions on $SU(2)$ and can be characterized as the 'maximal' unital commutative $\star$-algebra $\mathcal{A}$, generated by elements which we can denote as $\alpha, \nu, \alpha^*, \nu^*$ satisfying $\alpha^*\alpha + \nu^*\nu = 1$. This algebra can be generalized and deformed into a non-commutative one by replacing some relations with the following ones:

$$\alpha\alpha^* - \alpha^*\alpha = (2q - q^2)\nu^*\nu \qquad \nu^*\nu - \nu\nu^* = 0,$$

and

$$\nu\alpha - \alpha\nu = q\nu\alpha \qquad \nu^*\alpha - \alpha\nu^* = q\nu^*\alpha.$$

This algebra reduces to the previous commutative one when $q = 0$. In this respect this situation resembles the one on the phase-space where we consider 'deformation quantization' and the role of the parameter q is played by the Planck constant. Pursuing this analogy we may consider the formal product depending on the parameter q:

$$u \star_q v = uv + \sum_n q^n P_n(u, v),$$

where P_n are such that the product $\star_q$ is associative.

Since the commutator bracket

$$[u, v]_q = u \star_q v - v \star_q u$$

is a biderivation (as for any associative algebra) and satisfies the Jacobi identity we find that the 'quantum Poisson bracket' gives a Poisson bracket when restricted to 'first-order elements'

$$\{u, v\} = P_1(u, v) - P_1(v, u).$$

In general, we can write

$$\lim_{q \to 0} \frac{1}{q}[u, v]_q = \{u, v\} \tag{7.56}$$

From the defining commutation relations written by Woronowicz we get the corresponding quadratic Poisson brackets on the matrix elements of $SU(2)$:

$$\{\alpha, \bar{\alpha}\} = 2\bar{\nu}\nu, \quad \{\nu, \bar{\nu}\} = 0, \quad \{\nu, \alpha\} = \nu\alpha, \quad \{\bar{\nu}, \alpha\} = \bar{\nu}\alpha.$$

Passing to real coordinates, $\alpha = q_2 + ip_2$ and $\nu = q_1 + ip_1$, we get a purely imaginary bracket whose imaginary part is the following quadratic Poisson bracket

$$\begin{array}{lll} \{p_1, q_1\} = 0, & \{p_1, p_2\} = q_1 q_2, & \{p_1, q_1\} = -p_1 p_2, \\ \{q_1, p_2\} = q_1 q_2, & \{q_1, q_2\} = -q_1 p_2, & \{p_2, q_2\} = q_1^2 + p_1^2. \end{array}$$

The functions $q_1^2 + q_2^2 + p_1^2 + p_2^2$ is a Casimir function for this Lie algebra.

By performing a standard Poisson bracket reduction we find a bracket on S^3. If we identify this space with the group $SU(2)$ we get the Lie-Poisson structure on $SU(2)$:

The vector field

$$X = -q_1 \frac{\partial}{\partial p_1} + p_1 \frac{\partial}{\partial q_1} + q_2 \frac{\partial}{\partial p_2} - p_2 \frac{\partial}{\partial q_2}$$

selects a subalgebra of functions $\mathcal{F}$ by imposing the condition $\mathcal{L}_X f = 0, \forall f \in \mathcal{F}$. This reduced algebra can be regarded as the algebra generated by

$$u = -p_1^2 - q_1^2 + p_2^2 + q_2^2, \qquad \nu = 2(p_1 p_2 + q_1 q_2), \qquad z = 2(p_1 q_2 - q_1 p_2),$$

with brackets

$$\{v, u\} = 2(1 - u)z, \qquad \{u, z\} = 2(1 - u)v, \qquad \{z, v\} = 2(1 - u)u.$$

One finds that $u^2 + v^2 + z^2 = 1$ so that the reduced space of $SU(2)$ is the unit sphere S^2 and the reduced bracket vanishes at the North Pole ($u = 1, v = z = 0$).

It may be interesting to notice that the stereographic projection from the North Pole pulls-back the standard symplectic structure on $\mathbb{R}^2$ onto the one associated with this one on S^2-{North Pole}.

It is now possible to carry on the reduction at the non-commutative level. We identify the subalgebra $\mathcal{A}'_q \subset \mathcal{A}_q$ generated by the elements $u = \mathbb{I} - 2\nu^*\nu = \alpha^*\alpha - \nu^*\nu$, $w = 2\nu^*\alpha$ and $w^* = 2\alpha^*\nu$. We have $uu^* + w^*w = \mathbb{I}$ and the algebra $\mathcal{A}'_q$ admits a limit given by $\mathcal{A}'_0$ generated by the two-dimensional sphere S^2. The subalgebra $\mathcal{A}'_q$ can be considered as a quantum sphere.

The quantum Poisson bracket on S^2 is given by

$$[w, u] = (q^2 - 2q)(1 - u)w, \qquad [w^*, u] = -(q^2 - 2q)(1 - u)w^*,$$

and

$$[w, w^*] = -(2q^2 - 2q)(1 - u) + (4q - 6q^2 + 4q^3 - q^4)(1 - u)^2.$$

Passing to the classical limit we find, by setting $v = \mathrm{Re}(w)$, $z = -\mathrm{Im}(w)$:

$$\{v, u\} = 2(1 - u)z, \qquad \{u, z\} = 2(1 - u)u, \qquad \{z, v\} = 2(1 - u)u,$$

which coincides with the previous reduced Poisson bracket associated with the vector field X. In this case, the reduction procedure commutes with the 'quantum-classical' limit.

In this same setting it is now possible to consider a 'quantum dynamics' and the corresponding 'classical' one to see how they behave with respect to the reduction procedure.

On the algebra $\mathcal{A}_q$ we consider the dynamics defined by the Hamiltonian

$$H = \frac{1}{2}u = \frac{1}{2}(\mathbb{I} - 2\nu^*\nu) = \frac{1}{2}(\alpha^*\alpha - \nu^*\nu).$$

This choice ensures that our Hamiltonian defines a dynamics on $\mathcal{A}'_q$. The resulting equations of motion are

$$[H, \nu] = 0,\ [H, \nu^*] = 0,\ [H, \alpha] = (q^2 - 2q)\nu^*\nu\alpha,\ [H, \alpha^*] = -(q^2 - 2q)\nu^*\nu\alpha^*,$$

so that the dynamics written in the exponential form is

$$U(t) = \mathrm{e}^{it\,\mathrm{ad}_H}$$

and gives,

$$\nu(t) = \nu_0, \qquad \nu^*(t) = \nu^*(0)$$

$$\alpha(t) = \mathrm{e}^{it(q^2 - 2q)\nu^*\nu}\alpha_0, \qquad \alpha^*(t) = \mathrm{e}^{-it(q^2 - 2q)\nu^*\nu}\alpha_0^*.$$

Going to the 'classical limit' we find

$$H = \frac{1}{2}(q_2^2 + p_2^2 - q_1^2 - p_1^2),$$

with the associated vector field on S^3 given by [Li93]

$$\Gamma = 2(q_1^2 + p_1^2)\left(q_2\frac{\partial}{\partial p_2} - p_2\frac{\partial}{\partial q_2}\right),$$

and the corresponding solutions are given by

$$q_1(t) = q_1(0), \qquad p_1(t) = p_1(0)$$

$$p_2(t) = \cos\left(2t\left(q_1^2 + p_1^2\right)\right) p_2(0) + \sin\left(2t\left(q_1^2 + p_1^2\right)\right) q_2(0),$$

$$q_2(t) = -\sin\left(2t\left(q_1^2 + p_1^2\right)\right) p_2(0) + \cos\left(2t\left(q_1^2 + p_1^2\right)\right) q_2(0).$$

If we remember (7.56), this flow is actually the limit of the quantum flow when we take the limit of the deformation parameter $q \to 0$ and hence $q^2/q \to 0, q/q \to 1$. Indeed in this case $v^*v = q_1^2 + p_1^2$ and $\alpha = q_2 + ip_2$. As the Hamiltonian was chosen to be an element of $\mathcal{A}'_q$ we get a reduced dynamics given by

$$[H, w] = -\frac{1}{2}(q^2 - 2q)(1 - u)w, \qquad [H, w^*] = -\frac{1}{2}(q^2 - 2q)(1 - u)w^*.$$

The corresponding solutions for the endomorphism $\mathrm{e}^{it\,\mathrm{ad}_H}$ become

$$w(t) = \mathrm{e}^{-it\frac{1}{2}(q^2-2q)(1-u)}w(0), \qquad w^*(t) = \mathrm{e}^{it\frac{1}{2}(q^2-2q)(1-u)}w^*(0).$$

Passing to the classical limit we find the corresponding vector field on $\mathbb{R}^3$ tangent to S^2

$$\widetilde{\Gamma} = (1 - u)\left(z\frac{\partial}{\partial v} - v\frac{\partial}{\partial z}\right),$$

which is the reduced dynamics

$$\begin{aligned}
\frac{du}{dt} &= 0 \\
\frac{dv}{dt} &= 2(q_1^2 + p_1^2)(q_2 p_1 - p_2 q_1) = (1 - u)z, \\
\frac{dz}{dt} &= -2(q_1^2 + p_1^2)(p_1 p_2 + q_1 q_2) = -(1 - u)v.
\end{aligned} \tag{7.57}$$

By using the stereographic projection $S^2 \to \mathbb{R}^2$ given by $(x, y) = \frac{1}{1-u}(v, z)$ we find the associated vector field on $\mathbb{R}^2$

$$\Gamma(x, y) = \frac{2}{x^2 + y^2 + 1} \left(x \frac{\partial}{\partial y} - y \frac{\partial}{\partial x} \right).$$

This example is very instructive because it provides us with an example of reduced quantum dynamics that goes onto the corresponding reduced classical dynamics, i.e., reduction 'commutes' with 'dequantization'. Further details can be found in [Gr94].

7.5.4.1 Example: Deformed Oscillators

Another instance of a non-commutative algebra reduction is provided by the case of the deformed harmonic oscillator. Let us start thus by analyzing the case of deformed harmonic oscillators described in the Heisenberg picture. By including the deformation parameter in the picture we can deal with several situations at the same time, as we are going to see.

We consider a complex vector space V generated by $a, a^\dagger$. Out of V we construct the associative tensorial algebra $\mathcal{A} = \mathbb{C} \oplus V \oplus (V \otimes V) \oplus (V \otimes V \otimes V) \oplus \cdots$. A dynamics on V, say

$$\frac{d}{dt} a = -i\omega a, \quad \frac{d}{dt} a^\dagger = i\omega a^\dagger ,$$

defines a dynamics on $\mathcal{A}$ by extending it by using the Leibniz rule with respect to the tensor product

A bilateral ideal $I_{r,q}$ of $\mathcal{A}$, generated by the relation $a^\dagger a - qaa^\dagger + r = 0$, i.e., the most general element of $I_{r,q}$ has the form $A(a^\dagger a - qaa^\dagger + r)B$, with $A, B \in \mathcal{A}$, is also invariant under the previously defined equations of motion. It follows then that the dynamics defines a derivation, a 'reduced dynamics' on the quotient algebra $\mathcal{A}_{r,q} = \mathcal{A}/I_{r,q}$. When $q = 1$ and $r = 0$ the dynamics becomes a dynamics on a commuting algebra and therefore can be considered to be a classical dynamics. When $q = 1$ and $r = \hbar$ we get back the standard quantum dynamics of the harmonic oscillator. If we consider r to be a function of the 'number operator' defined as $n = a^\dagger a$ we obtain many of the proposed deformations of the harmonic oscillator existing in the literature. In particular, these deformations have been applied to the description of the magnetic dipole [Lo97].

It is clear now that this reduction procedure may be carried over to any realization or representation of the abstract algebra and the corresponding ideal $I_{r,q}$. In this example it is important that the starting dynamics is linear. The extension to the universal tensorial algebra gives a kind of abstract universal harmonic oscillator. The bilateral ideal we choose to quotient the tensor algebra is responsible for the physical identification of variables and may arise from a specific realization of the tensor algebra in terms of functions or operators.

7.5.5 Reduction in the Ehrenfest Formalism

This picture of Quantum Mechanics is not widely known but it arises in connection with the so-called Ehrenfest theorem which may be seen from the point of view of $\star$-products on phase space (see [Es04]). Some aspects of this picture have been considered by Weinberg [We89a] and more generally appear in the geometrical formulation of Quantum Mechanics [Ci90, Ci901, Ci91, Ci94].

We saw above how the Schrödinger picture assumes as a starting point the Hilbert space of states and derive the observable as real operators acting on this space of states. The Heisenberg picture starts from the observables, enlarged by means of complexification into a $\mathbb{C}^*$-algebra and derives the states as positive normalized linear functionals on the algebra of observables. In the Ehrenfest picture both spaces are considered jointly to define quadratic functions as

$$f_A(\psi) = \frac{1}{2}\langle\psi, A\psi\rangle. \tag{7.58}$$

In this way all operators are transformed into quadratic functions which are real-valued when the operators are Hermitian. The main advantage of this picture relies on the fact that we can define a Poisson bracket on the space of quadratic functions by setting

$$\{f_A, f_B\} := i f_{[A,B]}, \tag{7.59}$$

where $[A, B]$ stands for the commutator on the space of operators. By introducing an orthonormal basis in $\mathcal{H}$, say $\{\psi_k\}$, we may write the function f_A as

$$f_A(\psi) = \frac{1}{2}\sum_{jk} c_j c_k^* \langle\psi_j, A\psi_k\rangle, \qquad \psi = \sum_k c_k \psi_k$$

and the Poisson bracket then becomes

$$\{f_A, f_B\} = i\sum_k \left(\frac{\partial f_A}{\partial c_k}\frac{\partial f_B}{\partial c_k^*} - \frac{\partial f_A}{\partial c_k^*}\frac{\partial f_B}{\partial c_k}\right).$$

This bracket can be used to write the equations of motion in the form

$$i\frac{df_A}{dt} = \{f_H, f_A\},$$

where f_H is the function associated to the Hamiltonian operator.

While this way of writing the dynamics is very satisfactory because allows us to write the equations of motion in a 'classical way', one has lost the associative product of operators. Indeed, the point-wise product (somehow a natural one for the functions defined on a real differential manifold) of two quadratic functions will not be quadratic but a quartic function. To recover the associative product we can, however, get inspiration from the definition of the Poisson bracket (7.59) and introduce

$$(f_A \star f_B)(\psi) := f_{AB}(\psi) = \frac{1}{2}\langle \psi, AB\psi \rangle. \tag{7.60}$$

By inserting a resolution of the identity $\sum_j |\psi_j\rangle\langle\psi_j| = \mathbb{I}$ (since there is a numerable basis for $\mathcal{H}$) in between the two operators in AB, say

$$\langle \psi, A \sum_j |\psi_j\rangle\langle\psi_j|B\psi\rangle,$$

and writing the expression of ψ in terms of the basis elements $\psi = \sum_k c_k \psi_k$ we find a product

$$(f_A \star f_B)(\psi) = \sum_{jkl} c_j c_l^* \langle \psi_j, A\psi_k\rangle\langle\psi_k, B\psi_l\rangle,$$

which reproduces the associative product of operators but now it is no longer pointwise.

As a matter of fact the Poisson bracket defines derivations for this product, i.e.

$$\{f_A, f_B \star f_C\} = \{f_A, f_B\} \star f_C + f_B \star \{f_A, f_C\} \quad \forall f_A, f_B, f_C.$$

Therefore it is an instance of what Dirac calls a quantum Poisson bracket [Di58]. In the literature it is known as a Lie-Jordan bracket [Em84, La98].

Using both products, the Ehrenfest picture becomes equivalent to Schrödinger and Heisenberg ones.

Let us consider now how the expressions of the products are written in terms of a different bases, namely the basis of eigenstates of the position operator Q or the momentum operator P. We have thus two basis $\{|q\rangle\}$ and $\{|p\rangle\}$ satisfying $Q|q\rangle = q|q\rangle$ and $P|p\rangle = p|p\rangle$ and

$$\int_{-\infty}^{\infty} |q\rangle dq \langle q| = \mathbb{I} = \int_{-\infty}^{\infty} |p\rangle dp \langle p|.$$

Now the matrix elements $A_{kj} = \langle \psi_j, A\psi_k\rangle$ of the operators in the definition of the $\star$ product above become

$$A(q', q) = \langle q', Aq\rangle \quad \text{or} \quad A(p', p) = \langle p', Ap\rangle,$$

and the sum is replaced by an integral:

$$(f_A \star f_B)(\psi) = \int dq\, dq'\, dq'' c(q')c^*(q'')A(q', q)B(q, q''). \tag{7.61}$$

Thus this is a product of functions defined on $\mathbb{R}^n \times \mathbb{R}^n$ or $(\mathbb{R}^n)^* \times (\mathbb{R}^n)^*$, i.e., two copies of the configuration space or two copies of the momentum space. Following an idea of Dirac [Di45] one may get functions on $\mathbb{R}^n \times (\mathbb{R}^n)^*$ by using eigenstates of the position operator on the left and eigenstates of the momentum operator on the right:

$$A_l(q, p) = \langle q, Ap\rangle e^{-\frac{i}{\hbar}qp},$$

or also interchanging the roles of position and momentum:

$$A_l(p, q) = \langle p, Aq\rangle e^{\frac{i}{\hbar}qp}.$$

Without elaborating much on these aspects (we refer to [Ch05] for details) we simply state that the $\star$-product we have defined, when considered on phase space, becomes the standard Moyal product.

It is now clear that we may consider the reduction procedure in terms of non-commutative algebras when we consider the $\star$-product. We shall give a simple example where from a $\star$-product on $\mathbb{R}^4$ we get by means of a reduction procedure a $\star$-product on the dual of the Lie algebra of $SU(2)$. Further details connected with their use in non-commutative geometry can be found in [Gr02].

7.5.5.1 Example: Star Products on $\mathfrak{su}(2)$

We are going to show how it is possible to define star products on spaces such as $\mathfrak{su}(2)$ by using the reduction of the Moyal star product defined on a larger space ($\mathbb{R}^4$ in this case).

Let us then consider the coordinates $\{q_1, q_2, p_1, p_2\}$ for $\mathbb{R}^4$, $\{x, y, w\}$ for $\mathfrak{su}(2)$ and the mapping $\pi : \mathbb{R}^4 \to \mathbb{R}^3 \sim \mathfrak{su}(2)$ defined as:

$$f_1(q_1, q_2, p_1, p_2) = \pi^*(x) = \frac{1}{2}(q_1q_2 + p_1p_2)$$
$$f_2(q_1, q_2, p_1, p_2) = \pi^*(y) = \frac{1}{2}(q_1p_2 - q_2p_1)$$
$$f_3(q_1, q_2, p_1, p_2) = \pi^*w = \frac{1}{4}(q_1^2 + p_1^2 - q_2^2 - p_2^2)$$

It is useful to consider also the pull-back of the Casimir function of $\mathfrak{su}(2)$, $C = \frac{1}{2}(x^2 + y^2 + w^2)$, which becomes

$$\pi^*C = \frac{1}{32}\left(p_1^2 + q_1^2 + p_2^2 + q_2^2\right)^2.$$

To define a reduced star product on $\mathfrak{su}^*(2)$ we consider the Moyal star product on the functions of $\mathbb{R}^4$, and select a $\star$-subalgebra isomorphic to the $\star$-algebra of $\mathfrak{su}^*(2)$. To identify this subalgebra we need derivations of the $\star$-product that annihilate the algebra we are studying. We look then for a derivation H which is a derivation of both the point-wise algebra and the $\star$-algebra, to ensure that reduction commutes with the 'classical limit'. The commutative point-wise product condition will identify the quotient manifold, while the condition on the $\star$-product identifies a star product on functions defined on the quotient. We consider thus a vector field H on $\mathbb{R}^4$ satisfying

$$\mathcal{L}_H \pi^* x = 0 = \mathcal{L}_H \pi^* y = \mathcal{L}_H \pi^* w.$$

This condition characterizes the point-wise subalgebra of functions of $\mathbb{R}^4$ which are projectable on functions of $\mathbb{R}^3$. Such a vector field can be taken to be the Hamiltonian vector field associated to the Casimir function $\pi^* C$. It is simple to see that the Poisson subalgebra generated by the functions $\{\pi^* x, \pi^* y, \pi^* w, f_H\}$ where $f_H = q_1^2 + q_2^2 + p_1^2 + p_2^2$ is the Poisson commutant of the function f_H (see [Gr02]). And this set is an involutive Moyal subalgebra when we consider the Moyal product on them, i.e., for any functions F, G

$$\{f_H, F\} = 0 = \{f_H, G\} \Rightarrow \{f_H, F \star G\} = 0.$$

The star product on $\mathfrak{su}(2)$ is then defined as:

$$\pi^*(F \star_{\mathfrak{su}(2)} G) = \pi^* F \star \pi^* G.$$

As an example we can consider the product:

$$x_j \star_{\mathfrak{su}(2)} f(x_i) = \left(x_j - \frac{i\theta}{2} \epsilon_{jlm} x_l \frac{\partial}{\partial x_m} - \frac{\theta^2}{8} \left[\left(1 + x_k \frac{\partial}{\partial x_k} \right) \frac{\partial}{\partial x_j} - \frac{1}{2} x_j \frac{\partial}{\partial x_k} \frac{\partial}{\partial x_k} \right] \right) f(x_i).$$

The same procedure may be applied to obtain a reduced star product for all three dimensional Lie algebras (see [Gr02]) and to deal with a non-commutative differential calculus [Ma06].

Another comment is in order. The reduction procedures within Quantum Mechanics are most effective when they are formulated in a way such that the classical limit may be naturally considered in the chosen formalism. We believe that this may be considered as an indication that Quantum Mechanics should be formulated in a way that in some form it incorporates the so-called 'correspondence principle'.

References

[Ei61] Eisenhart, L.P.: Continuous Groups of Transformations. Dover, New York (1961)

[Fo59] Forsyth, A.R.: Theory of Differential Equations. Dover, New York (1959)

[Lie93] Lie, S., Scheffers G.: Vorlesungen über continuierliche Gruppen mit geometrischen und anderen Anwendungen. Teubner, Leipzig (1893) (Edited and revised by G. Scheffers)

[Mm85] Marmo, G., Saletan, E.J., Simoni, A., Vitale, B.: Dynamical Systems: A Differential Geometric Approach to Symmetry and Reduction. Wiley, Chichester (1985)

[Ma83] Marmo, G.: Function groups and reduction of Hamiltonian systems. Att. Accad. Sci. Torino Cl. Sci. Fis. Mat. Nat. **117**, 273–287 (1983)

[Ho85] Howe, R.: Dual pairs in physics: harmonic oscillators, photons, electrons and singletons. Lect. Appl. Math. **21**, 179–207 (1985)

[Ch07] Chaturvedi, S., Ercolessi, E., Marmo, G., Mukunda, N., Simon, R.: Ray-space Riccati evolution and geometric phases for n-level quantum systems. Pramana J. Phys. **63**, 317–327 (2007)

[OlPe81] Olshanetsky, M.A., Perelomov, A.M.: Classical integrable finite-dimensional systems related to Lie algebras. Phys Rep. **71**, 313–400 (1981)

[Ki76] Kirillov, A.A.: Elements of the Theory of Representations. Springer, Berlin (1976)

[Ki99] Kirillov, A.A.: Merits and demerits of the orbit method. Bull. Am. Math. Soc. **36**, 433–488 (1999)

[La91] Landi, G., Marmo, G., Sparano, G., Vilasi, G.: A generalized reduction procedure for dynamical systems. Mod. Phys. Lett. A **6**, 3445–3453 (1991)

[Ma92] Man'ko, V.I., Marmo, G.: Generalized reduction procedure and nonlinear non stationary dynamical systems. Mod. Phys. Lett. A **7**, 3411–3418 (1992)

[Ca07b] Cariñena, J.F., Clemente-Gallardo, J., Marmo, G.: Reduction procedures in classical and quantum mechanics. Int. J. Mod. Phys. **4**, 1363–1403 (2007)

[Ca09] Cariñena, J.F., Clemente-Gallardo, J., Marmo, G.: Towards a definition of quantum integrability. Int. J. Geom. Meth. Mod. Phys. **6**, 129–172 (2009)

[Ca06] Cariñena, J.F., Gracia, X., Marmo, G., Martínez, E., Muñoz-Lecanda, M., Román-Roy, N.: Geometric Hamilton-Jacobi theory. Int. J. Geom. Meth. Phys. **3**, 1417–1458 (2006)

[Am26] Amaldi, U., Levi-Civita, T.: Lezioni di Meccanica Razionale. N. Zanichelli, Bologna (1926)

[Pa57] Palais, R.S.: A global formulation of the Lie theory of transitive groups. Mem. Am. Math. Soc. **22**, 404 (1957)

[Ma05] Marmo, G., Parasecoli, E., Tulczyjew, W.: Space-time orientations and Maxwell's equations. Rep. Math. Phys. **56**, 209–248 (2005)

[Di31] Dirac, P.A.M.: Quantized singularities in the electromagnetic field. Proc. Roy. Soc. Lond. A **133**, 60–72 (1931)

[Po96] Poincaré, H.: Remarques sur une experience de M. Birkeland. Comp. Rend. Acad. Sci. **123**, 530–533 (1896)

[Ba80] Balachandran, A.P., Marmo, G., Skagerstam, B.S., Stern, A.: Magnetic monopoles with no strings. Nucl. Phys. B **152**, 385–396 (1980)

[Ba83] Balachandran, A.P., Marmo, G., Skagerstam, B.S., Stern, A.: Gauge Symmetries and Fiber Bundles: Applications to Particle Dynamics, Lecture Notes in Physics, vol 188. Springer, New York (1983)

[Ba91] Balachandran, A.P., Marmo, G., Skagerstam, B.S., Stern, A.: Classical Topology and Quantum States. World Scientific, River Edge (1991)

[La90] Landi, G., Marmo, G.: Algebraic Differential Calculus for Gauge Theories. Nucl. Phys. B (Proc.Suppl.) **18**, 171–206 (1990)

[Cu97] Cushman, R.H., Bates, L.M.: Global Aspects of Classical Integrable Systems. Birkhäuser, Basel (1997)

[Du80] Duistermaat, J.J.: On global action-angle variables. Commun. Pure Appl. Math. **33**, 687–706 (1980)

[Es04] Esposito, G., Marmo, G., Sudarshan, G.: From Classical to Quantum Mechanics: an introduction to the formalism. Cambridge University Press, Cambridge (2004)

[Ca94] Cariñena, J.F., Ibort, L.A., Marmo, G., Perelomov, A.M.: The Geometry of Poisson manifolds and Lie algebras. J. Phys. A Math. Gen. **27**, 7425–7449 (1994)

[Al98] Aldaya, V., Guerrero, J., Marmo, G.: Quantization on a Lie group: higher order polarizations. In: Gruber, B., Ramek, M. (eds.) Symmetries in Science X. Plenum Press, New York (1998)

[Av05] D'Avanzo, A., Marmo, G.: Reduction and unfolding: the Kepler problem. Int. J. Geom. Meth. Mod. Phys. **2**, 83–109 (2005)

[Zu93] Zhu, J.C., Klauder, J.R.: Classical symptoms of quantum illness. Am. J. Phys. **61**, 605–611 (1993)

[Av05b] D'Avanzo, A., Marmo, G., Valentino, A.: Reduction and unfolding for quantum systems: the hydrogen atom. Int. J. Geom. Meth. Mod. Phys. **2**, 1043–1062 (2005)

[Fe06] Feher, L., Pusztai, B.G.: A class of Calogero type reductions of free motion on a simple Lie group. Lett. Math. Phys. **79**, 263–277 (2007)

[Fe07] Feher, L., Pusztai, B.G.: Hamiltonian reductions of free particles under polar actions of compact Lie groups. Theor. Math. Phys. **155**, 646–658 (2008)

[Fe08] Feher, L., Pusztai, B.G.: On the self-adjointness of certain reduced Laplace-Beltrami operators. Rep. Math. Phys. **61**, 163–170 (2008)

[Wo87] Woronowicz, S.L.: Twisted $SU(2)$ group: an example of a non-commutative differential calculus. Pub. Res. Inst. Math. Sci. **23**, 117–181 (1987)

[Gr94] Grabowski, J., Landi, G., Marmo, G., Vilasi, G.: Generalized reduction procedure: symplectic and Poisson formalisms. Forts. Phys. **42**, 393–427 (1994)

[Li93] Lizzi, F., Marmo, G., Sparano, G., Vitale, P.: Dynamical aspects of Lie-Poisson structures. Mod. Phys. Lett. A **8**, 2973–2987 (1993)

[Lo97] López-Peña, R., Manko, V.I., Marmo, G.: Wigner problem for a precessing dipole. Phys. Rev. A **56**, 1126–1130 (1997)

[We89a] Weinberg, S.: Precision tests of quantum mechanics. Phys. Rev. Lett. **62**, 485–488 (1989)

[Ci90] Cirelli, R., Maniá, A., Pizzocchero, L.: Quantum mechanics as an infinite-dimensional Hamiltonian system with uncertainty structure. Int. J. Math. **31**, 2891–2897 (1990)

[Ci901] Cirelli, R., Maniá, A., Pizzocchero, L.: Quantum mechanics as an infinite-dimensional hamiltonian system with uncertainty structure II. J. Math. Phys. **31**, 2898–2903 (1990)

[Ci91] Cirelli, R., Maniá, A., Pizzocchero, L.: Quantum phase-space formulation of Schrödinger mechanics. Int. J. Mod Phys A **6**, 2133–2146 (1991)

[Ci94] Cirelli, R., Maniá, A., Pizzocchero, L.: A functional representation for non-commutative $\mathbb{C}^*$-algebras. Rev. Math. Phys. **6**, 675–697 (1994)

[Di58] Dirac, P.A.M.: The Principles of Quantum Mechanics. Clarendon Press, Oxford (1958)

[Em84] Emch, G.G.: Mathematical and Conceptual Foundations of 20th Century Physics. North-Holland, Amsterdam (1984)

[La98] Landsman. N.P.: Mathematical Topics Between Classical and Quantum Mechanics. Springer, New York (1998)

[Di45] Dirac, P.A.M.: On the analogy between classical and quantum mechanics. Rev. Mod. Phys. **17**, 195–199 (1945)

[Ch05] Chaturvedi, S., Ercolessi, E., Marmo, G., Morandi, G., Mukunda, N., Simon, R.: Wigner-Weyl correspondence in quantum mechanics for continuous and discrete time systems: a Dirac inspired view. J. Phys. A Math. Gen. **39**, 1405–1423 (2005)

[Gr02] Gracia-Bondia, J.M., Lizzi, F., Marmo, G., Vitale, P.: Infinitely many star products to play with. J. High Energy Phys. **4**, 025 (2002)

[Ma06] Marmo, G., Vitale, P., Zampini, A.: Noncommutative differential calculus for Moyal subalgebras. J. Geom. Phys. **55**, 611–622 (2006)

Chapter 8
Integrable and Superintegrable Systems

Complex models are rarely useful (unless for those writing their dissertations).
Vladimir Arnold, *On teaching mathematics*, 1997

8.1 Introduction: What Is Integrability?

There is no generally accepted definition of integrability that would include the various instances which are usually associated with the word "integrable". Occasionally the word 'solvable' is also used more or less as synonymous, but to emphasize the fact that the system need not be Hamiltonian. Any definition should, of course, include those systems that are usually termed completely integrable systems, or Liouville-Arnold integrable systems, moreover all systems that carry the qualification 'integrable' have the characteristic property that their solutions can, in principle, be constructed explicitly.

In this respect, also separability turns out to be instrumental for explicit integrability of the system. As a matter of fact, the decomposition of a matrix into semisimple plus a nilpotent part is a way to introduce a preliminary separation. Afterwards, the spectral decomposition of the semisimple part allows us to separate the system into pairwise commuting ones, to separate the nilpotent part into pairwise commuting ones, we restrict the analysis to those systems that are nilpotent of index two.

Decomposition into pairwise commuting subsystems formalizes the 'composition of independent motions' that is commonly used at the level of Newton's equations.

Spectral decomposition shows that one may search for particular coordinate systems in which the equations acquire a particular simple form. By using the insight coming from the fundamental theorem of algebra, one may reduce the analysis to two-dimensional vector spaces and decompose the matrix associated with the subsystem into a multiple of the identity plus a matrix proportional to σ_3 or one proportional to $i\sigma_2$ to deal with semisimple matrices, while the nilpotent part, with index two, may be represented by $\sigma_+ = \frac{1}{2}(\sigma_1 + i\sigma_2)$. The systems represented by means of the Pauli

J.F. Cariñena et al., *Geometry from Dynamics, Classical and Quantum*,
DOI 10.1007/978-94-017-9220-2_8

matrices as explained will have associated flows representing hyperbolic motions, circular motions or straight line motion respectively. These may be considered to be building blocks of generic systems with nilpotent part of index two. It should be clear that this decomposition is defined when we only use the linear group to achieve diagonalization. If we allow nonlinear coordinate transformations, then we can transform a semisimple system into a nilpotent one, which is the case for instance for the Harmonic Oscillator when we go to action-angle variables.

On the other hand the separability procedure is the one always attempted for solving Schrödinger equations or wave equations, usually under the name of separation of variables when we use the description in terms of differential operators. More generally, it is what we try to attempt when we deal with diagonalization of the Hamiltonian operator (let us remark that the Hamiltonian operator of Quantum Mechanics should be thought of as a vector field not as the Hamiltonian function of classical mechanics; as a matter of fact, if it is not Hermitian it need not be associated with a Hamiltonian function). Thus integrability in the previous meaning may be formalized by means of a decomposition of the vector field representing the dynamics as pairwise commuting vector fields, i.e., forming an Abelian algebra with coefficient which behaves like a numerical constant with respect to the vector fields entering the decomposition. Thus it imitates the low dimensions appearing in the spectral decomposition of the matrix representing the starting vector field.

We could require that the vector fields entering the decomposition should be maximally superintegrable, i.e. admitting a maximal number of constants of the motion. It should be stressed that this requirement may discriminate between different decompositions which would be allowed by the 'algebraic only' restriction. At the quantum level this amounts to searching for an Abelian algebra of pairwise commuting first-order differential operators (see Chap. 10) such that the Hamiltonian operator would be an element of the enveloping algebra generated by the Abelian Lie algebra.

The natural generalization originated by dropping the requirement of Abelianity of the algebra of vector fields entering the decomposition will be examined in the next chapter under the name of Lie–Scheffers systems. Again at the quantum level a Lie algebra of first-order differential operators would contain the Hamiltonian operator as an element of the enveloping algebra.

In conclusion, the various procedures to identify 'integrable systems' all seem to rely on the construction of Abelian algebras of vector fields which allow for the decomposition of the original dynamics.

As one-one tensor fields which are invariant under the dynamics allow, generically, construction of commuting vector fields out of the starting one, the so-called 'recursion operator method' uses this procedure. Often a recursion operator may be obtained by composing invariant covariant and contravariant $(0, 2)$-tensors, this would mean composing a symplectic structure with a Poisson structure. Some requirement of "integrability" or "flatness" of this tensor field (Nijenhuis condition) would guarantee that the algebra one generates starting with the dynamical vector field will be Abelian.

For scattering theory, the role of the connecting intertwining operator would be played by the Moeller or wave operator; again one constructs a change of coordinates connecting a "free system" with the interacting one. This would be the analogue of the construction of action-angle variables for systems with bounded orbits.

In general the 'comparison system' may be chosen in such a way that it is easily integrable; as we have seen, that is the case when this comparison system may be described in terms of matrices. The connecting map gives raise to the so-called Lax representation.

Of course very much as in the case of the construction of action-angle variables, the identification of maps connecting our system with one described by means of matrices is highly non trivial, even though quite powerful; the search for Lax representations of particular integrable systems has been more of an 'art' than a technique generating quite a few papers dealing with the construction of the map for relevant physical systems.

8.2 A First Approach to the Notion of Integrability: Systems with Bounded Trajectories

As it is customary in this book, to pin-point the mathematical structures involved, without introducing additional technicalities, we shall first consider linear systems in finite dimensions.

On a vector space V a dynamical linear system would have the form

$$\frac{d}{dt}v = Av,$$

with v a generic vector in V. With this linear system there is always associated another one on the dual linear space V^* and on any tensor space built out of the vector space V (recall the discussion in Sect. 2.2.5).

This equation is explicitly solved by the exponential map, as discussed at length in Sect. 2.2.2. Then we have $\Phi\colon \mathbb{R} \times V \to V$ given by $\Phi(t, x_0) = \exp(tA)x_0$. Thus for each initial condition x_0 we know its evolution $x(t) = \Phi(t, x_0)$ after a time t. From a minimalistic point of view this system would be 'integrable', nevertheless we do not feel satisfied, for instance we would not know how to answer simple questions such as:

1. Are there periodic solutions?
2. What happens when $t \to \pm\infty$?
3. If the system has some periodic orbits, are all the rest also periodic?
4. What are the symmetries and the constants of motion?

One has the expectation, or prejudice, that integrability should also carry along a simple and direct answer to all previous questions. In some sense we would like to have a kind of 'normal form' for integrable systems so that previous questions would be easily answerable.

In previous chapters we considered systems that allow for action-angle coordinates; these systems take the simple form

$$\frac{d}{dt}\varphi_j = \nu_j, \qquad \frac{d}{dt}\nu_j = 0, \qquad j = 1, ..., n.$$

The flow in these coordinates takes a 'linear form' given by

$$\varphi_j(t) = \varphi_j(0) + t\nu_j, \qquad \nu_j(t) = \nu_j(0).$$

In general this change of coordinates excludes some points of the carrier space because they do not define a global diffeomorphism. The placement and the 'extension' of these 'offending points' may change from system to system, making classification very difficult if not impossible.

The previous list of questions can be extended now to this situation adding a few more:

1. Does the existence of this change of coordinates depends on the Poisson structure or can it be defined independently of it?
2. Are the trajectories periodic for any initial condition?
3. If the system allows for many alternative invariant symplectic structures, are the action-angle variables uniquely defined?

In the following we would like to address these questions and try to show that even for simple systems the situation is rather composite.

8.2.1 Systems with Bounded Trajectories

In this section we would like to show to what extent a dynamical system may identify geometrical structures without an a priori assumption that it preserves a symplectic or a Poisson structure. We assume that our system is linear and has all orbits bounded.

We have already encountered this example in connection with the existence of invariant Hermitian structures, as a particular instance of Hamiltonian systems, and it was discussed in Sect. 6.2.3. Here we would like to take up again this example to exhibit some properties of integrable systems which would suggest a normal form independently of any invariant symplectic structure. Clearly, in finite-dimensional vector spaces, we can use any norm to define boundedness of the orbits. This condition on the boundedness of the orbits has rather stringent consequences on the structure of the flow. The following proposition, which relies only on linearity of the vector field and boundedness of the orbits, extracts the relevant ingredients entering the mathematical structures of this particular family of systems.

Proposition 8.1 *Any finite-dimensional linear system, admitting only bounded orbits, allows for action-angle variables and admits an infinite number of constants of motion and an infinite number of infinitesimal symmetries, moreover it admits an infinite number of alternative Hamiltonian descriptions.*

Exercise 8.1 Prove this proposition using the proof of Proposition 6.15 and the ideas sketched in the next paragraphs.

We shall see that even if two systems satisfy both requirements of this proposition, they need not be diffeomorphic, showing that an equivalence relation based on the diffeomorphism group is not adequate. In particular with such an equivalence relation we would find that systems with all orbits bounded would divide into an infinite number of equivalence classes of diffeomorphic systems. This result, surprising or not, shows that without many more qualifications, the form

$$\frac{d}{dt}\varphi_j = \nu_j; \qquad \frac{d}{dt}\nu_j = 0, \qquad j = 1, \ldots, n$$

is the best one may hope to achieve, if it exists at all.

8.2.1.1 Factorization Properties of the Flow

For any linear system, the flow $\phi_t = e^{tA}$, may be factorized[1] as $e^{tS}\, e^{tN}$ with N a nilpotent and S a semi-simple matrix such that $[S, N] = 0$. The requirement of boundedness of the trajectories, i.e., $||e^{tA}x(0)|| \leq K||x(0)||$ for any $x(0) \in V$ and any $t \in \mathbb{R}$ immediately requires $N = 0$ and S must have only imaginary eigenvalues.

If we make now the further assumption that eigenvalues have minimal degeneracy, we find that in the basis of eigenvectors for S, the matrix A acquires the form:

$$A = \begin{pmatrix} \begin{array}{|cc} 0 & \nu_1 \\ -\nu_1 & 0 \end{array} & & \\ & \ddots & \\ & & \begin{array}{|cc} 0 & \nu_n \\ -\nu_n & 0 \end{array} \end{pmatrix},$$

with $(\nu_1, \ldots, \nu_n) \in \mathbb{R}^n$. The representation of the vector field $\Gamma = X_A$ describing the dynamics gives:

$$\Gamma = \sum_{j=1}^{n} \nu_j \left(x_j \frac{\partial}{\partial y_j} - y_j \frac{\partial}{\partial x_j} \right), \quad \nu_j \in \mathbb{R}, \quad j = 1, \ldots, n.$$

If we use complex coordinates $z_j = x_j + iy_j$, $j = 1, \ldots, n$, we have:

[1] That is often quoted as 'the composition of independent motions'.

$$\Gamma = \sum_{j=1}^{n} i\nu_j \left(z_j \frac{\partial}{\partial z_j} - z_j^* \frac{\partial}{\partial z_j^*} \right),$$

and in terms of vector fields we deduce immediately (compare with the results on Sect. 1.2.7):

1. Any function of
$$|z_1|^2 = x_1^2 + y_1^2, \ldots, |z_n|^2 = x_n^2 + y_n^2. \tag{8.1}$$
is a constant of motion.
2. The module generated by vector fields:
$$X_j = x_j \frac{\partial}{\partial y_j} - y_j \frac{\partial}{\partial x_j}, \quad \Delta_j = x_j \frac{\partial}{\partial x_j} + y_j \frac{\partial}{\partial y_j},$$
with coefficients in the algebra of constants of motion consists of infinitesimal symmetries.
3. Many invariant symplectic structures are generated by (non-degenerate) linear combinations of
$$f_j(x_j^2 + y_j^2) dx_j \wedge dy_j. \tag{8.2}$$
4. Many invariant Poisson structures are generated by linear combinations of the bivector fields:
$$\Lambda_{jk} \left(\frac{\partial}{\partial x_j} \wedge \frac{\partial}{\partial y_k} \right), \quad \lambda_j X_j \wedge \Delta_j = \lambda_j (x_j^2 + y_j^2) \frac{\partial}{\partial y_j} \wedge \frac{\partial}{\partial x_j}, \quad \lambda_j, \Lambda_{jk} \in \mathbb{R}.$$

If we restrict our attention only to the semisimple part, the spectral decomposition of our matrix gives raise to a decomposition in terms of vector fields which are pairwise commuting; $X_S = a^j Y_j$ with $[Y_j, Y_k] = 0$, and $L_{Y_j} a_k = 0$ for any pair of indices j and k.

Clearly, each Y_j is maximally superintegrable if eigenvalues have minimal degeneracy. It also follows that each one of the vector fields entering the decomposition is a complete vector field and gives raise to a one-parameter subgroup of linear transformations. As we are considering diagonalization over the complex numbers, the real dimension of the vector space corresponding to each eigenvalue will be two. At the manifold level, to implement this property, we should introduce the notion of 'related' vector fields. Roughly speaking we are dealing with a map from the carrier space to another manifold of lower dimensions such that the algebra of functions on this manifold, when pulled back to the starting carrier space, will be an invariant subalgebra under the dynamical evolution. The restriction of the dynamical vector field to this subalgebra would play the role of the vector field Y_j. The algebraic properties emerging from the spectral decomposition of the semisimple part may be recasted into geometrical properties of integrable systems (see Sect. 8.3.1). It may be useful to recall that Marcus and Meyer [MM74] have shown that generic Hamiltonian systems are neither integrable nor ergodic.

8.2.1.2 Conditions on Frequencies

When further properties are satisfied by frequencies $\nu_1, \nu_2, \ldots, \nu_n$, additional properties emerge for the dynamics. For instance, if we assume that ν_j/ν_k is an irrational number for any pair of indices j, k, it follows that the closure of each trajectory starting with a generic initial condition $(x_1(0), x_2(0), \ldots, x_n(0))$ will be a torus of dimension n. Under these hypotheses it is clear that there will not be any additional constant of motion which is functionally independent from the algebra of functions which is generated by $\{x_j^2 + y_j^2 \mid j = 1, 2, \ldots, n\}$.

Infinitesimal symmetries however are linear combinations of the infinitesimal generators Δ_j, X_j. Obviously, the coefficients in any combination may be functions of the constants of motion.

Roughly speaking, the dynamical system has been split into a family of non-interacting systems, i.e., pairwise commuting. Each one $\Gamma_j = \nu_j X_j$ generates a one-parameter subgroup of the torus group of dimension n acting on the $2n$-dimensional vector space V.

The symmetries Δ_j pairwise commute among themselves and also commute with the action of the torus group. By using the differential forms $\theta_j = x_j dy_j - y_j dx_j$ and functions $f(x_1^2 + y_1^2, \ldots, x_n^2 + y_n^2)$ along with the infinitesimal symmetries Δ_j, we can generate a rather large family of invariant tensor fields for Γ.

If we use the torus action generated by X_j, all possible linear systems that share the same invariant foliation by n-dimensional tori emerge as one-parameter subgroups of toric transformations. When a given system has frequencies that are pairwise irrational, generic orbits are not closed and their closure coincides with the full torus. When particular initial conditions are chosen, say represented by vectors belonging to proper invariant subspaces, the corresponding orbits have a closure of dimension equal to half the real dimension of the minimal invariant subspace containing the chosen initial condition. To say it plainly, if the frequencies are pairwise irrational does not mean that the system has no closed trajectories. The invariance under dilations, generated by Δ_j, shows that on each plane from one periodic orbit we generate a new one.

8.2.1.3 Rationally Related Frequencies

We may assume now that some frequencies are rationally related, say $\nu_j = \nu n_j$ and $\nu_k = \nu n_k$, with the integers n_k, n_j relatively prime. In this case we have an additional constant of motion given by:

$$K = (z_j^*)^{n_k} (z_k)^{n_j} .$$

If some of the frequencies are rationally related we find additional constants of motion and the system is superintegrable. The given system is said to be maximally superintegrable when the number of independent constants of motion is maximal, i.e., $2n-1$ for a system of dimension $2n$. An important instance of a maximally superintegrable system, in action-angle variables is of the form:

$$\Gamma = \nu\left(\frac{\partial}{\partial\varphi_1} + \cdots + \frac{\partial}{\partial\varphi_n}\right),$$

then, not only $x_j^2 + y_j^2$ are constants of motion but the differences $\nu_j - \nu_k$ provide additional constants of motion.

When the frequencies satisfy $\nu_j = \nu m_j$ for all j and all pairs (m_j, m_k) are pairs of mutually prime integer numbers, any sequence of pairwise mutually prime integers, say $(m_1, m_2, \ldots, m_n)$ will identify a linear dynamical system which is maximally superintegrable, however two systems associated with two sequences of mutually prime integers which are also relatively prime as sequences, cannot represent diffeomorphic systems.

Thus even for maximally superintegrable systems, the classification under the diffeomorphism group would give an infinite number of equivalence classes.

8.3 The Geometrization of the Notion of Integrability

We have derived many of the previous properties for a linear system Γ with the only requirement that orbits must be bounded. For a generic vector field the objects we may be able to identify are:

1. The family of constants of motion under dynamical evolution: $f(x(t)) = f(x(0))$ for any $t \in \mathbb{R}$ and any initial condition $x(0)$.
   When a constant of motion is required to be at least continuous, this family captures the essence of the closure of the orbit for a generic initial condition.
2. The family of vector fields in the kernel of the map $\Phi = (f_1, \ldots, f_m)\colon M \to \mathbb{R}^m$, where M denotes the carrier space of the dynamics Γ, usually a smooth finite-dimensional manifold, and the functions $f_1, \ldots, f_m$ are functionally independent constants of motion.

If we consider differentiable constants of motion we may consider the algebra of vector fields defined by:

$$\mathcal{L}_X f = 0,$$

for any f smooth enough such that $\mathcal{L}_\Gamma f = 0$. This algebra of vector fields is a Lie module[2] with coefficients constants of motion of Γ and was called the bicommutant in Sect. 3.2.4 where a stability theorem was proved, Lemma 3.7, see also [IM12]).

[2] That is a module $\mathcal{M}$ over an algebra $\mathcal{A}$ with a Lie bracket which is $\mathcal{A}$-linear.

We might also consider the Lie module of vector fields Y such that $\mathcal{L}_X \mathcal{L}_Y f = 0$ for any constant of motion and all vector fields X such that $\mathcal{L}_X f = 0$, which is a natural way of defining the normalizer of the family of vector fields X. Actually the vector fields Y will take constants of motion into constants of motion.

The two Lie modules define a short exact sequence of Lie modules:

$$0 \to \mathfrak{X}^0 \to \mathfrak{X}^N \to \mathfrak{X}^Q \to 0,$$

where $\mathfrak{X}^0$ is the Lie module of vector fields defined by $\mathcal{L}_X f = 0$. The Lie module $\mathfrak{X}^N$ is the normalizer of the previous Lie module and $\mathfrak{X}^Q$ is the quotient of the previous two.

We should stress that this sequence is canonically associated with any vector field Γ. The Lie module $\mathfrak{X}^Q$ is essentially related to the algebra of vector fields defined on the manifold associated with the constants of motion for Γ. With reference to the previous example of linear systems with all orbits bounded, it would be generated by the real spectrum of constants of motion built out of $z_j^* z_j$ and $(z_j^{n_k})^* z_k^{n_j}$.

We could define an integrable system to be one for which the Lie modules $\mathfrak{X}^Q$ and $\mathfrak{X}^0$ have the same dimension, which would be superintegrable when $\dim \mathfrak{X}^0 < \dim \mathfrak{X}^Q$. Maximally superintegrable systems will happen when $\mathfrak{X}^0$ is generated by Γ itself and $\mathfrak{X}^Q$ has dimension $2n - 1$. This picture has the advantage of disregarding the parametrization of the dynamics.

The connection between infinitesimal symmetries and constants of motion requires preliminarily to associate vector fields and 1-forms. To associate vector fields with 1-forms we need a (0, 2)-tensor field. In the linear case it has been possible to construct the elementary closed 2-forms $f_j(x_j^2 + y_j^2) dx_j \wedge dy_j$, Eq. (8.2), by using the requirement of simplicity of eigenvalues. For generic vector fields this would correspond to the requirement that there exist projection maps, submersions: $\phi_j : M \to \mathbb{R}_j^2$, such that Γ is ϕ_j-related to a vector field Γ_j on $\mathbb{R}_j^2$ possessing a constant of motion and preserving the volume. In this way the pull-back of the volume form on $\mathbb{R}_j^2$ and h_j would provide us with the analogue of $f_j(x_j^2 + y_j^2) dx_j \wedge dy_j$.

In summary, the existence of an invariant non-degenerate closed 2-form, would allow us to create a correspondence between a subalgebra of vector fields in $\mathfrak{X}^0$ and closed 1-forms generated by $df_1, \ldots, df_n$.

Clearly it is conceivable that many such pairings are possible and therefore many alternative Hamiltonian descriptions could exist. If we write the factorization of Γ in the form:

$$\Gamma = \Lambda(dH),$$

it is now clear that any diffeomorphism which is also a symmetry for Γ would provide a new Hamiltonian description when it is not canonical:

$$\phi_* \Gamma = \Gamma = \phi_*(\Lambda) d(\phi^* H).$$

We have already commented this situation in the case of Hamiltonian linear systems (Sect. 5.2.6). In what follows we shall further analyze what can be said when we start from the normal form:

$$\Gamma = \nu_j \frac{\partial}{\partial \varphi_j}.$$

8.3.1 The Geometrical Notion of Integrability and the Erlangen Programme

When we move to the geometrical picture of our dynamical system, it is rather natural to consider all vector fields obtained by means of the diffeomorphism group acting on the starting one to represent the same abstract dynamical system. This group may be reduced, 'broken', to a smaller subgroup, by requiring that it preserves some additional geometrical structure, say a symplectic structure. In this way we would obtain the group of "canonical transformations" to go from one Hamiltonian vector field to another equivalent to it.

After the Erlangen Programme written by Felix Klein with the collaboration of Sophus Lie (Lie visited Klein for 2 months, just before the programme was written) it is by now accepted that a "geometry" or a "geometrical structure" on a manifold M amounts to selecting a subgroup of the group of diffeomorphisms of M [Kl72, Kl92].

In Physics this correspondence is quite familiar, for instance we have the Poincaré group for the Minkowski space-time in Special Relativity, the group $\mathrm{Diff}(\mathbb{R}^4)$ in general relativity, the group of symplectomorphisms in Hamiltonian dynamics, contact transformations, the unitary group in Quantum Mechanics and so on. In all previous cases, the subgroup of the group of diffeomorphisms determined by a geometry characterizes the geometry itself, for instance, if (M_1, ω_1) and (M_2, ω_2) are two symplectic manifolds and there is a group isomorphism $\Phi\colon \mathrm{Sp}(M_1, \omega_1) \to \mathrm{Sp}(M_2, \omega_2)$ of the corresponding groups of canonical transformations, i.e., symplectic diffeomorphisms defined by each one, then the two symplectic manifolds are (up to a conformal constant) symplectically equivalent, i.e., there exists a diffeomorphism $\varphi\colon M_1 \to M_2$ such that $\varphi^*\omega_2 = c\omega_1$ [Ba86, BG88]. Similar results were established by Grabowski for Poisson and Jacobi manifolds [Gr00]. Thus the automorphisms determined by some geometrical structures are essentially inner as it happens for the group of unitary transformations of a Hilbert space.

In this section we would like to identify the appropriate subgroup of $\mathrm{Diff}(M)$ associated with an integrable or a superintegrable system Γ.

Thus if we have an integrable Hamiltonian dynamical system Γ defined on a symplectic manifold (M, ω) the naive thought that the subgroup of the group of diffeomorphisms determined by it should be a subgroup of the group of canonical transformations of ω is immediately shown to be inadequate because integrable systems always admit alternative Hamiltonian descriptions and we would not know which canonical transformations to consider.

An example of the application of the Erlangen programme is readily shown by the geometrization of linear structures described in Sect. 3.3. Let us recall that a linear structure is characterized by a tensorial object (and the associated invariance subgroup), i.e., a complete vector field Δ such that:

1. There exists only one point, say $m_0 \in M$, such that $\Delta(m_0) = 0$.
2. The eigenvalue problem: $\mathcal{L}_\Delta f = 0 \cdot f$, $f \in \mathcal{F}(M)$ has only trivial solutions, i.e., $f =$ constant if M is connected.
3. The eigenvalue problem: $\mathcal{L}_\Delta f = f$, $f \in \mathcal{F}(M)$ has $\dim M = n$ independent solutions f_k such that $df_1 \wedge \cdots \wedge df_n \neq 0$.

Let us remark that the completeness condition on Δ allows us to 'generate' all of M starting with a properly chosen transversal codimension-one submanifold in a neighborhood of m_0 considered as a set of 'initial conditions'. Any such vector field Δ identifies a subgroup of Diff(M) by requiring $\varphi_*(\Delta) = \Delta$. Thus the subgroup $GL(M, \Delta)$ of Diff(M) of diffeomorphisms φ preserving Δ is exactly the group of linear isomorphisms $GL(n, \mathbb{R})$ with $n = \dim M$ where we use the global chart provided by the functions f_j to identify M with $\mathbb{R}^n$. In infinitesimal terms, vector fields generators of linear transformations are solutions of $[X, \Delta] = 0$. Moreover if we have two linear structures (M_1, Δ_1) and (M_2, Δ_2) defined in two manifolds M_1 and M_2, and there is a group isomorphisms $\Psi \colon GL(M_1, \Delta_1) \to GL(M_2, \Delta_2)$, then there exists a diffeomorphism $\psi \colon M_1 \to M_2$ such that $\psi_* \Delta_1 = \Delta_2$.

As it was shown in Sect. 3.3 the notion of linear structure can be weakened by replacing conditions 1–3 above. Specifically we may ask that the set of points in M satisfying $\Delta(m) = 0$ define a smooth submanifold Z of M of dimension k instead of condition 1. Condition 2 will be rephrased by asking that functions f such that $\mathcal{L}_\Delta f = 0$ define an Abelian algebra whose spectrum is diffeomorphic to Z and, finally, condition 3 will be substituted by demanding that $\mathcal{L}_\Delta f = f$, define $n - k$ functionally independent fibrewise-linear functions. Such a partial linear structure, as we have already seen, is actually equivalent to a vector bundle structure on the manifold M and the corresponding vector bundle autormophisms are selected by the condition $\varphi_* \Delta = \Delta$.

Another example in the same vein is concerned with possible tangent and cotangent bundle structures on a manifold M.

We recall (see Sect. 5.5.1) that a tangent bundle structure on M is identified by a pair (Δ, S), where Δ defines a partial linear structure on M, i.e., a vector bundle structure, and S is a $(1, 1)$-tensor field [FL89] such that $\ker S = \operatorname{Im} S$ (which implies $S^2 = 0$), $S(\Delta) = 0$ and $d_S^2 = 0$ where the 'twisted' differential d_S is defined by

$$d_S f(X) = df(S(x)). \tag{8.3}$$

The vector field Δ is required to define a vector bundle structure on M, i.e., a partial linear structure, that is to satisfy the modified conditions (1), (2) and (3) above with $2k = \dim M$. The vector bundle structure identified by functions in (3) becomes the tangent bundle structure of a manifold Q, so $M = TQ$. Notice that the

(1, 1)-tensor S is the soldering form of TQ. In natural bundle coordinates (q, v), these tensor fields take the form:

$$S = dq \otimes \frac{\partial}{\partial v}, \qquad \Delta = v\frac{\partial}{\partial v}.$$

The subgroup of Diff(M) identified by $\varphi_*\Delta = \Delta$ and $\varphi_* S = S$ is the group of tangent bundle automorphisms. A single manifold can be equipped with alternative tangent bundle structures by considering diffeomorphisms $\varphi \colon M \to M$ such that $\varphi_* S = S$ but not preserving Δ, we consider Δ_1 and Δ_2 given as $\Delta_1 = \Delta$ and $\Delta_2 = \varphi_*\Delta$. Then M becomes a double vector bundle if $[\Delta_1, \Delta_2] = 0$ and M will carry two tangent bundle structures TQ_1 and TQ_2.

Similarly one may define a cotangent bundle structure by means of a pair (Δ, θ) where again, Δ is a partial linear structure on M, and θ is a particular 1-form such that $d\theta$ is a symplectic form, $i_\Delta d\theta = \theta$, and requiring that solutions of $\mathcal{L}_\Delta f = 0$ be pairwise in involution and define a maximal Abelian subalgebra with respect to the Poisson bracket defined by $\omega = d\theta$. Notice that is equivalent to asking that the submanifold Q defined by $\Delta(m) = 0$ is Lagrangian with respect to $d\theta$. Thus $M = T^*Q$ and the canonical Liouville 1-form of T^*Q is just the 1-form θ above. Again as in the case of tangent bundle structures, alternative cotangent bundle structures on M can be constructed by choosing diffeomorphisms φ such that $\varphi^*\theta \neq \theta$.

Thus we will try to study the geometry of an integrable system Γ by determining its associated subgroup of the group of diffeomorphisms of the manifold by writing the system in "normal form" instead of determining such a subgroup from the subgroup of diffeomorphisms determined by some geometrical structure determined by it. In particular we shall consider systems with orbits possessing a compact closure, even though systems possessing unbounded orbits are relevant in scattering theory for instance.

8.4 A Normal Form for an Integrable System

8.4.1 Integrability and Alternative Hamiltonian Descriptions

We have encountered in various occasions throughout this book systems exhibiting action-angle variables. Let us review now the context in which action-angle variables are exhibited in most cases and which is related to the notion of complete integrability of a Hamiltonian vector field.

Definition 8.2 *A vector field Γ on a $2n$-dimensional symplectic manifold (M, ω) such that $i_\Gamma\omega = dH$, is said to be completely integrable if there exist n functionally independent first integrals $f_1, \ldots, f_n$, $df_1 \wedge \cdots \wedge df_n \neq 0$, such that:*

$$\{f_j, f_k\} = 0, \quad \{H, f_k\} = 0, \quad \forall j, k = 1, 2, \ldots, n\,. \tag{8.4}$$

The independence condition $df_1 \wedge \cdots \wedge df_n \neq 0$ may hold on some open dense submanifold of M, or in a weaker form, on some open invariant submanifold. We will find this situation for instance when considering scattering problems where we prefer to remove closed regions composed of bounded orbits from phase space.

When the system Γ possesses more than n first integrals, we recall that the system is said to be superintegrable. In this case, we have $f_1, \ldots, f_{n+k}, df_1 \wedge \cdots \wedge df_{n+k} \neq 0$, and $\{H, f_j\} = 0$, for all $j = 1, 2, \ldots, n+k$. In the particular case of $n+k = 2n-1$ the system is said to be maximally superintegrable.

To study the subgroup of diffeomorphisms of M appropriate for an integrable system, it is convenient to have a "normal form". As it was discussed in the introduction, in searching for normal forms, it is quite natural to ask which transformations are allowed on the system. Thus for Hamiltonian systems, it would be natural to use canonical transformations, that is to consider transformations of the systems belonging to the closed subgroup of symplectic diffeomorphisms of the group of diffeomorphisms of our manifold M. However, any integrable system admits alternative Hamiltonian descriptions, i.e., there are (ω_a, H_a), $a = 1, 2$, such that

$$i_\Gamma \omega_1 = dH_1, \quad i_\Gamma \omega_2 = dH_2.$$

In this case, which canonical transformations should we use, symplectic diffeomorphisms with respect to ω_1 or with respect to ω_2?

In this respect it is amusing to recall a quarrel between Levi–Civita and Birkhoff around this. Indeed in his 1934 paper "A general survey of the theory of adiabatic invariants" [Le34] Levi–Civita feels the need to write a section entitled "Birkhoff's severity against canonical variables and methods. Apology for a milder attitude" (p. 430). From the point of view of Birkhoff, one should consider the orbit of vector fields obtained by acting on Γ with Diff(M). All the vector fields in the same orbit share the same properties. Thus, it makes sense to restrict our attention to just one representative. Let us elaborate on this point in the particular case of linear systems.

We notice that a Hamiltonian vector field admits a factorization in terms of a Poisson tensor and an exact one-form:

$$\Gamma = \Lambda(dH). \tag{8.5}$$

Clearly, any vector field in the orbit of Γ will also be decomposable as above. Moreover, if we consider diffeomorphisms $\varphi \in \mathrm{Diff}(M)$ such that $\varphi_* \Gamma = \Gamma$, and apply it to the decomposition above (8.5), we find:

$$\Gamma = \varphi_*(\Lambda) d(\varphi_* H).$$

We conclude that any non-canonical transformation that is a symmetry for Γ, produces alternative Hamiltonian descriptions. In infinitesimal form: if X is an infinitesimal symmetry for the dynamics Γ, $[\Gamma, X] = 0$, we get:

$$\mathcal{L}_X(i_\Gamma \omega) = d\mathcal{L}_X H,$$

then

$$i_\Gamma(\mathcal{L}_X\omega) = d(\mathcal{L}_X H),$$

and the previous equation provides an alternative Hamiltonian description for Γ if $\mathcal{L}_X\omega$ is nondegenerate. In relation with alternative descriptions, we notice that there are additional ways to generate alternative Hamiltonian descriptions. Let us consider for instance a $(1, 1)$-tensor T such that:

$$\mathcal{L}_\Gamma T = 0. \tag{8.6}$$

Let us define as before Eq. (8.3), the twisted differential:

$$d_T f(X) = df(TX).$$

Then from any constant of motion F we will have the closed 2-form:

$$\omega_{T,F} = dd_T F,$$

now if $\omega_{T,F}$ is nondegenerate, it provides an alternative Hamiltonian description for Γ. In fact notice that:

$$i_\Gamma\omega_{T,F} = \mathcal{L}_\Gamma d_T F - d(i_\Gamma d_T F),$$

but $\mathcal{L}_\Gamma d_T F = 0$ because of the invariance condition (8.6) and the fact that F is a constant of motion. Hence $-dF(T\Gamma)$ is a Hamiltonian function for Γ with respect to the symplectic structure $\omega_{T,F}$.

8.4.2 Integrability and Normal Forms

Because of the previous discussion we should expect, therefore, that for vector fields with a large group of symmetries we will find always alternative Hamiltonian descriptions. However we should stress that there are alternative Hamiltonian descriptions which are not generated by diffeomorphisms.

In conclusion, we should accept any diffeomorphism to reduce a vector field Γ to its normal form. Thus, if the orbit through Γ contains a completely integrable system we can concentrate our attention on the standard normal form we are familiar with when we construct action-angle variables, i.e., we could consider the form:

$$\Gamma = \nu_j(I)\frac{\partial}{\partial\varphi_j}.$$

We should therefore study a normal form for integrable systems as emerging from the following conditions:

Definition 8.3 *Given a vector field* Γ *we will say that* $\Gamma = \nu^j(f_1, \ldots, f_n)X_j$ *is a normal form for* Γ *(and that* Γ *is integrable) if:*

i. *There exist n functionally independent first integrals* $f_1, \ldots, f_n$*, such that* $df_1 \wedge \cdots df_n \neq 0$.
ii. *There exist n complete vector fields* $X_1, \ldots, X_n$ *pairwise commuting* $[X_j, X_l] = 0$ *and independent* $X_1 \wedge \cdots X_n \neq 0$*, and,*
iii. $\mathcal{L}_{X_j} f_l = 0$ *for all* $j, l = 1, \ldots, n$.

We should notice that we have dropped the requirement for Γ to be Hamiltonian and consequently that the 'frequencies' ν^j are derivatives of the Hamiltonian function.

The usual Liouville–Arnold's theorem becomes now a particular way to find functions (coordinates) which reduce Γ to normal form.

A few remarks are in order here.

1. All integrable systems have the same normal form, then what distinguishes one system from another if any such distinction exists?
2. Which aspects of the normal form above for a given integrable system are able to discriminate integrable from superintegrable systems?

In connection with the first query, we immediately notice that many interesting aspects on the qualitative structure of the orbits of the system are to be extracted from the normal form because we know that usually specific integrable systems need not be diffeomorphic among them. We may rephrase our questions by investigating how many different orbits exist in $\mathfrak{X}(M)$ under the diffeomorphism group when each orbit is required to contain at least one element that is completely integrable.

To have an idea of the variety of situations we might be facing we shall investigate some variations on the theme of harmonic oscillators.

8.4.2.1 Hamiltonian Linear Systems

We may consider as a particular instance of the analysis performed in the previous sections the isotropic harmonic oscillator. Thus, let us consider $M = \mathbb{R}^{2n} = \mathbb{C}^n$ and the system Γ defined in cartesian coordinates (x_j, y_j) by (recall Sect. 1.2.2):

$$\frac{d}{dt}x_j = \omega y_j, \quad \frac{d}{dt}y_j = -\omega x_j,$$

then,

$$\Gamma = \sum_j \omega \left(y_j \frac{\partial}{\partial x_j} - x_j \frac{\partial}{\partial y_j} \right).$$

Introducing complex coordinates $z_j = y_j + ix_j$, we obtain:

$$\frac{d}{dt} z_j = i\omega z_j, \quad \frac{d}{dt} \bar{z}_j = -i\omega \bar{z}_j,$$

The algebra of first integrals is generated by the quadratic forms $z_l \bar{z}_m$ and because Γ is proportional to the vector field defined by the complex structure on $\mathbb{C}^n$, we conclude that its group of linear symmetries is $GL(n, \mathbb{C})$. For a given factorization $\Gamma = \Lambda(dH)$, the homogeneous space $GL(n, \mathbb{C})/GL(n, \mathbb{C}) \cap \mathrm{Sp}(2n, \mathbb{R})$ parameterizes alternative Hamiltonian descriptions, however not all of them.

What will happen then for a generic linear systems? Given a generic linear system, represented by the matrix A, it has a decomposition:

$$A = \Lambda \cdot H, \tag{8.7}$$

with Λ a non-degenerate skew-symmetric matrix and H a symmetric matrix if and only if $\mathrm{Tr} A^{2k+1} = 0$, $k = 0, 1, 2, \ldots$ (see Sect. 4.2.3). For nongeneric matrices conditions guaranteeing the existence of the factorization (8.7) are more cumbersome and we refer to the paper [GM93] for a full discussion.

For instance, when A is generic all linear symmetries are generated by powers of A. The non-canonical ones are given by even powers, therefore:

$$T = e^{\lambda A^{2k}}, \quad k = 1, 2, \ldots,$$

will be a non-canonical symmetry for any value of λ.

For instance, two alternative descriptions for the isotropic harmonic oscillator in $\mathbb{R}^4$ are given by:

$$\Lambda_1 = \frac{\partial}{\partial p_1} \wedge \frac{\partial}{\partial q_1} + \frac{\partial}{\partial p_2} \wedge \frac{\partial}{\partial q_2}, \qquad H_1 = \frac{1}{2}\omega(p_1^2 + p_2^2 + q_1^2 + q_2^2),$$
$$\Lambda_2 = \frac{\partial}{\partial p_1} \wedge \frac{\partial}{\partial q_2} + \frac{\partial}{\partial p_2} \wedge \frac{\partial}{\partial q_1}, \qquad H_2 = \omega(p_1 p_2 + q_1 q_2).$$

A particular invariant $(1, 1)$-tensor field is defined by:

$$T = dq_1 \otimes \frac{\partial}{\partial q_2} + dq_2 \otimes \frac{\partial}{\partial q_1} + dp_1 \otimes \frac{\partial}{\partial p_2} + dp_2 \otimes \frac{\partial}{\partial p_1}.$$

We may then consider the 2-form $dd_T F$ with $F = \frac{1}{4}(p_1^2 + p_2^2 + q_1^2 + q_2^2)^2$ and get:

$$\begin{aligned} dd_T F =& d(p_1^2 + p_2^2 + q_1^2 + q_2^2) \wedge d(p_1 p_2 + q_1 q_2) \\ &+ 2(p_1^2 + p_2^2 + q_1^2 + q_2^2)(dq_2 \wedge dq_1 + dp_1 \wedge dp_2). \end{aligned}$$

We finally remark that the selection of a specific decomposition with the Hamiltonian being positive definite gives a group of canonical symmetries which is the unitary group:

$$GL(n,\mathbb{C}) \cap O(2n,\mathbb{R}) = U(n),$$

therefore the system may be thought of as a 'quantum-like' system [EM10] (see Sect. 6.2.1).

8.4.3 The Group of Diffeomorphisms of an Integrable System

If Γ is an integrable system possessing a normal form like in Definition 8.3, then because $X_1, \ldots, X_n$ are complete and pairwise commuting we can define an action of the Abelian group $\mathbb{R}^n$ onto M. Moreover, because $\mathcal{L}_{X_j} \nu^k = 0$, we could redefine the vector field X_j to $Y_j = \nu^j X_j$ (no summation on j) and we would still have pairwise commuting vector fields $[Y_j, Y_k] = 0$ for all $j, k = 1, \ldots, n$. Notice that the completeness condition will not be spoiled by redefining the vector fields in this form, i.e., the vector fields Y_k will be complete (but with a different parametrization) and we would define an alternative action of $\mathbb{R}^n$ on M.

Thus what seems to matter is not the particular action of $\mathbb{R}^n$ but rather the integral leaves of the involutive distribution generated by $X_1, \ldots, X_n$. In those cases where the leaves are compact, say tori, we could require the choice of an action of $\mathbb{R}^n$ that factors to an action of $\mathbb{T}^n = \mathbb{R}^n/\mathbb{Z}^n$. Moreover we could select a particular basis of vector fields such that $X_1, \ldots, X_n$ each generate the action of a closed subgroup. With these particular prescriptions we decompose our dynamical vector field Γ. Let us denote the particularly chosen generators of closed subgroups by $Z_1, \ldots, Z_n$, then:

$$\Gamma = \omega^j Z_j .$$

Thus when the closure of generic orbits of Γ are n-dimensional, the system does not have additional first integrals. If for some initial conditions the closure of the orbits does not coincide with the full torus, there are additional invariant relations. When for any initial condition, the closure of the orbits is some $k < n$ dimensional torus then the system has additional first integrals and it is superintegrable.

When the closure is one dimensional for all initial conditions, the system is maximally superintegrable.

It is now clear that the geometry of integrable and superintegrable systems with bounded flow is associated with a toroidal generalized bundle, i.e., projections on the manifold M which have fibers diffeomorphic to tori of dimension going from one to n. i.e. exactly the situation that happens if M is a compact $2n$-dimensional symplectic manifold and the dynamics is invariant under the action of a torus group $\mathbb{T}^n$. Then because of Atiyah's convexity theorem, the momentum map $J : M \to \mathfrak{t}^*$ is a surjective map onto a convex polytope $P \subset \mathfrak{t}^*$. The fibers of J are invariant tori that on points on the interior of P are n-dimensional. The fibers corresponding to the boundary of the polytope are lower dimensional tori [At82].

Then the associated subgroup of diffeomorphisms of M determined by Γ is the subgroup of bundle automorphisms of such generalized toroidal bundle. As we will see later on, in connection with specific integrable or superintegrable systems, the most important obstruction to their being diffeomorphic is the energy-period theorem, that actually puts a restriction on the nature of the toroidal bundle of the system and in consequence on its group of diffeomorphisms.

8.4.4 Oscillators and Nonlinear Oscillators

To illustrate the previous situation we give now a few examples.

We may consider the isotropic Harmonic oscillator with two degrees of freedom. Say $M = \mathbb{R}^4$, $\omega = \sum_a dp_a \wedge dq_a$, and $H_0 = \frac{1}{2}\sum_a (p_a^2 + q_a^2)$. In this case on $\mathbb{R}_0^4 = \mathbb{R}^4 - \{0\}$, an open dense submanifold, the dynamics generates orbits that coincide with their closure and they are one-dimensional. The toroidal bundle is provided by:

$$S^1 \to \mathbb{R}_0^4 \to S^2 \times \mathbb{R}^+ .$$

We have $\mathbb{R}_0^4 \cong S^3 \times \mathbb{R}^+$ and the dynamics induces the Hopf fibration $S^1 \to S^3 \to S^2$ (see Sect. 5.4.5). The subgroup of diffeomorphisms of $\mathbb{R}_0^4$ which is selected by the fibration is the group of projectable diffeomorphisms. Clearly this large group of symmetries, when applied to a chosen Hamiltonian description, will generate many more alternative descriptions.

It should be remarked that the alternative Hamiltonian description provided by $dp_1 \wedge dq_1 - dp_2 \wedge dq_2$, and

$$H = \frac{1}{2}(p_1^2 + q_1^2) - \frac{1}{2}(p_2^2 + q_2^2),$$

cannot be derived from the standard one with the positive definite Hamiltonian H_0 because the diffeomorphism would preserve the signature of H (because it cannot map compact energy levels, the ones of the Hamiltonian H_0, into non-compact ones, the ones of H).

The system is actually superintegrable, indeed the quotient manifold under the action of the dynamics is three dimensional instead of two dimensional.

This example generalizes to any finite dimension and we have again for the n-dimensional isotropic harmonic oscillator the fibration:

$$S^1 \to \mathbb{R}_0^{2n} \to \mathbb{C}P^{n-1} \times \mathbb{R}^+ .$$

Again the symmetry group for the dynamics is the group of diffeomorphisms projectable under the previous projection, hence it is diffeomorphic to the central extension of the group of diffeomorphisms $\mathrm{Diff}(\mathbb{C}P^{n-1} \times \mathbb{R}^+)$ by $U(1)$.

The 1-form

$$d\tau = \frac{1}{n} \sum_k \left(\frac{p_k dq_k - q_k dp_k}{\omega(p_k^2 + q_k^2)} \right),$$

has the property that $i_\Gamma d\tau = 1$. Any closed 2-form on the quotient manifold and any function on the quotient manifold without critical points on the invariant open dense submanifold specified by $p_k^2 + q_k^2 \neq 0$, $k = 1, 2, \ldots, n$, give raise to an alternative Hamiltonian description.

This example is a normal form for maximally superintegrable systems with one-dimensional closed orbits and constant period.

In higher dimensions we can consider the Hamiltonian $H = \sum_a \omega_a H_a$, with $H_a = \frac{1}{2}(p_a^2 + q_a^2)$, $\omega_a \in \mathbb{R}$. The subgroup associated with the Hamiltonian vector fields Γ_a, $i_{\Gamma_a}\omega = dH_a$, are closed subgroups. When all frequencies are pairwise irrational, i.e., ω_a/ω_b is irrational, the closure of the generic orbit of Γ is the full torus, in this case there cannot be additional constants of motion. When some of the frequencies are pairwise rational, the closure of a generic orbit is a torus of lower dimensions. A particular example where the closure of the orbits goes from a one-dimensional torus to an n-dimensional one, depending on the initial conditions, is provided by:

$$H = \sum_a \pm (H_a)^2.$$

In this case we may also find invariant relations for particular values of the initial conditions.

This example gives raise to the so-called nonlinear oscillators and has been considered in quantum mechanics to give interesting consequences at the level of Planck's distribution law and for alternative commutation relations [MM97, LM97].

8.4.5 Obstructions to the Equivalence of Integrable Systems

If we inquire about the obstructions to the existence of diffeomorphisms conjugating two integrable systems, the energy-period theorem provides the first and most important one. We refer to the literature [Ne02, Go69] for various versions of this theorem, however a simple argument may be given as follows. On the carrier space of our dynamical system let us consider the 2-form $dp_k \wedge dq^k - dH \wedge dt$. Consider now the evolution $t \mapsto t + \tau$, $p_k(t) = p_k(t + \tau)$, $q^k(t) = q^k(t + \tau)$ and $H(q(t), p(t)) = H(q(t'), p(t'))$ for all $t' \in \mathbb{R}$. Because the evolution is canonical, we find $dH \wedge d\tau = 0$ on the submanifold on which the period is a differentiable function of (q^k, p_k). It follows that the period τ and the Hamiltonian H are functionally dependent.

It is well-known that the period of a dynamical system is an attribute of the vector field which is invariant under diffeomorphisms. It follows that if two Hamiltonian systems have different periods, they cannot be connected via diffeomorphisms. For instance the isotropic Harmonic oscillator and the Kepler problem cannot be connected by a diffeomorphism. Indeed the map connecting solutions of the Harmonic oscillator and those of the Kepler problem, the Kustaanhneimo–Stiefel map, as we discussed in Sect. 7.5.3, is a map defined on each energy level, i.e., for those orbits that have all the same period. The map changes from one energy-level to another (see for instance [DM05, DM05b]).

A more simple example is provided by $H = \frac{1}{2}(p^2 + q^2)$ and $H' = (p^2 + q^2)^2$. For the first one, the frequency is independent of the energy, while for the second one it depends on the initial conditions. The two systems cannot be diffeomorphic. This circumstance was the main motivation to introduce the classification of dynamical systems up to parametrization, i.e., up to conformal factors [IM98].

We hope we have made clear that the geometrical picture we have derived is the best one can do because each individual integrable system will give raise to infinitely many different situations which cannot be classified in a meaningful way otherwise (i.e., identifying a finite or a countable family of equivalence classes).

8.5 Lax Representation

The origin of Lax representation is to be found in the theory of nonlinear partial differential equations. The Schrödinger operator

$$L = -\frac{\partial^2}{\partial x^2} + u(x, t),$$

has the remarkable property that for u rapidly decreasing at infinity, the spectrum of L does not change with time if u satisfies the KdV equation [La68]. Lax suggested that the operator $L(t)$ must be similar to a fixed one, say $L(0)$, therefore $L(t) = U(t)L(0)U(t)^{-1}$. This implies that $\frac{\partial L}{\partial t} = [A, L]$ where

$$A = \frac{dU}{dt}U^{-1}.$$

There are no algorithmic procedures to construct Lax representations for "integrable" dynamical systems, in general they are constructed case by case with some educated guesswork.

The so-called Lax method for the study of integrable systems is particularly useful when dealing with integrable systems with an infinite number of degrees of freedom. The tangent bundle and the cotangent bundle picture are not immediately available without further assumptions on the structure of the "configuration space", an infinite-dimensional manifold. Normally the study of these systems require an extended use of the theory of partial differential equations, therefore it is out of the scope of our present treatment. Nevertheless the main difference with the finite-dimensional situation is mostly technical rather than conceptual, therefore we shall restrict our considerations to the finite-dimensional case.

The essential aspects of the Lax method rely on the following simple proposition:

Proposition 8.4 *If $A(t)$ and $B(t)$ are two families of $N \times N$ matrices, with $A(t)$ differentiable and $B(t)$ bounded, assuming that they satisfy the differential equation:*

$$\frac{dA}{dt} = [B, A], \quad A(0) = A_0, \tag{8.8}$$

then all matrices $A(t)$ have the same spectrum, with identical multiplicity (both geometrical and algebraic). We say that the map $t \mapsto A(t)$ is isospectral.

Furthermore if ξ_k is an eigenvector of A_0 belonging to the eigenvalue λ_0, the vector $U(t)\xi_k$ is an eigenvector of $A(t)$ with the same eigenvalue when $U(t)$ is the solution of the equation:

$$\frac{dU(t)}{dt} = B(t)U(t), \quad U(0) = \mathbb{I}, \quad U(t) \neq 0. \tag{8.9}$$

For the proof we notice that eigenvalues of $A(t)$ are completely determined by knowing $\mathrm{Tr}\,(A - a\mathbb{I})^k$ for a suitable choice of a and a sufficient number of choices of the power k. Using the Leibnitz rule:

$$\begin{aligned}
\frac{d}{dt}\mathrm{Tr}\,(A - a\mathbb{I})^k &= \sum_{j=0}^{k} \mathrm{Tr}\left[(A - a\mathbb{I})^j \frac{dA}{dt} \mathrm{Tr}\,(A - a\mathbb{I})^{k-j-1}\right] \\
&= \sum_{j=0}^{k} \mathrm{Tr}\left[(A - a\mathbb{I})^j [B, A] \mathrm{Tr}\,(A - a\mathbb{I})^{k-j-1}\right] = 0
\end{aligned}$$

where for the last step we have used the cyclicity property of the trace, that is, $\mathrm{Tr}\,(MN) = \mathrm{Tr}\,(NM)$, and the derivation property of $[B, \cdot]$.

Therefore if the equation of motion $\dot{A} = [B, A]$ is satisfied, we have found that the quantities:

$$f_m = \frac{1}{m}\mathrm{Tr}\,A^m(t), \quad m \in \mathbb{Z},$$

are constants of motion.

We notice that also

$$f_0 = \log|\det A(t)|$$

is a constant of motion.

The statement concerning eigenvectors and eigenvalues is shown by using:

$$A(t) = U(t)A_0U(t)^{-1},$$

with $U(t)$ solving the Eq. (8.9). Because $\tilde{A}(0) = A_0 = A(0)$, the uniqueness of the solution tells us that $\tilde{A}(t) = A(t)$.

We see that from $A_0\xi_k = \lambda_k\xi_k$ setting $\xi_k(t) = U(t)\xi_k$, we find:

$$A(t)\xi_k(t) = U(t)A_0U(t)^{-1}U(t)\xi_k = U(t)A_0\xi_k = \lambda_k U(t)\xi_k = \lambda_k\xi_k(t),$$

as we have stated.

Now, as a corollary, we have:

Corollary 8.5 *Let $dx/dt = Y(x)$ be a dynamical system on some manifold M and $\varphi(t, x_0)$ the solution with initial condition x_0. If there exist two families of matrices $A(x)$, $B(x)$, such that*

$$\frac{d}{dt}A(\varphi(t, x_0)) = [B(\varphi(t, x_0)), A(\varphi(t, x_0))],$$

for all x_0, then all eigenvalues of $A(x)$ are constants of motion for our dynamical system.

Remark 8.1 When the matrix $B(t)$ is skew-symmetric then matrices $U(t)$ are unitary.

Remark 8.2 If we want to recover the notion of integrability in the sense of Liouville–Arnold, we have also to show that the obtained constants of motion are in involution and functionally independent.

It is quite clear that the Lax method may be considered to be a generalization of the Heisenberg form of the equations of motion in Quantum Mechanics. Here the equations are not restricted to observables, i.e., Hermitian operators:

$$i\frac{dA}{dt} = [H, A],$$

whose exponential is given by unitary operators. In the Lax situation matrices are allowed to be generic matrices in the general linear group.

To find Lax families for a given dynamical system is rather non-trivial. Many examples, although in infinite dimensions, arise from geodetical motion on Lie groups (not restricted to be finite-dimensional). For simplicity we consider a Lie group G realized as a group of matrices.

Geodetical motions are described by the equations of motion

$$\frac{d}{dt}\left(g^{-1}\frac{dg}{dt}\right) = 0;$$

or, equivalently: $d/dt\left((dg/dt)\,g^{-1}\right) = 0$.

If $B_1, \ldots, B_N$ is a basis for the Lie algebra of the group G, the previous second-order differential equation is equivalent to the family of first-order differential equations:

$$\left(g^{-1}\frac{dg}{dt}\right) = B_j, \qquad j = 1, \ldots, N.$$

It is not difficult to show that the second-order differential equations admit a Lagrangian description by means of the Lagrangian:

$$L = \frac{1}{2}\mathrm{Tr}(g^{-1}\dot{g})^2 = \frac{1}{2}\mathrm{Tr}(\dot{g}g^{-1})^2,$$

with possible alternative Lagrangians

$$L_K = \frac{1}{2}\mathrm{Tr}K(g^{-1}\dot{g})^2,$$

with K any invertible matrix.

By using the momentum map associated with the left or right action of the group on itself, we find Lax-type equations on the dual of the Lie algebra expressed in terms of Poisson brackets.

Going from $\mathfrak{g}^*$ (the dual of the Lie algebra) to $\mathfrak{g}$ (the Lie algebra) by means of an invariant pairing, we obtain equations of motion in the matrix Lax-type.

Many general aspects of dynamical systems associated with differential equations on Lie groups will be considered in Chap. 9. Here we simply comment that by replacing a matrix with a $(1, 1)$-tensor field we find Lax-type equations when an invariant $(1, 1)$-tensor field can be factorized by means of a symplectic structure and a Poisson tensor. This approach is closely related to recursion operators and biHamiltonian systems.

8.5.1 The Toda Model

The Toda model describes the motion of N particles on a line interacting among themselves with a repulsive force decreasing exponentially with the distance. To get more symmetrical formulae, it is convenient to set:

$$q_0 = -\infty, \quad q_{N+1} = +\infty,$$

with $q_0, q_1, \ldots, q_N, q_{N+1}$ the position coordinates. Hamilton equations are given by:

$$\dot{q}_k = p_k; \quad \dot{p}_k = e^{-(q_k - q_{k-1})} - e^{-(q_{k+1} - q_k)}, \quad k = 1, \ldots, N.$$

These equations admit alternative Hamiltonian descriptions:

The first one (the standard):

$$\omega = \sum_{k=1}^{N} dq_k \wedge dp_k, \quad H(q,p) = \sum_{k=1}^{N} \left(\frac{1}{2} p_k^2 + e^{-(q_{k+1}-q_k)} \right),$$

but there is another one:

$$\Omega = \sum_{k=1}^{N} \left(e^{-(q_{k+1}+q_k)} dq_k \wedge d_{k+1} - p_k dq_k \wedge dp_k \right) + \frac{1}{2} \sum_{j,k=1}^{N} \epsilon(j-k) dp_k \wedge dp_j,$$

with

$$K(q,p) = \sum_{k=1}^{N} \left(\frac{1}{3} p_k^3 + (p_{k+1} + p_k) e^{-(q_{k+1}+q_k)} \right), \quad \epsilon(0) = 0, \quad \epsilon(p) = \frac{p}{|p|} \quad (p \neq 0).$$

It is possible to compose one symplectic structure with the inverse of the other to define a $(1,1)$-tensor field $S = \Omega^{-1} \circ \omega$, we find:

$$S = \begin{bmatrix} B & A \\ -E & B \end{bmatrix},$$

with

$$A_{ij} = \delta_{i+1,j} e^{-(q_{i+1}-q_i)} - \delta_{i,j+1} e^{-(q_{j+1}-q_j)} \quad B_{ij} = p_i \delta_{i,j}; \quad E_{i,j} = \epsilon(j-i).$$

Constant of the motion are given by $\mathrm{Tr} S^k$. We find for a few of them:

$$K_1 = \mathrm{Tr} S = 2 \sum_{k=1}^{N} p_k;$$

$$K_2 = \frac{1}{2} \mathrm{Tr} S^2 = \sum_{k=1}^{N} (p_k^2 + 2e^{-(q_{k+1}-q_k)});$$

$$K_3 = \frac{1}{3} \mathrm{Tr} S^3 = \sum_{k=1}^{N} \left(\frac{2}{3} p_k^3 + (p_{k+1} + p_k) e^{-(q_{k+1}-q_k)} \right);$$

$$K_4 = \frac{1}{4} \mathrm{Tr} S^4$$
$$= \sum_{k=1}^{N} \left(\frac{1}{2} p_k^4 + (p_k^2 + p_k p_{k+1} + p_{k+1}^2) e^{-(q_{k+1}-q_k)} + e^{-2(q_{k+1}-q_k)} + 2e^{-(q_{k+2}-q_k)} \right).$$

By setting:

$$D_{ij} = (\delta_{i,j} - \delta_{j,i+1})e^{-(q_{i+1}-q_i)} + (\delta_{i,j} - \delta_{j,i-1})e^{-(q_i-q_{i-1})}$$

we find the equations of motion in the Lax-type equations:

$$\frac{dA}{dt} = -[B, D], \quad \frac{dB}{dt} = \frac{1}{2}[E, D],$$

an alternative Lax representation is provided by:

$$\frac{dL}{dt} = [L, V],$$

with:

$$L_{ij} = \frac{1}{2}\left(p_i\delta_{i,j} + \delta_{j,i+1}e^{-(q_{i+1}-q_i)/2} + \delta_{i,j+1}e^{-(q_{j+1}-q_j)/2}\right),$$

and

$$V_{ij} = \frac{1}{2}\left(\delta_{i,j}e^{-(q_{i+1}-q_i)/2} + \delta_{i,j+1}e^{-(q_{j+1}-q_j)/2}\right),$$

with L symmetric and V skew-symmetric.

8.6 The Calogero System: Inverse Scattering

8.6.1 The Integrability of the Calogero-Moser System

In Sect. 1.2.12 we derived the Calogero-Moser system from a free systems by using different reduction techniques. In doing so, we know that our system is "integrable", however it is not obvious how the properties of the Calogero-Moser system are related with the geometric notion of integrability that we have been discussing in this chapter. In this section we will approach again the Calogero-Moser system from a different perspective enhancing some of the integrability properties that are more appealing from the present discussion.

We consider N mass points, assumed to have all the same mass which we put equal to one for convenience, moving on a line. They interact with each other via a repulsive force proportional to the third power of the inverse of their mutual distance. Let us denote by q_n the position of the nth point mass and by p_n the corresponding momentum, the Hamiltonian describing the system will be:

$$H(q, p) = \frac{1}{2}\sum_{n=1}^{N} p_n^2 + V(q), \qquad V(q) = \sum_{1\leq n<m\leq N} \frac{1}{(q_n - q_m)^2}.$$

The associated second-order equations of motion will be:

$$\frac{d^2 q_n}{dt^2} = -\frac{\partial V}{\partial q_n} = 2\sum_{k\neq n} \frac{q_k - q_n}{(q_k - q_n)^4}, \qquad n = 1, \ldots, N, \tag{8.10}$$

and, $p_n = \dot{q}_n$.

A Lax-type description of Eq. (8.10) is provided by the following two matrix-valued functions:

$$A(q, p)_{m,n} = \delta_{m,n} p_n - i\frac{\delta_{m,n} - 1}{q_m - q_n}, \tag{8.11}$$

$$B(q, p)_{m,n} = i\delta_{m.n} d_n + i\frac{\delta_{m,n} - 1}{(q_n - q_m)^2}, \quad d_n = \sum_{k\neq n} \frac{1}{(q_n - q_k)^2}. \tag{8.12}$$

We have already seen that, for $N = 2, 3$, this system may be obtained as a suitable reduction of a free system on the space of Hermitian matrices. In particular, this shows that the motion here is not bounded contrary to the analysis at the beginning of this chapter, see Sect. 8.2, and therefore not contained in a compact submanifold. As a result we cannot define action-angle variables for this system and the adequate geometrical notion of integrability goes beyond the methods of toric manifolds discussed so far.

It is possible to obtain the general Calogero system by considering the free motion on Hermitian matrices and performing a reduction as we did for $N = 2$. However here we would like to explain a different method which would apply when the condition of boundedness of e^{tA} does not apply.

8.6.2 Inverse Scattering: A Simple Example

Definition 8.6 *Given a second-order dynamics on a vector space of dimension N, we will say that the Cauchy data $x(0)$, $\dot{x}(0)$ define a scattering state (or scattering trajectory) if the two limits:*

$$\alpha_n^{\pm} = \lim_{t\to\pm\infty} \dot{x}_n(t); \quad \beta_n^{\pm} = \lim_{t\to\pm\infty} (x_n(t) - t\alpha_n),$$

exist.

The inverse scattering method relies on the fact that, under suitable conditions, the transformations:

$$(p, x) \mapsto (\alpha^-, \beta^-); \qquad (p, x) \mapsto (\alpha^+, \beta^+),$$

are symplectic transformations with respect to the standard symplectic structure on $T^*\mathbb{R}^N$, or with respect to the standard symplectic structure on $T^*\mathbb{R}^{3N}$ if we are considering N particles moving in Euclidean three dimensional space $\mathbb{R}^3$, and the functions α_k are first integrals in involution pairwise or, in other words, the system is asymptotically free.

Let us consider $H(x, p)$ the Hamiltonian of the system we are considering. Let us denote by $\varphi_t : \xi = (x, p) \to \varphi_t(\xi)$ the flow associated with H and by $\varphi_t^0 : \xi = (x, p) \to \varphi_t^0(\xi)$ the flow associated to a comparison Hamiltonian H^0. We often take as comparison Hamiltonian a 'free' motion system $H^0(x, p) = \frac{1}{2} p^T M p$ where M is a definite positive matrix (called the mass matrix of the system).

The asymptotic behavior of asymptotic trajectories express the fact that the limit:

$$\Psi(\xi) := \lim_{t\to\infty} (\varphi_{-t}^0 \circ \varphi_t)(\xi)$$

exists uniformly for $\xi = (x, p)$ in any compact set (with similar considerations applying for $t \to -\infty$).

The transformation Ψ turns out to be symplectic because it is the limit of symplectic transformations. More properly we should verify that also $T(\varphi_{-t}^0 \circ \varphi_t)$ converges and this should be verified using the explicit form of the derivative. This condition will be obtained on each instance under appropriate hypotheses on the speed of convergence to zero of $H - H^0$ when $||x|| \to \infty$.

By construction Ψ maps the Cauchy data onto 'scattering data' (α, β); in this way α and β are differentiable functions of the coordinates (x, p) and are themselves canonical variables because Ψ is a symplectic transformation.

In particular the functions α_k are N functions pairwise in involution and are functionally independent. They are defined only in the region of phase space corresponding to points for which scattering states exist in the asymptotic sense either for $t \to +\infty$ or $t \to -\infty$. The fact that our functions α_k are constants of motion follows from the observation that in any scattering process:

$$\varphi_s^0 \circ \Psi = \Psi \circ \varphi_s,$$

and therefore $H = H^0 \circ \Psi$. Then, being $(\alpha_1, \ldots, \alpha_N)$ a set of first integrals for the flow associated with H^0, we deduce that any $\alpha_k \circ \Psi$ is a constant of motion for the flow associated to H.

Notice that any system whose trajectories have the asymptotic behavior we described earlier has a carries space which is not compact, i.e., the assumption on the boundedness of the flow is not satisfied. The inverse scattering method may be considered to be a completion to the action-angle variables representation of integrable systems when the flow is not bounded.

8.6.3 Scattering States for the Calogero System

For the Calogero system, using the method of inverse scattering, we find for $t \to \infty$:

$$q_n(t) = \alpha_n^+ t + \beta_n^+ + O(t^{-1}), \quad p_n(t) = \alpha_n^+ + O(t^{-1}),$$

and for $t \to -\infty$:

$$q_n(t) = \alpha^-_{N-n+1} t + \beta^-_{N-n+1} + O(t^{-1}), \quad p_n(t) = \alpha^-_{N-n+1} + O(t^{-1}).$$

These statements follow from the form of the matrices in the Lax representation form, Eq. (8.11). In fact because

$$\lim_{t\to\pm\infty} p_n(t) = \alpha_n^\pm, \quad \lim_{t\to\pm\infty} ||q_n(t) - q_m(t)|| = +\infty, \quad n \neq m,$$

then, because of Eq. (8.11):

$$\lim_{t\to\pm\infty} A_{m,n}(q(t), p(t)) = \lim_{t\to\pm\infty} \left[\delta_{m,n} p_n(t) - i \frac{\delta_{m,n} - 1}{q_m(t) - q_n(t)} \right] = \delta_{m,n} \alpha_n^\pm .$$

Because the eigenvalues of the matrix A are first integrals, they will coincide with the eigenvalues of the limit matrix, thus:

$$\alpha_n^\pm = \lambda^\pm_{\pi(n)} = \lambda_{\pi^\pm(n)}, \quad \forall n,$$

where $n \mapsto \pi^\pm(n)$ is a permutation. It follows that the functions α_n^- will differ from the α_k^+ at most for a permutation.

The limit of $p_n(t)$ and $||q_n(t) - q_m(t)||$ follows from the following considerations. From the equations of motion we have:

$$\ddot{q}_N - \ddot{q}_1 = 2 \sum_{n<N} \frac{1}{(q_N - q_m)^3} + 2 \sum_{1<m} \frac{1}{(q_m - q_1)^3}. \tag{8.13}$$

The right-hand side of (8.13) is strictly positive for t finite. Indeed, material points must keep their relative order on the line, due to the conservation of energy and the singularity of the potential energy when two point masses coincide.

Integrating Eq. (8.13) between two finite times t_1, t_2 and using the fact that conservation of energy implies that each one of the $\dot{q}_n(t)$ is equibounded as functions of t, we have:

$$\int_{-\infty}^{+\infty} (q_N(t) - q_n(t))^{-3} dt < +\infty, \quad n < N,$$

$$\int_{-\infty}^{+\infty} (q_n(t) - q_1(t))^{-3} dt < +\infty, \quad n > 1.$$

Using now:

$$\ddot{q}_{N-1} - \ddot{q}_2 = 2 \sum_{2<n<N-1} \frac{1}{(q_{N-1} - q_n)^3} + 2 \sum_{2<m<N} \frac{1}{(q_m - q_2)^3} - \frac{2}{(q_N - q_{N-1})^3} - \frac{2}{(q_2 - q_1)^3},$$

and taking into account that the last two terms in the previous equation, even though not positive, admit a bounded integral to derive:

$$\int_{-\infty}^{+\infty} (q_{N-1}(t) - q_n(t))^{-3} dt < +\infty, \quad n < N-1,$$

$$\int_{-\infty}^{+\infty} (q_n(t) - q_2(t))^{-3} dt < +\infty, \quad n > 2.$$

Thus, by induction we get:

$$\int_{-\infty}^{+\infty} (q_n(t) - q_m(t))^{-3} dt < +\infty, \quad \forall n \neq m.$$

We finally deduce that:

$$\dot{q}_n(T) = \dot{q}_n(0) + \int_0^T \sum_{n \neq m} (q_n(t) - q_m(t))^{-3} dt$$

and therefore the limits for $\dot{q}_n(t)$ exist and are finite for $t \to \pm\infty$ and the asymptotic behavior is proved.

To show that α_n and β_n are symplectic coordinates and that their commutation relations are the required ones is more cumbersome, and to rely on the reduction procedure is much more convenient.

References

[MM74] Markus, L., Meyer, K.R.: Generic Hamiltonian dynamical systems are neither integrable nor ergodic. Mem. Am. Math. Soc. **144**, 1–52 (1974)

[IM12] Ibort, A., Marmo, G., Rodríguez, M.A., Tempesta, P.: Nilpotent Integrability (Preprint) (2012).

[Kl72] Klein, F.: A comparative review of recent researches in geometry. Univ. of Erlangen (1872)

[Kl92] Klein, F.: A comparative review of recent researches in geometry. Bull. New York Math. Soc. **2**, 215–249 (1892). arXiv: 0807.3161 v1[math.H0]

[Ba86] Banyaga, A.: On isomorphic classical diffeomorphism groups I. Proc. Am. Math. Soc. **98**, 113–118 (1986)

[BG88] Batlle, C., Gomis, J., Pons, J.M., Román-Roy, N.: Lagrangian and Hamiltonian constraints for second-order singular Lagrangians. J. Phys. A: Math. Gen. **21**, 2693–2703 (1988)

[Gr00] Grabowski, J.: Isomorphisms of Poisson and Jacobi brackets. Poisson geometry. Banach Center Publ. **51**, 79–85 (2000)

[FL89] de Filippo, S., Landi, G., Marmo, G., Vilasi, G.: Tensor fields defining a tangent bundle structure. Ann. Inst. H. Poincaré **50**, 205–218 (1989)

[Le34] Levi-Civita, T.: A general survey of the theory of adiabatic invariants. J. Math. Phys. **13**, 18–40 (1934)

[GM93] Giordano, M., Marmo, G., Rubano, C.: The inverse problem in the Hamiltonian formalism: integrability of linear Hamiltonian fields. Inverse Prob. **9**, 443–467 (1993)

[EM10] Ercolessi, E., Marmo, G., Morandi, G.: From the equations of motion to the canonical commutation relations. Riv. Nuovo Cim. **33**, 401–590 (2010)

[At82] Atiyah, M.: Convexity and commuting Hamiltonians. Bull. London Math. Soc. **14**, 1–15 (1982)

[MM97] Man'ko, V.I., Marmo, G., Sudarshan, E.C.G., Zaccaria, F.: f-oscillators and nonlinear coherent states. Phys. Scripta **55**, 528–541 (1997)

[LM97] López-Peña, R., Manko, V.I., Marmo, G.: Wigner problem for a precessing dipole. Phys. Rev. A **56**, 1126–1130 (1997)

[Ne02] Nekhoroshev, N.N.: Generalizations of Gordon theorem. Regular Chaotic Dyn. **7**, 239–247 (2002)

[Go69] Gordon, W.B.: On the relation between period and energy in periodic dynamical systems. J. Math. Mech. **19**, 111–114 (1969)

[DM05] D'Avanzo, A., Marmo, G.: Reduction and unfolding: the Kepler problem. Int. J. Geom. Meth. Mod. Phys. **2**, 83–109 (2005)

[DM05b] D'Avanzo, A., Marmo, G., Valentino, A.: Reduction and unfolding for quantum systems: the hydrogen atom. Int. J. Geom. Meth. Mod. Phys. **2**, 1043–1062 (2005)

[IM98] Ibort, A., Marmo, G.: A new look at completely integrable systems and double Lie groups. Contemp. Math. **219**, 159–172 (1998)

[La68] Lax, P.D.: Integrals of nonlinear equations of evolution and solitary waves. Comm. Pure Appl. Math. **XXI**, 467–90 (1968)

Chapter 9
Lie–Scheffers Systems

If only I knew how to get mathematicians interested in transformation groups and their applications to differential equations. I am certain, absolutely certain, that these theories will some time in the future be recognized as fundamental. When I wish such a recognition sooner, it is partly because then I could accomplish ten times more.

Sophus Lie, *Letter to Adolf Mayer,* 1884.

9.1 The Inhomogeneous Linear Equation Revisited

In 1893, Lie and Scheffers [Lie93] presented a result that has a deep implication regarding the notion of integrability that we are describing, but that has been almost unnoticed since then (for two modern general references see [Ca00, Ca07b] and [CL11]). We will start by analysing the simplest case of linear (systems of) differential equations.

To start with we can consider the inhomogeneous linear differential equation

$$\dot{x} = c_0(t) + c_1(t)\, x, \tag{9.1}$$

which is well known to admit a solution in terms of two quadratures:

$$x(t) = \exp\left(\int_0^t c_1(t')\, dt'\right)\left[C + \int_0^t \exp\left(-\int_0^{t'} c_1(t'')\, dt''\right) c_0(t')\, dt'\right]. \tag{9.2}$$

It contains two simpler cases when either c_0 or c_1 identically vanish.

The first remarkable point is that there is a superposition rule for such a differential equation which is given by the function:

$$\phi(x_1, x_2; k) = x_1 + k\,(x_2 - x_1), \tag{9.3}$$

J.F. Cariñena et al., *Geometry from Dynamics, Classical and Quantum*,
DOI 10.1007/978-94-017-9220-2_9

i.e., the differential equation is such that its general solution is given by

$$x(t) = \phi(x_1(t), x_2(t); k) = x_1(t) + k\,(x_2(t) - x_1(t)),$$

where $x_1(t)$ and $x_2(t)$ are two generic, but not proportional, particular solutions.

A second interesting property is that its solutions are integral curves of the t-dependent vector field in $\mathbb{R}$

$$X = (c_0(t) + c_1(t)x)\,\frac{\partial}{\partial x}, \tag{9.4}$$

which is a linear combination with t-dependent coefficients, $X(t) = c_0(t)\,X_0 + c_1(t)\,X_1$, of the two vector fields in $\mathbb{R}$

$$X_0 = \frac{\partial}{\partial x}, \qquad X_1 = x\,\frac{\partial}{\partial x}, \tag{9.5}$$

with the following commutation relation:

$$[X_0, X_1] = X_0, \tag{9.6}$$

i.e., they are closing on a two-dimensional real Lie algebra isomorphic to the Lie algebra of the affine group in one dimension, $\mathfrak{a}(1, \mathbb{R})$. The t-dependent vector field in $\mathbb{R}$ for the two simpler particular cases, either when $c_0 = 0$ or $c_1 = 0$, reduces to a multiple by a t-dependent coefficient of the vector field X_1 or the vector field X_0, respectively. These cases admit simpler superposition rules, involving one particular solution: $\phi(x; k) = k\,x$, when $c_0 = 0$ and $\phi(x; k) = x + k$ when $c_1 = 0$.

Another useful property is that if we consider the set of transformations

$$x(t) \mapsto \bar{x}(t) = \alpha(t)x(t) + \beta(t), \qquad \alpha(t) \neq 0, \forall t, \tag{9.7}$$

we see that $x(t)$ is a solution of the given equation if and only if $\bar{x}(t)$ is a solution of a similar equation

$$\dot{\bar{x}} = c_0'(t) + c_1'(t)\bar{x}, \tag{9.8}$$

with

$$c_0' = \alpha\,c_0 + \dot{\beta} - \beta\,c_1 - \beta\,\frac{\dot{\alpha}}{\alpha}, \qquad c_1' = c_1 + \frac{\dot{\alpha}}{\alpha}.$$

The mentioned set of transformations is a group $\mathcal{G}$ isomorphic to the group of curves in the affine group G in one dimension and is usually called the structure preserving group [Sc02, Nd08]. This way we obtain a group action of $\mathcal{G}$ on the set of inhomogeneous linear differential equations, in the sense that the composition of the transformations determined by first $(\alpha_1(t), \beta_1(t))$ and second $(\alpha_2(t), \beta_2(t))$ is the same as the transformation corresponding to the product $(\alpha_2(t)\alpha_1(t), \alpha_2(t)\beta_1(t) + \beta_2(t))$.

Furthermore, one can see that there is an element $(\alpha(t), \beta(t))$ such that the solution for a given initial condition $x(0) = x_0$ is

$$x(t) = \alpha(t)\, x_0 + \beta(t). \tag{9.9}$$

In fact, then

$$\alpha(t) = \exp\left(\int_0^t c_1(t')\, dt'\right),$$

$$\beta(t) = \exp\left(\int_0^t c_1(t')\, dt'\right) \int_0^t \exp\left(-\int_0^{t'} c_1(t'')\, dt''\right) c_0(t')\, dt'. \tag{9.10}$$

It is also to be remarked that the action of the group $\mathcal{G}$ can be used to reduce the given equation to one of the simpler cases. For instance, if $x_1(t)$ is a non-vanishing solution of the given equation, then under the transformation $x(t) = x_1(t) + \bar{x}(t)$, as

$$\dot{x}_1(t) + \dot{\bar{x}}(t) = c_0(t) + c_1(t)(x_1(t) + \bar{x}(t)),$$

the new equation becomes $\dot{\bar{x}} = c_1(t)\bar{x}$, which is one of the two above mentioned simpler cases.

9.2 Inhomogeneous Linear Systems

We recall that in Chap. 2, Sect. 2.2 it was discussed in detail systems of differential equations on vector spaces E associated with linear maps. They have associated one-parameter subgroups of linear automorphisms of E. So, consider the system

$$\frac{dx}{dt} = A\, x, \qquad x \in E, \tag{9.11}$$

with A being a given endomorphism of E. This system defines a flow ϕ_t which can be considered as a curve $g(t)$ on the general linear group $GL(n, \mathbb{R})$ given by $g(t) = \phi_t = e^{At}$. Such flow satisfies (Eq. (2.5)):

$$\frac{d\phi_t}{dt} = A\, \phi_t, \tag{9.12}$$

the one-parameter group property, $\phi_{t_1} \circ \phi_{t_2} = \phi_{t_1+t_2}$, and the initial condition $\phi_0 = I$. Note that choosing as initial condition each element of the canonical basis of $\mathbb{R}^n$ we find that the columns of $\phi_t = e^{At}$ are a set of linearly independent solutions of the given system. Any other such a set defines a t-dependent matrix $P(t)$ which is also a solution of the matrix differential equation

$$\frac{dP}{dt} = AP,$$

but for which $P(0)$ is a regular matrix such that $P(0) \neq I$. Then $P(t) = \exp(tA)P(0)$ and the important point is that A can be recovered from P as $A = (dP/dt)\, P^{-1}$. This implies that the flow ϕ_t can be determined from a fundamental set of solutions, i.e. when the $n \times n$ matrix X whose columns are the vectors defining the solutions $\{x_{(1)}(t), \dots, x_{(n)}(t)\}$,

$$X(t) = (x_{(1)}(t), \dots, x_{(n)}(t)),$$

is an invertible matrix, then the equation

$$X(t) = e^{tA} X(0),$$

shows that the evolution operator e^{tA} is determined as $e^{tA} = X(t)X(0)^{-1}$, i.e., the fundamental system of solutions allows us to find the flow of our first-order differential equation system. Moreover, the relation $x(t) = e^{At}\, x(0)$ for a solution tells us that the general solution can be written as a linear combination of n linearly independent solutions. In other words, the function $\Phi : \mathbb{R}^{n(n+1)} \to \mathbb{R}^n$ given by

$$\Phi(u_1, \dots, u_n, c_1, \dots, c_n) = c_1\, u_1 + \cdots + c_n\, u_n, \tag{9.13}$$

is such that, if $\{x_{(1)}(t), \dots, x_{(n)}(t)\}$ is a set of linearly independent solutions of the system, then $x(t) = \Phi(x_{(1)}(t), \dots, x_{(n)}(t), c_1, \dots, c_n)$ gives us the general solution. This superposition rule (9.13) for solutions is the same for all homogeneous linear systems, no matter of the choice of the coefficients $A^i{}_j$.

From here it is clear that with any linear differential equation on $\mathbb{R}^n$ we can associate an equation on $GL(n, \mathbb{R})$ by setting

$$\dot{g} = Ag, \tag{9.14}$$

or in other words, that is an equation for a curve in the group of automorphisms of E:

$$\frac{dg}{dt} g^{-1} = A. \tag{9.15}$$

We recall that the matrix A is an element of the Lie algebra $\mathfrak{gl}(n, \mathbb{R})$.

Moreover, this way of finding the resolvent can also be used for time-dependent systems

$$\frac{dx^i}{dt} = \sum_{j=1}^{n} A^i{}_j(t)\, x^j, \qquad i = 1, \dots, n\,. \tag{9.16}$$

or in a shorter form

$$\frac{dx}{dt} = A(t)\, x, \tag{9.17}$$

as was shown in Sect. 2.2.4. The general theory of systems of differential equations asserts that there exists a curve $g(t) = \phi_t$ in $GL(n, \mathbb{R})$ such that $x(t) = \phi_t(x(0))$, but now $\phi_{t_1} \circ \phi_{t_2} = \phi_{t_1+t_2}$ no longer holds, but only the flow Eq. (9.12) and the initial condition $\phi_0 = \text{id}$.

Now, if the matrix A is not constant, but we suppose that the matrix

$$B(t) = \int_0^t A(\zeta)\, d\zeta, \tag{9.18}$$

that satisfies $B(0) = 0$ and

$$\frac{dB(t)}{dt} = A(t), \tag{9.19}$$

is such that $A(t)\, B(t) = B(t)\, A(t)$, what, for instance, happens if $A(\zeta_1)A(\zeta_2) = A(\zeta_2)A(\zeta_1)$, for all couples of values ζ_1 and ζ_2, then, the commutativity of $A(t)$ and $B(t)$ implies that

$$\frac{dB^m(t)}{dt} = m\ A(t)\, B^{m-1}(t), \tag{9.20}$$

and then,

$$\int_0^t A(\zeta)\, B^{m-1}(\zeta)\, d\zeta = \frac{1}{m}\, B^m(t).$$

Therefore we find that the flow ϕ_t is given by

$$\phi_t = \exp B(t), \tag{9.21}$$

because

$$\frac{d}{dt}\left(\sum_{k=0}^{\infty} \frac{1}{k!} B^k(t)\right) = A(t)\left(\sum_{k=1}^{\infty} \frac{1}{(k-1)!} B^{k-1}(t)\right) = A(t)\left(\sum_{k=0}^{\infty} \frac{1}{k!} B^k(t)\right),$$

where use has been made of (9.20), and then

$$\frac{d}{dt} \exp(B(t))x_0 = A(t)\, (\exp(B(t))x_0)\, .$$

In the generic case of arbitrary dependence of A in the parameter t, we cannot explicitly write the automorphism ϕ_t but normally is written using the time ordered exponential as explained in Sect. 2.2.4, Eq. (2.43)

The linear superposition property for non-autonomous systems is still valid and then we can take advantage of the knowledge of a fundamental set of solutions to determine ϕ_t. In fact if $X(t)$ is the matrix determined by the set of n linearly independent solutions of the equation, then from

$$X(t) = \phi_t\, X(0)$$

we get $\phi_t = X(t)X(0)^{-1}$.

We obtain in this way a curve $g(t) = \phi_t$ in $GL(n, \mathbb{R})$, which is the solution of the time-dependent equation

$$\frac{dg}{dt}\, g^{-1} = A(t),$$

with $A(t)$ being a curve in $\mathfrak{gl}(n, \mathbb{R})$, such that $g(0) = I$.

The point we want to remark is that if the evolution preserves some geometrical structure, then

$$A(t) = \frac{d\phi_t}{dt} \circ \phi_t^{-1}$$

lies in certain subalgebra of $GL(n, \mathbb{R})$ and the computation of ϕ_t becomes simpler.

As in the preceding case of the linear equation, if we define a transformation

$$Y(t) = B(t)X(t), \tag{9.22}$$

where $B(t)$ is an invertible matrix, the new vectorial function $Y(t)$ satisfies a similar system

$$\frac{dy^i}{dt} = \sum_{j=1}^{n} A'^i(t)_j\, ya^j, \qquad i = 1, \ldots, n,$$

where

$$A'(t) = \dot{B}(t)\, B(t)^{-1} + B(t)\, A(t)\, B(t)^{-1}, \tag{9.23}$$

and use can be made of this property to transform a given system into a simpler one, or to obtain information on the given system (see later on). It is worth noting that this shows that $GL(n, \mathbb{R})$ is contained in the structure preserving group.

The critical fact is that the general solution is determined by a linear operator from a set of fundamental solutions, and that there is a linear superposition principle.

It is very natural to ask what happens when the vector field is nonlinear. The answer is that, at least in some cases, there is a kind of nonlinear superposition rule, as it was proved by Lie [Lie93]. This nonlinear superposition rule is simply a generalization of the previous construction to those cases where the action of the group is not linear and $\mathbb{R}^n$ is replaced for a manifold M. A simple example would be the inhomogenous linear system

$$\frac{dx^i}{dt}=\sum_{j=1}^{n}A^i{}_j(t)\,x^j+a^i(t)\,,\quad i=1,\ldots,n. \tag{9.24}$$

for which there is also a superposition rule involving now $n+1$ solutions,

$$\Phi(u_1,\ldots,u_{n+1},c_1,\ldots,c_n)=u_1+C_1(u_2-u_1)+\cdots+c_n(u_{n+1}-u_1) \tag{9.25}$$

and then the general solution can be obtained from a fundamental set of $n+1$ particular solutions as

$$\begin{aligned}x(t)&=\Phi(x_{(1)}(t),\ldots,x_{(n+1)}(t),c_1,\ldots,c_n)\\&=x_{(1)}(t)+c_1(x_{(2)}(t)-x_{(1)})(t)+\ldots+c_n(x_{(n+1)}(t)-x_{(1)}(t)).\end{aligned}$$

The properties of the general solution of a homogeneous linear system of first-order differential equations on $\mathbb{R}^n$ can be studied by means of the transformation procedure pointed out before.

Given a linear system $\dot{x}=A(t)x$, Eq. (9.11), we first can prove that the general solution is of the form $x(t)=H(t)K$, where K is a constant column matrix and $H(t)$ is a square matrix whose columns are n linearly independent for each value of t particular solutions of $\dot{x}=A(t)x$. Such a matrix $H(t)$ is necessarily non-singular, and is called a fundamental matrix.

In fact, suppose we make a change of variables by

$$\bar{x}=\alpha(t)x, \tag{9.26}$$

where α denotes a t-dependent invertible matrix. As indicated above, in terms of the new variable the original equation reads $\dot{\bar{x}}=\bar{A}(t)\bar{x}$, where

$$\bar{A}=\dot{\alpha}\,\alpha^{-1}+\alpha\,A\,\alpha^{-1}. \tag{9.27}$$

Let us first prove how we can choose α in such a way that $\bar{A}=0$. As far as uniqueness is concerned, note that α_1 and α_2 are two such matrices if and only if there exists a constant invertible matrix P such that $\alpha_2=P\alpha_1$. In fact, if $\dot{\alpha}_1+\alpha_1A=0$ then for each constant matrix P, $\alpha_2=P\alpha_1$ satisfies $\dot{\alpha}_2+\alpha_2A=0$. Conversely, if $\dot{\alpha}_1+\alpha_1A=0$ and $\dot{\alpha}_2+\alpha_2A=0$, then, as $d\alpha_1^{-1}/dt=A\alpha_1^{-1}$,

$$\frac{d}{dt}(\alpha_2\,\alpha_1^{-1})=\dot{\alpha}_2\alpha_1^{-1}+\alpha_2\frac{d}{dt}\alpha_1^{-1}=-\alpha_2\,A\,\alpha_1^{-1}+\alpha_2\,A\,\alpha_1^{-1}=0,$$

and therefore the matrix $P=\alpha_2\,\alpha_1^{-1}$ is constant.

Suppose now that n particular solutions of (9.11), $x_{(1)},\ldots,x_{(n)}$, are known that are linearly independent for each value of t, and define the matrix $H(t)$ such that

$$H^i{}_j(t)=x^i_{(j)}(t)\,,\qquad i,j=1,\ldots,n. \tag{9.28}$$

This matrix is such that

$$\frac{dH^i{}_j}{dt} = \frac{dx^i_{(j)}}{dt} = \sum_{k=1}^n A^i{}_k x^k_{(j)} = \sum_{k=1}^n A^i{}_k H^k{}_j = (AH)^i{}_j$$

or in matrix form,

$$\frac{dH}{dt} = AH.$$

As the matrix H is invertible, taking time derivative in $HH^{-1} = I$ we obtain

$$\frac{dH}{dt}H^{-1} + H\frac{dH^{-1}}{dt} = 0 \Longrightarrow \frac{dH^{-1}}{dt} = -H^{-1}\frac{dH}{dt}H^{-1} = -H^{-1}A,$$

and consequently $\alpha = H^{-1}$ satisfies $\dot{\alpha} + \alpha A = 0$. Therefore using (9.26) with $\alpha = H^{-1}$ produces $\bar{A} = 0$ and then the general solution is $\bar{x} = K = \bar{x}(t_0)$, from where we recover the form

$$x(t) = H(t)\bar{x}(t_0) = H(t)K = H(t)H^{-1}(t_0)x_0. \tag{9.29}$$

for the general solution of (9.11).

Any other transformation (9.26) for which $\bar{A}$ is zero must be obtained by right multiplication of the considered H by a constant matrix. This corresponds to a change of basis for the linear space of solutions of the system: a matrix H' determined by another basis of the space of solutions for the system is related with H by right-multiplication with a constant invertible matrix: $H'(t) = H(t)P$.

Let us now consider the above mentioned inhomogeneous linear system

$$\dot{x}^i = \sum_{j=1}^n A^i{}_j(t)x^j + b^i(t)\,, \qquad i = 1, \ldots, n. \tag{9.30}$$

or written in matrix form

$$\dot{x} = A(t)x + b. \tag{9.31}$$

In full similarity with the preceding case, if we introduce the change of variable

$$\bar{x} = \alpha(t)x + \beta(t), \tag{9.32}$$

where $\alpha(t)$ denotes a t-dependent invertible matrix and $\beta(t)$ a column matrix, with inverse transformation

$$x = \alpha^{-1}(t)(\bar{x} - \beta(t)), \tag{9.33}$$

then $\bar{x}$ satisfies

$$\dot{\bar{x}} = \bar{A}(t)\bar{x} + \bar{b}(t), \tag{9.34}$$

where

$$\bar{A} = \dot{\alpha}\,\alpha^{-1} + \alpha\,A\,\alpha^{-1}, \qquad \bar{b}(t) = \alpha b + \dot{\beta} - \dot{\alpha}\,\alpha^{-1}\beta - \alpha\,A\,\alpha^{-1}\beta = \alpha b + \dot{\beta} - \bar{A}\beta. \tag{9.35}$$

Suppose that we know n linearly independent solutions of the associated homogeneous system, construct the matrix H and choose $\alpha = H^{-1}$ as before. Then, as $\dot{\alpha} + \alpha A = 0$, the preceding equations reduce to

$$\bar{A} = 0, \qquad \bar{b} = H^{-1}b + \dot{\beta}.$$

Therefore, in order to get $\bar{b} = 0$ the column matrix $\beta(t)$ must be chosen such that $H^{-1}b + \dot{\beta} = 0$, which gives raise to the solution for β

$$\beta(t) = \beta(t_0) - \int_{t_0}^{t} H^{-1}(t')b(t')\,dt'.$$

With such choices for α and β, (9.34) turns out to be $\dot{\bar{x}} = 0$, with solution $\bar{x}(t) = \bar{x}(t_0)$, i.e.

$$x(t) = H(t)(\bar{x}(t_0) - \beta(t)) = H(t)\left(\bar{x}(t_0) - \beta(t_0) + \int_{t_0}^{t} H^{-1}(t')b(t')\,dt'\right),$$

which can be rewritten as

$$x(t) = H(t)H^{-1}(t_0)x(t_0) + \int_{t_0}^{t} H(t)H^{-1}(t')b(t')\,dt'. \tag{9.36}$$

and if we introduce the Green matrix $\mathcal{G}(t, t') = H(t)H^{-1}(t')$ the preceding equation becomes:

$$x(t) = \mathcal{G}(t, t_0)x(t_0) + \int_{t_0}^{t} \mathcal{G}(t, t')b(t')\,dt'. \tag{9.37}$$

Note that the Green matrix $\mathcal{G}(t, t')$ does not depend on the choice for the set of independent solutions of the homogeneous system, because for any other choice $H'(t) = H(t)P$, with P being a constant matrix and then $\mathcal{G}'(t, t') = H'(t)H'^{-1}(t') = H(t)H^{-1}(t') = \mathcal{G}(t, t')$. Furthermore, the Green matrix $\mathcal{G}(t, t')$ is such that

$$\frac{\partial \mathcal{G}(t, t')}{\partial t} = \dot{H}(t)H^{-1}(t') = A(t)H(t)H^{-1}(t') = A\mathcal{G}(t, t'),$$

and $\mathcal{G}(t,t) = I$. Conversely, given a matrix satisfying these two properties, the solution of (9.31) is given by (9.37), because taking derivatives with respect to t in this expression we see that

$$\frac{dx}{dt} = \frac{\partial \mathcal{G}(t,t_0)}{\partial t} x_0 + \mathcal{G}(t,t)b(t) + \int_{t_0}^{t} \frac{\partial \mathcal{G}(t,t')}{\partial t} b(t')\,dt'$$

i.e.,

$$\frac{dx}{dt} = A(t)\mathcal{G}(t,t_0)x(t_0) + A(t)\int_{t_0}^{t} \mathcal{G}(t,t')b(t')\,dt' + b(t)$$

and therefore $x(t)$ given by (9.37) satisfies (9.31).

9.3 Non-linear Superposition Rule

We can now look for a group theoretical approach to the problem. First of all, system (9.17) can be seen as the one determining the integral curves of the vector field

$$X_A = \sum_{i,j=1}^{n} A^i{}_j(t)\, x^j \,\frac{\partial}{\partial x^i},$$

while system (9.24) determines the integral curves of

$$X_{(A,a)} = \sum_{i,j=1}^{n} A^i{}_j(t)\, x^j \,\frac{\partial}{\partial x^i} + \sum_{i=1}^{n} a^i(t)\,\frac{\partial}{\partial x^i}.$$

The vector field X_A can be written as a linear combination

$$X_A = \sum_{i,j=1}^{n} A^i{}_j(t)\, X_{ij}$$

of the vector fields

$$X_{ij} = x^j \,\frac{\partial}{\partial x^i}, \qquad i,j = 1,\ldots,n, \tag{9.38}$$

and the vector field $X_{(A,a)}$ as a linear combination

$$X_{(A,a)} = \sum_{i,j=1}^{n} A^i{}_j(t)\, X_{ij} + \sum_{i=1}^{n} a^i(t)\, X_i, \tag{9.39}$$

with X_i the vector fields

$$X_i = \frac{\partial}{\partial x^i}, \qquad i = 1, \dots, n. \tag{9.40}$$

Notice that

$$[X_{ij}, X_{kl}] = \left[x^j \frac{\partial}{\partial x^i}, x^l \frac{\partial}{\partial x^k}\right] = \delta_{il}\, x^j \frac{\partial}{\partial x^k} - \delta_{kj}\, x^l \frac{\partial}{\partial x^i},$$

i.e.,

$$[X_{ij}, X_{kl}] = \delta_{il}\, X_{kj} - \delta_{kj}\, X_{il}, \tag{9.41}$$

which means that the vector fields $\{X_{ij} \mid i, j = 1, \dots, n\}$, which are the only ones appearing in the case of an homogeneous system, close on an n^2-dimensional Lie algebra isomorphic to the $\mathfrak{gl}(n, \mathbb{R})$-algebra. It suffices to compare these commutation relations with those of the $\mathfrak{gl}(n, \mathbb{R})$-algebra. The latter is generated by the matrices E_{ij} with elements $(E_{ij})_{kl} = \delta_{ik}\delta_{jl}$, which satisfy

$$[E_{ij}, E_{kl}] = \delta_{jk} E_{il} - \delta_{il} E_{kj}. \tag{9.42}$$

Moreover,

$$[X_i, X_k] = 0, \qquad \forall\, i, k = 1, \dots, n, \tag{9.43}$$

and

$$[X_{ij}, X_k] = -\delta_{kj}\, X_i, \qquad \forall\, i, j, k = 1, \dots, n. \tag{9.44}$$

Therefore, the Lie algebra generated by $\{X_{ij}, X_k \mid i, j, k = 1, \dots, n\}$ is isomorphic to the $(n^2 + n)$-dimensional Lie algebra of the affine group.

The generalisation of such a property is obvious and it was stated by Lie as follows [Lie93]:

Theorem 9.1 (Lie–Scheffers theorem) *Given a system*

$$\frac{dx^i}{dt} = X^i(x, t), \qquad i = 1, \dots, n, \tag{9.45}$$

the general solution can be expressed in terms of m fundamental solutions if there are r vector fields in $\mathbb{R}^n$, Y_1, ..., Y_r, *such that the vector field X,*

$$X = \sum_{i=1}^{n} X^i(x, t) \frac{\partial}{\partial x^i} \tag{9.46}$$

can be expressed as a linear combination

$$X = b_1(t) Y_1 + \cdots + b_r(t) Y_r \tag{9.47}$$

and furthermore the vector fields Y_α, close on a finite-dimensional real (or complex) Lie algebra, with dimension r, i.e., there exist r^3 real numbers $c_{\alpha\beta}{}^\gamma$ such that

$$[Y_\alpha, Y_\beta] = \sum_{\gamma=1}^{r} c_{\alpha\beta}{}^\gamma\, Y_\gamma. \tag{9.48}$$

Moreover, in this case $mn \geq r$.

The examples of homogeneous and inhomogeneous linear systems mentioned above are particular instances of this result. In the first case $m = n$, while in the second one $m = n + 1$. In the most general case m is arbitrary and the superposition rule will be a function $\Phi : \mathbb{R}^{n(m+1)} \to \mathbb{R}$, $x = \Phi(x_{(1)}, \ldots, x_{(m)}; k_1, \ldots, k_n)$, such that the general solution is

$$x(t) = \Phi(x_{(1)}(t), \ldots, x_{(m)(t)}; k_1, \ldots, k_n)\,,$$

with $\{x_{(a)}(t) \mid a = 1, \ldots, m\}$ being a set of particular solutions of the system and where $k_1, \ldots, k_n$ are real numbers. That is a generalisation of linear superposition rules for homogeneous linear systems (for which $m = n$ and $x = \Phi(x_{(1)}, \ldots, x_{(n)}; k_1, \ldots, k_n) = k_1\, x_{(1)} + \cdots + k_n\, x_{(n)}$) but the function Φ is nonlinear in the more general case.

As another example we can consider the Bernoulli differential equation

$$\dot{y} = b_1(t)\, y + b_2(t)\, y^n, \qquad n \neq 0, 1, \tag{9.49}$$

whose solutions are the integral curves of the t-dependent vector field

$$Y = b_1(t) Y_1 + b_2(t)\, Y_2 \tag{9.50}$$

with

$$Y_1 = y\,\frac{\partial}{\partial y}, \qquad Y_2 = y^n\,\frac{\partial}{\partial y}, \tag{9.51}$$

satisfying the commutation relations

$$[Y_1, Y_2] = (n-1) Y_2, \tag{9.52}$$

and therefore Y_2 and $\dfrac{1}{1-n}\, Y_1$ satisfy the same commutation relations as the vector fields X_0 and X_1 given in (9.5). This equation is therefore of Lie–Scheffers type with associated Lie algebra $\mathfrak{a}(1, \mathbb{R})$. There is only one action, up to conjugation by a diffeomorphism, of $\mathfrak{a}(1, \mathbb{R})$ on the real line, and consequently there is a diffeomorphism ϕ of $\mathbb{R}$ such that $y = \phi(x)$ and then $y = \phi(x)$ must be such that

$$\frac{1}{1-n}\, y = x\,\frac{dy}{dx}, \qquad \frac{dy}{dx} = y^n,$$

therefore x and y must be related as follows:

$$x = \frac{y^{1-n}}{1-n}, \tag{9.53}$$

the change used for reduction of the Bernoulli's equation to the inhomogeneous linear one, the factor $1/(1-n)$ being irrelevant.

As the general solution of the linear equation $\dot{x} = a_1(t)\, x + a_0(t)$ can be written as a linear combination of two solutions x_1 and x_2, $x(t) = k\, x_1(t) + (1-k)\, x_2(t)$, the general solution of Bernoulli equation can be written as a function of two particular solutions as follows:

$$y = [k\, y_1^{1-n} + (1-k)\, y_2^{1-n}]^{1/(1-n)} \tag{9.54}$$

which is a superposition rule of solutions but it is not an affine superposition anymore.

Another relevant example is to be studied in a later section, the Riccati equation, of a high relevance in classical and quantum physics.

We remark that in geometric terms, systems of first-order differential equations appear as local expressions determining the integral curves of a vector field in a manifold M. Correspondingly, we call Lie–Scheffers system a t-dependent vector field on a manifold M for which there exists a superposition rule, which now can be seen as a map

$$\Phi : M^{m+1} = M \times \cdots \times M \to M.$$

Explicit examples will be given in the next sections.

9.4 Related Maps

Many of the properties of the systems that we are going to consider here have a common framework. The idea is just to consider maps between differentiable manifolds $A : M_1 \to M_2$, $B : M_3 \to M_4$, $\alpha : M_1 \to M_3$ and $\beta : M_2 \to M_4$, in such a way that $B \circ \alpha = \beta \circ A$, i.e., the following diagram be commutative:

$$\begin{array}{ccc} M_1 & \xrightarrow{A} & M_2 \\ \alpha \downarrow & & \downarrow \beta \\ M_3 & \xrightarrow{B} & M_4 \end{array}$$

That is for instance the case when $M_1 = M_2$, $M_3 = M_4$ and a Lie group G acts both on M_1 and on M_3 and $A = \Phi_{1g}$, $B = \Phi_{2g}$, with $g \in G$, in such a way that the diagramme is commutative for any $g \in G$, i.e., $\Phi_{2g} \circ \alpha = \beta \circ \Phi_{1g}$, and then we will say that the maps α and β interchange the actions of G on M_1 and M_3:

$$\begin{array}{ccc} M_1 & \xrightarrow{\Phi_{1g}} & M_1 \\ {\scriptstyle\alpha}\downarrow & & \downarrow{\scriptstyle\beta} \\ M_3 & \xrightarrow{\Phi_{2g}} & M_3 \end{array}$$

The simplest case is when $G = \mathbb{R}$ and A and B can be considered as the flows of vector fields in M_1 and M_3 respectively, i.e., $A = \Phi_{1t}$ and $B = \Phi_{2t}$.

When $\alpha = \beta = F$ the map F is said to be equivariant. The equivariance condition means $\Phi_{2g} \circ F = F \circ \Phi_{1g}$.

Another example is when $M_2 = TM_1$, $M_4 = TM_3$ and $\beta = T\alpha$. Then A and B can be chosen as being sections for the corresponding tangent projections, i.e., vector fields, and in this case, $B \circ \alpha = T\alpha \circ A$, we will say that the vector fields A and B are α-related.

$$\begin{array}{ccc} M_1 & \xrightarrow{A} & TM_1 \\ {\scriptstyle\alpha}\downarrow & & \downarrow{\scriptstyle T\alpha} \\ M_3 & \xrightarrow{B} & TM_3 \end{array}$$

In particular, when $M_1 = M_3 = M$, $M_2 = M_4 = TM$, $A = \Gamma_1$ and $B = \Gamma_2$ are vector fields, and $\alpha = \phi$, the vector Γ_1 is ϕ-related with Γ_2, i.e., $\Gamma_2 \circ \phi = T\phi \circ \Gamma_1$, if $\Gamma_1(\phi^* f) = \phi^*(\Gamma_2 f), \forall f \in C^\infty(M)$:

$$\begin{array}{ccc} M & \xrightarrow{\Gamma_1} & TM \\ {\scriptstyle\phi}\downarrow & & \downarrow{\scriptstyle T\phi} \\ M & \xrightarrow{\Gamma_2} & TM \end{array}$$

In this case the subalgebra $\phi^*(M) \subset C^\infty(M)$ is invariant under Γ_1.

The problem of 'normal forms' for linear maps studied in Sect. 3.6 is once again of this kind. Now $M_1 = M_2 = E$ and $M_3 = M_4 = E'$ are linear spaces. An equivalence between inear maps from E to E' is defined as follows: α and β are equivalent if there are invertible maps A and B relating α and β.

$$\begin{array}{ccc} E & \xrightarrow{A} & E \\ {\scriptstyle\alpha}\downarrow & & \downarrow{\scriptstyle\beta} \\ E' & \xrightarrow{B} & E' \end{array}$$

When E and E' coincide and $A = B$ we will say that α and β are of the same conjugacy class. The diagonalization of an endomorphism is just one of this class with β being a diagonal matrix.

When we have related vector fields Γ_1 and Γ_2 with α being an differentiable and injective, the integrability of Γ_2 in the sense that we are able to determine the flow of Γ_2 by algebraic manipulations and quadratures, implies that Γ_1 is also integrable. In a similar way, if α is a submersion the integrability of Γ_1 implies the integrability of Γ_2. Note that γ is an integral curve of Γ_1 if and only if $\alpha \circ \gamma$ is an integral curve of Γ_2.

The framework of related maps is very general and by particularizing to some specific cases, i.e., by restricting the possible choices for the ingredients, we find interesting examples.

9.5 Lie–Scheffers Systems on Lie Groups and Homogeneous Spaces

Let now $\Phi\colon G \times M \to M$ be a left action of a Lie group G of dimension r on a differentiable manifold M (see Sect. 3.2.5 for more details). If $g \in G$ and $m \in M$, we denote $\Phi_m : G \to M$ and $\Phi_g : M \to M$ the maps defined by $\Phi_m(g) = \Phi_g(m) = \Phi(g, m)$. If a is an element of the Lie algebra of G, $a \in \mathfrak{g}$, then the fundamental vector field X_a is defined by

$$(X_a f)(m) = \frac{d}{dt} f(\exp(-t\, a)m)_{|t=0}, \tag{9.55}$$

which, therefore, it can also be defined by

$$X_a(m) = T\Phi_m(e)(-a). \tag{9.56}$$

Let us choose a basis $\{e_\alpha \mid \alpha = 1, \dots, r\}$ of the Lie algebra of the Lie group G with defining relations

$$[e_\alpha, e_\beta] = \sum_{\gamma=1}^{r} c_{\alpha\beta}{}^{\gamma}\, e_\gamma,$$

and denote X_α the corresponding fundamental vector fields, $X_\alpha = X_{e_\alpha}$, i.e., defined by

$$(X_\alpha f)(m) = \frac{d}{dt} f(\exp(-te_\alpha)m)_{|t=0}.$$

We recall that such fundamental vector fields satisfy

$$[X_\alpha, X_\beta] = \sum_{\gamma=1}^{r} c_{\alpha\beta}{}^{\gamma}\, X_\gamma.$$

Another important property is that given two actions of the same group on two different manifolds and a map $F : M_1 \to M_2$, the fundamental vector fields in the manifold M_1 are F-projectable on the corresponding ones on the second manifold iff the map F is equivariant. Note that F is equivariant, i.e., $\Phi_{2g} \circ F = F \circ \Phi_{1g}$, if and only if $F \circ \Phi_{1m} = \Phi_{2F(m)}$, and therefore, having in mind that

$$F_{*m}(X_a(m)) = F_{*m}(\Phi_{1m*e}(-a)) = (F \circ \Phi_{1m})_{*e}(-a)$$

we see that if F is equivariant,

$$F_{*m}(X_a(m)) = (\Phi_{2F(m)})_{*e}(-a)$$

i.e., the fundamental vector fields are F-related. One can also prove using the exponential map that if G is connected the converse property is also true.

It is well known that a Lie group acts on itself in a transitive and free way both on the left and on the right

$$L_g : G \to G\,, \qquad R_g : G \to G, \qquad g' \mapsto g\,g'\,, \qquad g' \mapsto g'\,g.$$

The diffeomorphisms L_g and R_g allow us to define the left- and right-invariant vector fields determined by their values in the neutral element $X_g^L = L_{g*e}X_e$ and $X_g^R = R_{g*e}X_e$. We use the shorter notation $X_\alpha^L = X_{e_\alpha}^L$ and $X_\alpha^R = X_{e_\alpha}^R$, $\alpha = 1, \ldots, r$, i.e., $(X_\alpha^L)_g = L_{g*e}(e_\alpha)$, $(X_\alpha^R)_g = R_{g*e}(e_\alpha)$. They generate, respectively, two Lie algebras

$$[X_\alpha^L, X_\beta^L] = \sum_{\gamma=1}^{r} c_{\alpha\beta}{}^{\gamma}\, X_\gamma^L\,, \quad [X_\alpha^R, X_\beta^R] = -\sum_{\gamma=1}^{r} c_{\alpha\beta}{}^{\gamma}\, X_\gamma^R. \tag{9.57}$$

We can also define a map $\psi : G \to G$ such that $\psi(g) = g^{-1}$, which is an anti-homomorphism because $\psi(g_2 g_1) = \psi(g_1)\psi(g_2)$, and moreover, from the relations $L_g \circ \psi = \psi \circ R_{g^{-1}}$, $R_g \circ \psi = \psi \circ L_{g^{-1}}$, and the corresponding ones,

$$L_{g*} \circ \psi_* = \psi_* \circ R_{g^{-1}*}, \quad R_{g*} \circ \psi_* = \psi_* \circ L_{g^{-1}*},$$

we see that if X is a left-invariant vector field, then $Y = \psi_*(X)$ is such that

$$R_{g*}(Y) = \left(R_{g*} \circ \psi_*\right)(X) = \left(\psi_* \circ L_{g^{-1}*}\right)(X) = \psi_*(X) = Y.$$

i.e., $Y = \psi_*(X)$ is right-invariant.

One can check that, as a consequence of the relation $R_{g_2} \circ \Phi_{m=g_1} = R_{g_1 g_2}$,

$$R_{g_2*g_1}(X_a(g_1)) = R_{g_2*g_1}(\Phi_{g_1*e}(-a)) = R_{g_1 g_2 *e}(-a) = X_a(g_1 g_2)$$

and therefore the fundamental vector fields of the left action are right-invariant vector fields in the Lie group G.

We remark that if we consider the left action of G on itself by left translations, $\Phi(g, g') = gg'$, then the fundamental vector fields X_a are right invariant because

$$(X_a)(g) = \Phi_{g*e}(-a) = R_{g*e}(-a) = -(X_a^R)(g),$$

where X_a^R is the right-invariant vector field in G determined by its value at the neutral element $(X_a^R)(e) = a$.

The commutation relations (9.57) mean that we can consider a Lie–Scheffers system on G given by

$$\bar{X}(g, t) = -\sum_{\alpha=1}^{r} b_\alpha(t)\, X_\alpha^R(g), \tag{9.58}$$

whose integral curves are the solutions $g(t)$ of

$$\dot{g} = -\sum_{\alpha=1}^{r} b_\alpha(t)\, X_\alpha^R(g),$$

where $\dot{g}(t)$ denotes the vector field along the curve $g(t)$ defined by the tangent vector at each point of the curve.

When applying $R_{g^{-1}(t)*g(t)}$ to both sides we obtain the equation

$$\dot{g}(t)\, g^{-1}(t) = -\sum_{\alpha=1}^{r} b_\alpha(t) e_\alpha \in T_e G, \tag{9.59}$$

where $\dot{g}(t)\, g^{-1}(t)$ is a short notation for $R_{g^{-1}(t)*g(t)}\dot{g}(t)$.

This equation can be solved using a generalization of a method developed by Wei and Norman for the linear case [WN63, WN64] for finding the time evolution operator for a linear systems of type $dU(t)/dt = H(t)U(t)$, with $U(0) = I$, see also [Ca98]. We should also mention that there exist alternative methods for solving equation (9.59) by reducing the problem to a simpler one.

Both procedures are based on the following property [Ca01, CR02]: If $g(t), g_1(t)$ and $g_2(t)$ are differentiable curves in G such that $g(t) = g_1(t)g_2(t)$, $\forall t \in \mathbb{R}$, then,

$$R_{g(t)^{-1}*g(t)}(\dot{g}(t)) = R_{g_1(t)^{-1}*g_1(t)}(\dot{g}_1(t)) + \mathrm{Ad}\,(g_1(t))\left\{R_{g_2(t)^{-1}*g_2(t)}(\dot{g}_2(t))\right\}. \tag{9.60}$$

The generalization of this property to several factors is as follows. Let now $g(t)$ be a curve in G which is given by the product of other l curves $g(t) = g_1(t)g_2(t)\cdots g_l(t) = \prod_{i=1}^{l} g_i(t)$. Then, denoting $h_s(t) = \prod_{i=s+1}^{l} g_i(t)$, for $s \in \{1, \ldots, l-1\}$, and applying (9.60) to $g(t) = g_1(t)\, h_1(t)$ we have

$$R_{g(t)^{-1}*g(t)}(\dot{g}(t)) = R_{g_1(t)^{-1}*g_1(t)}(\dot{g}_1(t)) + \mathrm{Ad}\,(g_1(t))\left\{R_{h_1(t)^{-1}*h_1(t)}(\dot{h}_1(t))\right\}.$$

Simply iterating, and using that $\mathrm{Ad}\,(gg') = \mathrm{Ad}\,(g)\mathrm{Ad}\,(g')$ for all $g, g' \in G$ we obtain

$$
\begin{aligned}
R_{g(t)^{-1}*g(t)}(\dot{g}(t)) &= R_{g_1(t)^{-1}*g_1(t)}(\dot{g}_1(t)) + \mathrm{Ad}\,(g_1(t))\left\{R_{g_2(t)^{-1}*g_2(t)}(\dot{g}_2(t))\right\} \\
&\quad + \cdots + \mathrm{Ad}\left(\prod_{i=1}^{l-1} g_i(t)\right)\left\{R_{g_l(t)^{-1}*g_l(t)}(\dot{g}_l(t))\right\} \\
&= \sum_{i=1}^{l} \mathrm{Ad}\left(\prod_{j<i} g_j(t)\right)\left\{R_{g_i(t)^{-1}*g_i(t)}(\dot{g}_i(t))\right\} \\
&= \sum_{i=1}^{l}\left(\prod_{j<i} \mathrm{Ad}\,(g_j(t))\right)\left\{R_{g_i(t)^{-1}*g_i(t)}(\dot{g}_i(t))\right\}, \qquad (9.61)
\end{aligned}
$$

where it has been taken $g_0(t) = e$ for all t.

The generalized Wei-Norman method (see [Ca98] and references therein) consists of writing the solution $g(t)$ of (9.59) in terms of its second kind canonical coordinates with respect to a basis $\{e_1, \ldots, e_r\}$ of the Lie algebra $\mathfrak{g}$, for each value of t, i.e.

$$
g(t) = \prod_{\alpha=1}^{r} \exp(-v_\alpha(t)e_\alpha) = \exp(-v_1(t)e_1)\cdots\exp(-v_r(t)e_r),
$$

and transforming the differential equation (9.59) into a differential equation system for the $v_\alpha(t)$, with initial conditions $v_\alpha(0) = 0$ for all $\alpha = 1, \ldots, r$. The minus signs in the exponentials have been introduced for computational convenience. Then, we use the result (9.61), taking $l = r = \dim G$ and $g_\alpha(t) = \exp(-v_\alpha(t)e_\alpha)$ for all α. Now, since $R_{g_\alpha(t)^{-1}*g_\alpha(t)}(\dot{g}_\alpha(t)) = -\dot{v}_\alpha(t)e_\alpha$, we see that (9.61) reduces to

$$
\begin{aligned}
R_{g(t)^{-1}*g(t)}(\dot{g}(t)) &= -\sum_{\alpha=1}^{r} \dot{v}_\alpha \left(\prod_{\beta<\alpha} \mathrm{Ad}\,(\exp(-v_\beta(t)e_\beta))\right) e_\alpha \\
&= -\sum_{\alpha=1}^{r} \dot{v}_\alpha \left(\prod_{\beta<\alpha} \exp(-v_\beta(t)\mathrm{ad}\,(e_\beta))\right) e_\alpha,
\end{aligned}
$$

where the identity $\mathrm{Ad}\,(\exp(a)) = \exp(\mathrm{ad}\,(a))$, has been used for all $a \in \mathfrak{g}$. Substituting in Eq. (9.59) we obtain the fundamental expression of the Wei-Norman method

$$
\sum_{\alpha=1}^{r} \dot{v}_\alpha \left(\prod_{\beta<\alpha} \exp(-v_\beta(t)\mathrm{ad}\,(e_\beta))\right) e_\alpha = \sum_{\alpha=1}^{r} b_\alpha(t)e_\alpha, \qquad (9.62)
$$

with $v_\alpha(0) = 0$, $\alpha = 1, \ldots, r$. The resulting differential equation system for the functions $v_\alpha(t)$ is integrable by quadratures if the Lie algebra is solvable [WN63, WN64], and in particular, for nilpotent Lie algebras.

Note also that the point that the t-dependent vector field (9.58) is right-invariant means that if $g(t)$ is the integral curve of such a vector field starting from e, then $\bar{g}(t) = g(t)\, g_0$ is the integral curve of (9.58) starting from g_0, and therefore the superposition rule involves only one solution, and is given by the following function $\phi : G \times G \to G$:

$$\phi(g, g_0) = g\, g_0.$$

Assume that a transitive action $\Phi : G \times M \to M$ of the Lie group G on a manifold M is given, i.e., M is a homogeneous space for G. Let x_0 be an arbitrary but fixed point in M and denote by G_{x_0} the stability or isotopy subgroup of $x_0 \in M$. This will provide us an identification of G/G_{x_0} with M given by $j(gG_{x_0}) = \Phi(g, x_0)$ (Sect. 3.2.5, Eq. 3.14). The important point is (see e.g., [Ca01]) that the natural projection map $\tau : G \to G/H$ is equivariant, with respect to the left action of G on itself by left translations and the natural left-action on G/H, and consequently, the fundamental vector fields corresponding to the two actions are τ-related. Therefore, the right-invariant vector fields X_α^R, which are the fundamental vector fields corresponding to the natural left action of G on itself, are τ-projectable and the τ-related vector fields in M are the fundamental vector fields $-X_\alpha = -X_{a_\alpha}$ corresponding to the natural left action of G on M, $\tau_{*g} X_\alpha^R(g) = -X_\alpha(gH)$. In this way system (9.58) will have an associated Lie–Scheffers system on M:

$$X = \sum_{\alpha=1}^{r} b_\alpha(t)\, X_\alpha.$$

The converse property is true in the following sense: Given such a Lie–Scheffers system in a manifold M defined by complete vector fields X_α and with associated Lie algebra $\mathfrak{g}$, we can see these as fundamental vector fields relative to an action given by integrating the vector fields. Then, the restriction to an orbit will provide a homogeneous space of the above type. The choice of a point x_0 in the homogeneous space allows us to identify the homogeneous space M with G/H, where H is the stability group of x_0. Different choices for x_0 will lead to conjugate subgroups [Ca01].

Hence we reduce the problem of integrating the given Lie–Scheffers system in M to studying the curve $g(t)$ on the Lie group itself such that $g(0) = e$ and is an integral curve of the t-dependent right-invariant vector field

$$X^G(g, t) = -\sum_{\alpha=1}^{r} b_\alpha(t)\, X_\alpha^R(g)$$

i.e., a solution $g(t)$ of

$$\dot{g} = -\sum_{\alpha=1}^{r} b_\alpha(t)\, X_\alpha^R(g).$$

Using the right translation $R_{g^{-1}}$ we find

$$\frac{dg}{dt}\, g^{-1} = -\sum_{\alpha=1}^{r} b_\alpha(t)\, e_\alpha\,. \tag{9.63}$$

Then, the given t-dependent vector field X on M is related to a t-dependent vector field in the group, X^G, and the integral curves of this latter vector field provide us the integral curves of the initial system.

The relatedness property of the t-dependent vector fields X^G and X, for each fixed value of t, is displayed in the next diagram:

$$\begin{array}{ccc} TG & \xrightarrow{T\Phi_{x_0}} & TM \\ {\scriptstyle X^G}\uparrow & & \uparrow{\scriptstyle X} \\ G & \xrightarrow{\Phi_{x_0}} & M \end{array}$$

Even if the Eq. (9.63) may be difficult to solve directly, it can be explicitly solved when a fundamental set of particular solutions of the system of differential equations is known (see e.g., [Ca98]) and this property leads to a superposition rule allowing us to find the general solution of the system in terms of any fundamental set of solutions by using a (nonlinear) superposition rule.

In this way the solution of the given system for any initial condition is reduced to only one higher dimensional system of first-order linear equations with initial condition $g(0) = e$. Once the curve $g(t)$ has been found, the solution of the given system determined by the initial condition $x(0) = x_0$ is $\Phi(g(t), x_0)$.

A set $x_1, \ldots, x_m$, of solutions is said to be a fundamental system of solutions, if

$$\begin{aligned} x_1(t) &= \Phi(g(t), x_1(0)) = F(g(t), x_1(0)) \\ \ldots &= \ldots\ldots \\ x_m(t) &= \Phi(g(t), x_m(0)) = F(g(t), x_m(0)) \end{aligned} \tag{9.64}$$

is a minimal set allowing us to solve for $g(t)$ via the implicit function Theorem. If this can be done we get

$$g(t) = G(x_1(t), \ldots, x_m(t); x_1(0), \ldots, x_m(0)),$$

and then any other solution can be written as

$$x(t) - F(G(x_1(t), \ldots, x_m(t); x_1(0), \ldots, x_m(0)), x(0)) = 0.$$

Therefore the left hand-side of this relation defines a constant of motion.

From the operational point of view, starting with the action $\Phi : G \times M \to M$ we should find the minimal integer number m such that the isotopy group of the action of G on the product $M \times \cdots \times M$ (m times), extended from Φ by $\Phi^m(g, x_1, \ldots, x_m) = (\Phi(g, x_1), \ldots, \Phi(g, x_m))$, reduces to the neutral element for a generic point. When expressed in terms of vector fields, that means that the corresponding diagonal extensions of the fundamental vector fields X_α do not vanish in a generic point. The solution then is found by adding a new component and looking for constants of motion.

9.6 Some Examples of Lie–Scheffers Systems

In this section we present some specific examples illustrating how the theory works. The procedure is first illustrated with one interesting example for the simplest case $n = 1$ and afterwards we analyse other examples arising in classical and Quantum Mechanics.

9.6.1 Riccati Equation

For the simplest case $n = 1$, according to Lie's Theorem we should look for a finite-dimensional real Lie algebra of differential operators

$$X_\alpha = f_\alpha(x)\frac{\partial}{\partial x}.$$

It can be shown that the only finite-dimensional Lie algebra that can be found from vector fields in one real variable are $\mathfrak{sl}(2, \mathbb{R})$ and its subalgebras. The uniquely defined (up to a change of variables) realization of $\mathfrak{sl}(2, \mathbb{R})$ is given by

$$X_0 = \frac{\partial}{\partial x}, \qquad X_1 = x\,\frac{\partial}{\partial x}, \qquad X_2 = x^2\,\frac{\partial}{\partial x},$$

with commutation relations

$$[X_0, X_1] = X_0\,, \qquad [X_1, X_2] = X_2\,, \qquad [X_2, X_0] = -2\,X_1\,.$$

More explicitly (see [Ca99]), let us consider the action of $G = SL(2, \mathbb{R})$ on $M = \overline{\mathbb{R}} = \mathbb{R} \cup \infty$, the real line completed with a point at the infinity, given by

$$\Phi(A, x) = \frac{\alpha x + \beta}{\gamma x + \delta},$$

when

$$A = \begin{pmatrix} \alpha & \beta \\ \gamma & \delta \end{pmatrix}$$

with the understanding that $\Phi(A, \infty) = \alpha/\gamma$ and $\Phi\,(-\delta/\gamma) = \infty$.

If we consider $x_0 = \infty$, then the stability subgroup G_{x_0} is the affine group in one dimension, made up by the matrices

$$A = \begin{pmatrix} \alpha & \beta \\ 0 & 1/\alpha \end{pmatrix}$$

with $\alpha \neq 0$. On the other side, if $x_0 = 0$ the stability group is that of matrices

$$A = \begin{pmatrix} \alpha & 0 \\ \gamma & 1/\alpha \end{pmatrix},$$

which is conjugate to the previous one.

The actions of the one-parameter subgroups of $SL(2, \mathbb{R})$ are, respectively, given by

$$x \mapsto x + t, \quad x \mapsto e^t x, \quad x \mapsto \frac{x}{1 - tx},$$

and they give raise to the corresponding fundamental vector fields X_0, X_1 and X_2, given above.

The Lie algebra $\mathfrak{sl}(2, \mathbb{R})$ of the Lie group $SL(2, \mathbb{R})$ is generated by the traceless matrices

$$e_0 = \begin{pmatrix} 0 & 1 \\ 0 & 0 \end{pmatrix}, \quad e_1 = \frac{1}{2}\begin{pmatrix} 1 & 0 \\ 0 & -1 \end{pmatrix}, \quad e_2 = \begin{pmatrix} 0 & 0 \\ -1 & 0 \end{pmatrix} \tag{9.65}$$

with commutation relations

$$[e_1, e_0] = e_0\,, \quad [e_1, e_2] = -e_2\,, \quad [e_2, e_0] = 2\,e_1\,.$$

The time-dependent vector field on $SL(2, \mathbb{R})$ given by

$$\bar{X} = [b_0(t)X_0^R + b_1(t)X_1^R + b_2(t)X_2^R],$$

gives raise on $M = SL(2, \mathbb{R})/A_1 \approx \overline{\mathbb{R}}$ to the t-dependent vector field

$$X(t) = -\left[b_0(t)\frac{\partial}{\partial x} + b_1(t)x\frac{\partial}{\partial x} + b_2(t)x^2\frac{\partial}{\partial x}\right],$$

whose integral curves are solutions of the Riccati equation

$$\dot{x} + b_0(t) + b_1(t)x + b_2(t)x^2 = 0. \tag{9.66}$$

In this case one can prove that $m = r = 3$, i.e., the superposition rule involves three different solutions, and the superposition principle comes from the relation

$$\frac{x - x_1}{x - x_2} : \frac{x_3 - x_1}{x_3 - x_2} = k, \tag{9.67}$$

or in other words, the general solution of (9.66) is

$$x(t) = \frac{x_1(t)(x_3(t) - x_2(t)) + k\, x_2(t)(x_1(t) - x_3(t))}{(x_3(t) - x_2(t)) + k\, (x_1(t) - x_3(t))}. \tag{9.68}$$

The value $k = \infty$ must be accepted, otherwise we do not obtain the solution x_2.

The solutions are obtained from the curve $g(t)$ in $SL(2, \mathbb{R})$ such that $g(0) = e$ and is a solution of the equation

$$\dot{g}g^{-1} = b_0(t)\, e_0 + e_1(t)a_1 + b_2(t)\, e_2,$$

as follows, $x(t) = \Phi(g(t), x(0))$.

Let us check now that the number m of fundamental solutions is $m = 3$ and determine the superposition rule: First we note that the determinant

$$\begin{vmatrix} 1 & \xi_1 & \xi_1^2 \\ 1 & \xi_2 & \xi_2^2 \\ 1 & \xi_3 & \xi_3^2 \end{vmatrix} = (\xi_1 - \xi_2)(\xi_2 - \xi_3)(\xi_3 - \xi_1)$$

is generically different from zero, while the system given by the two equations $a+b\,\xi_1+c\,\xi_1^2 = 0$, $a+b\,\xi_2+c\,\xi_2^2 = 0$ always has a non trivial solution. Consequently, $m = 3$ in this case.

Recall that for obtaining the general solution we should define the vector fields

$$\begin{aligned} V_0 &= \frac{\partial}{\partial x} + \frac{\partial}{\partial x_{(1)}} + \frac{\partial}{\partial x_{(2)}} + \frac{\partial}{\partial x_{(3)}}, \\ V_1 &= x\frac{\partial}{\partial x} + x_{(1)}\frac{\partial}{\partial x_{(1)}} + x_{(2)}\frac{\partial}{\partial x_{(2)}} + x_{(3)}\frac{\partial}{\partial x_{(3)}}, \\ V_2 &= x^2\frac{\partial}{\partial x} + x_{(1)}{}^2\frac{\partial}{\partial x_{(1)}} + x_{(2)}{}^2\frac{\partial}{\partial x_{(2)}} + x_{(3)}{}^2\frac{\partial}{\partial x_{(3)}} \end{aligned} \tag{9.69}$$

and look for a solution of the system

$$V_0 f = V_1 f = V_2 f = 0.$$

This system of partial differential equations is integrable, because the vector fields V_0, V_1 and V_2 close a Lie algebra, and therefore they define an integrable distribution.

The first equation $V_0 f = 0$ tells us that the function f depends only on the differences $u_1 = x - x_{(1)}$, $u_2 = x_{(1)} - x_{(2)}$ and $u_3 = x_{(2)} - x_{(3)}$, because the characteristic system is

$$\frac{dx}{1} = \frac{dx_{(1)}}{1} = \frac{dx_{(2)}}{1} = \frac{dx_{(3)}}{1},$$

and it has as first integrals the differences $u_1 = x - x_{(1)}$, $u_2 = x_{(1)} - x_{(2)}$ and $u_3 = x_{(2)} - x_{(3)}$. Now, if $f(x, x_{(1)}, x_{(2)}, x_{(3)}) = \varphi(u_1, u_2, u_3)$, the condition $V_1 f = 0$ is written

$$u_1 \frac{\partial \varphi}{\partial u_1} + u_2 \frac{\partial \varphi}{\partial u_2} + u_3 \frac{\partial \varphi}{\partial u_3} = 0,$$

i.e., the function φ should be homogeneous of degree zero, and therefore it only can depend on the quotients $v_1 = u_1/u_2$ and $v_2 = u_3/u_2$, $\varphi(u_1, u_2, u_3) = \phi(v_1, v_2)$. Finally, the condition $V_2 f = 0$ can be written in these coordinates, after a long computation, as

$$v_1(v_1 + 1) \frac{\partial \phi}{\partial v_1} - v_2(v_2 + 1) \frac{\partial \phi}{\partial v_2} = 0.$$

The corresponding characteristic system is

$$\frac{dv_1}{v_1(v_1 + 1)} = -\frac{dv_2}{v_2(v_2 + 1)},$$

and taking into account that

$$\int \frac{dx}{x(x+1)} = \log \frac{x}{x+1},$$

we obtain that the constant of motion f should be a function of

$$\zeta = \frac{v_1}{v_1 + 1} \frac{v_2}{v_2 + 1},$$

and therefore,

$$\frac{(x - x_{(1)})(x_{(2)} - x_{(3)})}{(x - x_{(2)})(x_{(1)} - x_{(3)})} = k$$

provides the nonlinear superposition rule giving $x(t)$ as a function of three independent solutions

$$x = \frac{(x_{(1)} - x_{(3)})x_{(2)}k + x_{(1)}(x_{(3)} - x_{(2)})}{x_{(1)} - x_{(3)})k + (x_{(3)} - x_{(2)})}.$$

Let us remark that for the $n = 1$ case there is only one nonlinear differential equation family satisfying Lie's theorem: the Riccati equation. Of course, proper subalgebras of $\mathfrak{sl}(2, \mathbb{R})$ lead to the linear inhomogeneous equation when $b_2 = 0$ or Lie homogeneous equation when $b_2 = b_3 = 0$ (remember also the Bernoulli equation studied before). However, for $n = 2$ in addition to $SL(3, \mathbb{R})$, $O(3, 1)$ and $O(2, 2)$, we can realize families of Lie algebras with arbitrary large Abelian ideals.

The importance of Lie–Scheffers systems in Supersymmetric Quantum Mechanics (see for instance [Co95] and references therein) is based on the above mentioned fact that Riccati equation can be considered as a Lie–Scheffers system with group $SL(2,\mathbb{R})$. Recall, for instance, that the condition for the determination of the superpotential W in the factorization of a Hamiltonian H in such a way that

$$H - c = \left(-\frac{d}{dx} + W\right)\left(\frac{d}{dx} + W\right),$$

is a Riccati equation. Actually, if

$$H = -\frac{d^2}{dx^2} + V(x),$$

then W can be found from V as a solution of the Riccati equation

$$W' = W^2 - (V - c).$$

Moreover, a similar equation plays a relevant role in the search for Shape Invariant potentials using the so-called Infeld and Hull Factorization method [In51] (see also [CR00] for a modern approach). Consequently, the Riccati equation plays a relevant role in the theory of intertwining operators and shape invariance in Quantum Mechanics (see e.g., [CRF01] and [CR00b]).

Finally, as indicated in Sect. 2.3.3, a linear system in two variables can be reduced to a Riccati equation. Therefore as a homogeneous linear second-order differential equation,

$$\frac{d^2z}{dx^2} + b(x)\frac{dz}{dx} + c(x)z = 0, \tag{9.70}$$

can be written in the form

$$\begin{cases} z' = v \\ v' = -b(x)v - c(x)\,z; \end{cases} \tag{9.71}$$

the quotient v/z satisfies a Riccati equation.

In other words, the Riccati equation is particularly important because it appears as a consequence of the reduction theory when taking into account that dilations are symmetries of linear second-order differential equations [Ca98]. Actually, the homogeneous linear second-order differential equation (9.70) admits as an infinitesimal symmetry the vector field $X = z\,\partial/\partial z$ generating dilations in the variable z, which is defined for $z \neq 0$. According to Lie theory we should change the coordinate z to a new one, $u = \varphi(z)$, such that $X = \partial/\partial u$. This change is determined by the equation $Xu = 1$, which leads to $u = \log|z|$, i.e., $|z| = e^u$. In both cases of regions with $z > 0$ or $z < 0$ we have:

$$\frac{dz}{dx} = z\frac{du}{dx}, \quad \text{and} \quad \frac{d^2z}{dx^2} = z\left(\frac{du}{dx}\right)^2 + z\,\frac{d^2u}{dx^2},$$

so Eq. (9.70) becomes

$$\frac{d^2u}{dx^2} + b(x)\frac{du}{dx} + \left(\frac{du}{dx}\right)^2 + c(x) = 0,$$

and as the dependent variable does not explicitly appear, the order can be lowered by introducing the new variable $w = du/dx$. We arrive to the following Riccati equation for w,

$$\frac{dw}{dx} = -w^2 - b(x)w - c(x). \tag{9.72}$$

Notice that $w = z^{-1}dz/dx$, which corresponds to $w = v/z$ in (9.71), and that this relation together with (9.72) is equivalent to the original second-order equation. In the particular case of the one-dimensional time-independent Schrödinger equation

$$-\frac{d^2\phi}{dx^2} + (V(x) - \epsilon)\phi = 0,$$

the reduced Riccati equation for $W = \phi^{-1}d\phi/dx$ is

$$W' = -W^2 + (V(x) - \epsilon), \tag{9.73}$$

which is the equation that W must satisfy in the previously mentioned factorization of $H = -d^2/dx^2 + V(x)$.

As another example of Lie–Scheffers system, not an equation but with $n = 2$, we can apply the previous procedure to rotation group $SO(3, \mathbb{R})$ acting transitively on the sphere $M = S^2$. We start by considering the linear action of $SO(3, \mathbb{R})$ on $\mathbb{R}^3$. The fundamental vector fields are:

$$X_1 = x_2\,\frac{\partial}{\partial x_3} - x_3\,\frac{\partial}{\partial x_2}, \quad X_2 = x_3\,\frac{\partial}{\partial x_1} - x_1\,\frac{\partial}{\partial x_3}, \quad X_3 = x_1\,\frac{\partial}{\partial x_2} - x_2\,\frac{\partial}{\partial x_1},$$

satisfying the commutation relations

$$[X_i, X_j] = -\epsilon_{ijk}X_k.$$

The function $x_1^2 + x_2^2 + x_3^2$ is invariant under such fundamental vector fields, i.e., $X_i(x_1^2 + x_2^2 + x_3^2) = 0$, $i = 1, 2, 3$, and therefore the restrictions of the fundamental vector fields on the sphere of radius one, $M = S^2$, are tangent to the sphere.

There we will consider a t-dependent vector field Γ on $\mathbb{R}^3$ of the form,

$$\Gamma = b_1(t)X_1 + b_2(t)X_2 + b_3(t)X_3,$$

and its restriction to M. The integral curves of such restriction are the integral curves of γ starting from a point of the sphere.

Note that the vector field in $\mathbb{R}^3$ is a linear vector field associated with a skew-symmetric matrix matrix.

A fundamental system of solutions is made by two solutions. In fact, if $Y(t) = (y_1(t), y_2(t), y_3(t))$ and $Z(t) = (z_1(t), z_2(t), z_3(t))$ are two solutions, i.e., curves on S^2, there is only one rotation $R(t)$ such that $Y(t) = R(t)Y(0)$ and $Z(t) = R(t)Z(0)$. For instance, we can consider the case in which $Y(0)$ and $Z(0)$ are orthogonal, because if

$$\langle Y(0), Z(0)\rangle = 0,$$

as the matrix defining the evolution is orthogonal, we find that

$$\langle Y(t), Z(t)\rangle = 0.$$

We then find for a generic solution,

$$X(t) = \langle Y(0), X(0)\rangle Y(t) + \langle Z(0), X(0)\rangle Z(t) + \langle Y(0)\wedge Z(0), X(0)\rangle (Y(t)\wedge Z(t)).$$

9.6.2 Euler Equations

We recall that given a Lie group G, each $g \in G$ defines an innner automorphism $i(g) : G \to G$ by $i(g)g' = gg'g^{-1}$, and therefore, as $i(g_2\, g_1) = i(g_2) \circ i(g_1)$, there is an action of G on itself defined by $\phi(g_2, g_1) = i(g_2)(g_1)$. This action allows us to define a linear representation of G with its Lie algebra as representation space, as follows: $\mathrm{Ad}\,(g) = i(g)_*$. Then it provides us a linear representation of its Lie algebra, called adjoint representation, and denoted ad, i.e.

$$\mathrm{Ad}\,(\exp(a)) = \exp(\mathrm{ad}\,(a)), \qquad a \in \mathfrak{g}.$$

As an instance we can consider $G = GL(E)$, where E is a linear space. Now let B_0 be an arbitrary but fixed endomorphism of E, choose another endomorphism A and define the curve in $\mathrm{End}(E)$,

$$B(t) = e^{At}\, B_0\, e^{-At}. \tag{9.74}$$

Such a curve satisfies $B(0) = B_0$ and

$$\frac{dB}{dt} = [A, B] = \mathrm{ad}\,(A)B.$$

That is an equation for a curve in $\mathrm{End}(E)$ which will be called, by similarity with the quantum case, the Heisenberg description of the dynamical system. The solution starting from B_0 is given by (9.74).

Remark that $\mathrm{ad}\,(A)$ is an inner derivation of the Lie algebra $\mathrm{End}(E)$ of the endomorphisms of E. This motivates our interest in characterizing derivations of such a Lie algebra that are inner derivations and study the evolution equations in $\mathrm{End}(E)$ of the type $\dot{B} = D(B)$.

More generally, the adjoint representation of a Lie group G provides a natural realization of G in the linear space of the Lie algebra $\mathfrak{g}$. In the associated action of G on $\mathfrak{g}$ the fundamental vector field defined by $a \in \mathfrak{g}$ is given by $X_a = \mathrm{ad}\,(a)$, i.e., $X_a(b) = [a, b]$. Consequently, the integral curves of the vector field X_a are given by $\dot{c} = [a, c]$. that is a Lax-like equation, a very special kind of equations that play an important rôle in the study of integrable systems [La68, CI85, CM94]. It is well known that the trace of the elements c in the curve is constant along the curve, because the trace of a commutator is always zero. Furthermore, when G is a subgroup of the general linear group, i.e., the elements of G are invertible matrices, then the elements of its Lie algebra $\mathfrak{g}$ are also matrices, but in general they are singular.

It is to be remarked that the different powers c^k of the curve solution of the equation $\dot{c} = [a, c]$ satisfy the same equation as c itself, and therefore their traces are also constant along the curve. In other words, not only the trace of c but also all the coefficients of its characteristic polynomial are constants of motion [CI83].

The case of Lie–Scheffers systems is when instead of having a fundamental vector field $\mathrm{ad}\,(a)$ defined by an element $a \in \mathfrak{g}$ we have a linear combination, with functions of t as coefficients,

$$X = \sum_{\alpha=1}^{r} b_\alpha(t)\, \mathrm{ad}\,(a_\alpha),$$

of such vector fields. In this case, according to the recipe established in the general theory of Lie systems for finding the general solution, we should look for the curve $g(t)$ in G, starting from the neutral element, $g(0) = e$ and satisfying the equation

$$\dot{g}(t)\, g^{-1}(t) = \sum_{\alpha=1}^{r} b_\alpha(t)\, \mathrm{ad}\,(a_\alpha).$$

Once such a curve has been found, the solution determined by the initial condition $c(0)$ is given by

$$c(t) = \mathrm{Ad}\,(g(t))c(0),$$

and in the particular case in which G is a subgroup of the general linear group it reduces to

$$c(t) = g(t)\, c(0)\, g^{-1}(t)\,.$$

9.6.3 SODE Lie–Scheffers Systems

Newtonian equations of motion have the following form:

$$\ddot{x}^i = F^i(x, \dot{x}), \qquad i = 1, \ldots, n. \tag{9.75}$$

An associated system of $2n$ first-order differential equations is:

$$\begin{cases} \dot{x}^i &= v^i \\ \dot{v}^i &= F^i(x, v) \end{cases} \tag{9.76}$$

which can be understood as the system determining the integral curves of a vector field in $T\mathbb{R}^n$:

$$X = v^i \frac{\partial}{\partial x} + F^i(x, v) \frac{\partial}{\partial v^i}.$$

We can say that the system of SODEs (9.75) is a SODE Lie–Scheffers system if the system of first-order differential equations (9.76) obtained by adding the new variables $v^i = \dot{x}^i$, with $i = 1, \ldots, n$, to system (9.75), is a Lie–Scheffers system.

One of the simplest examples is today called the Milne-Pinney equation:

$$\ddot{x} = -\omega^2(t)x + \frac{k}{x^3}, \qquad x \in \mathbb{R}_+ = (0, \infty), \tag{9.77}$$

where k is a constant which in the interesting cases in physics is $k < 0$. Such an equation was introduced by a Ukrainian mathematician of the nineteenth century, V.P. Ermakov, as a way of looking for a first integral for the time-dependent harmonic oscillator [Er80] (the case $k = 0$ corresponds to the harmonic oscillator and the interval $(0, \infty)$ is replaced by the real line). The Milne-Pinney equation describes the time-evolution of an isotonic oscillator [Cal69, Pe90] (also-called pseudo-oscillator), i.e., an oscillator with an additional inverse quadratic potential. This oscillator shares with the harmonic one the property of having a period independent of the energy [Ch05], i.e., they are isochronous systems, and in the quantum case they have an equispaced spectrum [ACMP07].

As usual, we can relate the Milne-Pinney equation to a system of first-order differential equations defined in $T\mathbb{R}_+$ by introducing a new auxiliary variable $v \equiv \dot{x}$. Such a system is given by

$$\begin{cases} \dot{x} = v \\ \dot{v} = -\omega^2(t)x + \dfrac{k}{x^3} \end{cases}$$

where $x \in \mathbb{R}_+$ and $(x, v) \in T_x\mathbb{R}_+$. The associated t-dependent vector field over $T\mathbb{R}_+$,

$$X = v\frac{\partial}{\partial x} + \left(-\omega^2(t)x + \frac{k}{x^3}\right)\frac{\partial}{\partial v},$$

is a Lie–Scheffers system because X can be written as

$$X = L_2 - \omega^2(t)L_1,$$

where the vector fields L_1 and L_2 are given by

$$L_1 = x\frac{\partial}{\partial v}, \qquad L_2 = \frac{k}{x^3}\frac{\partial}{\partial v} + v\frac{\partial}{\partial x},$$

and are such that

$$[L_1, L_2] = 2L_3, \quad [L_3, L_2] = -L_2, \quad [L_3, L_1] = L_1$$

with

$$L_3 = \frac{1}{2}\left(x\frac{\partial}{\partial x} - v\frac{\partial}{\partial v}\right),$$

i.e., they span a 3-dimensional real Lie algebra $\mathfrak{g}$ isomorphic to $\mathfrak{sl}(2, \mathbb{R})$. More details can be found in [CL08, CLr08] and [CdLS12]. Other interesting examples, such as the second-order Riccati equation, have recently been studied [CL11].

9.6.4 Schrödinger–Pauli Equation

As an example of Lie–Scheffers systems in Quantum Mechanics, we can consider the non-relativistic dynamics of a spin 1/2 particle, when only the spinorial part is considered [CGM01]. The dynamics of such a particle in a time-dependent magnetic field $\mathbf{B}$ is described by the so-called Schrödinger–Pauli equation:

$$i\hbar\frac{d\,|\psi\rangle}{dt} = H|\psi\rangle = -\mu\,\mathbf{B}\cdot\mathbf{S}\,|\psi\rangle,$$

with the coefficient μ being proportional to the Bohr magneton, $\mathbf{B} = (B^1, B^2, B^3)$ the t-dependent magnetic field, and $S_i = \frac{\hbar}{2}\sigma_i$. The matrices $-i\sigma^1$, $-i\sigma^2$ and $-i\sigma^3$ generate the real Lie algebra of traceless skew-Hermitian 2×2 matrices, the Lie algebra of the group $SU(2, \mathbb{C})$ and therefore of $SO(3, \mathbb{R})$.

As a consequence of the theory developed in this chapter, in order to find the general solution of the evolution equation, it suffices to determine the curve $R(t)$ in $SO(3, \mathbb{R})$ such that $\dot{R}R^{-1} = B^3M_3 + B^2M_2 + B^1M_1$ and $R(0) = I$, where

$$M_1 = \begin{pmatrix} 0 & 0 & 0 \\ 0 & 0 & -1 \\ 0 & 1 & 0 \end{pmatrix}, \quad M_2 = \begin{pmatrix} 0 & 0 & 1 \\ 0 & 0 & 0 \\ -1 & 0 & 0 \end{pmatrix}, \quad M_3 = \begin{pmatrix} 0 & -1 & 0 \\ 1 & 0 & 0 \\ 0 & 0 & 0 \end{pmatrix}.$$

Such a curve gives us the general solution for the dynamics

$$|\psi(t)\rangle = \bar{R}(t)|\psi(0)\rangle,$$

where $\bar{R}$ is an element in $SU(2,\mathbb{C})$ corresponding to R.

9.6.5 Smorodinsky–Winternitz Oscillator

As another example consider the differential equation of an n-dimensional Smorodinsky–Winternitz oscillator of the form [WSUP67, CLS13]

$$\begin{cases} \dot{x}_i = p_i, \\ \dot{p}_i = -\varpi^2(t)x_i + \dfrac{k}{x_i^3}, \end{cases} \qquad i = 1, \ldots, n,$$

which describes the integral curves of the t-dependent vector field on $T^*\mathbb{R}^n$

$$X_t = \sum_{i=1}^{n} \left[p_i \frac{\partial}{\partial x_i} + \left(-\varpi^2(t)x_i + \frac{k}{x_i^3} \right) \frac{\partial}{\partial p_i} \right].$$

Note that X_t can be written as $X_t = X_2 + \varpi^2(t) X_1$ with X_1, X_2 and $X_3 = -[X_1, X_2]$ being given by

$$X_1 = -\sum_{i=1}^{n} x_i \frac{\partial}{\partial p_i}, \quad X_2 = \sum_{i=1}^{n} \left(p_i \frac{\partial}{\partial x_i} + \frac{k}{x_i^3} \frac{\partial}{\partial p_i} \right), \quad X_3 = \sum_{i=1}^{n} \frac{1}{2} \left(x_i \frac{\partial}{\partial x_i} - p_i \frac{\partial}{\partial p_i} \right).$$

Therefore, X_t is a Lie–Scheffers system, because X_1, X_2 and X_3 close on a $\mathfrak{sl}(2,\mathbb{R})$ algebra:

$$[X_1, X_2] = -2X_3, \qquad [X_1, X_3] = -X_1, \qquad [X_2, X_3] = X_2.$$

Moreover, the preceding vector fields are Hamiltonian vector fields with respect to the usual symplectic form $\omega_0 = \sum_{i=1}^{n} dx^i \wedge dp_i$ with Hamiltonian functions

$$h_1 = \frac{1}{2} \sum_{i=1}^{n} x_i^2, \qquad h_2 = \frac{1}{2} \sum_{i=1}^{n} \left(p_i^2 + \frac{k}{x_i^2} \right), \qquad h_3 = \frac{1}{2} \sum_{i=1}^{n} x_i p_i,$$

which satisfy that

$$\{h_1, h_2\} = 2h_3, \qquad \{h_1, h_3\} = h_1, \qquad \{h_2, h_3\} = -h_2.$$

Consequently, every curve h_t that takes values in the three-dimensional Lie algebra $(W, \{\cdot, \cdot\})$ spanned by h_1, h_2 and h_3 gives raise to a Lie–Scheffers system in $T^*\mathbb{R}^n$ which is Hamiltonian with respect to the symplectic structure ω_0 in such a way that the t-dependent vector field is given by

$$X_t = X_2 + \varpi^2(t)X_1 = \widehat{\omega}_0^{-1}(dh_2 + \varpi^2(t)dh_1),$$

i.e., the Hamiltonian is $h_t = h_2 + \varpi^2(t)h_1$.

This suggests that we consider Lie–Scheffers systems admitting a similar Hamiltonian description.

9.7 Hamiltonian Systems of Lie–Scheffers Type

An important case of Lie–Scheffers systems is when (M, Ω) is a symplectic manifold and the vector fields in M arising in the expression of the t-dependent vector field describing a Lie–Scheffers system are Hamiltonian vector fields closing on a finite-dimensional real Lie algebra $\mathfrak{g}$ [CLS13]. When these vector fields are complete, they are fundamental vector fields of a symplectic action of a Lie group G with Lie algebra $\mathfrak{g}$ on the symplectic manifold (M, Ω).

The corresponding Hamiltonian functions h_α, defined by $i(X_\alpha)\Omega = -dh_\alpha$, do not close in general on the same Lie algebra $\mathfrak{g}$ when the Poisson bracket is considered, but we can only assure that

$$d\left(\{h_\alpha, h_\beta\} - h_{[\alpha,\beta]}\right) = 0\,,$$

i.e., that there exist real constants $\lambda_{\alpha\beta}$ such that

$$\{h_\alpha, h_\beta\} - h_{[\alpha,\beta]} = \lambda_{\alpha\beta}.$$

Therefore, the functions h_α span a Lie algebra extension of the original one.

The situation in Quantum Mechanics is quite similar: we have analyzed the geometry of a Hilbert space $\mathcal{H}$ when considered as a real manifold with a global chart in Sect. 6.4.4. The tangent space $T_\phi\mathcal{H}_\mathbb{R}$ at any point $\phi \in \mathcal{H}$ can be identified with $\mathcal{H}_\mathbb{R}$ itself, the isomorphism associating $\psi \in \mathcal{H}_\mathbb{R}$ with the vector $X_\psi \in T_\phi\mathcal{H}_\mathbb{R}$ being given by:

$$(X_\psi f)(\phi) := \frac{d}{dt} f(\phi + t\psi)\,|_{t=0}, \qquad \forall f \in C^\infty(\mathcal{H}).$$

We also recall that the Hilbert space $\mathcal{H}_{\mathbb{R}}$ as a real manifold is endowed with a symplectic 2-form Ω defined by

$$\Omega_\phi(\dot{\psi}, \dot{\psi}') = 2\,\mathrm{Im}\langle\psi|\psi'\rangle.$$

A vector field is just a map $A\colon \mathcal{H}_{\mathbb{R}} \to \mathcal{H}_{\mathbb{R}}$; in particular a linear operator A on $\mathcal{H}$ is a special kind of vector field. Given a smooth function $a\colon \mathcal{H} \to \mathbb{R}$, its differential da_ϕ at $\phi \in \mathcal{H}$ is an element of the (real) dual $\mathcal{H}'_{\mathbb{R}}$ given by

$$\langle da_\phi, \psi\rangle := \left(\frac{d}{dt}a(\phi + t\psi)\right)_{|t=0}.$$

Actually, the skew-self-adjoint linear operators $-iA$ in $\mathcal{H}$, that is, A is a self-adjoint operator, define Hamiltonian vector fields, the Hamiltonian function of $-iA$ being $a(\phi) = \frac{1}{2}\langle\phi, A\phi\rangle$. Therefore, the Schrödinger equation plays the rôle of Hamilton equations, because it determines the integral curves of the vector field $-iH$, where H is the Hamiltonian of the system [BCG91].

In particular, the theory of Lie–Scheffers systems applies in the previous framework of Quantum Mechanics when we have a t-dependent quantum Hamiltonian that can be written as a linear combination, with t-dependent coefficients, of Hamiltonians H_α, with $\alpha = 1, \dots, r$, closing on a finite-dimensional real Lie algebra under the commutator bracket. However, note that as indicated above this Lie algebra does not necessarily coincide with that of the corresponding classical problem, but it may be a Lie algebra extension of the latter.

For the illustration of these classical and quantum Lie–Scheffers systems of Hamiltonian type we consider an important example: the time-dependent classical and quantum quadratic Hamiltonians.

The configuration space for the classical system is the real line $\mathbb{R}$, the corresponding phase space $T^*\mathbb{R}$ being endowed with its canonical symplectic structure $\omega_0 = dq \wedge dp$, and the time-dependent classical Hamiltonian function is

$$H = \alpha(t)\,\frac{p^2}{2} + \beta(t)\,\frac{qp}{2} + \gamma(t)\,\frac{q^2}{2} + \delta(t)\,p + \epsilon(t)\,q. \tag{9.78}$$

The dynamical vector field solution of the dynamical equation $i(\Gamma_H)\,\omega_0 = dH$ is given by

$$\Gamma_H = \left(\alpha(t)\,p + \frac{1}{2}\beta(t)\,q + \delta(t)\right)\frac{\partial}{\partial q} - \left(\frac{1}{2}\beta(t)\,p + \gamma(t)\,q + \epsilon(t)\right)\frac{\partial}{\partial p}, \tag{9.79}$$

which can be rewritten as

$$\Gamma_H = \alpha(t)\,X_1 + \beta(t)\,X_2 + \gamma(t)\,X_3 - \delta(t)\,X_4 + \epsilon(t)\,X_5,$$

where the vector fields

$$X_1 = p\frac{\partial}{\partial q},\ X_2 = \frac{1}{2}\left(q\frac{\partial}{\partial q} - p\frac{\partial}{\partial p}\right),\ X_3 = -q\frac{\partial}{\partial p},\ X_4 = -\frac{\partial}{\partial q},\ X_5 = -\frac{\partial}{\partial p},$$

satisfy the following commutation relations:

$$\begin{aligned} &[X_1, X_2] = X_1, \quad [X_1, X_3] = 2X_2, \quad [X_1, X_4] = 0, \quad [X_1, X_5] = -X_4, \\ &[X_2, X_3] = X_3, \quad [X_2, X_4] = -\frac{1}{2}X_4, \quad [X_2, X_5] = \frac{1}{2}X_5, \\ &[X_3, X_4] = X_5, \quad [X_3, X_5] = 0, \quad [X_4, X_5] = 0, \end{aligned} \tag{9.80}$$

and therefore they close on a five-dimensional real Lie algebra. Consider the abstract, five-dimensional, Lie algebra $\mathfrak{g}$ such that in a basis $\{a_1, a_2, a_3, a_4, a_5\}$, the Lie products are analogous to that of (9.80). Then, the Lie algebra $\mathfrak{g}$ is a semi-direct sum of the Abelian two-dimensional Lie algebra generated by $\{a_4, a_5\}$ with the $\mathfrak{sl}(2, \mathbb{R})$ Lie algebra generated by $\{a_1, a_2, a_3\}$, i.e., $\mathfrak{g} = \mathbb{R}^2 \rtimes \mathfrak{sl}(2, \mathbb{R})$. The corresponding Lie group will be the semi-direct product $G = T_2 \rtimes SL(2, \mathbb{R})$ relative to the linear action of $SL(2, \mathbb{R})$ on the two-dimensional translation algebra. When computing the flows of the previous vector fields X_α, we see that they correspond to the affine action of G on $\mathbb{R}^2$, and therefore, the vector fields X_α can be regarded as fundamental fields associated to the previous basis of the Lie algebra with respect to the affine action.

In order to find the integral curves of the time-dependent vector field (9.79), we can solve first the corresponding equation in the Lie group G and then use the affine action of G on $\mathbb{R}^2$. We should find the curve $g(t)$ in G such that

$$\dot{g}g^{-1} = -\sum_{i=1}^{5} b_i(t)a_i, \qquad g(0) = e, \tag{9.81}$$

with $b_1(t) = \alpha(t)$, $b_2(t) = \beta(t)$, $b_3(t) = \gamma(t)$, $b_4(t) = -\delta(t)$, and $b_5(t) = \epsilon(t)$. The explicit calculation can be carried out by using the generalized Wei-Norman method [WN63, WN64], i.e., writing $g(t)$ in terms of a set of second class canonical coordinates, for instance,

$$g(t) = \exp(-v_4(t)a_4)\exp(-v_5(t)a_5)\exp(-v_1(t)a_1)\exp(-v_2(t)a_2)\exp(-v_3(t)a_3),$$

and then, equation (9.81) leads to the system

$$\begin{aligned} &\dot{v}_1 = b_1 + b_2 v_1 + b_3 v_1^2, \quad \dot{v}_2 = b_2 + 2b_3 v_1, \quad \dot{v}_3 = e^{v_2} b_3, \\ &\dot{v}_4 = b_4 + \frac{1}{2}b_2 v_4 + b_1 v_5, \quad \dot{v}_5 = b_5 - b_3 v_4 - \frac{1}{2}b_2 v_5, \end{aligned}$$

with initial conditions $v_1(0) = \cdots = v_5(0) = 0$.

For some specific choices of the functions $\alpha(t), \ldots, \epsilon(t)$, the problem becomes simpler and it may be enough to consider a subgroup, instead of the whole Lie group G, to deal with the arising system. For instance, the only non-vanishing coefficients of the classical Hamiltonian

$$H = \frac{p^2}{2m} + f(t)\,q,$$

are $\alpha(t) = 1/m$ and $\epsilon(t) = f(t)$, and then the problem is reduced to one in a three-dimensional subalgebra, generated by $\{X_1, X_4, X_5\}$. The associated Lie group will be the subgroup of G generated by $\{a_1, a_4, a_5\}$. The main point is that such a subgroup is solvable and therefore the problem can be integrated by quadratures.

Another remarkable property is that the Hamiltonian functions h_α corresponding to the Hamiltonian vector fields $X_1, \ldots, X_5$, defined by $i(X_\alpha)\omega = -dh_\alpha$, i.e.

$$h_1(q,p) = -\frac{p^2}{2}, \quad h_2(q,p) = -\frac{1}{2}q\,p, \quad h_3(q,p) = -\frac{q^2}{2}, \quad h_4(q,p) = p, \quad h_5(q,p) = -q,$$

have almost the same Poisson bracket relations as the vector fields X_α, but they do not coincide because of $\{h_4, h_5\} = 1$, instead of $[X_4, X_5] = 0$. In other words, they close on a Lie algebra which is a central extension of $\mathbb{R}^2 \rtimes \mathfrak{sl}(2, \mathbb{R})$ by a one-dimensional algebra.

Let us now consider the quantum case [Wo80], with applications in a number of physical problems, as for instance, the quantum motion of charged particles subject to time-dependent electromagnetic fields (see, e.g., [FMM94]), and connects with the theory of exact invariants developed by Lewis and Riesenfeld (see [LR69] and references therein).

A generic time-dependent quadratic quantum Hamiltonian is given by

$$H = \alpha(t)\,\frac{P^2}{2} + \beta(t)\,\frac{QP + PQ}{4} + \gamma(t)\,\frac{Q^2}{2} + \delta(t)P + \epsilon(t)\,Q + \phi(t)I, \tag{9.82}$$

where Q and P are the position and momentum operators satisfying the commutation relation

$$[Q, P] = i\,I.$$

The previous Hamiltonian can be written as a sum with t-dependent coefficients

$$H = \alpha(t)H_1 + \beta(t)H_2 + \gamma(t)H_3 - \delta(t)H_4 + \epsilon(t)H_5 - \phi(t)H_6,$$

of the Hamiltonians

$$H_1 = \frac{P^2}{2}, \ H_2 = \frac{1}{4}(QP + PQ), \ H_3 = \frac{Q^2}{2}, \ H_4 = -P, \ H_5 = Q, \ H_6 = -I,$$

which satisfy the commutation relations

$$[iH_1, iH_2] = iH_1\,, \quad [iH_1, iH_3] = 2iH_2\,, \quad [iH_1, iH_5] = -iH_4\,, \quad [iH_2, iH_3] = iH_3\,,$$
$$[iH_2, iH_4] = -\frac{i}{2}\,H_4\,, \quad [iH_2, iH_5] = \frac{i}{2}\,H_5\,, \quad [iH_3, iH_4] = iH_5\,, \quad [iH_4, iH_5] = iH_6\,,$$

and $[iH_1, iH_4] = [iH_3, iH_5] = [iH_\alpha, iH_6] = 0$, $\alpha = 1, \dots, 5$. That is, the skew-self-adjoint operators iH_α generate a six-dimensional real Lie algebra which is a central extension of the Lie algebra arising in the classical case, $\mathbb{R}^2 \rtimes \mathfrak{sl}(2, \mathbb{R})$, by a one-dimensional Lie algebra. It can be identified as the semi-direct sum of the Heisenberg–Weyl Lie algebra $\mathfrak{h}(3)$, which is an ideal in the total Lie algebra, with the Lie subalgebra $\mathfrak{sl}(2, \mathbb{R})$, i.e. $\mathfrak{h}(3) \rtimes \mathfrak{sl}(2, \mathbb{R})$. Sometimes this Lie algebra is referred to as the extended symplectic Lie algebra $\mathfrak{hsp}(2, \mathbb{R}) = \mathfrak{h}(3) \rtimes \mathfrak{sp}(2, \mathbb{R})$. The corresponding Lie group is the semi-direct product $H(3) \odot SL(2, \mathbb{R})$ of the Heisenberg–Weyl group $H(3)$ with $SL(2, \mathbb{R})$, see also [Wo80].

The time-evolution of a quantum system can be described in terms of the evolution operator $U(t)$ which satisfies the Schrödinger equation (see, e.g., [CDL77])

$$i\frac{dU}{dt} = H(t)U\,, \qquad U(0) = \mathrm{Id}\,,$$

where $H(t)$ is the Hamiltonian of the system. In our current case, the Hamiltonian is given by (9.82), and therefore the time-evolution of the system is given by an equation of the type

$$\dot{g}g^{-1} = -\sum_{\alpha=1}^{6} b_\alpha(t)\, a_\alpha, \qquad g(0) = e, \tag{9.83}$$

with the identification of $g(t)$ with $U(t)$, e with Id, iH_α with a_α for $\alpha \in \{1, \dots, 6\}$ and the time-dependent coefficients $b_\alpha(t)$ are given by

$$\begin{aligned} &b_1(t) = \alpha(t)\,,\quad b_2(t) = \beta(t)\,,\quad b_3(t) = \gamma(t),\\ &b_4(t) = -\delta(t)\,,\quad b_5(t) = \epsilon(t)\,,\quad b_6(t) = -\phi(t). \end{aligned}$$

We would like to remark that time-dependent quantum Hamiltonians are seldom studied, because it is generally difficult to find their time evolution. However, in the case where the system could be treated as a Lie–Scheffers system in a certain Lie group, the calculation of the evolution operator is reduced to the problem of integrating the system appearing after application of the Wei-Norman method. In the case where the associated Lie group is solvable, the integration can be made by quadratures, leading to an exact solution of the problem.

The calculation of the solution of (9.83) can be carried out by using the generalized Wei–Norman method, i.e. writing $g(t)$ in terms of a set of second class canonical coordinates. We take, for instance, the factorization

$$\begin{aligned} g(t) = &\exp(-v_4(t)a_4)\exp(-v_5(t)a_5)\exp(-v_6(t)a_6)\\ &\times \exp(-v_1(t)a_1)\exp(-v_2(t)a_2)\exp(-v_3(t)a_3) \end{aligned}$$

and therefore, the Eq. (9.83) leads in this case to the system

$$\dot{v}_1 = b_1 + b_2 v_1 + b_3 v_1^2, \quad \dot{v}_2 = b_2 + 2b_3 v_1, \quad \dot{v}_3 = e^{v_2} b_3,$$
$$\dot{v}_4 = b_4 + \frac{1}{2} b_2 v_4 + b_1 v_5, \quad \dot{v}_5 = b_5 - b_3 v_4 - \frac{1}{2} b_2 v_5,$$
$$\dot{v}_6 = b_6 + b_5 v_4 - \frac{1}{2} b_3 v_4^2 + \frac{1}{2} b_1 v_5^2,$$

with initial conditions $v_1(0) = \cdots = v_6(0) = 0$.

Analogously to what happened in the classical case, special choices of the time-dependent coefficient functions $\alpha(t), \ldots, \phi(t)$, may lead to problems for which the associated Lie algebra is a subalgebra of that of the complete system, and similarly for the Lie groups involved. For example, we could consider as well the quantum Hamiltonian linear in the positions

$$H = \frac{P^2}{2m} + f(t)\, Q,$$

which in the notation of (9.82) has the only non-vanishing coefficients $\alpha(t) = 1/m$ and $\epsilon(t) = f(t)$. This problem can be regarded as a Lie–Scheffers system associated to the four-dimensional Lie algebra generated by $\{iH_1, iH_4, iH_5, iH_6\}$, which is also solvable, and hence the problem can be solved by quadratures.

9.8 A Generalization of Lie–Scheffers Systems

The theory of Lie–Scheffers systems has been generalized to deal with more general cases, as for instance the Abel equation:

$$\dot{x} = A_0(t) + A_1(t)x + A_2(t)x^2 + A_3(t)x^3.$$

Note that then the linear space

$$V_{\text{Abel}}(\mathbb{R}) = \langle \frac{\partial}{\partial x}, x\frac{\partial}{\partial x}, x^2\frac{\partial}{\partial x}, x^3\frac{\partial}{\partial x} \rangle$$

is not a finite-dimensional real Lie algebra, because

$$\left[x^2\frac{\partial}{\partial x}, x^3\frac{\partial}{\partial x}\right] = x^4\frac{\partial}{\partial x}.$$

Note also that the vector fields of the Lie algebra in the hypothesis of the Lie–Scheffers Theorem play a double role, because on one side they generate a Lie algebra with an associated Lie group G, and on the other one, they explicitly appear in the dynamics.

The important fact is that the group $\mathcal{G}$ of curves in G transforms an element of the Lie family of systems into another one, and this fact may be used to simplify the problem. For instance, the transformed vector field may be included in a Lie subalgebra of the given Lie algebra and therefore it is a simpler 'Lie–Scheffers system'.

This fact admits the following generalization:

- The dynamics is, as before, $X = \sum_0^r b_\alpha X_\alpha$, but the vector fields X_α do not close a finite-dimensional Lie subalgebra of vector fields anymore but only span a linear space
- There is a Lie subalgebra W of (complete) vector fields $\{Y_a \mid a = 1, \dots, l\}$ such that

$$[Y_a, X_\alpha] = \sum_{\beta=1}^{r} c_{a\alpha}{}^{\beta} X_\beta.$$

In this case the group $\mathcal{G}$ of curves in the group corresponding to the Lie algebra also transforms each element of the Lie family of systems into another one. Appropriate curves can lead to simpler systems, for instance when the image of the given dynamics is included in a finite-dimensional Lie algebra, reducing the system to a Lie–Scheffers system

As an instance, for the Abel equation [CLR11] we can consider the Lie algebra

$$W_{\text{Abel}}(\mathbb{R}) = \text{span}\left\{\frac{\partial}{\partial x}, x\frac{\partial}{\partial x}\right\}.$$

which leaves invariant $V_{\text{Abel}}(\mathbb{R})$ and then the subgroup of transformations

$$x(t) = \alpha(t)\,\bar{x}(t) + \beta(t), \qquad \alpha(t) \neq 0,$$

is a structure preserving group and we can take advantage of such transformations to transform a given equation into another simpler one.

Under such a transformation the given Abel equation becomes a new Abel equation

$$\dot{\bar{x}} = \bar{A}_0(t) + \bar{A}_1(t)\bar{x} + \bar{A}_2(t)\bar{x}^2 + \bar{A}_3(t)\bar{x}^3,$$

with

$$\begin{aligned}
\bar{A}_3(t) &= A_3(t)\alpha^2(t),\\
\bar{A}_2(t) &= \alpha(t)(3A_3(t)\beta(t) + A_2(t)),\\
\bar{A}_1(t) &= 3A_3(t)\beta^2(t) + 2A_2(t)\beta(t) + A_1(t) - \dot{\alpha}(t)\alpha^{-1}(t),\\
\bar{A}_0(t) &= \alpha^{-1}(t)\left(A_3(t)\beta^3(t) + A_2(t)\beta^2(t) + A_1(t)\beta(t) + A_0(t) - \dot{\beta}(t)\right).
\end{aligned}$$

The action of the group of curves in the affine group produces orbits, i.e., equivalence classes of Abel equations, with the same properties of integrability or existence of superposition rules. For instance, Abel equations of an orbit containing an integrable by quadratures Riccati equation are also integrable by quadratures. The orbits are characterized by different values of invariant functions.

One must determine the Lie algebras contained in $V_{\text{Abel}}(\mathbb{R})$ reachable by the set of transformations (the coefficient of X_3 cannot be transformed to be zero). It is possible to check that the only subalgebra of dimension three is the one of the Riccati equation which is not reachable by the considered set of transformations.

There is a one-parameter family of two-dimensional subalgebras of $V_{\mathrm{Abel}}(\mathbb{R})$ reachable from a proper Abel equation, those generated by

$$(-2\mu^3+3\mu x^2+x^3)\partial_x,\qquad (\mu+x)\partial_x,\quad \mu\in\mathbb{R}.$$

These are integrable by two quadratures. For instance the case $\mu=0$ corresponds to Bernoulli equation. Finally, the Lie algebras generated by constant vector fields

$$(d_0+d_1x+d_2x^2+x^3)\partial_x,\qquad d_i\in\mathbb{R}$$

lead to separable differential equations, therefore integrable by one quadrature.

One can proceed in a similar way with the Gambier equation [CGL13], which can be described as the coupling of two Riccati equations:

$$\begin{cases}\dfrac{dy}{dt} = -\,y^2+a_1y+a_2,\\[4pt] \dfrac{dx}{dt} = a_0x^2+nyx+\sigma,\end{cases}$$

where n is an integer, σ is a constant, which can be scaled to 1 unless it happens to be 0, and a_0, a_1, a_2 are certain functions depending on time.

If $n\neq 0$, we can eliminate y between the two equations above, which gives raise to the one referred to as second-order Gambier equation:

$$\frac{d^2x}{dt^2}=\frac{n-1}{xn}\left(\frac{dx}{dt}\right)^2+a_0\frac{(n+2)}{n}x\frac{dx}{dt}+a_1\frac{dx}{dt}-\sigma\frac{(n-2)}{nx}\frac{dx}{dt}$$
$$-\frac{a_0^2}{n}x^3+\left(\frac{da_0}{dt}-a_0a_1\right)x^2+\left(a_2n-2a_0\frac{\sigma}{n}\right)x-a_1\sigma-\frac{\sigma^2}{nx}.$$

Particular examples are the second-order Riccati equation ($n=1$ and $\sigma=0$):

$$\frac{d^2x}{dt^2}=(a_1+3a_0x)\frac{dx}{dt}-a_0^2x^3+\left(\frac{da_0}{dt}-a_0a_1\right)x^2+a_2x,$$

and the second-order Kummer–Schwarz equation ($n=-2$, $a_1=\sigma=0$):

$$\frac{d^2x}{dt^2}=\frac{3}{2x}\left(\frac{dx}{dt}\right)^2-2c_0x^3+2\omega(t)x.$$

Coming back to the general case, the vector fields on $T\mathbb{R}_0$, with $\mathbb{R}_0\equiv\mathbb{R}-\{0\}$, involved are:

$$Y_1=v\frac{\partial}{\partial x},\qquad Y_2=\frac{v^2}{x}\frac{\partial}{\partial v},\qquad Y_3=xv\frac{\partial}{\partial v},\qquad Y_4=v\frac{\partial}{\partial v},\qquad Y_5=\frac{v}{x}\frac{\partial}{\partial v},$$
$$Y_6=x^3\frac{\partial}{\partial v},\qquad Y_7=x^2\frac{\partial}{\partial v},\qquad Y_8=x\frac{\partial}{\partial v},\qquad Y_9=\frac{\partial}{\partial v},\qquad Y_{10}=\frac{1}{x}\frac{\partial}{\partial v},$$

to be completed with the vector field

$$Y_{11} = x\frac{\partial}{\partial x}.$$

Then,

$$X = \sum_{\alpha=1}^{11} b_\alpha\left(a_0, \frac{da_0}{dt}, a_1, a_2, \sigma, n\right) Y_\alpha,$$

where $b_{11} = 0$ and

$$b_1 = 1, \quad b_2 = \frac{n-1}{n}, \quad b_3 = a_0\frac{n+2}{n}, \quad b_4 = a_1, \quad b_5 = -\sigma\frac{n-2}{n},$$

$$b_6 = -\frac{a_0^2}{n}, \quad b_7 = \frac{da_0}{dt} - a_0 a_1, \quad b_8 = a_2 n - 2a_0\frac{\sigma}{n}, \quad b_9 = -a_1\sigma, \quad b_{10} = -\frac{\sigma^2}{n}.$$

Here the important Lie algebra W is the one generated by Y_4, Y_8 and Y_{11}:

$$[Y_4, Y_8] = -Y_8, \qquad [Y_4, Y_{11}] = 0, \qquad [Y_8, Y_{11}] = -Y_8.$$

More details can be found in [CGL13].

References

[Lie93] Lie, S., Scheffers, G.: Vorlesungen über continuierliche Gruppen mit geometrischen und anderen Anwendungen. Teubner, Leipzig (1893) (Edited and revised by G. Scheffers)

[Ca00] Cariñena, J.F., Grabowski, J., Marmo, G.: Lie-Scheffers Systems: A Geometric Approach. Bibliopolis, Napoli (2000)

[Ca07b] Cariñena, J.F., Grabowski, J., Marmo, G.: Superposition rules, Lie theorem, and partial differential equations. Rep. Math. Phys. **60**, 237–258 (2007)

[CL11] Cariñena, JF., de Lucas, J.: Lie systems: theory, generalisations, and applications. Dissertationes Mathematicae 479, Institute of Mathematics, Polish Academy of Sciences, Warszawa (2011)

[Sc02] Schwarz, F.: Equivalence classes, symmetries and solutions of linear third-order differential equations. Computing **69**, 141–162 (2002)

[Nd08] Ndogmo, J.C.: A method for the equivalence group and its infinitesimal generators. J. Phys. A: Math. Theor. **41**, 102001 (2008)

[WN63] Wei, J., Norman, E.: Lie algebraic solution of linear differential equations. J. Math. Phys. **4**, 575–581 (1963)

[WN64] Wei, J., Norman, E.: On global representations of the solutions of linear differential equations as a product of exponentials. Proc. Amer. Math. Soc. **15**, 327–334 (1964)

[Ca98] Cariñena, J.F., Marmo, G., Nasarre, J.: The nonlinear superposition principle and the Wei-Norman method. Int. J. Mod. Phys. **A13**, 3601–3627 (1998)

[Ca01] Cariñena, J.F., Grabowski, J., Ramos, A.: Reduction of time-dependent systems admitting a superposition principle. Act. Appl. Math. **66**, 67–87 (2001)

[CR02] Cariñena, J.F., Ramos, A.: A new geometric approach to Lie systems and physical applications. Acta Appl. Math. **70**, 43–69 (2002)

[Ca99] Cariñena, J.F., Ramos, A.: Integrability of the Riccati equation from a group theoretical viewpoint. Int. J. Modern Phys. **A14**, 1935–1951 (1999)

[Co95] Cooper, F., Khare, A., Sukhatme, U.: Supersymmetry and quantum mechanics. Phys. Rep. **251**, 267–385 (1995)

[In51] Infeld, L., Hull, T.E.: The factorization method. Rev. Mod. Phys. **23**, 21–68 (1951)

[CR00] Cariñena, J.F., Ramos, A.: Riccati equation, factorization method and shape invariance. Rev. Math. Phys. **12**, 1279–1304 (2000)

[CRF01] Cariñena, J.F., Fernández, D.J., Ramos, A.: Group theoretical approach to the intertwined Hamiltonians. Ann. Phys. **292**, 42–66 (2001)

[CR00b] Cariñena, J.F., Ramos, A.: The partnership of potentials in Quantum Mechanics and shape invariance. Mod. Phys. Lett. **A15**, 1079–1088 (2000)

[La68] Lax, P.D.: Integrals of nonlinear equations of evolution and solitary waves. Comm. Pure Appl. Math. **21**, 467–490 (1968)

[CI85] Cariñena, J.F., Ibort, L.A.: Geometrical setting for Lax equations associated to dynamical systems. Phys. Lett. **107A**, 356–358 (1985)

[CM94] Cariñena, J.F., Martínez, E.: A new geometric setting for Lax equations. Int. J. Mod. Phys. **A9**, 4973–4986 (1994)

[CI83] Cariñena, J.F., Ibort, L.A.: Non-Noether constants of motion. J. Phys. A: Math. Gen. **16**, 1–7 (1983)

[Er80] Ermakov, V.P.: Second-order differential equations. Conditions of complete integrability, Univ. Isz. Kiev Series III **9**, 1–25 (1880) (translation by A.O. Harin)

[Cal69] Calogero, F.: Solution of a three body problem in one dimension. J. Math. Phys. **10**, 2191–2196 (1969)

[Pe90] Perelomov, A.M.: Integrable Systems of Classical Mechanics and Lie Algebras. Birkhauser, Basel (1990)

[Ch05] Chalykh, O.A., Vesselov, A.P.: A remark on rational isochronous potentials. J. Nonlin. Math. Phys. **12**(Suppl. 1), 179–183 (2005)

[ACMP07] Asorey, M., Cariñena, J.F., Marmo, G., Perelomov, A.M.: Isoperiodic classical systems and their quantum counterparts. Ann. Phys. **322**, 1444–1465 (2007)

[CL08] Cariñena, J.F., de Lucas, J.: A nonlinear superposition rule for solutions of the Milne-Pinney equation. Phys. Lett. **A372**, 5385–5389 (2008)

[CLr08] Cariñena, J.F., de Lucas, J., Rañada, M.F.: Recent applications of the theory of Lie systems in Ermakov systems. SIGMA **4**, 031 (2008)

[CdLS12] Cariñena, J.F., de Lucas, J., Sardón, C.: A new Lie systems approach to second-order Riccati equation. Int. J. Geom. Methods Mod. Phys. **9**, 1260007 (2012)

[CL11] Cariñena, J.F., de Lucas, J.: Superposition rules and second-order Riccati equations. J. Geom. Mech. **3**, 1–22 (2011)

[CGM01] Cariñena, J.F., Grabowski, J., Marmo, G.: Some applications in physics of differential equation systems admitting a superposition rule. Rep. Math. Phys. **48**, 47–58 (2001)

[WSUP67] Winternitz, P., Smorodinskii, Y.A., Uhlir, M., Fris, J.: Symmetry groups in classical and Quantum Mechanics. Sov. J. Nucl. Phys. **4**, 444–450 (1967)

[CLS13] Cariñena, J.F., de Lucas, J., Sardón, C.: Lie-Hamilton systems: theory and applications. Int. J. Geom. Methods Mod. Phys. **10**, 1350047 (2013)

[BCG91] Boya, L.J., Cariñena, J.F., Gracia-Bondía, J.M.: Symplectic structure of the Aharonov-Anandan geometric phase. Phys. Lett. **161A**, 30–34 (1991)

[Wo80] Wolf, K.B.: On time-dependent quadratic quantum Hamiltonians. SIAM J. Appl. Math. **40**, 419–431 (1980)

[FMM94] Fernández, D.J., Mielnik, B.: Controlling quantum motion. J. Math. Phys. **35**, 2083–2104 (1994)

[LR69] Lewis Jr, H.R., Riesenfeld, W.B.: An exact quantum theory of the time-dependent harmonic oscillator and of a charged particle in a time-dependent electromagnetic field. J. Math. Phys. **10**, 1458–1473 (1969)

[CDL77] Cohen-Tannoudji, C., Diu, B., Laloë, F.: Mécanique Quantique, vol. I. Hermann, Paris (1977)

[CLR11] Cariñena, J.F., de Lucas, J., Rañada, M.F.: A geometric approach to integrability of Abel differential equations. Int. J. Theor. Phys. **50**, 2114–2124 (2011)

[CGL13] Cariñena, J.F., Guha, P., de Lucas, J.: A quasi-Lie schemes approach to second-order Gambier equations. SIGMA **9**, 026 (2013)

Chapter 10
Appendices

Appendix A: Glossary of Mathematical Terms

In this appendix we summarize the most fundamental and common mathematical notions that appear everywhere on the book refereeing the reader to the quoted literature for further information.

A.1 Glossary of Algebraic Terms

Groups

Definition 10.1 A group G is a set together with a binary composition law $\circ \colon G \times G \to G$ such that it has a neutral element e ($e \circ g = g \circ e = g$), is associative $(g \circ h) \circ k = g \circ (h \circ k)$, and every element $g \in G$ has an inverse g^{-1} ($g \circ g^{-1} = g^{-1} \circ g = e$), for all $g, h, k \in G$.

A group G is Abelian or commutative if $g \circ h = h \circ g$ for all $g, h \in G$. It is frequent to omit the symbol $\circ$ (unless there is risk of confusion) when writing the composition $g \circ h$ of two elements $g, h \in G$ that will written in what follows as gh.

Examples of groups are provided by the collection $S(X)$ of all bijective maps $\alpha \colon X \to X$ of an arbitrary set X with the binary operation the composition law of maps. If X is finite with n elements $S(X)$ is called the group of permutations of order n and denoted by S_n.

A set S with a composition law $*$ verifying only the associativity and the existence of a neutral element is called a semigroup.[1]

[1] A fundamental algebraic structure extending the notion of group (even though it will not be used in this book) is the notion of groupoid that we will skip here.

J.F. Cariñena et al., *Geometry from Dynamics, Classical and Quantum*,
DOI 10.1007/978-94-017-9220-2_10

Definition 10.2 A subgroup H of a group G is a subset $H \subset G$ such that $hk \in H$ and $h^{-1} \in H$ for any $h, k \in H$. A subgroup H of G is called normal if $ghg^{-1} \in H$ for any $g \in G$ and $h \in H$.

Given a group G and a subgroup H, the quotient G/H denotes the family of subsets $gH = \{gh \mid h \in H\}$, $g \in G$, called (right) cosets. If H is a normal subgroup of G the set G/H becomes a group with the induced composition law $(gH) \circ (g'H) = (gg')H$.

Definition 10.3 Given two groups G, G', a group homomorphism from G to G' is a map $\varphi\colon G \to G'$ such that $\varphi(gg') = \varphi(g)\varphi(g')$, for all $g, g' \in G$.

The set $\{g \in G \mid \varphi(g) = e'\}$ is called the kernel of the homomorphism φ and is always a normal subgroup of G. It will be denoted by $\ker \varphi$. The range of the homomorphism φ is a subgroup of G' and it will be denoted by Im φ.

Definition 10.4 A group homomorphism $\varphi\colon G \to G'$ is called a monomorphism if it is injective; epimorphism if it is surjective and isomorphism if it is bijective. If φ if a group isomorphism the inverse φ^{-1} is a group homomorphism too, hence a group isomorphism. The set of all group isomorphisms $\varphi\colon G \to G$ is called the group of automorphisms of G and denoted Aut(G).

Given a group homomorphism $\varphi\colon G \to G'$, if we denote by π the canonical projection $\pi : G \to G/\ker G$ given by $\pi(g) = g \ker \varphi$, by $\bar{\varphi}\colon G/\ker \varphi \to \mathrm{Im}\,\varphi$, the map given by $\overline{\varphi}(g \ker \varphi) = \varphi(g)$, and by $i\colon \mathrm{Im}\,\varphi \to G'$ the canonical inclusion, then π is a group epimorphism, $\overline{\varphi}$ is a group isomorphism, i is a group monomorphism, and $\varphi = i \circ \overline{\varphi} \circ \pi$ (First theorem of group isomorphisms).

Rings and Fields

Definition 10.5 A ring $\mathcal{R}$ is an Abelian group (with the group operation denoted additively as '+'), equipped with a further operation, denoted multiplicatively, which is associative, $(ab)c = a(bc)$, and satisfies the distributive laws $(a+b)c = ac+bc$ and $a(b+c) = ab+ac$, for all $a, b, c \in \mathcal{R}$.

A subring $\mathcal{Y}$ of a ring $\mathcal{R}$ is a subgroup $\mathcal{Y} \subset \mathcal{R}$ of the group structure such that $xy \in \mathcal{Y}$ for any $x, y \in \mathcal{Y}$.

The ring $\mathcal{R}$ is commutative if $ab = ba$, $\forall a, b \in \mathcal{R}$. A unit element in the ring $\mathcal{R}$ is an (unique) element $1 \in \mathcal{R}$ satisfying: $1a = a1$, for all $a \in \mathcal{R}$. The symbol 0 stands for the neutral element of the (Abelian) group structure of $\mathcal{R}$.

Given a ring $\mathcal{R}$ we can always consider that it has unit element because of the following construction. Suppose that $\mathcal{R}$ doesn't have a unit, then consider the set[2] $\overline{\mathcal{R}} = \mathcal{R} \times \mathbb{Z}$ of pairs (x, n) where $n \in \mathbb{Z}$, with the composition laws: $(x, n)+(y, m) = (x+y, n+m)$ and $(x, n)\cdot(y, m) = (xy+mx+ny, nm)$ where $x, y \in \mathcal{R}, n, m \in \mathbb{Z}$,

[2] We may use any ring with unit instead of $\mathbb{Z}$ like $\mathbb{R}$ or $\mathbb{C}$ as it will be done in the case of algebras (see later).

and $nx = x + \cdots + x$, n times. The set $\overline{\mathcal{R}}$ with the previous operations is a ring with unit element $(0, 1)$. The subring of elements of the form $(x, 0)$ is isomorphic (see below) to the original ring $\mathcal{R}$. In what follows we will consider that all rings have a unit 1.

A familiar example of a ring is provided by the set $\mathbb{Z}$ of the integers. Other relevant examples are:

(i) The ring $\mathbb{R}[x]$ of the polynomials with coefficients in $\mathbb{R}$.

(ii) The set of all real-valued functions $f : X \to \mathbb{R}$ on a set X is a ring, denoted as $\mathcal{F}(X)$, under the usual definition of pointwise addition and multiplication of functions.

(iii) The collection $M_n(\mathbb{Z})$ of all $n \times n$ matrices with integer entries. Similarly the matrices $M_n(\mathcal{R})$ with entries in a ring $\mathcal{R}$ with the standard addition and multiplication of matrices $(A \cdot B)_{jk} = \sum_{l=1}^{n} A_{jl} B_{lk}$.

(iv) Given an Abelian group $(A, +)$ the family of all group homomorphisms $\varphi : A \to A$, is a ring denoted by $\mathrm{End}(A)$ with the addition $(\varphi + \psi)(a) = \varphi(a) + \psi(a)$, and the composition of maps $(\varphi \circ \psi)(a) = \varphi(\psi(a))$, for all $a \in A$. The unit element is the identity map.

(v) The set of all the differential operators (Appendix G) with constant (real) coefficients. The Fourier transform exhibits an isomorphism with the ring of polynomials in 'momentum space'.

Definition 10.6 Given a ring $\mathcal{R}$ a left (right) ideal $\mathcal{J}$ is a subgroup $\mathcal{J} \subset \mathcal{R}$ of the Abelian group structure such that $ax \in \mathcal{J}$ $(xa \in \mathcal{R})$ for all $a \in \mathcal{R}$, $x \in \mathcal{J}$. An Abelian subgroup which is both a left and a right ideal is called a bilateral ideal. If $\mathcal{R}$ is a commutative ring any left or right ideal is a bilateral ideal (or ideal for short).

The subgroups $\{0\}$ and $\mathcal{R}$ are ideals, called improper ideals of the ring $\mathcal{R}$.

The abelian subgroup $p\mathbb{Z}$ of integers multiples of p is an ideal of $\mathbb{Z}$. Given a polynomial $p(x)$ its multiples are again a ideal on $\mathbb{R}[x]$. Given a set X and a subset $Y \subset X$ the set of functions vanishing at Y is an ideal on $\mathcal{F}(X)$.

Given an ideal $\mathcal{J}$ of the ring $\mathcal{R}$, the quotient Abelian group $\mathcal{R}/\mathcal{J}$ is again a ring with the induced product $(x + \mathcal{J})(y + \mathcal{J}) = xy + \mathcal{J}$.

Given a subset $S \subset \mathcal{R}$ the set of finite linear combinations $\sum_k a_k x_k$ $(\sum_k x_k a_k)$ of elements $x_k \in S$ and $a_k \in \mathcal{R}$ is a left (right) ideal, called the left (right) ideal generated by S and denoted by $\mathcal{R}S$ $(S\mathcal{R})$. In what follows we will omit 'left' or 'right' when referring to ideals if the definitions and properties stated hold when substituting left for right and vice versa.

The intersection of two ideals is again and ideal. The ideal generated by S is the smallest ideal containing it, that is, the intersection of all ideals containing it.

Definition 10.7 An ideal $\mathcal{J}$ of the ring $\mathcal{R}$ is called maximal if is proper and there doesn't exist an ideal $\mathcal{I}$ such that $\mathcal{J} \subset \mathcal{I} \subset \mathcal{R}$ (all inclusions strict). Equivalently, $\mathcal{J}$ is a maximal ideal if for any $x \notin \mathcal{J}$ the ideal generated by $\mathcal{J} \cup \{x\}$ is $\mathcal{R}$.

Definition 10.8 Given two rings $\mathcal{R}_a$, $a = 1, 2$, a ring homomorphism Φ is a group homomorphism $\Phi : \mathcal{R}_1 \to \mathcal{R}_2$ of the corresponding Abelian groups, such that

$\Phi(xy) = \Phi(x)\Phi(y)$ for all $x, y \in \mathcal{R}_1$. Given a ring homomorphism Φ the kernel of Φ as a group homomorphism is an ideal of $\mathcal{R}_1$ and it will be denoted by $\ker \Phi$.

A ring homomorphism will be called mono-, epi- or isomorphism if the corresponding group homomorphism is mono-, epi- or isomorphism.

Definition 10.9 A division ring is a ring $\mathbb{K}$ such that its non zero elements form a group under multiplication. A field is a commutative division ring.

The only ideals on a field $\mathbb{K}$ are $\{0\}$ and $\mathbb{K}$, and this property characterizes fields. Standard examples of fields are the set $\mathbb{Q}$ of the rational numbers and the sets $\mathbb{R}$ and $\mathbb{C}$ of the real and complex numbers.

If $\mathcal{J}$ is a maximal ideal of $\mathcal{R}$, then $\mathbb{R}/\mathcal{J}$ is a field an conversely. For instance if p is a prime number the ideal $p\mathbb{Z}$ is maximal and the congruence classes of integers mod p, $\mathbb{Z}_p = \mathbb{Z}/p\mathbb{Z}$ is a field.

The ideal $\mathcal{J}_x$ of all real-valued functions on a set X vanishing at the point $x \in X$ is maximal. The quotient $\mathcal{F}(X)/\mathcal{J}_x$ is just the field $\mathbb{R}$.

Definition 10.10 The family of all maximal ideals of the ring $\mathcal{R}$ is called the maximal spectrum (or simply, the spectrum) of $\mathcal{R}$. It will be denoted by $\mathrm{Spec}(\mathcal{R})$.

Modules, Linear Spaces and Algebras

Definition 10.11 A left $\mathcal{R}$-module $\mathcal{M}$ over a ring $\mathcal{R}$ is an Abelian group $(\mathcal{M}, +)$ together with a product $\circ\colon \mathcal{R} \times \mathcal{M} \to \mathcal{M}$, $(x, m) \mapsto x \circ m$, such that the map $\alpha_x\colon \mathcal{M} \to \mathcal{M}$ defined by $\alpha_x(m) = x \circ m$, for all $x \in \mathcal{R}$ and $m \in \mathcal{M}$, is a ring homomorphism from $\mathcal{R}$ to the ring of endomorphisms $\mathrm{End}(\mathcal{M})$ of $\mathcal{M}$. It is also said that $\mathcal{R}$ is represented in the ring of endomorphisms of $\mathcal{M}$ as an Abelian group. As usual we will use a multiplicative notation for $\circ$.

Right $\mathcal{R}$-modules are defined in a similar way. A set $\mathcal{M}$ which is a left and a right $\mathcal{R}$-module will be called and $\mathcal{R}$-bimodule. In what follows, unless stated otherwise, we will just call 'module' a left $\mathcal{R}$-module.

An ideal $\mathcal{J}$ of a ring $\mathcal{R}$ is a left (right) $\mathcal{R}$-module with the operation $x \circ a = xa$ ($a \circ x = ax$) for all $a \in \mathcal{J}$ and $x \in \mathcal{R}$. The Cartesian product $\mathcal{J}^n = \mathcal{J} \times \cdots \times \mathcal{J}$ is a left $\mathcal{R}$-module with the operation $x(a_1, \ldots, a_n) = (xa_1, \ldots, xa_n)$.

Let $\mathcal{M} = \mathcal{F}(X, \mathbb{R}^n)$ be the space of functions on the set X with values in $\mathbb{R}^n$. Then $\mathcal{M}$ is a left (right) $\mathcal{F}(X)$-module with the operation $f \circ (\sigma_1, \ldots, \sigma_n) = (f\sigma_1, \ldots, f\sigma_n)$, with f any real-valued function on X and $\sigma = (\sigma_1, \ldots, \sigma_n)$ any element in $\mathcal{M}$.

A submodule of the module $\mathcal{M}$ is a subgroup (necessarily Abelian) $\mathcal{N} \subset \mathcal{M}$ such that $xn \in \mathcal{N}$ for any $x \in \mathcal{R}$ and $n \in \mathcal{N}$. The quotient (Abelian) group $\mathcal{M}/\mathcal{N}$ with the operation $x(m + \mathcal{N}) = xm + \mathcal{N}$ is a module.

The intersection of two submodules of a module is again a module. Given a set $S \subset \mathcal{M}$ the module generated by S is the smallest module in $\mathcal{M}$ containing S, or equivalently the set of all finite linear combinations $\sum_k x_k m_k$, with $x_k \in \mathcal{R}$ and $m_k \in \mathcal{M}$. The set S will be called a generating system if $\mathcal{M}$ is the module generated by S. The module $\mathcal{M}$ is finitely generated if it possess a finite generating set.

We will say that $\mathcal{M}$ is a free module with $\mathcal{B}$ as a basis if every $m \in \mathcal{M}$ can be written uniquely as a finite sum: $m = \sum_k x_k m_k$, $m_k \in \mathcal{B}$, $x_k \in \mathcal{R}$. Given any set S the module $\mathcal{M}/S)$ defined as the family of finite formal linear combinations $\sum_k x_k s_k$, $s_k \in S$ and $x_k \in \mathcal{R}$, is called the free module generated by S. Clearly S is a basis for $\mathcal{M}/S)$. A finitely generated free $\mathbb{R}$-module has a finite basis. Basis need not be unique.

Definition 10.12 A homomorphism of modules $\Phi\colon \mathcal{M}_1 \to \mathcal{M}_2$ is a group homomorphism such that $\Phi(xm) = x\Phi(m)$ for all $x \in \mathcal{R}$, $m \in \mathcal{M}_1$. A homomorphism of modules is a mono-, epi- or isomorphism if is so as a group homomorphism.

Definition 10.13 A linear space (or vector space) E is a left $\mathbb{K}$-module where $\mathbb{K}$ is a field, tipically $\mathbb{R}$ or $\mathbb{C}$, in which case it will be called a real or complex (linear) space respectively.

Any linear space is a $\mathbb{K}$-bimodule with the right $\mathbb{K}$-module structure given by the obvious operation $v \circ \lambda = \lambda v$, $\lambda \in \mathbb{K}$, $v \in E$. Any linear space is a free module, hence it has a basis, sometimes called a Hamel basis if its cardinal is non-finite. The dimension of a linear space is the cardinal of any basis on it. A linear subspace V of a linear space E is a submodule of E.

If E, F are linear spaces, a module homomorphism $\Phi\colon E \to F$ is called a linear map. The space of linear maps from E to itself is a linear space and a ring denoted by $\mathrm{End}(E)$. The overall structure of $\mathrm{End}(E)$ is called an algebra (see Definition 10.14) and, accordingly, it is called the algebra of endomorphisms of E. The group of all isomorphisms from E into itself is called the general linear group of the space E and is denoted by $GL(E)$.

If $\Phi\colon E \to F$ is a linear map, there is a canonical linear isomorphism $\overline{\Phi}\colon E/\ker\Phi \to \Phi(E) \subset F$ which constitutes the extension to linear spaces of the first isomorphism theorem for Abelian groups.

Associative Algebras and Lie Algebras

Definition 10.14 An algebra is a ring that is at the same time a vector space $(\mathcal{A}, +, \circ)$ and such that: $\lambda(u \circ v) = (\lambda u) \circ v = u \circ (\lambda v)$ and $\forall \lambda \in \mathbb{K}$.

Any ring $\mathcal{R}$ can be made into an algebra over the field $\mathbb{K}$ by considering the family of finite linear combinations of the form $\sum_k \lambda_k x_k$, $\lambda_k \in \mathbb{K}$ and $x_k \in \mathcal{R}$, with the obvious addition and multiplication operations. Such algebra is denoted by $\mathcal{R} \otimes \mathbb{K}$. The notion of subrings, ideals, ring homomorphisms, and all other related concepts extend without change to the class of algebras. Thus, in what follows we will just discuss algebras.

All examples of rings discussed before are algebras or can be made into algebras by using the construction before. For finite-dimensional algebras, that is, possessing a finite-dimensional basis, say $\mathcal{B} = \{e_1, \ldots, e_n\}$, we can write, $e_i \circ e_j = \sum_{k=1}^n c_{ij}^k e_k$, $c_{ijk} \in \mathbb{K}$. The n^3 constants c_{ij}^k are called the structure constants of the algebra relative to the basis $\mathcal{B}$.

A map $\Phi\colon \mathcal{A}_1 \to \mathcal{A}_2$ is an algebra homomorphism if it is a linear map and a ring homomorphism. The group of algebra isomorphisms from $\mathcal{A}$ into itself is the called the automorphisms group of $\mathcal{A}$, sometimes denoted by $\mathrm{Aut}(\mathcal{A})$.

The algebras we will be dealing most in this book are associative and Lie algebras.

Definition 10.15 An algebra $(\mathcal{A}, +, \circ)$ is associative if $(u \circ v) \circ w = u \circ (v \circ w)$ for all $u, v, w \in \mathcal{A}$. An associative algebra $\mathcal{A}$ is commutative if $u \circ v = v \circ u$ for all $u, v \in \mathcal{A}$.

A Lie algebra is an algebra $(\mathcal{L}, +, [\cdot, \cdot])$ such that: $[u, v] = -[v, u]$, and $[u[v, w]] + [v, [w, u]] + [w, [u, v]] = 0$ (Jacobi identity), for all $u, v, w \in \mathcal{L}$.

With any associative algebra $\mathcal{A}$ we can associate a Lie algebra by defining a new product as, $[u, v] = u \circ v - v \circ u$ called the commutator. It is straightforward to check that this product does indeed satisfy the Jacobi identity. Obviously, if the algebra $\mathcal{A}$ is commutative, the commutator always vanish and the Lie algebra is trivial.

Typical examples of Lie algebras are provided by the linear transformations $\mathrm{End}(E)$ of a vector space E with the Lie algebra product defined by the commutator of linear maps. Ado's theorem [Ja79] states that any finite-dimensional Lie algebra is isomorphic to a Lie subalgebra of the Lie algebra $\mathrm{End}(E)$ with E a finite-dimensional linear space.

Definition 10.16 Given an algebra $\mathcal{A}$, a derivation is a linear map $D\colon \mathcal{A} \to \mathcal{A}$ such that: $D(a \circ b) = a \circ D(b) + D(a) \circ b$ (Leibnitz' identity), for all $a, b \in \mathcal{A}$. The set of derivations of $\mathcal{A}$ will be denoted as $\mathrm{Der}(\mathcal{A})$.

The set of derivations of $\mathcal{A}$ can be made into a Lie algebra by defining the product of derivations as,[3]

$$[D_1, D_2](a) = D_1(D_2(a)) - D_2(D_1(a)), \qquad \forall D_1, D_2 \in \mathrm{Der}(\mathcal{A}), a \in \mathcal{A}, \quad (10.1)$$

We can consider derivations on the algebra of derivations, i.e., linear maps, $\mathfrak{D}\colon \mathrm{Der}(\mathcal{A}) \to \mathrm{Der}(\mathcal{A})$, satisfying,

$$\mathfrak{D}([D_1, D_2]) = [\mathfrak{D}(D_1), D_2] + [D_1, \mathfrak{D}(D_2)]. \quad (10.2)$$

Any derivation $D \in \mathrm{Der}(\mathcal{A})$ defines a derivation $\mathfrak{D}_D$, called inner, of the Lie algebra of derivations: $\mathfrak{D}_D(D_1) = [D, D_1]$. With this definition, the Jacobi identity is nothing but the assertion that $\mathfrak{D}_D$ is a derivation on the Lie algebra $\mathrm{Der}(\mathcal{A})$ for any D.

[3] Notice that the composition of derivations is not a derivation in general.

Graded Algebras, Graded Lie Algebras and Graded Derivations

Definition 10.17 An algebra $\mathcal{A}$ is a graded algebra if, as a linear space, $\mathcal{A}$ is a graded linear space over the Abelian group H, that is, $\mathcal{A} = \bigoplus_{r \in H} \mathcal{A}^r$, and $\mathcal{A}^r \circ \mathcal{A}^s \subseteq \mathcal{A}^{r+s}$.

A Lie algebra $\mathcal{L}$ is a graded Lie algebra if $\mathcal{L} = \bigoplus_{r \in H} \mathcal{L}_r$ is a graded algebra over the Abelian group H, i.e., $[\mathcal{L}_r, \mathcal{L}_s] \subseteq \mathcal{L}_{r+s}$, it is graded commutative: $[x, y] = -(-1)^{rs} [y, x]$, for all $x \in \mathcal{L}_r$, $y \in \mathcal{L}_s$, and satisfies the graded Jacobi identity:

$$(-1)^{r_1 r_3} [x_1, [x_2, x_3]] + (-1)^{r_2 r_1} [x_2, [x_3, x_1]] + (-1)^{r_3 r_2} [x_3, [x_1, x_2]] = 0, \quad x_i \in \mathcal{L}_{r_i}, i = 1, 2, 3 .$$

Definition 10.18 Let $\mathcal{A} = \oplus_r \mathcal{A}^r$ be a graded algebra. A linear map $D \colon \mathcal{A} \to A$ is said to be a (graded) derivation of degree k if $D(\mathcal{A}^r) \subset D(\mathcal{A}^{r+k})$ and, for any pair of homogeneous elements, $\alpha \in \mathcal{A}^r$ and $\beta \in \mathcal{A}^s$, $D(\alpha \circ \beta) = (D\alpha) \circ \beta + (-1)^{kr} \alpha \circ (D\beta)$.

It should be remarked that, if D_1 and D_2 are derivations of a graded Lie algebra of degrees k_1 and k_2 respectively, then the graded commutator defined by,

$$[D_1, D_2] = D_1 \circ D_2 - (-1)^{k_1 k_2} D_2 \circ D_1 \tag{10.3}$$

is a derivation of degree $k_1 + k_2$. On the other hand, for $H = \mathbb{Z}$, it can be shown that, if $\mathcal{A}^1$ generates $\mathcal{A}^r$ for all $r \leq 1$, then any derivation of $\mathcal{A}$ will be determined by giving the images of the elements of degree zero and one.

A.2. Topology: A Brief Dictionary

Topological Spaces: Hausdorff and Separable Spaces

Definition 10.19 A topological space is a set X and a family of subsets $\mathcal{O}$ of X containing $\emptyset$ and X, and such that if O_α, $\alpha \in A$ is any collection of subsets in $\mathcal{O}$, then $\bigcup_\alpha O_\alpha$ belongs to $\mathcal{O}$, and the intersection of any two elements in the family belongs to the family. The family $\mathcal{O}$ is called a topology (for the space X) and subsets O in the family $\mathcal{O}$ are called open sets. The open sets $\emptyset$, X are called improper open subsets. A proper open set is an open set different from $\emptyset$ and X.

Given a point $x \in X$, a subset $N \subset X$ is said to be a neighborhood of x it it contains and open set O that contains x, i.e., $x \in O \subset N$. Given a subset $S \subset X$, a point $x \in S$ is said to be in the interior of S if S is a neighborhood of x. The interior of S is the union of all its interior points. The interior of S will be denoted by $\mathrm{Int}(X)$.

Given two topological spaces X and Y its Cartesian product $X \times Y$ is a topological space too, its topology defined by the family of arbitrary unions of sets of the form $O_X \times O_Y$ where O_X is open in X and O_Y open in Y. This topology is called the product topology (of the topologies in X and Y).

The set of real numbers $\mathbb{R}$ with the topology obtained by considering arbitrary unions of open intervals, including the improper subsets, is a topological space. We will call this topology the natural topology in $\mathbb{R}$ and unless stated otherwise this will be the topology that will be considered on it. The Cartesian product $\mathbb{R}^n$ of n copies of $\mathbb{R}$ is a topological space with the product topology (again the only topology that will be considered). In particular the identification of $\mathbb{C}$ with $\mathbb{R}^2$ by means of $z = x + iy$ allows us to consider **C** as a topological space.

Definition 10.20 A topological space X is called Hausdorff, also said to satisfy the T_2 separation property, if given two different points $x \neq y$, there exists neighborhoods N_x of x and N_y of y such that $N_x \bigcap N_y = \emptyset$.

The topological space $\mathbb{R}^n$ is Hausdorff.

Definition 10.21 Let $(X, \mathcal{O})$ be a topological space. A subset $F \subset X$ is said to be closed if $X \backslash F$ is open. Notice that the improper subsets $\emptyset$ and X are always open and closed in any topology.

It is clear that the intersection of an arbitrary family of closed sets is closed. Given a subset $S \subset X$, the intersection of all closed sets containing S is called the closure of S and is by construction the smallest closed set containing S. The closure of S will be denoted by $\overline{S}$. The set $\overline{S}\backslash \text{Int}(S)$ is called the boundary of S and denoted by ∂S.

Definition 10.22 A subset $S \subset X$ of the topological space X is said to be dense if $\overline{S} = X$. A topological space satisfies the second countability axiom, also-called separable, if it contains a countable dense subset.

Continuous Maps and Homeomorphisms

Definition 10.23 A map $\varphi \colon X \to Y$ from the topological space X to the topological space Y is continuous if $\varphi^{-1}(O)$ is open in X for any open set $O \subset Y$.

The composition $\psi \circ \varphi$ of two continuous maps $\varphi \colon X \to Y$, $\psi \colon Y \to Z$ is continuos.

Definition 10.24 A bijective continuous map $\varphi \colon X \to Y$ such that the inverse map $\varphi^{-1} \colon Y \to X$ is continuous too, is called a homeomorphism. The family of all homeomorphisms from X into itself form a group.

If $f, h \colon X \to \mathbb{R}$ are real-valued continuous functions, then the pointwise sum $(f + h)(x) = f(x) + h(x)$, the pointwise product $(f \cdot h)(x) = f(x)h(x)$, and the function $(\lambda \cdot f)(x) = \lambda(f(x))$ where $\lambda \in \mathbb{R}$, are continuous. The set of real-valued continous functions on X is an algebra denoted by $C(X)$ (in particular a linear space over the field $\mathbb{R}$).

In a similar way the set of all continuous complex-valued functions $f \colon X \to \mathbb{C}$ is an algebra. Notice that the real and imaginary parts u, v of f given respectively as

$u(x) = \text{Re}(f(x))$, $v(x) = \text{Im}(f(x))$, are such that $f = u + iv$ and f is continuous iff u and v are continuous. The space of complex-valued continuous functions on X is a complex algebra, that is an algebra which is a complex linear space. The linear complex linear space of complex-valued continuous functions on X is the complexification of the real space of real-valued continuous functions on X and it will be denoted either as $C(X; \mathbb{C})$ or $C(X)^{\mathbb{C}}$.

Fundamental examples of topological spaces are linear spaces E equipped with a topology such that the addition and multiplication maps $+: E \times E \to E$ and $\cdot \mathbb{C} \times E \to E$ are continuous. Such linear spaces will be called topological linear spaces.

Compactness

A covering of a subset S of a topological space X is a family $\mathcal{C}$ of subsets $C_\alpha \subset X$, labelled by a set $\alpha \in A$ such that $S \subset \bigcup_\alpha C_\alpha$. The covering $\mathcal{C}$ is called open if all its elements are open sets. Given a covering $\mathcal{C}$ of S a sub covering is a subfamily $\mathcal{C}'$ of $\mathcal{C}$ (that is there is a subset $A' \subset A$ such that C_α belongs to $\mathcal{C}'$ if $\alpha \in A' \subset A$) which is still a covering of S.

Definition 10.25 A subset $K \subset X$ of the topological space X is compact if given an open covering $\mathcal{C}$ of K there exists a finite sub covering.

Bolzano's theorem states that any real-valued continuous function f on a compact space X reaches its maximum and minimum, in other words, $f(X) = [a, b]$ where $a = \min f$ and $b = \max f$.

Heine-Borel theorem states that a subset of R^n is compact iff it is closed and bounded (see later on Appendix A.2, Metric spaces).

Given a covering $\mathcal{C}$ of a given subset $S \subset X$, a covering $\mathcal{C}'$ of S such that for any set $C'_{\alpha'}$ in $\mathcal{C}'$ there exists a set C_α in $\mathcal{C}$ such that $C'_{\alpha'} \subset C_\alpha$ is called a refinement of $\mathcal{C}$.

Definition 10.26 A subset S of the topological space X is called paracompact if given any open covering $\mathcal{C}$ of it, there exists a refinement $\mathcal{C}'$ which is locally finite, that is by that that given any point $x \in X$ there exists only a finite number of sets in $\mathcal{C}'$ containing x.

Subspaces, Conectedness

Given a subspace $S \subset X$ of the topological space X, the family of sets $S \cap O$ with O an open set in X defines a topology on S called the trace topology. The subset $S \subset X$ equipped with the trace topology is said to be a topological subspace of X.

Definition 10.27 A subset S of a topological space is said to be connected if it doesn't have proper subsets which are open and closed simultaneously with respect to the trace topology.

A continuous curve on the topological space X is a continuous map $\gamma: I \to X$. A subset $S \subset X$ is said to be arc-connected if given any two points $x, y \in S$ there exists a continuous curve $\gamma: [a, b] \to S$ such that $\gamma(a) = x$ and $\gamma(b) = y$. Then we will say that the continuous curve γ joins x and y. If S is open and arc-connected, it is connected.

Let S be topological subspace, then two points $x, y \in S$ are in the same connected component of S if there exists a continuous curve joining x and y. The relation 'being in the same connected component' is an equivalence relation and a connected component of S is any of the corresponding equivalence classes. A connected component of S is open, hence connected.

The Fundamental Group of a Topological Space

Definition 10.28 Given $x, y \in X$, two continuous curves $\gamma_1, \gamma_2: [0, 1] \to X$ are homotopic if there exists a continuous map $H: [0, 1] \times [0, 1] \to X$ such that $H(0, t) = \gamma_1(t)$, $H(1, t) = \gamma_2(t)$ for all $0 \leq t \leq 1$, and $H(s, 0) = x$, $H(s, 1) = y$ for all $0 \leq s \leq 1$. The points x, y are called the end-points of the curves. The relation γ is homotopic to γ' is an equivalence relation in the set of all continuos maps form $[0, 1]$ to X with fixed end-points x, y.

Given a point $x \in X$, the family of homotopy equivalence classes of continuous curves with coinciding end-points x is a group with respect to the composition law:

$$(\gamma_1 \star \gamma_2) = \begin{cases} \gamma_2(2t), & \text{if } 0 \leq t < 1/2 \\ \gamma_1(2t - 1), & \text{if } 1/2 \leq t < 1 \end{cases}$$

This group will be called the fundamental group of the topological space at the point x and denoted by $\pi_1(X, x)$.

If x, y lie in the same connected components of X the corresponding fundamental groups are isomorphic. If the space X is arc-connected then the fundamental groups at all points are isomorphic. The equivalence class of homotopy groups defined by $\pi_1(X, x)$,x an arbitrary point in X will be called the fundamental group of X, the first homotopy group of X of the Poincarŕe group of X, and will be denoted simply as $\pi_1(X)$.

Definition 10.29 A connected and arc-connected topological space X will be said to be simply connected if $\pi_1(X) = 0$ or, in other wordis, if any closed continuous curve is homotopic to a constant continuous curve.

Metric Spaces: Completeness

Definition 10.30 A metric space is a set M and a map $d: M \times M \to \mathbb{R}^+$ where $\mathbb{R}+$ denotes the set of non-negative real numbers, such that $d(x, x) = 0$ iff $x = 0$,

$d(x, y) = d/y, x)$ and $d(x, y) \leq d(x, z) + d(z, y)$ ("triangle inequality"), for all $x, y, z \in M$. Then the map d is called a metric (or distance) on M.

Given a pont x and a positive real number r the subset $\{y \in M \mid d(x.y) < r$ is called the open ball or radius r and center x in M, and it will be denoted as $B_r(x)$. We construct a topology on a metric space (M, d) by defining open sets as arbitrary union of open balls or, equivalently, sets O such that for any $x \in O$, there exists $r > 0$ such that $B_r(x) \subset O$. Such topology is called the metric topology.

Two metrics d_1, d_2 on M are equivalent if the identity map $i\colon M toM$ is continuous with respect to both metric topologies in M. Equivalently, d_1 is equivalent to d_2 if there exists two positive real numbers $B > A > 0$ such that for any pair of points $x, y \in M$, we have $Ad_1(x, y) \leq d_2(x, y) \leq Bd_1(x, y)$.

The set $\mathbb{R}^n$ is a metric space with respect to the metrics $d_p(x, y = (\sum_{i=1}^n |x^i.y^i|^p)^{1/p}$ for all $p \geq 1$. Moreover we define $d_\infty(x, y) = \max\{|x^i - y^i| \mid i = 1, \ldots, n\}$. The metrics d_p, $1 \leq p \leq \infty$ are equivalent.

Definition 10.31 A sequence $|\{x_n\}, n = 1, 2, \ldots$, is said to be convergent if there exists $\lim x_n$, that is, there exists $x \in M$ such that for every $\epsilon > 0$, there exists $N \in \mathbb{N}$ such that if $n \geq N$, then $d(x_n, x) < \epsilon$. A sequence $|\{x_n\}, n = 1, 2, \ldots$, is said to be Cauchy, if given $\epsilon > 0$, there exists $N \in \mathbb{N}$ such that if $n, m \geq N$, then $d(x_n, x_m) < \epsilon$. It is immediate to show that any convergent sequence is Cauchy.

Definition 10.32 A metric space (M, d) is complete if any Cauchy sequence is convergent. We will also say that the metric topology is complete.

The metric space $\mathbb{R}^n$ with respect to any metric d_p is complete.

The space of real-valued continuous functions on a compact space is a complete metric space with respect to the metric $d_\infty(f, h) = \max\{|f(x)| \mid x \in X\}$.

Let $X = \mathbb{R}^n$ and consider the subset of continuous real-valued functions such that the following integral is convergent:

$$d_p(f, h) = \left(\int_{\mathbb{R}^n} |f(x) - h(x)|^p d^n x \right)^{1/p} < \infty$$

The function d_p defines a metric on the corresponding subspace, however it is not complete for any $1 \leq p < \infty$.

A.3. A Concise Account of Differential Calculus

Banach and Hilbert Spaces

Definition 10.33 Let E be a real or complex linear space. A norm on E is a function $|| \cdot ||\colon E \to \mathbb{R}^+$ such that $||x|| = 0$ iff $x = 0$, $||\lambda x|| = |\lambda| ||x||$, and $||x + y|| \leq ||x|| + ||y||$ for all $\lambda \in \mathbb{C}$, $x, y \in E$.

An inner product on a complex linear space E is a map $\langle\cdot,\cdot\rangle E\times E\to\mathbb{C}$ such that $\langle x,x\rangle=0$ iff $x=0$, $\langle x,\lambda y+\mu z\rangle=\lambda\langle x,y\rangle+\mu\langle x,z\rangle$ and $\langle x,y\rangle=\overline{\langle y,x\rangle}$ for all $\lambda,\mu\in\mathbb{C}$, $x,y,z\in E$.

It is obvious that if $||\cdot||$ is a norm, the function $d(x,y)=||x-y||$ defines a metric on E. The topology on E defined by the metric d is called the norm topology, and the linear space E equipped with this topology is a topological vector space.

If $\langle\cdot,\cdot\rangle$ is an inner product then it is always satisfied the following fundamental Cauchy-Schwartz inequality: $|\langle x,y\rangle|\leq||x||||y||$ for all $x,y\in E$. Then it is clear that if $\langle\cdot,\cdot\rangle$ is an inner product the function $||x||=+\sqrt{\langle x,x\rangle}$ defines a norm that will be called the norm associated to the inner product $\langle\cdot,\cdot\rangle$.

Definition 10.34 A complex linear space E equipped with a norm will be called a normed space. A Banach space is normed space E which is complete with respect to the norm topology.

Similarly, a complex linear space E equipped with an inner product is called a pre-Hibert space. A Hilbert space is a pre-Hilbert space $\mathcal{H}$ which is complete with respect to the topology defined by the norm associated to the inner product. Every Hilbert space is a Banach space with respect to the norm associated to the inner product.

Examples of Banach spaces are provided by the completion of the spaces of continuous functions on $\mathbb{R}^m$ with respect to the norm $||f||_p=d_p(f,0)$, $1\leq p\leq\infty$. Such spaces are denoted by $L^p(\mathbb{R}^n)$. In the particular instance of $p=2$, the norm $||\cdot||_2$ is associated to the inner product $\langle f,h\rangle_2=\int\overline{f(x)}h(x)d^nx$ and $L^2(\mathbb{R}^2)$ is a Hilbert space with respect to this product.

The natural extension of $\mathbb{C}^n$ to infinite dimensions is the space $l^2(\mathbb{C})$ consisting on infinite sequences $z=\{z_k\}$, $k=1,2,\ldots$. of complex numbers such that $\sum_k|z_k|^2<\infty$. It is a Hilbert space with respect to the inner product $\langle z,w\rangle=\sum_k\bar{z}_kw_k$.

Definition 10.35 Let E be a Banach space. A linear map $T\colon E\to E$ is said to be continuous (or bounded) if it is continuous with respect to the norm topology. The space of bounded linear maps from E to E form an algebra. The function $||T||=\sup\{||Tx||\mid||x||=1\}$ defines a norm on the complex linear space $\mathcal{B}(E)$ of bounded linear maps from E to E sometimes called the operator norm.

The space $\mathcal{B}(E)$ is a Banach space with the operator norm.

Given a Hilbert space $\mathcal{H}$ and a bounded linear operator T, its adjoint operator $T^\dagger$ is the (unique) operator defined by $\langle T^\dagger x,y\rangle=\langle x,Ty\rangle$, for all $x,y\in\mathcal{H}$.

Definition 10.36 Let $\mathcal{H}$ be a Hilbert space. An operator $U\colon\mathcal{H}\to\mathcal{H}$ is said to be unitary if $\langle Ux,Uy\rangle=\langle x,y\rangle$ for all $x,y\in\mathcal{H}$. Notice that an unitary operator is necessarily bounded and $||U||=1$.

A bounded operator $T\colon\mathcal{H}\to\mathcal{H}$ is called self-adjoint (or Hermitean) if $\langle Tx,y\rangle=\langle x,Ty\rangle$, or all $x,y\in\mathcal{H}$, that is $T=T^\dagger$.

Definition 10.37 Let T be a (non-necessarily bounded) linear operator defined on a subspace $\mathcal{D} \subset \mathcal{H}$. We will say that λ is in the spectrum of T if the operator $T - \lambda\mathbb{I}$ is not invertible. The spectrum of T will be denoted by $\sigma(T)$. A complex number λ will be said to be an eigenvalue of T if there exists a non-zero vector $x \in \mathcal{D}$ such that $Tx = \lambda x$. If λ is an eigvenvalue of T then $\lambda \in \sigma(T)$.

Theorem 10.38 (Easy spectral theorem) *Given a bounded self-adjoint operator T on a Hilbert space $\mathcal{T}$, then $\sigma(T) \subset \mathbb{R}$ is compact. Moreover, there exists a Borel measure E on $\mathbb{R}$ with values in the space of projectors on $\mathcal{H}$ such that $T = \int \lambda E(d\lambda)$.*

A similar theorem holds for unbounded (normal) self-adjoint operators.

$\mathbb{C}^*$-Algebras

Definition 10.39 A Banach algebra $\mathcal{B}$ is an algebra with a complete norm $||\cdot||$ such that $||xy|| \leq ||x||||y||$ for all $x, y \in \mathcal{B}$.

Standard examples of Banach algebras are provided by the space $C(K)$ of continuous functions on a compact space K and the norm the supremum norm $||\cdot||_\infty$. Morevoer the space $\mathcal{B}(E)$ of bounded operators on a Banach space is a Banach algebra. In particular the space of finite-dimensional square matrices $M_n(\mathbb{C})$ is a Banach algebra.

We can associate with each Banach algebra $\mathcal{B}$ the space Spec($\mathcal{B}$) of closed proper maximal ideals on it called the maximal spectrum (or just the spectrum) of $\mathcal{B}$.

If $\mathcal{B}$ is a commutative Banach algebra with unit, then maximal ideals are in one-to-one correspondence with the kernel of algebra morphisms $\rho\colon \mathcal{B} \to \mathbb{C}$. The spectrum of $\mathcal{B}$ is compact in a natural topology and there is a natural map from $\mathcal{B}$ into the space of continuous functions $C(\mathrm{Spec}(\mathcal{B}))$, called the Gelf'and transform, defined by $x \in \mathcal{B} \mapsto \widehat{x}$, with $\widehat{x}(\rho) = \rho(x)$ which is a homeomorphism.

Definition 10.40 A $\star$-algebra is a complex algebra $\mathcal{A}$ possessing a complex anti-linear map $\star\colon \mathcal{A} \to \mathcal{A}$, $x \mapsto x^\star$, such that $\star^2 = \mathbb{I}$, and $(xy)^\star = y^\star x^\star$, for all $x, y \in \mathcal{A}$.

A C^*-algebra is a $\star$-algebra $\mathcal{A}$ carrying a complete norm $||\cdot||$ such that $(\mathcal{A}, ||\cdot||)$ is a Banach algebra and $||x^\star x|| = ||x||^2$ for all $x \in \mathcal{A}$.

The space of continuous complex-valued functions on a compact topological space with the map $f^\star(x) = \overline{f(x)}$ and the supremum norm is a commutative C^*-algebra. Given a complex Hilbert space $\mathcal{H}$ the algebra of all bounded operators $\mathcal{B}(\mathcal{H})$ is a C^*-algebra with the operation $\star\colon T \mapsto T^\dagger$. The GNS (Gel'fand-Naimark-Segal) Theorem, establishes that given a C^*-algebra, there exists a complex Hilbert space $\mathcal{H}$ and an invective morphism of C^*-algbras $\pi : \mathcal{A} \to \mathcal{B}(\mathcal{H})$, that is, any C^*-algebra can be faithfully represented as a C^*-subalgebra of the C^*-algebra of bounded operators on a Hilbert space.

As a particular instance of the GNS theorem we obtain that any finite-dimensional C^*-algebra can be represented faithfully as an algebra of finite-dimensional complex matrices where $A^\star$ is the Hermitean conjugate of the matrix A.

Differential Calculus on Banach Spaces

Most of the standard results of calculus in several variable stand in the general setting of Banach spaces (see for instance [La85]). We will succintly review them.

Definition 10.41 Let $f : O \subset E \to F$ be a from the open set O in the Banach space E to the Banach space F. Let x be a point in O. We will say that f is differentiable at x if there exists a continuous linear map $L : E \to F$ such that the following limit exists and is zero:

$$\lim_{h \to 0} \frac{||f(x+h) - f(x) - L(h)||}{||h||} = 0.$$

In such case we will say that L is the derivative (or differential) of f at x and will be denoted either as $f'(x), df(x)$ of $Df(x)$. If f is differentiable for any $x \in O$ we will say that f is differentiable in O. In such case we have a map $df : O \to \mathcal{B}(E, F)$ from the open set O into the Banach space $\mathcal{B}(E, F)$ of bounded linear operators from E to F. If the map df is continuous we will say that f is of class C^1 on O.

We will say that f is of class C^k, $k > 1$ on O if $D^{k-1}f$ is differentiable for all $x \in O$ and $D^k f = D(D^{k-1}f)$ is continuous on O. The collection of functions of class C^k on O with values in F will be denoted $C^k(O, F)$. We will say that a function f is of class C^∞ or smooth if it is of class C^k for all k.

Clearly the composition addition and multiplication by complex numbers of functions of class C^k are again of class C^k and the standard rules apply: $D(f+g) = Df + Dg$, $D(\lambda f) = \lambda Df$ and if $f, h : O \subset E \to A$ with A a normed algebra, then $D(f \cdot h) = Df \cdot h + f \cdot Dh$ (Leibnitz' rule).

Standard examples are the spaces $C^k(O, \mathbb{R})$ (denoted simply by $C^k(O)$) with $O \subset \mathbb{R}^n$. In such cases even if $C^k(O)$ is a linear space, the topologies induced by the norms $|| \cdot ||_p$, $1 \leq p < \infty$, are not complete (even if O is compact).

There is a natural topology on $C^\infty(O)$ which is being discussed in the main text.

We will establish without proofs the main theorems of differential calculus on Banach spaces:

Theorem 10.42 (Inverse function theorem) *Let E, F be a Banach spaces, $O \subset E$ and open set, and $f : O \to F$ a function of class C^1 such that the differential $Df(x_0) : E \to F$ at the point $x_0 \in O$ is a topological isomorphism, that is, $Df(x_0)$ is invertible and continuous and its inverse is continuous too. Then there exists an open neighborhood V of x_0 in O such that the restriction of f to V is a diffeomorphism onto its image, i.e., the map $f : V \to f(V)$ is invertible and its inverse $f^{-1} : f(V) \to V$ is C^1.*

Theorem 10.43 (Implicit function theorem) *Let E, F be Banach spaces, E is the direct sum of the Banach spaces E_1 and E_2 and $\pi_a : E \to E_a$, $a = 1, 2$, denote the corresponding projections. Let $\Phi : O \to F$, $O \subset E$ an open set, be a function of class C^1 such that $\Phi(x_0, y_0) = 0$ at the point $(x_0, y_0) \in O$, $x_0 \in E_1$,*

$y_0 \in E_2$. We will suppose that the differential $D\Phi(x_0, y_0)\colon E \to F$ is surjective and, if we denote by $\Phi_{x_0} : \pi_2(O) \to F$ the map $\Phi_{x_0}(y) = \Phi(x_0, y)$, that the differential $D\Phi_{x_0}(y_0)\colon E_2 \to F$ is a topological isomorphism. Then, there exists an open neighborhood W of x_0 in $\pi_1(O) \subset E_1$ and a C^1 map $\psi\colon W \to \pi_2(O) \subset E_2$, such that $y_0 = \psi(x_0)$ and $\Phi(x, \psi(x)) = 0$ for all $x \in W$. The function ψ is called the implicit function determined by Φ at the point (x_0, y_0).

We have the following corollary in finite dimensions.

Corollary 10.44 *Let $\Phi\colon O \subset \mathbb{R}^{n+m} \to \mathbb{R}^m$ be a C^1 function, and $\mathcal{S} = \{\mathbf{x} \in O \mid \Phi(x) = \mathbf{0}\}$. Suppose that the components φ^i of $\Phi = (\varphi^1, \ldots, \varphi^m)$ are functionally independent on $\mathcal{S}$, then for every point $\mathbf{x} \in \mathcal{S}$ one can choose a subset $(x^{i_1}, \ldots, x^{i_n})$ of the coordinates $(x^1, \ldots, x^{n+m})$ and associate with any point $\mathbf{x} \in \mathcal{S}$ an open neighborhood $U_\mathbf{x}$ in $\mathcal{S}$ and an open set $V_\mathbf{x}$ in $\mathbb{R}^m$ with coordinates $(x^{i_1}, \ldots, x^{i_n})$ that can be mapped difeomorphically one onto each other.*

Appendix B: Tensor Algebra

The Tensor Algebra of a Linear Space

Let E and F be linear spaces (we shall consider only real or complex linear spaces). The tensor product of the linear spaces E and F, denoted by $E \otimes F$, can be simply defined as the vector space generated by elements of the form $u \otimes w$,[4] $u \in E$ and $w \in F$ with relations $(u+u')\otimes w = u\otimes w + u'\otimes w$, $u\otimes(w+w') = u\otimes w + u\otimes w'$, $(\lambda u) \otimes w = u \otimes (\lambda w)$, for all $u, u' \in E$, $w, w' \in F$ and $\lambda \in \mathbb{K}$. In other words, $E \otimes F$ is the quotient space of the vector space generated by all elements of the form $u \otimes w$ by the subspace whose elements have the form $(u+u')\otimes w - u\otimes w - u'\otimes w$, $u\otimes(w+w') - u\otimes w - u\otimes w'$ or, $(\lambda u)\otimes w - u\otimes(\lambda w)$. The tensor product $E \otimes F$ is sometimes called the algebraic tensor product of E and F.

If the vector spaces involved in the tensor product are finite-dimensional, the tensor product of both is still finite dimensional. It is easy to check that if $\{u_i\}$, $i = 1, \ldots n$, is a basis for E and $\{w_k\}$, $k = 1, \ldots, m$, is a basis for F, then $\{u_i \otimes w_k\}$, $i = 1, \ldots, n$, $k = 1, \ldots, m$ is a basis for $E \otimes F$, hence $\dim E \otimes F = \dim E \dim F$, and no ambiguities arise when equipping it with a norm.

However, in the infinite-dimensional case, the construction of norms on the algebraic tensor product of two normed spaces becomes an extremely subtle question. We will just mention here that if E, F are separable Hilbert spaces, there is a canonical norm on $E \otimes F$ associated to the inner product defined by $\langle u \otimes w, u' \otimes w'\rangle = \langle u, u'\rangle_E \langle w, w'\rangle_F$ for all $u \otimes w, u' \otimes w' \in E \otimes F$. The completion of the algebraic tensor product $E \otimes F$ with respect to the norm associated to the inner product before is denoted by $E\widehat{\otimes}F$.

---

[4] The symbol '$\otimes$' is conventional.

There are natural isomorphisms $E \otimes F \cong F \otimes E$ and $(E \otimes F) \otimes G \cong E \otimes (F \otimes G)$, for any three linear spaces E, F and G.

The first natural space associated to a linear space E is the linear space of $\mathbb{K}$-linear maps from E to $\mathbb{K}$, denoted E^* and called the algebraic dual of E. If the space E is an infinite-dimensional topological space, the natural notion of dual space is the space of continuous linear maps from E to $\mathbb{K}$ which is called the topological dual and denoted by E'. Of course in finite dimensions $E^* = E'$ and if E is a Hilbert space, because of Riesz duality theorem, $E' \cong E$. In what follow we will just consider finite-dimensional spaces and then $E^{**} \cong E$.

Using as building blocks E and E^* it is possible to construct the linear space,

$$T_s^r(E) = E \otimes \cdots \otimes E \otimes E^* \otimes \cdots \otimes E^*,$$

with r factors E and s factors E^*. Elements $T \in T_s^r(E)$ are called homogeneous tensors of type (r, s). It is also said that T is an r-contravariant, s-covariant homogeneous tensor. Notice that if $\{u_i\}, i = 1, \ldots, n$ is a basis for E and $\{u^k\}, k = 1, \ldots, n$ denotes its dual basis, i.e., the basis in E^* defined by the formula: $u^k(u_j) = \delta^k_j$, than

$$T = T^{j_1,\ldots,j_r}_{k_1,\ldots,k_s} u_{j_1} \otimes \cdots \otimes u_{j_r} \otimes u^{k_1} \otimes \cdots \otimes u^{k_s}$$

(where Einstein's convention of summation over repeated indices at different levels is used from now on). Notice that the space of linear maps from E to E is isomorphic to $E^* \otimes E$, that is, to the space of homogeneous tensors of type $(1, 1)$ on E. Similarly the linear space of linear maps from E^* to E is isomorphic to $E \otimes E$ and so on.

The direct sum,

$$T^\bullet_\bullet(E) = \bigoplus_{r,s \geq 0} T_s^r(E), \tag{10.4}$$

carries a graded associative algebra structure under the tensor multiplication $\otimes$ and is called the tensor algebra over E.

The group $GL(E)$ can be realized (or represented) on $T(E)$ as follows: let $\varphi \colon E \to E$ be a linear isomorphism, then we define the linear map $T_r^s(\varphi) \colon T_r^s(E) \to T_r^s(E)$, by $T_r^s(\varphi)(u_{j_1} \otimes \cdots \otimes u_{j_r} \otimes u^{k_1} \otimes \cdots \otimes u^{k_s}) = \varphi(u_{j_1}) \otimes \cdots \otimes \varphi(u_{j_r}) \otimes (\varphi^{-1})^*(u^{k_1}) \otimes \cdots \otimes (\varphi^{-1})^*(u^{k_s})$, for any basis element in $T_r^s(E)$ and where $\varphi^*(\alpha)(u) = \alpha(\varphi(u))$, for all $u \in E$ and $\alpha \in E^*$. We will call $T_r^s\varphi$ the natural (or tensorial) extension of φ to the space of homogeneous tensors of type (r, s). It is a simple exercise to chek that $T_r^s(\varphi \circ \psi) = T_r^s(\varphi) \circ T_r^s(\psi)$ for all $\varphi, \psi \in GL(E)$. In what follows, unless there is risk of confusion, we denote $T_r^s\varphi$ simply as φ. It is easy to check that $\varphi(T \otimes S) = \varphi(T) \otimes \varphi(S)$, for all tensors $T, S \in T^\bullet_\bullet(E)$.

The Exterior Algebras of Forms and Multivectors

The tensor algebra over E, has two distinguished (graded) subalgebras: the algebras $T_\bullet(E) = \bigoplus_s T_s^0(E)$ and $T^\bullet(E) = \bigoplus_r T_0^r(E)$ of fully covariant and contravariant tensors on E, respectively. Additional subalgebras can be considered by considering complete symmetric or antisymmetric tensors in $T_s^0(E)$ or $T_0^r(E)$ for instance.

Defining the operators $\mathcal{S}_n$ (symmetrizer) and $\mathcal{A}_n$ (anti-symmetrizer) on $T_n^0(E)$ by the formulas:

$$\mathcal{S}_n(u_1 \otimes \cdots \otimes u_n) = \frac{1}{n!} \sum_{\pi \in S_n} u_{\pi(1)} \otimes \cdots \otimes u_{\pi(n)} \tag{10.5}$$

and

$$\mathcal{A}_n(u_1 \otimes \cdots \otimes u_n) = \frac{1}{n!} \sum_{\pi \in S_n} \sigma(\pi) u_{\pi(1)} \otimes \cdots \otimes u_{\pi(n)}, \tag{10.6}$$

(where the π's are permutations of n objects that define the symmetric group S_n, and $\sigma(\pi)$ is the signature of π). The range $\mathcal{S}_r(T_n^0(E))$ of the linear map $\mathcal{S}$ are the completely symmetric covariant tensors of order n in E and $\mathcal{S}_\bullet(E) = \bigoplus_{n\geq 0} \mathcal{S}_n(T_n^0(E))$ is called the covariant symmetric algebra over E.

If, instead, we use the anti-symmetrizer $\mathcal{A}_n$ we obtain the space $\mathcal{A}_n(T_k^0(E))$ of completely anti-symmetric covariant tensors of order k, commonly denoted by $\Lambda^k(E)$. Clearly, the dimension of $\Lambda^k(E)$ is $\binom{\dim E}{k}$, and $\Lambda^k(E) = 0$ if $k > \dim E$. Then the direct sum $\bigoplus_{k=0}^{\dim E} \Lambda^k(E)$ is called the exterior algebra of E, has dimension $2^{\dim E}$, and is denoted as $\Lambda^\bullet(E)$.

Similarly, starting with $T_0^r(E)$ we get the algebras $\mathcal{S}^\bullet(E) = \bigoplus_r S^r(E)$ where $S^r(E) = \mathcal{S}_r(T_0^r(E))$, and $V^\bullet(E) = \bigoplus_r V^r(E)$ where $V^r(E) = \mathcal{A}_r(T_0^r(E))$. Elements on $V^\bullet(E)$ are called multivectors and elements on $\Lambda^\bullet(E)$ are called forms. If a multivector u belongs to $A^r(E)$ it is said to be of degree r and the degree is denoted also by $|u|$ (similar notions hold for linear forms).

Both $\Lambda^\bullet(E)$ and $V^\bullet(E)$ inherit a graded associative algebra structure from the graded algebra structure of the tensor algebra. The corresponding products are given by $\alpha \wedge \beta = \mathcal{A}(\alpha \otimes \beta)$ in the case of forms and $V \wedge U = \mathcal{A}(V \otimes U)$ in the case of multivectors.

Finally, we will point it out that a tensor in $T_s^r(E)$ can be represented as a multilinear mapping, $T : E^* \otimes \cdots \otimes E^* \otimes E \otimes \cdots \otimes E \to \mathbb{K}$, defined as $T(u^{j_1} \otimes \cdots u^{j_r} \otimes u_{k_1} \otimes \cdots \otimes u_{k_s}) = T_{k_1,\ldots,k_s}^{j_1,\ldots,j_r}$. Thus the tensor T defines a multilinear map that is precisely s-multilinear in E and r-multilinear in E^*.

Pull-Back and Push-Forward of Forms and Multi-vectors

Even if only the group $GL(E)$ acts in the tensor algebra, we can define in a natural way the pull-back of forms and the push-forward of multi-vectors[5] along a linear map $\varphi\colon E \to F$.

Thus let α be a k-form on F. Define $\varphi^*\alpha$ as the k-form on E by the formula:

$$\varphi^*\alpha(v_1, \ldots, v_k) = \alpha(\varphi(v_1), \ldots, \varphi(v_k)), \qquad \forall v_1, \ldots, v_k \in E.$$

It is easy to check that $(\varphi \circ \psi)^* = \psi^* \circ \varphi^*$, and $\varphi^*(\alpha \wedge \beta) = \varphi^*(\alpha) \wedge \alpha^*(\beta)$.

Similarly we define the push-forward of a k-multi-vector V on E along φ as the k-multi-vector $\varphi_* V$ on F defined as:

$$\varphi_* V(\alpha_1, \ldots, \alpha_k) = V(\varphi^*(\alpha_1), \ldots, \varphi^*(\alpha_k)), \qquad \forall \alpha_1, \ldots, \alpha_k \in F^*.$$

Now we have $(\varphi \circ \psi)_* = \varphi_* \circ \psi_*$.

The Algebra of Polynomials Over a Linear Space

Concentrating now on the subalgebra of r-covariant tensors, we see that evaluation of $P \in T_0^r\,(E)$ 'along the diagonal', i.e., for arguments $u_1 = \cdots = u_r = u$, provides us with functions of the form,

$$P\,(u) = p_{i_1 \ldots i_r} u^{i_1} \ldots u^{i_r} \tag{10.7}$$

with distinguished homogeneity properties with respect to the given linear structure in E.

To begin with, linear functions correspond to tensors $\alpha \in T_0^1(E)$ (commonly denote as T^*E) that maps $\alpha\colon E \to \mathbb{R}$. Linear maps are continuous homogenous functions of degree one.

Tensors $b \in T_0^2(E)$, are bilinear maps $b\colon E \times E \to \mathbb{R}$ and define continuous quadratic forms $Q(u) = b(u, u)$ which are homogeneous of degree 2, $Q(\lambda\, u) = \lambda^2 Q(u)$. Moreover, the symmetric part of the bilinear map b is recovered from the expression:

$$\frac{1}{2}\left(Q\,(x + y) - Q\,(x) - Q\,(y)\right) \tag{10.8}$$

We can repeat the same procedure to obtain cubic forms from trilinear maps $t\colon E \times E \times E \to \mathbb{R}$ as $C(u) = t(u, u, u)$. Cubic forms are clearly homogeneous of degree 3: $C\,(\lambda u) = \lambda^3 C\,(u)$, and the totally symmetric part of the trilinear form t can be

[5] The definitions extend without change to covariant tensors and contravariant tensors respectively.

recovered from C by a formula similar to Eq. (10.8). In general, multilinear maps m of degree r will define 'm'-tic forms or homogeneous polynomials of degree r on E. It is an interesting exercise to recover the totally symmetric part of m from the corresponding homogeneous polynomial.

We denote by $\mathcal{P}^r$ the space of homogeneous polynomial of degree r in E. They form a linear subspace on the space of all polynomials $\mathcal{P}(E)$,

$$\mathcal{P}(E) = \mathcal{P}^0 \oplus \mathcal{P}^1 \oplus \mathcal{P}^2 \oplus \cdots \oplus \mathcal{P}^r \oplus \cdots = \bigoplus_{r \geq 0} \mathcal{P}^r \tag{10.9}$$

The space of polynomials $\mathcal{P}(E)$ is also a subspace of the linear space of all real-valued functions on E. This subspace is also a commutative algebra because the product of two polynomial functions is again a polynomial function and, as a vector space, is an infinite-dimensional real space. On the contrary, as an algebra, it is generated by linear functions because any polynomial function is a finite sum of finite products of the linear functions x^i. Thus the commutative algebra $\mathcal{P}$ is finitely generated.

Appendix C: Smooth Manifolds: A Standard Approach

In the main text we proceed by introducing first exterior differential calculus on finite-dimensional linear spaces and, afterwards we jumped to the notion of a differentiable space as the space defined by the quotient of the algebra of smooth functions on some $\mathbb{R}^n$ by a closed ideal. This way of proceeding was a fast track towards the possibility of using dynamics on manifolds without expending too much time in discussing the theory of smooth manifolds in abstract terms. Actually, as it was discussed any smooth manifold can be described in this way, and then the use of local coordinates and all the apparatus of exterior differential calculus comes directly from the $\mathbb{R}^n$ in which our manifold is embedded.

However, in order to help readers that want to 'taste' a more orthodox introduction to smooth manifolds, we have summarized the main aspects of the theory in this Appendix. Actually, we decided to discuss them again as submanifolds in some $\mathbb{R}^n$ in order to make easier the connection with the formalism of differentiable spaces used in the main text.

At this stage, capitalizing on what has been done in Sects. 2.3 and 2.4 of Chap. 2, we can try to extend the notions already established to the case in which differential calculus needs to be done on 'surfaces' that are (with all the technical qualifications that will given below) subsets of $\mathbb{R}^n$ that are 'curved' in some sense (i.e., for the time being, that cannot be endowed with a linear space structure). These are the *submanifolds* of (or immersed in) $\mathbb{R}^n$. In this connection we may quote again a famous theorem by Whitney stating that *any* finite-dimensional manifold can be viewed as a submanifold of $\mathbb{R}^n$ for n large enough.

Theorem 10.45 (Whitney's embedding theorem) [Whit44] *For $n > 0$, every paracompact hausdorff n-manifold embeds in $\mathbb{R}^{2n}$. Moreover it immerses in $\mathbb{R}^{2n-1}$ if $n > 1$.*

We will limit ourselves in what follows to what we will call 'regular' submanifolds, i.e., those that can be defined as surface levels of functions (again with some regularity properties to be specified), although most of what we will say applies also to more general cases.

Regular Submanifolds of $\mathbb{R}^n$ as Level Surfaces of Functions

We will introduce first the notion of a regular level surface.

Consider a map $\mathbf{F} : \mathbb{R}^n \to \mathbb{R}^k, k = n-m$, with $m \leq n-1$, and $\mathbf{F} = (f^1, \dots, f^k)$, with $f^i : \mathbb{R}^n \to \mathbb{R}$, $i = 1, \dots, k = n - m$, smooth functions. The subset of $\mathbb{R}^n$ defined by $\mathcal{S} = \mathbf{F}^{-1}(c)$ with $c \in \mathbb{R}^k$ in the range of $\mathbf{F}$ with the topology inherited from $\mathbb{R}^n$, is called a level surface of $\mathbf{F}$. The level surface $\mathcal{S} = \mathbf{F}^{-1}(c)$ will be called a regular level surface if c is a regular value of $\mathbf{F}$, that is the Jacobian matrix of f:

$$Jf(x) = \frac{\partial(f^1, \dots, f^k)}{\partial(x^i, \dots, x^n)}, \tag{10.10}$$

must be of maximal rank k on any $x \in \mathcal{S}$. Equivalently, the values $\epsilon^{j_1 \dots j_k}_{i_1 \dots i_k} \frac{\partial f^1}{\partial x^{j_1}} \cdots \frac{\partial f^k}{\partial x^{j_k}}$ of the minor determinants of rank k of the Jacobian are not all zero, or because

$$df^1 \wedge \cdots \wedge df^k = \frac{1}{k!} \frac{\partial(f^1, \dots, f^k)}{\partial(x^i, \dots, x^n)} dx^{i_1} \wedge \cdots \wedge dx^{i_k}, \tag{10.11}$$

the functions f^i are functionally independent along $\mathcal{S}$, that is:

$$(df^1 \wedge \cdots \wedge df^k) \mid_{\mathcal{S}} \neq 0. \tag{10.12}$$

Under these conditions, the Implicit Function Theorem, Theorem 10.43 and its Corollary 10.44, assures that for every point $\mathbf{x} \in \mathcal{S}$ we can choose a subset $(x^{i_1}, \dots, x^{i_m})$ of $m = n - k$ of the coordinates $(x^1, \dots, x^n)$ and associate with a point $\mathbf{x} \in \mathcal{S}$ an open neighborhood $U_{\mathbf{x}}$ in $\mathcal{S}$ and an open set $V_{\mathbf{x}}$ in $\mathbb{R}^m$ with coordinates $(x^{i_1}, \dots, x^{i_m})$ that can be mapped diffeomorphically one onto each other. We say then that $\mathcal{S}$ can be 'locally modeled on $\mathbb{R}^m$' and as m independent coordinates suffice to parametrize the neighborhood of every point, this justifies saying that $\mathcal{S}$ is an m-dimensional surface immersed in $\mathbb{R}^n$.

Example 10.1 Let be $n = 3$ and $k = 1$ with $\mathbf{x} = (x^1, x^2, x^3)$ and F such that $F(\mathbf{x}) = \|\mathbf{x}\|^2 - 1$. Then, the zero level set of F, $\mathcal{S} = F^{-1}(0)$, is the two-sphere S^2, which can be covered by the six open sets $U_i^{\pm} = \{\mathbf{x} \in S^2 \mid \pm x^i > 0\}$, on each one of which points on S^2 are parametrized by a pair of their coordinates in $\mathbb{R}^3$. For example, on $U_1^\pm$ we have the maps:

$$\phi_1^\pm : U_1^+ \to D_1 = \{(x^2, x^3) \in \mathbb{R}^2 \mid (x^2)^2 + (x^3)^2 < 1\} ,$$

via $\phi_1^\pm(x^1, x^2, x^3) = (x^2, x^3)$, that can be inverted as:

$$(\phi_1^\pm)^{-1} : D_1 \to U_1^+ , \qquad (\phi_1^\pm)^{-1}(x^2, x^3) = (\pm\sqrt{1 - \ (x^2)^2 + (x^3)^2}, x^2, x^3) .$$

It is left as an exercise to define the remaining homeomorphisms $\phi_i^\pm$ $(i = 2, 3)$ between half-spheres without the boundary and open disks in some $\mathbb{R}^2$.

This literal use of the Implicit Function Theorem is admittedly not the most efficient way of modeling open subsets of S^2 on open sets of $\mathbb{R}^2$. A different and maybe more elegant way is offered by the stereographic projections from the two poles onto $\mathbb{C} \cong \mathbb{R}^2$, namely,

(i) Stereographic (N): Here $U_N = \{S^2 - (0, 0, 1)\}$ and $\phi_N : U_N \to \mathbb{C}$ is defined by:

$$\phi_N(x^1, x^2, x^3) = z = x + i\, y = \frac{x^1 + i\, x^2}{1 - x^3} , \tag{10.13}$$

that can be inverted as:

$$x^1 = \frac{\bar{z} + z}{|z|^2 + 1} , \qquad x^2 = i\, \frac{\bar{z} - z}{|z|^2 + 1} , \qquad x^3 = \frac{|z|^2 - 1}{|z|^2 + 1} \tag{10.14}$$

(ii) Stereographic (S): Here $U_S = \{S^2 - (0, 0, -1)\}$ and $\phi_S : U_S \to \mathbb{C}$ is given by

$$\phi_S(x^1, x^2, x^3) = z = x + i\, y = \frac{x^1 + i\, x^2}{1 + x^3} , \tag{10.15}$$

that can be inverted as:

$$x^1 = \frac{\bar{z} + z}{1 + |z|^2} , \qquad x^2 = i\, \frac{\bar{z} - z}{1 + |z|^2} , \qquad x^3 = \frac{1 - |z|^2}{1 + |z|^2} . \tag{10.16}$$

Remark 10.1 As S^2 is compact it cannot, for obvious topological reasons, be mapped homeomorphically onto a single open set in $\mathbb{R}^2$. That is the reason why we need at least two open sets, as in the stereographic projections, to represent the whole of S^2.

C.1. Charts and Atlases: Submanifolds

What the Implicit Function Theorem is telling us is that we can cover our surface $\mathcal{S}$ with a collection $\{U_i\}$ of open sets each one of which can be mapped homeomorphically on (i.e., 'modeled on') some open subset of $\mathbb{R}^m$ via an associated collection $\{\phi_i\}$ of maps. When this happens, we say that each pair (U_i, ϕ_i) is a chart for $\mathcal{S}$, and the open set U_i is said to be the domain of the chart.

It can happen that some points of $\mathcal{S}$ belong to the intersection of the domains of two charts, $\{U_i, \phi_i\}$ and $\{U_j, \phi_j\}$ say, and hence, they have different parametrizations in $U_i \cap U_j$. The problem then arises of how to compare these different descriptions.

Consider again the example of the previous Subsection, namely the two-sphere S^2, and, e.g., the two charts: $\{U_1^+, \phi_1^+\}$, and $\{U_3^+, \phi_3^+\}$. Hence, $U_1^+ \cap U_3^+ = \{\mathbf{x} \in S^2,\ x^1, x^3 > 0\}$ and D_1 and D_3 is the unit open disks in the planes $\{x^2, x^3\}$ and $\{x^1, x^2\}$ respectively. So, $\phi_1^+(x^1, x^2, x^3) = (x^2, x^3)$, $(\phi_1^+)^{-1}(x^2, x^3) = (\sqrt{1 - (x^2)^2 + (x^3)^2}, x^2, x^3)$, while $\phi_3^+(x^1, x^2, x^3) = (x^1, x^2)$ and $(\phi_3^+)^{-1}(x^2, x^3) = (x^1, x^2, \sqrt{1 - (x^1)^2 + (x^2)^2})$. Combining them together we find that on $U_1^+ \cap U_3^+$, $\phi_3^+ \circ (\phi_1^+)^{-1} : D_1 \to D_3$ is given by

$$(x^2, x^3) \mapsto \left(\sqrt{1 - (x^2)^2 + (x^3)^2}, x^2\right), \tag{10.17}$$

and similarly, $\phi_1^+ \circ (\phi_3^+)^{-1} : D_3 \to D_1$ is defined by

$$(x^1, x^2) \mapsto \left(x^2, \sqrt{1 - (x^1)^2 + (x^2)^2}\right). \tag{10.18}$$

Quite similarly, in the case of the stereographic projections, denoting the two charts as (U_i, ϕ_i), $i = N, S$, $U_{N(S)}$ is the sphere without the $N(S)$ pole, and $U_N \cap U_S$ projects onto the punctured plane $\mathbb{C} - \{0\} \cong \mathbb{R}^2 - \{\mathbf{0}\}$. Denoting by z and z' the projections associated with ϕ_N and ϕ_S respectively, then:

$$\phi_S \circ \phi_N^{-1} : z \mapsto z' = \frac{1}{\bar{z}}, \tag{10.19}$$

(the bar denoting complex conjugation). This is also a C^∞ diffeomorphism of the punctured plane onto itself. In particular, it maps diffeomorphically the exterior of the unit disk onto its punctured interior.

The functions $\phi_3^+ \circ (\phi_1^+)^{-1}$ and $\phi_1^+ \circ (\phi_3^+)^{-1}$, or $\phi_S \circ \phi_N^{-1}$ and $\phi_N \circ \phi_S^{-1}$, (that are the inverse of each other) or, more generally, $\phi_i \circ \phi_j^{-1} : \phi_j(U_i \cap U_j) \to \phi_i(U_i \cap U_j)$ on a general surface $\mathcal{S}$, are called transition functions, and we see on the example of the sphere that they are all smooth diffeomorphisms. We state here without proof that the same holds true in general whenever the conditions of the Inverse Function Theorem are fulfilled. We will abstract the situation depicted above to get the following definition.

Definition 10.46 Let $\mathcal{S}$ be a topological space. A collection of open sets and maps $\{(U_i, \phi_i)\}$ such that $\bigcup_i U_i = \mathcal{S}$, i.e., the family $\{U_i\}$ is a open covering of $\mathcal{S}$, $\phi_i : U_i \to \phi_i(U_i) \subseteq \mathbb{R}^m$ is a homeomorphism onto its image, and the maps ϕ_i's are such that $\phi_i \circ \phi_j^{-1} : \phi_j(U_i \cap U_j) \to \phi_i(U_i \cap U_j))$ are smooth for all i, j, is called an smooth atlas[6] (or just an atlas for short). A chart is compatible with an atlas when the union of the atlas with the chart is an atlas. Two different atlases are said to be compatible iff their union is once again an atlas. Given an atlas and considering the union of all compatible atlas we will find a compatible maximal atlas. A maximal atlas on a topological space is called a differentiable structure.

Exercise 10.2 Show that the two atlases for the two-sphere that have been introduced before are compatible.

We get immediately:

Proposition 10.47 *A regular level surface can be equipped with a differentiable structure.*

Definition 10.48 The regular level surface $\mathcal{S}$ equipped with its differentiable structure is said to be a regular submanifold of $\mathbb{R}^n$.

Again, the previous statements can be made into a general definition.

Definition 10.49 A (smooth) manifold M modeled on the real Banach space E (also-called a Banach manifold) is a Hausdorff and second countable topological space with an (smooth) atlas with charts (U_i, ϕ_i), $\phi : U_i \to E$ defining its differentiable structure. If E is finite dimensional, say $\mathbb{R}^n$, then we say that M has dimension n.

When $\mathcal{S}$ is a subset of $\mathbb{R}^n$ for some n, together with it goes a natural map, the so-called canonical identification (or inclusion) map, $i : \mathcal{S} \to \mathbb{R}^n$ by which points on $\mathcal{S}$ are identified with points in $\mathbb{R}^n$, $i : \mathbf{x} \in \mathcal{S} \mapsto \mathbf{x} \in \mathbb{R}^n$. We will make a systematic use of the identification map in what follows. It can also be shown that the induced topology is the weakest topology under which the identification map is a continuous map.

From now on we will work on the class of smooth manifolds, even though all the ideas can be particularized to the situation where the manifolds are just regular level surfaces of $\mathbb{R}^n$.

C.2. Tangent and Cotangent Bundle: Orientability

What we want to show briefly here is that the existence of a differentiable structure allows us to extend all of the machinery of differential and exterior calculus from linear spaces as it was developed in Sect. 2.4 to manifolds, and that the fulfillment

6 C^r-atlas are defined similarly.

of the compatibility conditions on charts ensures the internal consistency of the extensions.

We say that a curve $c\colon I \to M$, where I is an interval in $\mathbb{R}$, is a (smooth) differentiable curve if, for any chart (U, ϕ), the curve: $\widetilde{c} = \phi \circ c : I \to \mathbb{R}^m$ is a (smooth) differentiable curve in $\mathbb{R}^m$. If (U', ϕ') is another chart with $U \cap U' \neq \emptyset$, the fact that $\phi' \circ \phi^{-1}$ is a C^∞-diffeomorphism ensures that the equivalent image $\widetilde{c}' = \phi' \circ c \equiv (\phi' \circ \phi^{-1}) \circ \widetilde{c}$, enjoys the same properties, so the notion of differentiability of a curve is established once a given differentiable structure has been established on M. In what follows we will consider that curves are always smooth and that 0 is in the interior of their domains.

Quite similarly, a function $f\colon M \to \mathbb{R}$ is said to be (smooth) differentiable if the function $\widetilde{f} : (\phi^{-1})^* f = f \circ \phi^{-1} : \phi(U) \to \mathbb{R}$ is (smooth) differentiable for any chart (U, ϕ), and once again the compatibility conditions ensure the consistency of the definition, i.e., that it depends only on the differentiable structure and not on the particular chart that has been used for the definition of $\widetilde{f}$.

In this way we can define the associative commutative algebra $\mathcal{F}(M)$ of smooth functions on M (sometimes denoted as $C^\infty(M)$, even though we will stick most of the time to the previous notation). The sum and the product of two functions f_1 and f_2 are the functions defined by $(f_1 + f_2)(x) = f_1(x) + f_2(x)$ and $(f_1\, f_2)(x) = f_1(x)\, f_2(x)$, and the product by a scalar λ by $(\lambda\, f)(x) = \lambda\, f(x)$ for all $x \in M$ (and clearly all of them are smooth if f_1 and f_2 are smooth).

Given the point $x \in M$, we may define an equivalence relation on the family of smooth functions as follows: we will say that the functions f and g are equivalent at x if there exists an open neighborhood U of x such that $f\mid_U = g\mid_U$. The equivalence classes of functions at x are called germs of functions at x and will be denoted as $[f]_x$. The space $\mathcal{F}_x$ of germs of functions at x inherits the structure of a commutative associative algebra from $\mathcal{F}(M)$ by means of $[f]_x + [g]_x = [f + g]_x$ and $[f]_x[g]_x = [fg]_x$. The algebra $\mathcal{F}_x$ has a natural maximal ideal given by all germs of functions vanishing at x that will be denoted by $\mathfrak{m}_x$.

Proceeding further, we can, much as in the linear case (see Sect. 2.3.7), define the equivalence of curves at a point $x \in M$ as follows: Two differentiable curves c_1 and c_2 through the point x, i.e., $c_1(0) = c_2(0) = x$, are tangent at x if in a given chart (U, ϕ) (hence in all), with $x \in U$, they satisfy $\frac{d}{dt}(\phi \circ c_1)|_{t=0} = \frac{d}{dt}(\phi \circ c_2)|_{t=0}$. The relation '$c_1$ is equivalent to c_2 if they are tangent at x' is indeed an equivalence relation and that it does not depend on the choice of the chart.

Let us denote by $[c]_x$ an equivalence class of curves at x, $c = c(t)$ being any one of its representatives with $c(0) = x$. Each equivalence class $[c]_x$ defines linear map on the algebra of germs at x, by means of $[c]_x([f]_x) = \frac{d}{dt}(f \circ c(t))\mid_{t=0}$. Hence the space of equivalence classes of curves at x becomes a linear space that will be denoted by T_xM and its elements, the classes $[c]_x$ will be denoted by v_x or simply $v \in T_xM$ and will be called tangent vectors to the manifold M at the point x. The space T_xM consequently will be called the tangent space to M at x.

If M is a manifold of dimension n, given x an a chart (U, ϕ) such that $x \in U$, there is a natural basis of T_xM induced from the standard basis $\{e_i\}$ in $\mathbb{R}^n$, by means of the family of curves $c_i(t) = \phi^{-1}(\phi(x) + te_i)$, $t \in (-\epsilon, \epsilon)$ for ϵ small enough. Then the tangent vector corresponding to the class $[c_i]_x$ will be denoted as $\frac{\partial}{\partial x^i}\mid_x$. Notice that given any tangent vector $v = [c]_x$ in $T_x M$, then because of $\frac{d}{dt}\phi \circ c(t)\mid_{t=0} = \sum_{i=0}^{n} v^i e_i$, we get that $v = v^i \frac{\partial}{\partial x^i}$.

It is clear that $[c]_x$ determines a unique vector $\mathbf{v}$ in $\mathbb{R}^m$ (the common tangent at $\phi(x)$ of all the image curves). In turn, any vector $\mathbf{v} \in \mathbb{R}^m$ determines the curve $c : t \mapsto \phi^{-1}(\phi(x) + \mathbf{v}t)$ in $\mathcal{S}$, and hence an equivalence class. We denote this bijection by $T_x\phi$:

$$T_x\phi : \{[c]_x\} \longleftrightarrow \mathbb{R}^m \quad by \qquad [c]_x \longleftrightarrow \mathbf{v} = \frac{d}{dt}(\phi \circ c)|_{t=0} \tag{10.20}$$

Exercise 10.3 Determine how $T_x\phi$ transforms under a change of charts and prove that

$$T_x\phi' = \frac{d}{dt}(\phi' \circ \phi^{-1})\, T_x\phi\,. \tag{10.21}$$

Much like as in the linear case on can endow the set $\{[c]_x\}$ with a real linear structure defined by

$$\alpha[c_1]_x + \beta[c_2]_x = (T_x\phi)^{-1}\left(\alpha\frac{d}{dt}(\phi \circ c_1)|_{t=0} + \beta\frac{d}{dt}(\phi \circ c_2)|_{t=0}\right) \tag{10.22}$$

where $\alpha, \beta \in \mathbb{R}$. Again, it is left as an exercise to show that the right-hand side does not depend on the choice of the chart. Then,

Definition 10.50 The real vector space $T_x\mathcal{S} = \{[c]_x\}$ is called the tangent space to $\mathcal{S}$ at the point x.

Besides considering real-valued functions on $\mathcal{S}$, we will need in the following also to consider, more generally, maps between manifolds. Let then $\mathcal{S}$ and Σ be (sub)manifolds with $\dim(\mathcal{S}) = m$ and $\dim(\Sigma) = k$, and consider a map between them $F : \mathcal{S} \to \Sigma$.

Let (U, ϕ) be a chart on $\mathcal{S}$ containing the point x_0 and (V, ψ) a chart of Σ such that $F(x_0) \in V$. The function F is a differentiable map in x_0 if the map $\psi \circ F \circ \phi^{-1} : \phi(U) \to \psi(V)$ is differentiable at the point $\phi(x_0)$. The function is differentiable in an open set of $\mathcal{S}$ when it is differentiable in each point of the open set. The map F induces in an obvious way a map $T_{x_0}(F)$ called the derivative, or the differential of F between the tangent spaces at $x_0 \in \mathcal{S}$ and $F(x_0) \in \Sigma$, i.e.

$$T_{x_0}(F) : T_{x_0}\mathcal{S} \to T_{F(x_0)}\Sigma \tag{10.23}$$

by means of

$$[c]_{x_0} \mapsto [F \circ c]_{F(x_0)} \tag{10.24}$$

From the definition (see above) of $T_{x_0}(\phi)$ (and of $T_{F(x_0)}(\psi)$) one obtains immediately,

$$T_{x_0}(F) = (T_{F(x_0)}(\psi))^{-1} \circ D(\psi \circ F \circ \phi^{-1}) \circ T_{x_0}(\phi) \tag{10.25}$$

where D is the differential of a function of $\mathbb{R}^m$ into $\mathbb{R}^n$, from where it is evident (but that could have been inferred directly from the definition) that $T_{x_0}(F)$ is a linear map between tangent spaces. Of course $D(\psi \circ F \circ \phi^{-1})$, the 'local representative' of F, is nothing but the Jacobian matrix of the 'local representative' $\psi \circ F \circ \phi^{-1}$ of F on the charts.

Let us go back now to tangent spaces. Tangent vectors are also-called "contravariant vectors' to distinguish them from covariant ones (see below).

Noting that, given $\mathbf{u}_{x_0} = [c]_{x_0} \in T_{x_0}\mathcal{S}$ and $f \in \mathcal{F}(\mathcal{S})$, the composition map $f \circ c$ can be defined entirely in terms of its image on the range of a chart, i.e., that $f \circ c \equiv f \circ \phi^{-1} \circ \phi \circ c \equiv \widetilde{f} \circ \widetilde{c}$, the proof that tangent vectors at x_0 act as derivations at the point x_0 on the algebra of smooth functions (and vice versa) can be extended once more from the context of linear spaces to that of (sub)manifolds without essentially no significant changes, and we do not repeat it here. Then, if the curve c is a representative of the vector $\mathbf{u}_{x_0}$ at the point x_0,

$$\frac{d}{dt}(f \circ c) = \frac{d}{dt}(\widetilde{f} \circ \widetilde{c}) \tag{10.26}$$

and if $\{\mathbf{e}_i \mid i = 1, \ldots, m\}$ is a basis of unit vectors in $\mathbb{R}^m$ (and hence also a basis for $T_{\phi(x_0)}\mathbb{R}^m$), the set of inverse images

$$[c_i]_{x_0} = (T_{x_0}\phi)^{-1}(\mathbf{e}_i)\,, \qquad i = 1, \ldots, m, \tag{10.27}$$

in turn defines a basis in $T_{x_0}\mathcal{S}$, more explicitly $[c_i]_{x_0} = [\phi^{-1}(\phi(x_0) + \mathbf{e}_i t)]_{x_0}$, and hence

$$\frac{d}{dt}(f \circ c_i)|_{t=0} = \left(\frac{\partial \widetilde{f}}{\partial x^i}\right)_{|\phi(x_0)}, \tag{10.28}$$

where $(x^1, \ldots, x^m)$ are the local coordinates associated with the chart (U, ϕ) and

$$\mathbf{u}_{x_0} f = \frac{\partial \widetilde{f}}{\partial x^i}\frac{d\widetilde{c}_i}{dt}\,.$$

It is then natural to introduce, in the domain of a chart, the notation

$$\left(\frac{\partial}{\partial x^i}\right)_{|x_0} = [c_i]_{x_0} = (T_{x_0}\phi)^{-1}(\mathbf{e}_i)\,. \tag{10.29}$$

Every tangent vector $\mathbf{u}_{x_0}$ at the point x_0 is then represented in the form

$$\mathbf{u}_{x_0} = u^i \left(\frac{\partial}{\partial x^i}\right)_{|x_0} \tag{10.30}$$

(and $T_{x_0}\phi(\mathbf{u}_{x_0}) = u^i\,\mathbf{e}_i$). In general, just as in the linear case,

$$\mathbf{u}_{x_0}(f) = u^i \left(\frac{\partial \tilde{f}}{\partial x^i}\right)_{|\phi(x_0)} \tag{10.31}$$

(and in particular $u^i = \mathbf{u}_{x_0}(x^i)$).

Note that the vector $\mathbf{u}_{x_0}$ can be seen as a linear map $\mathbf{u}_{x_0} : \mathcal{F}(\mathcal{S}) \to \mathcal{F}(\mathcal{S})$ such that satisfies the Leibnitz rule,

$$\mathbf{u}_{x_0}(f_1\, f_2) = f_1(x_0)\,\mathbf{u}_{x_0} f_2 + f_2(x_0)\,\mathbf{u}_{x_0} f_1 \tag{10.32}$$

Exercise 10.4 Prove that, if $(x'^1, \ldots, x'^m)$ are the local coordinates associated with a different chart (U', ϕ'), then,

$$\left(\frac{\partial}{\partial x'^i}\right)_{|x_0} = \left(\frac{\partial x^k}{\partial x'^i}\right)_{|\phi'(x_0)} \left(\frac{\partial}{\partial x^k}\right)_{|\phi(x_0)} . \tag{10.33}$$

and hence that

$$u'^i = \left(\frac{\partial x'^i}{\partial x^k}\right)_{|\phi(x_0)} u^k . \tag{10.34}$$

Remark 10.2 We are employing here an apparently cumbersome notation. This is to stress the fact that, while local bases can be defined rather easily at every point of a (sub)manifold, there is no 'natural' or 'canonical' relationship between them at different points, and that unless some additional structure such as some notion of 'parallel transport' of vectors on the manifold has been established, what we will do only in a later section. However, in all the cases in which no ambiguities can arise, we use the simplified notation $\{(\partial/\partial x^i)_{|x_0}\}$ for the natural bases in the domain of a chart.

We next go from the tangent to the cotangent space at a point much in the same way as we went in the linear case, namely:

Definition 10.51 The cotangent space $T^*_{x_0}\mathcal{S}$ at the point $x_0 \in \mathcal{S}$ is the dual space of $T_{x_0}\mathcal{S}$. Any $\omega \in T^*_{x_0}\mathcal{S}$ is called covariant vector, or covector for short, at the point x_0, and is a linear functional on $T_{x_0}\mathcal{S}$.

Definition 10.52 The differential of a function $f \in \mathcal{F}(\mathcal{S})$ at the point x_0 is the unique covector $df(x_0)$ such that

$$df(x_0)(\mathbf{u}) = \mathbf{u}(f)\,, \qquad \forall \mathbf{u} \in T_{x_0}\mathcal{S}\,. \tag{10.35}$$

Now, if $\{(\partial/\partial x^i)_{|x_0} \mid i = 1, \ldots, m\}$ is the basis of $T_{x_0}\mathcal{S}$ associated to a given chart and we choose for f a coordinate function, e.g., $f : x \to x^i$, then, the previous definition gives

$$dx^i(x_0)\ ((\partial/\partial x^j)_{|x_0}) = \delta^i_j\ , \tag{10.36}$$

and hence: $\{dx^i(x_0) \mid i = 1, \ldots, m\}$ is a basis of $T^*_{x_0}\mathcal{S}$ dual to the natural basis in $T_{x_0}\mathcal{S}$ that is called the cobasis associate to the given chart. Any covector ω at the point x_0 is locally expressed in a unique way in the form $\omega = \omega_i\, dx^i(x_0)$ and $\omega(\mathbf{u}) = \omega_i\, u^i$, for any pair of elements $\omega \in T^*_{x_0}(\mathcal{S})$ and $\mathbf{u} \in T_{x_0}(\mathcal{S})$. In particular,

$$df(x_0) = \left(\frac{\partial \widetilde{f}}{\partial x^j}\right)_{|\phi(x_0)} dx^j(x_0)\,, \tag{10.37}$$

for any $f \in \mathcal{F}(\mathcal{S})$. Under a change of chart the associated bases are related by

$$dx'^i(x_0) = \left(\frac{\partial x'^i}{\partial x^k}\right)_{|\phi(x_0)} dx^k(x_0) \tag{10.38}$$

and therefore the relation among coordinates of a covector in both bases is

$$\omega'_i = \left(\frac{\partial x^k}{\partial x'^i}\right)_{|\phi'(x_0)} \omega_k\ . \tag{10.39}$$

The transformation law explains the nomenclature that has been introduced for vectors and covectors, because the matrix appearing in the preceding relation is the inverse of that of relation (10.34).

Vector fields and 1-forms are defined here too as smooth assignments on $\mathcal{S}$ of tangent and cotangent vectors respectively. In the domain of a chart that amounts to promoting the u^i's and/or the ω_i's to smooth functions, transforming appropriately when moving from a chart to an overlapping one. Out of vector fields and forms we will be able eventually to construct tensor fields of arbitrary rank as well.

The unions

$$T\mathcal{S} = \bigcup_{x\in\mathcal{S}} T_x\mathcal{S}\,, \qquad T^*\mathcal{S} = \bigcup_{x\in\mathcal{S}} T^*_x\mathcal{S} \tag{10.40}$$

are called the tangent and cotangent bundle respectively. They have been defined here only as set-theoretic unions. It can be shown that there is a canonical way to endow both of them with a differentiable structure inherited from the one on $\mathcal{S}$.

Pull-Back of Covariant Tensors and Forms from $\mathbb{R}^n$ to Submanifolds

The operation of taking the pull-back of a map has been already defined mainly in connection with functions in the linear context. We want to extend it here to manifolds and to general $\binom{0}{r}$-type (i.e., totally covariant) tensor fields. Let $F: \mathcal{S} \rightarrow \Sigma$ be a differentiable (smooth, actually) map between (sub)manifolds. In analogy with the linear case, the pull-back (to $\mathcal{S}$) of a function $g : \Sigma \rightarrow \mathbb{R}$ can be defined as $F^*g = g \circ F$. Let now $t \in T_r^0(\Sigma)$ be a totally covariant tensor field of rank r on Σ. Then,

Definition 10.53 The pull-back by F on tensors of type $(0, r)$ is the map:

$$F^*: T_r^0(\Sigma) \rightarrow T_r^0(\mathcal{S}) \tag{10.41}$$

defined by:

$$(F^*t)_x(\mathbf{u}_1, \ldots, \mathbf{u}_r) = t_{F(x)}(T_x F(\mathbf{u}_1), \ldots, T_x F(\mathbf{u}_r)) \tag{10.42}$$

where $\mathbf{u}_1, \ldots, \mathbf{u}_r \in T_x\mathcal{S}$.

Note that $(F^*t)_x$ has the same symmetry properties under interchange of indices as t_x, and therefore, the pull-back of symmetric or skew-symmetric tensors are also symmetric or skew-symmetric. In particular, forms of arbitrary rank may be pulled-back, a possibility of which we will make large use, as well as other covariant tensors.

In particular, consider the case in which $\Sigma \equiv \mathbb{R}^n$ and F is the identification map $F = i : \mathcal{S} \hookrightarrow \mathbb{R}^n$. Assume that a Riemannian metric (see below Appendix E) $\widetilde{g}$ has been assigned on $\mathbb{R}^n$ (for example the standard Euclidean metric). Then,

$$g = i^*\widetilde{g} \tag{10.43}$$

is a Riemannian metric on $\mathcal{S}$. So we obtain that if $\mathcal{S}$ is a regular immersed submanifold of $\mathbb{R}^n$ for some n it can be always endowed with a Riemannian metric.

Example 10.5 Consider $\mathbb{R}^3$ (with Cartesian coordinates (x, y, z)) with the Euclidean metric $\widetilde{g} = (dx)^2 + (dy)^2 + (dz)^2$. If we choose polar coordinates on the two-sphere S^2 (actually the two-sphere with the poles removed, which is homeomorphic to a cylinder), given by $x = \sin\theta\, \cos\phi,\ y = \sin\theta\, \sin\phi,\ z = \cos\theta, \theta \in (0, \pi), \phi \in [0, 2\pi]$, then $\mathcal{S}^2$ inherits[7] the Riemannian metric:

$$g = (d\theta)^2 + \sin^2\theta\, (d\phi)^2\,, \tag{10.44}$$

i.e., the Riemannian structure is defined by,

---

[7] See below, page 647.

$$g_{\theta\theta} = 1\,, g_{\phi\phi} = \sin^2\theta, g_{\theta\phi} = g_{\phi\theta} = 0. \tag{10.45}$$

Note however that the 1-forms $d\theta$ and $d\phi$ are well defined only on the cylinder (i.e., on the sphere without the poles) and are not globally defined on the sphere (see also the discussion in Remark 2.7 above). This has to do with the fact that the sphere S^2 is not 'parallelizable'.

Orientability and Orientations

Consider $\mathbb{R}^n$ to start with. Any two coordinate systems $(x^i), (x'^i), i = 1, \ldots, n$ on $\mathbb{R}^n$ (or on any open subset of it) will be said to define the same orientation iff their Jacobian determinant: $J = \frac{\partial(x'^1,\ldots,x'^n)}{\partial(x^1,\ldots,x^n)}$ is everywhere positive. As the sign of the Jacobian can be changed by just changing the sign of one of the coordinates, this implies that $\mathbb{R}^n$ and its open subsets can have only two different orientations. Comparing the standard volume-forms associated with $\Omega = dx^1 \wedge \cdots \wedge dx^n$ and $\Omega' = dx'^1 \wedge \cdots \wedge dx'^n$, we know that: $\Omega' = J \cdot \Omega$, so the two volume-forms will be proportional (as they should) with an everywhere positive proportionality factor.

The obvious way to extend this notion to submanifolds of $\mathbb{R}^n$ is given by the following:

Definition 10.54 A submanifold $\mathcal{S}$ of $\mathbb{R}^n$ will be orientable iff it can be given an oriented atlas, i.e., an atlas $\{(U_i, \phi_i)\}$ such that for any pair of overlapping charts, $(U_i, \phi_i), (U_j, \phi_j)$, such that $U_i \cap U_j \neq \emptyset$, $\det[D(\phi_i \circ \phi_j^{-1})] > 0$ holds, where $D(\phi_i \circ \phi_j^{-1})$ is (cfr. Sect. 5.4.5.) the Jacobian matrix of the map: $\phi_i \circ \phi_j^{-1} : \phi_j(U_j) \to \phi_i(U_i)$.

For example, on the two-sphere S^2, the atlas defined by the stereographic projection (see Sect. 1.1.1) has: $D(\phi_S \circ (\phi_N)^{-1}) = -1/|z|^2$ and is not an oriented atlas. However,the atlas defined using the covering with the six (open) half-spheres $U_i^{\pm}$ has, e.g.: $D(\phi_3^+ \circ (\phi_1^+)^{-1}) = x^3/\sqrt{1-(x^2)^2-(x^3)^2} > 0$ and so do the other Jacobians. The atlas is therefore an oriented one and S^2 is orientable. This carries over to the general case, namely it can be shown that all regular submanifolds (obtained as level sets of functions) of $\mathbb{R}^n$ are orientable.

Orientability has to do with the global existence of volume-forms (i.e., of nowhere vanishing forms of maximal rank). Actually, it can be proved [Ab78, Ab88] that:

Proposition 10.55 *A manifold $\mathcal{S}$ admits a volume-form iff it is orientable.*

Exterior Differential Calculus

We have seen that there are manifolds that arise as submanifolds of some $\mathbb{R}^n$, and we have mentioned also Whitney's theorem [Whit44], Theorem 10.45, according to which any manifold may be seen as a sumbanifold of $\mathbb{R}^n$ when n is large enough.

We may now take advantage of the intrinsic differential calculus introduced so far on $\mathbb{R}^n$ to induce a differential calculus on submanifolds.

From the conceptual point of view, this means that we are not required to go over again through the definition of vector fields and 1-forms as we have already done it in Sect. 2.4. The main idea here is to capitalize on our presentation in Sect. 2.4.2, i.e., expressing the exterior derivative and the Lie derivative in terms of appropriateli chosen vector fields and 1-forms.

Let us go back to the bare essentials involved in our presentation of the external differential calculus on $\mathbb{R}^n$. We have expressed vector fields and 1-forms as,

$$X = f^j \frac{\partial}{\partial x^j}; \; df = \frac{\partial f}{\partial x^j} dx^j \tag{10.46}$$

When the x^j's are globally defined so are the vector fields $\partial/\partial x^j$ and 1-forms dx^j, $j = 1, \ldots, n$. They provide then us with a basis of the modules of vector fields and 1-forms respectively, the coefficients being functions on $\mathbb{R}^n$.

From the point of view of the differential equations associated with vector fields we notice that by integrating the differential equations associated with each one of the $\partial/\partial x^j$'s we obtain a global action of the Abelian translation group $\mathbb{R}^n$. The corresponding 1-forms dx^j provide us with the dual version, in terms of covariant fields, of the same action, because they give raise to a one-to-one map between our manifold E (the vector space manifold we are dealing with) and $\mathbb{R}^n$. This one-to-one map allows us to 'export' on E all the structures that are available on $\mathbb{R}^n$.

When the submanifold of E we are considering is 'curved' in some way, for instance if it is a sphere, it is reasonable to expect that a global action of the Abelian translation group will not exixt on it. Our strategy to define an external differential calculus will be to replace the $\partial/\partial x^j$'s with other vector fields 'adapted' to the submanifold, i.e., a basis of them tangent to the submanifold itself. At the same time we will consider an 'adapted' basis of 1-forms.

The most simple example is provided by the submanifold of $\mathbb{R}^2$, defined by the equation,

$$a^2x^2 + b^2y^2 = 1 \tag{10.47}$$

which represents an ellipse.

An adapted basis of vector fields and 1-forms on $\mathbb{R}^2 - \{\mathbf{0}\}$ is given by,

$$R = \frac{a}{b}x\frac{\partial}{\partial y} - \frac{b}{a}y\frac{\partial}{\partial x}, \qquad \Delta = x\frac{\partial}{x\partial} + y\frac{\partial}{\partial y} \tag{10.48}$$

and,

$$\theta = ab\frac{xdy - ydx}{a^2x^2 + b^2y^2}, \qquad \alpha = \frac{a^2xdx + b^2ydy}{a^2x^2 + b^2y^2} \tag{10.49}$$

We have excluded the origin from $\mathbb{R}^2$ because R and Δ are zero and θ and α would not be defined there.

We find,

$$\begin{aligned} \theta(R) &= 1, \quad & \theta(\Delta) &= 0 \\ \alpha(R) &= 0, \quad & \alpha(\Delta) &= 1 \end{aligned} \tag{10.50}$$

These relations allow us to write a (1, 1) tensor field that represents the identity map on the tangent or on the cotangent bundle as,

$$T_{\mathbb{I}} = \theta \otimes R + \alpha \otimes \Delta \equiv \mathbb{I} \tag{10.51}$$

Any vector field, say X, may be written in the form,

$$X = T_{\mathbb{I}}(X) = \theta(X)\,R + \alpha(X)\,\Delta \tag{10.52}$$

and similarly for any 1-form β,

$$\beta = T_{\mathbb{I}}(\beta) = \beta(R)\theta + \beta(\Delta)\alpha \tag{10.53}$$

which, for exact 1-forms, becomes,

$$df = \theta\mathcal{L}_R(f) + \alpha\mathcal{L}_\Delta(f) \tag{10.54}$$

which replaces the expression:

$$df = dx\frac{\partial f}{\partial x} + dy\frac{\partial f}{\partial y} \tag{10.55}$$

Stated otherwise, we are writing the exterior differential d in the adapted basis as,

$$d = \theta \wedge \mathcal{L}_R + \alpha \wedge \mathcal{L}_\Delta \tag{10.56}$$

On a generic 1-form, say, $\beta = g\,df$, we have,

$$\begin{aligned} d(g\,df) &= (\theta\mathcal{L}_R(g) + \alpha\mathcal{L}_\Delta(g)) \wedge (\theta\mathcal{L}_R(f) + \alpha\mathcal{L}_\Delta(g)) \\ &= (\theta \wedge \alpha)(\mathcal{L}_R(g)\,\mathcal{L}_\Delta(f) - \mathcal{L}_R(f)\,\mathcal{L}_\Delta(g)) \end{aligned} \tag{10.57}$$

If: $i_S : S^1 \hookrightarrow \mathbb{R}^2$ is the identification map immerging the ellipse in $\mathbb{R}^2$, we see immediately that from: $i_S^*\alpha = 0$ there follows:

$$i_S^* d_{\mathbb{R}^2} f = d_{S^1} i_S^* f \tag{10.58}$$

where we have written $d_{\mathbb{R}^2}$ and d_{S^1} to emphasize that they are the exterior differentials on $\mathbb{R}^2$ and S^1 respectively.

A generalization of our simple example to many dimensions would require an 'adapted basis' of vector fields, say,

$$\{R_1, R_2, \ldots, R_k; Y_1, Y_2, \ldots, Y_{n-k}\} \tag{10.59}$$

and corresponding 1-forms,

$$\{\theta^1, \theta^2, \ldots, \theta^k; \alpha^1, \alpha^2, \ldots, \alpha^{n-k}\} \tag{10.60}$$

such that,

$$\begin{array}{ll} \theta^j(R_l) = \delta^j_l, & \alpha^m(R_l) = 0 \\ \theta^j(Y_m) = 0, & \alpha^m(Y_n) = \delta^m_n \end{array} \tag{10.61}$$

The corresponding decomposition of the identity,

$$T_{\mathbb{I}} = \sum_{l=1}^{k} \theta^l \otimes R_l + \sum_{j=1}^{n-k} \alpha^j \otimes Y_j \tag{10.62}$$

allows us to decompose 1-forms and vector fields as,

$$\beta = \sum_{l=1}^{k} \beta(R_l)\,\theta^l + \sum_{j=1}^{n-k} \beta(Y_j)\,\alpha^j \tag{10.63}$$

and,

$$X = \sum_{l=1}^{k} \theta^l(X)\,R_l + \sum_{j=1}^{n-k} \alpha^j(X)\,Y_j \tag{10.64}$$

Of course, these decompositions extend to any tensor field. On exact 1-forms, we find,

$$df = \sum_{l=1}^{k} \left(\mathcal{L}_{R_l}(f)\right)\theta^l + \sum_{j=1}^{n-k} \left(\mathcal{L}_{Y_j}(f)\right)\alpha^j \tag{10.65}$$

We have not said anything on the commutation relations: $\left[R_l, R_j\right]$, $[R_l, Y_n]$ and $[Y_m, Y_n]$ among the vector fields of the adapted basis, as well as on the corresponding relations on 1-forms, that must be of the form,

$$d\theta^l = c^l_{mn}\theta^m \wedge \theta^n + a^l_{jm}\theta^j \wedge \alpha^m + b^l_{rs}\alpha^r \wedge \alpha^s \tag{10.66}$$

with similar relations for the $d\alpha^j$'s.

Qualifying these relations identifies special bases. Clearly, the Abelian situation we started from in $\mathbb{R}^n$ in terms of the $\partial/\partial x^j$'s and dx^j's uses commuting vector fields and exact 1-forms. In a generic situation, for generic submanifolds, it is not granted that we can be able to find globally an adapted basis. For instance, the two-dimensional sphere does not admit, for topological reasons, a global basis of 1-forms (nor of vector fields).

When such a basis exists, the submanifold is said to be *parallelizable*. Thus, in general, a manifold will be said to be parallelizable if it admits of a global basis of 1-forms and vector fields considered as modules with coefficients smooth functions. A large class of parallelizable manifolds is provided by the Lie groups.

The exterior differential calculus we have considered here may be thought of as arising from the Abelian Lie group structure of $\mathbb{R}^n$. Generalizations are provided by replacing the Abelian vector group $\mathbb{R}^n$ with a generic Lie group G.

When the (sub)manifold we are considering is not parallelizable, but is a homogeneous space of a Lie group G, it is still possible to take advantage of the exterior differential calculus associated with G by considering the algebra of functions on the submanifold as a subalgebra of functions in $\mathcal{F}(G)$. We shall make sense of these statements by considering the parallelizable submanifold $S^3 \subset \mathbb{R}^4$ (the group manifold of $SU(2)$) and the submanifold $S^2 \subset \mathbb{R}^3$, which is not parallelizable but may be considered a homogeneous space of $SU(2)$.

Differential Calculus on $S^3 \subset \mathbb{R}^4$

We recall that a convenient parametrization of $\mathbb{R}^4$ is provided by the complex, Hermitian matrices,

$$h = \begin{pmatrix} x_0 + x_3 & x_1 - ix_2 \\ x_1 + ix_2 & x_0 - x_3 \end{pmatrix} \tag{10.67}$$

which, in terms of Pauli matrices $\boldsymbol{\sigma} = (\sigma^1, \sigma^2, \sigma^3)$ given by:

$$\sigma^0 = \begin{pmatrix} 1 & 0 \\ 0 & 1 \end{pmatrix}, \sigma^1 = \begin{pmatrix} 0 & 1 \\ 1 & 0 \end{pmatrix}, \sigma^2 = \begin{pmatrix} 0 & -i \\ i & 0 \end{pmatrix}, \sigma^3 = \begin{pmatrix} 1 & 0 \\ 0 & -1 \end{pmatrix}, \tag{10.68}$$

may be written as

$$h = x_0 \sigma^0 + \mathbf{x} \cdot \boldsymbol{\sigma} \tag{10.69}$$

With this choice of parametrization by Hermitian matrices we find,

$$\det(h) = x_0^2 - x_1^2 - x_2^2 - x_3^2 \tag{10.70}$$

which makes it appropriate when we deal with Special Relativity.

If we consider a different parametrization using *anti*-Hermitian matrices,

$$g = \begin{pmatrix} x_0 + ix_3 & ix_1 + x_2 \\ ix_1 - x_2 & x_0 - ix_3 \end{pmatrix} = x_0\sigma^0 + i\mathbf{x} \cdot \boldsymbol{\sigma} \tag{10.71}$$

we find,

$$\det(g) = x_0^2 + x_1^2 + x_2^2 + x_3^2 \tag{10.72}$$

In this parametrization matrices with $\det(g) = 1$ represent the unit sphere $S^3 \subset \mathbb{R}^4$. We may set $g = Rs$, with: $s \in SU(2) \approx S^3$, to find $\det(g) = R^2$. Hence,

$$s = \frac{1}{\sqrt{x_0^2 + x_1^2 + x_2^2 + x_3^2}} \begin{pmatrix} x_0 + ix_3 & ix_1 + x_2 \\ ix_1 - x_2 & x_0 - ix_3 \end{pmatrix} \tag{10.73}$$

We notice also that: $g^\dagger = Rs^\dagger = Rs^{-1}$.

To find a basis of 1-forms adapted to S^3 we may set,

$$g^{-1}dg = \theta^0\sigma_0 + i\theta^a\sigma_a \tag{10.74}$$

$(a = 1, 2, 3)$ and we find,

$$R^{-1}s^{-1}d(Rs) = \left(R^{-1}dR\right)\sigma_0 + s^{-1}ds \tag{10.75}$$

The basis of 1-forms is given specifically by,

$$\theta^0 = \frac{x_0\,dx_0 + x_1\,dx_1 + x_2\,dx_2 + x_3\,dx_3}{x_0^2 + x_1^2 + x_2^2 + x_3^2} \tag{10.76}$$

$$\theta^1 = \frac{x_0\,dx_1 - x_1\,dx_0 + x_3\,dx_2 - x_2\,dx_3}{x_0^2 + x_1^2 + x_2^2 + x_3^2} \tag{10.77}$$

$$\theta^2 = \frac{x_0\,dx_2 - x_2\,dx_0 + x_1\,dx_3 - x_3\,dx_1}{x_0^2 + x_1^2 + x_2^2 + x_3^2} \tag{10.78}$$

$$\theta^3 = \frac{x_0\,dx_3 - x_3\,dx_0 + x_2\,dx_1 - x_1\,dx_2}{x_0^2 + x_1^2 + x_2^2 + x_3^2} \tag{10.79}$$

The corresponding basis of vector fields is given by,

$$Y_0 = x_\mu \frac{\partial}{\partial x_\mu}, \qquad (\mu = 0, 1, 2, 3) \tag{10.80}$$

$$Y_1 = x_0 \frac{\partial}{\partial x_1} - x_1 \frac{\partial}{\partial x_0} + x_3 \frac{\partial}{\partial x_2} - x_2 \frac{\partial}{\partial x_3} \tag{10.81}$$

$$Y_2 = x_0 \frac{\partial}{\partial x_2} - x_2 \frac{\partial}{\partial x_0} + x_1 \frac{\partial}{\partial x_3} - x_3 \frac{\partial}{\partial x_1} \tag{10.82}$$

$$Y_3 = x_0 \frac{\partial}{\partial x_3} - x_3 \frac{\partial}{\partial x_0} + x_2 \frac{\partial}{\partial x_1} - x_1 \frac{\partial}{\partial x_2} \tag{10.83}$$

Again, the adapted decomposition of the identity is given by,

$$T_{\mathbb{I}} = \theta^0 \otimes Y_0 + \theta^a \otimes Y_a \tag{10.84}$$

The exterior differential of a function is simply given by,

$$df = \left(\mathcal{L}_{Y_0}(f)\right)\theta^0 + \left(\mathcal{L}_{Y_a}(f)\right)\theta^a \tag{10.85}$$

It can be extended to any form by using the mutual relations between the Lie, inner and exterior derivatives.

As an example, we write explicitly the exterior derivative of a 1-form α. We consider first: $\alpha = \alpha(Y_0)\theta^0 + \alpha(Y_a)\theta^a$, and then,

$$d\alpha = (d\alpha(Y_0)) \wedge \theta^0 + \alpha(Y_0)\, d\theta^0 + (d\alpha(Y_a)) \wedge \theta^a + \alpha(Y_a)\, d\theta^a \tag{10.86}$$

We notice that, with respect to the use of bases with closed 1-forms, we get here additional terms due to the fact that the basis is 'non-holonomic', this meaning that: $d\theta^a \neq 0$.

In our specific case: $d\theta^0 = 0$ and: $d\theta^a + (1/2)\,\varepsilon^a_{bc}\theta^b \wedge \theta^c = 0$, and we get,

$$\begin{aligned} d\alpha &= \left(\mathcal{L}_{Y_a}(\alpha(Y_0)) - \mathcal{L}_{Y_0}(\alpha(Y_a))\right)\theta^a \wedge \theta^0 \\ &\quad + \left(\mathcal{L}_{Y_b}(\alpha(Y_c)) - \mathcal{L}_{Y_c}(\alpha(Y_b)) - \varepsilon^a_{bc}\alpha(Y_a)\right)\theta^b \wedge \theta^c \end{aligned} \tag{10.87}$$

A volume-form can be written as,

$$\Omega = \theta^0 \wedge \theta^1 \wedge \theta^2 \wedge \theta^3 \tag{10.88}$$

while the Euclidean metric tensor will be,

$$g = \theta^0 \otimes \theta^0 + \theta^1 \otimes \theta^1 + \theta^2 \otimes \theta^2 + \theta^3 \otimes \theta^3 \tag{10.89}$$

It is also possible to derive $dx_0, \ldots, dx_3$ by noticing that,

$$dg = dx_0\, \sigma_0 + i\, d\mathbf{x} \cdot \boldsymbol{\sigma} \tag{10.90}$$

and therefore,

$$dx_0 = \frac{1}{2}\operatorname{Tr}(dg\,\sigma_0) = \frac{1}{2}\operatorname{Tr}\left\{g\left(\theta^0\sigma_0 + i\theta^a\sigma_a\right)\sigma_0\right\} \tag{10.91}$$

and,

$$dx_a = \frac{1}{2}\operatorname{Tr}\left(g\left(\theta^0\sigma_0 + i\theta^a\sigma_a\right)\sigma_a\right). \tag{10.92}$$

Also, and more directly, using Eq. (10.84) (cfr. also Eq. 10.85),

$$dx_\mu = T_{\mathbb{I}}\left(dx_\mu\right) = \left(\mathcal{L}_{Y_0}\left(x_\mu\right)\right)\theta^0 + \left(\mathcal{L}_{Y_a}\left(x_\mu\right)\right)\theta^a, \qquad \mu = 0, 1, 2, 3 \tag{10.93}$$

Differential Calculus on $S^2 \subset \mathbb{R}^3$

For $S^2 \subset \mathbb{R}^3$ it is not possible to find a global adapted basis, because S^2 is not parallelizable. For instance, the 1-forms $d\theta$ and $d\phi$ used in Example (10.5) cannot be smooth on the whole of S^2. As a matter of fact, they are well defined only on $(-\pi, \pi) \times [0, 2\pi]$.

To deal with a smooth exterior differential calculus with globally defined fields we may consider the projection map,

$$\begin{aligned} \pi : SU(2) \equiv S^3 &\longrightarrow S^2 \subset \mathbb{R}^3 \\ s &\mapsto s^{-1}\sigma_3 s = \mathbf{x}\cdot\boldsymbol{\sigma} \end{aligned} \tag{10.94}$$

with $\det\left(s^{-1}\sigma_3 s\right) = -1 = -\left(x_1^2 + x_2^2 + x_3^2\right)$ and

$$x_i = \frac{1}{2}\operatorname{Tr}\left\{\sigma_i s^{-1}\sigma_3 s\right\}, \qquad i = 1, 2, 3 \tag{10.95}$$

The algebra of functions on S^2 may be pulled back to a subalgebra of $\mathcal{F}(SU(2))$ simply replacing $f(\mathbf{x})$ with $f\left(s^{-1}\sigma_3 s\right)$.

The vector fields Y_1, Y_2 and Y_3 act on this subalgebra as derivations. They are left-invariant and generate right translations, commuting therefore with the infinitesimal generator of the left action $s \mapsto e^{i\varphi\sigma_3}s$, giving,

$$s^{-1}\sigma_3 s = \left(e^{i\varphi\sigma_3}s\right)^{-1}\sigma_3\left(e^{i\varphi\sigma_3}s\right). \tag{10.96}$$

The dual 1-forms are well defined on S^3 but not on S^2, because the projections on S^2 of Y_1, Y_2 and Y_3 are not independent. In this way we may write,

$$d(\pi^* f) = \left(\mathcal{L}_{Y_1} f\right)\theta^1 + \left(\mathcal{L}_{Y_2} f\right)\theta^2 + \left(\mathcal{L}_{Y_3} f\right)\theta^3 \tag{10.97}$$

which may be expressed then in terms 1-forms that are globally defined on S^3, but cannot be expressed in terms globally defined 1-forms on S^2.

Therefore, the use of Y_1, Y_2, Y_3 and $\theta^1, \theta^2, \theta^3$ allows for an exterior differential calculus on tensorial objects on S^2 by considering them as objects in the subalgebra of tensors on S^3.

Appendix D: Differential Concomitants: Nijenhuis, Schouten and Other Brackets

Along the main text many computations with tensors involve the use of various brackets with strong algebraic flavor. We will summarize in this appendix the definitions and main properties of some of them for the ease of the reader. Mostly, we follow the conventions in [Ab88], trying to avoid the proliferation of factorial factors.

The Lie Derivative and the Lie Bracket

As in the main text M will denote a smooth manifold, $\mathfrak{X}(M)$ will denote the Lie algebra of smooth vector fields on M and $\Omega^{\cdot}(M)$ the algebra of differential forms on M. The smooth tensors of type (p, q) over M will be denoted as $\mathcal{T}_q^p(M)$ and the bigraded tensorial algebra over M with the tensor product $\otimes$, $\mathcal{T}(M)$.

Given two tensors t_1 of type (p_1, q_1) and t_2 of type (p_2, q_2), we will define the contraction of t_1 with t_2, $p_1 \leq q_2$ and $p_2 \leq q_1$, as the tensor $i_{t_1} t_2$ of type $(0, q_2 - p_1 + q_1 - p_2)$ obtained by contracting the contravariant indices of t_1 with the first p_1 covariant indices of t_2, and the contravariant indices of t_2 with the first p_2 contravariant indices of t_1.

In particular the contraction of a vector field X with a k-form α is the $k-1$-form $i_X\alpha$:

$$i_X\alpha(X_1, \dots, X_{k-1}) = \alpha(X, X_1, \dots, X_{k-1}) .$$

The contraction of A of type $(1, 1)$ with a k form α will be the k-form:

$$i_A\alpha(X_1, \dots, X_k) = \sum_{i=1}^{k} \alpha(X_1, \dots, A(X_i), \dots, X_k).$$

We have already shown that in the algebra of differential forms $\Omega^\bullet(M)$ we have the differential operator d of degree 1 satisfying (see Appendix G):

1. $d^2 = 0$.
2. $d(\alpha \wedge \beta) = d\alpha \wedge \beta + (-1)^k \alpha \wedge d\beta$, for all $\alpha, \beta \in \Omega^\bullet(M)$, $\alpha \in \Omega^k(M)$.

The Lie derivative is the degree 0 operator on the algebra $\Omega^\bullet(M)$ defined by using Cartan's formula:

$$\mathcal{L}_X\alpha = i_X d\alpha + di_X\alpha.$$

The Lie derivative satisfies the following properties:

1. $\mathcal{L}_X d\alpha = d\mathcal{L}_X\alpha$.
2. $\mathcal{L}_X i_Y\alpha = i_Y\mathcal{L}_X\alpha + i_{[X,Y]}\alpha$.
3. $\mathcal{L}_{fX}\alpha = f\mathcal{L}_X\alpha + df \wedge i_X\alpha$.
4. $\mathcal{L}_{[X,Y]}\alpha = \mathcal{L}_X\mathcal{L}_Y\alpha - \mathcal{L}_Y\mathcal{L}_X\alpha$.
5. $\mathcal{L}_X(\alpha \wedge \beta) = (\mathcal{L}_X\alpha) \wedge \beta + \alpha \wedge (\mathcal{L}_X\beta)$.

The Lie derivative of a vector field Y along the direction of X is precisely the Lie bracket $[X, Y]$, that is: $\mathcal{L}_X Y = [X, Y]$. Hence, we denote the Lie derivative of any contravariant skew symmetric tensor A as:

$$\mathcal{L}_X A = [X, A].$$

Then it is easy to check that for any skew symmetric contravariant tensor A, we have

$$i_{\mathcal{L}_X A}\alpha = i_X d(i_A\alpha) - i_A di_X\alpha.$$

The Nijenhuis Bracket

We may consider if it possible to extend the bracket $[X, A]$ defined above for a vector field and a contravariant tensor to all skew symmetric contravariant tensors in such a way that it coincides with the Lie derivative on vector fields and making it into the graded analog of a Lie algebra. The answer is positive and was studies by Nijenhuis [Ni55].

Let A, B be skew symmetric homogeneous contravariant tensors of degrees q and p respectively. Then we define $[A, B]$ to be a $q + p - 1$ homogeneous skew symmetric contravariant tensor such that when contracted with any $q+p-1$–closed form we get:

$$i_{[A,B]}\alpha = (-1)[(q+1)pi_A di_B\alpha + (-1)^q i_B di_A\alpha,$$

moreover,

$$[A, B] = (-1)^{qp}[B, A],$$

and, it satisfies the graded Jacobi identity:

$$(-1)^{qp}[A,[B,C]] + (-1)^{pr}[C,[A,B]] + (-1)^{rq}[B,[C,A]] = 0$$

with C homogeneous of degree r.

The space of skew symmetric contrariant tensors on M becomes a graded Lie algebra with degree the degree of the tensors minus 1 (that is the graded degree of a vector field is 0) [Ni64].

The Schouten-Nijenhuis Bracket

Finally, it is very useful a bracket defined on $(1, 1)$ tensors and that gives a $(1, 2)$ tensor, skew symmetric in the covariant part. This tensor is actually the "differential concomitant" defined for $(1, q)$ tensors, skew symmetric in the covariant part. In other words, we may define a bracket on vector-valued forms on M [Ni66].

Thus if T, R are two $(1, 1)$ tensors, we define $[T, R]$ as a 2-form with values in vector fields as:

$$\begin{aligned}[T,R](X,Y) = \frac{1}{2}(&[TX,RY] + [RX,TY] + (TR+RT)[X,Y] \\ &-T([X,RY]+[RX,Y]) - R([X,TY]+[TX,Y])).\end{aligned}$$

it follows from the previous definition that $[T, R] = [R, T]$. If $T = R$, the previous expression reduces to:

$$[T,T](X,Y) = [TX,TY] + T^2[X,Y] - T[X,TY] - T[TX,Y],$$

which is just the well known torsion operator used to characterize the integrability of a almost complex structure.

The extension of this bracket to the graded algebra of vector-valued forms provident with a graded Lie algebra structure.

Appendix E: Covariant Calculus

The Covariant Derivative

The exterior derivative calculus we have been developing up to now is applied to skew–symmetric forms built out of functions, $f, g, \ldots$, and the exterior differential operator d. We have $f, dg, h\,dg, dg \wedge df$, and so on. Along with d we have developed the Lie derivative along a vector field X, $\mathcal{L}_X$, and the contraction with respect to X,

i_X. These operations, which are graded derivations, satisfy the relation,

$$i_X d + d i_X = \mathcal{L}_X \tag{10.98}$$

This calculus, however, does not enable us to differentiate general tensors fields. We assume here again that our carrier space is an affine space M of dimension n. We have seen that for a function f, df is a covariant vector field which in a given coordinate system is expressed by,

$$df = \sum_{i=1}^{n} \frac{\partial f}{\partial x^i} dx^i \tag{10.99}$$

In other words, the partial derivatives of f with respect to x^i are the components of a covariant vector field. This process of generating new vector fields by taking the partial derivatives of the components of a tensor field with respect to a coordinate system cannot be extended to tensor fields of higher orders. Let us consider a contravariant vector field X whose local components in two coordinate systems x^i and $\tilde{x}^i$ are A^i and $\tilde{A}^i$ respectively, so that we have,

$$X = \sum_{i=1}^{n} A^i \frac{\partial}{\partial x^i} = \sum_{i=1}^{n} \tilde{A}^i \frac{\partial}{\partial \widetilde{x}^i} \tag{10.100}$$

or,

$$\widetilde{A}^i = \sum_{j=1}^{n} \frac{\partial \widetilde{x}^i}{\partial x^j} A^j \tag{10.101}$$

Partial differentiation gives,

$$\frac{\partial \widetilde{A}^i}{\partial \widetilde{x}^k} = \sum_{j,l=1}^{n} \frac{\partial \widetilde{x}^i}{\partial x^j} \frac{\partial x^l}{\partial \widetilde{x}^k} \frac{\partial A^j}{\partial x^l} + \sum_{j,l=1}^{n} \frac{\partial^2 \widetilde{x}^i}{\partial x^j \partial x^l} \frac{\partial x^l}{\partial \widetilde{x}^k} A^j \tag{10.102}$$

which, because of the second terms on the right-hand side, shows that the $\partial A^i/\partial x^j$'s are not the components of a tensor field of type $(1, 1)$.

To remedy this situation we introduce the notion of affine connection.

Definition 10.56 An affine connection is given by a set of linear differential 1–forms $\omega^i{}_j$, with $i, j = 1, \ldots, n$, in each coordinate system, in such a way that when going from one coordinate system to another we have,

$$dP^i{}_j + \sum_{k=1}^{n} \tilde{\omega}^i{}_k P^k{}_j = \sum_{k=1}^{n} \omega^k{}_j P^i{}_k \tag{10.103}$$

where,

$$P^i{}_j = \frac{\partial \tilde{x}^i}{\partial x^j} \tag{10.104}$$

Notice that Eq. (10.101) can be rewritten as,

$$\tilde{A}^i = \sum_{j=1}^{n} P^i{}_j \, A^j \tag{10.105}$$

It is also remarkable that such set of $\omega^i{}_j$ is a matrix–valued 1-form.

We have to show that this definition is consistent. By this we mean that if we consider three coordinate systems, x^i, $\tilde{x}^i$, $\tilde{\tilde{x}}^i$, the relation amongst x^i, and $\tilde{\tilde{x}}^i$ follows from those of x^i with $\tilde{x}^i$, and those of $\tilde{x}^i$ with $\tilde{\tilde{x}}^i$.

Indeed from,

$$d\left(\frac{\partial \tilde{x}^i}{\partial x^j}\right) + \sum_{k=1}^{n} \tilde{\omega}^i{}_k \frac{\partial \tilde{x}^k}{\partial x^j} = \sum_{k=1}^{n} \frac{\partial \tilde{x}^i}{\partial x^k} \omega^k{}_j \tag{10.106}$$

and,

$$d\left(\frac{\partial \tilde{\tilde{x}}^k}{\partial x^j}\right) + \sum_{k=1}^{n} \tilde{\tilde{\omega}}^i{}_k \frac{\partial \tilde{\tilde{x}}^k}{\partial \tilde{x}^j} = \sum_{k=1}^{n} \frac{\partial \tilde{\tilde{x}}^i}{\partial \tilde{x}^k} \tilde{\omega}^k{}_j \tag{10.107}$$

we get,

$$\begin{aligned}
& d\left(\tfrac{\partial \tilde{\tilde{x}}^i}{\partial x^j}\right) + \textstyle\sum_{k=1}^{n} \tilde{\tilde{\omega}}^i{}_k \tfrac{\partial \tilde{\tilde{x}}^k}{\partial x^j} = d\left(\textstyle\sum_{k=1}^{n} \tfrac{\partial \tilde{\tilde{x}}^i}{\partial \tilde{x}^k} \tfrac{\partial \tilde{x}^k}{\partial x^j}\right) + \textstyle\sum_{k,l=1}^{n} \tilde{\tilde{\omega}}^i{}_l \tfrac{\partial \tilde{\tilde{x}}^k}{\partial \tilde{x}^k} \tfrac{\partial \tilde{x}^k}{\partial x^j} \\
&= \textstyle\sum_{k=1}^{n} \tfrac{\partial \tilde{x}^k}{\partial x^j} \left[d\left(\tfrac{\partial \tilde{\tilde{x}}^i}{\partial \tilde{x}^k}\right) + \textstyle\sum_{l=1}^{n} \tilde{\tilde{\omega}}^i{}_l \tfrac{\partial \tilde{\tilde{x}}^l}{\partial \tilde{x}^k}\right] + \textstyle\sum_{k=1}^{n} \tfrac{\partial \tilde{\tilde{x}}^i}{\partial \tilde{x}^k} d\left(\tfrac{\partial \tilde{x}^k}{\partial x^j}\right) \\
&= \textstyle\sum_{k,l=1}^{n} \tfrac{\partial \tilde{x}^k}{\partial x^j} \tfrac{\partial \tilde{\tilde{x}}^i}{\partial \tilde{x}^l} \tilde{\omega}^l{}_k + \textstyle\sum_{k=1}^{n} \tfrac{\partial \tilde{\tilde{x}}^i}{\partial \tilde{x}^k} d\left(\tfrac{\partial \tilde{\tilde{x}}^k}{\partial x^j}\right) \\
&= \textstyle\sum_{k=1}^{n} \tfrac{\partial \tilde{\tilde{x}}^i}{\partial \tilde{x}^k} \left[\textstyle\sum_{l=1}^{n} \tilde{\omega}^k{}_l \tfrac{\partial \tilde{x}^l}{\partial \tilde{x}^j} + d\left(\tfrac{\partial \tilde{\tilde{x}}^k}{\partial x^j}\right)\right] \\
&= \textstyle\sum_{k,l=1}^{n} \tfrac{\partial \tilde{\tilde{x}}^k}{\partial \tilde{x}^k} \tfrac{\partial \tilde{x}^k}{\partial x^l} \omega^l{}_j = \textstyle\sum_{l=1}^{n} \tfrac{\partial \tilde{\tilde{x}}^i}{\partial x^l} \omega^l{}_j
\end{aligned} \tag{10.108}$$

which proves the consistency of the condition.

Condition (10.103) can be put into an equivalent form using the quantities,

$$Q^i{}_k = \frac{\partial x^i}{\partial \tilde{x}^k} \tag{10.109}$$

Indeed, matrix–valued functions Q and P are inverses to each other, i.e.,

$$\sum_{j=1}^{n} P^i{}_j Q^j{}_k = \sum_{j=1}^{n} Q^i{}_j P^j{}_k = \delta^i_k \tag{10.110}$$

Differentiating any one of these equations we find the equivalent condition,

$$dQ^i{}_j + \sum_{k=1}^{n} \omega^i{}_k Q^k{}_j = \sum_{k=1}^{n} Q^i{}_k \tilde{\omega}^k{}_j \tag{10.111}$$

In fact,

$$\sum_{j=1}^{n} dQ^k{}_j P^j{}_i = \sum_{j,l=1}^{n} Q^k{}_j \left(-\tilde{\omega}^j{}_l P^l{}_i + P^j{}_l \omega^l{}_i\right) = 0 \tag{10.112}$$

and then,

$$\sum_{j=1}^{n} dQ^k{}_j P^j{}_i - \sum_{j,l=1}^{n} Q^k{}_j \tilde{\omega}^j{}_l P^l{}_i + \omega^k{}_i = 0 \tag{10.113}$$

which implies that,

$$dQ^k{}_i + \sum_{j=1}^{n} \omega^k{}_j Q^j{}_i = \sum_{j=1}^{n} Q^k{}_j \tilde{\omega}^j{}_i \tag{10.114}$$

Connections exist on any affine space. This can be easily shown by considering a second-order dynamics associated with a non-degenerate quadratic in velocities. This will be seen in a later section.

We assume that on our affine space an affine connection has been selected and try to study the properties of M arising from this connection.

By differentiating (10.101) and using (10.103) we get,

$$d\tilde{A}^i = \sum_{j=1}^{n} P^i{}_j dA^j + \sum_{j=1}^{n} dP^i{}_j A^j = \sum_{j=1}^{n} P^i{}_j dA^j + \sum_{j,k=1}^{n} (P^i{}_k \omega^k{}_j - \tilde{\omega}^i{}_k P^k{}_j) A^j \tag{10.115}$$

which can be rewritten as,

$$d\tilde{A}^i + \sum_{k=1}^{n} \tilde{\omega}^i{}_k \tilde{A}^k = \sum_{j=1}^{n} P^i{}_j \left(dA^j + \sum_{k=1}^{n} \omega^j{}_k A^k\right) \tag{10.116}$$

We define,

$$DA^i = dA^i + \sum_{k=1}^{n} \omega^i{}_k A^k = \sum_{j=1}^{n} A^i{}_{;j} dx^j \tag{10.117}$$

where the $A^i{}_{;j}$'s turn out to be the components of the (1,1) tensor field,

$$A^i{}_{;j} = \frac{\partial A^i}{\partial x^j} + \sum_k \Gamma^i_{kj} A^k; \quad \omega^i{}_k = \sum_j \Gamma^i_{kj} dx^j \tag{10.118}$$

because DA^i transform as the components of a vector field.

Writing (10.116) in the form,

$$D\tilde{A}^i = \sum_{j=1}^{n} P^i{}_j (DA^j) \tag{10.119}$$

we see that the DA^j's are linear differential forms transforming like contravariant vectors. They are said to define the covariant differential of the vector field X.

More generally, let, for instance, $A^{ij}{}_k$ be the components of a tensor field of type (2, 1), relative to coordinates (x^i). Under a change of coordinates they transform in the following way,

$$\tilde{A}^{ij}{}_k = \sum_{l,m=1}^{n} P^i{}_l P^j{}_m Q^r{}_k A^{lm}{}_r \tag{10.120}$$

Making use of (10.103) and (10.115) we get in a similar way as above,

$$D\tilde{A}^{ij}{}_k = \sum_{l,m=1}^{n} P^i{}_l P^j{}_m Q^r{}_k DA^{lm}{}_r \tag{10.121}$$

where,

$$DA^{ij}{}_k = \sum_{l=1}^{n} DA^{ij}{}_{k;l} dx^l = dA^{ij}{}_k + \sum_{l=1}^{n} (\omega^i{}_l A^{lj}{}_k + \omega^j{}_l A^{il}{}_k - \omega^l{}_k A^{ij}{}_l) \tag{10.122}$$

It follows that the components $A^{ij}{}_{k;l}$ define a tensor field of type (2, 2). We declare $DA^{ij}{}_k$ to be the covariant differential of the tensor field $A^{ij}{}_k$.

Now, it should be clear how these considerations extend to a tensor field of any type. In particular, for a scalar function f we define $Df = df$.

Under tensor multiplication and contraction of tensor fields we have, for instance,

$$D\left(\sum_{l=1}^{n} A^{ijk}{}_{l}\, B^{l}{}_{i}\right) = \sum_{l=1}^{n}(DA^{ijk}{}_{l})B^{l}{}_{i} + \sum_{l=1}^{n} A^{ijk}{}_{l}(DB^{l}{}_{i}) \tag{10.123}$$

as it may be verified by direct computation. In a given coordinate system we may introduce the quantities Γ^{i}_{jk} via the relation,

$$\omega^{i}{}_{k} = \sum_{j=1}^{n} \Gamma^{i}_{jk}\, dx^{j} \tag{10.124}$$

and then (10.117) becomes,

$$DA^{i} = dA^{i} + \sum_{j=1}^{n} \Gamma^{i}_{jk}\, dx^{j}\, A^{k} \tag{10.125}$$

Conditions (10.103) can be written as,

$$\Gamma^{i}_{jk} = \sum_{l=1}^{n} \frac{\partial^{2}\widetilde{x}l}{\partial x^{j}\partial x^{k}} \frac{\partial x^{i}}{\partial \widetilde{x}^{l}} + \sum_{l,m,q=1}^{n} \widetilde{\Gamma}^{l}_{mq} \frac{\partial \widetilde{x}^{m}}{\partial x^{j}} \frac{\partial \widetilde{x}^{q}}{\partial x^{k}} \frac{\partial \widetilde{x}^{i}}{\partial x^{l}} \tag{10.126}$$

The numbers Γ^{i}_{jk} are called the *Christoffel symbols* of the connection.

Second Order Differential Equations Associated with a Connection

Definition 10.57 Let $\gamma : I \subset \mathbb{R} \to M$ be a smooth curve with parameter t. Then, a vector field along γ is a map $V : I \to TM$ such that $V(t) \in T_{\gamma(t)}M$.

For instance, if $W \in \mathfrak{X}(M)$ we can define $V = W \circ \gamma$. Notice that not every vector field along γ can be factorized in such way. For instance, if there are two values t_1 and t_2 in I such that $\gamma(t_1) = \gamma(t_2)$, vector fields along γ such that $V(t_1) \neq V(t_2)$ cannot be factorized in such a way. However, the factorization is possible if we restrict ourselves to a small enough neighborhood of a t_0.

Another way of constructing a vector field along γ is by considering a vector field in I and defining the image under the tangent map $T\gamma$. As a very important example of it, we can consider the vector field in I given by d/dt, because then the corresponding vector field along γ is the tangent vector in each point of the curve, $\dot{\gamma}$, defined by,

$$\dot{\gamma} : f \mapsto \frac{d}{dt} f(\gamma(t)) = \sum_{i=1}^{n} \frac{\partial f}{\partial x^i} \frac{dx^i}{dt} \tag{10.127}$$

Definition 10.58 Let $\gamma : I \subset \mathbb{R} \to M$ be a smooth curve with parameter t and V a vector field along γ with coordinates $V^i(t)$ in a particular coordinate system. Then, the vector field along γ given in the same coordinate system by,

$$\frac{DV^i}{dt} = \langle DV^i(V(t), \dot{\gamma}(t) \rangle, \tag{10.128}$$

is called absolute derivative or covariant derivative along γ of the vector field along γ, $V : I \to TM$.

Definition 10.59 Let $\gamma : I \subset \mathbb{R} \to M$ and V be as in the preceding definition. Then V is said to be a covariantly constant vector field along γ if,

$$\frac{DV^i}{dt} = 0 \tag{10.129}$$

Analytically the condition can be written as,

$$\frac{DV^i}{dt} = \frac{dV^i}{dt} + \sum_{j=1}^{n} \Gamma^i_{kj} V^k \frac{dx^j}{dt} = 0 \tag{10.130}$$

In particular, if $\gamma : \mathbb{R} \to M$ is a smooth curve, then the tangent vector field along γ associated with the curve γ, $\dot{\gamma}$, is parallel along γ if $\frac{D\dot{\gamma}^i}{dt} = 0$ along γ.

Definition 10.60 A parametrized curve is said to be auto-parallel if its tangent vectors are parallel along γ.

The preceding equations (10.130) can be considered as a differential equation system for $V^i(t)$, therefore there exists one and only one solution $V^i(t)$ corresponding to any initial values $V^i(t_0)$. We interpret this geometrically by saying that given a curve $\gamma : I \to \mathbb{R}^n$ and a vector $V^i(t_0)$ at the point $\gamma(t_0)$, the solution of the differential equation system (10.130) provides a vector field along γ (for t enough close to t_0), $V(t)$, and we usually say that the vector $V^i(t_0)$ is parallelely transported along γ. Moreover, as the differential equation system is linear, the vector field along γ obtained by parallel transport from the vector $V(t_0) + W(t_0)$, is the sum of those obtained from $V(t_0)$ and $W(t_0)$, respectively, and for any real number α, the vector field along γ obtained from $\alpha V(t_0)$ is $\alpha V(t)$.

By setting $V^i = \frac{dx^i}{dt}$ we get the second-order differential equation for autoparallel curves (also-called *geodesics*),

$$\frac{d^2x^i}{dt^2} + \sum_{j,k=1}^{n} \Gamma^i_{jk} \frac{dx^j}{dt} \frac{dx^k}{dt} = 0 \tag{10.131}$$

In this sense, geodesics are the generalization of straight lines, because for affinely parametrized straight lines, its tangent vector is parallelely transported along the straight line.

It is also remarkable that the equation is invariant under affine reparametrization; in other words, for any couple a, b, of real numbers, with $a \neq 0$, if γ is a geodesic curve then $\bar{\gamma}$ defined by $\bar{\gamma}(t) = \gamma(at + b)$ is a geodesic too.

There is one, and only one, solution $x^i(t)$ corresponding to any given initial values $x^i(t_0)$ and $\frac{dx^i}{dt}(t_0)$, which is the geodesic characterized by the mentioned initial conditions.

It is clear from the preceding formula that any second-order differential equation with *forces* quadratic in the velocities defines an affine connection.

Definition 10.61 Let $\gamma : I \subset \mathbb{R} \to M$ be a smooth curve with parameter t. Then, a covariant vector field along γ, or a 1–form along γ, is a map $\alpha : I \to TM$ such that $\alpha(t) \in T^*_{\gamma(t)}M$.

Notice that a 1–form along a curve γ, α, can be contracted with a vector field along the same curve, V, giving raise to a map $f : I \to \mathbb{R}$ by,

$$f(t) = \langle \alpha(t), V(t) \rangle. \tag{10.132}$$

Then the covariant derivative along γ can be defined by using the relation,

$$\frac{d}{dt}\langle \alpha(t), V(t) \rangle = \left\langle \frac{D}{dt}\alpha(t), V(t) \right\rangle + \left\langle \alpha(t), \frac{DV}{dt} \right\rangle, \tag{10.133}$$

and we will obtain,

$$\frac{D\alpha_i}{dt} = \frac{d\alpha_i}{dt} - \sum_{j,k=1}^{n} \Gamma^j_{ik}\alpha_j \frac{dx^k}{dt} \tag{10.134}$$

and then we will say that the 1–form along the curve γ, α, is constant along γ, or parallel along γ, if

$$\frac{d\alpha_i}{dt} - \sum_{j,k=1}^{n} \Gamma^j_{ik}\alpha_j \frac{dx^k}{dt} = 0\,, \tag{10.135}$$

and therefore to define $D\alpha_i$ by

$$D\alpha_i = d\alpha_i - \sum_{j,k=1}^{n} \Gamma^j_{ik}\alpha_j dx^k\,. \tag{10.136}$$

The components of $D\alpha_i$ are given by

$$D\alpha_i = \sum_{j=1}^{n} \alpha_{i;j}\, dx^j \,. \tag{10.137}$$

E.2. Torsion and Curvature

The covariant differential of a tensor field can be introduced in a similar way. Given a tensor field, the covariant differential gives a new tensor field with one more covariant index. For instance, we obtain $A^i{}_{jk;l}$ out of $A^i{}_{jk}$. Applying this process again, we see that the components $A^i{}_{jk;l;m}$ defined by,

$$DA^i{}_{jk;l} = \sum_{m=1}^{n} A^i{}_{jk;l;m}\, dx^m \tag{10.138}$$

are those of a tensor field of type (1, 4).

It will be of importance to find a formula for the difference $A^i{}_{jk;l;m} - A^i{}_{jk;m;l}$. Consider first a function f. We find,

$$f_{;i} = \frac{\partial f}{\partial x^i} \tag{10.139}$$

and,

$$Df_{;i} = \sum_{j=1}^{n} \frac{\partial^2 f}{\partial x^i \partial x^j} dx^j - \sum_{j,k=1}^{n} \Gamma^k_{ij} \frac{\partial f}{\partial x^k}\, dx^j \tag{10.140}$$

and then,

$$f_{;i;j} = \frac{\partial^2 f}{\partial x^i \partial x^j} dx^j - \Gamma^k_{ij} \frac{\partial f}{\partial x^k} \tag{10.141}$$

It follows that,

$$f_{;i;j} - f_{;j;i} = \sum_{k=1}^{n} (\Gamma^k_{ji} - \Gamma^k_{ij}) \frac{\partial f}{\partial x^k} \tag{10.142}$$

Defining,

$$T^k_{ij} = \Gamma^k_{ji} - \Gamma^k_{ij} \tag{10.143}$$

and having in mind (10.126), we see that the components T^k_{ij} transform as the coordinates of a (1, 2) tensor, because of the first term in the right-hand side of (10.126), which destroys the tensorial character, is symmetric in the interchange of indices i and j, and therefore the T^k_{ij}'s transform as the components of a tensor of type (1, 2). It depends alone of the affine connection and it is called the torsion tensor. It vanishes if and only if the components of the Christoffel symbols are symmetric under the interchange of the two lower indices.

We can also see that for any function f the components,

$$f_{;i;j} - f_{;j;i} = \sum_{k=1}^{n} T^k_{ij} \frac{\partial f}{\partial x^k} \tag{10.144}$$

are those of a tensor field of type (0, 2), which is the contraction of the torsion tensor and the differential of f.

Next we consider a contravariant vector field with components A^i. We find, by definition,

$$A^i_{;j} = \frac{\partial A^i}{\partial x^j} + \sum_{l=1}^{n} \Gamma^i_{jl} A^l \tag{10.145}$$

and,

$$A^i_{;j;k} = \sum_{l=1}^{n} \left(\frac{\partial \Gamma^i_{jl}}{\partial x^k} + \sum_{m=1}^{n} \Gamma^i_{km} \Gamma^m_{jl} \right) A^l - \sum_{l=1}^{n} \Gamma^l_{kj} A^i{}_{;l} + \quad (j \Leftrightarrow k) \tag{10.146}$$

It follows that,

$$A^i_{;j;k} - A^i_{;k;j} = \sum_{l=1}^{n} \left(T^l_{jk} A^i_{;l} + R^i_{jlk} A^l \right) \tag{10.147}$$

where,

$$R^i_{jlk} = \frac{\partial \Gamma^i_{kj}}{\partial xl} - \frac{\partial \Gamma^i_{lj}}{\partial x^k} + \sum_{m=1}^{n} \left(\Gamma^i_{lm} \Gamma^m_{kj} - \Gamma^i_{km} \Gamma^m_{lj} \right) \tag{10.148}$$

The R^i_{jkl}'s are the components of a tensor field of type (1, 3), skew–symmetric in its last two indices. It is called the *curvature tensor*.

Remark 10.3 The derivation of similar formulae for general tensor fields does not give raise to new tensors out of the affine connection.

Up to now we have been working in coordinates (x^i) and we have used a basis of 1–forms given by (dx^i) and a basis of vector fields given by $\partial/\partial x^i$.

In general we could use a different frame at each point, say, we could use $X_1, X_2, \ldots, X_n$, independent vector fields, which in previous coordinate basis would be,

$$X_i = \sum_{k=1}^{n} A^k{}_i \frac{\partial}{\partial x^k} \tag{10.149}$$

implying,

$$\det ||A^k{}_j|| \neq 0 \tag{10.150}$$

because of independence. By introducing the matrix $B_k{}^j$ such that,

$$\sum_{k=1}^{n} B_k{}^j A^k{}_i = \delta_i^j \tag{10.151}$$

we may introduce the dual basis of 1–forms given by,

$$\theta^j = \sum_{k=1}^{n} B_k{}^j \, dx^k \tag{10.152}$$

In a different coordinate system,

$$X_i = \sum_{k=1}^{n} \tilde{A}^k{}_i \frac{\partial}{\partial \tilde{x}^k} = \sum_{k=1}^{n} A^k{}_i \frac{\partial}{\partial x^k} \tag{10.153}$$

and,

$$\theta^j = \sum_{k=1}^{n} \tilde{B}_k{}^j \, d\tilde{x}^k = \sum_{k=1}^{n} B_k{}^j \, dx^k \tag{10.154}$$

By considering the $A^k{}_i$ as independent variables (introducing the space of frames at each point of M) we have that,

$$DA^k{}_i = dA^k{}_i + \sum_{l=1}^{n} \omega^k{}_l A^l{}_i \tag{10.155}$$

considered as linear differential frames in $(x^i, A^k{}_j)$. We also get,

$$D\tilde{A}^k{}_i = \sum_{l=1}^{n} \frac{\partial \tilde{x}^k}{\partial x^l} DA^l{}_i \tag{10.156}$$

which using the preceding expressions gives,

$$\sum_{j=1}^{n} \tilde{B}_k{}^j \, D\tilde{A}^j{}_i = \sum_{j=1}^{n} B_j{}^k \, DA^j{}_i \tag{10.157}$$

They represent differential forms in the space of frames.
We set,

$$\theta_i^k = \sum_{j=1}^{n} \tilde{B}_j{}^k \, D\tilde{A}^j{}_i = \sum_{j=1}^{n} B_j{}^k \, DA^j{}_i \tag{10.158}$$

These $n + n^2$ linear differential 1–forms, θ^i and θ_i^k are linearly independent and we find,

$$\theta^i = \sum_{j=1}^{n} B_j{}^i \, dx^j\,, \quad \theta_i^k = \sum_{j=1}^{n} B_j{}^k \left(dA^j{}_i + \sum_{l=1}^{n} \omega^i{}_l A^l{}_i \right) \tag{10.159}$$

or,

$$dx^i = \sum_{j=1}^{n} A^i{}_j dx^j\,, \quad \sum_{j=1}^{n} A^k{}_j \theta_i^j = dA^k{}_i + \sum_{l=1}^{n} \omega^k{}_l A^l{}_i \tag{10.160}$$

Applying exterior differentiation on these equations and simplifying we get,

$$dx^i = \sum_{j,k=1}^{n} T^i_{jk} \, dx^j \wedge dx^k \tag{10.161}$$

$$d\omega^k{}_j - \sum_{l=1}^{n} \omega^l{}_j \wedge \omega^k{}_l = \sum_{l,m=1}^{n} \frac{1}{2} R^k_{jlm} dx^l \wedge dx^m \tag{10.162}$$

or,

$$d0^i - \sum_{k=1}^{n} \theta^k \wedge \theta_k^i = \frac{1}{2} \sum_{l,m=1}^{n} P^i_{lm} \theta^l \wedge \theta^m \tag{10.163}$$

$$d\theta_j^i - \sum_{l=1}^{n} \theta_j^l \wedge \theta_l^i = \frac{1}{2} \sum_{l,m=1}^{n} S^i_{jlm} \theta^l \wedge \theta^m \tag{10.164}$$

where,

$$\frac{1}{2}P^i_{lm} = \sum_{k,p,q=1}^{n} B_k{}^i A^p{}_l A^q{}_m T^k_{pq} \ ; \ S^i_{jlm} = \sum_{k,p,q,r=1}^{n} B_k{}^i A^p{}_k A^q{}_l A^r{}_m R^k{}_{pqr} \tag{10.165}$$

Theses equations are called *equations of structure*.
When we use a holonomic frame, i.e.,

$$A^k{}_l = B_l{}^k = \delta^k_l \tag{10.166}$$

we find,

$$-\sum_{k=1}^{n} dx^k \wedge \omega^j{}_k = \sum_{l,m=1}^{n} T^j_{lm} dx^l \wedge dx^m \tag{10.167}$$

and,

$$d\omega^j{}_i - \sum_{l=1}^{n} \omega^l{}_i \wedge \omega^j{}_l = \frac{1}{2} \sum_{l,m=1}^{n} R^j_{ilm} du^l \wedge du^m \tag{10.168}$$

It is sometimes convenient to introduce the exterior quadratic differential forms,

$$\Theta^j = \frac{1}{2} \sum_{l,m=1}^{n} P^j_{lm} \theta^l \wedge \theta^m \ , \quad \Theta^j_i = \frac{1}{2} \sum_{l,m=1}^{n} S^j_{ilm} \theta^l \wedge \theta^m \tag{10.169}$$

and write the previous equations in the form,

$$d\theta^j - \sum_{k=1}^{n} \theta^k \wedge \theta^j_k = \Theta^j \ , \quad \Theta^j_i = \sum_{k=1}^{n} d\theta^j_i - \sum_{k=1}^{n} \theta^k_i \wedge \theta^j_k \tag{10.170}$$

Exterior differentiation of this system gives,

$$d\Theta^j - \sum_{k=1}^{n} \Theta^k \wedge \theta^j_k - \sum_{k=1}^{n} \theta^k \wedge \Theta^j_k = 0 \tag{10.171}$$

$$d\Theta^j_i + \sum_{k=1}^{n} \Theta^k_i \wedge \theta^j_k - \sum_{k=1}^{n} \theta^k_i \wedge \Theta^j_k = 0 \tag{10.172}$$

which are called Bianchi identities.

Summarizing, an affine connection on M gives raise to n^2+n linearly independent 1–forms θ^i and θ^k_i on the space of frames on M whose exterior derivatives satisfy equations (10.171, 10.172). Viceversa, any set of n^2+n linearly independent 1–

forms on the space of frames whose exterior derivatives satisfy equations (10.171, 10.172) gives raise to an affine connection on M.

Riemannian Connections

Given a Riemannian metric g, it is possible to define a scalar product of two vectors with the same origin as the contraction of their tensor product with the metric at the point. If g_{ij} are the components of the metric tensor with respect to the natural frame of a coordinate system (x^i) and A^i and B^i are the components of the vectors X and Y the scalar product is given by,

$$X \cdot Y = \sum_{i,j=1}^{n} g_{ij} A^i B^j \tag{10.173}$$

The length of a vector X is the positive square root of the product $X^2 = X \cdot X$.

A Riemannian metric also enables us to define the arc length of a parametrized curve. If $\gamma : \mathbb{R} \to M$ is a curve with components $x^i(t) = \gamma^i(t)$, we have,

$$ds^2 = g_{ij}\, dx^i\, dx^j \tag{10.174}$$

and,

$$s = \int_{t_0}^{t_1} \sqrt{g_{ij} \frac{dx^i}{dt} \frac{dx^j}{dt}}\, dt \tag{10.175}$$

defines the arc length between the points $\gamma(t_0)$ and $\gamma(t_1)$.

The angle between vectors is also given by,

$$\cos\theta = \frac{X \cdot Y}{\sqrt{X \cdot X}\sqrt{Y \cdot Y}} \tag{10.176}$$

Definition 10.62 A map $F : M \to M'$ between Riemannian manifolds is said to be isometric if it preserves distances between points, and therefore it also preserves angles between curves.

One of the main results of Riemannian geometry was the discovery of the dominant role played by the Levi–Civita connection.

E.3. The Levi–Civita Connection

Theorem 10.63 *On a Riemannian manifold there is one affine connection with the following properties*
(i) The torsion tensor is zero
(ii) The scalar product of two vectors remains unchanged when they are displaced parallelely along a curve.

One such torsionless connection is called a metric connection.

Proof 10.1 This theorem can be easily proved in coordinates (x^i). We consider a parametrized curve, $x^i = x^i(t)$ and along it the vector fields with components $A^i(t)$ and $B^i(t)$. The conditions for parallelism along the curve are,

$$\frac{dA^i}{dt} + \sum_{j,k=1}^{n} \Gamma^i_{kj} A^k \frac{dx^j}{dt} = 0 \tag{10.177}$$

and,

$$\frac{dB^i}{dt} + \sum_{j,k=1}^{n} \Gamma^i_{kj} B^k \frac{dx^j}{dt} = 0 \tag{10.178}$$

for $i = 1, \ldots, n.$□

Property *(ii)* in the statement of the theorem is equivalent to,

$$\frac{d}{dt} \sum_{i,j=1}^{n} g_{ij} A^i B^j = 0 \tag{10.179}$$

Whenever these conditions are satisfied we find,

$$\frac{d}{dt} \sum_{i,j=1}^{n} g_{ij} A^i B^j = \sum_{i,j=1}^{n} \left(\frac{dg_{ij}}{dt} - \sum_{l=1}^{n} g_{il}\, \Gamma^l_{kj} \frac{dx^k}{dt} - \sum_{l=1}^{n} g_{lj}\, \Gamma^l_{ik} \frac{dx^k}{dt} \right) A^i B^j = 0 \tag{10.180}$$

with the requirement that this holds for all parametrized curves and all vector fields along the curve, therefore,

$$\frac{\partial g_{ij}}{\partial x^k} = \sum_{l=1}^{n} \left(g_{il} \Gamma^l_{jk} + g_{lj} \Gamma^l_{ik} \right) , \quad i, j, k = 1, \ldots, n. \tag{10.181}$$

On the other side, the vanishing of the torsion can be expressed by,

$$\Gamma^i_{jk} = \Gamma^i_{kj} \qquad i, j, k = 1, \ldots, n \tag{10.182}$$

or by introducing $\Gamma_{kij} = \sum_{l=1}^{n} g_{li}$, by: $\Gamma_{kij} = \Gamma_{jik}$. With these Christoffel symbols of first kind, (10.181) can be rewritten as,

$$\frac{\partial g_{ij}}{\partial x^k} = \sum_{l=1}^{n} (\Gamma_{jik} + \Gamma_{ijk}) \tag{10.183}$$

Permuting cyclically n i, j, k we get,

$$\begin{aligned} \frac{\partial g_{jk}}{\partial x^i} &= \Gamma_{jki} + \Gamma_{kji} \\ \frac{\partial g_{ki}}{\partial x^j} &= \Gamma_{kij} + \Gamma_{ikj} , \quad i, j, k = 1, \ldots, n \end{aligned} \tag{10.184}$$

Since the Γ_{ijk}'s are symmetric in the first and third indices, these two equations together with previous one give,

$$\Gamma_{ijk} = \frac{1}{2} \left(\frac{\partial g_{ij}}{\partial x^k} + \frac{\partial g_{jk}}{\partial x^i} - \frac{\partial g_{ik}}{\partial x^j} \right) \tag{10.185}$$

or,

$$\Gamma^i_{jk} = \frac{1}{2} \sum_{l=1}^{n} g^{il} \left(\frac{\partial g_{kl}}{\partial x^j} + \frac{\partial g_{jl}}{\partial x^k} - \frac{\partial g_{jk}}{\partial x^l} \right) \tag{10.186}$$

when g^{ij} are defined by,

$$\sum_{j=1}^{n} g_{ij} g^{jk} = \delta_i^k \tag{10.187}$$

It follows that the affine connection is completely determined. The components Γ_{ijk} and Γ^i_{jk} are called respectively the Christoffel symbols of the first and second kind, respectively. This particular connection associated with a metric tensor is called the connection of Levi–Civita.

Remark 10.4 If we consider other quadratic forms constructed out of two vector fields and a (0,2)–type tensor, say,

$$\sum_{i,j=1}^{n} \alpha_{ij} A^i B^j \tag{10.188}$$

we may say that a connection is an α–connection if,

$$\frac{d}{dt} \sum_{i,j=1}^{n} \alpha_{ij} A^i B^j = 0 \tag{10.189}$$

For instance, α could be a symplectic structure, a non positive definite product, an Hermitean structure and so on.

When the connection is a metric connection there are further properties on the Riemann tensor. In terms of the structure forms, and using the metric tensor to lower some indices, we find $\theta_i, \quad \theta_{ij}, \quad \Theta_{ij}, \quad S_{ijkl}$ out of $\theta^i, \quad \theta_i^j, \quad \Theta_i^j, \quad S_{ikl}^j$ respectively.

Then the following properties hold,

$$\theta_{ij} + \theta_{ji} = 0\,, \quad \Theta_{ij} + \Theta_{ji} = 0 \tag{10.190}$$

and θ_i, θ_{ij} (with $i < j$) are linearly independent.

These forms satisfy the equations,

$$d\theta_i = \sum_{j=1}^{n} \theta_j \wedge \theta_{ji} \tag{10.191}$$

$$d\theta_{ij} = \sum_{k=1}^{n} \theta_{ik} \wedge \theta_{kj} + \Theta_{ij} \tag{10.192}$$

$$d\Theta_{ij} + \sum_{k=1}^{n} \Theta_{ik} \wedge \theta_{kj} - \sum_{k=1}^{n} \theta_{ik} \wedge \Theta_{kj} = 0 \tag{10.193}$$

and the functions S_{ijkl} have the following symmetry properties,

$$S_{ijkl} = -S_{jikl} = -S_{ijlk} \tag{10.194}$$

$$S_{ijkl} = -S_{klij} \tag{10.195}$$

$$S_{ijkl} + S_{iklj} + S_{iljk} = 0 \tag{10.196}$$

E.4. Properties of Connections and Comparison with Other Approaches

A linear space M can be endowed with two o more linear connections. Notice that if $\Gamma^{(1)}$ and $\Gamma^{(2)}$ are the Christoffel symbols of two such connections, then the difference symbols $\Gamma^{(1)} - \Gamma^{(2)}$ transform under changes of coordinates as the components of a (1,2) tensor field, even if the symbols of each connection do not transform as a tensor field due to the presence of the first term in the right-hand side of (10.126). However, when taking the differences $\Gamma^{(1)} - \Gamma^{(2)}$, the first terms cancel and therefore the components of the difference $\Gamma^{(1)} - \Gamma^{(2)}$ transform as the components of a (1,2) tensor field. Conversely, given a connection with Christoffel symbols $\Gamma^{(1)}$ and a tensor field L of type (1,1) we can define a new set of Christoffel symbols $\Gamma^{(2)}$ by writing in any coordinates,

$$\Gamma^{(2)i}_{jk} = \Gamma^{(1)i}_{jk} + L^i_{jk} \tag{10.197}$$

because if the $\Gamma^{(1)i}_{jk}$ satisfy the property of changing as (10.126) when going from one set to another of coordinates, the same is true for $\Gamma^{(2)i}_{jk}$, because L being a (1, 1)–tensor field, its coordinates transform as,

$$L^i_{jk} = \sum_{m,p,q=1}^{n} \tilde{L}^m_{pq} \frac{\partial \tilde{x}^p}{\partial x^j} \frac{\partial \tilde{x}^q}{\partial x^k} \frac{\partial x^i}{\partial \tilde{x}^m} \tag{10.198}$$

Now, given two affine connections, any convex combination gives a new connection, because the symbols $\Gamma^i_{jk} = \lambda \Gamma^{(1)i}_{jk} + (1-\lambda)\Gamma^{(2)i}_{jk}$, with $0 \leq \lambda \leq 1$ transform properly when going from one coordinate system to another. If we change the lower indices in the symbols we also obtain a new connection because the first term in (10.126) is symmetric under such interchange of indices. Therefore, if $\Gamma^{(1)i}_{jk}$ are the symbols of a connection, then $\Gamma^{(2)i}_{jk} = \Gamma^{(1)i}_{kj}$ is a new connection in which the torsion changes the sign and the convex combination of both with $\lambda = 1/2$ gives raise to a symmetric connection. We notice that geodesics are indifferent to torsion.

Let us now consider a curve $\sigma : I \to M$, with $0 \in I$, and a vector $u \in T_{\sigma(0)}M$. We have seen before that we can transport this vector all along the curve by parallel transport and we obtain in this way a curve in TM, $\sigma^h : I \to TM$, which projects on σ. The curve so obtained will be called the horizontal lift of σ.

The same can be done for vectors. So, given a vector $u \in T_x M$, we can lift it to a vector tangent to TM at w, for every $w \in T_x M$, just by choosing a curve $\sigma : I \to M$ such that $\sigma(0) = x$ and $\dot{\sigma}(0) = u$ and taking the tangent vector to the horizontal lift σ^h of such a curve starting at $w \in T_x M$ at $t = 0$. The vector so obtained is said to be horizontal and the lift so defined does not depend on the choice of the curve σ.

In fact, if $W(t)$ is a vector field along σ such that $W(0) = w$, then the condition of the curve σ^h of being obtained by parallel transport means that,

$$\dot{W}^i(t) + \sum_{j,k=1}^{n} \Gamma^i_{jk}(x) W^j u^k = 0 \tag{10.199}$$

and therefore the vector tangent to σ^h at t_0 is,

$$\sum_{i=1}^{n} \left(\dot{\sigma}^i(0) \frac{\partial}{\partial x^i} + \dot{W}^i(0) \frac{\partial}{\partial v^i} \right) = \sum_{k=1}^{n} u^k \left(\frac{\partial}{\partial x^k} - \sum_{i,j=1}^{n} \Gamma^i_{jk}(x) w^j \frac{\partial}{\partial v^i} \right) \tag{10.200}$$

We see in this expression for the lifting of a vector that the lifting of a vector depends linearly on it. Moreover, the lifting of a vector is zero if and only if the vector is null. In other words, the linear map of lifting is injective.

The preceding expression for the lifting of a vector displays clearly that we can choose a basis of horizontal vector fields formed by,

$$H_i = \frac{\partial}{\partial x^i} - \sum_{j,k=1}^{n} \Gamma^k_{ji}(x) v^j \frac{\partial}{\partial v^k}\,, \quad i = 1, \dots, n \tag{10.201}$$

These fields determine a distribution in TM which in general is not integrable because,

$$\begin{aligned}
&\left[\frac{\partial}{\partial x^i} - \sum_{k,l=1}^{n} \Gamma^k_{li} v^l \frac{\partial}{\partial v^k}, \frac{\partial}{\partial x^j} - \sum_{h,g=1}^{n} \Gamma^h_{gj} v^g \frac{\partial}{\partial v^h}\right] \\
&= - \sum_{h,g=1}^{n} \frac{\partial \Gamma^h_{gj}}{\partial x^i} v^g \frac{\partial}{\partial v^h} + \sum_{k,l=1}^{n} \frac{\partial \Gamma^k_{li}}{\partial x^j} v^l \frac{\partial}{\partial v^k} \\
&+ \sum_{k,l,h=1}^{n} v^l \Gamma^k_{li} \Gamma^h_{kj} \frac{\partial}{\partial v^h} - \sum_{j,k,l,g=1}^{n} v^g \Gamma^l_{gj} \Gamma^k_{li} \frac{\partial}{\partial v^k} \\
&= \sum_{j,k,l,g=1}^{n} v^l \left[\frac{\partial \Gamma^k_{li}}{\partial x^j} - \frac{\partial \Gamma^k_{lj}}{\partial x^i} + \Gamma^g_{li} \Gamma^k_{gj} - \Gamma^l_{gj} \Gamma^k_{li}\right] \frac{\partial}{\partial v^k}
\end{aligned} \tag{10.202}$$

which can be rewritten as,

$$\left[\frac{\partial}{\partial x^i} - \sum_{k,l=1}^{n} \Gamma^k_{li} v^l \frac{\partial}{\partial v^k}, \frac{\partial}{\partial x^j} - \sum_{g,h=1}^{n} \Gamma^h_{gj} v^g \frac{\partial}{\partial v^h}\right] = \sum_{k,l=1}^{n} R^k{}_{lji} v^l \frac{\partial}{\partial v^k} \tag{10.203}$$

and therefore, the horizontal distribution is integrable if and only if the curvature of the connection vanishes.

The horizontal distribution can also be defined as the kernel of the 1-forms,

$$\eta^i = dv^i + \sum_{j,k=1}^{n} \Gamma^i_{jk} v^j dx^k \tag{10.204}$$

because,

$$\eta^i(H_j) = (dv^i + \sum_{l,k=1}^{n} \Gamma^i_{lk} v^l dx^k)(\frac{\partial}{\partial x^j} - \sum_{l,k=1}^{n} \Gamma^k_{lj}(x) v^l \frac{\partial}{\partial v^k}) = \sum_{m=1}^{n} \Gamma^i_{mj} v^m - \sum_{k=1}^{n} \Gamma^i_{kj} v^k = 0 \tag{10.205}$$

Conversely, if the vector field,

$$X = \sum_{i=1}^{n} \left(A^i \frac{\partial}{\partial x^i} + B^i \frac{\partial}{\partial v^i} \right) \tag{10.206}$$

is annihilated by the 1–forms η^i, then as,

$$\eta^i(X) = B^i + \sum_{j,k=1}^{n} \Gamma^i_{jk} v^j A^k \tag{10.207}$$

we see that X is of the form,

$$X = \sum_{i=1}^{n} A^i \left(\frac{\partial}{\partial x^i} - \sum_{j,k=1}^{n} \Gamma^k_{ji} v^j \frac{\partial}{\partial v^k} \right) \tag{10.208}$$

and then it lies in the horizontal distribution. Another very remarkable property of the linear connections we are considering is that if V^h is the horizontal lift of a vector field V in the base manifold, then $[\Delta, V^h] = 0$, i.e., the lifting so defined is homogeneous of degree zero in velocities.

Let us now mention the relationship between the theory of connections as explained here, which is based in the ideas of Levi–Civita, and other alternative but equivalent approaches.

We recall that the affine connection provide a method of transporting a vector at a point $x \in M$ to another point along any chosen curve σ. That means that if $y \in M$ is another point of the given curve the affine connection provides a map: $\mathcal{Y}_{y,x} : T_x M \to T_y M$, where mention is not made to the curve σ for the sake of notational simplicity. Such a map $\mathcal{Y}$ satisfies that for any triplet of points in $\sigma(I)$, $\mathcal{Y}_{y,z} \circ \mathcal{Y}_{z,x} = \mathcal{Y}_{y,x}$. Obviously: $\mathcal{Y}_{x,x} = Id_{T_x M}$ (along a degenerate curve reduced to a point), and: $\mathcal{Y}_{x,y} = \left(\mathcal{Y}_{y,x}\right)^{-1}$, when the curves we consider are the opposite one to another.

We remark that as the transport rule is defined by solving the ordinary differential equation system,

$$\frac{dV^i}{dt} + \sum_{j,k=1}^{n} \Gamma^i_{jk} V^j \frac{d\sigma^k}{dt} = 0 \tag{10.209}$$

which is linear , the map $\mathcal{Y}_{y,x}$ so defined is linear . Moreover, when we consider two curves σ and σ' which are related by a reparametrization, i.e., there exists a diffeomorphism $h : I' \to I$, such that $\sigma' = \sigma \circ h$, we see that $\dot{\sigma}' = \dot{\sigma}\, \dot{h}$ and taking into account that V becomes $V' = V \circ h$, we also see that,

$$\frac{dV'^i}{dt'} = \dot{h}\frac{dV^i}{dt} \tag{10.210}$$

and therefore the maps $\mathcal{Y}_{x,y}$ obtained for the two curves coincide. In other words, the transport rule is defined for each path, i.e., a curve freed from its parametrization.

Conversely, the affine connection can be introduced by giving an appropriate rule of parallel transport in M, i.e., a rule of parallel transport is given for each path in M. Additional smoothness conditions have to be imposed.

So, the rule allows us to say that a vector field along a path is a parallel field when it may be obtained by parallel el transport of a given vector at some point of the curve.

The rule of parallel transport can be used to define the absolute o covariant derivative along a curve σ of a vector field along σ, as follows:

If $V(t) \in \mathfrak{X}(\sigma)$, we can define $DV/Dt \in \mathfrak{X}(\sigma)$ by,

$$\frac{DV}{Dt} = \lim_{\delta \to 0} \frac{1}{\delta}\left[V(t+\delta)_{\parallel} - V(t)\right] \tag{10.211}$$

where $V(t+\delta)_{\parallel}$ is the vector in $T_{\sigma(t)}$ obtained by transporting parallelely $V(t+\delta)$ from $\sigma(t+\delta)$ to $\sigma(t)$,

$$V(t+\delta)_{\parallel} = \mathcal{Y}_{\sigma(t),\sigma(t+\delta)} V(t+\delta) \tag{10.212}$$

With this definition we can easily check that,

$$\frac{D}{Dt}(U+V) = \frac{DU}{Dt} + \frac{DV}{Dt}, \quad \frac{D}{Dt} fV = f\frac{DV}{Dt} + \frac{df}{dt}V, \; \forall f \in \mathcal{F}(I) \tag{10.213}$$

and that $DV/Dt = 0$ if and only if V is parallel along the curve σ.

We will assume, as an additional requirement on our transport rule the two following properties:

PT1. The value of the absolute derivative along a curve σ at a point $\sigma(t)$ depends only on $\dot{\sigma}(t)$: if σ and σ' are curves such that $\sigma(0) = \sigma'(0) = x$ and $\dot{\sigma}(0) = \dot{\sigma}'(0) = v$, and if V is a vector field defined in a neighborhood of $x \in M$, then the absolute derivatives of V at the point x along σ and σ' are equal.

PT2. For any vector field V defined in a neighborhood of $x \in M$, the map $T_xM \to T_xM$ assigning to each $u \in T_xM$ the absolute derivative of V along any curve whose tangent vector is at x in u is linear.

With these two properties for the rule of parallel transport we can define the concept of covariant derivative along a vector $u \in T_xM$ of a vector field defined in a neighborhood of $x \in M$ as follows:

If $V \in \mathfrak{X}(M)$ y $v \in T_xM$, with $v \neq 0$, let us choose a curve $\sigma : I \to M$ such that $\sigma(0) = x$ and $\dot{\sigma}(0) = v$, and then, by assumption, $(DV/dt)(0)$ does not depend on the choice of the curve σ, but only on x and v, and we will denote this vector $\nabla_v V$. When $v = 0$, by definition, $\nabla_0 V = 0$, for any vector field V.

Given two different vector fields $U, V \in \mathfrak{X}(M)$, we can define a new one to be denoted $\nabla_U V \in \mathfrak{X}(M)$ by means of,

$$(\nabla_U V)(x) = \nabla_{U_x} V \tag{10.214}$$

In this way we have found a map $\nabla : \mathfrak{X}(M) \times \mathfrak{X}(M) \to \mathfrak{X}(M)$ given by $\nabla(U, V) = \nabla_U V$ such that the following properties hold,

$$(i)\ \nabla_{U+V} W = \nabla_U W + \nabla_V W \tag{10.215}$$

$$(ii)\ \nabla_U (V + W) = \nabla_U V + \nabla_U W \tag{10.216}$$

$$(iii)\ \nabla_{fU} V = f\,\nabla_U V,\ \forall f \in \mathcal{F}(M) \tag{10.217}$$

$$(iv)\ \nabla_U (fV) = (Uf)V + f\,\nabla_U V \tag{10.218}$$

In this approach to connections the Christoffel symbols arise directly from ∇. In fact, if x^i are arbitrary coordinates, then:

$$\nabla_{\partial/\partial x^i}\left(\frac{\partial}{\partial x^j}\right) = \sum_{k=1}^{n} \Gamma^k_{ji} \frac{\partial}{\partial x^k} \tag{10.219}$$

Therefore, in arbitrary coordinates, $\nabla_U V$ will be expressed by:

$$\nabla_U V = \sum_{i,k=1}^{n} U^k \left(\frac{\partial V^i}{\partial x^k} + \sum_{j=1} \Gamma^i_{jk}(x) V^j\right) \frac{\partial}{\partial x^i} \tag{10.220}$$

because if,

$$U = \sum_{i=1}^{n} U^i \frac{\partial}{\partial x^i}, \quad \sum_{i=1}^{n} V^i \frac{\partial}{\partial x^i} \tag{10.221}$$

then,

$$\begin{aligned} &\nabla_{\sum_{i=1}^n U^i \partial/\partial x^i} \sum_{j=1}^n V^j \frac{\partial}{\partial x^j} \\ &= \sum_{i,j=1}^n U^i \left(\frac{\partial V^j}{\partial x^i}\frac{\partial}{\partial x^j} + V^j \nabla_{\partial/\partial x^i}\frac{\partial}{\partial x^j}\right) \end{aligned} \tag{10.222}$$

where use is made of the preceding properties, and therefore, using (10.219) we find (10.220).

From this expression we find that,

$$\nabla_v V \left(\frac{D(V \circ \sigma)}{dt}\right)_{t=0} = \sum_{i,k=1} v^k \left(\frac{\partial V^i}{\partial x^k} + \Gamma^i_{jk}(x) V^j(x)\right) \frac{\partial}{\partial x^i} \tag{10.223}$$

If v is a vector in the point $x \in M$, the covariant derivative of $\nabla_v \alpha$ of a 1-form α can be defined in a similar way by means of $(D(\alpha \circ \gamma)/Dt)_{t=0}$, γ being a curve such that $\gamma(0) = x$ and $\dot{\gamma}(0) = v$. This definition does not depend on the choice of the curve. If U is a vector field in M, we can define $\nabla_U \alpha$ by,

$$(\nabla_U \alpha)(x) = \nabla_{U_x} \alpha \tag{10.224}$$

We find in this way a map $\nabla : \mathfrak{X}(M) \times \Omega^1(M) \to \Omega^1(M)$ such that,

$$(i)\ \nabla_{U+V}\alpha = \nabla_U \alpha + \nabla_V \alpha \tag{10.225}$$

$$(ii) \nabla_U(\alpha + \beta) = \nabla_U \alpha + \nabla_U \beta \tag{10.226}$$

$$(iii) \nabla_{fU}\alpha = f\, \nabla_U \alpha,\ \forall f \in \mathcal{F}(M) \tag{10.227}$$

$$(iv) \nabla_U(f\alpha) = (Uf)\alpha + f\, \nabla_U \alpha \tag{10.228}$$

Finally, the relation,

$$\frac{d}{dt}\langle \alpha, V \rangle = \left\langle \frac{D\alpha}{Dt}, V \right\rangle + \left\langle \alpha, \frac{DV}{Dt} \right\rangle, \tag{10.229}$$

translates to,

$$W\langle \alpha, V \rangle = \langle \nabla_W \alpha, V \rangle + \langle \alpha, \nabla_W V \rangle. \tag{10.230}$$

As far as the torsion and curvature is concerned, we should remark that,

$$T(X, Y) = \nabla_X Y - \nabla_Y X - [X, Y], \quad X, Y \in \mathfrak{X}(M) \tag{10.231}$$

is such that $T(X, Y) = -T(Y, X)$ and for each function f in M,

$$T(fX, Y) = f\, T(X, Y) \tag{10.232}$$

because,

$$\begin{aligned} T(fX, Y) =& f\, \nabla_X Y - (Yf)X - f\, \nabla_Y X - [fX, Y] \\ =& f\, \nabla_X Y - (Yf)X - f\, \nabla_Y X - f[X, Y] + (Yf)\, X = f\, T(X, Y) \end{aligned} \tag{10.233}$$

Consequently, T defines a $(1, 2)$ tensor field. In local coordinates, the components of T are given by,

$$T\left(\frac{\partial}{\partial x^i}, \frac{\partial}{\partial x^j}\right) = \nabla_{\frac{\partial}{\partial x^i}}\left(\frac{\partial}{\partial x^j}\right) - \nabla_{\frac{\partial}{\partial x^j}}\left(\frac{\partial}{\partial x^i}\right) = \sum_{k=1}^{n}(\Gamma^k_{ji} - \Gamma^k_{ij})\frac{\partial}{\partial x^k} \quad (10.234)$$

and then T is but the torsion tensor we introduced before.

In a similar way, if $R(X, Y)Z$, is defined by,

$$R(X, Y)Z = \nabla_X(\nabla_Y Z) - \nabla_Y(\nabla_X Z) - \nabla_{[X,Y]}Z = 0 \quad (10.235)$$

then it satisfies $R(X, Y)Z = -R(Y, X)Z$, together with,

$$R(fX, Y)Z = f\,R(X, Y)Z = R(X, Y)(fZ) \quad (10.236)$$

$\forall f \in \mathcal{F}(M)$, because,

$$R(fX, Y)Z = f\nabla_X(\nabla_Y Z) - (Yf)\nabla_X Z - f\nabla_{[X,Y]}Z + (Yf)\nabla_X Z = fR(X, Y)Z \quad (10.237)$$

and,

$$\begin{aligned} R(X, Y)(fZ) = \nabla_X((Yf)Z + f\nabla_Y Z) - \nabla_Y((Xf)Z + f\nabla_X Z) \\ -f\nabla_{[X,Y]}Z - (([X, Y])f)Z = f\,R(X, Y)Z \end{aligned} \quad (10.238)$$

R can then be seen as a map from $\mathfrak{X}(M) \times \mathfrak{X}(M) \times \mathfrak{X}(M)$ into $\mathfrak{X}(M)$ which is $\mathcal{F}(M)$-multilinear and therefore it defines a tensor of type $(1, 3)$. Its components $R^a{}_{bcd}$ in a coordinate system are,

$$R\left(\frac{\partial}{\partial x^k}, \frac{\partial}{\partial x^i}\right)\frac{\partial}{\partial x^j} = \sum_{i=1}^{n} R^i{}_{jkl}\frac{\partial}{\partial x^i} \quad (10.239)$$

and then,

$$R^i{}_{jkl} = \frac{\partial \Gamma^i_{jl}}{\partial x^k} - \frac{\partial \Gamma^i_{jk}}{\partial x^l} + \sum_{m=1}^{n}(\Gamma^m_{jl}\,\Gamma^i_{mk} - \Gamma^m_{jk}\,\Gamma^i_{ml}) \quad (10.240)$$

Therefore, the tensor R is the curvature tensor introduced before.

If the connection is torsionless, $T = 0$, when applying ∇_X on both sides of the equation $T(X, Y) = 0$ we will obtain,

$$\nabla_X(\nabla_Y Z) - \nabla_X(\nabla_Z Y) - \nabla_X([Y, Z]) = 0 \quad (10.241)$$

and permuting cyclically the vector fields X, Y, Z and adding the corresponding expressions we find that,

$$\nabla_X(\nabla_Y Z) - \nabla_X(\nabla_Z Y) - \nabla_X([Y, Z]) + \nabla_Y(\nabla_Z X) - \nabla_Y(\nabla_X Z) - \nabla_Y([Z, X]) + \nabla_Z(\nabla_X Y) - \nabla_Z(\nabla_Y X) - \nabla_Z([X, Y]) = 0 \tag{10.242}$$

We can now use the relations $\nabla_X(\nabla_Y Z) - \nabla_X(\nabla_Z Y) + R(X, Y)Z + \nabla_{[X,Y]}Z$ and analogously for the other cyclic permutations and then we get,

$$R(X, Y)Z + \nabla_{[X,Y]}Z - \nabla_Z([X, Y]) + R(Y, Z)X + \nabla_{[Y,Z]}X - \nabla_X([Y, Z]) + R(Z, X)Y + \nabla_{[Z,X]}Y - \nabla_Y([Z, X]) = 0 \tag{10.243}$$

In the torsionless case we are considering the terms in which R does not appear cancel, because of Jacobi identity and we obtain the relation,

$$R(X, Y)Z + R(Y, Z)X + R(Z, X)Y = 0 \tag{10.244}$$

which is known as *Ricci* or *first Bianchi identity*.

Koszul's approach to the theory of connections was based in relations (10.215) to (10.218). In other words, we can consider a connnection as a map $\nabla : \mathfrak{X}(M) \times \mathfrak{X}(M) \to \mathfrak{X}(M)$ given by $\nabla(U, V) = \nabla_U V$ satisfying properties (10.215) to (10.218). In this case the connection 1-forms can be recovered by,

$$\omega^i{}_j(V) = \left\langle dx^i, \nabla_V \frac{\partial}{\partial x^j} \right\rangle, \tag{10.245}$$

and Christoffel symbols are then defined by (10.124).

In the case in which (M, g) is a Riemannian manifold, given a connection in the Koszul approach by means of the ∇ map, we can see that the connection is a metric connection if and only if for any three vector fields $X, Y, Z \in \mathfrak{X}(M)$, the following relation holds,

$$X(Y \cdot Z) = (\nabla_X Y) \cdot Z + Y \cdot (\nabla_X Z) \tag{10.246}$$

In fact, if ∇ satisfies the mentioned property, then given a curve in M, $\gamma : I \to M$, starting from $x = \gamma(0)$, for any couple of vectors $u, v \in T_x M$, let U, V be the vector fields along γ obtained by parallel transport of u and v respectively. If $f : I \to \mathbb{R}$ is the function given by $f(t) = U(t) \cdot V(t)$, then,

$$\frac{df}{dt} = T_{\gamma(t)} f = (\nabla_T U(t)) \cdot V(t) + U(t) \cdot (\nabla_T V(t)) \tag{10.247}$$

where T is a vector field in M extending to $\frac{d\gamma}{dt}$, and as U and V are constant along the curve, we see that $\frac{df}{dt} = 0$.

Conversely, let assume that the connection is metric. Given any three vector fields, $X, Y, Z \in \mathfrak{X}(M)$, for any point $x \in M$, let $\gamma : I \to M$ be the geodesic curve starting from x and with tangent vector X_x. If $\{u_1, \ldots, u_n\}$ is an orthonormal basis of $T_x M$ and $U_i \in \mathfrak{X}(\gamma)$ denote the vector fields along γ obtained by parallel transport of u_i, as by assumption the connection is metric, the set of vector fields along γ, $U_1, \ldots, U_n$, determine an orthonormal basis of the tangent spaces at each point of the curve. Let the coordinate expressions of the vector fields along γ, $Y \circ \gamma$ and $Z \circ \gamma$, be,

$$Y(t) = \sum_{i=1}^{n} Y^i(t) U_i, \qquad Z(t) = \sum_{i=1}^{n} Z^i(t) U_i \tag{10.248}$$

and then $Y(t) \cdot Z(t) = \sum_{i,j=1}^{n} g_{ij}\, Y^i(t) Z^j(t)$, where g_{ij} are constant. Let T denote as before a vector field in M extending $\frac{d\gamma}{dt}$, and then,

$$T_{\gamma(t)}(Y(t) \cdot Z(t)) = \sum_{i,j=1}^{n} g_{ij} \frac{d}{dt} \left(Y^i(t) Z^j(t) \right) \tag{10.249}$$

The vector fields along γ U_i are constant, and then,

$$\nabla_T Y = \left(\frac{d}{dt} \sum_{j=1}^{n} Y^j(t) \right) U_j, \quad \nabla_T Z = \left(\frac{d}{dt} \sum_{j=1}^{n} Z^j(t) \right) U_j \tag{10.250}$$

Therefore,

$$(\nabla_T Y \cdot Z) + (Y \cdot \nabla_T Z) = \sum_{i,j=1}^{n} g_{ij} \frac{d}{dt} \left(Y^i(t) Z^j(t) \right) \tag{10.251}$$

Particularizing for the value $t = 0$, and having in mind that by construction of γ, $T(0) = X_x$, we obtain the relation (10.246).

The explicit expression of the torsionless metric connection of Levi–Civita is given by the so-called 'Koszul formula',

$$\begin{aligned} 2(\nabla_X Y \cdot Z) &= X(Y \cdot Z) + Y(Z \cdot X) - Z(X \cdot Y) \\ &\quad - (X \cdot [Y, Z]) + (Y \cdot [Z, X]) + (Z \cdot [X, Y]) \end{aligned} \tag{10.252}$$

In a local coordinate system x^i in M, we can see that the Koszul formula yields,

$$2 \left(\nabla_{\partial/\partial x^i} \left(\frac{\partial}{\partial x^j} \right) \right) \cdot \frac{\partial}{\partial x^k} = \frac{\partial g_{jk}}{\partial x^i} + \frac{\partial g_{ik}}{\partial x^j} - \frac{\partial g_{ij}}{\partial x^k} \tag{10.253}$$

and using the definition of Christoffel symbols,

$$\left(\nabla_{\partial/\partial x^i}\left(\frac{\partial}{\partial x^j}\right)\right)\cdot\frac{\partial}{\partial x^k}=\sum_{l=1}^{n}\Gamma^l_{\ ji}g_{lk} \tag{10.254}$$

Therefore,

$$\Gamma^k_{ji}=\frac{1}{2}\sum_{l=1}^{n}g^{kl}\left(\frac{\partial g_{jl}}{\partial x^i}+\frac{\partial g_{il}}{\partial x^j}-\frac{\partial g_{ij}}{\partial x^l}\right) \tag{10.255}$$

Similarly, given a skew–symmetric in the lower indices tensor of type (1, 2), T, there is only one metric connection with such a torsion tensor T which is given by the generalization of the Koszul formula,

$$\begin{aligned}2\,(\nabla_X Y)\cdot Z &= X(Y\cdot Z)+Y(Z\cdot X)-Z(X\cdot Y)-X\cdot[Y,Z]+Y\cdot[Z,X]\\ &\quad +Z\cdot[X,Y]+X\cdot T(Y,Z)-Y\cdot T(Z,X)+Z\cdot T(X,Y)\end{aligned} \tag{10.256}$$

On the other hand, we have been able to associate to the connection an horizontal distribution satisfying an additional property of homogeneity. If we suppose that, conversely, we give as an input an horizontal distribution, with appropriate smoothness conditions, i.e., we choose in every tangent vector to M a n-dimensional subspace complementary to the vertical subspace, we can use it to lift vector fields in M to *horizontal vector fields* in TM, i.e., the lift of the vector field $V\in\mathfrak{X}(M)$ is the unique horizontal vector field such that $T\tau(V^h)=V$. More specifically, in a given coordinate system, the horizontal lifting of $\partial/\partial q^i$ is,

$$\left(\frac{\partial}{\partial q^i}\right)^h=\frac{\partial}{\partial q^i}-\sum_{j=1}^{n}\Gamma^j_i\frac{\partial}{\partial v^j} \tag{10.257}$$

If the distribution verifies the mentioned homogeneity condition $[\Delta, V^h]=0$, from the relation,

$$\left[\sum_{k=1}^{n}v^k\frac{\partial}{\partial v^k},\frac{\partial}{\partial q^i}-\sum_{j=1}^{n}\Gamma^j_i\frac{\partial}{\partial v^j}\right]=\sum_{j,k=1}^{n}\left(-v^k\frac{\partial\Gamma^j_i}{\partial v^k}+\Gamma^k_i\right)\frac{\partial}{\partial v^j}=0 \tag{10.258}$$

from where we see that,

$$\sum_{k=1}^{n}v^k\frac{\partial\Gamma^j_i}{\partial v^k}=\Gamma^j_i\,,\quad i,j=1,\ldots,n \tag{10.259}$$

and therefore the functions Γ_i^j are homogeneous of degree one. If they are smooth at $v = 0$, they should be linear, i.e., there will be of the form,

$$\Gamma_i^j = \sum_{k=1}^{n} \Gamma_{ki}^j v^k \tag{10.260}$$

Once we have been able to lift vectors, we can also lift curves in the base, i.e., given a curve in the base there exists one curve in TM starting from each vector and such that the tangent vector to the curve is horizontal. With these horizontal lifts of curves we can do the parallel transport from a vector at a point of M along a curve in M just by using the horizontal lift of the curve.

Given a symmetric linear connection on a affine space M, one can define a vector field Γ on TM whose value at $v \in TM$ is its horizontal lift to $T_v(TM)$. In local coordinates the expression of this vector field is,

$$\Gamma = \sum_{i=1}^{n} v^i \left(\frac{\partial}{\partial q^i} - \sum_{j,k} \Gamma_{jk}^i \frac{\partial}{\partial v^j} v^k \right) \tag{10.261}$$

This vector field is called the *geodesic spray* of the connection and its integral curves are just the natural lift of geodesics in M, which are determined by (10.131). Notice that, by construction, the natural lift of a geodesics is a horizontal curve. As we see, the geodesic spray is a second-order differential equation vector field satisfying an additional condition of degree–one homogeneity which corresponds to the affine reparametrization property of geodesics. More generally, a second-order differential equation vector field is said to be a spray if the set of its integral curves is invariant under any affine reparametrization: these curves are called geodesics of the spray. Given a spray Γ, there is a symmetric connection whose geodesic spray is Γ; and conversely, the connection is fully determined by its geodesic spray, that can also be used to construct the exponential map $\exp : T_x M \to M$. A Riemannian structure on M determines one symmetric connection, the Levi-Civita connection, and consequently a Riemannian spray.

Riemannian and Pseudo-Riemannian Metrics on Linear Vector Spaces

Let again E be an n-dimensional vector space equipped with coordinates $x^1, \ldots, x^n$, and let $g \in S_2(E) = S(T_2^0(E))$ be a fully covariant symmetric tensor field of rank two. In coordinates,

$$g = g_{ij} dx^i \otimes dx^j \tag{10.262}$$

where: $g_{ij} = g_{ji}$ and $g_{ij} \in \mathcal{F}(E)$, i.e., the g_{ij}'s are smooth functions on E. The tensor g induces the maps:

$$g : \mathfrak{X}(E) \times \mathfrak{X}(E) \to \mathcal{F}(E) \tag{10.263}$$

where $\mathfrak{X}(E)$ is the set of smooth sections of the tangent bundle,[8] via: $(X, Y) \mapsto g(X, Y)$, and:

$$\widehat{g} : \mathfrak{X}(E) \to \Omega^k(E) = \mathfrak{X}^*(E) \tag{10.264}$$

by means of,

$$X = X^i \frac{\partial}{\partial x^i} \mapsto \widehat{g}(X) = X_i \, dx^i; \quad X_i = g_{ij} X^j \tag{10.265}$$

Of course: $g(X, Y) = g_{ij} X^i Y^j = i_X \widehat{g}(Y) = i_Y \widehat{g}(X)$. The tensor g will be said to be non-degenerate if $\widehat{g}(X) = 0$ when and only when $X = 0$, i.e., if,

$$g(X, Y) = 0, \ \forall Y \longrightarrow X = 0 \tag{10.266}$$

Definition 10.64 A fully covariant symmetric and non-degenerate second-rank tensor field g will be said to define a pseudo-riemannian metric on E. If the stronger condition,

$$g(X, X) = 0 \leftrightarrow X = 0 \tag{10.267}$$

holds, the metric will be said to be Riemannian.

Example 10.6 In $\mathbb{R}^n$, the tensor:

$$g_{ij} = \delta_{ij} dx^i \otimes dx^j \tag{10.268}$$

defines the (Riemannian) Euclidean metric associated with a given coordinate system.

Example 10.7 Let $E = \mathbb{R}^4$ with coordinates: x^0, x^1, x^2, x^3. The Lorentz metric on $\mathbb{R}^4$ is the pseudo-riemannian metric,

$$g_{ij} = dx^0 \otimes dx^0 - \sum_{i=1}^{3} dx^i \otimes dx^i \tag{10.269}$$

Remark 10.5 If g is a Riemannian metric, then:

---

[8] i.e. The set of vector fields on E.

Proposition 10.65 *$g(X, X)$ has constant sign on E.*

So, if that is the case we may always assume, without loss of generality: $g(X, X) > 0$ for all $X \neq 0$.

Remark 10.6 If g is nondegenerate, then the map of Eq. (10.265) becomes actually a bijection. In other words, and again in local coordinates, this implies that the matrix $||g_{ij}||$ is invertible. Defining then the inverse matrix (with contravariant indices) $||g^{ij}||$, such that,

$$g^{ij} g_{jk} = \delta^i_k \tag{10.270}$$

we can also define the map,

$$\widetilde{g} = \hat{g}^{-1} : \mathfrak{X}^*(E) \to \mathfrak{X}(E) \tag{10.271}$$

via,

$$\omega = \omega_i \, dx^i \mapsto \hat{g}^{-1}(\omega) = \omega^i \frac{\partial}{\partial x^i}; \qquad \omega^i = g^{ij} \omega_j \tag{10.272}$$

Hence, g (or $\widehat{g}$ and/or $\widetilde{g}$) can be used to "raise" and "lower" indices. We can associate with $||g^{ij}||$ the fully contravariant symmetric second-rank tensor $\widetilde{g}$ defined by,

$$\widetilde{g} = g^{ij} \frac{\partial}{\partial x^i} \otimes \frac{\partial}{\partial x^j} \tag{10.273}$$

The right-hand side of Eq. (10.263), being bilinear, symmetric and nondegenerate, will define for us (pointwise) the *scalar product* of vector fields (which need not be positive if the metric is not Riemannian, of course), also denoted as $(X, Y)_g$, or simply as (X, Y) for short if the metric g has been selected once and for all. Quite similarly, the scalar product of two one-forms: $\omega = \omega_i dx^i$ and: $\eta = \eta_i dx^i$ will be pointwise defined as,

$$(\omega, \eta)_g = g^{ij} \omega_i \eta_j \tag{10.274}$$

Let's denote now by G the determinant of the matrix $||g_{ij}||$,

$$G = \det ||g_{ij}|| = \varepsilon^{i_1 \ldots i_n}_{1 \ldots n} g_{1 i_1} \cdots g_{n i_n} \tag{10.275}$$

Under a change of coordinates: $x^i \mapsto \overline{x}^i$,

$$g_{ij} \mapsto \overline{g}_{ij} = g_{hk} \frac{\partial x^h}{\partial \overline{x}^i} \frac{\partial x^k}{\partial \overline{x}^j} \tag{10.276}$$

and hence,

$$G \mapsto \overline{G} = G[\frac{\partial(x^1, \ldots, x^n)}{\partial(\overline{x}^1, \ldots, \overline{x}^n)}]^2 \quad (10.277)$$

(proof left as an exercise). Therefore, if we limit ourselves to transformations with a positive jacobian (orientation-preserving changes of coordinates), then,

$$\sqrt{|G|} \mapsto \sqrt{|\overline{G}|} = \sqrt{|G|}\frac{\partial(x^1, \ldots, x^n)}{\partial(\overline{x}^1, \ldots, \overline{x}^n)} \quad (10.278)$$

i.e., $\sqrt{|G|}$ transforms (with respect to transformations with a positive jacobian) as the coefficient of a form of maximal rank. Being nowhere vanishing:

Proposition 10.66 *If g is a (pseudo) Riemannian metric, then,*

$$\Omega = \sqrt{|G|} dx^1 \wedge \cdots \wedge dx^n \quad (10.279)$$

is a volume-form on E.

The volume-form can be written also as,

$$\Omega = \frac{1}{n!}\Omega_{i_1 \cdots i_n} dx^{i_1} \wedge \cdots \wedge dx^{i_n} \quad (10.280)$$

where,

$$\Omega_{i_1 \cdots i_n} = \sqrt{|G|}\varepsilon^{1 \cdots n}_{i_1 \cdots i_n} \quad (10.281)$$

are the totally antisymmetric components of Ω.

Appendix F: Cohomology Theories of Lie Groups and Lie Algebras

In many instances, some of them found in this book, there appear functions defined on groups that must satisfy equations deeply related to the group structure. Such equations are often related to the cohomological structure of the group and their understanding provides the clues to solve the problems from which they emerge.

Groups often appear as represented on certain modules or acting on given spaces, we will be interested in cohomology theories for groups G with coefficients in G-modules $\mathcal{M}$ (for instance the algebra of smooth functions on a manifold where the group G acts, or the module of symmetric tensors over G, or the Lie algebra of G itself, etc.)

We will be mainly interested in Lie groups. We always have the de Rham cohomology of the group as a smooth manifold on one side, where the group structure is disregarded, and the so-called Eilenberg-Maclane cohomology on the other, where G is just an algebraic group. We will discuss in this appendix the relation among both notions and the relation with a few other cohomology theories that appear naturally along the road.

In this context, both Lie algebras and associative algebras arise naturally, and their cohomology theories will be also succintly discussed as well as their relation to deformation theory.

F.1. Eilenberg-MacLane Cohomology

Let G be an algebraic group and $\mathcal{M}$ a G-module (that is, there is a morphism $\rho\colon G \to \mathrm{Aut}(\mathcal{M})$ or, in other words, G acts on $\mathcal{M}$). We can define the set of n-cochains of G with coefficients on $\mathcal{M}$ as the set of normalized maps $f\colon G \times \overset{n}{\cdots} \times G \to \mathcal{M}$, $f(g_1, \ldots, g_n) = 0$ if $g_i = e$ for some index i. We will denote the Abelian group of n-cochains by $C^n(G, \mathcal{M})$ (the group operation induced pointwise from the Abelian group structure of $\mathcal{M}$ as a module. We can form in this way the Abelian group $C^\bullet(G, \mathcal{M}) = \bigoplus_{n\geq 0} C^n(G, \mathcal{M})$, with $C^0(G, \mathcal{M}) = \mathcal{M}$. We may define now the graded homomorphism of degree 1, $\delta_n\colon C^n(G, \mathcal{M}) \to C^{n+1}(G, \mathcal{M})$, as follows:

$$\begin{aligned}\delta_n f(g_1, \ldots, g_{n+1}) = &\rho(g_1) f(g_2, \ldots, g_{n+1}) \\ &+\sum_{i=1}^{n}(-1)^i f(g_1, \ldots, g_i g_{i+1}, \ldots, g_{n+1}) \\ &+(-1)^{n+1} f(g_1, \ldots, g_n), \end{aligned} \tag{10.282}$$

for all $g_i \in G, i = 1, \ldots, n+1$. A direct computation shows that $\delta_{n+1} \circ \delta_n = 0$. The extension δ to $C^\bullet(G, \mathcal{M})$ of the homomorphisms δ_n, defined by $\delta\mid_{C^n(G,\mathcal{M})} = \delta_n$, satisfies $\delta^2 = 0$. Hence we have defined a cochain complex $(C^\bullet(G, \mathcal{M}), \delta)$ whose cohomology $H^\bullet(G, \mathcal{M}) = \bigoplus_{n\geq 0} H^n(G, A)$ is a graded Abelian group whose n-th factor, called the n th Eilenberg-MacLane cohomology group $H^n(G, \mathcal{M})$ is given by [ML75]:

$$H^n(G, \mathcal{M}) = \ker \delta_n / \mathrm{Im}\delta_{n-1}\,.$$

If we call n-cocycles the n-cochains f such that $\delta_n f = 0$ and we denote them by $Z^n(G, \mathcal{M})$ and we called n-coboundaries the n-cochains f such that there exists a $n-1$-cochain h such that $f = \delta_{n-1} h$ and we denote them by $B^n(G; \mathcal{M})$, then $B^n(G, \mathcal{M})$ is a subgroup of $Z^n(G, \mathcal{M})$ and $H^n(G, \mathcal{M}) = Z^n(G, \mathcal{M})/B^n(G, \mathcal{M})$.

Mackey-Moore and Bargmann-Mostow Cohomologies

Let G be now a topological group. Then we consider $\mathcal{M}$ to be a topological module and the representation ρ of G on $\mathcal{M}$ will be assumed to be strongly continuous, i.e., $\rho(g)$ is a homeomorphism for each $g \in G$ and $\rho(g)x$ is continuous for each $x \in \mathcal{M}$. In that case we will say that $\mathcal{M}$ is a topological G-module. We denote by $C_b^n(G, \mathcal{M})$ the space of Borel measurable n-cochains in G with values in $\mathcal{M}$. Then it is obvious that the Eilenberg-MacLane cohomology operator preserves the Borelian cochains, then it induces a cohomology operator in $C_b^\bullet(G, \mathcal{M}) = \bigoplus_{n\geq 0} C_b^n(G, \mathcal{M})$. The cohomology of the cochain complex $(C_b^\bullet(G, \mathcal{M}), \delta)$ is called the Mackey-Moore cohomology [Mo64, Mac57], it will be denoted by $H_b^\bullet(G, \mathcal{M})$ and it is useful when dealing with topological groups not too wild, for instance Polish groups (that is, topological groups which are separable, metrizable and complete). In such cases, for instance, $H_b^2(G, \mathcal{M})$ measures the space of topological extensions of G by $\mathcal{M}$ (they are actually the same if G and $\mathcal{M}$ are locally compact [Mac76]).

Even if it is not known like that in the literature, we will call the Bargmann-Mostow cohomology [Ba54, Ho77], the cohomology defined on a topological group G by considering the cochain complex $(C_0^\bullet(G, \mathcal{M}), \delta)$ of continuous cochains on G with values in $\mathcal{M}$, where $\mathcal{M}$ is as before a topological G-module and δ is the restriction of the Eilenberg-MacLane cohomology operator to continuous cochains. We will denote such cohomology as $H_c^\bullet(G, \mathcal{M})$ and it is trivial to see that if G is discrete, then:

$$H^\bullet(G, \mathcal{M}) = H_b^\bullet(G, \mathcal{M}) = H_c^\bullet(G, \mathcal{M}).$$

In general there are just natural morphisms between the three previous cohomologies (the sequence is not exact):

$$H_c^\bullet(G, \mathcal{M}) \to H_b^\bullet(G, \mathcal{M}) \to H^\bullet(G, \mathcal{M}).$$

However the most appealing cohomology studied by Bargmann is not the continuous cohomology but a local cohomology that we describe now. Let $C_{\text{loc}}^n(G, \mathcal{M})$ be the space of germs of continuous n-cochains from G to $\mathcal{M}$, that is $[f] \in C_{\text{loc}}^n(G, \mathcal{M})$ is the equivalence class containing the continuous n-cochain f with respect to the equivalence relation, '$f \equiv h$ if there is a neighborhood U of e such that $f\mid_U = h\mid_U$'. Now it is easy to check that the Eilenberg-MacLane cohomology operator is compatible with the preceding equivalence relation, thus it induces an operator, denoted with the same symbol, in $C_{\text{loc}}^\bullet(G, \mathcal{M}) = \bigoplus_{n\geq 0} C_{\text{loc}}^n(G, \mathcal{M})$. We denote by $H_{\text{loc}}^\bullet(G, \mathcal{M})$ the corresponding cohomology. It is remarkable that if G is a Lie group, on each cocycle class $[f]$ there exists a smooth representative f [Ba54]. Thus if G is a Lie group, we may consider $H_{\text{loc}}^\bullet(G, \mathcal{M})$ as the cohomology of the cochain complex of germs of smooth cochains from G to $\mathcal{M}$.

F.2. Smooth Cohomologies on Lie groups and de Rham Cohomology

Under the name of smooth cohomologies we put together all those cohomologies defined on Lie groups that use explicitly the smooth structure of the group. These cohomologies play an important role in various places along the main text, as they characterize either properties of the action of Lie groups on dynamical systems, or they 'measure' the obstructions for some properties of the group to hold true. In any case we are interested in differentiable objects and that is the main reason we restrict ourselves to consider them.

We have already seen that $H^\bullet_{\mathrm{loc}}(G, \mathcal{M})$ is basically a smooth cohomology (and as we will see later on, it coincides with the cohomology of the Lie algebra of G). A natural extension of $H^\bullet_{\mathrm{loc}}(G, \mathcal{M})$ is to consider the cohomology defined on $\mathcal{M}$-valued smooth cochains on G. This cohomology will be denoted as $H^\bullet_{\mathrm{diff}}(G, \mathcal{M})$ and is related to the structure of a maximal subgroup as we will discuss at the end of this appendix.

From a completely different perspective, that is, not considering the Eilenberg-MacLane cochain complex on G, but starting afresh by considering that G is a smooth manifold, we may define the de Rham cohomology of G with values in $\mathcal{M}$. Let $\Omega^p(G, \mathcal{M})$ be the set of $\mathcal{M}$-valued smooth p–forms on G. In the graded algebra $\Omega^*(G, \mathcal{M}) = \bigoplus_{p\geq 0} \Omega^p(G, \mathcal{M})$ we define the differental operator of degree 1 as [Ch48] :

$$
\begin{aligned}
d\alpha_g(X_1, \ldots, X_{k+1}) &= \sum_{i=1}^{k+1} (-1)^{i+1} \frac{d}{ds} \rho(\phi_s^{(i)}(g)) \left((\phi_{-s}^{(i)})^* \alpha_g(X_1, \ldots, \hat{X}_i, \ldots, X_{k+1}) \right) \\
&+ \sum_{i<j} (-1)^{i+j} \alpha_g([X_i, X_j], X_1, \ldots, \hat{X}, \ldots, \hat{X}_j, \ldots, X_{k+1}), \qquad (10.283)
\end{aligned}
$$

where $X_i \in \mathfrak{X}(G)$, $i = 1, \ldots, k+1$ and $\phi_s^{(i)}$ is the local flow of the vector field X_i at the point $g \in G$. It can be shown that $d^2 = 0$ (because it is the natural extension of the exterior differential in G), and its associated cohomology, denoted $H^\bullet_{\mathrm{dR}}(G, \mathcal{M})$, will be called the de Rham cohomology of G with values in $\mathcal{M}$.

If $\mathcal{M} = \mathbb{R}$ with the trivial G-action, $H^\bullet_{\mathrm{dR}}(G, \mathbb{R})$ becomes the standard de Rham cohomology of G, isomorphic to the singular (topological) cohomology of G, $H^\bullet(G)$.

The complex $\Omega^\bullet(G, \mathcal{M})$ has various interesting subcomplexes: the subcomplex of equivariant forms, the subcomplex of left- (or right-) invariant forms and that of bi-invariant forms.

First of all notice that the group G acts on itself by left- (right-) translations, then we may consider the subcomplex of left- (right-) invariant forms denoted by $\Omega^\bullet_L(G, \mathcal{M})$ ($\Omega^\bullet_R(G, \mathcal{M})$).

The group G acts also on $\mathcal{M}$, then we may consider equivariant forms, that is forms α satisfying $g^*\alpha = g \cdot \alpha$, where $g \cdot \alpha$ means $\rho(g)(\alpha(X_1, \ldots, X_k))$ for any vector fields X_i on G. Notice that if the form α is equivariant, then:

$$\frac{d}{ds}\rho(\phi_s^{(i)}(g))\left((\phi_{-s}^{(i)})^*\alpha_g(X_1,\ldots,\hat{X}_i,\ldots,X_{k+1})\right)=0,$$

and formula (10.283) becomes simply:

$$d\alpha_g(X_1,\ldots,X_{k+1})=\sum_{i<j}(-1)^{i+j}\alpha_g([X_i,X_j],X_1,\ldots,\hat{X},\ldots,\hat{X}_j,\ldots,X_{k+1}), \tag{10.284}$$

It is clear that d maps equivariant forms into equivariant forms, thus we have defined a sub complex $E^\bullet(G,\mathcal{M})=\bigoplus_{k\geq 0}E^k(G,\mathcal{M})$ where $E^k(G,\mathcal{M})$, $k\geq 1$, $E^0=\mathcal{M}^G$, denotes the set of equivariant k-forms. The cohomology of the equivariant subcomplex will be denoted by $H_E^\bullet(G,\mathcal{M})$.

The cohomology operator d restricted to the subcomplex of left-invariant forms $\Omega_L^\bullet(G,\mathcal{M})$ becomes:

$$\begin{aligned} d\alpha(X_1,\ldots,X_{k+1})=&\sum_{i=1}^{k+1}(-1)^{i+1}\hat{\rho}(\xi)\left(\alpha(X_1,\ldots,\hat{X}_i,\ldots,X_{k+1})\right)\\ &+\sum_{i<j}(-1)^{i+j}\alpha([X_i,X_j],X_1,\ldots,\hat{X},\ldots,\hat{X}_j,\ldots,X_{k+1}), \end{aligned} \tag{10.285}$$

where $\xi_i=X_i(e)$. The cohomology defined in the subcomplex of left-invariant forms coincide with the cohomology of the Lie algebra of G (see later Appendix F.3) and if the group G is compact, connected and $\mathcal{M}=\mathbb{R}$, then all the cohomologies we have described coincide. The cohomology defined over the subcomplex $\Omega_L^\bullet(G,\mathcal{M})$ will be denoted by $H_L^\bullet(G,\mathcal{M})$.

A relevant question at this point is how the smooth cohomologies $H_{\text{loc}}^\bullet(G,\mathcal{M})$ and $H_{\text{diff}}^\bullet(G,\mathcal{M})$ are related with the de Rham like cohomologies $H_{\text{dR}}^\bullet(G,\mathcal{M})$ and $H_L^\bullet(G,\mathcal{M})$. We will see in the Appendix F.3 that $H_{\text{dR}}^\bullet(G,\mathcal{M})$ is the same as $H_L^\bullet(G,\mathcal{M})$, but $H_{\text{diff}}^\bullet(G,\mathcal{M})$ is related to $H_L^\bullet(G,\mathcal{M})$ by means of a maximal compact subgroup of G.

F.3. Chevalley Cohomology of a Lie Algebra

Let us consider now a Lie algebra $\mathfrak{g}$ and a representation $\hat{\rho}$ of $\mathfrak{g}$ on a module $\mathcal{M}$, that is, a Lie algebras homomorphism $\hat{\rho}\colon\mathfrak{g}\to\operatorname{End}(\mathcal{M})$. We consider now the graded algebra of $\mathcal{M}$–valued skew-symmetric forms on $\mathfrak{g}$, $\Lambda^\bullet(\mathfrak{g},\mathcal{M})=\bigoplus_{k\geq 0}\Lambda^k(\mathfrak{g},\mathcal{M})$, with

$$\Lambda^k(\mathfrak{g},\mathcal{M})=\Lambda^k(\mathfrak{g})\otimes\mathcal{M}=\{c\colon\mathfrak{g}\times\overset{k}{\cdots}\times\mathfrak{g}\mid c\ \text{skew-symmetric}\}.$$

Let us define the graded operator of degree one $d_C\colon \Lambda^k(\mathfrak{g}, \mathcal{M}) \to \Lambda^{k+1}(\mathfrak{g}, \mathcal{M})$ as:

$$d_C c(\xi_1, \ldots, \xi_{k+1}) = \sum_{i=1}^{k+1}(-1)^{i+1}\hat{\rho}(\xi_i)c(\xi_1, \ldots, \hat{\xi}_i, \ldots, \xi_{k+1}) + \sum_{i<j}(-1)^{i+j}c([\xi_i, \xi_j], \xi_1, \ldots, \hat{\xi}, \ldots, \hat{\xi}_j, \ldots, \xi_{k+1}), \tag{10.286}$$

for all $\xi_i \in \mathfrak{g}$, $i = 1, \ldots, k+1$. It can be checked that $d_C^2 = 0$ and the cohomology defined by d_C is called the Chevalley cohomology of $\mathfrak{g}$ associated to the representation $\hat{\rho}$ with values in A and denoted by $H_C^\bullet(\mathfrak{g}, \mathcal{M})$ [Ch48].

If $\mathfrak{g}$ is the Lie algebra of a Lie group G and $\hat{\rho}$ is the representation of $\mathfrak{g}$ defined by a representation $\rho\colon G \to \mathrm{Aut}(\mathcal{M})$, then it is obvious that we may identify the cochains of the complex $\Lambda^\bullet(\mathfrak{g}, \mathcal{M})$ with the left-invariant cochains on G with values on $\mathcal{M}$, $\Omega_L^\bullet(G, \mathcal{M})$. Then, the Chevalley operator d_C is just the restriction of the exterior differential d on G and ξ^L is the unique left-invariant vector field on G determined by the Lie algebra element ξ. Then $H_L^*(G.\mathcal{M}) \cong H_C^\bullet(\mathfrak{g}, \mathcal{M})$.

Again, it is not hard to show that $H_{\mathrm{loc}}^\bullet(G, \mathcal{M})$ is isomorphic to $H_C^\bullet(\mathfrak{g}, \mathcal{M})$ because each cochain $f \in H_{\mathrm{loc}}^k(G, \mathcal{M})$ defines a unique left-invariant k-form on G by simply taking the differential at e of all factors. Then we get $H_{\mathrm{loc}}^\bullet(G, \mathcal{M}) \cong H_C^\bullet(\mathfrak{g}, \mathcal{M})$ [Ba54, Ho80].

On the other hand, the global case is more complicated. Basically, the situation we are facing is as follows. If G is a non-compact Lie group and K is a maximal compact subgroup, then K is a strong retraction of G and the de Rham cohomology of G and K are the same. Then it remains to analyze the Eilenberg-MacLane cohomology coming from the quotient G/K. We may define relative cohomology groups $H^n(G, K, \mathcal{M})$ and it can be shown that $H^n(G, K, \mathcal{M}) \cong H^n_{\mathrm{diff}}(G, \mathcal{M})$ [Ho80].

Along the main text, only the cohomologies $H_C^\bullet(\mathfrak{g}, \mathcal{M})$ and $H^*(G.\mathcal{M})$ will be used. In general, unless it is necessary to be more specific, we will omit the subindexes L, C and dR when writing the corresponding cohomology groups.

F.4. Cohomology Theory of Associative Algebras

In this section we wil introduce the basic notions of cohomology theory for associative algebras and its relation with the deformation theory for associative algebras, introduced by Gerstenhaber [G64].

Let $\mathcal{A}$ be an associative algebra over a field $\mathbb{K}$ and $\mathcal{M}$ be a $\mathcal{A}$-bimodule, i.e., $\mathcal{M}$ is a module that is the carrier space for a linear representation Ψ of $\mathcal{A}$ and a linear anti-representation Ψ' of $\mathcal{A}$ that commute. The action of an element $a \in \mathcal{A}$ on an element $m \in \mathcal{M}$ will be denoted either am or ma depending if we are consider the left- or right action.

A n-cochain on $\mathcal{A}$ with values in $\mathcal{M}$, is a n-linear map $\alpha\colon \mathcal{A} \times \cdots \times \mathcal{A} \to \mathcal{M}$ (n times). The set of such n-cochains, which can be regarded as an additive group, will be denoted $C^n(\mathcal{A}, \mathcal{M})$, and for every $n \in \mathbb{N}$ we introduce the coboundary operator (compare with Eq. 10.282) $\delta_n\colon C^n(\mathcal{A}, \mathcal{M}) \to C^{n+1}(\mathcal{A}, \mathcal{M})$, by means of [GH46]:

$$(\delta_n \alpha)(a_1, \ldots, a_{n+1}) := a_1 \alpha(a_2, \ldots, a_{n+1}) + \sum_{i=1}^{n} (-1)^i \alpha(a_1, \ldots, a_i a_{i+1}, \ldots, a_{n+1}) + (-1)^{n+1} \alpha(a_1, \ldots, \ldots, a_n) a_{n+1} .$$

For instance, when $n = 1$ we obtain $(\delta\alpha_1)(a_1, a_2) = a_1\alpha_1(a_2) - \alpha_1(a_1a_2) + \alpha_1(a_1)a_2$, and for $n = 2$,

$$(\delta\alpha_2)(a_1, a_2, a_3) = a_1\alpha_2(a_2, a_3) - \alpha_2(a_1a_2, a_3) + \alpha_2(a_1, a_2a_3) - \alpha_2(a_1, a_2)a_3 .$$

It is now an easy but cumbersome task to check that the linear maps δ_n satisfy $\delta_{n+1} \circ \delta_n = 0$.

We can form now the cochain complex $C^\bullet(\mathcal{A}, \mathcal{M}) = \bigoplus_{n \geq 0} C^n(\mathcal{A}, \mathcal{M})$ and the natural extension of the operators δ_n to it, will allow us to define the corresponding cohomology $H^\bullet(\mathcal{A}, \mathcal{M}) = \bigoplus_{n \geq 0} H^n(\mathcal{A}, \mathcal{M})$, called the Hochschild cohomology of the associative algebra $\mathcal{A}$ with coefficients in $\mathcal{M}$, whose nth cohomology groups is defined as $H^n(\mathcal{A}, \mathcal{M}) = Z^n(\mathcal{A}, \mathcal{M})/B^n(\mathcal{A}, \mathcal{M})$, with $Z^n(\mathcal{A}, \mathcal{M})$ the group of n-cocycles, $\delta_n\alpha = 0$, and $B^n(\mathcal{A}, \mathcal{M})$ the subgroup of n-coboundaries, $\alpha = \delta_{n-1}\beta$ with $\beta \in C^{n-1}(\mathcal{A}, \mathcal{M})$.

A simplest example happens when $\mathcal{M}$ is the additive group of $\mathcal{A}$ itself, and then the $\mathcal{A}$-module structure is given by left and right multiplication. In this case, if $F\colon \mathcal{A} \times \mathcal{A} \to \mathcal{A}$ is a bilinear map, it defines a 2-cochain and then

$$(\delta F)(a, b, c) = aF_1(b, c) - F_1(ab, c) + F(a, bc) - F_1(a, b)c .$$

and therefore $\delta F = 0$ reduces to the condition (10.288) below.

F.5. Deformation of Associative Algebras

Let $R = \mathbb{K}[\lambda]$ denote the ring of power series in one variable λ, and $K = K(\lambda)$ the field of power series of R. If V is the underlying linear space of $\mathcal{A}$, let V_K denote the linear space obtained from V by extending the coefficient domain from $\mathbb{K}$ to K i.e., $V_K = V \otimes_{\mathbb{K}} K$. Let suppose that $f_\lambda : V_K \times V_K \to V_K$ is a bilinear function expressible as

$$f_\lambda(a, b) = F_0(a, b) + \lambda\, F_1(a, b) + \lambda^2 F_2(a, b) + \cdots , \tag{10.287}$$

where $F_0(a, b) = a * b$ and F_k are bilinear functions over V.

The map f_λ can be used to define an algebra structure $\mathcal{A}_\lambda$ on V_K. The conditions for f_λ to be associative, i.e.

$$f_\lambda(a, f_\lambda(b,c)) = f_\lambda(f_\lambda(a,b),c) ,$$

are

$$\sum_{j=0}^{k} F_j(a, F_{k-j}(b,c)) = \sum_{j=0}^{k} F_j(F_{k-j}(a,b),c) ,$$

for any k, with the convention $F_0(a,b) = a * b$. Notice that when $k = 0$, it reduces to the associativity condition on $\mathcal{A}$. For $k = 1$, the condition reads:

$$a * F_1(b,c) + F_1(a, b*c) = F_1(a*b, c) + F_1(a,b) * c , \tag{10.288}$$

which can be written as $\delta F_1 = 0$, i.e., in terms of Hochschild cohomology, the function F_1, which is called infinitesimal deformation, is an element of the group $Z^2(\mathcal{A},\mathcal{A})$ of 2-cocycles of $\mathcal{A}$ with coefficients on $\mathcal{A}$.

Coming back to the new deformed algebra, even if the original algebra was Abelian, the deformed one is, in general, no longer commutative, but

$$f_\lambda(a,b) - f_\lambda(b,a) = \lambda[F_1(a,b) - F_1(b,a)] + \vartheta(\lambda^2) ,$$

where $\vartheta(\lambda^2)$ is of degree higher than one in λ, i.e.

$$\lim_{\lambda\to 0} \frac{1}{\lambda}\left(\vartheta(\lambda^2)\right) = 0 .$$

This allows us to define a new bilinear map in $\mathcal{A}$, to be denoted $\{\cdot,\cdot\}$ by

$$\{a,b\} = \lim_{\lambda\to 0}\frac{1}{\lambda}[f_\lambda(a,b) - f_\lambda(b,a)] = F_1(a,b) - F_1(b,a) ,$$

which, by construction, is a Lie algebra structure. Moreover,

$$\{a, b*c\} = F_1(a, b*c) - F_1(b*c, a) , \quad b*\{a,c\} = b*(F_1(a,c) - F_1(c,a)),$$

and

$$\{a,b\}*c = (F_1(a,b) - F_1(b,a))*c ,$$

we see that

$$\begin{aligned}\{a, b * c\} - b * \{a, c\} - \{a, b\} * c &= F_1(a, b * c) - F_1(b * c, a) \\ &\quad -b * [F_1(a, c) - F_1(c, a)] \\ &\quad -[F_1(a, b) - F_1(b, a)] * c ,\end{aligned}$$

and therefore, using that

$$\begin{aligned}b * [F_1(a, c) - F_1(c, a)] = F_1(b * a, c) - F_1(b, a * c) + F_1(b, a) * c \\ - F_1(b * c, a) + F_1(b, c * a) - F_1(b, c) * a\end{aligned}$$

we find that

$$\begin{aligned}\{a, b * c\} - b * \{a, c\} - \{a, b\} * c = & F_1(a, b * c) - F_1(b * c, a) \\ & - [F_1(b * a, c) - F_1(b, a * c) + F_1(b, a) * c \\ & - F_1(b * c, a) + F_1(b, c * a) - F_1(b, c) * a] \\ & - F_1(a, b) * c + F_1(b, a) * c ,\end{aligned}$$

and simplifying terms and replacing

$$F_1(a, b * c) - F_1(a, b) * c = F_1(a * b, c) - a * F_1(b, c) ,$$

we will arrive to

$$\begin{aligned}\{a, b * c\} - b * \{a, c\} - \{a, b\} * c = & F_1(a * b, c) - a * F_1(b, c) \\ & - F_1(b * a, c) + F_1(b, a * c) \\ & - F_1(b, c * a) + F_1(b, c) * a ,\end{aligned}$$

which can be reordered as

$$\begin{aligned}\{a, b * c\} - b * \{a, c\} - \{a, b\} * c = & F_1(a * b, c) - F_1(b * a, c) + F_1(b, c) * a \\ & -a * F_1(b, c) + F_1(b, a * c) - F_1(b, c * a) ,\end{aligned}$$

which clearly shows that if the original algebra is commutative, then

$$\{a, b * c\} - b * \{a, c\} - \{a, b\} * c = 0 ,$$

i.e., the map $\{a, \cdot\}$ is a derivation in the commutative and associative algebra $(\mathcal{A}, *)$. This leads us to introduce the general concept of Poisson bracket and Poisson algebra.

Definition 10.67 A Poisson algebra is a set $\mathcal{A}$ endowed with a commutative associative algebra $(\mathcal{A}, +, \cdot, *)$, and a Lie algebra structure defined by the composition law $\{\cdot, \cdot\}$ such that

$$\{a, b * c\} = b * \{a, c\} + c * \{a, b\} .$$

The element $\{a, b\}$ is called the Poisson bracket of the elements $a \in \mathcal{A}$ and $b \in \mathcal{A}$.

This means that the composition law $\{\cdot, \cdot\}$ defines a Lie algebra structure in the commutative and associative algebra $(\mathcal{A}, +, *)$

$$\{a, b + \lambda c\} = \{a, b\} + \lambda\{a, c\}, \quad \{a, b\} + \{b, a\} = 0 ,$$

and for any triple of elements of $\mathcal{A}$

$$\{a, \{b, c\}\} + \{b, \{c, a\}\} + \{c, \{a, b\}\} = 0 .$$

Given a Poisson algebra $(\mathcal{A}, +, \cdot, *, \{\cdot, \cdot\})$, a Poisson subalgebra is a subset $\mathcal{B}$ which is invariant under the composition law, and, therefore, when endowed with the composition lawss which are the restrictions onto $\mathcal{B}$ of the corresponding ones in $\mathcal{A}$ is a Poisson subalgebra. A subalgebra $\mathcal{I}$ is said to be a Poisson ideal when

$$\mathcal{A}\mathcal{I} \subset \mathcal{I}, \qquad \{\mathcal{A}, \mathcal{I}\} \subset \mathcal{I} .$$

In such a case, the equivalence relacion defined by $\mathcal{I}$ is compatible with the composition laws and then $\mathcal{A}/\mathcal{I}$ can be endowed with a Poisson algebra structure.

A linear map $D : \mathcal{A} \to \mathcal{A}$ is said to be a derivation of the Poisson algebra when

$$D(a * b) = (DA) * b + a * (Db) , \quad D\{a, b\} = \{Da, b\} + \{a, Db\} ,$$

and the set of all derivations of a Poisson algebra is a Lie algebra when the following composition law is defined:

$$[D_1, D_2] = D_1 \circ D_2 - D_2 \circ D_1 .$$

In particular, for each element $a \in \mathcal{A}$, the map $D_a : \mathcal{A} \to \mathcal{A}$ defined by

$$D_a(b) = \{a, b\} ,$$

is a derivation. Such derivations are called inner derivations.

The elements $a \in \mathcal{A}$ of the Poisson algebra such that

$$\{a, b\} = 0 , \quad \forall b \in \mathcal{A} ,$$

are called Casimir elements of the Poisson algebra.

The most interesting case is when $\mathcal{A}$ is the algebra of differentiable functions on a vector space, or, even more generally, on a differentiable manifold.

Let us now consider the case in which $\mathfrak{g}$ is a Lie algebra rather than an associative one. In similarity what we did with associative algebras, let $\mathfrak{g}$ be a Lie algebra and $\mathfrak{a}$

be a $\mathfrak{g}$-module, i.e., $\mathfrak{a}$ is a module that is the carrier space for a linear representation Ψ of $\mathfrak{g}$, i.e., $\Psi\colon \mathfrak{g} \to \mathrm{End}\,\mathfrak{a}$ is such that

$$\Psi(a)\Psi(b) - \Psi(b)\Psi(a) = \Psi([a,b]) \,,$$

we will call n-cochain to a n-linear alternating mapping from $\mathfrak{g} \times \ldots \times \mathfrak{g}$ (n times) into $\mathfrak{a}$. We denote by $C^n(\mathfrak{g},\mathfrak{a})$ the space of n-cochains. For every $n \in \mathbb{N}$ we define $\delta_n : C^n(\mathfrak{g},\mathfrak{a}) \to C^{n+1}(\mathfrak{g},\mathfrak{a})$ by Cariñena and Ibort [CI88]

$$\begin{aligned}(\delta_n\alpha)(a_1,\ldots,a_{n+1}) := &\sum_{i=1}^{n+1}(-1)^{i+1}\Psi(a_i)\alpha(a_1,\ldots,\widehat{a}_i,\ldots,a_{n+1})\\ &+\sum_{i<j}(-1)^{i+j}\alpha([a_i,a_j],a_1,\ldots,\widehat{a}_i,\ldots,\widehat{a}_j,\ldots,a_{n+1})\,,\end{aligned}$$

where $\widehat{a}_i$ denotes, as usual, that the element a_i is omitted.

In particular, if $\alpha : \mathfrak{g} \to \mathfrak{a}$ is a linear map, then

$$\delta\alpha(a_1,a_2) = \Psi(a_1)\alpha(a_2) - \Psi(a_2)\alpha(a_1) - \alpha([a_1,a_2])\,,$$

and if $\beta \in C^2(\mathfrak{g},\mathfrak{a})$,

$$\begin{aligned}\delta\beta(a_1,a_2,a_3) = &\Psi(a_1)\beta(a_2,a_3) - \Psi(a_2)\beta(a_1,a_3) + \Psi(a_3)\beta(a_1,a_2)\\ &-\beta([a_1,a_2],a_3) + \beta([a_1,a_3],a_2) - \beta([a_2,a_3],a_1)\,.\end{aligned}$$

The linear maps δ_n satisfy

$$\delta_{n+1}\circ\delta_n = 0\,.$$

The proof is a simple but cumbersome checking.

The linear operator δ on $C(\mathfrak{g},\mathfrak{a}) := \bigoplus_{n=0}^{\infty} C^n(\mathfrak{g},\mathfrak{a})$ whose restriction to each $C^n(\mathfrak{g},\mathfrak{a})$ is δ_n, satisfies $\delta^2 = 0$. We will then denote

$$\begin{aligned}B^n(\mathfrak{g},\mathfrak{a}) &:= \{\alpha \in C^n(\mathfrak{g},\mathfrak{a}) \mid \exists\beta \in C^{n-1}(\mathfrak{g},\mathfrak{a})\ \text{such that}\ \alpha = \delta\beta\} = \mathrm{Im}\,\delta_{n-1}\,,\\ Z^n(\mathfrak{g},\mathfrak{a}) &:= \{\alpha \in C^n(\mathfrak{g},\mathfrak{a}) \mid \delta\alpha = 0\} = \ker\delta_n\,.\end{aligned}$$

The elements of Z^n are called n-cocycles, and those of B^n are called n-coboundaries. Since $\delta^2 = 0$, we have $B^n \subset Z^n$. The n-th cohomology group $H^n(\mathfrak{g},\mathfrak{a})$ is defined as

$$H^n(\mathfrak{g},\mathfrak{a}) := \frac{Z^n(\mathfrak{g},\mathfrak{a})}{B^n(\mathfrak{g},\mathfrak{a})}\,,$$

and we will define $B^0(\mathfrak{g},\mathfrak{a}) = 0$, by convention.

In particular, a linear map $\alpha : \mathfrak{g} \to \mathfrak{a}$ is a coboundary if there exists $a \in \mathfrak{a}$ such that

$$\alpha(a_1) = \Psi(a_1)a ,$$

and $\beta \in C^2(\mathfrak{g}, \mathfrak{a})$ is a cocycle (δ-closed) when

$$\Psi(a_1)\beta(a_2, a_3) - \Psi(a_2)\beta(a_1, a_3) + \Psi(a_3)\beta(a_1, a_2) - \beta([a_1, a_2], a_3) + \beta([a_1, a_3], a_2) - \beta([a_2, a_3], a_1) = 0 ,$$

and a coboundary (δ-exact) when there exist a linear map $\alpha : \mathfrak{g} \to \mathfrak{a}$ such that

$$\beta(a_1, a_2) = \Psi(a_1)\alpha(a_2) - \Psi(a_2)\alpha(a_1) - \alpha([a_1, a_2]) .$$

As a first example, let $\mathfrak{a} := \mathbb{R}$, $\Psi(a) := 0 , \ \forall a \in \mathfrak{g}$. We can also replace $\mathbb{R}$ for any trivial $\mathfrak{g}$-module. In this case, the operator δ reduces to

$$(\delta\alpha)(a_1, a_2, \ldots, a_{n+1}) = \sum_{i<j}(-1)^{i+j}\alpha([a_i, a_j], a_1, \ldots, \widehat{a}_i, \ldots, \widehat{a}_j, \ldots, a_{n+1}) .$$

In particular, if $\alpha \in C^1(\mathfrak{g}, \mathbb{R})$, then $(\delta\alpha)(a_1, a_2) = -\alpha([a_1, a_2])$. Thus $Z^1(\mathfrak{g}, \mathbb{R})$ is the set

$$Z^1(\mathfrak{g}, \mathbb{R}) = \{\alpha \in \mathfrak{g}^* \mid \alpha([a_1, a_2]) = 0, \ \forall a_1, a_2 \in \mathfrak{g}\} ,$$

and $B^1(\mathfrak{g}, \mathbb{R}) = 0$, by convention. If $\beta \in C^2(\mathfrak{g}, \mathbb{R})$, then β is a 2-cocycle if

$$\beta([a_1, a_2], a_3) + \beta([a_2, a_3], a_1) + \beta([a_3, a_1], a_2) = 0 ,$$

and $\beta \in B^2(\mathfrak{g}, \mathbb{R})$ if there exists a linear map $\tau : \mathfrak{g} \to \mathbb{R}$ such that

$$\beta(a_1, a_2) = \tau([a_1, a_2]) .$$

As a second example we can consider the case in which $\mathfrak{g}$ is the Lie algebra of vector fields, $\mathfrak{g} = \mathfrak{X}(M)$, and the cohomology corresponding to its action on the set of functions in M. In other words, we consider $\mathfrak{a} := C^\infty(M)$ and define $\Psi(X)f := \mathcal{L}_X f$, which is a linear representation of $\mathfrak{g}$, because of

$$[\Psi(X) \circ \Psi(Y) - \Psi(Y \circ \Psi(X)]f = (\mathcal{L}_X\mathcal{L}_Y - \mathcal{L}_Y\mathcal{L}_X)f = \mathcal{L}_{[X,Y]}f = \Psi([X, Y])f.$$

The δ-operator takes the following form: if $\alpha \in C^n(\mathfrak{g}, C^\infty(M))$,

$$(\delta_n\alpha)(X_1,\ldots,X_{n+1}) = \sum_{i=1}^{n+1}(-1)^{i+1}\mathcal{L}_{X_i}\alpha(X_1,\ldots,\widehat{X}_i,\ldots,X_{n+1})$$
$$+\sum_{i<j}(-1)^{i+j}\alpha([X_i,X_j],X_1,\ldots,\widehat{X}_i,\ldots,\widehat{X}_j,\ldots,X_{n+1}).$$

In particular, if $\alpha \in C^1(\mathfrak{g}, C^\infty(M))$,

$$(\delta\alpha)(X,Y) = \mathcal{L}_X\alpha(Y) - \mathcal{L}_Y\alpha(X) - \alpha([X,Y]).$$

The elements of $B^1(\mathfrak{g}, C^\infty(M))$ are those α for which $\exists\beta \in \Omega^p(M)$ with

$$\alpha(X) = \mathcal{L}_X\beta,$$

while the elements of $Z^1(\mathfrak{g}, C^\infty(M))$ are linear maps $\alpha : \mathfrak{g} \to C^\infty(M)$ satisfying

$$\mathcal{L}_X\alpha(Y) - \mathcal{L}_Y\alpha(X) = \alpha([X,Y]).$$

We have seen that the divergence of a vector field is the generalization of the trace and that the set of all divergence-free vector fields is an infinite-dimensional Lie Algebra. We want to remark that the association of $\mathfrak{X}(E)$ with ${}^F(E)$ given by $X \mapsto \operatorname{div} X$, is a 1-cocycle, i.e.,

$$\mathcal{L}_X\operatorname{div} Y - \mathcal{L}_Y\operatorname{div} X = \operatorname{divx}\,[X,Y], \tag{10.289}$$

and therefore, the set ker div $\subset \mathfrak{X}(E)$ is the Lie subalgebra which generalizes $\mathfrak{isl}(n)$. We notice that this time it is not possible to decompose a vector field like in (4.29).

However, it is possible to consider a new vector space $E \times \mathbb{R}$, and a new volume

$$\widetilde{\Omega} = dx^1 \wedge dx^2 \wedge \cdots \wedge dx^n \wedge ds = \Omega \wedge ds \tag{10.290}$$

such that we can associate to any vector field X a new vector field

$$\widetilde{X} = X + (\operatorname{div} X)\, s\frac{\partial}{\partial s},$$

which is divergence-free with respect to $\widetilde{\Omega}$. In fact,

$$\mathcal{L}_{\widetilde{X}}\widetilde{\Omega} = (\mathcal{L}_X\Omega)\wedge ds - \Omega\wedge d((\operatorname{div} X)s) = (\mathcal{L}_X\Omega)\wedge ds - (\operatorname{div} X)\Omega\wedge ds$$
$$= -d((\operatorname{div} X)\widetilde{\Omega})) = 0.$$

Coming back to the theory of deformations, when the algebra $\mathcal{A}$ is a Lie algebra $\mathfrak{g}$ rather than an associative one, and we consider the one-parameter deformation given by (10.287), the conditions for $f_\lambda = [\cdot,\cdot]_\lambda$ to define a Lie algebra structure,

$$f_\lambda(a,b) = -f_\lambda(b,a)\,,$$
$$f_\lambda(f_\lambda(a,b),c) + f_\lambda(f_\lambda(b,c),a) + f_\lambda(f_\lambda(c,a),b) = 0\,,$$

are translated to

$$F_k(a,b) = -F_k(b,a)\,,$$
$$\sum_{j=0}^{k} F_j(F_{k-j}(a,b),c) + \sum_{j=0}^{k} F_j(F_{k-j}(b,c),a) + \sum_{j=0}^{k} F_j(F_{k-j}(c,a),b) = 0\,.$$

For $k = 0$ these conditions only reduce to the conditions for $mathfrakg$ to be a Lie algebra, with respect to the composition law given by F_0, i.e., $[a,b] = F_0(a,b)$, while for $k = 1$, we get

$$F_1([a,b],c) + F_1([b,c],a) - F_1([a,c],b) - [a, F_1(b,c)] + [b, F_1(a,c)] - [c, F_1(a,b)] = 0\,, \tag{10.291}$$

together with $F_1(a,b) + F_1(b,a) = 0$, or in other words, when considering $\mathfrak{g}$ as a $\mathfrak{g}$-module by means of the adjoint representation of $\mathfrak{g}$ and F_1 as an element of $C^2(\mathfrak{g},\mathfrak{g})$, the conditions only mean that F_1 is again a 2-cocycle, $\delta F_1 = 0$.

For $k = 2$ we have

$$F_0(F_2(a,b),c) + F_0(F_2(b,c),a) + F_0(F_2(c,a),b) + F_1(F_1(a,b),c) + F_1(F_1(b,c),a) + F_1(F_1(c,a),b) + F_2(F_0(a,b),c) + F_2(F_0(b,c),a) + F_2(F_0(c,a),b) = 0\,.$$

Very often we consider deformations in which $F_k = 0$ for $k \geq 2$, i.e. (10.287) reduces to

$$[a,b]_\lambda = [a,b] + t\,F_1(a,b)\,, \tag{10.292}$$

and in this case the condition for $k = 2$ is

$$F_1(F_1(a,b),c) + F_1(F_1(b,c),a) + F_1(F_1(c,a),b) = 0\,,$$

i.e., $[a,b]_1 = F_1(a,b)$ also defines a Lie algebra structure. The condition (10.291) can be written as $\delta F_1 = 0$, with δ denoting the coboundary operator in the complex with coefficients in the adjoint representation.

In fact, in the case we are considering, $\mathfrak{g}$ can be considered as a $\mathfrak{g}$-module by means of the adjoint representation, i.e., $\Psi : \mathfrak{g} \to \operatorname{End}\mathfrak{g}$ is given by $\Psi(a)(b) = [a,b]$, and then, if $\beta \in C^n(\mathfrak{g},\mathfrak{g})$, then

$$\delta\beta(a_1,\ldots,a_{n+1}) = \sum_{i=1}^{n+1}(-1)^{i+1}[a_i,\beta(a_1,\ldots,\widehat{a_i},\ldots,a_{n+1})] + \sum_{i<j}(-1)^{i+j}\beta([a_i,a_j],a_1,\ldots,\widehat{a_i},\ldots,\widehat{a_j},\ldots,a_{n+1}) .$$

In particular, a 1-cochain is given by a linear map $A:\mathfrak{g}\to\mathfrak{g}$. The coboundary of such 1-cochain is

$$\delta A(a_1,a_2) = [a_1,A(a_2)] - [a_2,A(a_1)] - A([a_1,a_2]) .$$

Note that the linear map A is a derivation of the Lie algebra $\mathfrak{g}$ if and only if $\delta A = 0$.

The coboundary of a 2-cochain $\zeta:\mathfrak{g}\times\mathfrak{g}\to\mathfrak{g}$ is

$$\delta\zeta(a_1,a_2,a_3) = [a_1,\zeta(a_2,a_3)] - \zeta([a_1,a_2],a_3) + [a_2,\zeta(a_3,a_1)] - \zeta([a_2,a_3],a_1) + [a_3,\zeta(a_1,a_2)] - \zeta([a_3,a_1],a_2) .$$

Then, the Jacobi identity in the Lie algebra can be written $\delta\zeta = 0$ where ζ is just the bilinear map defining the composition law in the Lie algebra, $\zeta(a,b) = [a,b]$.

The skew-symmetric bilinear map $F_1:\mathfrak{g}\times\mathfrak{g}\to\mathfrak{g}$ appearing in (10.287) defines a 2-cochain and

$$(\delta F_1)(a,b,c) = [a,F_1(b,c)] - [b,F_1(a,c)] + [c,F_1(a,b)] - F_1([a,b],c) + F_1([a,c],b) - F_1([b,c],a) .$$

Therefore the condition (10.291) can be written as $\delta F_1 = 0$.

Definition 10.68 A deformation T_λ is said to be trivial if there exists a linear operator A such that $T_\lambda = I + \lambda A$ and

$$T_\lambda[a,b]_\lambda = [T_\lambda a, T_\lambda b] , \qquad \forall\, a,b\in\mathfrak{g} ,$$

where $[a,b]_\lambda = [a,b]+\lambda[a,b]_1$, i.e., F_1 in (10.292) is given by $F_1(a,b) = [a,b]_1$.

Taking into account that

$$T_\lambda[a,b]_\lambda = (I+\lambda A)([a,b]+\lambda\,[a,b]_1) = [a,b]+\lambda\,(A[a,b]+[a,b]_1)+\lambda^2 A[a,b]_1 ,$$

and

$$[T_\lambda a, T_\lambda b] = [a+\lambda\, Aa, b+\lambda\, Ab] = [a,b]+\lambda([Aa,b]+[a,Ab])+\lambda^2[Aa,Ab] ,$$

we see that T_λ is a trivial deformation if and only if

$$[a,b]_1 = [Aa,b]+[a,Ab] - A[a,b] , \qquad (10.293)$$

$$A[a,b]_1 = [Aa,Ab] . \qquad (10.294)$$

The first condition can be expressed in terms of the coboundary operator in the complex with coefficients in the adjoint representation as follows:

$$F_1 = \delta A \ .$$

We should also remark that using in the second relation the expression for $[a, b]_1$ given in the first one we obtain the following relation for the linear operator A:

$$[Aa, Ab] - A[Aa, b] - A[a, Ab] + A^2[a, b] = 0 \ .$$

A linear map A satisfying this condition is called a Nijenhuis map.

In general, the Nijenhuis torsion of A is defined as the (1,2)-tensor such that

$$N_A(a, b) = [Aa, Ab] - A[Aa, b] - A[a, Ab] + A^2[a, b] \ .$$

The linear map A is a Nijenhuis map when its Nijenhuis torsion vanishes.

When such a condition is satisfied, the bilinear map $[a, b]_1$ defines an alternative Lie bracket, which is ususally denoted

$$[a, b]_A = [Aa, b] + [a, Ab] - A[a, b] \ ,$$

and then the preceding condition can we written as

$$A([a, b]_A) = [Aa, Ab] \ ,$$

i.e. $A : (\mathfrak{g}, [\cdot, \cdot]_A) \to (\mathfrak{g}, [\cdot, \cdot]_A)$ is a homomorphism of Lie algebras.

In fact, in order to prove the Jacobi identity for $[\cdot, \cdot]_A$, it suffices to note that for any three elements, $a_1, a_2, a_3 \in \mathfrak{g}$,

$$\begin{aligned}
&[a_1, [a_2, a_3]_A]_A + [a_3, [a_1, a_2]_A]_A + [a_2, [a_3, a_1]_A]_A \\
&= [A(a_1), [A(a_2), a_3]] + [A(a_1), [a_2, A(a_3)]] - [A(a_1), A([a_2, a_3])] + [A(a_3), [A(a_1), a_2]] \\
&\quad +[A(a_3), [a_1, A(a_2)]] - [A(a_3), A([a_1, a_2])] + [A(a_2), [A(a_3), a_1]] \\
&\quad +[A(a_2), [a_3, A(a_1)]] - [A(a_2), A([a_3, a_1])] + [a_1, A([A(a_2), a_3])] \\
&\quad +A([a_2, A(a_3)]) - A^2([a_2, a_3])] + [a_3, A([A(a_1), a_2])] + A([a_1, A(a_2)]) \\
&\quad -A^2([a_1, a_2])] + [a_2, A([A(a_3), a_1])] + A([a_3, A(a_1)]) - A^2([a_3, a_1])]
\end{aligned}$$

and using the Jacobi identity for $[\cdot, \cdot]$, we finally get

$$\begin{aligned}
&[a_1, [a_2, a_3]_A]_A + [a_3, [a_1, a_2]_A]_A + [a_2, [a_3, a_1]_A]_A \\
&\qquad = [a_3, N_A(a_1, a_2)] + [a_2, N_A)(a_3, a_1)] + [a_1, N_A(a_2, a_3)] \ .
\end{aligned}$$

Therefore, we see that if A is a Nijenhuis map, $N_A = 0$, then $[\cdot,\cdot]_A$ satisfies Jacobi identity and therefore it defines a new Lie algebra bracket. This is a sufficient, but not necessary, condition for δA to define a new Lie algebra bracket.

As indicated above, we remark that the vanishing of the Nijenhuis torsion of A, $N_A = 0$ also implies that $A : (\mathfrak{g}, [\cdot,\cdot]_A) \to (\mathfrak{g}, [\cdot,\cdot])$ is a Lie algebra homomorphism, because

$$A([a_1, a_2]_A) - [A(a_1), A(a_2)] = -N_A(a_1, a_2) = 0\,.$$

In summary, the knowledge of a Nijenhuis map A allows us to define a new Lie algebra structure on $\mathfrak{g}$, such that the map A is a homomorphism of Lie algebras $A : (\mathfrak{g}, [\cdot,\cdot]_A) \to (\mathfrak{g}, [\cdot,\cdot])$.

A particularly important case is that of $\mathfrak{g} = \mathfrak{X}(M)$, the Lie algebra of vector fields on a manifold M. Then the linear maps are given by $(1, 1)$-tensor fields T in M. Given a $(1, 1)$-tensor field T, the Nijenhuis torsion of T, N_T is defined by

$$N_T(X, Y) = T([T(X), Y] + [X, T(Y)]) - T^2([X, Y]) - [T(X), T(Y)]\,,$$

for any pair of vector fields $X, Y \in \mathfrak{X}(M)$.

A Nijenhuis structure on a manifold M is a $(1, 1)$-tensor field T with vanishing Nijenhuis torsion,

$$N_T(X, Y) = 0\,,$$

i.e.

$$T([T(X), Y] + [X, T(Y)]) - T^2([X, Y]) - [T(X), T(Y)] = 0\,.$$

Note that, from the relations

$$(\mathcal{L}_{T(X)}T)(Y) = [T(X), T(Y)] - T([T(X), Y])\,,$$

and

$$(T \circ \mathcal{L}_X T)(Y) = T([X, T(Y)] - T^2([X, Y])\,,$$

we see that the condition for the $(1, 1)$-tensor field T to be a Nijenhuis tensor can be written as

$$T \circ \mathcal{L}_X T = \mathcal{L}_{T(X)} T\,, \qquad \forall X \in \mathfrak{X}(M)\,.$$

A Nijenhuis structure allows us to define an alternative Lie algebra structure on $\mathfrak{X}(M)$ with the new Lie algebra bracket

$$[X, Y]_T = [T(X), Y] + [X, T(Y)] - T([X, Y])\,.$$

Moreover, as a consequence of the vanishing of N_T, the linear map

$$T : (\mathfrak{X}(M), [\cdot, \cdot]_T) \to (\mathfrak{X}(M), [\cdot, \cdot])$$

is a Lie algebra homomorphism.

F.6. Poisson Algebras and Deformation Quantization

As an instance of an abstract associative algebra, given a $n \times m$ matrix J, we can consider an associative algebra in the set of $m \times n$ matrices by means of a new composition law

$$A * B = AJB.$$

When $m = n$ the matrix J is a square matrix and the composition law $A * B$ is an alternative associative composition law in the set of square $n \times n$ matrices.

We can also consider not only one alternative structure but all a family depending on a parameter λ, and the composition law will be written $A *_\lambda B$. This problem was recently studied [CGM00] and it may be relevant in the study of alternative quantum descriptions for a given system. For instance, in the linear space $\mathcal{M}_n$ of $n \times n$ matrices the matrix J can be replaced by $\exp(\lambda K)$ with λ being an arbitrary parameter, and K any $n \times n$ matrix, and then

$$A *_\lambda B = A\, e^{\lambda K}\, B\ .$$

Consequently, the Lie algebra structure associated to this new associative algebra structure is

$$[A, B]_\lambda = A *_\lambda B - B *_\lambda A\ ,$$

which also satisfies the compatibility condition

$$[A, B *_\lambda C]_\lambda = [A, B]_\lambda *_\lambda C + B *_\lambda [A, C]_\lambda\ .$$

Note that if we define the map of the algebra in itself given by

$$\phi_\lambda(A) = e^{\frac{1}{2}\lambda K}\, A\, e^{\frac{1}{2}\lambda K}\ ,$$

we see that ϕ_λ is a linear map $\phi_\lambda : \mathcal{M}_n \to \mathcal{M}_n$ such that

$$\phi_\lambda(A *_\lambda B) = e^{\frac{1}{2}\lambda K}\, A\, e^{\lambda K}\, B e^{\frac{1}{2}\lambda K}\ ,$$

and then

$$\phi_\lambda(A *_\lambda B) = \phi_\lambda(A)\,\phi_\lambda(B)\,,$$

i.e., $\phi_\lambda : (\mathcal{M}_n, *_\lambda) \to (\mathcal{M}_n, \cdot)$ is a homomorphism of associative algebras, and, in an analogous way, as

$$\phi_\lambda([A, B]_\lambda) = [\phi_\lambda(A), \phi_\lambda(B)]\,,$$

ϕ_λ is also a Lie algebra homomorphism- The image of the neutral element is $\phi_\lambda(1) = e^{\lambda K}$, and the neutral element for this new product is $1_\lambda = e^{-\lambda K}$. Note that $\phi_\lambda(e^{-\lambda K}) = 1$.

Many of the algebras we will use are algebras of functions on some set. If we consider a linear space E, the set of real functions defined in E is an algebra with the usual point-wise composition laws. But the important remark is that it is possible to show [RR84] that as far as a new associative local algebra structure $(E, *)$, there will exist a real function h in E such that

$$f * g = h\, f\, g \Longleftrightarrow (f * g)(x) = h(x)\, f(x)\, g(x)\,.$$

Therefore we will fix our attention in non-local algebras of functions for which the expression of the product is given by an integral kernel depending on three variables:

$$(f * g)(x) = \int_E K(x, y, z)\, f(y) g(z)\, dy\, dz\,.$$

Of course, in order to this product be associative the values of the kernels are not arbitrary but they should satisfy the following relation

$$\int_E K(x, y, z)\, K(z, t, u)\, dz = \int_E K(x, z, u)\, K(z, y, t)\, dz\,.$$

In fact,

$$\begin{aligned}[(f * g) * h](x) &= \int K(x, y, z)(f * g)(y) h(z)\, dy\, dz \\ &= \int K(x, y, z)\, K(y, t, u) f(t)\, g(u)\, h(z)\, dy\, dz\, dt\, du\,,\end{aligned}$$

which, with the change of variables $y \mapsto z, z \mapsto u, t \mapsto y, u \mapsto t$, becomes

$$[(f * g) * h](x) = \int K(x, z, u)\, K(z, y, t) f(y)\, g(t)\, h(u)\, dy\, dz\, dt\, du\,,$$

while

$$[f * (g * h)](x) = \int K(x, y, z) f(y)(g * h)(z)\, dy\, dz$$
$$= \int K(x, y, z)\, K(z, t, u) f(y)\, g(t)\, h(u)\, dy\, dz\, dt\, du\,,$$

and therefore the mentioned relation follows.

This new product will be in general non commutative unless the following condition be satisfied:

$$K(x, y, z) = K(x, z, y)\,.$$

A way of constructing an associative algebra structure is by transporting the structure from an associative algebra by means of an invertible map. This mechanism has been very often used in the formulation of Quantum Mechanics in phase space. The idea is to associate to a selfadjont operator A in a Hilbert space $\mathcal{H}$ a function in the phase space, f_A, which is usually called the symbol of the operator A, this correspondence being one-to-one. This allows us to define a new composition law in the space of such functions by

$$f_A * f_B = f_{AB}\,.$$

These twisted products can depend on a parameter, as indicated before. For instance, a remarkable example is the one giving raise to the so-called Moyal quantization, which is defined on functions of a phase space as follows:

$$(f * g) = \exp\,[-\lambda\, D]\,(f\, g)\,,$$

where $D(fg)$ is given by the rule

$$D(fg) = \frac{\partial f}{\partial q^i}\,\frac{\partial g}{\partial p_j} - \frac{\partial g}{\partial q^i}\,\frac{\partial f}{\partial p_j} = \{f, g\}\,.$$

Here λ is a parameter which in physical applications is related to the Planck constant $\hbar$.

In the most general case, a deformation of the associative and commutative pointwise product of functions will be given by

$$f *_\lambda g = f\, g + \lambda\, F_1(f, g) + \lambda^2 F_2(f, g) + \cdots\,.$$

The new deformed product is non-commutative, and the lowest degree term in the difference of the deformed product defines a skew-symmetric bilinear function,

$$\{f, g\} = \lim_{\lambda \to 0} \frac{1}{\lambda}[f *_\lambda g - g *_\lambda f] = F_1(f, g) - F_1(g, f)\,.$$

The new composition law is associative if the functions F_i satisfy some properties which will be later established in full generality. For instance,

$$f_1\, F_1(f_2, f_3) + F_1(f_1, f_2\, f_3) = F_1(f_1\, f_2, f_3) + F_1(f_1, f_2) f_3\,.$$

In case in which this deformed algebra is non-commutative, we can construct the corresponding Lie algebra and take the limit when λ goes to zero, obtaining in this way a Lie algebra. It can also be shown that the bracket we have introduced satisfies the relation

$$\{f, g_1\, g_2\} = g_1\, \{f, g_2\} + g_2\, \{f, g_1\}\,,$$

for any triple of functions, i.e., our bracket satisfies the properties of a Poisson bracket. See the next section after the digression for more details. As a consequence we have arrived in this way to a new structure which will play a relevant rôle as characterizing the possible deformations of the algebra of functions. Next section is a digression of a mathematical nature which can be skipped in a first lecture.

Appendix G: Differential Operators

G.1. Local Differential Operators

If U is an open subset of $\mathbb{R}^n$ and D_k denotes the differential operator $D_k = -i\,\partial/\partial x_k$, we define, for every multi-index $\alpha = (\alpha_1, \ldots, \alpha_n)$, $\alpha_k \in \mathbb{N}$, $k = 1, \ldots, n$: $D^\alpha = D_1^{\alpha_1} \cdots D_n^{\alpha_n}$.

Definition 10.69 A differential operator P of order r is a linear map on the space of complex-valued functions on U of the form $Pu = \sum_{|\alpha| \le r} a_\alpha\, D^\alpha u$, $u \in C^\infty(U; \mathbb{C})$, whose coefficients $a_\alpha \in C^\infty(U; \mathbb{C})$, and where $|\alpha|$ denotes $|\alpha| = \alpha_1 + \cdots + \alpha_r$. The symbol of P is the function on $U \times \mathbb{R}^n$ given by $p(x, \xi) = \sum_{|\alpha| \le r} a_\alpha(x)\, \xi^\alpha$, where $\xi^\alpha \equiv \xi_1^{\alpha_1} \cdots \xi_n^{\alpha_n}$. The principal symbol of P is the function on $U \times \mathbb{R}^n$ defined by $p_r(x, \xi) = \sum_{|\alpha| = r} a_\alpha(x)\, \xi^\alpha$.

Notice that if P is a differential operator of order r, then

$$\left(e^{-itf}\, P e^{itf}\right) u = \sum_{k=0}^{r} t^{r-k}\, P_k u\,,$$

with P_k being a differential operator of order k which does not depend on t. For instance, P_0 is just multiplication by the function $p_0(x) = P(x, \partial f/\partial x)$. In fact, from

$$\left(e^{-itf}D^\alpha e^{itf}\right)u = \left(e^{-itf}D_1 e^{itf}\right)^{\alpha_1}\cdots\left(e^{-itf}D_n e^{itf}\right)^{\alpha_n}u$$

and $\left(e^{-itf}D_k e^{itf}\right)u = D_k u + t\,\frac{\partial f}{\partial x_k}$, we see that

$$\left(e^{-itf}D^\alpha e^{itf}\right)u = \left(D_1 + t\,\frac{\partial f}{\partial x_1}\right)^{\alpha_1}\cdots\left(D_n + t\,\frac{\partial f}{\partial x_n}\right)^{\alpha_n}u\,.$$

Consequently, $\left(e^{-itf}D^\alpha e^{itf}\right)u$ takes the above mentioned form and the coefficient of t^r is the product $(\partial f/\partial x_1)^{\alpha_1}\cdots(\partial f/\partial x_n)^{\alpha_n}$. In particular, we can write $\left(e^{-i\,\xi\cdot x}Pe^{i\,\xi\cdot x}\right)u = p(x,\xi)\,u$.

As an immediate consequence, if P and Q are differential operators with principal symbols $p_r(x,\xi)$ and $q_s(x,\xi)$, then the principal symbol of PQ is the product $p_r(x,\xi)\,q_s(x,\xi)$.

Let us observe that if $\widehat{u}$ denotes the Fourier transform of the function u, i.e.

$$u(x) = \frac{1}{(2\pi)^n}\int e^{ix\cdot\xi}\widehat{u}(\psi)\,d\xi\,,$$

then

$$Pu(x) = \frac{1}{(2\pi)^n}\int e^{ix\cdot\xi}\,p(x,\xi)\widehat{u}(\psi)\,d\xi\,,$$

whenever the integral is meaningful.

Note however that we are using the explicit form of the differential operators D_k which depend very much of the choice for the coordinates or, in other terms, when seen as generators of translations, they depend on the choice of a linear structure on $\mathbb{R}^n$. There are however alternative ways of defining differential operators on $\mathbb{R}^n$ which do not depend on such choices.

Our first approach consists of considering the associative algebra of functions in $\mathbb{R}^n$ together with the Lie algebra $\mathfrak{X}(\mathbb{R}^n)$ of vector fields in $\mathbb{R}^n$. Then we can endow the linear space $\mathfrak{X}(\mathbb{R}^n)\oplus C^\infty(\mathbb{R}^n)$ with a Lie algebra structure by defining

$$[(X_1,f_1),(X_2,f_2)] = ([X_1,X_2], X_1(f_2) - X_2(f_1))\,,$$

usually called the 'holomorph' of $C^\infty(\mathbb{R}^n)$ and then, the enveloping algebra of such Lie algebra is the associative algebra of differential operators in $\mathbb{R}^n$, that is the quotient algebra of the associative algebra generated by elements (X, f) quotiented by the bilateral ideal generated by the relations $(X, f)\otimes(Y, g) - (Y, g)\otimes(X, f) - [(X, f), (Y, g)]$ for all $(X, f), (Y, g)$.

The differential operators of order zero are the only differential operators that are not zero on constant functions and correspond to functions. The set $\mathcal{D}(\mathbb{R}^n)$ of differential operators can be written as $\mathcal{D}(\mathbb{R}^n) = C^\infty(\mathbb{R}^n) \oplus \mathcal{D}_c(\mathbb{R}^n)$, where $\mathcal{D}_c(\mathbb{R}^n)$ denotes the set of differential operators that are zero on constant functions.

The preceding algebraic characterization allows for a definition of differential operators on any subamnifold $M \subset \mathbb{R}^n$. It would be enough to restrict ourselves to consider only vector fields which are tangent to M instead of $\mathfrak{X}(\mathbb{R}^n)$, to use $\mathcal{F}(M)$ instead of $C^\infty(\mathbb{R}^n)$ and consider the holomorph of $\mathcal{F}(M)$. The corresponding enveloping algebra provides us the algebra of differential operators in the submanifold M. In other words, considering $\mathcal{F}(M)$ as an Abelian Lie algebra, $\mathcal{D}_1(M)$ is the algebra of derivations $\mathfrak{X}(M)$ and $\mathcal{D}(M)$ becomes what is known in the literature as the 'holomorph' of $\mathcal{F}(M)$. In this way the algebra of differential operators becomes the enveloping algebra of the holomorph of $\mathcal{F}(M)$.

There is still another intrinsic way of defining the concept of differential operator. The idea again consists in replacing $\mathbb{R}^n$ by its algebra of differentiable functions and the observation that the differential operators D_k are linear maps satisfying the following property (see [Al91, Gr04]):

$$\left[D_k, m_f\right] = m_{D_k(f)},$$

where m_f is the multiplication operation by f, i.e., the operation of degree zero $m_f(g) = fg$, with $f, g \in C^\infty(U)$. It follows that if P is a differential operator of degree r of the form indicated, then

$$[P, m_f] = \sum_{|\sigma| \leq k} a_\sigma \left[D_k, m_f\right],$$

is of degree at most $r-1$. Iterating the previous formula for a set of $r+1$ functions $f_0, \cdots, f_r \in C^\infty(U)$, one finds that

$$[\ldots[[P, m_{f_0}], m_{f_1}], \ldots, m_{f_r}] = 0,$$

i.e., differential operators of degree no greater than r are such that for $r+1$ functions $f_j \in C^\infty(U)$ they satisfy the preceding identity. In particular the operators m_f are differential operators of order zero and the usual vector fields are differential operators of order one.

Then we may consider the algebra $\mathcal{F}(M)$ of differentiable functions in M as differential operators of order zero and then the differential operators P of degree no greater than $r > 0$ are $\mathbb{R}$-linear maps from $\mathcal{F}(M)$ into itself such that

$$[\cdots[P, m_{f_0}], m_{f_1}], \cdots, m_{f_r}] = 0.$$

Usually one adds the locallity requirement that P does not increase the support, i.e., $\mathrm{supp}(Pf) \subset \mathrm{supp}(f), \forall f \in \mathcal{F}(M)$.

The set of differential operators of degree not greater than r is both a left $\mathcal{F}(M)$-module and a right $\mathcal{F}(M)$-module and both actions commute, therefore, it is a bimodule. We denote $\mathcal{D}_r(M)$ the set of differential operators of order at most r. Note that $\mathcal{D}_p(M) \subset \mathcal{D}_q(M)$ when $p \leq q$, and that for any couple of nonnegative integer numbers,

$$\mathcal{D}_p(M) \circ \mathcal{D}_q(M) \subset \mathcal{D}_{p+q}(M) , \qquad [\mathcal{D}_p(M), \mathcal{D}_q](M)] \subset \mathcal{D}_{p+q-1}(M) . \quad (10.295)$$

This shows that $\mathcal{D}_1(M)$ is a Lie subalgebra, while $\mathcal{D}_0(M)$ is a subalgebra, and the following sequence of Lie algebras is exact:

$$0 \to \mathcal{D}_0(M) \to \mathcal{D}_1(M) \to \operatorname{Der} \mathcal{F}(M) \to 0$$

i.e., $\mathcal{D}_1(M)$ is a Lie algebra extension of $\mathcal{D}_{1c}(M) = \mathfrak{X}(M) = \operatorname{Der} \mathcal{F}(M)$ by $\mathcal{D}_0(M)$.

If we denote $\mathfrak{m}_x$ the ideal of functions vanishing on x, then differential operators of degree not greater than k are alternatively characterized by the following property: $P \in \mathcal{D}_k(M)$ is a $\mathbb{R}$-linear map $P : \mathcal{F}(M) \to \mathcal{F}(M)$ such that the image of $\mathfrak{m}_x^{k+1}$ lies in $\mathfrak{m}_x$. In fact, given $P \in \mathcal{D}_k(M)$, then for any pair of functions $f \in \mathfrak{m}_x$ and $g \in \mathfrak{m}_x^k$,

$$P(fg) = f\, P(g) + [P, m_f](g) ,$$

and on the right-hand side both terms are in $\mathfrak{m}_x$, therefore $P(\mathfrak{m}_x^{k+1}) \subset \mathfrak{m}_x$. Conversely, asssume that P is a linear map such that $P(\mathfrak{m}_x^{k+1}) \subset \mathfrak{m}_x$, and consider first the case $k = 0$. Then, for any $g \in \mathcal{F}(M)$ and $x \in M$, using the identity

$$(P - P(1))g = P(g - g(x)) + P(g(x)) - P(1)(g - g(x)) - P(1)g(x),$$

and taking into account that $P(g(x)) = P(g(x)\,1) = g(x)P(1) = P(1)\,g(x)$, it reduces to:

$$(P - P(1))g = P(g - g(x)) - P(1)(g - g(x)) ,$$

and consequently $(P - P(1))g \in \mathfrak{m}_x, \forall x \in M$, i.e., $P = P(1)$. If k is different from zero, the condition $P(\mathfrak{m}_x^{k+1}) \subset \mathfrak{m}_x$ says that if $f \in \mathcal{F}(M)$ and $g \in \mathfrak{m}_x^k$, then we define $\bar{f} = f - f(x)$, and we can write

$$[P, m_f]g = P(fg) - f P(g) = P(\bar{f}\, g) - \bar{f}\, P(g) ,$$

and both terms on the right-hand side are in $\mathfrak{m}_x$.

The relations (10.295) also show that the set of differential operators on M, denoted as $\mathcal{D}(M)$, can be given the structure of a graded associative algebra and it is also a bi-module over $\mathcal{F}$.

Differential Operators on Vector Bundles

Given two vector bundles over the same base, $\pi_a\colon E_a \to M$, $a = 1, 2$, the corresponding spaces of sections $\mathcal{E}_1$ and $\mathcal{E}_2$ are both $\mathcal{F}(M)$-modules. The concept of differential operator is given by recurrence. A $\mathbb{R}$-linear map $P\colon \Gamma(E_1) \to \Gamma(E_2)$ is said to be a differential operator of order zero. Therefore $\mathcal{D}_0(E_1, E_2) = \mathrm{Hom}(E_1, E_2)$. If $k \geq 1$, we say that such a P is a differential operator of order no greater than k if for any function $f \in \mathcal{F}(M)$ the commutator $[P, m_f]$ is a differential operator of order no greater than $k - 1$, where m_f denotes as usual the map of multiplying a section by the function f.

The set of differential operators of order $\leq k$ is a $\mathcal{F}(M)$-module, to be denoted $\mathcal{D}_k(E_1, E_2)$ and similarly, $\mathcal{D}(E_1, E_2) = \bigcup_k \mathcal{D}_k(E_1, E_2)$.

When E_1 and E_2 are the trivial bundle $\pi_a\colon M \times \mathbb{R} \to M$, $a = 1, 2$, then $\mathcal{E}_1 = \mathcal{E}_2 = \mathcal{F}(M)$ and we recover the spaces $\mathcal{D}$ we mentioned before.

There are other equivalent definitions of differential operators. For instance, $P \in \mathcal{D}_k(E_1, E_2)$, for $k > 0$, if and only if P is a linear map $P\colon \Gamma(E_1) \to \Gamma(E_2)$, and for every $k + 1$ functions $f_i \in \mathcal{F}(M)$, $i = 1, \ldots, k + 1$, such that $f_i(x) = 0$, and every section $s : M \to E_1$, we have $P(f_1 \cdots f_{k+1} s)(x) = 0$. Or in another equivalent form, if for every function f such that $f(x) = 0$, and every section $s : M \to E_1$, we have that $P(f^{k+1} s)(x) = 0$.

Example 10.8 1. Let M be a differentiable manifold and consider the vector bundles $E_1 = \Lambda^r(T^*M)$, $E_2 = \Lambda^{r+1}(T^*M)$. Then $\Gamma(E_1)$ is the set of r-forms in M while $\Gamma(E_2)$ is the set of $(r + 1)$-forms. The exterior dfferential acting on r-forms d_r is a differential operator of order 1.

Note that given multi-indexes $I = (i_1, \ldots, i_r)$, with $1 \leq i_1 < \ldots < i_r \leq \dim M$, and $J = (j_1, \ldots, j_{r+1})$ with $1 \leq j_1 < \cdots < j_{r+1} \leq n$, then the operator d_r is given by

$$d\left(\sum_I f_I dx_{i_1} \wedge \cdots \wedge dx_{i_r}\right) = i \sum_J (-1)^k (D_k f_{j_1 \cdots \hat{j}_k \cdots j_{r+1}}) dx_{j_1} \wedge \cdots \wedge dx_{j_{r+1}}.$$

Consider now $E_1 = E_2 = \Lambda^\bullet(T^*M)$, then the exterior differential d is differential operator of order 1.

2. Let be M be a differentiable manifold and $E_1 = \Lambda^r(T^*M)$ (with $r > 1$), $E_2 = \Lambda^{r-1}(T^*M)$ and $X \in \mathfrak{X}(M)$ an arbitrary but fixed vector field. Contraction with such vector field defines a differential operator i_X of order 0 such that

$$i_X \left(f_{i_1 \cdots i_r} dx_{i_1} \wedge \cdots \wedge dx_{i_r}\right) = \sum_k (-1)^{k+1} f_{i_1 \cdots i_r} dx_{i_1} \wedge \cdots \wedge \widehat{dx_k} \wedge \cdots \wedge dx_{i_r}.$$

As before, we can define now a differential operator i_X of order 0 acting on sections of the exterior bundle $\Lambda^\bullet(T^*M)$.

3. In the same vein, the Lie derivative $\mathcal{L}_X$ is a differential operator of order 1 given by Cartan's formula: $\mathcal{L}_X \alpha = i_X d\alpha + d i_X \alpha$.
4. Pauli and Dirac operator.
   In non-relativistic mechanics the description of Stern-Gerlach experiment requires wave functions which have values in $\mathbb{C}^2$ to take into account the spin degrees of freedom. From our general point of view, these wave functions, spinorial functions, may be considered as sections of a vector bundle with typical fibre $\mathbb{C}^2$. Thus the Hamiltonian operator and all other observables should be operators acting on vector-valued wave functions to be able to extract expectation values out of measurements performed while the system is in these states.
   A convenient way to take into account the electron spin, is to consider the Hamilton operator represented by the following operator, called the Pauli Hamiltonian:

$$H = \frac{1}{2m}(\boldsymbol{\sigma} \cdot \mathbf{P})^2 + U(x, y, z)\,\sigma_0\,,$$

where $\sigma_0, \sigma_1, \sigma_2, \sigma_3$, are Pauli matrices (Eq. 10.68).

The vector-valued operator $\mathbf{P}$ given by

$$\mathbf{P} = \mathbf{p} + \frac{e}{c}\mathbf{A},$$

with $\mathbf{p}$ the standard momentum, and $\mathbf{A}$ being the vector potential of the external electromagnetic field B in which the electron is moving.

We thus have:

$$\boldsymbol{\sigma} \cdot \mathbf{p} = \frac{\hbar}{i}\begin{pmatrix} \dfrac{\partial}{\partial x_3} & \dfrac{\partial}{\partial x_1} - i\dfrac{\partial}{\partial x_2} \\ \dfrac{\partial}{\partial x_1} + i\dfrac{\partial}{\partial x_2} & -\dfrac{\partial}{\partial x_3} \end{pmatrix}, \qquad \sigma \cdot \mathbf{A} = A_1\sigma_1 + A_2\sigma_2 + A_3\sigma_3.$$

These operators turn out to be finite-dimensional matrices whose entries are differential operators. They act on sections of vector bundles or on the tensor product of the Hilbert space of complex-valued square integrable functions times the finite-dimensional Hilbert space describing the 'internal structure', in our case the spin degrees of freedom.
Similarly, for the relativistically invariant Dirac equation one replaces $\boldsymbol{\sigma} \cdot \mathbf{p}$ with the quantity

$$\gamma^\mu p_\mu = \gamma^0 p_0 + \gamma^1 p_1 + \gamma^2 p_2 + \gamma^3 p_3\,,$$

where $(p^0, \mathbf{p})$ is the four-momentum and $(\gamma^0, \gamma^1, \gamma^2, \gamma^3)$ is a four-vector whose components are matrices. These matrices $(\gamma^0, \boldsymbol{\gamma})$ transform like a vector, while the four-momentum transforms like a covector and then the total 'scalar product' is Poincaré invariant. Here the γ-matrices may be represented as

$$\gamma^0 = \begin{pmatrix} 1 & 0 \\ 0 & 1 \end{pmatrix}, \quad \gamma^i = \begin{pmatrix} 0 & \sigma^i \\ -\sigma^i & 0 \end{pmatrix}, \quad i = 1, 2, 3.$$

and the Dirac equation acquires the form

$$\left(\gamma^0 \frac{\partial}{\partial x^o} + \gamma^1 \frac{\partial}{\partial x^1} + \gamma^2 \frac{\partial}{\partial x^2} + \gamma^3 \frac{\partial}{\partial x^3}\right) \psi = im\psi \,.$$

Remark 10.7 The matrices γ^μ are generators of a Clifford algebra and have the following structure constants:

$$(\gamma^0)^2 = 1 \,, \quad (\gamma^j)^2 = -1 \,, \quad j = 1, 2, 3$$

$$\gamma^0\gamma^j + \gamma^j\gamma^0 = 0 \,, \quad \gamma^j\gamma^k + \gamma^k\gamma^j = 0 \,, \; j \neq k \,.$$

Any product of a finite number of matrices may be expressed in the form

$$a_0 \, 1 + \sum_j a_j \, \gamma^j + \sum_{j<k} a_{jk} \, \gamma^j\gamma^k + \sum_{j<k<l} a_{jkl} \, \gamma^j\gamma^k\gamma^l + a_{0123} \, \gamma^0\gamma^1\gamma^2\gamma^3 \,.$$

By considering the left- and right- multiplication by functions on $\mathbb{R}^4$ we can construct differential operators with coefficients given by the matrices generated by the γ's.

G.2. The Codifferential and the Laplace-Beltrami Operator

The Hodge-Star Operator

As a preliminary, we would like to extend the notion of the scalar product from 1-forms to forms of higher rank. So, let us consider a manifold M with a (pseudo-) Riemannian metric g whose volume form is given by Ω.

Definition 10.70 The scalar product of two p-forms: $\alpha = \frac{1}{p!}\alpha_{i_1 \ldots i_p} dx^{i_1} \wedge \ldots \wedge dx^{i_p}$ and: $\beta = \frac{1}{p!}\beta_{i_1 \cdots i_p} dx^{i_1} \wedge \cdots \wedge dx^{i_p}$ is given by

$$(\alpha, \beta) = \frac{1}{p!} \alpha_{i_1 \ldots i_p} \beta^{i_1 \ldots i_p} \tag{10.296}$$

where,

$$\beta^{i_1 \ldots i_p} = g^{i_1 j_1} \ldots g^{i_p j_p} \beta_{j_1 \cdots j_p} \tag{10.297}$$

In analogy with Eq. (10.297), the totally contravariant components of the volume-form can be defined as:

$$\Omega^{i_1,\ldots,i_n} = g^{i_1 j_1} \ldots g^{i_n j_n} \Omega_{j_1,\ldots,j_n}. \tag{10.298}$$

Let β be a p-form. Then there exists a unique $(n-p)$-form $*\beta$ such that,

$$\alpha \wedge *\beta = (\alpha, \beta)\, \Omega, \qquad \forall \alpha \in \Omega^p(M). \tag{10.299}$$

The existence of $*\beta$ follows from the explicit construction,

$$*\, \beta = \frac{1}{(n-p)!} (*\beta)_{j_1 \cdots j_{n-p}} dx^{j_1} \wedge \cdots \wedge dx^{j_{n-p}}, \tag{10.300}$$

where,

$$(*\beta)_{j_1 \cdots j_{n-p}} = \frac{1}{p!} \Omega_{i_1 \cdots i_p, j_1 \cdots j_{n-p}} \beta^{i_1 \cdots i_p}. \tag{10.301}$$

The uniqueness follows in turn from the fact, whose proof is immediate, that if α is a p-form and γ an $(n-p)$-form, then, $\alpha \wedge \gamma = 0\ \forall \alpha$ implies $\gamma = 0$.

Definition 10.71 With the conventions before, the Hodge star operator is the linear isomorphism,

$$* : \Omega^p(M) \to \Omega^{n-p}(M) \tag{10.302}$$

that associates with every p-form β the $(n-p)$-form $*\beta$.

Proposition 10.72 *The Hodge star operator satisfies the properties,*

1. $\alpha \wedge (*\beta) = \beta \wedge (*\alpha)$.
2. $(*\alpha, *\beta) = sgn(G)(\alpha, \beta)$.
3. $* \circ * = (-)^{p(n-p)} sgn(G) Id_{\Omega^p}$ *on p–forms.*
4. $*\, \Omega = sgn\,(G)$.

where $G = \det(g_{ij})$.

Exercise 10.9 The proof is left as an exercise on tensor algebra.

Example 10.10 Let us consider $M = \mathbb{R}^3$ with the standard Euclidean metric. As $(-1)^{p(n-p)} = 1$ and $G = 1$ then $\Omega = dx^1 \wedge dx^2 \wedge dx^3$ and:

1. $*f = f\Omega$
2. If $\alpha = \alpha_i\, dx^i$, then $*\alpha = \frac{1}{2}\varepsilon_{ijk}\, \alpha^i\, dx^j \wedge dx^k$.
3. If $\omega = \frac{1}{2}\omega_{ij}\, dx^i \wedge dx^j$, then $*\omega = \frac{1}{2}\varepsilon_{ijk}\, \omega^{ij}\, dx^k$.
4. $*(h\, dx^1 \wedge dx^2 \wedge dx^3) = h$.

The Laplace-Beltrami Operator

The star operator and the exterior derivative can be combined to define a new differential operator:

Definition 10.73 The codifferential δ is the differential operator whose action on p-forms is given by,

$$\delta = (-1)^{n(p+1)+1}\mathrm{sgn}(G) * d* \tag{10.303}$$

Remark 10.8 The codifferential maps then p-forms into $(p-1)$-forms, thus it is a linear map of degree -1 with respect to the natural grading of the exterior algebra $\Omega^\bullet(M)$ of forms.

Moreover δ is a differential operator of order 1 as it is easy to check. However, at variance with the exterior differential d, it is not a derivation on forms. Indeed, as can be checked easily, δ does not obey the Leibnitz rule.

Proposition 10.74 *The codifferential δ operator satisfies the following properties:*

1. $\delta f = 0$ *on functions.*
2. $\delta^2 = \delta \circ \delta = 0.$
3. $*\delta\, d = d\,\delta\, *; \quad *\, d\,\delta = \delta\, d\, *.$
4. $*\,\delta\,\omega = (-1)^{n+p+1} d\, *\, \omega; \quad \omega \in \Omega^p.$
5. $d * \delta = \delta * d = 0.$
6. $\delta * \omega = (-)^{n+p+1} * d\omega; \quad \omega \in \Omega^p.$

The proof of all of the above properties is elementary, and is left as an exercise.

The exterior differential d is 'natural' with respect to diffeomorphisms, i.e.: $\phi^* d = d\phi^*$ for any $\phi \in \mathrm{Diff}\,(M)$, a similar discussion was not done for the codifferential. We can pose now the same problem for the codifferential. Let's use here a more complete notation, i.e. let's denote by δ_g the codifferential associated with a given metric g, and let's also denote with $\langle\cdot,\cdot\rangle_g$ the global scalar product associated with the same metric. Then:

Proposition 10.75 *If $\phi \in \mathrm{Diff}\,(M)$, then, $\phi^*\delta_g = \delta_{\phi^* g}\phi^*$.*

Indeed, it is easy to show that:

$$\phi^*(\langle\delta_g\alpha, \beta\rangle_g) = \langle\phi^*\delta_g\alpha, \phi^*\beta\rangle_{\phi^* g},$$

but, on the other hand,

$$\phi^*(\langle\delta_g\alpha, \beta\rangle_g) = \phi^*(\langle\alpha, d\beta\rangle_g) = \langle\phi^*\alpha, d\phi^*\beta\rangle_{\phi^* g} = \langle\delta_{\phi^* g}\phi^*\alpha, \phi^*\beta\rangle_{\phi^* g}.$$

Equating the right-hand side 's we obtain the desired result.

It follows from the previous proposition that the codifferential δ is 'natural' only with respect to the diffeomorphisms that leave the metric invariant, i.e., such that: $\phi^* g = g$. These are the isometries, and hence the codifferential is 'natural' with respect to isometries.

Example 10.11 If $F = \frac{1}{2} F_{ij} dx^i \wedge dx^j$ is a 2-form, then it is easy to prove that,

$$\delta F = (\delta F)_i \, dx^i \tag{10.304}$$

where,

$$(\delta F)_i = -\frac{1}{\sqrt{|G|}} g_{ij} \frac{\partial}{\partial x^k} \sqrt{|G|} F^{jk} \tag{10.305}$$

A further differential operator that, as we shall shortly, is a convenient generalization to forms of arbitrary rank of the more familiar Laplacian can be constructed by combining the actions of d and δ, and precisely:

Definition 10.76 The Hodge operator is the second-order differential operator $\Delta \colon \Omega^p \to \Omega^p$ defined by,

$$\Delta = (d + \delta)^2 \equiv d\delta + \delta d = d * d * + * d * d. \tag{10.306}$$

Restricted to functions, the Hodge operator becomes the familiar Laplace-Beltrami operator that, in local coordinates takes the explicit form:

$$\Delta f = -\frac{1}{\sqrt{|G|}} \frac{\partial}{\partial x^i} \sqrt{|G|} g^{ij} \frac{\partial}{\partial x^j} f. \tag{10.307}$$

Vector Analysis in the Presence of a (Pseudo-)Riemannian Metric

Let g be a (pseudo-) Riemannian metric on M. Then the gradient of a function $f \in \mathcal{F}(M)$ is the vector field,

$$\text{grad}(f) = \nabla f = \hat{g}^{-1}(df) \tag{10.308}$$

or, explicitly, $\text{grad}(f) = (\text{grad}(f))^i \frac{\partial}{\partial x^i}$, with $(\text{grad}(f))^i = g^{ij} \frac{\partial f}{\partial x^j}$.

Let now Ω be the volume-form associated with g (actually we could do the same with any other volume-form) and let X be a vector field. We recall that the *divergence* of the vector field X is the unique function div (X) (or $\nabla \cdot X$) such that:

$$\mathcal{L}_X \Omega = \text{div}(X) \, \Omega \tag{10.309}$$

and that if Ω is the volume-form associated with an Euclidean metric in $\mathbb{R}^n$, $\mathrm{div}(X)$ becomes the familiar expression,

$$\mathrm{div}(X) = \sum_i (\partial X^i/\partial x^i). \tag{10.310}$$

Denoting now by α_X the 1-form associated with X by means of g, i.e. $\alpha_X = \widehat{g}(X)$, it is a simple exercise to show that,

$$i_X\Omega = *\alpha_X \tag{10.311}$$

and we have furthermore:

$$\mathrm{div}X = -\delta\alpha_X. \tag{10.312}$$

Indeed, using the Cartan identity we get:

$$\mathrm{div}\,(X)\,\Omega = d\,(i_X\Omega) = d * \alpha_X \tag{10.313}$$

and Eq. (10.312) follows then by taking the 'star' of both sides and using Eq. (10.303).

Exercise 10.12 Prove that, in local coordinates,

$$\mathrm{div}\,(X) = \frac{1}{\sqrt{G}}\frac{\partial}{\partial x^i}\left(\sqrt{G}X^i\right)$$

and that, introducing the Christoffel symbols,

$$\Gamma^i_{jk} = \frac{1}{2}g^{il}\left\{\frac{\partial g_{lj}}{\partial x^k} + \frac{\partial g_{lk}}{\partial x^j} - \frac{\partial g_{jk}}{\partial x^l}\right\}$$

one has,

$$\mathrm{div}\,X = \frac{\partial X^i}{\partial x^i} + \Gamma^i_{ik}X^k.$$

Let's calculate now the function $\mathrm{div}(\mathrm{grad}(f))$. As it is obvious that $\alpha_{\mathrm{grad}(f)} = df$. Then we obtain,

$$-\,\mathrm{div}(\mathrm{grad}(f)) = \delta df = \Delta f, \tag{10.314}$$

where Δ is the Laplace-Beltrami operator Eq. (10.307).

Example 10.13 (i) In $\mathbb{R}^3$ with the Euclidean metric,

$$\Delta f = -\nabla^2 f \tag{10.315}$$

with ∇^2 the usual Laplacian: $\nabla^2 = \sum_i \partial^2/\partial(x^i)^2$. The minus sign in the above equation makes it clear that the Laplace-Beltrami operator is a positive (semi)definite elliptic operator in this case.

(ii) In $\mathbb{R}^4$ with the Lorentz metric,

$$\Delta f = -\Box^2 f \tag{10.316}$$

with $\Box^2$ the D'Alambertian (or wave operator),

$$\Box^2 = g^{ij}\frac{\partial^2}{\partial x^i \partial x^j} = \frac{\partial^2}{\partial (x^0)^2} - \sum_{i=1}^{3} \frac{\partial^2}{\partial (x^i)^2} \tag{10.317}$$

Let's specialize now the previous discussion to M a linear space of dimension 3. If X is a vector field, $*d\alpha_X$ will be a 1-form, and because $(-1)^{p(n-p)} = 1$. Then the rotational of a vector field X is the vector field curl X defined by,

$$\text{curl } X = \hat{g}^{-1}(*d\alpha_X).$$

Exercise 10.14 Prove that, in coordinates,

$$* \, d\alpha_X = \frac{1}{2}\sqrt{|G|}\varepsilon_{ijk} g^{jl} g^{km} \left\{ \frac{\partial X_l}{\partial x^m} - \frac{\partial X_m}{\partial x^l} \right\} dx^i \tag{10.318}$$

and that,

$$\text{curl } X = \sqrt{|G|}\left(\left(\frac{\partial X_2}{\partial x^1} - \frac{\partial X_1}{\partial x^2} \right) \frac{\partial}{\partial x^3} + \text{cyclic permutations} \right) \tag{10.319}$$

It is easy to check that:

$$\alpha_{\text{curl } X} = \widehat{g}(\text{curl } X) = *d\alpha_X,$$

and hence,

$$\text{div}(\text{curl } X) = -\delta * d\alpha_X \,,$$

as well as,

$$\hat{g}^{-1}(\delta * df) = \text{curl}(\text{grad}(f)).$$

The familiar identities: 'div · curl = 0' and 'curl · grad = 0' of elementary vector analysis in $\mathbb{R}^3$ all follow then from $\delta * d = 0$ (ultimately, from $d^2 = 0$).

Let now X have a vanishing curl. As both $\widehat{g}$ and the star operator are isomorphisms,

$$\text{curl}\, X = 0 \Longleftrightarrow d\alpha_X = 0,$$

and, by Poincare''s Lemma (which holds globally in the present context) $\alpha_X = df$ for some $f \in \mathcal{F}(E)$. Similarly: div $X = 0$ implies $\delta\alpha_X = 0$ and hence, $d{*}\alpha_X = 0$. Again by Poincare''s Lemma: ${*}\alpha_X = d\alpha_Y$ for some $Y \in \mathfrak{X}(E)$ and hence, $\alpha_X = {*}d\alpha_Y$. It follows then that the familiar theorems of vector analysis,

$$\text{curl}\, X = 0 \Longleftrightarrow X = \text{grad}(f)$$

and,

$$\text{div}\, X = 0 \Longleftrightarrow X = \text{curl}\, Y$$

are both consequences of the Poincare' Lemma.

Exercise 10.15 Observing that $\alpha_{fX} = f\alpha_X$ and using, $d(fg) = g\,df + f\,dg$, and $d\alpha_{fX} = fd\alpha_X + df \wedge \alpha_X$, prove that:

$$\text{grad}(fg) = f\,\text{grad}(g) + g\,\text{grad}(f),$$

and

$$\text{curl}(fX) = f\,\text{curl}\, X + \text{grad}(f) \times X,$$

where the cross-product of any two vector fields X and Y , $X \times Y$, is defined via,

$$\alpha_{X\times Y} = {*}(\alpha_X \wedge \alpha_Y),$$

or,

$$X \times Y = \widehat{g}^{-1}({*}(\alpha_X \wedge \alpha_Y)).$$

References

[Ja79] Jacobson, N.: Lie Algebras, No. 10. Dover Publications, New York (1979)

[La85] E.A. Lacomba, A.B. Bugdor.: Circuits, Chain Complexes, and Symplectic Geometry. H. Poincaré (1985)

[Whit44] Whitney, H.: The self intersections of a smooth n-manifold in $2n$-space. Ann. Math. **45**, 220–246 (1944)

[Ab78] Abraham, R., Marsden, J.E.: Foundations of Mechanics, 2nd edn. Benjamin, Reading (1978)

[Ab88] Abraham R., Marsden, J.E., Ratiu, T.: Manifolds, Tensor Analysis and Applications, 2nd edn. Springer, New York (1988)

[Ni55] Nijenhuis, A.: Jacobi-type identities for bilinear differential concomitants of certain tensor fields I, II. Indag. Math. **17**, 390–397 (1955); ibid. Indag. Math. **17**, 397–403 (1955)

[Ni64] Nijenhuis, A., Richardson, R.W.: Cohomology and deformations of algebraic structures. Bull. Amer. Math. Soc. **70**, 406–411 (1964)

[Ni66] Nijenhuis, A., Richardson, R.W.: Cohomology and deformations in graded Lie algebras. Bull. Amer. Math. Soc. **72**, 1–29 (1966)

[ML75] MacLane, S.: Homology. Graduate Texts of Math, vol. 114, p. 975. Springer, New York (1975)

[Mo64] Moore, C.: Extensions and low dimensional cohomology theory of locally compact groups, I, II. Trans. Amer. Math. Soc. **113**, 4063 (1964); ibid. Trans. Am. Math. Soc. **136**, 64–86 (1966)

[Mac57] Mackey, G.W.: Les ensembles Borliens et les extensions des groupes. J. Math. Pures Appl. **36**, 171–178 (1957)

[Mac76] Mackey, G.W.: Induced representations of groups and quantum mechanics. Pubbl. della Classe de Scienze della Scuola Normale Superiore di Pisa (Published jointly with Benjamin Inc. and Editore Boringhiori) (1976)

[Ba54] Bargmann, V.: On unitary ray representations of continuous groups. Ann. Math. **59**, 1–46 (1954)

[Ho77] Houard, J.C.: On invariance groups and Lagrangian theories. J. Math. Phys. **18**, 502–516 (1977)

[Ch48] Chevalley, C., Eilenberg, S.: Cohomology theory of Lie groups and algebras. Trans. Amer. Math. Soc. **63**, 85–124 (1948)

[Ho80] Houard, J.C.: An integral formula for cocycles of Lie groups. Ann. Inst. H. Poincaré **32A**, 221–247 (1980)

[G64] Gerstenhaber, M.: On the deformation of rings and algebras. Ann. Math. **79**, 59–103 (1964)

[GH46] Hochschild, G.: On the cohomology theory for associative algebras. Ann. Math. **47**, 568–79 (1946)

[CI88] Cariñena, J.F., Ibort, L.A.: Noncanonical groups of transformations, anomalies and cohomology. J. Math. Phys. **29**, 541–545 (1988)

[CGM00] Cariñena, J.F., Grabowski, J., Marmo, G.: Quantum bi-Hamiltonian systems. Int. J. Mod. Phys. A **15**, 4797–4810 (2000)

[RR84] Rubio, R.: Algèbres asociatives locales sur l'espace de sections d'un fibré en droites. C.R. Acad. Sci. Paris **299**, 699–701 (1984)

[Al91] Alekseevskij, D.V., Vinogradov, A.M., Lyčagin, V.V.E.: Geometry I: Basic ideas and concepts of differential geometry. In: Gamkrelidze, R.V. (ed.). Springer (1991)

[Gr04] Grabowski, J., Poncin, N.: Automorphisms of quantum and classical Poisson algebras. Compos. Math. **140**, 511–527 (2004); ibid., Lie algebraic characterization of manifolds. Cent. Eur. J. Math. **2**, 811–825 (2004)

Index

J.F. Cariñena et al., *Geometry from Dynamics, Classical and Quantum*,
DOI 10.1007/978-94-017-9220-2

9789401792196